MEDICAL MICROBIOLOGY

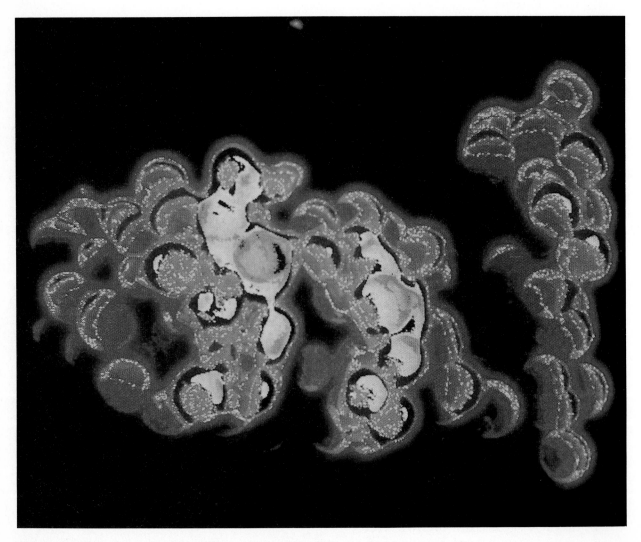

Photomicrograph of *Staphylococcus aureus*, by Frieder Michler.

MEDICAL
MICROBIOLOGY

PATRICK R. MURRAY, Ph.D.
Professor, Department of Pathology,
Washington University School of Medicine;
Director, Clinical Microbiology Laboratory,
Barnes Hospital,
St. Louis, Missouri

W. LAWRENCE DREW, M.D., Ph.D.
Director, Clinical Microbiology
   and Infectious Diseases,
Mount Zion Hospital and Medical Center,
San Francisco, California

GEORGE S. KOBAYASHI, Ph.D.
Professor, Department of Medicine,
Washington University School of Medicine;
Associate Director,
Clinical Microbiology Laboratory,
Barnes Hospital,
St. Louis, Missouri

JOHN H. THOMPSON, JR., Ph.D.
Emeritus Professor of Laboratory Medicine,
Mayo Medical School,
Mayo Clinic and Foundation,
Rochester, Minnesota

With **640** illustrations

THE C. V. MOSBY COMPANY
ST. LOUIS • BALTIMORE • PHILADELPHIA • TORONTO 1990

Editor Stephanie Bircher Manning
Developmental Editor Elaine Steinborn
Assistant Editors Anne Gunter and Jo Salway
Project Managers John A. Rogers and Carol Sullivan Wiseman
Book Design John Rokusek and Candace Conner
Cover Design Candace Conner
Illustration Program Donald O'Connor

Cover Photo © Frieder Michler—Peter Arnold Inc.

Printed in the United States of America

The C. V. Mosby Company
11830 Westline Industrial Drive, St. Louis, Missouri 63146

Library of Congress Cataloging in Publication Data

Medical microbiology/Patrick R. Murray . . . [et al.].
    p. cm.
  Includes bibliographical references.
  ISBN 0-8016-3586-1
  1. Medical microbiology.   I. Murray, Patrick R.
  [DNLM: 1. Microbiology.   2. Parasitology.   QW 4 M4862]
QR46.M4683   1990
616'.01—dc20
DNLM/DLC
for Library of Congress   89-13294
                               CIP

TSI/VH/VH 9 8 7 6 5 4 3 2 1

Contributing Authors

HENRY D. ISENBERG, Ph.D.
Chief, Division of Microbiology,
Long Island Campus for Albert Einstein College
    of Medicine,
New Hyde Park, New York;
Professor of Laboratory Medicine,
Albert Einstein College of Medicine,
New York, New York

KEN S. ROSENTHAL, Ph.D.
Associate Professor of Microbiology and Immunology,
Northeastern Ohio Universities College of Medicine,
Rootstown, Ohio

JAMES W. SMITH, M.D.
Chief, Infectious Disease Section,
Department of Veteran's Affairs Medical Center;
Professor, Internal Medicine,
University of Texas Southwestern Medical Center—
    Dallas,
Dallas, Texas

JON B. SUZUKI, D.D.S., Ph.D.
Professor, Microbiology and Periodontics,
Department of Surgery,
Division of Oral Maxillofacial Surgery,
University of Maryland and The Johns Hopkins
    University School of Medicine,
Baltimore, Maryland

ELLA M. SWIERKOSZ, Ph.D.
Associate Professor of Pediatric/Adolescent Medicine,
St. Louis University School of Medicine;
Director, Diagnostic Microbiology and Virology,
Cardinal Glennon Children's Hospital,
St. Louis, Missouri

RICHARD B. THOMSON, Jr., Ph.D.
Assistant Professor of Microbiology,
Northeastern Ohio Universities College of Medicine,
Rootstown, Ohio;
Director, Microbiology,
Akron City Hospital;
Director, Virology,
Children's Hospital Medical Center of Akron,
Akron, Ohio

To our families
for their patience and understanding
while we labored with this endeavor.

Preface

Microbiology, at present, is enjoying a period of unequaled popularity. Every year new diseases are discovered (e.g., Legionnaires' disease, cat scratch disease, Lyme disease, AIDS), or old diseases find a new host or clinical presentation (e.g., cryptosporidiosis, infant botulism). With the application of sophisticated technologic skills in molecular biology, biochemistry, immunology, and electron microscopy, the structural organization of these pathogens and the pathogenesis of their associated diseases can be clearly delineated. Use of well-defined epidemiologic principles combined with these technologic skills enables us to determine the prevalence of these diseases. One could say we are blessed with an abundance of information.

Unfortunately, our education is built upon layers and layers of information. Although it would be wonderful to know all things about all subjects, this is impossible. The dilemma of the student, and too often the educator, is to sort through the plethora of information and glean what is considered to be the important principles. For the medical student, this is the information required for understanding the pathogen, its associated diseases, and the pathogenesis of these diseases. The challenge for the educator is to present this information clearly and unambiguously, which requires the use of a textbook that is comprehensive but not an encyclopedia of all known facts.

Once could question the wisdom of writing a textbook of medical microbiology. A large number already exist, and—with the continual expansion of our knowledge—most textbooks rapidly become outdated. However, we believed an unmet need remained. Our experience as educators had convinced us that the right textbook, one that contains a balanced and comprehensive presentation of *medically important* information, did not exist. There was a real need for a textbook that offers appropriate weight and balance to the four major areas of medical microbiology— bacteriology, mycology, parasitology, and virology— yet is concise in presentation, focusing at all times on what the student truly needs to know.

This textbook of medical microbiology has been prepared for students who are either in medical or dental training or at the advanced undergraduate level. The information was carefully selected to help students understand the unique as well as the fundamental properties of microorganisms, and their relationship to the diseases caused by these microorganisms. We have avoided constructing a textbook that attempts to present all aspects of cellular biology, genetics, and immunology, because these sciences have evolved from minor components taught within microbiology to separate disciplines with their own textbooks.

In addition to the careful selection of information for this text, we have tried to follow a consistent format. Microorganisms have been divided into the four major classes—bacteriology, mycology, parasitology, and virology; the classes subdivided into the major genera and groups; and the groups presented with appropriate discussion of their physiology and structure, epidemiology, associated clinical syndromes, laboratory diagnosis, and treatment, prevention, and control of disease. We have supplemented this presentation with numerous outlines, tables, and illustrations. We believe this approach helps orient the student, emphasizes important ideas, and graphically depicts otherwise difficult concepts.

It is difficult for one individual to describe effectively all aspects of medical microbiology. Although we each have our areas of specialization, we also recognize our limitations. It was tempting to prepare a multi-authored text, thus sharing the burden of writing. However, we believe such a text would suffer from a lack of internal consistency, which would confuse the student. Thus we avoided this temptation except for those subjects for which expertise clearly exists elsewhere, and we supplemented our presentations with the advice and review of our knowledgeable colleagues. In particular we are indebted to the considerable assistance of Henry Isenberg (author of Chapters 2 and 3), James Smith (Chapter 4), Ella Swierkosz (Chapters 25 and 26), Jon Suzuki (Chapter 27), Eric Spitzer (contributions in mycology), and Tom Smith, Ken Rosenthal, and Tom Thomson (contributions in virology). In addition, no book would be possible without the patient assistance and appropriate prodding of a dedicated editorial staff. This book may never have come to fruition without the help of Stephanie Bircher Manning, Elaine Steinborn, and Anne

Gunter—no amount of thanks could compensate their tireless efforts. We would also like to thank Don O'Connor for transforming our ideas from concepts to artistic reality, and John Rogers for his meticulous attention to detail during the final stages of this publication.

Finally, we welcome comments from students and educators. Were we successful in clearly presenting the information relevant to an understanding of medicine? Ultimately, the success of this textbook will be measured by the understanding and appreciation of medical microbiology that it stimulates in the student. If we have been successful in expanding the student's knowledge of microorganisms and their pathogenic potential, then we have accomplished our goals.

Patrick R. Murray
W. Lawrence Drew
George S. Kobayashi
John H. Thompson, Jr.

Contents

BACTERIOLOGY

PART **I**

CHAPTER 1

Classification of Microorganisms

At one time the living world was subdivided into plants and animals. However, with the development of microscopes, the existence of organisms invisible to the unaided eye was discovered. It was rapidly appreciated that these microorganisms were neither plants nor animals, and in 1866 Haekel proposed a new kingdom, the **Protista,** that encompassed bacteria, fungi, protozoa, and algae.

The most numerous group of the Protista are the bacteria. Their taxonomy is complex and subject to frequent changes as we expand our knowledge of their physiologic and genetic properties. Despite this problem, bacteria can be subdivided by their staining properties with the Gram stain and by their cell morphology. Tables 1-1 to 1-3 summarize the current taxonomy classification of the medically important bacteria.

Another group of organisms was soon discovered—the viruses. Unlike members of the plant, animal, and Protista kingdoms, viruses are obligate intracellular parasites, unable to survive or replicate independently. Viruses have a simple structure, with their genetic information encoded in either DNA or RNA (but not both), packaged in a protein shell (capsid), and—for some viruses—enclosed in a membrane or envelope. Viral particles are generally very small, with individual particles observed only by electron microscopy. However, some viruses (e.g., *Poxvirus*) can approach the size of a small bacterium. A detailed discussion of viral classification and physiologic properties is presented in Chapters 38 and 39.

With refinements in our ability to study the biology of the Protista, it was soon recognized that these microorganisms could be subdivided into two groups based on cell structure: **eucaryotic** cells and **procaryotic** cells (Box 1-1).

EUCARYOTIC CELL STRUCTURE

Eucaryotic cells are structurally more complex than procaryotic cells (Table 1-4), containing a variety of membrane-enclosed organelles (Figure 1-1). A more detailed discussion of the structure of eucaryotic cells is presented in the introductory chapters for

Table 1-1	Medically Important Gram-Positive Bacteria	
Shape of Cell	**Family**	**Genus**
Cocci	Micrococcaceae	*Staphylococcus*
		Micrococcus
	Streptococcaceae	*Streptococcus*
		Enterococcus
	Miscellaneous	*Peptostreptococcus*
		Peptococcus
		Aerococcus
Bacilli	Bacillaceae	*Bacillus*
		Clostridium
	Propionibacteriaceae	*Propionibacterium*
		Eubacterium
		Corynebacterium
	Lactobacillaceae	*Lactobacillus*
	Uncertain affiliation	*Listeria*
		Erysipelothrix
	Actinomycetaceae	*Actinomyces*
		Bifidobacterium
	Nocardiaceae	*Nocardia*
	Mycobacteriaceae	*Mycobacteria*

Table 1-2 Medically Important Gram-Negative Bacteria

Shape of Cell	Family	Genus
Cocci	Neisseriaceae	*Neisseria*
		Branhamella
		Acinetobacter
		Kingella
		Moraxella
	Veillonellaceae	*Veillonella*
Bacilli	Enterobacteriaceae	*Escherichia*
		Edwardsiella
		Citrobacter
		Salmonella
		Shigella
		Klebsiella
		Enterobacter
		Serratia
		Proteus
		Providencia
		Morganella
		Yersinia
	Vibrionaceae	*Vibrio*
		Aeromonas
		Plesiomonas
	Spirillaceae	*Campylobacter*
	Pseudomonadaceae	*Pseudomonas*
		Xanthomonas
	Pasteurellaceae	*Pasteurella*
		Haemophilus
		Actinobacillus
	Miscellaneous	*Brucella*
		Bordetella
		Francisella
		Cardiobacterium
		Eikenella
		Flavobacterium
		Calymmatobacterium
		Streptobacillus
		Spririllum
		Cat scratch bacillus
	Legionellaceae	*Legionella*
	Bacteroidaceae	*Bacteroides*
		Fusobacterium

Table 1-3 Medically Important Miscellaneous Bacteria

Family or Order	Genus
Spirochetales	*Treponema*
	Borrelia
	Leptospira
Chlamydiaceae	*Chlamydia*
Mycoplasmataceae	*Mycoplasma*
	Ureaplasma
Rickettsiaceae	*Rickettsia*
	Coxiella
	Rochalimaea
	Ehrlichia

tein (histones). The chromosomes in turn are surrounded by a two-layer membrane, of which the outer membrane is contiguous with the endoplasmic reticulum.

Endoplasmic Reticulum

The endoplasmic reticulum extends throughout the cell cytoplasm and is subdivided into two types: rough and smooth. The rough endoplasmic reticulum is covered with ribosomes, which are used for protein synthesis. Transport of proteins through the cell cytoplasm is inside the channels in the endoplasmic reticulum. One specialized structure of the smooth endoplasmic reticulum is the **Golgi complex.** Proteins from the rough endoplasmic reticulum migrate to the Golgi complex, where they are packaged in vesicles. The vesicles are then transported to the cell surface, where they fuse with the cell membrane and release the proteins.

mycology (Chapter 28) and parasitology (Chapter 32). This section is intended to summarize the most important structures found in eucaryotic cells and their function.

Nucleus

The genetic information of the cell, DNA, is organized into multiple chromosomes covered with pro-

Box 1-1 Classification of Protista of Medical Interest

Eucaryotic Organisms

Algae
Protozoa
Fungi

Procaryotic Organisms

Bacteria

Table 1-4 Comparison of Eucaryotic and Procaryotic Cells

	Eucaryote	Procaryote
Nuclear structure		
DNA	Multiple chromosomes associated with histone proteins	Single, naked, circular chromosome
Nuclear membrane	Present	Absent
Cytoplasmic structure		
Endoplasmic reticulum	Present	Absent
Mitochondria	Present	Absent
Ribosomes	80S	70S
Lysosomes	Present	Absent
Plasma membrane	Present; contains sterols	Present; no sterols except in *Mycoplasma*
Cell wall	Absent or composed of cellulose or chitin	Complex structure with peptidoglycan layer, protein, and lipids
Capsule	Absent	Frequently present

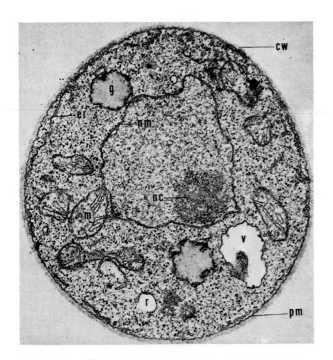

FIGURE 1-1 Electron micrograph of eucaryotic cell, the fungus *Cryptococcus neoformans*. Note the complexity of this cell, with a well-defined nucleus containing a nucleolus *(nc)* and bounded by a nuclear membrane *(nm)*. The cytoplasm contains ribosomes *(r)*, mitochondria *(m)*, vacuoles *(v)*, storage granules *(g)*, and endoplasmic reticulum *(er)*. The cell is bounded by a plasma membrane *(pm)* and a cell wall *(cw)*. (From Emmons CW et al: Medical mycology, ed 2, Philadelphia, 1970, Lea & Febiger.)

Mitochondria

The membrane-enclosed mitochondria are reminiscent of bacteria—they contain their own DNA and synthetic machinery and are capable of self-replication. Membranes in the mitochondria are the site of the respiratory electron transport system, the primary source of energy in the cell.

Lysosomes

Hydrolytic enzymes for the degradation of macromolecules and microorganisms are present in these membrane-enclosed structures.

Plasma Membrane

The plasma membrane, a lipoprotein structure, encloses the cell cytoplasm and regulates transport of macromolecules into and out of the cell.

Cell Wall

The cell wall forms rigid outer barrier that, when present in eucaryotic cells, is most commonly composed of polysaccharides such as cellulose or chitin.

PROCARYOTIC CELL STRUCTURE

Procaryotic cells are smaller and generally less complex than eucaryotic organisms (Figure 1-2). This section is intended to provide an overview of the most

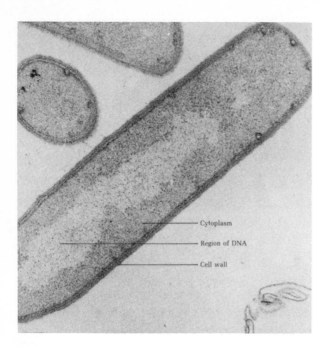

Cytoplasm

Region of DNA

Cell wall

FIGURE 1-2   Electron micrograph of a procaryotic cell, a *Bacillus* species. Note the rather simple structure in comparison with eukaryotic cells. The DNA is not organized in a nucleus, and the endoplasmic reticulum system is absent. However, the complexity of the procaryotic cell wall is obvious at even this magnification. (From Volk WA, Benjamin DC, Kadner RJ, and Parsons JT: Essentials of medical microbiology, ed 3, Philadelphia, 1986, JB Lippincott.)

important structures present in procaryotic cells, the physiologic and genetic properties of bacteria are discussed in Chapters 2 and 3.

Nucleus (Nuclear or Chromatin Body)

In contrast with eucaryotic cells, the genetic material of bacteria and other procaryotic cells is diffuse, organized into a single, naked, circular chromosome. A nuclear membrane and mitotic apparatus to facilitate separation of the chromosome during replication are not present. The chromosome is attached to a mesosome embedded in the cytoplasmic membrane. This structure is important for the replication of the chromosome.

Cell Membrane

The cell membrane is composed of phospholipids and proteins and, in contrast with eucaryotic cells,

does not contain sterols (except for *Mycoplasma*). Because procaryotic cells lack mitochondria and a complex membranous network like the endoplasmic reticulum within the cell cytoplasm, the electron transport enzymes are located in the cell membrane. The membrane also serves as an osmotic barrier for the cell, contains transport systems for solutes, and regulates transport of cell products to the extracellular environment.

Cell Wall

The cell wall in procaryotic cells is extremely complex. This rigid structure protects the cell from rupture caused by the high osmotic pressure inside the bacterial cell. Additionally, the cell wall is the site of many of the antigenic determinants that characterize and differentiate microorganisms. Endotoxin activity associated with certain groups of bacteria is also associated with the cell wall.

Bacteria have historically been subdivided by their reactions with the Gram stain. Although both gram-positive and gram-negative bacteria have cell walls, their differential staining properties are in large part attributed to the structure of the cell wall (Figure 1-3).

The basic structure of the cell wall of gram-positive bacteria is a thick (15 to 80 nm) peptidoglycan layer composed of chains of alternating subunits of *N*-acetylglucosamine and *N*-acetylmuramic acid. The rigidity of the structure is achieved by cross-linking tetrapeptide chains attached to the *N*-acetylmuramic acid subunit with oligopeptide bridges. Species-specific cell wall antigens are defined by the protein chains and bridges. Unique to many gram-negative and some gram-positive bacteria is the presence of diaminopimelic acid (a precursor of L-lysine) in some tetrapeptide chains. All gram-positive cell walls also contain teichoic acid bound to the cytoplasmic membrane; most gram-positive cells also have teichoic acids bound to the peptidoglycan layer. Additional antigens may be bound to the cell wall.

The structure of gram-negative cell walls is more complex (Figures 1-3 and 1-4). The **peptidoglycan layer** is thinner, only 1 to 2 nm. Outside the peptidoglycan layer is a phospholipid outer membrane (absent in gram-positive bacteria). The area between the outer membrane and the cytoplasmic membrane is called the periplasmic space. The outer membrane prevents loss of periplasmic proteins and forms a pro-

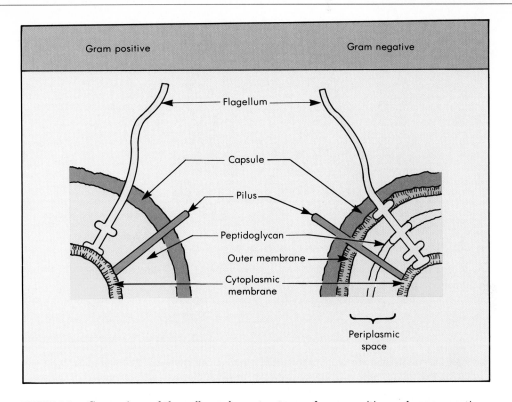

FIGURE 1-3   Comparison of the cell envelope structures of gram-positive and gram-negative bacteria. Gram-positive bacteria have a relatively simple cell wall, with a thick peptidoglycan layer surrounded by a polysaccharide capsule. The cell envelope of gram-negative bacteria is more complex, with a thin peptidoglycan layer sandwiched between two membranes. The space between the membranes is called the periplasmic space. (Redrawn from Ingraham JL, Maaioe O, and Neidhardt FC: Growth of the bacterial cell, Sunderland, Mass., 1983, Sinauer Associates, Inc.)

tective barrier preventing exposure of the bacteria to hydrolytic enzymes and toxic substances such as bile in the gastrointestinal tract. Membrane proteins **(porins)** are also present in the outer membrane and serve to regulate transport through membrane pores. This regulatory process is in part responsible for the resistance of gram-negative bacteria to antibiotics. A lipoprotein layer is present in the gram-negative cell wall that stabilizes the outer membrane and cross-links it to diaminopimelic acids in the peptidoglycan layer.

Embedded in the outside layer of the outer membrane is **lipopolysaccharide (LPS).** This molecule is the most significant structure in gram-negative bacteria. The lipid component (lipid A; **endotoxin**) in LPS is responsible for the toxic properties of this group of bacteria. The polysaccharide component consists of a core common to all gram-negative bacilli

and a variable terminal segment that is exposed on the outer surface of the bacteria and is the major surface antigen. The variable polysaccharide has been exploited for the serologic differentiation of bacterial isolates, and the genetic material that codes for the common core polysaccharide has been used as a target for molecular probes aimed at detecting gram-negative bacteria in clinical specimens.

Capsule

Many bacteria are enclosed with a polysaccharide covering. This can either be organized in a well-defined capsule (e.g., *Streptococcus pneumoniae, Klebsiella pneumoniae*) or in a diffuse layer surrounding the cell, commonly referred to as a slime layer (*Staphylococcus epidermidis*). This polysaccharide layer can be important in preventing bacterial phago-

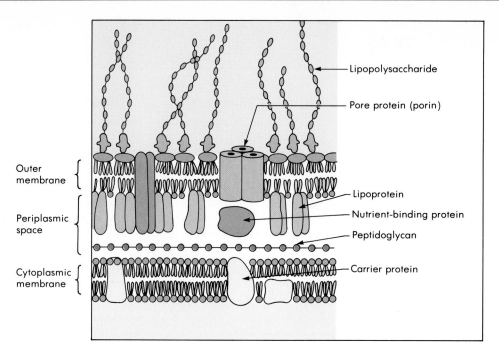

FIGURE 1-4   Molecular structure of the gram-negative cell envelope. The innermost layer surrounding the cell cytoplasm is the cytoplasmic membrane, which has embedded in the membrane carrier proteins to facilitate transport of solutes through the membrane barrier. Outside the cytoplasmic membrane is a thin peptidoglycan layer and an outer membrane. The area between the cytoplasmic membrane and the outer membrane is referred to as the periplasmic space and is filled with hydrated peptidoglycan and binding proteins for nutrients. Within the outer membrane additional porin proteins and other membrane proteins are present. A lipoprotein layer binds the outer membrane to the peptidoglycan layer. Finally, a complex lipopolysaccharide is attached to the outer layer of the outer membrane. (Redrawn from Lugtenberg B and Van Alphen LV: Biochim Biophys Acta 51:737, 1983.)

cytosis and in aiding the adherence of bacteria to tissues (e.g., a heart valve) or artificial devices (e.g., intravenous catheters).

Pili and Flagella

The pili (fimbriae) are short, hairlike structures that can aid in adherence of bacteria to host target cells or facilitate exchange of DNA during bacterial conjugation. The flagella are longer proteins important in bacterial motility. The number and distribution of flagella can be important in the taxonomic classification of isolates. Organisms with a single flagellum at one or both ends are called **monotrichous;** organisms with many flagella at one or both ends, **lophotrichous;** and those with flagella distributed over the entire bacterial surface, **peritrichous.**

Spores

Spore formation is a means by which some organisms are able to survive extremely harsh environmental conditions. The genetic material of the cell is concentrated and then surrounded by a protective coat, rendering the cell impervious to desiccation, heat, and many chemical agents. The bacterium in this state is metabolically inert and can remain stable for months to years. When exposed to favorable conditions, germination can occur, with the production of a single cell that subsequently can undergo normal replication. It should be obvious that the complete eradication of disease caused by spore-forming organisms is difficult or impossible. The two major groups of organisms that form spores are the aerobic genus *Bacillus* and the anaerobic genus *Clostridium*.

MICROBIAL INTERACTIONS

Medical microbiology is the study of the interactions between animals (primarily humans) and microorganisms such as viruses, bacteria, fungi, and protozoa. Although the primary interest is in diseases precipitated by these interactions, it must also be appreciated that microorganisms play a critical role in human survival. The normal population of endogenous organisms participates in the metabolism of food products, provides essential growth factors, protects against infection with highly virulent microorganisms, and stimulates and refines the immune response. In the absence of these organisms, life as we know it would be impossible.

The microbial flora in and on the human body is in a continual state of flux determined by such varied factors as age, diet, and health. It is important to recognize that changes in our physical well-being can drastically disrupt the delicate balance that is maintained among the heterogeneous organisms that coexist within us. For example, hospitalization can lead to replacement of normally avirulent organisms in the oropharynx with potentially invasive gram-negative bacilli. Although it is recognized that respiratory assistance devices or the indiscriminant use of antibiotics can precipitate this alteration, the complete explanation is not known.

Two important points must be emphasized: (1) care should be taken to maintain the normal balance of microbes, and (2) an important distinction exists between colonization (also called infection) with a pathogenic organism and disease. Indeed, most human diseases are caused by organisms that are a normal part of the human microbial flora. Diseases caused by these organisms develop only when the organisms enter a site where they are not normally found and the defensive mechanisms of the human body are unable to contain them. For example, *Streptococcus pneumoniae, Staphylococcus aureus,* and many gram-negative bacilli can be found as part of the normal oropharyngeal flora. However, when these organisms are introduced into the lower respiratory tract and the body is unable to contain them, bronchopulmonary disease develops.

An understanding of medical microbiology requires knowledge not only of the different classes of microbes but also of their propensity for causing disease. Some organisms will not cause disease except in immunocompromised patients with conditions that favor the growth of the organism. An example of infection caused by one of these opportunistic pathogens is *Staphylococcus epidermidis* infection of an intravenous catheter in a hospitalized patient. At the other end of the spectrum, some organisms are always associated with disease (e.g., *Mycobacterium tuberculosis, Shigella, Neisseria gonorrhoeae*). Between these two extremes are the majority of organisms associated with disease, the facultative pathogens (e.g., *Staphylococcus aureus, Escherichia coli*). The factors responsible for bacterial virulence will be discussed in subsequent chapters.

CHAPTER 2

Bacterial Metabolism

A knowledge of microbial physiology, biochemistry, and nutrition is required to understand the effects of the microbial world on human health. In addition, it must be remembered that our limited knowledge of microorganisms has been obtained with pure cultures of various representatives, raised under unnatural conditions on convenient, almost universal nutrients, and under stable conditions of temperature and atmosphere. Conclusions about microorganisms in natural situations are therefore interpolations of these findings.

Microorganisms are composed of the same carbohydrates, amino acids, and nucleic acids that make up plants and animals—even the human body. Furthermore, the pathways that lead to the conversion of nutrients into usable compounds and the finely tuned efforts to use energy conservatively to sustain life exist in all living creatures, including microorganisms. However, differences in degree do exist. Microorganisms can use substrates that no animal or plant can use for their carbon energy and nitrogen requirements. They also tolerate extremes of temperature as long as water is present in liquid form, and some can exist at extremes of pH that no other life form can tolerate.

ENERGY AND NUTRIENT REQUIREMENTS

One way of defining life is to consider the energy sources required for the synthesis of essential compounds by a living organism. Thus the metabolism of an organism must be interpreted from two distinct yet interrelated aspects—energy and nutrient requirements. Microorganisms, more so than all other forms of life, are profoundly affected by these circumstances. Their ability to handle changes in the environment depends on immediate physiologic responses because they lack the specialized cells present in higher forms of life that preserve homeostasis. The immediate physiologic capabilities are expressed by the vast array of energy sources that microorganisms can use and by their biosynthetic potential, an attribute not necessarily expressed when an organism is cultivated in a chemically undefined broth in the routine laboratory.

Microbial energy requirements may be classified into three major categories: chemotrophy, phototrophy, and paratrophy (Table 2-1). **Chemotrophy** is the derivation of biologic energy from reactions taking place without light. In bacteria two types of chemotrophy prevail: chemoorganotrophy and chemolithotrophy. **Chemoorganotrophy,** shared with all animal forms, indicates that the energy is derived from the oxidation or fermentation of exogenous organic compounds. In addition, some bacteria can use energy liberated by the oxidation of exogenous inorganic compounds **(chemolithotrophy),** a process in which bacteria can produce all their constituents by reducing carbon dioxide with the energy liberated during inorganic oxidation.

In **phototrophy,** energy is provided by photochemical reactions. The photoorganotrophic process, limited to certain groups of anaerobic prokaryotes, involves exogenous organic hydrogen donors. Photolithotrophic organisms depend on inorganic hydrogen donors and include the photolysis of water.

In **paratrophy,** energy must be provided by the

Table 2-1 Microbial Energy Requirements

	Carbon Source	Energy Source
Chemothroph		
Chemoorganotroph	Organic compounds	Oxidation-reduction reactions
Chemolithotroph	CO_2	Oxidation-reduction reactions
Phototroph		
Photoorganotroph	Organic compounds	Light
Photolithotroph	CO_2	Light
Paratroph	Host cell	Host cell

host cell, whether animal, plant, or microorganism. The various viruses are in this category.

The capacity of an organism to synthesize essential metabolites or, conversely, the need of an organism to be supplied with already synthesized molecules that it requires but cannot manufacture, may also be divided into four major groups: heterotrophy, autotrophy, mesotrophy, and hypotrophy (Box 2-1). **Heterotrophy** is the need of any organism for an exogenous supply of one or more metabolites essential for its survival; examples of metabolites are vitamins and growth factors. **Autotrophy** implies that the organism can synthesize all its essential metabolites. In the obligatively autotrophic organisms, all metabolic needs are served by the ability of the organism to reduce oxidized inorganic nutrients; **mesotrophic** organisms, on the other hand, require one or more reduced inorganic nutrients. Finally, **hypotrophy** describes the intracellular parasite's need for the host's metabolic processes to meet its nutritional requirements.

The most common microorganisms that can tolerate the intimate human biosphere are chemoorganotrophic heterotrophs. However, the array of metabolic capabilities displayed by all colonizing and potentially disease-involved microorganisms is considerable. Therefore the prevention or treatment of infectious disease requires an understanding of the factors that control microbial proliferation.

BACTERIAL GROWTH

Growth (i.e., the increase in size and division of any microorganism or cell) has been the main indicator of microbial viability. In fact, we do not know if an organism is alive unless it multiplies in a laboratory medium that provides all the material essential for the manufacture of cell structures. Many bacteria but by no means all have been practically the only microorganisms studied intensively with regard to growth because of difficulties unique to other types of organisms:

1. Viruses require suitable host cell systems. This imposes restrictions related to the laboratory strain of tissue culture cells, embryonated eggs, or animal model infected by the particular virus.
2. The complexity of fungal growth in differentiation has prevented the formulation of universal conclusions.
3. Most protozoa, especially those significant in human disease, have not been adapted to laboratory cultivation, and we lack information concerning many aspects of their growth cycle.

Bacteria are known to multiply by **binary fission,** the division of a single bacterium into two daughter bacteria, in a suitable environment. The time required to accomplish this division, or doubling, is the **generation time.** Growth of many bacteria is enhanced by a nutrient-rich laboratory medium to obtain a generation time as short as possible and thereby rapidly gain pertinent information for diagnosis and treatment of patients.

Box 2-1 Microbial Requirements for Metabolites

Heterotrophy	Mesotrophy
Autotrophy	Hypotrophy

Balanced Growth

When bacteria are inoculated into a nutrient broth (infusion medium), each grows in size and eventually divides into two daughter cells. The growth and division of these daughter bacteria continue until some essential nutrient is exhausted, the amount of oxygen is diminished, the hydrogen ion concentration has increased (pH has dropped) to interfering levels, or toxic end products have accumulated. This is **balanced growth,** because all activities of the population occur from cell division to cell division (i.e., all constituents are duplicated within that period). For fast-growing bacteria in ordinary laboratory media this period is typically 30 minutes or less. Estimation of the time required to reach this stage (generation time, or doubling time) can be made in ways ranging from actual bacterial counts at specific time intervals to the determination of the rate of increase of a specific enzymatic activity of the bacterium. During balanced growth, the mean generation time can be established as the time required for doubling the number, mass, or any other constituent of the bacterial population. All of these variables are related to the **growth rate constant,** the factor that expresses the average number of times the cells will divide in a given time period.

The rate of synthesizing cells constituents and the accumulation of end products mirror the complexity of the nutrient environment. In the usual laboratory media many different compounds are available for transport and immediate incorporation into the cell mass. The demands on the biosynthetic capability of the bacterium are therefore minimal (e.g., there is no need to synthesize amino acids, purines, pyrimidines, vitamins). The result is rapid growth and multiplication. When the same bacterium is transferred to a chemically defined medium containing minerals and a single reduced carbon and energy source, much slower multiplication occurs during balanced growth of those organisms that need no additional growth-stimulating factors. A further increase in the generation time can be achieved by substituting carbon and energy sources that differ from the commonly used glucose source and require the production of special transport systems into the cell.

This difference in nutrient milieu is also reflected in the number of bacteria or in the total bacterial mass each type of medium can support. If we permit an organism to proliferate in a given medium until all growth ceases—usually 18 to 24 hours—we find that the nutrient-rich laboratory broth may have allowed proliferation to a density of 10^{10} organisms per milliliter corresponding to several hundred micrograms of bacterial dry weight. Less complete diets can reduce that number by several magnitudes.

Studies of bacterial growth on various media have revealed that identifying characteristics are not necessarily fixed. The following have been observed:

1. **Size.** The size of the individual bacterium is not characteristic of the species. Great plasticity has been observed even under controlled laboratory conditions.
2. **Shape.** Long filaments of bacteria can be produced by agents such as ultraviolet light and cell wall active antibiotic (e.g., penicillin) applied at subinhibitory concentrations. Even other elements of shape that are thought to be characteristic of certain bacteria can be grossly distorted under such circumstances.
3. **Morphology.** Bacteria that grow faster are large and contain large quantities of RNA, whereas those that grow more slowly contain greater amounts of plasma membrane and cell wall per bacterium. Thus the synthesis of the bacterial envelope may limit fast-growing organisms and account for the larger size and decreased surface-to-volume ratio.

Only the rate of protein synthesis reflecting ribosomal RNA control appears to be related to the rate of bacterial proliferation under the conditions of balanced growth.

Unbalanced Growth

It is unlikely that balanced growth occurs often under the constantly changing, natural environmental conditions. Simulation of these conditions in the laboratory has shown that, when nutrients are added to growing cultures, the bacteria immediately shift to a faster mode of growth. Depleting an environment of nutrients (shift down), results in a slowing of bacterial growth. Despite the relative lack of RNA in slow-growing bacteria, RNA increases almost immediately during the shift to faster growth, followed by an increase in protein synthesis and finally an increase in DNA. Increased rates of multiplication occur as soon as the cells reach the dimensions characteristic of fast growth.

Unbalanced bacterial growth is especially significant in situations that call for control of microorganisms in human environments and during therapy, (e.g., application of disinfectants, antiseptics, antibiotics). Studies of bacteria undergoing physically or chemically induced unbalanced growth indicate that not all control efforts result in bacterial death. For example, other effects may be seen:

1. Some biosynthetic pathways proceed even though others are curtailed or inhibited by these measures.
2. DNA synthesis in bacteria occurs in a cycle that is completed in each individual and may occur even in the absence of RNA or protein synthesis.
3. Agents may interfere with RNA synthesis and produce a gradual decline in protein synthesis proportional to the decay of messenger RNA.
4. Protein synthesis may be inhibited by bacteriostatic antibiotics, leading eventually to a halt in RNA production.

Growth Cycles

After 18 to 24 hours of cultivation in the laboratory under ideal conditions of nutrition, oxygen availability, and buffering, etc., a considerable biomass of bacteria accumulates. Transfer of an extremely small aliquot of such a culture to an identical fresh medium causes bacterial multiplication to resume after a **lag phase;** then **exponential growth** occurs for several hours, and finally a **stationary growth** phase ensues (Figure 2-1). The bacterial density at the various phases of this growth cycle are usually estimated by nephelometric or turbidimetric measurements that reflect the number of organisms in the suspension. It should be obvious that these measurements do not distinguish between the number of viable and nonviable organisms present. During the phase of exponential growth, there is a good correlation between bacterial density and viable particles. However, this relationship does not exist during the stationary growth phase when many of the measured particles are nonviable organisms.

The following differences are also noted during the different phases of growth.

Lag Phase When bacteria from a stationary phase of growth are inoculated into a fresh, nutritionally enriched medium, a lag phase of growth occurs, during which the metabolic activity of the organism is increased in preparation for bacterial division. An increase in bacterial RNA and protein synthesis is

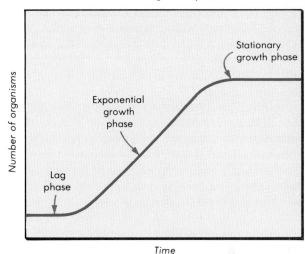

Bacterial growth cycle

FIGURE 2-1   Bacterial growth cycle.

noted, as is an increase in the size of the bacterial particles. Although it is expected that this shift in metabolic activity would require a period of time, there is no explanation for the very prolonged lag phase experienced with some microorganisms.

Exponential Growth Phase Exponential growth, the next step in bacterial proliferation, represents the peak of growth activity in a culture medium. The bacterial metabolism functions as a well-integrated system as long as nutrients remain in adequate concentrations and end products or toxins allow for the smooth performance of all functions. During this stage bacteria are especially susceptible to agents such as antibiotics, probably because the physiologic efficiency provides more frequent opportunities for interaction with the antibiotic and the bacterial envelope barriers are more easily traversed.

Stationary Growth Phase Depending on the bacterial species and the specific nutrient environment, the stationary phase gradually ensues. As essential nutrients—especially carbon and nitrogen sources—are depleted, growth (as indicated by increased size) is altered. Multiplication and DNA synthesis, on the other hand, continue for some time. The individual bacteria are quite small at this juncture.

In the undefined nutrient media most commonly used in the laboratory, the decline in available nutrients is even more gradual, suggesting a sequential use of medium constituents and a possible selection of those variants within the population that have some tolerance for less favorable conditions. In addition, some cannibalistic variants in the population may be able to use the dead remnants of their own kind for nutrients.

Although many aspects of bacterial growth require further study and the events in a single bacterium are almost entirely unknown, the growth and proliferation that are observed reflect a well-integrated system of biochemical activities. Thousands of reactions may be involved in achieving growth and multiplication, but a sense of orderly sequence emerges.

BACTERIAL METABOLISM

If we imagine an idealized bacterium growing on simple mineral fare and glucose, we can conceive how these compounds enter the bacterial interior and undergo processing under the regulation of three broad categories of enzymes:

1. Enzymes that break glucose into usable units
2. Enzymes that construct cellular building blocks from these units
3. Enzymes that arrange the building blocks into bacterial organelles

The first group of enzymes obtains energy from the conversion of glucose into smaller components that serve as building blocks for further activities. The next group of enzymes constructs these building blocks (e.g., amino acids, nucleotides, amino sugars, lipids). Although the carbon skeleton is derived from glucose, other elements, such as nitrogen, sulfur, phosphorus and essential trace metals, originate from an external mineral pool and must be incorporated into appropriate compounds. This type of biosynthesis uses the energy released in the degradation of glucose. The third group of enzymes then arranges the various building blocks into the organelles of the bacterium by constructing macromolecules according to the instructions found in the chromosomal and, when present, plasmid DNA. When enough macromolecules have been prepared to serve two daughter cells, division occurs.

In view of the competition for nutrients that must exist under natural conditions, an exacting regulatory mechanism that controls enzyme quantity, as well as rate of enzyme function, must exist. To appreciate how microorganisms can achieve this end in minimal time, an overview of microbial metabolism follows.

Intermediate Metabolism

Microorganisms can degrade almost any organic compound and convert its carbon skeleton into usable constituents. For example, *Pseudomonas putida* can grow on octane, camphor, and naphthalene as the sole source of carbon and energy. Each compound is attacked by enzymes that are not required in the decomposition of the other compounds. Other bacteria and certain fungi can use atmospheric nitrogen to meet their needs for protein synthesis, whereas a large number of microorganisms use ammonia or nitrate for this purpose. The specialized metabolism of the photosynthetic and chemolithoautotrophic bacteria, while of great significance in nature, is mentioned here only to complete the spectrum of activities microorganisms can perform.

Even with the variety of substrates attacked, the metabolic sequences that take place are common to all living forms. This unity of biochemistry applies not only to degradative or catabolic approaches, but also with even the biosynthetic pathways and the building of macromolecules. All living creatures share this common metabolic denominator and differ only in the number and complexity of genetic instructions.

Microbial metabolism can be approached from many perspectives. Microorganisms can use various carbon sources to synthesize their constituent parts and regulate their metabolic processes. They must therefore respond appropriately to the available starting materials. A bacterium may be capable of auxoautotrophy (multiplying in a mineral solution with a single reduced carbon source) but does not demonstrate this ability when it is grown on mixtures of organic carbon and nitrogen sources that are available in laboratory culture medium or infected tissues. Synthesis of the appropriate enzymes required for growth is conservative (i.e., only the enzymes required for utilization of available nutrients are produced). How microorganisms can sense, integrate, and control this information is discussed later.

The function of metabolism is to maintain and synthesize cell components, which consist of macromolecules. These cell components account for approximately 90% of the dry weight of microorganisms, leaving low molecular weight compounds such as amino acids, monosaccharides, coenzymes, nucleotides, and inorganic salts as the metabolic pool. This dynamic collection of low molecular weight compounds serves as the reservoir for the organism's major metabolic function—the synthesis of macromolecular machinery to impose order on all vital processes.

Each macromolecule is composed of characteristic constituents that have been assembled according to genetic instructions. However, the construction of each macromolecule requires that each subunit be activated or energized to take its proper place in the sequence. The required energy is ultimately derived from adenosine triphosphate (ATP), often through

intermediate molecules, just as in higher forms of life. Macromolecules are assembled in accordance with the universal order of biology: The assemblage of the nucleic acids and proteins is template-directed; that of polysaccharides and lipids depends on enzyme specificities. Therefore metabolic activities must both provide the building blocks and also generate the energy to perform all tasks.

Many different substrates can serve as starting materials for the various bacteria. However, each special substrate feeds into the major pathways of metabolism. Glucose can be used by almost all microorganisms, and consideration of its use to provide

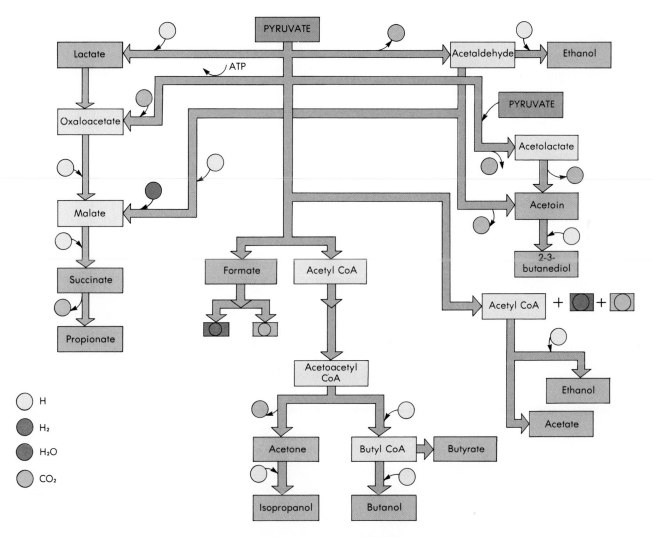

FIGURE 2-2   Diversity of pyruvate metabolism demonstrated by microorganisms. This diversity of metabolic products has been exploited in the clinical laboratory as phenotypic markers for the differential classification of microbial isolates.

materials and energy for biosynthesis is appropriate. Glucose metabolites also serve to support obligatively and facultatively anaerobic microorganisms by acting as the hydrogen receptor in a series of balanced oxidation-reduction reactions. The different approaches to the use of glucose result from the presence or absence of enzymes involved in specific reaction sequences. However, these approaches all share the enzymes that convert the triosephosphates to pyruvate, making this sequence the pivotal reaction for providing carbon skeletons and energy.

Microorganisms share with other living forms the Embden-Meyerhof-Paras glycolytic pathway, which converts glucose to pyruvate, and the pentose phosphate cycle. Not all microorganisms use these approaches, but many may employ these cascades singly or in combination to degrade glucose. Other pathways used include the Entner-Doudoroff pathway, the phosphoketolase pathway, the methylglyoxal bypass, and the tricarboxylic acid cycle.

The glycolytic pathway to pyruvate and pyruvate metabolism provides the basis for a number of microbial reactions (Figure 2-2). Homolactic acid bacteria such as streptococci and some lactobacilli can produce lactate from pyruvate, whereas heterofermentative bacteria such as other lactobacilli, Enterobacteriaceae, and clostridia produce a mixture of compounds in addition to lactic acid. Examples of these reactions include production of ethanol by diverse microorganisms such as yeasts, *Acetobacter, Zymomonas,* and *Erwinia*. Acetoin can be produced by *Klebsiella* and *Enterobacter*. Some yeasts can form acetoin directly from pyruvate and reduce it to butanediol. Acetyl-CoA and formate can be formed from pyruvate, with organisms such as *Escherichia coli* converting formate to hydrogen and carbon dioxide and other members of Enterobacteriaceae reducing acetyl CoA to acetate and/or ethanol. Acetyl CoA can also be converted to butyrate, butanol, or isopropanol by clostridia. Other anaerobic bacteria can also reduce lactate or pyruvate to succinate or propionate.

Microorganisms must possess mechanisms to carefully maintain available energy, but they do not have the elaborate mechanisms of higher forms of life. Microorganisms must be able to generate energy by turning glycolysis on and then turning the sequence off whenever excess ATP is available. Control is provided by phosphofructokinase and pyruvate kinase (Figure 2-3). These reactions differ from the usual

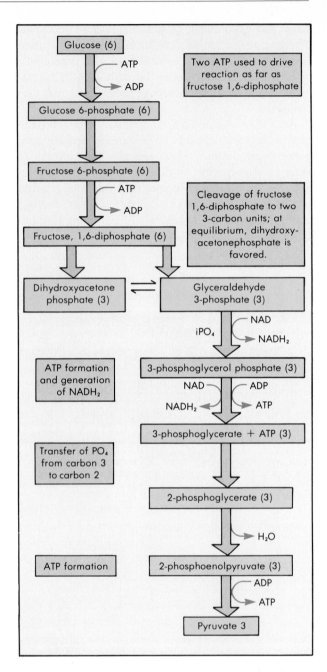

FIGURE 2-3   Embden-Meyerhof-Paras glycolytic pathway. Phosphofructokinase catalyzes the conversion of fructose-6-phosphate to 1,6-diphosphate. The product of this reaction activates pyruvate kinase, the enzyme that reduces 2-phosphoenolpyruvate to pyruvate with the generation of ATP. This reaction in turn inhibits phosphofructokinase activity. (Redrawn from Montgomery R et al: Biochemistry: a case-oriented approach, ed 4, St Louis, 1983, The CV Mosby Co.)

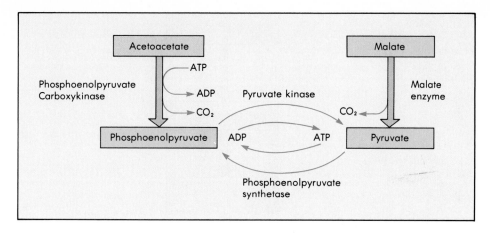

FIGURE 2-4   Enzymatic conversion of 4-carbon compounds to 3-carbon pyruvate.

energy-linked control that is encountered with catabolic processes not regulated by the end product and also the negative feedback effects that operate in biosynthesis. In this amphibolic (both anabolic and catabolic) pathway, precursors must be activated (e.g., the first or early intermediate metabolite stimulates the final event in the metabolic cascade).

An appreciable number of microorganisms capable of using three- or four-carbon compounds must be able to synthesize glucose from these sources. This is a formidable task because the Embden-Meyerhof-Parnas glycolytic pathway cannot be reversed readily. The high energy requirements of hexokinase, phosphofructokinase, and pyruvate kinase must be circumvented. Studies with mutants of *E. coli* indicate that two enzymes enable growth with four-carbon compounds (Figure 2-4). One is phosphoenolpyruvate carboxykinase, which catalyzes the conversion of acetoacetate to phosphoenolpyruvate and CO_2. The second enzyme, malate enzyme, directs the decarboxylation of malate to pyruvate. The pyruvate can then be activated by phosphoenolpyruvate synthetase, an enzyme induced during growth on the three-carbon compounds. Studies with bacterial mutants have indicated the presence of fructose diphosphatase, the enzyme that hydrolyzes the phosphate on carbon-1. The lack of this enzyme does not permit *E. coli* to grow on acetate, succinate, or glycerol unless hexoses are present. With these enzymes and the reversible sequences of the Embden-Meyerhof-Parnas pathway, bacteria can perform adequate gluconeogenesis.

The main thrust of the glycolytic Embden-Meyerhof-Parnas pathway from glucose to pyruvate is the provision of three-carbon skeletons, the essential starting point of all metabolic manipulations that result in macromolecules. Even in photosynthesis, 3-phosphoglycerate provides this apparently universal three-carbon starting point. Therefore, regardless of the carbon source configuration, all living creatures must possess the means to convert such a source to a three-carbon building block that can be manipulated properly and efficiently into its own structural integrity.

Pentose-Phosphate Cycle

The pentose-phosphate pathway of glucose utilization is present in many living organisms, including some microorganisms (Figure 2-5). The initial step of this cycle involves oxidative decarboxylation of glucose to a pentose (ribulose-5-phosphate), which is then anaerobically rearranged. The pathway provides carbon skeletons for the nucleotides, various carbohydrate moieties, and energy to generate ATP as follows:

1. It fulfills the need for NADPH and nucleotide synthesis. As with other aerobic organisms, bacteria can convert the pentoses in equilibrium with one another (e.g., ribose-5-phosphate, arabinose-5-phosphate, and xylulose-5-phosphate) into fructose-6-phosphate and glyceraldehyde 3-phosphate through the reversible action of transaldolases and transketolases. Glucose can be converted into

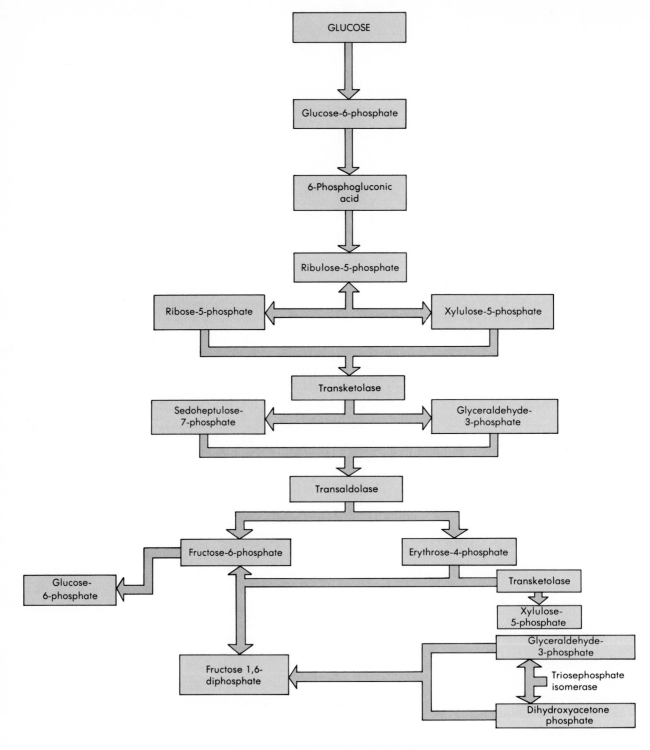

FIGURE 2-5   Pentose-phosphate cycle.

three carbon dioxide molecules, and glyceralde-hyde can enter the tricarboxylic acid cycle via pyruvate. This cycle is especially important in *Brucella abortus,* which lacks the enzymes of the gly-colytic pathway but can nevertheless metabolize glucose.

2. The pentose phosphate cycle also permits aerobic growth without any participation by the Krebs cycle enzymes. Glyceraldehyde 3-phosphate can be converted to dihydroxyacetone phosphate by triosephosphate isomerase. The product of this interaction can condense with glyceraldehyde 3-phosphate to yield fructose-1, 6-diphosphate by reversing aldolase. Fructose can then be reintro-duced into the pentose phosphate cascade after dephosphorylation and isomerization to glucose-6-phosphate.

3. This cycle is an important means whereby bacte-ria can use external pentoses as carbon and energy sources. After pentose phosphorylation, the compound enters the cycle and can be direct-ed into the various metabolic pathways, including the Embden-Meyerhof-Parnas cycle.

4. The pentose phosphate cycle is an efficient means of obtaining energy. Its overall reaction may be summarized as follows:

6 Glucose-6-phosphate + 12 NADP$^+$ + 6 H$_2$O →
5 Glucose-6-phosphate + 6 CO$_2$ + 12 NADPH +
12 H$^+$ + P$_i$ (inorganic) [P$_i$]

Since oxidation of NADPH yields 3 ATP molecules, the cycle produces 36 energy-rich phosphates per glucose molecule (one phosphate is expended in phosphorylation of glucose for a net yield of 35).

5. In addition to its role in energy metabolism, the NADPH generated is important in fatty acid syn-thesis and the diversion of many cycle intermedi-ates toward nucleotide synthesis.

Entner-Doudoroff Pathway

A number of gram-negative bacteria (e.g., some species of *Pseudomonas, Alcaligenes, Rhizopium, Thiobacillus, Xanthomonas,* and the gram-positive bacterium *Enterococcus faecalis* substitute the an-aerobic Entner-Doudoroff pathway for glycolysis (Fig-ure 2-6). This is a strictly microbial manipulation of glucose-6-phosphate and involves the production of 6-phosphogluconic acid. However, instead of the decarboxylation that occurs in the pentose phosphate cycle, a dehydratase removes water in the presence of ferrous iron (Fe^{2+}) and forms 2-keto-3-deoxy-6-

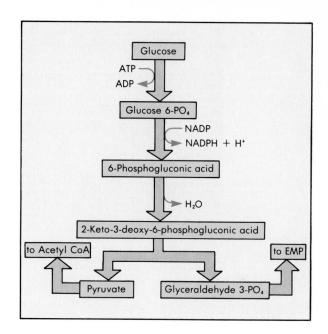

FIGURE 2-6   Entner-Doudoroff pathway. (Redrawn from Montgomery R et al: Biochemistry: a case-oriented approach, ed 4, St Louis, 1983, The CV Mosby Co.)

phosphogluconic acid. This compound is then split by an aldolase into glyceraldehyde 3-phosphate and pyruvate. These compounds can then be metabolized as usual.

Many bacteria use this alternative pathway not to degrade hexoses but to metabolize gluconate, man-nonate, glucuronate, and related compounds. *Pseu-domonas fluorescens* and *Pseudomonas putida,* which are significant in hospital-acquired infections and capable of flourishing in restricted nutrient con-centrations, contain glucose dehydrogenase, the enzyme that converts glucose to gluconic acid, in their membrane. They can carry out this process extracellularly, gaining an ecologic advantage over organisms forced to transport the hexose preferential-ly.

Phosphoketolase Pathway

The phosphoketolase pathway (Figure 2-7) is another strictly microbial means of glucose fermenta-tion that yields lactate, acetate or ethanol, and carbon dioxide. Pentoses may also be attacked via this mech-anism, and lactate and acetate are produced. The bacteria that must use this pathway lack the phos-phofructokinase, aldolase, and triosephosphate isom-

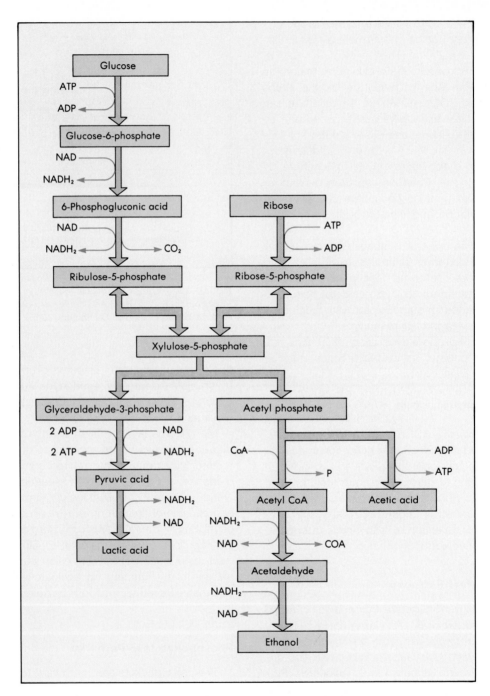

FIGURE 2-7   Phosphoketolase pathway.

erase of the glycolytic cycle. They produce their end products by converting glucose-6-phosphate to 6-phosphogluconic acid with reduced NAD rather than NADP. Another molecule of this nucleotide is reduced in the decarboxylation of the acid to xylulose-5-phosphate. A phosphoketolase in the presence of thiamine pyrophosphate, inorganic phosphate, magnesium, iron, and a thiol compound splits xylulose-5-phosphate into 3-phosphoglyceraldehyde and acetyl phosphate. The glyceraldehyde is converted to pyruvate and reduced to lactate. A hexose yields reduced NADH, two molecules in the decarboxylation to the pentose, and an additional two molecules in the reaction that leads from the glyceraldehyde to pyruvate. Two protons of NAD.H are consumed in reducing pyruvate to lactate, while four remain to participate in acetaldehyde and ethanol conversions.

Tricarboxylic Acid Cycle (TCA)

The TCA cycle (Figure 2-8) is the most important metabolic pathway present in all living forms capable of efficient oxygen consumption. It fulfills a dual function, providing energy and supplying intermediate substances for biosynthesis. If one considers acetyl-CoA and pyruvate together with this cycle, practically all of the significant basic biosynthetic processes (e.g., amino acid and fatty acid synthesis) originate here, especially when the role of acetyl-CoA in lipid degradation and synthesis is included. Thus, with respect to the amino acids (Figure 2-9), pyruvate gives rise to alanine, through α-ketoisovalerate to valine, and by way of α-ketoisocaproate to leucine. Oxaloacetate is the starting point for aspartic acid, which in bacteria is converted to the important bacterial constituent diaminopimelic acid. Although

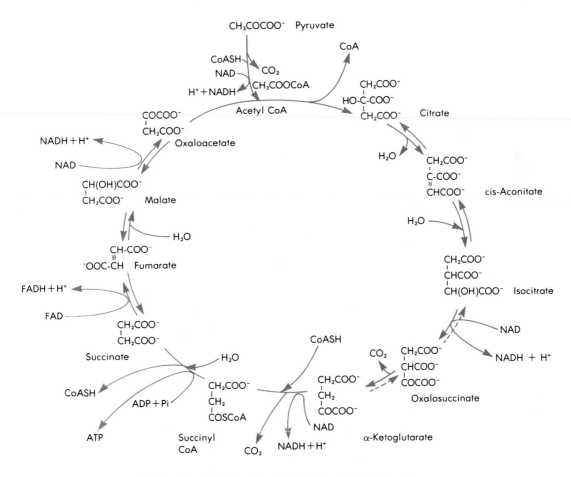

FIGURE 2-8   Tricarboxylic acid (TCA) cycle. (Redrawn from Montgomery R et al: Biochemistry: a case-oriented approach, ed 4, St Louis, 1983, The CV Mosby Co.)

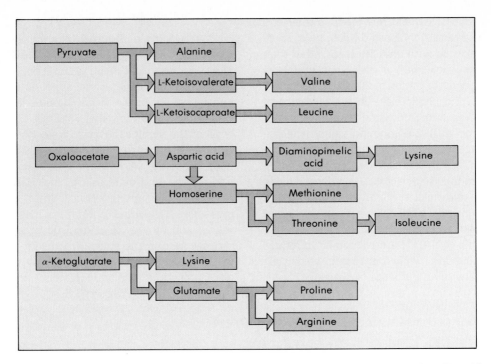

FIGURE 2-9   Examples of amino acids derived from intermediates of the tricarboxylic acid cycle.

present as such in most bacteria, this compound is converted by some organisms to lysine. Aspartic acid also serves as the precursor of homoserine, from which methionine and threonine are derived. Threonine itself can be converted to isoleucine. The α-ketoglutarate of the tricarboxylic acid cycle is the starting material for the fungal synthesis of lysine. Its most important derivative is glutamate, from which eventually proline and arginine are formed. To complete this picture of amino acid building block derivation, the various pentose phosphates provide histidine through imidazoleglycerophosphate. Serine (and through it glycine), cysteine, and cystine are synthesized from 3-phosphoglycerate. Serine reacting with indole glycerophosphate yields tryptophan. This amino acid can also be a derivative of chorismic acid, the condensation product of erythrose-4-phosphate and phosphoenolpyruvate. Chorismic acid, through the intermediate anthranilic acid, gives rise to indole glycerophosphate with which serine condenses. Chorismic acid, through prephenic acid, serves as the source of tyrosine and phenylalanine.

As all of this proceeds, the tricarboxylic acid cycle must provide energy to microorganisms depending on the physiologic demands of the moment. If *E. coli* is growing in a mineral medium with ammonium as its nitrogen source and glucose as the sole carbon source, the cycle provides building blocks while glycolysis furnishes energy as long as the exponential growth of the organism ensues. During the stationary phase of growth, the tricarboxylic acid cycle provides energy while glucose leads to the storage of glycogen in the bacterium.

The means of obtaining energy from the tricarboxylic acid cycle are also universal. Pyruvate, isocitrate, α-ketoglutarate, succinate, and malate are the substrates used to pass the energy contained in the carbon-to-carbon bond to liberate carbon dioxide and form water through oxidative phosphorylation. Classically, these steps involve the reduction of NAD or NADP, which is reoxidized by FAD. Reduced FAD is regenerated by the cytochrome chain, which through a series of reductions and oxidations of its central iron, finally reduces atmospheric oxygen to water. The complete oxidation of one glucose molecule yields 38 molecules of ATP, most of which are derived from the tricarboxylic acid cycle.

Many other oxidizing substrates are also metabolized through the cycle. As already indicated, fatty acids, degraded to acetyl-CoA, find entry in this manner; amino acids can reenter as energy sources through the same keto acids that served their synthesis; and intermediates of the cycle found in the environment are metabolized once they have gained entry into the organism.

Recycling Carbon

One important attribute of microorganisms is their ability to decompose most organic compounds, including those that occur naturally, as well as synthesized organic compounds. Advances in technology and medical science have profoundly affected microbial ecology, and microorganisms unknown a few years ago are now known to complicate the recovery of patients. Many can use unusual substrates.

Polymers of various composition are attacked by exoenzymes, which are excreted by bacteria and fungi. These include various carbohydrases that cleave starches, cellulose, and disaccharides. The latter may actually be transported into the organisms as such by specific permeases and metabolized there. Many microorganisms elaborate and excrete proteases. Some streptococci, staphylococci, certain members of Enterobacteriaceae, and many other organisms elaborate extracellular DNase and/or RNase.

In addition, special microbial enzymes attack certain amino acids. Thus serine and threonine form keto acids with their corresponding dehydratases, whereas the action of aspartase, present in many organisms, results in fumarate. In general, the conversion to keto acids can be enhanced by oxidation linked to the cytochrome system. Both L- and D-amino acid oxidases are found. As already shown, bacteria must possess this capability since D-amino acids, probably as a result of bacterial racemases, are incorporated into the peptidoglycan cell wall constituents. On the whole, these flavo-protein oxidases are nonspecific and capable of attacking as many as 10 different amino acids. They feed electrons directly into the cytochrome system.

Amino acids serve as significant carbon sources for anaerobic bacteria. Since these organisms lack the cytochrome sequence, oxidation by NADP-linked dehydrogenases constitutes the important metabolic sequence. Especially prevalent are alanine and glutamate dehydrogenases. Other anaerobic bacteria can ferment amino acids, which may result in the produc-

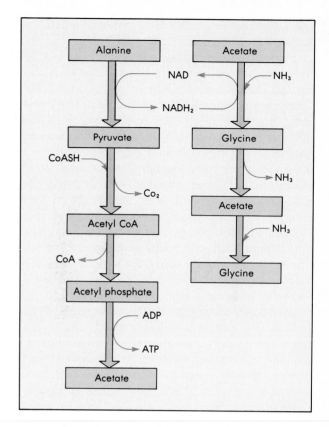

FIGURE 2-10   Stickland reactions.

tion of acetate, carbon dioxide, ammonia, and, in varying amounts, hydrogen gas. Thus alanine can be fermented via the acrylyl pathway, also useful in lactate fermentation, with acrylate as an intermediate substance. Other amino acids can also fulfill this function in anaerobic metabolism; it requires the production of special enzymes and coenzymes.

Mixtures of amino acids are fermented, especially by clostridia, via the Stickland reaction (Figure 2-10). One of the amino acids is oxidized while the second amino acid is reduced, resulting in the liberation of ammonia and carbon dioxide.

Microorganisms can also transaminate amino acids and do so readily for all amino acids with their keto acid counterparts. As in animals, ketoisovalerate and ketoisocaproate require special enzymes, but their final oxidation is through acetyl and propionyl-CoA. The latter compound can be used in the same manner as in higher forms, but different pathways exist in microorganisms. Conversion to acrylil-CoA and lactyl-CoA results finally in pyruvate, a pathway found in *E. coli* and other bacteria.

Purines and pyrimidines are decomposed readily. The glyoxylate pathway may be employed in an anaplerotic fashion (i.e., a metabolic pathway to replenish necessary intermediates drained off by biosynthesis). This pathway enables organisms to condense acetates to form succinate. It consists of a microbial shortcut of the tricarboxylic acid cycle, with which it shares some enzymes. Isocitrate can be acted on by isocitrate lyase to yield succinate and glyoxylate; the glyoxylate can combine with acetyl-CoA to form malate under the auspices of malate synthetase. Undoubtedly, the glyoxylate produced by decomposition of purines may be metabolized in this fashion. The bases are initially converted to urate, which is oxidatively decarboxylated to allantoin and transformed to allantoic acid.

Even water and soluble aliphatic hydrocarbons can serve as nutrients for some bacteria and fungi, including yeasts. Usually, their metabolism involves oxidation of a terminal methyl group to a primary alcohol. Several specific enzymes must be synthesized to accomplish this feat. The alcohol is then further oxidized to an aldehyde and corresponding acid, which is degraded by the usual β-oxidation of fatty acids.

Aromatic compounds are also not readily decomposed by higher forms. Mammals can degrade only phenylalanine and tyrosine. Tryptophan and the many other aromatic materials of living and nonliving origin accumulate in nature, to be decomposed by bacteria and fungi. The pathway of degradation of phenylalanine and tyrosine in mammals is shared by microorganisms. Most other aromatic compounds are changed by microorganisms to either catechol or protocatechuate, which is then cleaved oxidatively by the β-ketoadipic pathway. Protocatechuate is produced from such diverse precursors as benzoate, hydroxymandelate, quinate, vanillate, parahydroxybenzoate, and shikimate. Catechol results from the metabolism of anthracene, phenanthrene, naphthalene, salicylate, benzene, mandelate, tryptophan, and phenol, among others. Both orthocleavage and metacleavage of the aromatic ring have been noted. In both of these approaches reactions are catalyzed by various oxygenases that finally produce a 3-oxoadipate, which is a 6-carbon dicarboxylic acid that can be metabolized to succinate through the intervention of Coenzyme A.

Recycling Nitrogen

Ammonia in the external environment can be incorporated into intermediate substances by NADP-specific glutamate synthetase, dehydrogenase, and ketoglutarate of the tricarboxylic acid cycle and then via transaminases to other keto acids. A glutamine synthetase leads to the formation of glutamine but requires energy in the form of ATP. However, glutamine does serve as the amino nitrogen donor in such important reactions as the synthesis of purines, tryptophan, histidine, and carbamylphosphate, the precursor of pyrimidines. When the supply of ammonia is limited, glutamine synthetase and glutamate synthetase cooperate to form glutamate from the keto acid and ammonia; glutamine is formed and the amido group transferred to ketoglutarate.

In nature, competition for nitrogen compounds may be severe, especially between fungi and bacteria. Under these circumstances bacteria are favored until fungal metabolism can lower the environmental pH to levels that prevent bacterial growth.

The limiting role nitrogen plays in the economy of nature has considerable impact on the nutrition of all living forms. Plants use nitrate exclusively as their nitrogen source. Many bacteria and fungi can also use nitrate, converting it to nitrite, then ammonia, and even nitrogen gas.

Nitratase is an NADP or NAD-linked molybdoflavoprotein, whereas nitrite is reduced by a metalloflavoprotein that can use a variety of electron sources, including NADPH, hydrogen, or photoreduction. Ferredoxin plays a crucial role in this step, quite probably being reduced by the former agents (e.g. NADPH, H_2) and then catalyzing the reduction of nitrite to ammonia. A number of bacteria can carry this sequence still further, releasing nitrogen gas, a characteristic helpful in the classification of certain clinically significant bacteria. Other organisms stop at the nitrite step, another taxonomic property especially useful for the recognition of Enterobacteriaceae (Figure 2-11). Note that nitrite is considered a potentially carcinogenic compound because it is readily converted to nitrosamine. Thus we are subjected to carcinogen exposure by the activity of our normal intestinal microbiota acting on materials normally encountered in our food.

An entirely different but equally important dilemma in nature is solved by specialized bacterial activity. As stated earlier, plants require nitrate as their major source of nitrogen. However, practically all

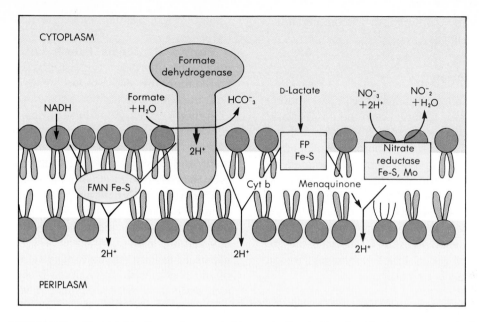

FIGURE 2-11   Proposed pathway for anaerobic respiration in *E. coli* with nitrate as terminal electron acceptor and NADH as electron donor. During anaerobic growth, formate is produced and formate dehydrogenase is induced. When lactate is the carbon source, pathway begins at FP-FeS complex. There are two types of cytochrome b involved in the scheme. *FP*, Flavoprotein; *FeS*, iron-sulfur proteins; *Mo*, molybdenum.

metabolic activities concerned with nitrogen ultimately produce ammonia, contributing to the famines that have plagued humans throughout history. Only some chemolithoautotrophic bacteria can reverse this process stoichimetrically, some producing nitrite from ammonia and others nitrate from nitrite.

The sequence of nitrate to nitrite or to nitrogen can enable certain bacteria to oxidize glucose without atmospheric oxygen. The free energy of nitrate reduction is almost comparable to the ATP gained by oxygen respiration. This capability enables some obligatively aerobic bacteria to grow, albeit minimally, under strictly anaerobic conditions.

The ability to incorporate atmospheric nitrogen demonstrates the seemingly limitless capacity of primitive organisms to survive, and it also underlines our dependence on microbial activities.

The nitrogenase complex responsible for the anaerobic reduction of nitrogen can also reduce compounds such as N_2O, cyanide, azide, and acetylene. Discovery of the reduction of acetylene to ethylene has explained some steps of the mechanism for nitro-

gen fixation. Nitrogenase is a complex of at least two enzymes: an iron-containing enzyme, azoferredoxin, and the much larger molybdenum and iron-containing molybdoferredoxin. Since atmospheric nitrogen is extremely stable, both ATP and the reduced form of the enzyme complex are required for fixation. Various metabolic intermediates serve to reduce these compounds, often depending on the nature of the prokaryote. In the absence of nitrogen, nitrogenase can produce hydrogen gas.

The best known example of nitrogen fixation occurs in the symbiotic relationship between bacteria and the root nodules of legumous plants, which together help to control atmospheric oxygen excess. In free-living aerobic bacteria such as *Azotobacter* species and *Mycobacterium flavum*, methods to uncouple oxidative respiration are used. In the legumes the bacteria that infect root hairs produce swollen, distorted forms called ''bacteroids'' that are provided by the plant with the necessary nutrients and protected against excess oxygen by leghemoglobin, a hemoglobin-like compound that supplies sufficient oxygen for the aerobic *Rhizobium* and the energy required by the organism without endangering nitrogenase activity.

Sulfur Assimilation

Most microorganisms fulfill their sulfur requirement from inorganic sulfate. Since there is a considerable charge difference between the inorganic and organic forms of this element, inorganic sulfate must be reduced to sulfite in order to be incorporated into biologically useful materials. This process is distinct from the bacterial use of sulfate as a terminal electron acceptor and proceeds in an entirely different fashion. In bacteria, O-acetylserine serves as the hydrogen sulfite acceptor, resulting in the formation of L-cysteine, acetate, and water. In eukaryotic microorganisms the sulfite acceptor is serine.

Biosynthetic Pathways

All microorganisms contain a dynamic pool of central intermediary metabolites that represent the starting point of all the organism's biosynthetic efforts. Although very small, it provides the materials from which nucleic acids, proteins, polysaccharides, and lipids are synthesized. This pool consists of phosphorylated carbohydrates, pyruvate, acetate, oxaloacetate, succinate, and α-ketoglutarate, along with minerals essential for the catalytic processes and structural integrity. Most of the biosynthetic pathways are branched. This presents a real advantage, because the number of biosynthetic efforts need not be as complex as the large number of important products would suggest. The branching pathways provide opportunities for several materials to be produced by proper modification of precursor compounds. In addition, the synthesized material may play an important role in cellular function or serve as a constituent of a macromolecule. The need of the organism dictates the use to which each synthesized material is put.

The biosynthesis of protoplasm proceeds along the universal lines that led Kluyver to postulate his concept of the unity of biochemistry. Many microorganisms, both procaryote and eucaryote, are able to synthesize all the required materials; other microorganisms, and of course animals, require complex nutrients. The pathways from precursor to finished building block proceed along universal routes for all forms capable of complete synthesis. Thus the synthesis of purines and pyrimidines requires phosphorylated ribose but employs this compound differently for each base.

Purine Synthesis Purine synthesis proceeds from 5-phosphoribosyl pyrophosphate to 5-phosphoribosylamine, the amino group donated by glutamine with the elimination of pyrophosphate. ATP and glycine then interact to produce glycineamide ribotide, which is converted with (1) further expenditure of energy, (2) introduction of an additional carbon fragment, and (3) amination by glutamine through formyl glycineamide ribotide to aminoimidazole ribotide.

Pyrimidine Synthesis Synthesis of pyrimidine is not initiated on pentose. Instead, carbamylphosphate, formed from carbon dioxide and the glutamine amino group with energy from ATP, condenses with aspartate to form carbamyl aspartic acid. This compound becomes the pyrimidine precursor dihydroorotic acid when water is eliminated, which is reduced to orotic acid and then condenses with 5-phosphoribosyl pyrophosphate to form the corresponding ribonucleotide. The elimination of carbon dioxide yields uridylic acid, which, as the high energy compound uridine triphosphate, can be transaminated to cytidine triphosphate. Both of these pyrimidine triphosphates can be incorporated into RNA.

Amino Acid Synthesis Of the required 20 amino acids 19 can be derived from five branched biosynthetic pathways, each usually identified by the first

Table 2-2 Amino Acid Families

Source	Family	Amino Acids
α-Ketoglutarate	Glutamate	Glutamate Glutamine Arginine Proline
Transamination of oxaloacetate	Aspartate	Aspartate Asparagine Methionine Threonine Isoleucine and lysine (in part)
Condensation between phosphoenolpyruvate and erythrose-4-phosphate	Aromatic	Phenylalanine Tyrosine Tryptophan
	Pyruvate	Alanine Valine Leucine
	Histidine	Histidine

amino acid derived from the non-amino acid precursors (Table 2-2).

The glutamate family (Figure 2-12), derived from α-ketoglutaric acid, is converted to L-glutamic acid by ammonia in the presence of reduced NADP. Glutamic acid is changed to glutamine in the presence of ammonia and ATP. By combining with acetyl-CoA, N-acetylglutamate is formed, which converts in the presence of NADP.H_2 and ATP to N-acetylglutamic semialdehyde. Transaminated, this compound forms N-acetylornithine. By eliminating the acetate group, the amino acid L-ornithine is formed. This acid, through condensation with carbamyl phosphate, forms L-citrulline. The cycle is completed by the transamination of aspartic acid to form fumaric acid with the synthesis of L-arginine. The last three reactions are, of course, familiar to students of biochemistry who have studied the formation of urea in humans. In addition, glutamic acid in the presence of NADP.H_2 and ATP can be converted into a glutamic semialdehyde that spontaneously forms an intermediate ring structure, which in the presence of additional reduced NADP forms the amino acid L-proline.

Oxaloacetic acid is transaminated to L-aspartic acid, which in the presence of ATP and ammonia can form the amide L-asparagine (Figure 2-13). Aspartic acid in the presence of NADP.H_2 and ATP can form aspartic semialdehyde, a compound that can contribute to the lysine pathway and can be converted in the presence of additional NADP.H_2 to homoserine. This compound leads to methionine or, in the presence of additional ATP, forms phosphohomoserine, which by the elimination of the phosphate produces L-threonine. Deamination of the threonine produces α-ketobutyric acid, which is readily converted to L-isoleucine. The aspartic semialdehyde mentioned previously leads to the synthesis of lysine by condensing with pyruvic acid to form dihydrodipicolinic acid, a heterocyclic ring structure. In the presence of NADP.H_2 this compound converts to tetrahydrodipicolinic acid, which forms an intermediate through the action of water and succinyl-CoA. This intermediate rearranges itself into L-diamino pimelic acid and, by the elimination of carbon dioxide, results in the production of L-lysine. This derivation of L-lysine from aspartic semialdehyde is characteristic of all prokaryotes, as well as certain Zygomycetes, green algae, and plants.

The compound dipicolinic acid, which is very important in spore formation, is formed in the aspartate pathway to lysine. Thus, in bacteria capable of forming spores, this pathway serves an important

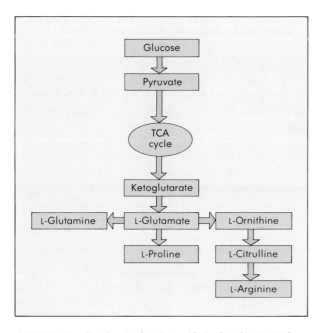

FIGURE 2-12   Synthesis of amino acids in the glutamate family.

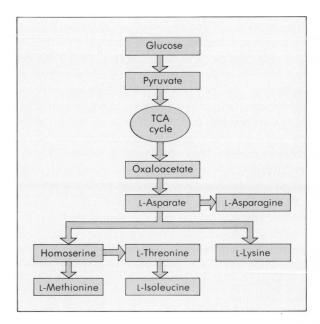

FIGURE 2-13   Synthesis of amino acids in the aspartate family.

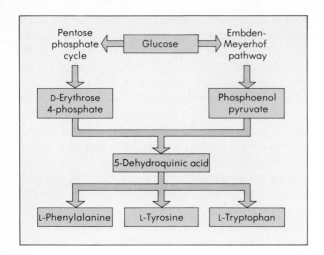

FIGURE 2-14   Synthesis of amino acids in the aromatic family.

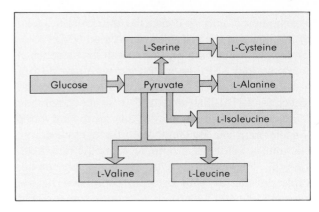

FIGURE 2-15   Synthesis of amino acids in the pyruvate family.

function. Bacteria also use diaminopimelic acid intermediates to produce cell wall peptidoglycans, adding a further role to the lysine branch of the aspartate family.

Synthesis of amino acids in the aromatic family (Figure 2-14) is initiated by the combination of erythrose-4-phosphate and phosphoenolpyruvate to form 5-dehydroquinic acid, an acid that is reduced, and eliminates water to form shikimic acid. In the presence of ATP it forms 5-phosphoshikimic acid. This intermediate condenses with phosphoenolpyruvic acid to form chorismic acid. Chorismic acid can be converted to anthranilate, which is the precursor of tryptophan, by condensing with 5-phosphoribosyl pyrophosphate to produce the intermediate phosphoribosyl anthranilic acid. By eliminating carbon dioxide and two water moieties, indole glycerol phosphate is formed. This compound is converted to L-tryptophan with the deamination of serine to 3-phosphoglyceraldehyde. Chorismic acid forms the precursor for L-phenylalanine and L-tyrosine (i.e., prephenic acid). Prephenic acid, in the presence of NAD and by eliminating CO_2, forms phenylpyruvic acid, which on transamination yields L-phenylalanine. By merely eliminating CO_2, prephenic acid forms parahydroxiphenylpyruvic acid, which can be transaminated to yield L-tyrosine.

The pyruvate family of amino acids (Figure 2-15) results from the oxidation of 3-phosphoglyceric acid to 3-phosphopyruvic acid, which can be transaminated to 3-phosphoserine. Elimination of the phosphate leads to L-serine, an important amino acid. Acetylation of serine and the addition of hydrogen sulfite leads to the formation of L-cysteine through an intermediate acetyl serine.

Pyruvate can be transaminated to yield L-alanine. It may also become an active acetaldehyde that condenses with α-ketobutyrate to form α-acetohydroxibutyric acid. This intermediate may be reduced to α, β-dihydroxi-β-methyl valeric acid, which by losing water forms α-keto-β-methyl valeric acid. Transamination of this compound yields L-isoleucine.

Pyruvate may condense with an active acetaldehyde originally formed from pyruvate to yield α-acetolactic acid, which can be reduced to α-β-dihydroxiisovaleric acid. By eliminating water, α-ketoisovaleric acid is formed and, on transamination, yields the amino acid L-valine. The intermediate may, however, condense with acetyl-CoA to form an intermediate that leads to α-ketoisocaproic acid by the elimination of carbon dioxide and hydrogen. Transamination of the α-ketoisocaproate leads to the formation of the amino acid L-leucine.

A rather complex reaction, which involves the condensation of phosphoribosyl pyrophosphate and glutamine in the presence of ATP, initiates the biosynthesis of L-histidine. Hydration of the intermediate leads to a series of intermediate compounds that finally, by transamination and reduction, yield the amino acid L-histidine. Although two of the atoms of the histidine imidazole ring are derived from the 5-phosphoribosyl pyrophosphate, the remaining three atoms are contributed by the purine nucleus of ATP and the additional nitrogen is derived from glutamine. This appears to be a most destructive use of a high-energy compound, but the intermediate formed is aminoimidazole carboxamide ribotide, which serves as an intermediate in purine synthesis. Thus the relationship between the biosynthesis of histidine and that of purine nucleotides is very close. This intermediate could conceivably serve either for needed purines or lead to the production of L-histidine.

SUMMARY

Kluyver enunciated the concept of the unity of biochemistry based on his observations, which showed the universal patterns of metabolisms existing in bacteria, plants, and animals. Andre Lwoff recognized that not only were the tools of life shared by all living forms, but the differences between higher and lower forms reflect only variances in the genetic instructions available. The metabolic and genetic ingenuity of the bacterial world underlies the survival of all life on this planet. On occasion, the bacterial actions may result in harm to other forms. Our understandable emphasis in medical microbiology on this harmful aspect of our usually beneficial relationship should not detract from the all-pervasive significance of these primitive organisms.

BIBLIOGRAPHY

Gottschalk G: Bacterial metabolism, New York, 1979, Springer-Verlag.

Gunsans IC and Stanier RY: The bacteria, vol II, Metabolism, New York, 1961, Academic Press.

Kluyver AJ: The chemical activities of microorganisms, London, England, 1931, University of London Press.

Lwoff A: Biological order, Cambridge, Mass., 1962, MIT Press.

Mandelstom J and McQuiley K: Biochemistry of bacterial growth, ed 2, Oxford, 1973, Blackwell Scientific Publications.

Norris JR and Richmond MH: Essays in microbiology, Chicago, 1978, John Wiley & Sons.

Stephenson M: Bacterial metabolism, London, England, 1943, Longmans, Green & Co.

Wilson G, Miles A, and Parker MT: Topley and Wilson's principles of bacteriology, virology and immunology, ed 7, Baltimore, 1982, Williams & Wilkins.

3

Bacterial Genetics

Microbiology was the last biologic discipline that retained the concepts of Lamarckian genetics (i.e., the environment plays a determining role in the organism's adaptation to external challenges). However, microbiology also provided the tools for the modern revolution in understanding genetics. Griffith, in 1928, described a phenomenon of pneumococcal reversion from a rough (R) type to the smooth (S) form by adding extracts of killed S forms of a different capsular antigenic composition to the living R forms in the peritoneum of mice. In 1944 Avery, McCleod, and McCarty demonstrated that this reversion from R to S forms resulted from the DNA coding for S forms contained in the extract and initiated the new era of molecular biology.

Bacterial genetics shares all the major features of eucaryotic organisms but on a somewhat simpler scale. A repetition of all the features discovered in the last 3 to 4 decades would be redundant; instead only those aspects pertinent to general appreciation and those peculiar to bacteria will be discussed.

TERMINOLOGY

The synthesis of proteins is primarily controlled by DNA. The design of each protein is transmitted from DNA to mRNA, which instructs the cellular machinery to assemble a particular protein. Proteins serve structural and enzymatic purposes and thereby determine the structure and metabolic function of any organism. The total genetic potential of DNA constitutes the **genotype** of the bacterium; what actually becomes manifest or discernible to the observer is the **phenotype.** Many capabilities of the genome remain latent; thus the ability to hydrolyze lactose, for exam-

ple, may not be evident until the need arises to synthesize that particular enzyme, β-galactosidase. Therefore the bacterial environment influences the phenotypic expression of the organism.

Bacteria carry their genome in a single, giant, circular loop of DNA called a bacterial **chromosome.** This chromosome is much longer than the organism and is folded into a compact package that can be visualized by electron microscopy after appropriate treatment. It differs from the eucaryotic chromosome because it lacks a nuclear membrane; it lies naked within the bacterial body, its acidic nature neutralized by basic amides.

The bacterial genome consists of 3,000 to 6,000 **genes**—distinct DNA sequences that specify the sequences of amino acids in a polypeptide chain. Each gene determines a particular kind of amino acid assembly. Its function can be altered or the entire gene deleted by **genetic mutation.** A number of special genes, probably a small minority, specify the structures of ribosomal and transfer RNAs, the cellular workhorses for the assembly of proteins.

In contrast to eucaryotes, bacterial genes have only one correct reading frame. The chain of nucleotides that constitute a gene is composed of groups of bases. Each set of three bases, known as a **codon,** specifies a particular amino acid. Since reading of the message can begin at any base, three different transcriptions would be possible. Bacterial transcription, however, is constrained to only one interpretation (i.e., the gene and the resultant polypeptide are colinear). Bacteria also do not possess sequences that intervene between genes, as are found in eucaryotes; they lack **introns.** The bacterial genome is thus composed of **exons,** primary transcript regions that are reflected in mRNA.

Each bacterium can be regarded as a **clone** (i.e., its progeny are produced vegetatively and are identical to the initial organism). Undirected spontaneous mutation was proved to be present in bacteria by Luria and Delbrück in 1943. Further studies indicated the **haploid** nature of bacteria; they produce no heterozygotes and do not manifest dominance. However, auxotrophic mutants which are defective in the synthesis of an essential metabolite, occur. They are detected by growth in the presence of the missing metabolite and their inability to proliferate in a milieu that supports the original, so-called **wild-type** representative. Expression of mutation is often delayed; bacteria grow to exhaust the previously formed enzymes before mutation can be expressed.

GENE EXCHANGE

Bacteria differ from eucaryotic diploid organisms by the lack of meiosis and subsequent zygote formation as a result of male and female haploid gamete fusion. The random mixing of more or less altered chromosomes that leads to the eucaryotic haploid gamete and in a sense ensures the individuality of offspring in higher forms is lacking in procaryotic organisms.

Gene exchange between bacteria is possible by three mechanisms: **transformation** (the addition of "foreign" DNA), **transduction** (the result of infection with a temperate phage), and **conjugation** (a quasisexual introduction of donor DNA into a recipient). The effect of bacterial gene transfer never leads to zygote formation. At best, a **merozygote** may be formed, in which part of the donor bacterium's genome **(exogenote)** is transmitted to an intact recipient. **Recombination** (i.e., replacement of resident genes by exogenote genes or addition of exogenote genes to the recipient gene pool) then takes place. Certain bacterial genetic information transfer can be described as infectious heredity, since transduction and conjugation transfer segments of DNA that are capable of autonomous reproduction in the recipient bacterium as viruses or bacteriophages (in the case of transduction) or as plasmids (following conjugal transfers). These exchanges produce additions to the recipient genome and may be integrated into the recipient chromosome or passed on to daughter bacteria as autonomously replicating units. The various methods of genetic transfer in bacteria could

be regarded as an evolutionary process that started with the very primitive and haphazard transformation, changed to a more deliberate delivery of donor information in the conjugative process, and finally evolved to the bacteriophage, which represents both an infectious and a genetic information exchange.

Transfer of genetic information among bacteria occurs most often between members of the same species; however, it does occur between totally unrelated organisms. In the latter instance transformation and transduction are inefficient. Conjugative plasmid transmission is independent of homology between donor and recipient chromosomes, allowing for the survival of donor genes. Some plasmids are readily moved between diverse bacteria (e.g., organisms such as *Escherichia coli, Neisseria* species and *Pseudomonas aeruginosa*). Other plasmids display a narrower host range.

Transformation

This mode of gene transfer is regarded mostly as a laboratory phenomenon, but naturally occurring transformation has been observed in cultures of *Haemophilus, Neisseria, Streptococcus, Staphylococcus, Bacillus* and *Acinetobacter*. Artificial transformation involves large DNA fragments (10^5 to 10^7 d). Uptake, requiring energy, is thought to take place in zones of cell wall synthesis. Several different mechanisms have been observed. *Streptococcus pneumoniae* and *Bacillus* species attack one strand of entering DNA with a membrane endonuclease and produce small (7 to 10 kb) segments that form a complex with recipient DNA, finally being covalently integrated. *Haemophilus influenzae* permits the entry of double-stranded DNA. Entry reflects the presence of nucleotides that are peculiar to this bacterium and are bound to a membrane protein. *Neisseria* require the presence of fimbriae, sometimes referred to as pili, for promoting transformation. Artificial transformation requires modification of the bacterial envelope, especially in members of the family Enterobacteriaceae. The presence of calcium chloride eases this penetration, but DNA uptake is very inefficient. The main value of the process is its use in genetic mapping.

Transduction

Genetic transfer by transduction involves bacteriophages or bacterial viruses. Bacteriophages, like all

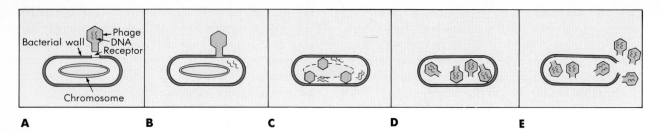

FIGURE 3-1   Bacterial transduction with release of mature bacteriophages following cell lyses. **A,** The phage tail combines with a specific receptor site in the bacterial cell wall. **B,** The phage DNA is injected into the bacterium. **C,** Replication of the bacterial chromosome is disrupted, and phage DNA codes for formation of phage components. **D,** The components are assembled into phage particles. **E,** The cell lyses and releases the mature phage particles.

viruses, differ from living organisms, whether bacteria, animal, or plants; they harbor only one type of nucleic acid, either DNA or RNA. Usually, bacteriophage DNA is introduced into a bacterium with the viral protein shell remaining outside the organism. The bacterial chromosome disintegrates, while the bacteriophage DNA instructs the bacterial machinery to synthesize bacteriophage components. These components are assembled into phage particles, and the bacterium undergoes lysis, releasing mature bacteriophages (Figure 3-1).

Bacteriophage infection can follow a different scenario called **lysogeny** (Figure 3-2). Under these conditions bacteriophage entry does not result in lysis. Instead the genome of the phage associates with the bacterial chromosome and is replicated as an integral part of the bacterial genome for many generations. Such bacteriophages, referred to as **temperate phages,** do at times revert to the virulent state, leading to lysis and the production of bacteriophages.

The temperate phage, in gaining its independence from the bacterial chromosome, may carry with it

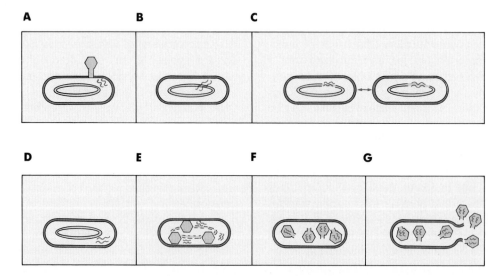

FIGURE 3-2   Lysogenic infection of bacterium with temperate bacteriophage. **A,** The phage infects a sensitive bacterium, and the phage DNA is injected. **B,** The phage DNA becomes integrated with the bacterial chromosome. **C,** The bacterium multiplies, apparently unaffected by the infection. It has been lysogenized. **D,** Occasionally, the phage DNA becomes detached from a bacterial chromosome and takes control (**E**). An individual cell (or by induction all the cells) produces phage components. **F,** The components are then later assembled into phage particles. **G,** Ultimately, the cell lyses and releases mature phage particles.

small pieces of donor bacterial DNA, which, when delivered to the next host cell, add a new attribute to the new host's capabilities. This genetic transfer is **transduction.** A wide range of bacteria can be transduced; the delivery of the donor DNA is safeguarded by the protein coat of the phage. Many bacterial characteristics are the result of transduction; the best example is the production of diphtheria toxin by a lysogenized *Corynebacterium diphtheriae.* When former host genes are included in the lysogenization of a new host, characteristics such as antibiotic resistance or beta-galactosidase production may be transduced. In the process of forming bacteriophages, the lysogenized host cell may include pieces of host DNA along with viral DNA into the viral protein coat. The chance delivery of such a particle may lead to the transduction of a segment of bacterial DNA. Transduction provides an additional tool in mapping the bacterial genome and has considerable importance in biotechnology.

Conjugation

Lederberg's demonstration that sexlike exchanges were possible between consenting bacteria was an unprecedented revelation that brought the procaryotes fully into the realm of modern biologic concepts. Fortunately, the study was undertaken with an old laboratory strain, *Escherichia coli* K12, the prolonged laboratory domestication of which had produced considerable surface changes that eased the process of conjugation. Lederberg succeeded in gaining a prototroph from the mating of two doubly auxotrophic strains, each different in its requirement for two different essential nutrients. The probability that back mutation could have produced such a prototroph approaches zero. It was demonstrated that contact between the cells was required and that the strains displayed polarity (i.e., some strains could only be donors while others acted only as recipients). Thus **fertility factor** positive (F^+) and fertility factor negative (F^-) varients were required, with the additional proviso that F^- strains must be viable for successful conjugation (Figure 3-3).

Further studies revealed that F^+ was determined by a plasmid that is transferred with high frequency but leads to chromosomal transfer at a significantly lower rate. Only part of the chromosome is transmitted usually, yielding merozygotes that are transient stages of the stable haploid bacterium. The F^+ plas-

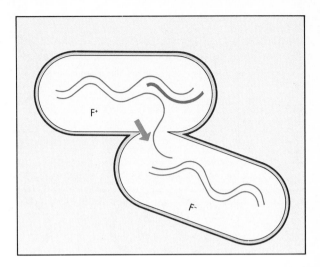

FIGURE 3-3   Bacterial conjugation.

mid can be integrated into the bacterial chromosome, producing an Hfr form, leading to increased chromosomal transfer to the recipient. Artificial interruption of mating with a blender, rapid agitation by pipetting, or strong vibrations permitted mapping of the genes along the bacterial circular chromosome. Once entry was achieved, the chromosome advanced at a constant rate, with 1% of the chromosome passing into the recipient per minute (15 μm or 50,000 bases at 37° C). Therefore distances on a chromosome can be measured in units of time, the entire process taking approximately 100 minutes.

The circular bacterial chromosome permits insertion of the F^+ plasmids at different sites, leading to different sequences transferred initially (Figure 3-4). The circular nature of the bacterial chromosome and of the plasmid was confirmed later with radioautographs. These studies led to exact mapping of the *E. coli* genes (Figure 3-5).

The mechanism of bacterial sex was illuminated with the use of bacteriophages specific for the F^+ pilus. The bacteriophages attach only to pilin, an 11,800 d protein that forms a hollow tube connecting the F^+ bacterium to the recipient F^- variant. Other adhesive projections such as fimbriae are not attacked by the phage. A specific outer membrane protein on the recipient bacterium is the specific site of the F^+ pilus attachment, a site that serves as the attachment receptor for other F^- specific phages. In F^+ bacteria this receptor no longer functions. At present, there is no acceptable evidence that even

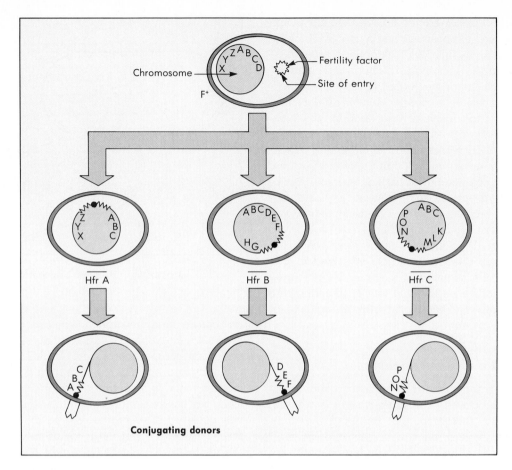

FIGURE 3-4   Different Hfr strains are produced by the insertion of the fertility F factor at different sites in the chromosome.

single strands of donor DNA pass through the pilus. Instead, the pilus may function as an attachment that recognizes a suitable recipient. It may also cause wall-to-wall contact between the pairs by retraction, leading to the formation of a cytoplasmic bridge. Evidence favors the need for cell-to-cell contact in conjugation, since surfactants that disrupt the pilus do not disaggregate mating bacteria or interfere with DNA transfer. Replication of the DNA usually starts at a specific site and moves along the molecule. During conjugative DNA transfer, only one strand of DNA enters the recipient while the other remains in the donor, but both strands are replicated immediately in each of the participants.

The genetic reservoir of bacteria in the chromosome allows for the survival of the organism but seems not to provide sufficient information to permit a bacterium to respond to many environmental challenges. This deficit is met by many bacteria through the presence of **plasmids,** circular DNA composed of 1,000 to 30,000 base pairs (the *E. coli* chromosome contains 4 million pairs). Plasmids play an especially significant role in the ability of bacteria to evade toxic substances and antibiotics. Of course, the fertility factor is a plasmid, and plasmids that communicate certain physiologic, adhesive, and protective coating advantages to recipients are known as well. Two classes of these bacterial organelles are known; the larger plasmids are often transmitted by conjugation, whereas some bacteria carry smaller versions of the extrachromosomal DNA that are nonconjugative but can be mobilized.

Plasmids and the bacterial chromosome can exchange genetic information by way of insertion

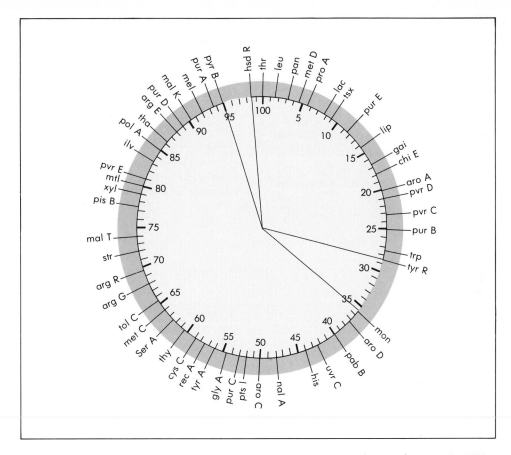

FIGURE 3-5   Chromosomal maps of *Escherichia coli* genes. The time (minutes) to transfer DNA segments from the origin of the first Hfr strain is represented by the numbers. (Redrawn from Bachmann BJ, Low KB and Taylor AL: Bacteriol Rev 40:116, 1976.)

sequences and transposons. **Insertion sequences (IS elements)** were discovered as bacterial gene inactivators after insertion in the middle of the gene structure. Removal of the element results in a recurrence of gene activity. Insertion sequences are found in many bacteria, but their function is still unknown. One end of the IS elements is an inverted repeat of the other end (Figure 3-6). Between the two ends one can encounter bases ranging from 100 to 1,000 nucleotides (usually less than 1,500). IS elements do not contain instructions for protein synthesis; thus their presence is noted only when interference with the chromosomal gene occurs.

Plasmids can also become part of the bacterial chromosome as **transposons.** The ends of transposons are also repeated bases, often several hundred nucleotides long. These ends may act as IS elements.

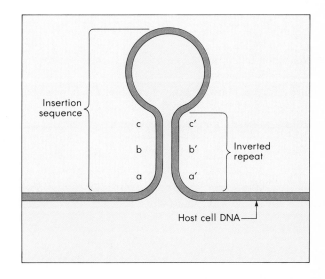

FIGURE 3-6   Insertion sequence in bacterial DNA.

Antibiotic-resistant plasmids especially are known to be transposed into the bacterial chromosome.

Although transposons and insertion sequences have been observed primarily in bacteria, transposable elements have also been suspected and even demonstrated in eukaryotes.

The **mutation rate** of bacteria is relatively constant when observed on the basis of cell division and is expressed as the number of mutants per number of cell divisions; it is not reflected in terms of the number of cells per unit of time. The **mutation frequency** is the total number of mutants in a population and represents the mutant pool before recent divisions.

BIBLIOGRAPHY

Davis B, Dulbecco R, Eisen, and Ginsburg HS: Microbiology, ed. 3, Cambridge, Mass. 1980, Harper & Row.

Lederberg J: Gene recombination and linked segregation in *Escherichia coli,* Genetics 32:502, 1947.

Ptashne M: A genetic switch: gene control and phage γ, Oxford, England, 1986, Cell Press and Blackwell Scientific Publications.

Wilson G, Miles A, and Parker MT: Topley and Wilson's principles of bacteriology, virology and immunology, ed 7, Baltimore, 1984, Williams & Wilkins.

CHAPTER 4

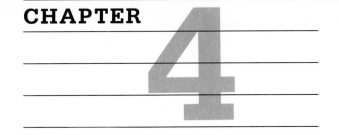

Chemotherapy of Bacterial Infections

An important year in chemotherapy of systemic bacterial infections was 1935. Although antiseptics had been applied topically to prevent growth of microorganisms, systemic bacterial infections did not respond to any existing agents. In 1935 the red azo dye protosil was shown to protect mice against systemic streptococcal infection and was curative in patients suffering from such infections. It was soon demonstrated that protosil was cleaved in the body to release *p*-aminobenzenesulfonamide, or sulfanilamide, and the chemotherapeutic activity was attributable to sulfanilamide. These observations with the first sulfa drug initiated a new era in medicine. Numerous derivatives of sulfanilamide were synthesized, and chemotherapy of systemic infections became possible.

Eventually, compounds (antibiotics) produced by microorganisms were discovered to inhibit the growth of other microorganisms. Fleming first noted that the mold *Penicillium* prevented the multiplication of staphylococci. A concentrate from a culture of this mold was prepared, and the remarkable activity

	Box 4-1 Terminology
Antibacterial spectrum	Range of activity of a compound against microorganisms. A broad-spectrum antibacterial drug can inhibit a wide variety of both gram-positive and gram-negative bacteria, whereas a narrow spectrum drug is active only against selected organisms.
Antimicrobial activity	Activity of a chemotherapeutic agent tested in the laboratory and expressed as the lowest concentration at which the drug inhibits multiplication of the microorganism (minimum inhibitory concentration, or MIC).
Bactericidal activity	Ability of a chemotherapeutic agent to kill a microorganism; expressed as the minimum bactericidal concentration (MBC).
Antibiotic combinations	Combinations of antibiotics may be used: (1) to broaden the antibacterial spectrum in presumed mixed infections pending culture results, (2) to prevent emergence of resistant organisms during therapy, and (3) for a synergistic killing effect.
Antibiotic synergism	Combination of two antibiotics (e.g, penicillin and streptomycin) that have enhanced bactercidal activity when tested together compared with each alone (Figure 4-1).
Antibiotic antagonism	Situation in which one antibiotic interferes with the killing action of another antibiotic.
β-lactamase	An enzyme that breaks the β-lactam ring in penicillins (penicillinase) or cephalosporins (cephalosporinase). Hydrolysis of the ring protects the bacteria from the antimicrobial activity of the antibiotic.

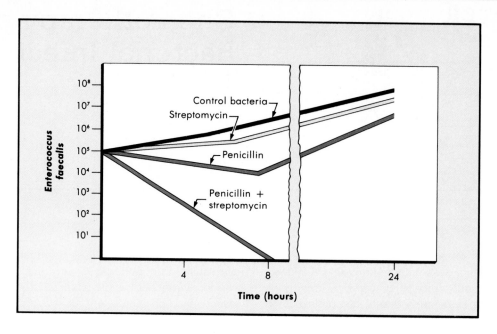

FIGURE 4-1   Combination of two antibiotics.

and lack of toxicity of the first antibiotic, penicillin, was demonstrated. Later, in the 1940s and 1950s, streptomycin and the tetracyclines were developed and were followed rapidly by additional aminoglycosides, semisynthetic penicillins, cephalosporins, quinolones and other antimicrobials. All greatly increased the range and effectiveness of antibacterial agents. Box 4-1 defines the terms appropriate to this discussion.

MECHANISMS OF ANTIBIOTIC ACTION

The basic mechanisms of antibiotic action are listed in Box 4-2.

Box 4-2   Five Basic Mechanisms of Antibiotic Action

---

1. Inhibition of cell wall synthesis
2. Alteration of cell membranes
3. Inhibition of protein synthesis
4. Inhibition of nucleic acid synthesis
5. Antimetabolic activity or competitive antagonism

Inhibition of Cell Wall Synthesis

The cross-linkage of precursors during synthesis of the bacterial cell wall is catalyzed by specific enzymes (e.g., transpeptidases and carboxypeptidases). These regulatory proteins are also called penicillin binding proteins (PBPs) because they are also bound by β-lactam antibiotics (e.g., penicillin). The rigid structure of the cell wall permits bacteria to maintain a very high internal osmotic pressure. However, when bacteria are exposed to penicillin and the antibiotic binds to the PBPs in the cell membrane (Figure 4-2), autolytic enzymes are released that degrade the preformed cell wall. Cell wall synthesis is also inhibited, resulting in bacterial cell death.

Alteration of Cell Membranes

The polymyxin class of antibiotics consists of cationic branched cyclic decapeptides that destroy the cytoplasmic membranes of susceptible bacteria (Figure 4-3). This detergent-like activity is prevented when the antibiotic is unable to penetrate through the outer cell wall to the inner cytoplasmic membrane. The antifungal polyene antibiotics (e.g., amphotericin B, nystatin) have a similar activity on cell membranes. These antibiotics are discussed further in Chapters 28 through 31.

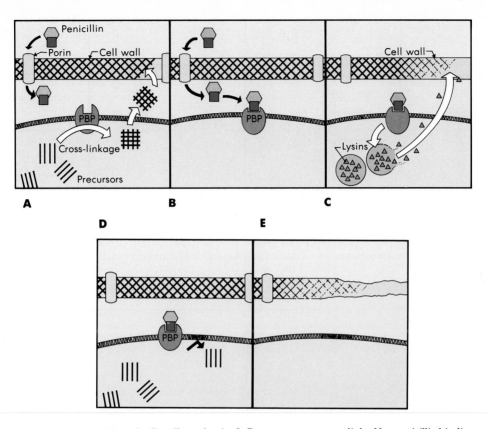

FIGURE 4-2 Inhibition of cell wall synthesis. **A,** Precursors are cross-linked by penicillin binding protein *(PBP)* and then added to cell wall. **B,** Penicillin enters the cell through porins and binds to PBP. **C,** Binding leads to release of autolysins, which break down preformed cell wall. **D,** After penicillin binds to PBP, PBP can no longer synthesize proteins essential to integrity of cell wall. **E,** Cell wall loses integrity and can no longer preserve osmotic pressure.

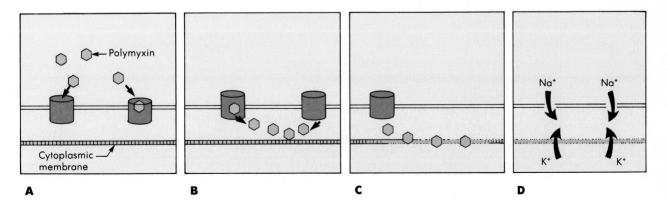

FIGURE 4-3 Alterations of cell membranes. **A,** Bacterial cell. **B,** Penetration of polymixin to inner cytoplasmic membrane. **C,** Detergentlike disruption of cytoplasmic membrane. **D,** Loss of cell integrity with subsequent cell death.

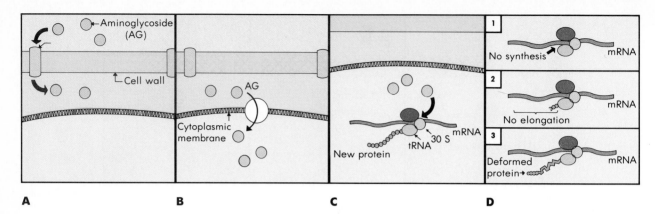

FIGURE 4-4   Inhibition of protein synthesis. **A,** Aminoglycoside *(AG)* enters bacteria through porins. **B,** AG actively transported across cytoplasmic membrane. **C,** AG binds to 30 S ribosomal subunit. **D,** As a consequence of binding: *(1)* failure to initiate protein synthesis; *(2)* failure of elongation of protein; and *(3)* misreading of tRNA, leading to deformed proteins.

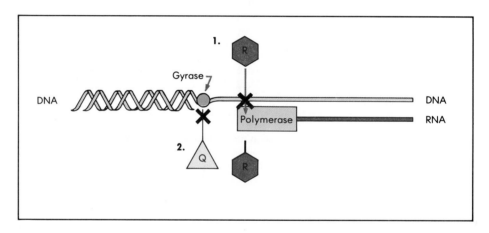

FIGURE 4-5   Inhibition of nucleic acid synthesis. *1,* Rifampin (R) binds to DNA-dependent RNA polymerase and inhibits RNA synthesis. *2,* Quinolone (Q) inhibits DNA gyrase to prevent super-coiling of DNA.

Inhibition of Protein Synthesis

Antibiotics such as tetracycline and aminoglycosides inhibit protein synthesis. After the antibiotics enter the cell and transverse the cell membrane, they bind to ribosomal subunits (Figure 4-4). As a consequence, some antibiotics inhibit mitochondrial protein synthesis, others stop elongation of nascent proteins, and the action of other antibiotics leads to deformation of proteins.

Inhibition of Nucleic Acid Synthesis

Some antimicrobial agents inhibit nucleic acid synthesis by either binding to RNA polymerase (e.g.,

rifampin) or inhibiting DNA gyrase (e.g., quinolone; Figure 4-5).

Antimetabolic Activity or Competitive Antagonism

Some antibacterial compounds act as antimetabolites (Figure 4-6). Sulfonamide competes with *p*-aminobenzoic acid, preventing synthesis of folic acid that is required by certain microorganisms. Because mammalian organisms do not synthesize folic acid (required as a vitamin), sulfonamides do not interfere with mammalian cell metabolism.

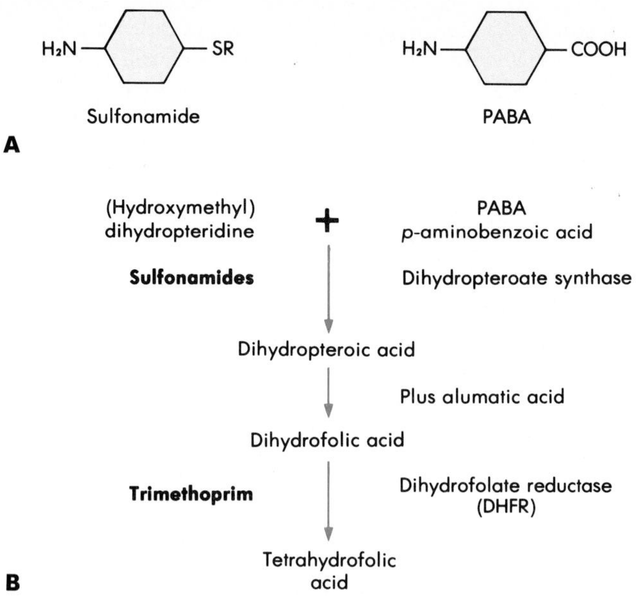

A

Sulfonamide   PABA

B

FIGURE 4-6   Antimetabolic activity or competitive antagonism. **A,** Sulfonamide, which resembles PABA, competitively inhibits one step in synthesis of folic acid, which is required by bacteria. **B,** Trimethoprim inhibits enzymatic action of DHFR.

ANTIMICROBIAL AGENTS

Antibiotics That Inhibit Cell Wall Synthesis

Penicillins Penicillin compounds are highly effective antibiotics with extremely low toxicity. The base compound is an organic acid with a β-lactam ring (Figure 4-7) obtained from culture of the mold *Penicillium chrysogenum*. If the mold is grown by a fermentation process, large amounts of a key intermediate, 6-aminopenicillanic, are produced. Biochemical substitution of this intermediate yields derivatives (Box 4-3) that have decreased acid lability and thus increased gastrointestinal absorption, resistance to destruction by penicillinase, or a widening of the spectrum so gram-negative organisms are susceptible to the compound.

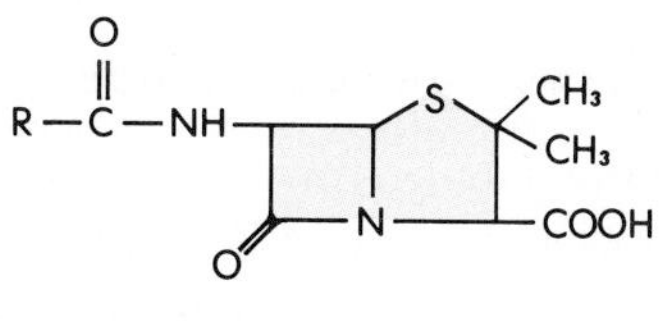

FIGURE 4-7   Structure of penicillin.

> **Box 4-3**   Penicillins With Improved Pharmacokinetic Properties or Enhanced Antimicrobial Spectrum
>
> **Acid-Stable Penicillins**
> Penicillin V, ampicillin, amoxicillin
>
> **Penicillinase-Resistant Penicillins**
> Nafcillin, oxacillin, methicillin, dicloxacillin, cloxacillin
>
> **Enhanced Spectrum of Activity**
> Carbenicillin, ticarcillin, mezlocillin, piperacillin

Penicillin G is incompletely absorbed because it is inactivated by gastric acid. Thus it is used mainly as an intravenous drug for serious infections with penicillin-sensitive organisms (e.g., streptococci, gonococcus). Penicillin V is more resistant to acid and is the preferred oral form for treatment of susceptible streptococci. Penicillinase-resistant penicillins such as nafcillin and cloxacillin are used to treat infections caused by penicillinase-producing staphylococci, including bacteremia, cellulitis, and osteomyelitis. Parenteral penicillins have been developed (e.g., carbenicillin, ticarcillin, piperacillin) that can be effective against gram-negative bacteria (e.g., *Enterobacter, Pseudomonas*) when administered in high dosages.

The toxicity of penicillin G is extremely low, but hypersensitivity to penicillin is commonly present (1% to 8% of the general population are allergic to the penicillins). The hypersensitivity reactions range from immediate anaphylactic reactions to late manifestations such as a skin rash. All penicillins share a basic antigenic site, so they all can cause reactions in hypersensitive patients. Other adverse effects (e.g., leukopenia, hepatitis, interstitial nephritis, diarrhea, platelet dysfunction) can occur, particularly with the more recently developed penicillins.

Cephalosporins The cephalosporins (Figure 4-8) are β-lactam antibiotics derived from 7-aminocephalosporanic acid, which was originally isolated for a *Cephalosporium* mold. The antibiotics have the same mechanism of action as the penicillins but have a wider antibacterial spectrum, are resistant to many β-lactamases, and have improved pharmacokinetic properties.

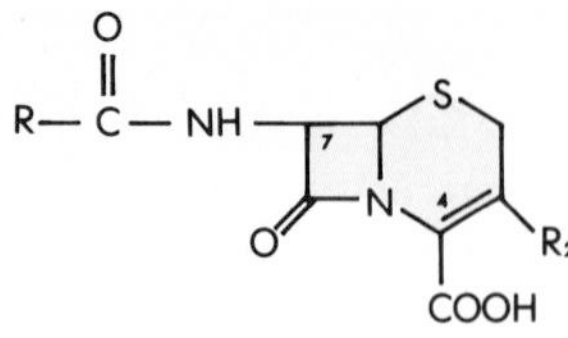

FIGURE 4-8   Structure of cephalosporin.

With biochemical modification of the basic cephalosporin molecule, significant improvements in antibiotic activity and pharmacokinetic properties have been realized. The activity of the first-generation cephalosporins *(Table 4-1)* was similar to that of ampicillin. Many of the second-generation antibiotics (e.g., cefaclor, cefuroxime) had expanded activity against *Haemophilus influenzae,* an important pediatric pathogen, and cefoxitin and cefotetan were active against *Bacteroides fragilis,* an important anaerobic pathogen. The third-generation cephalosporins further extended the antibacterial spectrum of cephalosporins to include virtually all Enterobacteriaceae and *Pseudomonas aeruginosa.*

Unfortunately, with these refinements the second- and third-generation antibiotics were less active against gram-positive cocci. Furthermore, all cephalosporins are ineffective against penicillin-resistant *Streptococcus pneumoniae,* methicillin-resistant *Staphylococcus, Enterococcus,* and *Listeria.* Additionally, organisms such as *Enterobacter, Serratia,* and *Pseudomonas* can develop resistance during therapy with the cephalosporins and then display cross-resistance to all β-lactam antibiotics.

The cephalosporins are used in many clinical situations because these antibiotics have a low level of toxicity. However, they should be used with caution in persons allergic to penicillin because they share cross-reactive antigens.

Other β-Lactam Antibiotics Several β-lactam antibiotics have slightly different biochemical structures from the penicillins and cephalosporins but have similar potent antibacterial activity. Imipenem is a carbapenem with excellent in vitro and in vivo activity for aerobic and anaerobic gram-positive and gram-negative bacteria. Aztreonam, a monobactam, is a narrow-spectrum antibiotic with activity specific for gram-negative bacilli (e.g., Enterobacteriaceae, *Pseudomonas*). Finally, β-lactamase inhibitors (e.g., clavulanic acid, sulbactam) have been combined with some penicillins (e.g., ampicillin, amoxicillin, ticarcillin) to treat infections caused by β-lactamase producing bacteria.

Other Antibiotics Vancomycin, obtained from an actinomycete, is a complex glycopeptide bactericidal against gram-positive bacteria. The antibiotic is poorly absorbed when administered orally; however, this property has proved useful for the treatment of gastrointestinal disease caused by *Staphylococcus aureus* or *Clostridium difficile.* The drug is administered intravenously for treatment of serious systemic infections in patients infected with methicillin-resistant staphylococci or with a history of allergy to the penicillins. Bacitracin, another cell wall–active antibiotic, is a mixture of polypeptides used topically for skin infections caused by gram-positive bacteria.

Antibiotics That Alter Cell Membranes

Polymyxins The polymyxins are basic peptides that act as cationic detergents to cause lysis of the lipoprotein cell membrane. Although polymyxin B and colistin are active against gram-negative bacte-

Table 4-1 Selected Cephalosporins	
Parenteral	**Oral**
FIRST GENERATION	
Activity Similar to Penicillins Like Ampicillin	
Cephalothin	Cephalexin
Cefazolin	Cephradine
	Cefadroxil
SECOND GENERATION	
Expanded Activity Against Some Enterobacteriaceae, *Haemophilus,* and Anaerobes; Decreased Activity Against Gram-Positive Cocci	
Cefuroxime	Cefaclor
Cefamandole	
Cefoxitin	
Cefotetan	
THIRD GENERATION	
Further Expanded Activity Against Gram-Negative Bacilli Including *Pseudomonas;* Generally Less Activity Against Gram-Positive Cocci	
Cefotaxime	
Ceftizoxime	
Ceftazidime	
Ceftriaxone	

Box 4-4 Aminoglycosides	
Streptomycin	Tobramycin
Kanamycin	Amikacin
Gentamicin	Netilmicin

---

ria, serious nephrotoxicity has limited their internal use. They are used chiefly to treat local infections such as external otitis, eye infections, and skin infections with sensitive organisms.

Antibiotics That Inhibit Protein Synthesis

Aminoglycosides The aminoglycoside antibiotics (aminocyclitols; Box 4-4) are bactericidal antibiotics commonly used to treat serious infections caused by many gram-negative bacilli and some gram-positive organisms. Streptococci and anaerobes are resistant to aminoglycosides.

These antibiotics inhibit bacterial protein synthesis by interfering with binding of bacterial tRNA to the 30 S ribosomal subunit. Resistance to the antibacterial action of aminoglycosides can develop one of three ways:

1. Mutation of the ribosome binding site
2. Enzymatic modification (e.g., acetylation, phosphorylation) of the antibiotic
3. Decreased uptake

Enzymatic modification is the most common form of resistance (see discussion on genetic resistance at the end of this chapter).

Because aminoglycosides are poorly absorbed after oral administration, they are injected intramuscularly or intravenously. Dosage must be modified in patients who have renal failure because the aminoglycosides are excreted unchanged by glomerular filtration into urine. The toxic concentration of aminoglycosides in serum is close to the levels required for therapeutic efficacy. Therefore it is important to monitor the serum levels closely to avoid nephrotoxicity, as well as dose-related changes in vestibular and auditory function.

Gentamicin and tobramycin have a broad spectrum of activity, with tobramycin being slightly more active against *Pseudomonas aeruginosa.* Netilmicin is reported to be less ototoxic than either gentamicin

or tobramycin, but netilmicin also has less antibacterial activity. All three aminoglycosides are used to treat systemic infections caused by susceptible gram-negative bacteria, including the Enterobacteriaceae and *Pseudomonas.* Because enzymatic modification of amikacin is rare, this aminoglycoside is used to treat infections caused by gram-negative bacteria that are resistant to other aminoglycosides. Streptomycin has been used for the treatment of tuberculosis, tularemia, and streptococcal endocarditis (when combined with a penicillin). Spectinomycin is an aminocyclitol that can be used to treat penicillin-resistant gonorrhea or penicillin-allergic patients with gonorrhea.

Tetracyclines The tetracycline antibiotics are broad-spectrum, bacteriostatic antibiotics that inhibit protein synthesis in bacteria by blocking the binding of tRNA to the 30 S ribosomal subunit. Tetracyclines are effective in treatment of *Mycoplasma pneumoniae,* cholera, rickettsial disease, brucellosis, chlamydial urethritis, as well as gonorrhea, uncomplicated urinary tract infections, and acne.

Although all tetracyclines are absorbed rapidly from the gastrointestinal tract, the drugs should be taken 30 minutes before meals or adequate serum levels will not be reached. Antacids should also be avoided for at least an hour afterward because divalent cations bind the tetracyclines and prevent absorption. Tetracyclines distribute widely in most fluids and tissues, localizing particularly in bones and teeth. Because the tetracyclines will cause permanent discoloration of teeth, these antibiotics should not be used in pregnant women or children less than 8 years of age. Additionally, tetracyclines do not accumulate in cerebrospinal fluid, so they cannot be used to treat meningitis or other central nervous system infections.

Chloramphenicol Chloramphenicol has a broad antibacterial spectrum similar to that of tetracycline but is considered the drug of choice only for treatment of typhoid fever. The reason for this is, in addition to interfering with bacterial protein synthesis, chloramphenicol disrupts protein synthesis in human bone marrow cells and can produce blood dyscrasias such as aplastic anemia (1 case per 24,000 treated patients).

Erythromycin Erythromycin, a macrolide antibiotic, is a bacteriostatic organic base used mainly to treat pulmonary infections caused by *Mycoplasma, Legionella,* and gram-positive organisms in patients allergic to penicillin. The antibiotic disrupts protein synthesis by binding to the 50 S ribosomal subunit. Bacterial resistance to erythromycin develops by modification of ribosomal proteins, which in turn prevents binding by the antibiotic.

Clindamycin Clindamycin also blocks protein synthesis by binding to the 50 S ribosome. It is active against staphylococci and anaerobic gram-negative bacilli but generally inactive against aerobic gram-negative bacteria. Clindamycin can be administered orally or intravenously with good penetration into tissues such as bone. Although intravenous administration of clindamycin is associated with relatively few side effects, oral administration can be responsible for gastrointestinal disturbances ranging from mild diarrhea to life-threatening pseudomembranous colitis.

Antibiotics That Inhibit Nucleic Acid Synthesis

Rifampin Rifampin, a semisynthetic derivative of rifamycin B produced by *Streptomyces mediterranei,* is bactericidal for *Mycobacterium tuberculosis* and is very active against aerobic gram-positive cocci. Because resistance can develop rapidly, rifampin is usually combined with one or more other effective antibiotics. The drug inhibits DNA-dependent RNA polymerase. Rifampin is administered orally, metabolized in the liver, and can be associated with increased hepatotoxicity of isoniazid (used for the treatment of tuberculosis), as well as increased metabolism of other drugs.

Quinolones The quinolones (Box 4-5) are synthetic chemotherapeutic agents that inhibit bacterial DNA gyrases or topoisomerases, which are required to supercoil strands of bacterial DNA into the bacterial cell. Nalidixic acid was used to treat urinary tract infections caused by a variety of gram-negative bacteria, but resistance to the drug developed rapidly. This drug has now been replaced by newer, more

active quinolones such as norfloxacin and ciprofloxacin. At present, norfloxacin is restricted to treatment of urinary tract infections because adequate serum levels cannot be maintained.

Metronidazole Metronidazole was originally introduced as an oral agent for treatment of *Trichomonas* vaginitis. It is also effective in treatment of amebiasis, giardiasis, and serious anaerobic bacterial infections (including *Bacteriodes fragilis*) but has no significant activity against aerobic or facultatively anaerobic bacteria. The antimicrobial properties of metronidazole appear to be mediated by a partially reduced intermediate, which results in DNA breakage. The drug diffuses well to all tissues, including the central nervous system.

Antibiotics With Antimetabolic Activity

Sulfonamides Sulfonamides are true antimetabolites; they block a specific step in the biosynthetic pathway of folic acid (see Figure 4-6). The combination trimethoprim and sulfamethoxazole blocks sequential steps in the pathway of folic acid synthesis. Sulfonamides are effective against a broad range of gram-positive and gram-negative organisms, such as *Norcardia, Chlamydia,* and some protozoa. Short-acting sulfonamides such as sulfisoxazole are among the drugs of choice for treatment of acute urinary tract infections caused by susceptible bacteria such as Escherichia coli. Trimethoprim-sulfamethoxazole is effective against a large variety of gram-positive and gram-negative microorganisms and is the drug of choice for acute and chronic urinary tract infections. The combination is active in infections caused by *Pneumocystis carinii,* bacterial infections of the lower respiratory tract, otitis media, and uncomplicated gonorrhea.

Short-acting sulfonamides (e.g., sulfisoxazole and sulfadiazine) are administered orally and are absorbed rapidly from the gastrointestinal tract. Absorption of trimethoprim-sulfamethoxazole is also rapid. Toxicity with the sulfonamides has been reported with most organ systems, including gastrointestinal effects, hepatitis, bone-marrow depression, and hypersensitivity reactions.

Nitrofurantoin Nitrofurantoin is an effective agent used for the treatment of urinary tract infections caused by gram-negative bacilli. It does not reach therapeutically effective concentrations in oth-

Box 4-5 Quinolones	
Nalidixic acid	Cinoxacin
Norfloxacin	Ofloxacin
Ciprofloxacin	

er body sites, so it cannot be used for treating nonurinary tract infections.

BACTERIAL RESISTANCE

Bacterial resistance to antibiotics may be present on a nongenetic basis or may develop on a genetic basis during therapy. Nongenetic resistance is most frequently attributable to the absence of targets for the drug in the bacteria. If the bacteria have no receptors that bind the drug or lack the metabolic pathway necessary for drug activity, the bacteria are intrinsically resistant (e.g., vancomycin or erythromycin for gram-negative bacilli). Inadequate permeability of a compound may also account for the ineffectiveness of tetracycline against some gram-negative bacteria (Figure 4-9, A). Certain microorganisms can escape the consequences of drug action by (1) synthesizing an enzyme that destroys the antibiotic (e.g., the β-lactamase that cleaves the β-lactam rings of penicillin and cephalosporin; Figure 4-9, B), (2) altering macromolecules to which the antibiotic binds (e.g., streptomycin binding to ribosomes in certain organisms; Figure 4-9, C), and (3) inactivation of antibiotic by a biochemical step, as with acetylation of aminoglycosides by gram-negative bacilli (Figure 4-9, D).

Genetic resistance may be chromosomal in origin or may be transmitted by extrachromosomal plasmids. Chromosomal resistance to several unrelated antibiotics can be transferred to susceptible organisms by cell-to-cell contact or conjugation. The bacteria contain extrachromosomal DNA or resistance (R) plasmids, which act like viruses without coats. These R plasmids are found in a variety of gram-negative bacilli, including *Shigella, Salmonella, Klebsiella, Vibrio, Pasteurella,* and *Escherichia coli.*

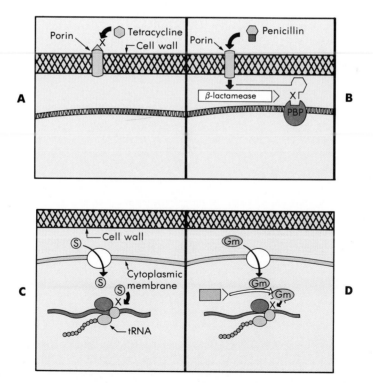

FIGURE 4-9   **A,** Change in porin so bacteria is impermeable to drug. **B,** Enzyme inactivates penicillin by opening β-lactam ring, so it cannot bind to PBP. **C,** There is a decrease in the affinity of streptomycin (S) for the target site on the ribosome. **D,** The enzyme acetylates gentamicin so that it will not bind to the ribosome.

BIBLIOGRAPHY

Bartlett, JG: Anti-anaerobic antibacterial agents, Lancet 2:478, 1982.

The choice of antimicrobial drugs, Med Lett Drugs Ther 26:19, 1984.

Holdiness MR: Clinical pharmacokinetics of the antituberculosis drugs, Clin Pharmacokinet 9:511, 1984.

Kucers A: Chloramphenicol, erythromycin, vancomycin, tetracyclines, Lancet 2:425, 1982.

Kunin CM, Tupasi T, and Craig WA: Use of antibiotics: a brief exposition of the problem and some tentative solutions, Ann Intern Med 79:555, 1973.

Murray BE and Moellering RC: Patterns and mechanisms of antibiotic resistance, Med Clin North Am 68:899, 1978.

Neu HD: Clinical uses of cephalosporins, Lancet 2:252, 1982.

Phillips I: Aminoglycosides, Lancet 2:311, 1982.

Pratt WB and Fekety R: The antimicrobial drugs, New York, 1986, Oxford University Press.

Rahal JJ Jr: Antibiotic combinations: the clinical relevance of synergy and antagonism, Medicine (Baltimore) 57:179, 1978.

Ramiez CA, Bran JL, Mejia CR, and Garcia JF: Open, prospective study of the clinical efficacy of ciprofloxacin, Antimicrob Agents Chemother 28:128, 1985.

Ruben RH and Swartz MN: Trimethoprim-sulfamethoxazole, N Engl J Med 303:426, 1980.

Smith JW: Proper use of antibiotics, Texas Med 73:37, 1977.

Stead WW and Dutt AK: Chemotherapy for tuberculosis today, Am Rev Respir Dis 125(suppl):94, 1982.

Tomasz A: Penicillin-binding proteins in bacteria, Ann Intern Med 96:502, 1982.

Staphylococcus

The family Micrococcaceae consists of four genera: *Planococcus, Stomatococcus, Micrococcus,* and *Staphylococcus. Planococcus,* a genus of motile, gram-positive cocci, has not been found in humans. *Stomatococcus* and *Micrococcus* can colonize humans, but these organisms are only rarely associated with disease. In contrast with the other Micrococcaceae, *Staphylococcus* is a significant human pathogen, causing a wide spectrum of diseases ranging from superficial cutaneous infections to life-threatening systemic maladies.

The name *Staphylococcus* is derived from the Greek term for grapelike cocci. This name is appropriate because the cellular arrangement of these gram-positive cocci resembles a cluster of grapes, although this is most characteristic for staphylococci grown on agar media. Organisms in clinical material are seen as single cells, pairs, or short chains.

Staphylococci are gram-positive cocci, 0.5 to 1.5 μm in diameter, nonmotile, facultatively anaerobic, catalase positive, and able to grow in a medium containing 10% sodium chloride and in a temperature range from 18° to 40° C. The organisms normally grow on the skin and mucous membranes of humans. The 19 species of *Staphylococcus* currently recognized in *Bergey's Manual of Systematic Bacteriology* are subdivided into four group based on DNA-DNA hybridization analysis: *S. epidermidis* group, *S. saprophyticus* group, *S. simulans* group, and *S. sciuri* group. *S. aureus* and four other species are not related to these groups. Of the large number of staphylococcal species, only three are commonly associated with human disease: *S. aureus, S. epidermidis,* and *S. saprophyticus.*

MICROBIAL PHYSIOLOGY AND STRUCTURE

Staphylococcal structure and function are outlined in Table 5-1 and Figure 5-1.

Capsule

The presence of a loose-fitting, polysaccharide capsule is only occasionally found on staphylococci cultured in vitro but is believed to be more commonly present in vivo. The capsule protects the bacteria from interaction with complement, antibodies, and phagocytic cells.

Peptidoglycan

The peptidoglycan layer, composed of peptide cross-linked glycan chains, is the major structural component of the staphylococcal cell wall (Figure 5-2). The glycan chains are built with approximately 10 alternating subunits of *N*-acetylmuramic acid and *N*-acetylglucosamine. Pentapeptide side chains are attached to the *N*-acetylmuramic acid subunits, the glycan chains are then cross-linked with peptide bridges betweeen the side chains. Lysozyme (muramidase) present in tears, saliva, and human leukocytes, monocytes, and macrophages can hydrolyze the linkage between the glycan subunits.

Protein A

The surface of most *S. aureus* strains is uniformly coated with protein A, a protein that is covalently

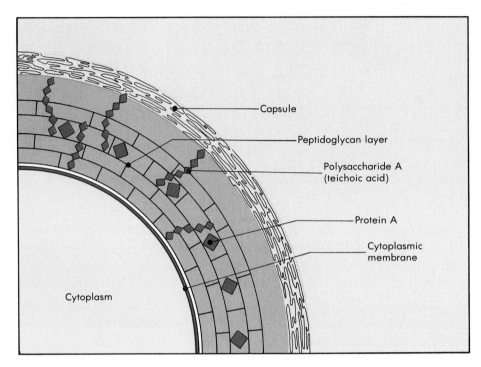

FIGURE 5-1   Staphylococcal cell wall structure.

Table 5-1   Staphylococcal Structure and Function

Structure	Function
Capsule	Inhibits opsonization and phagocytosis
	Protects from C′-mediated leukocyte destruction
Peptidoglycan	Osmotic stability
	Stimulates production of endogenous pyrogen
	Leukocyte chemoattractant
	Inhibits phagocytosis and chemotaxis
Protein A	Binds IgG1, IgG2, IgG4 Fc receptors
	Inhibits opsonization and phagocytosis
	Leukocyte chemoattractant
	Anticomplementary
Teichoic acid	Regulates cationic concentration at cell membrane
	Receptor for bacteriophages
	Attachment site for mucosal surface receptors
Cytoplasmic membrane	Osmotic barrier
	Regulates transport into and out of cell
	Site of biosynthetic and respiratory enzymes

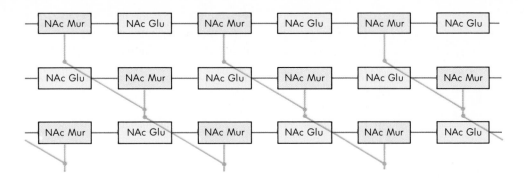

FIGURE 5-2   The peptidoglycan layer consists of three integral parts. The glycan chains are built with 10 to 12 alternating *N*-acetylglucosamine (NAc Glu) and *N*-acetylmuramic acid (NAc Mur) subunits jointed with β-1,4 glycosidic bonds. Vertical pentapeptide side chains are linked to the muramic acid subunits, and the side chains are in turn cross-linked with diagonal intrapeptide bridges. For example, the glycan chains in *S. aureus* are cross-linked with pentaglycine bridges attached to L-glycine in one pentapeptide chain and D-alanine in an adjacent chain.

linked to the peptidoglycan layer and that has the unique affinity for binding the Fc receptor of immunoglobulins IgG1, IgG2, and IgG4. This latter property, a pseudoimmune complex, has been exploited in some serologic tests where protein A–coated *S. aureus* is used as a nonspecific carrier of antibodies directed against other antigens.

Teichoic Acid

Teichoic acids are complex, phosphate-containing polysaccharides bound to both the peptidoglycan layer and cytoplasmic membrane. These polysaccharides are species specific; ribitol teichoic acid with *N*-acetylglucosamine residues (polysaccharide A) are present in *S. aureus,* and glycerol teichoic acid with glucosyl residues (polysaccharide B) are present in *S. epidermidis*. Although the teichoic acids are poor immunogens, a specific antibody response is stimulated when they are bound to peptidoglycan. The monitoring of this antibody response has been used to detect systemic staphylococcal disease.

Cytoplasmic Membrane

The cytoplasmic membrane is a complex of protein, lipids, and a small amount of carbohydrates that forms an osmotic barrier for the cell and provides an anchor site for the cellular biosynthetic and respiratory enzymes.

VIRULENCE FACTORS

Staphylococcal Toxins

At least five cytolytic, or membrane-damaging, toxins are produced by *S. aureus:* alpha, beta, delta, gamma, and leukocidin (Figure 5-3). These toxins have also been described as hemolysins, but the activities of the first four toxins are not restricted to red blood cells and leukocidin is unable to lyse erythrocytes. These cytotoxins are capable of lysing neutrophils, with the release of the lysosomal enzymes that subsequently damage the surrounding tissues. Other important staphylococcal toxins include exfoliative toxin, toxic shock syndrome toxin-1, and enterotoxins.

Alpha Toxin This toxin is cytotoxic for a number of cells, including erythrocytes, leukocytes, hepatocytes, platelets, human diploid fibroblasts, HeLa cells, and Ehrlich ascites carcinoma cells. The toxin also disrupts the smooth muscle in blood vessels. Species variation in susceptibility to alpha toxin is seen; for example, rabbit erythrocytes are 100 times more sensitive to alpha toxin than are human erythrocytes. Alpha toxin is a protein that is genetically encoded on a chromosomal, as well as plasmid, loci. Although the precise mechanism of toxin action is not known, the toxin appears to insert into hydrophobic regions of the cell membrane with the subsequent disruption of membrane integrity. The alpha toxin is believed to be

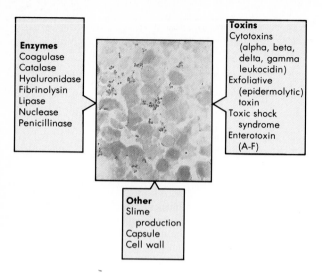

Enzymes	Toxins
Coagulase	Cytotoxins
Catalase	(alpha, beta,
Hyaluronidase	delta, gamma
Fibrinolysin	leukocidin)
Lipase	Exfoliative
Nuclease	(epidermolytic)
Penicillinase	toxin
	Toxic shock
	syndrome
	Enterotoxin
	(A-F)

Other
Slime
production
Capsule
Cell wall

FIGURE 5-3   Staphylococcal virulence factors.

an important mediator of tissue damage in staphylococcal disease.

Beta Toxin This toxin, also called sphingomyelinase C, is a heat-labile protein that is toxic for a variety of cells, including erythrocytes, leukocytes, and macrophages. This enzyme catalyzes the hydrolysis of membrane phospholipids in susceptible cells, with the amount of lysis proportional to the concentration of sphingomyelin exposed on the cell surface. Beta toxin, together with alpha toxin, is believed to be responsible for the tissue destruction and abscess formation characteristic of staphylococcal diseases.

Delta Toxin This is a thermostable, large, heterogeneous protein. The toxin has a wide spectrum of cytolytic activity, which is consistent with the belief that delta toxin disrupts cellular membranes by a detergent-like action.

Gamma Toxin This toxin is able to lyse a variety of species of erythrocytes, including human, sheep, and rabbit, as well as human lymphoblastic cells. Although two separate proteins are required for toxin

activity, the mode of action is not defined.

Leukocidin This toxin has an F component and an S component. Neither component alone has appreciable activity against the leukocyte membrane. However, the combination of the two molecules facilitates structural changes in the cell membrane and induces increased permeability. Leukocidin allows organisms to resist phagocytosis and subsequent bacterial cell death.

Exfoliative Toxin Staphylococcal scalded skin syndrome (SSSS), a spectrum of diseases characterized by exfoliative dermatitis, is mediated by the action of exfoliative toxin, also known as exfoliatin or epidermolytic toxin. The majority of stains associated with SSSS belong to phage group II, although disease has been reported with isolates of groups I and III (Table 5-2). Two distinct forms of exfoliative toxin (A and B) have been identified, but no clear relationship between specific phage groups and toxin types has been established. The exfoliative toxin gene is carried on nonchromosomal replicons.

The mechanism of toxin action is unknown. Ultrastructural studies have demonstrated that exposure to the toxin is followed by a disturbance of the adhesiveness of cells in the stratum granulosum layer of the outer epidermidis. Whether this is a result of enzymatic destruction of the intercellular cement or is due to nonenzymatic conformational changes has not been determined. The toxins are not associated with cytolysis or inflammation. Following exposure to the toxin, neutralizing antibodies develop and are protective in an experimental mouse model. This might explain why SSSS is seen only in young children and does not develop in older children or adults.

Toxic Shock Syndrome Toxin-1 Toxic shock syndrome, a staphylococcal disease charcterized by fever, hypotension, rash followed by desquamation, and involvement of multiple organ systems, is known to be mediated by the activity of a toxin. Toxic shock syndrome toxin-1 (TSST-1), formerly called pyrogenic

Table 5-2	Characteristics of Exfoliative Toxins	
Properties	Exfoliative Toxin A	Exfoliative Toxin B
Size	24,000 daltons	24,000 daltons
Temperature tolerance	Stable (100° C, 20 min)	Labile (60° C, 30 min)
EDTA treatment	Inactivated	No effect
DNA	Chromosomal	Plasmid

exotoxin C and enterotoxin F, is an exotoxin secreted during growth of the staphylococci and can reproduce most of the clinical manifestations of toxic shock syndrome in an experimental rabbit model (rash and desquamation are not seen). TSST-1 has not been found in staphylococcal isolated from all patients with toxic shock syndrome; however, most of these non-TSST-1 producing isolates are reported to product exterotoxin B. The role of this second toxin in toxic shock syndrome has not been adequately defined. The presence of TSST-1 toxin in species other than *S. aureus* remains controversial. Although the production of TSST-1 has been described for coagulase-negative staphylococci associated with toxic shock syndrome, this has not been confirmed by other investigators.

Enterotoxins Five serologically distinct staphylococcal enterotoxins (A through E) have been described (enterotoxin F is now called TSST-1). The enterotoxins are resistant to hydrolysis by gastric and jejunal enzymes and are stable to heating at 100° C for 30 minutes. These toxins are found in both *S. aureus* and *S. epidermidis,* with 30% to 50% of all *S. aureus* strains producing an enterotoxin. Although primarily associated with phage group III, other phage groups also produce the enterotoxins. Enterotoxin A is most commonly associated with disease. Enterotoxins C and D are associated with contaminated milk products, and enterotoxin B is associated with staphylococcal pseudomembranous enterocolitis. The mechanism of toxin activity is not understood because a satisfactory animal model is not available. The enterotoxins stimulate intestinal peristalsis and have a central nervous system effect, as manifested by the intense vomiting associated with this gastrointestinal disease.

Staphylococcal Enzymes

Coagulase *S. aureus* strains possess two forms of coagulase: bound (also called clumping factor) and free. Coagulase bound to the staphylococcal cell wall can directly convert fibrinogen to insoluble fibrin and cause the staphylococci to clump together. The cell-free coagulase accomplishes the same result by a somewhat different mechanism (Figure 5-4). Coagulase is used as a marker for virulence, separating *S. aureus* (coagulase positive) from the other staphylococcal species (coagulase negative). The role of coagulase in the pathogenesis of disease is speculative, but coagulase may cause the formation of a fibrin lay-

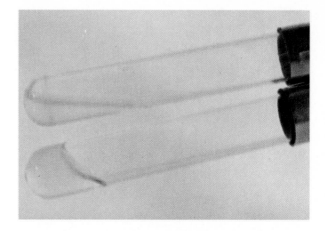

FIGURE 5-4 Coagulase reaction. Cell-free coagulase reacts with a globulin plasma factor (coagulase-reacting factor, or CRF) to form a thrombinlike factor (staphylothrombin). This factor, in turn, catalyzes the conversion of fibrinogen to insoluble fibrin. The test is performed by inoculating the staphylococci into a tube of plasma. The plasma will clot in 4 to 24 hours in a positive reaction. (From MacFaddin JF: Biochemical tests for identification of medical bacteria, ed 2 Baltimore, © 1980, William & Wilkins.)

er around a staphylococcal abscess, thus localizing the infection and protecting the organisms from phagocytosis.

Catalase All staphylococci produce catalase, an enzyme that catalyzes the conversion of hydrogen peroxide to water and oxygen. Catalase protects the organisms from the toxic hydrogen peroxide that accumulates during bacterial metabolism and that is released following phagocytosis.

Hyaluronidase This enzyme hydrolyzes hyaluronic acids, the acidic mucopolysaccharides present in the acellular matrix of connective tissue. Hyaluronidase facilitates the spread of *S. aureus* in tissues. More than 90% of *S. aureus* strains produce this enzyme.

Fibrinolysin This enzyme, also called staphylokinase, is produced by virtually all *S. aureus* strains and can dissolve fibrin clots. Staphylokinase is distinct from the fibrinolytic enzymes produced by streptococci.

Lipases All *S. aureus* and more than 30% of coagulase-negative staphylococci produce several different lipases. As their name implies, these enzymes hydrolyze lipids, which is essential for the survival of staphylococci in the sebaceous areas of the body. It is believed that these enzymes are required for invasion of staphylococci into cutaneous and subcutaneous

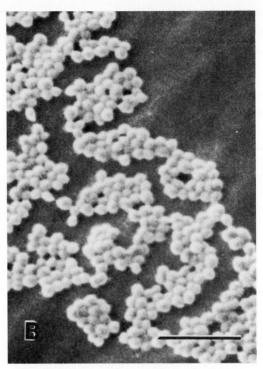

FIGURE 5-5   *S. epidermidis* and other coagulase-negative staphylococci are commonly associated with infections of prosthetic devices, shunts, and catheters. The ability of these organisms to form slime facilitates their colonization of plastic surfaces. Following attachment to the plastic, the organisms are able to erode the surface. This interaction with the surface and production of polysaccharide slime renders the organisms relatively resistant to natural immune mechanisms and antibiotic activity. In many patients the foreign body must be removed for an effective cure. **A,** Slime-producing strain forming an indistinct mat encased in the polysaccharide matrix. **B,** Non-slime-producing strain present as a mat of clearly defined cells without the adherent matrix. (From Christensen GD et al: Inf Immun 37:324, 1982.)

tissues and the formation of superficial skin infections (e.g., furuncles, or boils), and carbuncles).

Nuclease Another marker for *S. aureus* is the presence of a thermostable nuclease. The role of this enzyme in pathogenesis is unknown.

Penicillinase When penicillin was introduced for the management of bacterial infections, greater than 90% of staphylococci were susceptible. However, resistance quickly developed and was mediated by the production of penicillinase.

Other Virulence Factors

Slime Production This polysaccharide substance is produced by some strains of coagulase-negative staphylococci and facilitates their adherence to catheters and other synthetic material (e.g., graft, prosthetic valves and joints, and shunts) (Figure 5-5).

Slime also interferes with the cellular immune response by inhibiting the chemotaxis and phagocytosis of polymorphonuclear leukocytes and proliferation of mononuclear cells following mitogen exposure. Slime-producing strains are more virulent than are nonproducing strains if a foreign body (e.g., prosthetic device, catheter, shunt) is present.

Capsule and Cell Wall The capsule and cell wall components interfere with chemotaxis, opsonization, and phagocytosis. In addition, attachment to mucosal surfaces is mediated by the cell wall teichoic acids through their specific binding to fibronectin.

EPIDEMIOLOGY

Staphylococci are ubiquitous. Virtually all persons have coagulase-negative staphylococci on their skin surface, and transient colonization is common with *S. aureus,* particularly in warm, moist skin folds. *S. aureus* and coagulase-negative staphylococci are also found in the oropharynx, gastrointestinal tract, and urogenital tract. Colonization of neonates with *S. aureus* is common on the umbilical stump, skin surface, and perineal area. Carriage on colonization in older children and adults is more common in the anterior nasopharynx. Adherence to the mucosal epithelium is regulated by receptors for staphylococcal teichoic acids. Approximately 15% of normal healthy adults are persistent nasopharyngeal carriers of *S. aureus.* A higher incidence of carriage has been reported in hospitalized patients, medical personnel, individuals with eczematous skin diseases, and in individuals who regularly use needles illicitly (drug abusers) or for medical reasons (e.g., insulin-dependent diabetics, patients receiving allergy injections, or those undergoing hemodialysis).

Because staphylococci are carried on the skin surface and in the nasopharynx, shedding of the bacteria is commonplace and is responsible for many hospital-acquired infections. Staphylococci are susceptible to high temperature, as well as to disinfectants and antiseptic solutions. However, the organisms are capable of survival on dry surfaces for long periods of time. Transfer of the organisms to a susceptible individual can be either by direct contact or by means of fomites (contaminated inanimate objects). Therefore medical personnel must use proper handwashing techniques to prevent transfer of staphylococci from themselves to patients or between patients.

CLINICAL SYNDROMES

Staphylococcus aureus

S. aureus produces disease by either production of toxin or direct invasion and destruction of tissue. The clinical manifestations of some staphylococcal diseases are almost exclusively due to toxin activity (e.g., staphylococcal scalded skin syndrome, toxic shock syndrome, and staphylococcal food poisoning), whereas the other diseases involve proliferation of the organisms with abscess formation and tissue destruction (Figure 5-6).

Staphylococcal Scaled Skin Syndrome In 1878 Gottfried Ritter von Rittershain described 297 infants less than 1 month of age who had bullous exfoliative dermatitis. The disease, now called Ritter's disease or staphylococcal scalded skin syndrome (SSSS), was characterized by an abrupt onset with localized perioral erythema (redness and inflammation around the mouth) that spread to cover the entire body within 2 days. Under slight pressure the skin could be displaced (a positive Nikolsky sign). Soon afterward large bullae or cutaneous blisters formed, followed by desquamation of the epithelium. The blisters contained clear fluid with no organisms or leukocytes present (Figure 5-7). Recovery of intact epithelium occurred within 7 to 10 days.

This disease is mediated by an exfoliative (epidermolytic) toxin produced by specific strains of *S. aureus.* The disease is usually preceded by staphylococcal conjunctivitis (inflammation of the membrane covering the eye) or an upper respiratory tract infection. Growth of the organism leads to localized cutaneous inflammation, possibly resulting from the effect of alpha toxin on blood vessels, followed by the systemic distribution of exfoliative toxin via the bloodstream. Because the disease is toxin-mediated and the organism remains localized, abscess formation and tissue destruction are not observed. In addition, recovery is rapid and scarring is not seen because only the top layer of epidermis is sloughed. Mortality is low and when observed is due to secondary bacterial infection of the denuded skin areas. Neutralizing antibodies to the exfoliative toxin are produced and believed to be partially responsible for absence of disease in adults. However, whereas the toxin-producing strains of *S. aureus* can induce disease in newborn mice, disease cannot be established in adult mice. Thus other factors, such as changes in the skin sensitivity to the toxin, play a role in this disease.

A localized form of SSSS is bullous impetigo. Specific strains of toxin-producing *S. aureus* (e.g., phage type 71) are associated with superficial skin blisters. In contrast with the disseminated manifestations of SSSS, bullous impetigo is seen with localized blisters that are culture positive. Erythema does not extend beyond the borders of the blister, and the Nikolsky sign is not present. The disease is primarily restricted to infants and young children and is highly communicable.

SSSS is a disease seen almost exclusively in very young children and must be differentiated from toxic epidermal necrolysis (TEN), a disease observed primarily in older children and adults. TEN is caused by a hypersensitivity reaction to drugs.

Toxic Shock Syndrome (TSS) TSS was initially described in children, although it is now recognized

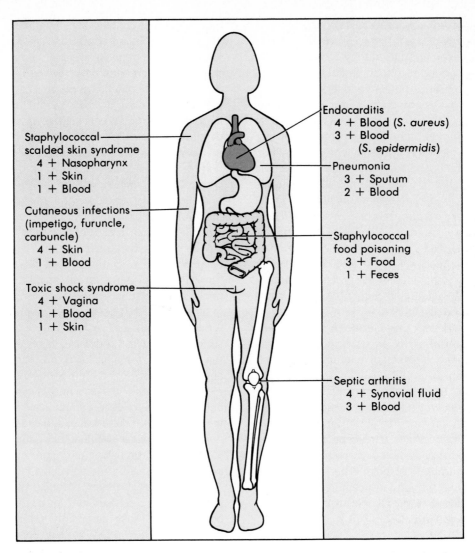

FIGURE 5-6   Staphylococcal diseases. Isolation of staphylococci from sites of infection. *1+*, <10% positive cultures; *2+*, 10% to 50% positive cultures; *3+*, 50% to 90% positive cultures; *4+*, >90% positive cultures.

as primarily a disease in menstruating women and others with localized staphylococcal infections. At present, 80% to 90% of patients with TSS are menstruating women. The disease is observed with an abrupt onset of fever, hypotension, and diffuse macular erythematous rash, as well as involvement of multiple organ systems (gastrointestinal, musculature, renal, hepatic, hematologic, central nervous system). The rash is followed by desquamation involving the entire skin surface, including the palms and soles.

The initial fatality rate was 5% to 10%, however, this has been decreased with a better understanding of the etiology and epidemiology of this disease.

Specific strains of *S. aureus,* those producing toxic shock syndrome toxin-1 (TSST-1) or a related toxin, are responsible for TSS. Vaginal carriage of toxin-producing strains has been reported in virtually all women with TSS but in less than 10% of healthy women. In the presence of hyperabsorbent tampons, these organisms can multiply rapidly and release tox-

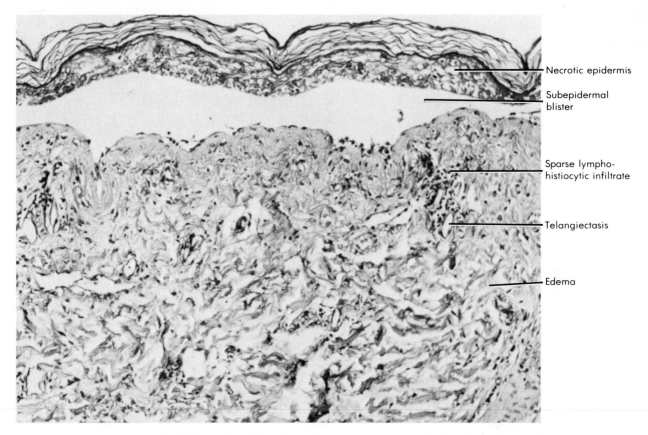

FIGURE 5-7   Staphylococcal scalded skin syndrome is characterized by intradermal separation of the skin through the stratum granulosum layer.

in for systemic distribution. Toxin production has also been associated with staphylococcal strains isolated in wounds in patients with TSS.

TSST-1 enhances by more than 1,000 times the susceptibility of rabbits to the lethal effect of endotoxin. This may explain the observed shock in TSS. Furthermore, the toxin can induce fever and dilation of peripheral blood vessels, with subsequent rash formation. The role of other staphylococcal toxins in this disease is not adequately defined.

Unless the patient is specifically treated with an effective antibiotic, the risk for recurrent disease is as high as 65%.

Staphylococcal Food Poisoning One of the most common food-borne illnesses, staphylococcal food poisoning is an intoxication rather than infection (Figure 5-8). Five serologically distinct enterotoxins (A through E) have been described, with enterotoxin

Table 5-3   Food Incriminated in 131 Staphylococcal Food Poisoning Outbreaks Reported to the Centers for Disease Control During a 5-Year Period

Food Product	Number of Outbreaks	Percentage of Total
Meat (ham, pork, beef)	50	38
Salad (potato, egg, other)	20	15
Baked foods	13	10
Poultry	13	10
Dairy (milk, cheese, butter)	4	3
Shellfish	2	2
Vegetables, fruits	2	2
Multiple sources, unknown	27	20

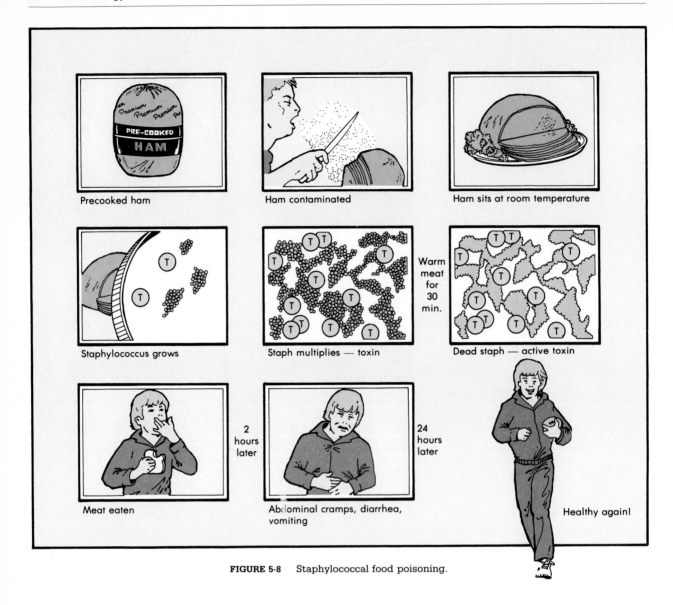

Precooked ham

Ham contaminated

Ham sits at room temperature

Staphylococcus grows

Staph multiplies — toxin

Warm meat for 30 min.

Dead staph — active toxin

Meat eaten

2 hours later

Abdominal cramps, diarrhea, vomiting

24 hours later

Healthy again!

FIGURE 5-8   Staphylococcal food poisoning.

A most commonly associated with food poisoning. Disease is due to ingestion of toxin-contaminated food rather than the organisms. The foods most commonly implicated are processed meats such as ham and salted pork, custard-filled pastries, potato salad, and ice cream (Table 5-3). Growth of *S. aureus* in salted meats is consistent with the ability of this organism to replicate selectively in nutrient media supplemented with up to 15% sodium chloride. In contrast with many other forms of food poisoning in which an animal reservoir is important, staphylococcal food poisoning is the result of contamination of the food by a human carrier. Although contamination can be avoided by excluding individuals with an obvious staphylococcal skin infection from preparing food, about one half of the carriers are asymptomatic with colonization most commonly in the nasopharynx. After the staphylococci have been introduced into the food, the food must remain at room temperature or warmer to permit the growth of the organisms and release of the toxin. The toxin is heat stable; thus cooking the contaminated food will kill the organisms but not inactivate the enterotoxin. Furthermore, the contaminated food will not appear or taste tainted.

Following ingestion of the food, the onset of disease is abrupt and rapid with a mean incubation peri-

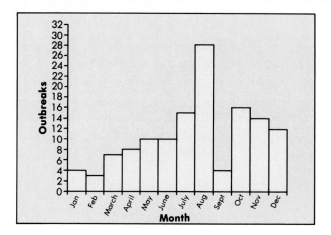

FIGURE 5-9   Staphylococcal food poisoning can be seen year-round but is most common during the summer and late year holidays (Thanksgiving and Christmas).

od of 4 hours, which is consistent with a disease mediated by preformed toxin. Further production of toxin by ingested staphylococci does not occur; so the course of disease is rapid, with symptoms lasting generally less than 24 hours. Staphylococcal food poisoning is characterized by severe vomiting, diarrhea, and abdominal pain or nausea. Sweating and headache may occur, but an elevated fever is not seen. Diarrhea is watery and nonbloody, and dehydration may result from significant fluid loss.

The toxin-producing organisms can be cultured from the contaminated food if the organisms are not killed during the food preparation (see Figure 5-8). Because the enterotoxins are heat stable, the food can be submitted to a public health facility for toxin testing. These tests are not performed in most hospital laboratories. Figure 5-9 illustrates the proportion of outbreaks of staphylococcal food poisoning during a 12-month period.

Treatment is symptomatic for relief of the abdominal cramping and diarrhea and replacement of fluids. Antibiotic therapy is not indicated. Neutralizing antibodies to the toxin can develop and be protective. However, a second episode of staphylococcal food poisoning can occur, with a second strain of staphylococcus organisms producing a serologically distinct enterotoxin.

Certain strains of *S. aureus* can also cause enterocolitis, manifested clinically by profuse watery diarrhea, fever, and dehydration. This is primarily observed in patients who have received broad-spec-

trum antibiotics, which suppress the normal colonic flora and permit the growth of *S. aureus*. The diagnosis of staphylococcal enterocolitis can be confirmed only after other infectious causes have been excluded (e.g., *Clostridium difficile* colitis). Abundant staphylococci are present, and the normal gram-negative flora are absent on Gram stains and culture of stool specimens. Fecal leukocytes are observed, and white plaques on the colonic mucosa with ulceration will be seen.

Cutaneous Infections   Localized pyogenic staphylococcal infections include impetigo, folliculitis, furuncles, and carbuncles. Impetigo is a superficial infection affecting mostly young children and is manifested primarily on the face and limbs. The initial presentation is a small macule (flattened red spot) that develops into a pus-filled vesicle (pustule) on an erythematous base (Figure 5-10). After the pustule ruptures, crusting will occur. Multiple vesicles at different stages of development are common. Impetigo is usually caused by group A Streptococcus, with *S. aureus* alone or in combination with group A Streptococcus responsible for 20% of the cases.

Folliculitis is a pyogenic infection localized to the hair follicle. The base of the follicle is raised and reddened, with a small collection of pus beneath the epidermis surface. If this occurs at the base of the eyelid, it is called a **stye**.

Furuncles, or boils, are an extension of folliculitis. Large, painful, raised nodules with an underlying collection of dead and necrotic tissue are characteristic. These can be drained spontaneously or with a surgical incision.

Carbuncles result from the coalescence of furuncles and extend to the deeper subcutaneous tissue. Multiple sinus tracts are usually present. In contrast with folliculitis and furuncles, chills and fevers are associated with carbuncles and may indicate systemic spread of the staphylococci. Bacteremia with secondary spread to other tissues is common with carbuncles.

Staphylococcal wound infections can also occur following surgery or a traumatic injury, with the introduction into the wound of organisms colonizing the skin. In an immunocompetent individual the staphylococci are not able to establish an infection unless a foreign body is present in the wound (e. g., stitches, splinter, dirt). Infections are characterized by edema, erythema, pain, and an accumulation of purulent material. If the wound is reopened and the foreign

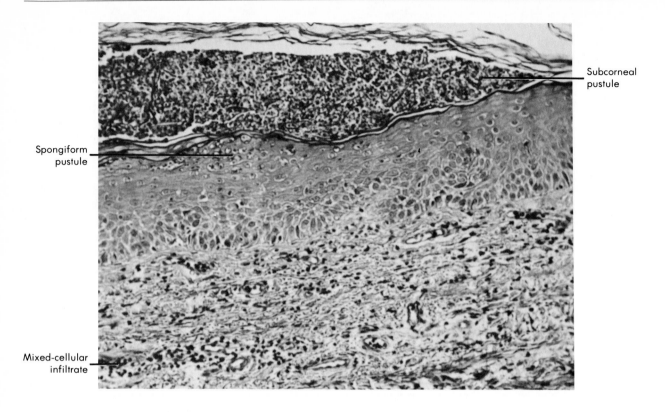

Subcorneal pustule

Spongiform pustule

Mixed-cellular infiltrate

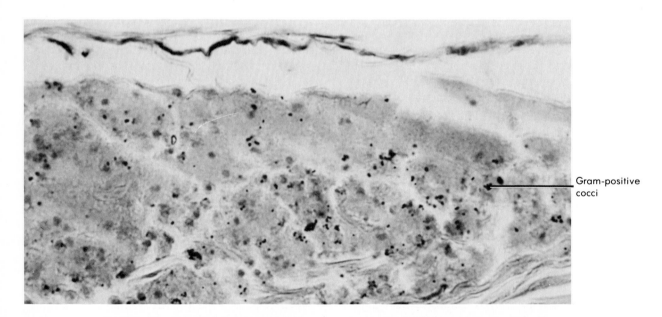

Gram-positive cocci

FIGURE 5-10   Staphylococcal impetigo. Note the invasion of polymorphonuclear (PMN) leuko-cytes and presence of bacteria. This is in contrast with staphylococcal scalded skin syndrome, where PMN leukocytes and bacteria are usually absent in the skin blisters.

matter removed with drainage of the purulence, the infection can be easily managed. If signs such as fever and malaise are observed, or if the wound does not clear with localized management, then antibiotic therapy directed against *S. aureus* is indicated.

Bacteremia and Endocarditis *S. aureus* is the most common gram-positive organism causing bacteremia. Whereas most bacteremias originate from an identifiable focus such as an infection of the skin, lungs, or other tissues, the primary source of infection is not recognized in approximately one third of all staphylococcal bacteremias. Most likely the infection spreads to the bloodstream from an innocuous-appearing skin infection. More than one half of *S. aureus* bacteremias and virtually all *S. epidermidis* bacteremias are acquired in the hospital following a surgical procedure or result from a contaminated intravascular catheter. *S. aureus* bacteremias, particularly long-dwelling episodes, are associated with dissemination to other body sites, including the heart.

S. aureus endocarditis is a serious disease, with a mortality rate approaching 50%. *S. aureus* endocarditis may initially be seen with nonspecific flu-like symptoms; however, the patient's condition will deteriorate rapidly with disruptions of cardiac output and peripheral evidence of septic embolization. Unless appropriate medical and surgical intervention is immediate, the patient's prognosis is poor. An exception to this rule is *S. aureus* endocarditis in parenteral drug abusers, whose disease normally involves the right side of the heart (tricuspid valve) rather than the left side. The initial symptoms may be mild, although fever, chills, and pleuritic chest pain caused by pulmonary emboli are generally present. Clinical cure of the endocarditis is the rule, although complications from secondary spread to other organs is common.

Pneumonia and Empyema *S. aureus* respiratory disease can develop following aspiration of oral secretions or hematogenous spread of the organism from a distal site. Aspiration pneumonia is primarily seen in the very young and the aged and in patients with cystic fibrosis, influenza, chronic obstructive pulmonary disease (COPD), or bronchiectasis. The clinical and radiologic presentations are not unique for this organism. Radiographic examination will reveal patchy infiltrates with consolidation or abscess formation, consistent with the ability of the organism to form localized abscesses. Hematogenous pneumo-nias are common in patients with endocarditis and in patients with long-dwelling bacteremias from contaminated intravascular catheters and access sites for hemodialysis.

Empyema will occur in 10% of patients with pneumonia, and *S. aureus* is responsible for one third of all cases of empyema. Because the organism is able to form walled-off areas of consolidation (loculation), drainage of the purulent material is sometimes difficult.

Osteomyelitis and Septic Arthritis *S. aureus* osteomyelitis can be a result of hematogenous infection or secondary infection resulting from trauma or an overlying staphylococcal infection. Hematogenous spread in children, generally from a cutaneous staphylococcal infection, usually involves the metaphyseal area of long bones, a highly vascularized area of bony growth. Hematogenous osteomyelitis in children is characterized by sudden onset of localized pain over the involved bone and high fever. Positive blood cultures are documented in about half of all infections. Hematogenous osteomyelitis in adults commonly occurs as vertebral osteomyelitis but rarely as an infection of the long bones. Intense back pain with fever is the initial symptom. Radiographic evidence of osteomyelitis is not seen until 2 to 3 weeks after initial signs (in both children and adults). Brodie's abscess is a sequestered focus of staphylococcal osteomyelitis of the metaphyseal area of a long bone in adults. Staphylococcal osteomyelitis following trauma or surgery is generally accompanied with evidence of inflammation and purulent drainage from the wound or sinus tract overlying the infected bone. With appropriate antibiotic therapy, and surgery when indicated, the cure rate for staphylococcal osteomyelitis is excellent.

S. aureus is the primary cause of septic arthritis in young children and in adults receiving intraarticular injections or with mechanically abnormal joints. *S. aureus* is replaced by *Neisseria gonorrhoeae* as the most common cause of septic arthritis in sexually active persons. Staphylococcal arthritis is characterized by a painful, erythematous, inflamed joint with purulence on aspiration. Infection is usually demonstrated in the large joints (e.g., shoulder, knee, hip, elbow). The prognosis is excellent in children, although in adults it is influenced by the underlying disease and the occurrence of secondary infectious complications.

Staphylococcus epidermidis

Endocarditis *S. epidermidis* and the related coagulase-negative staphylococci can infect native and prosthetic heart valves (Figure 5-11). Infections of native valves are believed due to the inoculation of organisms onto a damaged heart valve (e.g., congenital malformation, secondary to rheumatic heart disease). This form of staphylococcal endocarditis is relatively rare and more commonly associated with streptococci. In contrast, staphylococci are a major cause of artificial valve endocarditis. The organisms are introduced at the time of heart surgery, and the infection characteristically has an indolent course with clinical signs and symptoms not developing for as long as 1 year after surgery. Although infection can involve the heart valve, the more common occurrence is infection at the site where the valve is sewn to the heart tissue. Thus infection with abscess formation can lead to separation of the valve at the suture line, with mechanical heart failure. Septic embolization and persistent bacteremia are less common with staphylococcal prosthetic valve endocarditis than with other forms of endocarditis because of the nature and site of infection. Prognosis is guarded for this infection, and prompt medical and surgical management is critical.

Catheter and Shunt Infections From 20% to 65% of all infections of prosthetic devices, catheters, and shunts are caused by coagulase-negative staphylococci. This has become a major medical problem with the introduction of long-dwelling catheters for feeding and medical management of critically ill patients. *S. epidermidis* is particularly well suited for these infections by its ability to produce a polysaccharide slime that can bond it to catheters and shunts and protect it from antibiotics and inflammatory cells. Infections of cerebrospinal fluid shunts can lead to meningitis. Persistent bacteremia is generally observed because the organisms have continual access to the bloodstream. Immune complex–mediated glomerulonephritis occurs in patients with long-standing disease.

Prosthetic Joint Infections Infections of artificial joints, particularly the hip, can be caused by coagulase-negative staphylococci. Clinical manifestations are usually limited to localized pain and mechanical failure of the joint. Systemic signs such as fever and leukocytosis are not prominent, and blood cultures are usually noncontributory. Surgical replacement of the joint, together with antimicrobial therapy, is required. The risk of reinfection of the new joint is significantly increased.

Staphylococcus saprophyticus

Urinary Tract Infection Young, sexually active women are predisposed to developing symptomatic urinary tract infection with *S. saprophyticus.* The organism appears to be restricted to this population and is only infrequently found as an asymptomatic colonizer of the urinary tract. Infected women usually have dysuria (pain upon urination), pyuria (pus in urine), and large numbers of organisms in their urine. Rapid response to appropriate antibiotics and the absence of reinfections are common.

LABORATORY DIAGNOSIS

Microscopy

Staphylococcus is a gram-positive coccus that will form clusters of cocci when grown on agar media, but it is commonly seen as single cells or small groups of organisms when clinical material is examined by microscopy. The success of detecting the organism in a clinical specimen depends on the type of the infection (e. g., abscess, bacteremia, impetigo) and the quality of the material submitted for analysis. If abscess material is examined, the base of the abscess

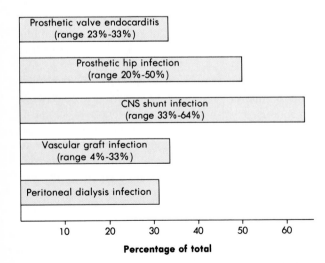

FIGURE 5-11   Role of *S. epidermidis* in infections associated with prosthetic devices.

should be scraped with a swab or curette. Large numbers of organisms should be observed in the actively growing location, but relatively few would be seen in the pus. Small numbers of organisms are generally present in bacteremic patients (an average of less than 1 org/ml of blood), so Gram stains of blood are unrewarding for most patients. Staphylococci will be seen in the nasopharynx of patients with staphylococcal scalded skin syndrome or in the vagina in patients with toxic shock syndrome, but these organisms cannot be differentiated from the staphylococci that normally colonize these sites. Large numbers of staphylococci will be seen in food implicated in staphylococcal food poisoning in more than 50% of documented outbreaks, but organisms are generally not seen in fecal specimens collected from ill patients.

Culture

Clinical specimens should be inoculated onto nutritionally enriched agar media supplemented with sheep blood. If a mixture of organisms is present in the specimen, *S. aureus* can be isolated on agar media supplemented with 7.5% sodium chloride (inhibitory for most other organisms) and mannitol (fermented by *S. aureus* but not other staphylococci). Staphylococci will grow rapidly on nonselective media both aerobically and anaerobically, with large, smooth colonies seen within 24 hours. Yellow pigmentation due to carotenoids is common with *S. aureus*, particularly when the cultures are incubated at room temperature. On sheep blood agar hemolysis caused by cytotoxins is seen with almost all isolates of *S. aureus* and relatively few strains of coagulase-negative staphylococci.

Serology

Attempts to detect staphylococcal structural antigens in blood or clinical specimens have been generally unsuccessful. However, antibodies to cell wall teichoic acids are present in many patients with prolonged *S. aureus* bacteremia with secondary foci of infection. Antibodies develop within 2 weeks of the onset of disease and are positive in virtually all patients with staphylococcal endocarditis. Although antibodies are not reliably detected in patients with sequestered foci of infection, such as osteomyelitis in adults, the presence of elevated antibody titers in a bacteremic patient is considered consistent with the need for a prolonged course of antimicrobial therapy. Antibody titers are measured by either counterimmunoelectrophoresis (CIE) or immunodiffusion. CIE is more sensitive but less specific, and at present the immunodiffusion assay is the preferred method for measuring teichoic acid antibodies.

Identification

Staphylococci can be separated from other gram-positive cocci by their microscopic and colonial morphology and biochemical properties (Table 5-4). *S. aureus*, *S. epidermidis*, and *S. saprophyticus* can be

Table 5-4   Differentiation of Staphylococcal Species

	S. aureus	*S. epidermidis*	*S. saprophyticus*
Colony color	**Yellow**-white	White	White
Hemolysis on blood agar	+	±	−
Anaerobic growth	+	+	±
Coagulase	+	−	−
Glucose fermentation	+	+	−
Mannitol fermentation	+	−	−
Heat-resistant endonuc lease	+	−	−
Protein A	+	−	−
Novobiocin	Susc.	Susc.	**Res.**
Teichoic acid			
Ribitol-*N*-acetylglucosamine	+	−	−
Glycerol-glucose	−	+	−
Glycerol-*N*-acetylglucosamine	−	−	+

differentiated by relatively simple biochemical tests. However, separation of *S. epidermidis* and *S. saprophyticus* from other coagulase-negative staphylococci is more complex and is not undertaken in most clinical laboratories.

Intraspecies characterization of isolates for epidemiologic purposes can be performed by analysis of antibiotic susceptibility patterns (antibiograms), biochemical profiles (biotyping), or susceptibility to bacteriphages (phage typing). Whereas the first two test procedures can be performed in most laboratories and are reproducible if the tests are carefully performed, phage typing is available only through a few research laboratories. Phage typing distinguishes staphylococcal strains by their pattern of susceptibility to lysis by an international collection of specific bacteriophages. Currently, five lytic groups are recognized (groups I through V). Particular strains are identified by their susceptibility to specific bacteriophages (e.g., phage type 80/81 is lysed by bacteriphages 80 and 81.

TREATMENT

Resistance quickly developed in staphylococci after penicillin was introduced, and today fewer than 10% of the strains are susceptible to this antibiotic. Resistance is mediated by production of penicillinase (β-lactamase), which hydrolyzes the β-lactam ring of penicillin. The genetic information encoding production of this enzyme is carried on transmissible plasmids, which facilitate the rapid dissemination of resistance among staphylococci.

With the problems with penicillin-resistant staphylococci, semisynthetic penicillins resistant to β-lactamase hydrolysis (e.g., methicillin, naficillin, oxacillin, dicloxacillin) were developed. Unfortunately, resistance to these antibiotics also followed, first in Europe, Scandanavia, and Japan and then more recently in the United States. Although the majority of patients with serious staphylococcal infections can be treated with these penicillins, approximately 10% to 15% of *S. aureus* strains and as many as 40% of coagulase-negative staphylococci are resistant to these penicillins. This is a particularly serious problem in large academic medical centers where resistant *S. aureus* have become well established. Although the mechanism of resistance is not definitively understood, evidence now indicates that the proteins present in the cell wall to which the penicil-

lins must bind for their effects (penicillin-binding proteins, or PBPs) are altered, which prevents binding. This phenomenon appears to be chromosomally mediated. Some strains of staphylococci also produce large quantities of β-lactamases that, when present in sufficient amounts, can hydrolyze these antibiotics. Resistance to these semisynthetic penicillins is also associated with resistance to other classes of antibiotics, including clindamycin, erythromycin, and the aminoglycosides. The detection of resistance to the semisynthetic penicillins is sometimes difficult because resistance is preferentially expressed at low temperatures (30° C).

Despite the propensity for staphylococci to develop resistance to antibiotics, virtually all strains are uniformly susceptible to vancomycin. This is the antibiotic of choice in treatment of disease caused by staphylococci resistant to β-lactam antibiotics.

PREVENTION AND CONTROL

Staphylococci are ubiquitous organisms present on the skin and mucous membranes. Introduction through breaks in the skin is frequently unavoidable. However, the number of organisms required to establish an infection (infectious dose) is generally large unless a foreign body is present in the wound (e.g., dirt, splinter, stitch). Proper cleansing of the wound and application of an appropriate disinfectant (e.g., germicidal soap, iodine solution, hexachlorophene) should prevent most infections in healthy individuals.

The spread of staphylococci from person-to-person is more difficult to manage. Surgical infections with organisms contaminating the operative site during the surgical procedure can be established by relatively few organisms because foreign bodies and devitalized tissue are present. Although it is unrealistic to sterilize the operating room personnel and environment, proper handwashing and the covering of exposed skin surfaces should minimize the risk of contamination during the operative procedure. The spread of methicillin-resistant organsism can also be difficult to control because asymptomatic nasopharyngeal carriage is the most frequent source of these organisms. Some success has been seen with chemoprophylaxis with the combination of vancomycin and rifampin.

BIBLIOGRAPHY

Christensen GD, Simpson WA, Bisno AL, and Beachey EH: Adherence of slime-producing strains of *staphylococcus epidermidis* to smooth surfaces, Inf Immun 37:318-326, 1982.

Christensen GD, Simpson WA, Bisno AL, and Beachey EH: Experimental foreign body infections in mice challenged with slime-producing *staphylococcus epidermidis,* Inf Immun 40:407-410, 1983.

Committee on Toxic Shock Syndrome. Report of toxic shock syndrome conference. I and II. Ann Intern Med 96: 1982.

Crass BA and Bergdoll MS: Involvement of coagulase-negative staphylococci in toxic shock syndrome, J Clin Microbiol 23:43-45, 1986.

Crass BA and Bergdoll MS: Toxin involvement in toxic shock syndrome, J Infect Dis 153:918-926, 1986.

Dobrin RS, Day NK, Quie PG, et al: The role of complement, immunoglobulin and bacterial antigen in coagulase-negative staphylococcal shunt nephritis, Am J Med 59:660-673, 1975.

Franson TR, Sheth NK, Rose HD, and Sohnle PG: Scanning electron microscopy of bacteria adherent to intravascular catheters, J Clin Microbiol 20:500-505, 1984.

Holmberg SD and Blake PA: Staphylococcal food poisoning in the United States: new facts and old misconceptions, JAMA 251:487-489, 1984.

Hovelius B and Mardh PA: *Staphylococcus saprophyticus* as a common cause of urinary tract infections, Rev Infect Dis 6:328-337, 1984.

Johnson GM, Lee DA, Regelmann WE, et al: Interference with granulocyte function by *staphylococcus epidermidis* slime, Inf Immun 54:13-20, 1986.

Kaplan MH and Tenenbaum MJ: *Staphylococcus aureus:* cellular biology and clinical appliation, Am J Med 72:248-258, 1982.

Latham RH, Running R, and Stamm WE: Urinary tract infections in young adult women caused by *staphylococcus saprophyticus,* JAMA 250:3063-3066, 1983.

Lennette EH, Balows A, Hausler WJ, and Shadomy HJ: Manual of clinical microbiology, ed 4, Washington, DC, 1985, American Society of Microbiology.

Lowy FD and Hammer SM: *Staphylococcus epidermidis* infections, Ann Intern Med 99:834-839, 1983.

MacFaddin JF: Biochemical tests for identification of medical bacteria. ed 2, Baltimore, 1980, Williams and Wilkins.

Mandell GL, Douglas RG, and Bennett JE: Principles and practice of infectious diseases, ed 2, New York, 1985, John Wiley & Sons.

Musher DM, Verbrugh HA, and Verhoef J: Suppression of phagocytosis and chemotaxis by cell wall components of *staphylococus aureus,* J Immunol 127:84-88, 1981.

Parsonnet J, Harrison AE, Spencer SP, et al: Nonproduction of toxic shock syndrome toxin 1 by coagulase-negative staphylococci, J Clin Microbiol 25:1370-1372, 1987.

Peterson PK, Wilkinson BJ, Kim Y et al: Influence of encapsulatum on staphylococcal opsonization and phagocytosis by human polymorphonuclear leukocytes, Inf Immun 19:943-949 1978.

Proctor RA: Fibronectin and the pathogenesis of infections, Rev Infect Dis 9(suppl. 4): S317-S430, 1987.

Rogolsky M: Nonenteric toxins of *staphylococcus aureus,* Microbiol Rev 43:320-360, 1979.

Schlievert PM: Toxic shock syndrome: an update, Lab

Sheagren JN: *Staphylococcus aureus:* the persistent pathogen, N Engl J Med 310:1368-1373, 1437-1442, 1984.

Sneath PHA, Mair NS, Sharpe ME, and Holt JG: Bergey's manual of systematic bacteriology, vol 2, Baltimore, 1986, Williams & Wilkins.

Verhoef J, Musher DM, Spika JS, et al: The effect of staphylococcal peptidoglycan on polymorphonuclear leukocytes *in vitro* and *in vivo.* Scand J Infect Dis Suppl 41:79-85, 1983.

CHAPTER 6

Streptococcaceae

The genus *Streptococcus* encompasses a diverse collection of species of gram-positive cocci that are commonly arranged in pairs or chains. Most species are facultative anaerobes, although atmospheric requirements may range from strictly anaerobic to capnophilic (growth dependent on carbon dioxide). The nutritional requirements are complex, necessitating the use of blood or serum-enriched medium for their isolation. Carbohydrates are fermented with the production of lactic acid and, unlike *Staphylococcus* species, the organisms are catalase negative.

The role of streptococci in human disease was appreciated very early (Box 6-1). However, the differentiation of streptococcal species and their role in human disease was slow to be defined precisely, resulting in at least four different schemes for classifying these organisms: clinical presentation (pyogenic, oral, enteric streptococci), serologic properties (Lancefield groupings A through H, K through V), hemolytic patterns (complete [β] hemolysis, incomplete [α] hemolysis, and no [γ] hemolysis), and biochemical (physiologic) properties (Table 6-1). The serologic classification scheme was developed by Lancefield in 1933 for differentiating pathogenic, β-hemolytic strains. Most β-hemolytic strains and some α- and nonhemolytic strains possess group-specific antigens, which are either cell wall carbohydrates or teichoic acids. These antigens can be readily detected by immunologic probes and have been useful for the rapid identification of the most common streptococcal pathogens (e.g., group A *Streptococcus*—streptococcal pharyngitis; group B *Streptococcus*—neonatal septicemia, pneumonia, and meningitis; group D *Streptococcus*—urinary tract infections

and endocarditis). Unfortunately, not all streptococci possess these group-specific cell wall antigens. Thus organisms such as *S. pneumoniae* and the numerous species of α- and nonhemolytic streptococci (collectively termed the viridans group of streptococci) must be identified by their physiologic properties.

Box 6-1 Streptococcal Disease	
S. pneumoniae	Pneumonia
	Sinusitis
	Otitis media
	Meningitis
	Bacteremia
S. pyogenes (group A)	Pharyngitis
	Pyoderma
	Rheumatic fever
	Glomerulonephritis
S. agalactiae (group B)	Neonatal meningitis
	Neonatal pneumonia
	Neonatal bacteremia
	Postpartum sepsis
Enterococcus (group D)	Urinary tract infection
	Intraabdominal abscess
	Wound infection
	Endocarditis
Viridans group	Dental caries
	Bacteremia
	Endocarditis
	Intraabdominal abscess

Table 6-1   Classification of Common Streptococcal Pathogens in Humans

Serologic Classification	Biochemical Classification	Hemolytic Pattern
A	*S. pyogenes*	β
B	*S. agalactiae*	β,α,γ
C	*S. equisimilis*	β
D	*Enterococcus faecalis*	α,γ,β
	E. faecium	α,γ,β
	S. bovis	α,γ
F	*S. milleri* group	γ,β,α
G	*S. milleri* group	γ,β,α
—	*S. pneumoniae*	α
Viridans	*S. salivarius*	α,γ
	S. sanguis	
	S. mutans	
	S. acidominimus	
	S. uberis	
	S. mitis	
	S. milleri group	

The different classification schemes for streptococci can be confusing. For some serologic categories there is a single species (e. g., the only group A *Streptococcus* of clinical significance is *S. pyogenes*. Some groups have multiple species (e. g., group D), and some species have multiple group-specific antigens (e.g., *S. milleri* groups can possess antigens for groups A, C, G, G, or K. Some species have no group-specific antigens (e. g., *S. pneumoniae*, viridans group of species), and some species have multiple hemolytic patterns (e. g., group B *S. agalactiae*). Despite the problems with these different classification schemes, it is unlikely that a uniform scheme will be developed.

STREPTOCOCCUS PNEUMONIAE

Streptococcus pneumoniae (also called pneumo-coccus and formerly Diplococcus pneumoniae) was initially isolated independently by Pasteur and Steinberg more than 100 years ago. Since that time research with this organism has increased our understanding of molecular genetics (Figure 6-1), antibiotic resistance, and vaccine related immunoprophylaxis. Unfortunately, pneumococcal disease is still a leading cause of morbidity and mortality.

Microbial Physiology and Structure

The pneumococcus is an encapsulated gram-positive coccus. The cells are 0.5 to 1.25 μm in diameter, oval or lancet-shaped, and arranged in pairs or short chains. Older cultures will decolorize readily and appear gram negative. Colonial morphology will vary. Encapsulated strains are generally large (1 to 3 mm on blood agar; smaller on chocolatized or heated blood agar), round, mucoid, and unpigmented; nonencapsulated strains are smaller and appear flat. All colonies will undergo autolysis with aging, the central portion of the colony will be depressed, and older colonies will dissolve leaving a "ghost" or remnant on the agar. Colonies will partially lyse blood agar (α-hemolysis) when incubated aerobically and may completely lyse the blood (β-hemolysis) when grown anaerobically.

The organism has fastidious nutritional requirements and is capable of growth only on enriched media supplemented with blood products (e.g., tryptic soy agar or brain heart infusion agar with blood). *S. pneumoniae* is able to ferment a number of carbohydrates, with lactic acid as the primary metabolic byproduct. In media with high glucose concentrations, *S. pneumoniae* grow poorly because lactic acid rapidly reaches toxic levels. Like all streptococci the organism lacks catalase. Unless an exogenous source of catalase is provided (e.g., from blood), the accumulation of hydrogen peroxide will inhibit growth of *S. pneumoniae*. The poor growth of isolates on chocolatized blood agar is the result of the heat-denaturation of catalase present in the blood.

Virulent strains of *S. pneumoniae* are covered with a complex polysaccharide capsule. The capsular polysaccharides are antigenically distinct and have been

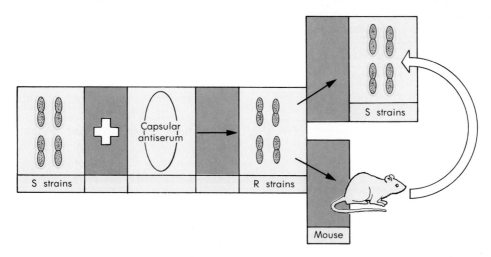

FIGURE 6-1   Genetic variation. Under the selective pressure of type-specific antiserum, rough variants of encapsulated strains of *S. pneumoniae* are selected. These R strains can revert back to encapsulated (S) strains by spontaneous back mutations. S strains can also be selected by passage through a mouse, where the virulent S strains will rapidly overgrow the avirulent R strains. Passage of pneumococci in mice is a means to maintain virulent strains in the laboratory. Avery, MacLeod, and McCarty demonstrated the DNA from a smooth strain could transform a different strain from rough to smooth. This observation formed the beginning of molecular genetics.

used for serologic classification of strains. Currently, 84 serotypes are recognized. Purified capsular material from the most common serotypes are used in a polyvalent vaccine. Antibodies directed against the capsules have also been used for diagnostic purposes.

The cell wall peptidoglycan layer of the pneumococcus is typical of gram-positive cocci, with alternating subunits of *N*-acetylglucosamine and *N*-acetylmuramic acid cross-linked by peptide bridges. The other major component of the cell wall is teichoic acid rich in galactosamine, phosphate, and choline. The presence of choline is unique to the cell wall of *S. pneumoniae* and plays an important regulatory role in cell wall hydrolysis. In the absence of choline the pneumococcal autolytic enzyme is unable to function, and cell division ceases. Two forms of teichoic acid exist in the pneumococcal cell wall: one exposed on the cell surface and a similar form covalently bound to the plasma membrane lipids. The exposed teichoic acid (also called **C-substance**) is species specific and unrelated to the group-specific carbohydrates described by Lancefield in β-hemolytic streptococci. The C-substance will precipitate a serum globulin fraction (**C-reactive protein**, or **CRP**) in the presence of calcium. CRP is present in low concentrations

in healthy individuals but is elevated in the patients with acute inflammatory diseases. The membrane lipoteichoic acid can cross-react with the Forssman surface antigens on mammalian cells. Other somatic antigens include the poorly characterized species-specific R-protein and the type-specific M-protein. Neither antigen protects against phagocytosis.

Pathogenesis

Capsule   The virulence of *S. pneumoniae* is directly associated with the capsule, which inhibits phagocytosis in the absence of specific antibodies (Box 6-2). Encapsulated (smooth) strains are able to cause disease in humans and experimental animals, whereas nonencapsulated (rough) strains are avirulent. Antibodies directed against the capsular polysaccharides protect against disease caused by immunologically related strains. The capsular polysaccharides are soluble and have been referred to as **specific soluble substance (SSS)**. Free polysaccharides can protect viable organisms from phagocytosis by binding with opsonic antibodies.

The role of toxins in the pathogenesis of pneumococcal disease has not been demonstrated. However, an undefined or presently unrecognized factor is very

---

Box 6-2 *Streptococcus pneumoniae* Virulence Factors

Capsule	Inhibits phagocytosis in absence of antibodies
Pneumolysin	Hemolysin, dermotoxic
Purpura-producing principle	Causes dermal hemorrhage
Neuraminidase	Spreading factor
Amidase	Autolysin important for cell division

---

likely responsible for the rapid onset of pneumoncoccal disease and the fulminant course observed in some groups of patients (e.g., patients with splenectomy). The following toxins and enzymes have been described.

Pneumolysin This is a temperature- and oxygen-labile hemolysin immunologically related to streptolysin 0, produced by β-hemolytic streptococci. Pneumolysin is responsible for the β-hemolysis observed when *S. pneumoniae* is grown anaerobically. The protein is dermatoxic and causes hemolytic anemia in experimental rabbit infections.

Purpura-Producing Principle This substance is released during cell autolysis and can cause dermal hemorrhage in experimental animals. A role in human disease has not been identified.

Neuraminidase This enzyme is active against cell glycoproteins and glycolipids and may play a role in the spread of pneumococci through infected tissues.

Autolysins The pneumococcal autolysin, amidase, hydrolyzes the peptidoglycan layer at the bond between *N*-acetylmuramic acid and the alanine residue on the peptide cross-bridge (Figure 6-2). If choline is absent from the cell wall teichoic acid, the amidase is inactive. Although its role in cell division is well defined, the importance of the amidase in pathogenesis is not known.

Epidemiology

S. pneumoniae is a common inhabitant of the throat and nasopharynx of healthy individuals. Car-

riage has been reported to range from 5% to 75%, but these percentages are significantly affected by the methods used to detect the organism. Intraperitoneal inoculation of mice with specimens is the most sensitive method for isolating *S. pneumoniae;* however, this technique is rarely used. Colonization is more common in children than in adults, with *S. pneumoniae* initially detected at about 6 months of age (Figure 6-3). Subsequently, the child is transiently colonized with other serotypes of the organism. The duration of carriage decreases with each successive serotype carried, in part related to the development of serotype-specific immunity. Acquisition of new serotypes occurs throughout the year, although carriage and associated disease is highest during the winter and spring months. The strains of pneumococci that cause disease are the same ones associated with carriage. When infection occurs, it is generally with a newly acquired serotype rather than one associated with prolonged carriage.

Pneumococcal disease originates from spread of organisms colonizing the nasopharynx and oropharynx to distal loci: lungs (pneumonia), paranasal sinuses (sinusitis), ears (otitis media), and meninges (meningitis) (Figure 6-4). Bacteremia, with subsequent spread to other body sites, can occur with all of these infections.

Clinical Syndromes

Pneumonia *S. pneumoniae* is the most common cause of bacterial pneumonia in the United States. Infections are caused by aspiration of the endogenous oral organisms. Although strains can be transferred from one person to another by droplets in a closed population, epidemics are rare.

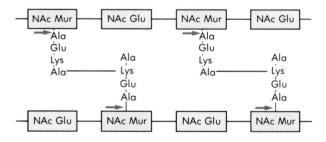

FIGURE 6-2 Pneumococcal cell wall. The peptidoglycan layer consists of alternating *N*-acetylglucosamine and *N*-acetylmuramic acid residues cross-linked with tetrapeptide bridges. The site of amidase activity is indicated by the arrows.

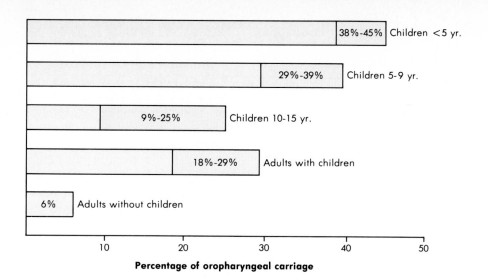

FIGURE 6-3 Carriage of *S. pneumoniae* is age-related. The highest incidence is observed in young children and adults with young children present in the home.

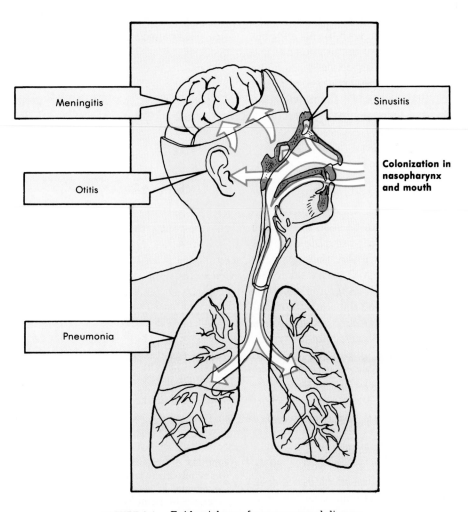

FIGURE 6-4 Epidemiology of pneumococcal disease.

Disease occurs when the natural defense mechanisms (epiglottal reflex, trapping of bacteria by the mucus-producing cells lining the bronchus, removal of organisms by the ciliated respiratory epithelium and cough reflex) are circumvented, permitting organisms colonizing the oropharynx to gain access into the lower airways. Pneumococcal disease is most commonly associated with an antecedent viral respiratory disease such as influenza or measles, or with other conditions that interfere with bacterial clearance, such as chronic pulmonary disease, alcoholism, congestive heart failure, diabetes mellitus, and chronic renal disease.

The pathogenesis of pneumococcal pneumonia is due to bacterial multiplication in the alveolar spaces. Following aspiration, the bacteria undergo rapid growth in nutrient-rich edema fluid. Erythrocytes, leaking from congested capillaries, accumulate in the alveoli, followed by the migration of first neutrophils and then alveolar macrophages, with the subsequent phagocytosis and destruction of the organisms. The clinical manifestations of pneumococcal pneumonia are abrupt in onset, with a severe shaking chill and sustained fever of 102° to 105° F. Commonly, the patient has symptoms of a viral respiratory infection 1 to 3 days before the initial onset. A productive cough with blood-tinged sputum is seen in most patients, and chest pain (pleurisy) is common. Because the disease is associated with aspiration, disease is generally localized in the lower lobes of the lungs—hence the name **lobar pneumonia**. However, a more generalized bronchopneumonia can be seen in children or the elderly. Recovery is usually rapid after the initiation of appropriate antimicrobial therapy, with complete radiologic resolution in 2 to 3 weeks. The overall mortality is 5%, although this is influenced by the serotype of the organism and the age and underlying disease of the patient. Mortality is significantly increased in disease caused by *S. pneumoniae* type 3, as well as in elderly patients or in patients for whom bacteremia is documented. Severe pneumococcal disease is also observed in patients with splenic dysfunction or splenectomy due to decreased bacterial clearance from the bloodstream and defective production of early antibodies.

Abscess formation is not commonly associated with pneumococcal pneumonia except with specific serotypes (e.g., serotype 3). Pleural effusions are seen in about 25% of patients with pneumococcal pneumonia, and empyema (purulent effusion) is a rare complication.

Sinusitis and Otitis Media *S. pneumoniae* is a common cause of acute infections of the paranasal sinuses and ear. Disease is usually preceded by a viral infection of the upper respiratory tract, leading to infiltration with polymorphonuclear leukocytes and obstruction of the sinuses and ear canal. Middle ear infection (otitis media) is primarily seen in young children; bacterial sinusitis can occur at all ages.

Meningitis Spread of *S. pneumoniae* into the central nervous system can follow bacteremia, infections of the ear or sinuses, or head trauma with communication between the subarachnoid space and the nasopharynx. Bacterial meningitis can occur at all ages, although it is primarily a pediatric disease. Pneumococcal meningitis is relatively uncommon in neonates; however, about 15% of meningitis in children and 30% to 50% of adult disease is caused by *S. pneumoniae*. Mortality and severe neurologic deficits are from four times to 20 times more common with disease caused by *S. pneumoniae* compared with the

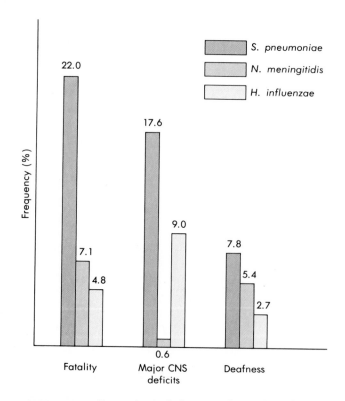

FIGURE 6-5   Comparison of the mortality and morbidity associated with bacterial meningitis caused by *Streptococcus pneumoniae, Neisseria meningitidis,* and *Haemophilus influenzae.* (Redrawn from Christensen G et al: Infect Immun 37:318, 1982.)

other common causes of bacterial meningitis (Figure 6-5).

Bacteremia Bacteremia will occur in 25% to 30% of patients with pneumococcal pneumonia and more than 80% of patients with meningitis. In contrast, bacteria are generally not present in the bloodstream of patients with sinusitis or otitis media. Endocarditis can occur in patients with normal or previously damaged heart valves. Destruction of the valve tissue is common.

Laboratory Diagnosis

Microscopy Examination of sputum by Gram stain is a rapid method for diagnosing pneumococcal disease. The organisms will appear as lancet-shaped diplococci surrounded by an unstained capsule. *S. pneumoniae* may appear gram negative because they can be easily over-decolorized during the staining procedure. If the Gram stain is consistent with *S. pneumoniae*, this can be confirmed with a quellung reaction (Figure 6-6).

Culture Specimens should be inoculated onto an enriched nutrient medium supplemented with blood. Recovery of *S. pneumoniae* in sputum cultures from patients with pneumonia is frequently difficult because the organism grows slowly because of its fastidious nutritional requirements and is rapidly over-grown by contaminating oral bacteria. A selective medium such as blood agar with 5 μg/ml gentamicin has been used with some success to isolate the organism from sputum specimens; however, some technical skill is still required to separate *S. pneumoniae* from the other α-hemolytic streptococci frequently present in the specimen. The definitive diagnosis of the organism responsible for sinusitis or otitis dictates that an aspirate from the infected focus be obtained (Figure 6-7). The isolation of *S. pneumoniae* from cerebrospinal fluid is usually without problems, unless antibiotic therapy is initiated before the specimen is collected. In the presence of even a single dose of antibiotics as many as one half of the cultures will be negative.

Identification Isolates of *S. pneumoniae* undergo rapid lysis when the autolysins are activated following exposure to bile. A presumptive identification can be made by placing a drop of bile on an isolated colony. Colonies of *S. pneumoniae* will be solubilized within a few minutes, whereas other α-hemolytic streptococci will be unchanged. This test can also be performed by adding bile to a broth culture of *S. pneumoniae* with the rapid lysis of the organisms and clearing of the broth. *S. pneumoniae* can also be identified by its susceptibility to **optochin** (ethylhydrocupreine dihydrochloride). The isolate is streaked onto a blood agar plate, and a disk saturated with optochin is placed in the middle of the inoculum. After overnight incubation a zone of inhibited bacterial growth will be seen around the disk. Additional biochemical and serologic tests can be performed for a definitive identification (Table 6-2).

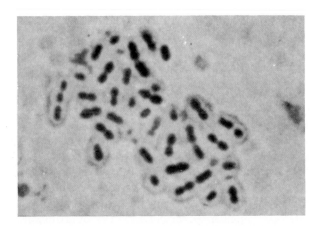

FIGURE 6-6 Quellung reaction. The polysaccharide capsule of *S. pneumoniae* can be demonstrated by mixing the bacteria with specific anticapsular antibodies and the examining the suspension microscopically. A positive reaction is seen by an increased refractiveness around the bacteria. Electron micrograph of *S. pneumoniae* type 1 after exposure to type 1 antibodies.

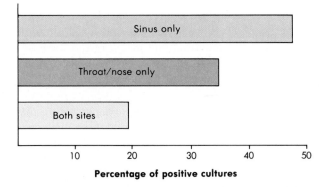

FIGURE 6-7 Correlation between cultures of sinus aspirates and throat or nasopharyngeal cultures. Throat and nasopharyngeal cultures are insensitive and nonspecific indicators of the etiology of sinusitis.

Table 6-2 Presumptive Identification of Streptococci

Organism	Susceptibility to:			Hydrolysis of:		Growth in:		Lysis by Bile
	Bacitracin	**Optochin**	**CAMP**	**Hippurate**	**Esculin**	**Bile**	**6.5% NaCl**	
β-HEMOLYTIC STREPTOCOCCI								
Group A	S	R	−	−	−	−	−	−
Group B	R†	R	+	+	−	−	+†	−
GROUP D STREPTOCOCCI								
Enterococcus*	R	R	−	−†	+	+	+	−
Nonenterococci	R	R	−	−	+	+	−	−
VIRIDANS GROUP	R†	R	−	−	−†	−†	−	−
S. PNEUMONIAE	R	S	−	−	−	−	−	+

*Although historically classified as a group D *Streptococcus*, *Enterococcus* is now recognized as a separate genus.
†Test variations can be seen.

Antigen Detection Soluble pneumococcal capsular polysaccharide can be rapidly detected in infected body fluids by immunologic assays such as counterimmunoelectrophoresis (CIE) or latex agglutination. The tests are very sensitive (able to detect 50 ng of antigen) and specific, although serotypes 7 and 14 are not detected by routine CIE procedures. The antigen detected in cerebrospinal fluid and serum is identical to the purified capsular polysaccharide; however, the antigen is hydrolyzed before it is excreted into the urine and has only partial immunologic identity.

Treatment

Before the availability of antibiotics, specific treatment was guided by the passive infusion of type-specific capsular antibodies. These opsonizing antibodies enhanced polymorphonuclear leukocyte–medicated phagocytosis and killing of the bacteria. However, this immunotherapy was discontinued with the advent of antimicrobial therapy. Penicillin rapidly became the treatment of choice for pneumococcal disease. For patients allergic to penicillin alternative effective agents have included the cephalosporins, erythromycin, and chloramphenicol (for meningitis). Resistance to tetracycline is well documented. In 1977 isolates of *S. pneumoniae* resistant to multiple antibiotics, including penicillin, were reported in South Africa. Although high-level resistance to penicillin (minimum inhibitory concentration, or MIC ≥ 2μg/ml) is still relatively uncommon, approximately 5% of all strains of *S. pneumoniae* isolated in the United States are considered to be moderately resistant (between MIC 0.1 and 1.0 μg/ml). Furthermore, a gradual but definite increase in MIC values (from a range of 0.006-0.008 to 0.03-0.05 μg/ml) has been noted with naturally occurring strains of *S. pneumoniae*. The overall mortality rate is higher with the moderately and highly resistant strains, and high doses of penicillin or an alternative antibiotic are required for successful treatment. Increased resistance to penicillins is not related to β-lactamase hydrolysis of the antibiotic. Rather, resistance is associated with a decreased affinity of the antibiotic for the penicillin-binding proteins present in the bacterial cell wall.

Prevention and Control

Because the pneumococcal capsule is the primary virulence factor associated with this organism, efforts to prevent or control the disease have focused on the development of an effective vaccine. The first vaccine was used in 1914 in South African gold miners, a population at high risk for pneumococcal disease. Over the ensuing years the vaccines have been refined and improved to the current vaccine, which contains 23 different capsular polysaccharides. Approximately 94% of all strains isolated from infected patients are either included in the vaccine or are serologically related to the vaccine serotypes. Longitudinal studies have documented that the introduction of the vaccine has not influenced the serotypes of *S. pneumoniae* that are associated with disease. The vaccine is immunogenic in normal adults, and the immunity is long lived. However, the effectiveness of the vaccine in patients at risk for pneumococcal disease (e.g., patients with asplenia, sickle cell disease, a hematologic malignancy, a renal transplant, or the elderly) is less satisfactory.

GROUP A *STREPTOCOCCUS*

Group A *Streptococcus,* also called *Streptococcus pyogenes,* is an important cause of bacterial pharyngitis, impetigo, and pyoderma. Additionally, the organism is responsible for nonsuppurative sequelae—acute rheumatic fever and glomerulonephritis.

Microbial Physiology and Structure

Group A streptococci are 0.5 to 1.0 μm spherical cocci (compared with the lancet-shaped diplococci of *S. pneumoniae*) that form short chains in clinical specimens and longer chains when grown in liquid media. Growth is optimal on enriched blood agar media but is inhibited if the medium contains a high concentration of glucose. After 24 hours of incubation, 1 to 2 mm white colonies with a large zone of β-hemolysis are observed. Encapsulated strains will appear mucoid on freshly prepared media, but on dry media they appear wrinkled (matt appearance). Nonencapsulated colonies are smaller and glossy.

The antigenic structure of group A streptococci is well-defined (Figure 6-8). The outermost layer of the cell is the capsule, which is composed of hyaluronic acid, identical to that found in connective tissue. For this reason the capsule is nonimmunogenic (again in contrast with *S. pneumoniae*). Although the capsule is present in actively growing cells, it rapidly diffuses into the extracellular space in nondividing cells.

The basic structural framework of the cell wall is the peptidoglycan layer, similar in composition to that found in other gram-positive bacteria. Within the cell wall are the group- and type-specific antigens of group A streptococci. The **group-specific carbohydrate** of group A streptococci, which is approximately 10% of the dry weight of the cell, is a dimer of *N*-acetylglucosamine and rhamnose. Three type-specific protein antigens have also been identified. The **M protein** is a major antigen associated with virulent streptococci. In the absence of the M protein the strains are not infectious. This protein is located on the end of the hairlike fimbriae that are anchored in the cell wall and extend through the capsule. Thus the M protein is exposed in encapsulated strains. Two other type-specific proteins are the T and R antigens. The **M and T (trypsin-resistant) proteins** are particularly important epidemiologic markers of

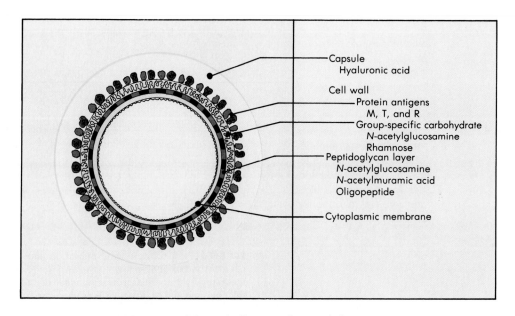

FIGURE 6-8   Schematic diagram of group A *Streptococcus.*

group A strains. The last important antigenic structure in group A streptococci is the **lipoteichoic acid** antigen. This also is associated with fimbriae and is exposed on the cell surface.

Pathogenesis

The virulence of group A streptococci is determined by a variety of structural molecules and elaborated toxins and enzymes (Box 6-3). Although the precise role each plays in pathogenesis is often not well defined, it is clear that M protein and lipoteichoic acid are particularly important in the establishment of infection, and the observed clinical manifestations can be directly attributed to molecules such as the erythrogenic toxins and streptolysins.

Capsule The hyaluronic acid capsule of group A streptococci is nonimmunogenic and protects the cells against phagocytosis. However, the major antiphagocytic structural component is the M protein.

M Protein In the absence of specific antibodies against the M protein, the cells are protected against phagocytosis. The M protein also prevents interaction with complement. The other structural proteins,

T and R, have no defined role in the virulence of group A streptococci.

Lipoteichoic Acid The ability of group A streptococci to bind to epithelial cells in the mouth and on the skin is mediated by lipoteichoic acid exposed on the cell surface (Figures 6-9 and 6-10).

Erythrogenic Toxin This toxin is produced by lysogenic strains of streptococci, similar to the diphtheria toxin produced in *Corynebacterium diphtheria*. Three immunologically distinct, heat-labile toxins (A, B, and C) have been described in group A streptococci and in rare strains of groups C and G streptococci. The toxin is responsible for the rash observed in scarlet fever, although it is unclear whether this is due to the direct effect of the toxin on the capillary bed or is secondary to a hypersensitivity reaction. Toxin injected intradermally will produce localized erythema at 24 hours in a susceptible individual **(Dick test).** Antitoxin injected intradermally in a patient with scarlet fever will produce localized blanching, indicating neutralization of the erythrogenic toxin **(Schultz-Charlton reaction).** Neither test is currently used for diagnostic purposes.

Streptolysin S and O Streptolysin S is an oxygen stable, nonimmunogenic cell-bound hemolysin capable of lysing erythrocytes, as well as leukocytes and platelets, following direct cell contact. Streptolysin S can also stimulate release of lysosomal contents after engulfment, with subsequent death of the phagocytic

Box 6-3	Group A *Streptococcus* Virulence Factors
Capsule	Nonimmunogenic
M protein	Antiphagocytic, anticomplementary
Lipoteichoic acid	Mediates adherence to epithelial cells
Erythrogenic toxins	Mediates scarlatiniform rash
Streptolysin S	Lyses leukocytes, platelets, and erythrocytes; stimulates release of lysosomal enzymes; nonimmunogenic
Streptolysin O	Lyses leukocytes, platelets, and erythrocytes; stimulates release of lysosomal enzymes; immunogenic
Streptokinases	Lyses blood clots, facilitates spread of bacteria in tissues
DNAse	Depolymerizes cell-free DNA in purulent material

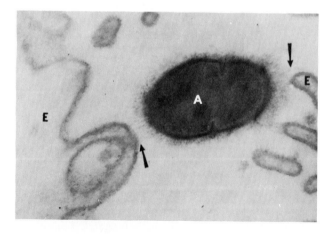

FIGURE 6-9   Electron micrograph of group **A** streptococcus *(A)* attached to pharyngeal epithelial cells *(E)*. Attachment is mediated by fimbriae (arrow) composed of streptococcal lipoteichoic acid and **M** protein. (Reproduced from The Journal of Experimental Medicine, 1976, 143:759, by copyright permission of The Rockefeller University Press.)

cell. Streptolysin O is inactivated reversibly by oxygen and irreversibly by cholesterol. Unlike streptolysin S, antibodies are readily formed against streptolysin O and are useful for documenting a recent infection **(ASO test).** This hemolysin will cross-react with similar oxygen-labile toxins produced by *S. pneumoniae* and *Clostridium* species. Streptolysin O is also capable of killing leukocytes by lysis of their cytoplasmic granules with release of hydrolytic enzymes.

Streptokinases At least two forms (A and B) have been described. These enzymes are capable of lysing blood clots and may be responsible for the rapid spread of group A streptococci in infected tissues.

DNAse Four immunologically distinct deoxyribonucleases (A through D) have been identified. These enzymes are not cytolytic but are capable of depolymerizing free DNA present in pus. This reduces the viscosity of the abscess material and facilitates spread of the organisms. Antibodies developed against DNAse B are an important marker of cutaneous group A streptococcal infections.

Other Enzymes Other enzymes have been described, including hyaluronidase ("spreading factor") and diphosphopyridine nucleotidase (DPNase), but their role in pathogenesis is unknown.

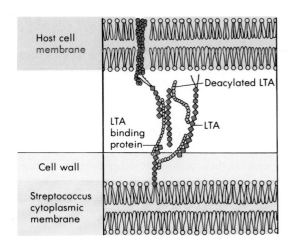

FIGURE 6-10 Schematic representation of attachment of group A streptococcus to an oral epithelial cell. The ionic interaction between negatively charged lipoteichoic acid (*LTA*) and positively charged protein (M protein) results in a fibrillar network. This exposes the glycolipid moiety of LTA to fibronectin receptors on the epithelial cell. (Modified from Ofek I et al: J Bacteriol 149:426, 1982.)

Epidemiology

Group A streptococci commonly colonize the oropharynx of healthy children and young adults. Although the incidence of carriage is reported to be 15% to 20%, these figures are misleading. Highly selective culture techniques are required to detect small numbers of organisms in oropharyngeal secretions. Colonization with group A streptococci is transient, regulated by the individual's ability to mount specific immunity to the M protein of the colonizing strain and the presence of competitive organisms in the oropharynx. Bacteria such as the α- and nonhemolytic streptococci are able to produce antibiotic-like substances called **bacteriocins,** which suppress the growth of group A streptococci. In general, group A streptococcal disease is caused by recently acquired strains that are able to establish an infection of the pharnyx or skin before specific antibodies are produced or competitive organisms are able to proliferate.

Suppurative Streptococcal Disease: Clinical Syndromes

Pharyngitis Group A *Streptococcus* is the major cause of bacterial pharyngitis, with group C and G occasionally involved. This is primarily a disease of children between the age of 5 to 15 years, but infants and adults are also susceptible. The pathogen is spread by person-to-person contact via respiratory droplets. Crowding, such as in classrooms and with play activities for children, increases the opportunity for spread of the organism.

Infection generally develops 2 to 4 days after exposure, with an abrupt onset of sore throat, fever, malaise, and headache. The posterior pharynx can appear erythematous with an exudate, and cervical lymphadenopathy can be prominent. Despite these clinical signs and symptoms, it is difficult to differentiate streptococcal pharyngitis from viral pharyngitis. For example, only about 50% of patients with "strep throat" will have pharyngeal or tonsillar exudates. Likewise, most young children with exudative pharyngitis will have viral disease. The specific diagnosis can be made only by bacteriologic or serologic tests.

Scarlet fever is a complication of streptococcal pharyngitis (Figure 6-11) seen when the infecting strain is lysogenized by a temperate bacteriophage that stimulates production of erythrogenic toxin.

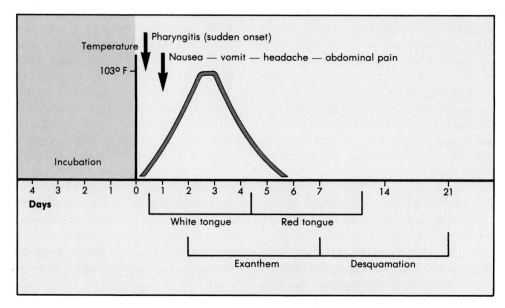

FIGURE 6-11   Evolution of signs and symptoms of scarlet fever. (Modified from Habif TP: Clinical dermatology: a color guide to diagnosis and therapy, St Louis, 1985, The CV Mosby Co.)

Within 1 to 2 days after the initial clinical symptoms of pharyngitis, a diffuse erythematous rash will initially appear on the upper chest and then spread to the extremities. The area around the mouth is generally spared (circumoral pallor), as are the palms and soles. The tongue will initially be covered with a yellowish-white coating that will later be shed revealing a red, raw surface ("strawberry tongue"). The rash, which blanches upon pressure, is best seen on the abdomen and in skin folds (Pastia's lines). The rash will disappear over the next 5 to 7 days and is followed by desquamation.

Suppurative complications of streptococcal pharyngitis rarely occur with the advent of antimicrobial therapy. However, abscesses of the peritonsillar and retropharyngeal areas can be observed, as well as disseminated infections to the brain, heart, bones, and joints.

Pyoderma Streptococcal skin infections most commonly occur in warm, moist environments during the summer months. Pyoderma is seen primarily in young children (2 to 5 years of age) with poor personal hygiene, Clinical disease is preceded by initial colonization of the skin with group A streptococci via direct contact with another infected child or fomite,

or transfer by an arthropod vector. Introduction into the subcutaneous tissues is by minor break in the skin integrity (e.g., scratch or insect bite). Group A streptococci are responsible for the majority of streptococcal skin infections, although group C and G have also been implicated. The strains of streptococci that cause skin infections are different from those that cause pharyngitis, although pyoderma serotypes can colonize the pharynx and establish a persistent carriage state.

Other Suppurative Diseases Group A streptococci have been associated with a variety of other suppurative infections, including cellulitis, puerperal sepsis, lymphangitis, pneumonia, and others. Although these infections can be seen occasionally, they have become exceedingly rare with the introduction of antibiotic therapy.

Nonsuppurative Streptococcal Disease: Clinical Syndromes

Rheumatic Fever Rheumatic fever is a nonsuppurative complication of group A streptococcal disease. It is characterized by inflammatory changes of the heart, joints, blood vessels, and subcutaneous tissues. Chronic, progressive damage to the heart valves may occur, although the specific mechanisms of tissue damage are unknown. A number of theories have been proposed, including (1) direct destruction of the tissue by the organism or a streptococcal enzyme (e.g., streptolysin), (2) serum sickness–like reaction mediated by complexes of antibodies and antigens, and (3) an autoimmune reaction. This latter explanation is currently favored because antibodies directed against heart tissue have been identified in patients with uncomplicated streptococcal disease and rheumatic heart disease. These antibodies can bind to cardiac and skeletal muscles, as well as to the smooth muscles in blood vessels.

Rheumatic fever is associated with specific serotypes of group A *Streptococcus* and only with upper respiratory infections. Cutaneous streptococcal infections do not initiate rheumatic fever. The epidemiology of the disease mimics streptococcal pharyngitis—it most commonly occurs in young school-age children, with no male or female predilection, and it occurs during the fall or winter months. Rheumatic fever usually follows severe streptococcal disease, although as many as one third of the patients will have had an asymptomatic infection with group A *Streptococcus*. Recurrence will occur with subsequent streptococcal infection if antibiotic prophylaxis is not used. The risk for recurrence will decrease with time.

Because no specific diagnostic test is available to identify patients with rheumatic fever, diagnosis is made by clinical parameters. The revised criteria of Jones are currently used (Table 6-3). Critical to the diagnosis is documentation of recent group A streptococcal disease by either culture or serologic testing.

Acute Glomerulonephritis The other nonsuppurative complication of streptococcal disease is acute glomerulonephritis, which is characterized by acute inflammation of the renal glomerulus, with edema, hypertension, hematuria, and proteinuria. Specific nephritogenic strains of group A streptococci are associated with this disease. The pharyngeal strains and pyodermal strains differ. The epidemiology of the disease is similar to the initial streptococcal infection (Table 6-4). The clinical diagnosis is based on the clinical presentation and evidence of a recent group A streptococcal infection. Young patients generally have an uneventful recovery, whereas the long term prognosis for adult patients is unclear. Progressive, irreversible loss of renal function has been reported in adults.

Laboratory Diagnosis

Microscopy A rapid, preliminary diagnosis of group A streptococcal pyoderma can be made with a Gram stain. Streptococci do not normally colonize the skin surface; thus the presence of gram-positive cocci in pairs and chains associated with leukocytes is significant. In contrast, streptococci are part of the normal oropharyngeal flora so their presence in a respiratory specimen collected from a patient with pharyngitis has a poor predictive value. However, if the organisms are associated with disrupted leukocytes (presumably due to the release of streptolysin), the Gram stain has been shown in experienced hands to be a rapid test for the diagnosis of streptococcal

---

Table 6-3   Jones Criteria for Diagnosing Rheumatic Fever

Major Manifestations	Minor Manifestations	Supporting Evidence
Carditis	Fever	Increased ASO antibodies
Polyarthritis	Arthralgia	Throat culture with
Chorea	Previous rheumatic	group A *Streptococcus*
Erythema marginatum	fever	Recent scarlet fever
Subcutaneous nodules		

Modified from Rodnan GP and Schumacher HR, editors: Primer on rheumatic diseases, ed. 8, Atlanta, 1983, The Arthritis Foundation.
The presence of two major criteria or of one major and two minor criteria indicates a high probability for the presence of rheumatic fever if supported by evidence of a preceding streptococcal infection.

Table 6-4 Epidemiologic Features of Streptococcal Glomerulonephritis

Feature	Pharyngitis-Associated Acute Glomerulonephritis	Pyoderma-Associated Acute Glomerulone-phritis
Seasonal occurrence	Winter and spring	Late summer and early fall
Geographic distribution	Common in temperate and cold climates	Common in hot or tropical climates
Age	School-age children	Preschool-age children
Familial occurrence	Common	Common
Attack rate after infection with nephritogenic strain	10%-15%	10%-15%
Carrier state	Pharynx (common)	Skin (rare)
Serologic types	Limited in pharynx	Limited to skin
Antistreptolysin 0 response	Common	Uncommon
Anti-DNAse B response	Common	Common

Modified from Wannamaker LW: Differences between streptococcal infections of the throat and of the skin, N Engl J Med 282:23, 1970.

pharyngitis. When properly performed, the Gram stain has been reported to be more reliable than diagnosis of streptococcal pharyngitis by clinical parameters alone. Unfortunately, the accurate interpretation of these stains requires technical experience and cannot be recommended for the inexperienced.

Culture The proper specimen must be collected for the isolation of group A streptococci. Specimens collected from posterior oropharyngeal sites from infected patients (e.g., tonsils, posterior pharynx, posterior tongue) yield quantitatively more group A streptococci than do specimens from the anterior areas of the mouth. This is logical, because there is a greater chance of isolating the pathogen when the site of infection is sampled. However, it should also be appreciated that the mouth, particularly saliva, is colonized with bacteria that inhibited the growth of group A streptococci. Contamination of even a well-collected specimen may obscure or suppress the growth of group A streptococci. The recovery of group A streptococci from skin infections is less of a problem. The crusted top of the lesion is raised, and the purulent material and base of the lesion cultured. Group A streptococci should be recovered in large numbers. Open, draining skin pustules should be avoided because these might be superinfected with staphylococci.

As discussed previously, streptococci have fastidious growth requirements. To suppress the oral bacterial flora, antibiotics (e.g., trimethoprim-sulfamethoxazole) have been added to blood agar plates. Although these selective plates have proved to be very useful, a delay in the growth of group A streptococci has been observed and necessitates prolonged incubation (2 to 3 days) of the cultures. It is also unclear what atmosphere of incubation should be used. Although both streptolysin O and S are expressed in an oxygen-free (anaerobic) atmosphere, group A streptococci will be confused under these conditions with other β-hemolytic streptococci present in the oropharynx. Because virtually all group A streptococci produce streptolysin S, cultures can be incubated in air.

Identification Historically, group A streptococci were identified by their susceptibility to **bacitracin.** A disk saturated with bacitracin is placed onto a plate seeded with group A streptococci and, after overnight incubation, a zone of inhibited growth is considered positive for group A streptococci. Some α-hemolytic strains are inhibited by bacitracin. If errors in interpretation of the hemolytic patterns are made, then these organisms would be misidentified. This is particularly a problem in small laboratories with limited experience. A bacitracin disk can be placed directly onto the primarily seeded culture (direct test). This procedure will identify approximately 50% of the positive cultures, so negative tests must be confirmed. Definitive identification of group A streptococci is made by demonstrating the group-specific carbohydrate. This was not routinely practical until the recent introduction of direct antigen detection tests.

Antigen Detection Specific identification of group A streptococci can be made directly by the detection of the group-specific antigen. Isolated colonies or the direct clinical specimen can be used. The group-specific antigen is extracted with nitrous acid or less commonly by enzymatic methods. The extract is then mixed with specific antibodies, either bound to latex particles or immobilized on a filter membrane (ELISA procedure). Agglutination of the latex particles or development of a positive indicator in the ELISA procedure represents a positive test. These assays have been shown to be very specific. The test sensitivity is more difficult to assay. Most evaluations have demonstrated that these immunologic assays will detect large numbers of organisms in the clinical specimen but are relatively insensitive if only a few organisms are present. Serologic assays have demonstrated that significant disease can be associated with relatively few cultured organisms, presumably due to the collection of an inadequate sample. Thus these rapid identification tests can be used for the initial screening of a specimen, and—if positive—no additional testing is required. If negative, the specimen should be cultured. If small numbers of organisms are isolated, then clinical and serologic tests can be used to differentiate between colonization and disease.

Antibody Detection Patients with group A streptococcal disease produce antibodies to a number of specific enzymes. Although antibodies against the M protein are produced and are important for immunity to develop, these antibodies are not measured because they appear late in the clinical course of disease and are type specific. Antibodies against streptolysin O **(ASO test)** are measured most commonly. The antibodies appear 3 to 4 weeks after the intitial exposure to the organism and persist. Measurement of these antibodies is particularly useful for documenting recent streptococcal pharyngitis in a patient with rheumatic fever or acute glomerulonephritis. An elevated ASO titer is not observed in patients with streptococcal pyoderma. It is thought that the streptolysin is inactivated by the lipids present on the skin. Other antibodies produced against streptococcal enzymes, particularly **DNAse B,** have been documented in patients with streptococcal pyoderma and pharyngitis.

Treatment

Group A streptococci are exquisitely sensitive to penicillin. For patients with a history of penicillin allergy, erythromycin can be used. Persistent oropharyngeal carriage of group A streptococci after a complete course of therapy can occur. This may represent poor compliance with the prescribed course of therapy, reinfection with a new strain, or persistent carriage in a sequestered focus. Antibiotic resistance has not been reported; thus an additional course of treatment can be initiated. If carriage persists, retreatment is not indicated because prolonged antibiotic therapy can disrupt the normal protective bacterial flora. Carriers have not been shown to be at increased risk for relapse infections or transmission of their organism to susceptible individuals. Antibiotic therapy in patients with pharyngitis will speed the relief of symptoms and, if initiated within 10 days of initial clinical disease, prevent rheumatic fever. Antibiotic therapy does not appear to influence the progression to acute glomerulonephritis.

Prevention and Control

Patients with a history of rheumatic fever require long-term use of antibiotic prophylaxis to prevent recurrent disease. In addition damage to the heart valve predisposes the patient to subsequent endocarditis. Antibiotic prophylaxis is required before the use of procedures that induce transient bacteremias (e.g., dental procedures). Specific antibiotic therapy will not alter the course of acute glomerulonephritis and is not indicated for prophylaxis, because recurrent disease is not observed.

GROUP B *STREPTOCOCCUS*

Group B *Streptococcus,* or *Streptococcus agalactiae,* was initially recognized as a cause of puerperal sepsis. Although the organism is still associated with this disease, it has more recently gained notoriety as a significant cause of septicemia, pneumonia, and meningitis in newborn children.

Microbial Physiology and Structure

Group B streptococci are gram-positive cocci (0.6 to 1.2 μm) that form short chains in clinical specimens and longer chains in culture. The organism grows well on nutritionally enriched medium as buttery colonies, larger than seen with group A streptococci, and surrounded by a narrow zone of β-hemoly-

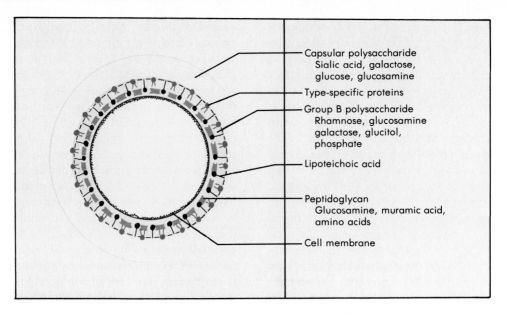

FIGURE 6-12   Schematic representation of group B *Streptococcus*. (Redrawn from Christensen KK, Christensen P and Ferrieri P: Neonatal group B streptococcal infections, vol. 35, Basel, Switzerland, 1985, Karger.)

sis. Some strains will be nonhemolytic or α-hemolytic.

The group-specific cell wall polysaccharide antigen is composed of rhamnose, *N*-acetylglucosamine, and galactose (Figure 6-12). Serologic cross-reactions have been observed between some group B and group G streptococci. Five immunologically distinct, type-specific antigens have been identified in the polysaccharide capsule: Ia, Ib, Ic, II, and III. Serotype Ic possesses the Ia polysaccharide, as well as a unique protein surface antigen, Ibc. These serotypes are important epidemiologic markers.

Pathogenesis

Antibodies developed against the type-specific capsular antigens in group B streptococci are protective. This in part explains the predilection of this organism for neonates. In the absence of type-specific maternal antibodies, the infant is at increased risk for infection. Bactericidal activity for group B streptococci also requires complement. If the level of complement is low in the neonate, then there is a greater likelihood of systemic spread of the organism in colonized infants.

Group B streptococci produce a number of enzymes, including deoxyribonucleases, hyaluroni-

dase, neuraminidase, proteases, hippuricase, and hemolysins. Although these enzymes are useful for identication of the organism, their role in the pathogenesis of infection is unknown.

Epidemiology

Group B streptococci commonly colonize the upper respiratory tract and vagina. Vaginal carriage has been reported in as many as 40% of pregnant women, although this is influenced by the time of sampling during the gestation period and culture techniques employed. Most women are transiently colonized with small numbers of orgnaisms, which confounds attempts to identify prospectively neonates at risk for infection.

Infection with subsequent development of disease in the neonate can occur in utero, at the time of birth, or during the first few months of life. Infections at or before birth are called early-onset disease (Table 6-5). Premature rupture of membranes, prolonged labor, preterm delivery, or maternal disease increases the risk of fetal infection in colonized women. Approximately 60% of infants born of colonized mothers will become colonized with their mother's serotype organism. The likelihood of colonization at the time of birth is increased if the mother is heavily colonized

Table 6-5 Epidemiology of Neonatal Group B Streptococcal Disease

Features	Early-Onset Disease	Late-Onset Disease
Age of onset	<7 days	1 week to 3 months
Obstetrical complications	Yes	Infrequent
Time of acquisition	In utero or at delivery	Postpartum
Clinical disease	Bacteremia, pneumonia, meningitis	Bacteremia, meningitis, osteomyelitis
Mortality rate	High	Low (<20%)
Streptococcal serotype	I, II, or III	III most common

Table 6-6 Correlation Between Maternal and Neonatal Colonization With Group B *Streptococcus*

Quantitative Perinatal Culture	Number of Women	Percentage of Infants	
		Negative Culture	Positive Culture
−	88	99	1
1+	55	82	18
2+	45	78	22
3+	107	36	64

From Boyer KM, et al: J Infect Dis 148:802, 1983. The proportion of children colonized with abundant group B streptococci, and at greatest risk for neonatal disease, is directly related to the level of maternal colonization at the time of delivery.

(Table 6-6). The incidence of neonatal disease is approximately 3 per 1,000 live births, with early-onset disease about twice as frequent as late-onset disease. Late-onset disease generally occurs in infants from 1 week to 3 months of life.

Early-onset disease is associated with all serotypes of group B *Streptococcus* in a proportion similar to maternal colonization. One exception would be infants with early-onset meningitis—80% of these infants are infected with serotype III. Serotype III is also responsible for more than 90% of late-onset infections.

Clinical Syndromes

Early-Onset Neonatal Disease Clinical symptoms of group B streptococcal disease acquired in utero or at the time of delivery generally will develop during the first 5 days of life. Early-onset disease, characterized by bacteremia, pneumonia, or meningitis, will appear indistinguishable from sepsis caused by other organisms. Mortality has been decreased with rapid diagnosis and better supportive care. However, 60% of infected, low birth weight, premature infants will die, and a significant proportion of survivors will have neurologic sequelae—including blindness, deafness, and severe mental retardation.

Late-Onset Neonatal Disease Disease in older infants is acquired from an exogenous source (e.g., mother, another infant). The predominant manifestation is bacteremia with meningitis, which again resembles disease caused by other bacterial pathogens. Although survival (>80%) is significantly better than with early-onset disease, neurologic complications in children with meningitis are common.

Postpartum Sepsis Postpartum group B streptococcal disease generally is seen as endometritis or a wound infection, with bacteremia frequently documented. Because child-bearing women are generally in good health, the prognosis is excellent when appropriate therapy is initiated. Secondary complications following bacteremia, such as endocarditis, meningitis, or osteomyelitis, have been rarely reported.

Laboratory Diagnosis

Culture Group B streptococci readily grow on a nutritionally enriched medium, producing large colonies after 24 hours of incubation. β-hemolysis may be either difficult to detect or absent, which will present a problem in the detection of the organism in mixed cultures. A selective medium, with antibiotics used to suppress the growth of other organisms, has been used with some success, and carriage of small numbers of organisms can be detected by inoculating specimens into an enrichment broth.

Antigen Detection Direct detection of the organism with antibodies prepared against the group-specific carbohydrate is useful for the rapid detection of group B streptococcal disease in neonates. A variety of methods have been used, including counterimmunoelectrophoresis, staphylococcal coagglutination, and latex agglutination. In the latter two methods antibodies developed against the group-specific antigen are attached to killed staphylococci or latex particles. These assay methods are sensitive, specific, and can be used with cerebrospinal fluid, urine, or serum.

Identification Definitive identification of group B streptococci is made by demonstrating the group-specific carbohydrate. This can now be performed with antisera prepared by a number of commercial manufacturers. In addition, specific biochemical reactions can be used to make a presumptive identification. One commonly used test is the demonstration of the **CAMP factor** (Figure 6-13).

Treatment

Group B streptococci are uniformly susceptible to penicillin G, which is the drug of choice. However, the minimum inhibitory concentration (MIC) is approximately 10 times greater than with group A streptococci. In addition, tolerance to penicillin (ability to inhibit but not to kill the organism) has been reported. For these reasons, a combination of penicillin plus an aminoglycoside is frequently used in serious infections. Resistance to erythromycin and tetracycline has also been observed.

Prevention and Control

Attempts to prevent neonatal disease have met with limited success. Although early-onset disease occurs in infants of colonized women, the incidence of colonization is high. Only a small proportion of

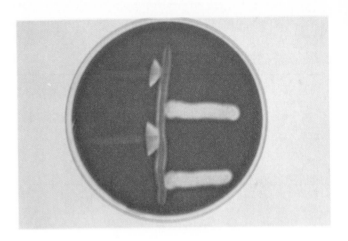

FIGURE 6-13   CAMP reaction with group B *Streptococcus*. Group B streptococci produce a diffusible, heat-stable protein (CAMP factor) that enhances β-hemolysis of *Staphylococcus aureus*. *S. aureus* produces sphingomyelinase C, which can bind to erythrocyte membranes. When exposed to the group B CAMP factor, the cells undergo hemolysis. Although most group B streptococci produce CAMP factor and stimulate enhanced hemolysis, it has also been found in some group C, F, and G streptococci. (From Howard BJ: Clinical and pathogenic microbiology, St Louis, 1987, The CV Mosby Co.)

these women will deliver infants who will become colonized and subsequently develop disease, and it is difficult to identify which infants are at greatest risk. Intrapartum antibiotic therapy reduces the incidence of neonatal disease, but the routine treatment of all colonized women may be impractical.

Passive immunization of high-risk babies by transfusion with blood containing type-specific antibodies has reduced morbidity and mortality associated with group B streptococcal disease. However, most whole blood has inadequate levels of protective antibodies. Future efforts to eliminate this disease will be directed toward the detection and immunization of antibody negative women of childbearing age.

GROUP D *STREPTOCOCCUS*

Group D streptococci consist of two subgroups: enterococcal species and nonenterococcal species. Historically, these subgroups were differentiated by their physiologic properties. Recently, DNA-DNA hybridization studies have demonstrated that the most common members of the enterococcal group were not closely related to the *Streptococcus* genus and have been reclassified in the new genus *Enterococcus*. The enterococci associated with human infections are *E. faecalis* and *E. faecium,* and the most commonly isolated nonenterococcal group D *Streptococcus* is *S. bovis*. For the purpose of clarity, the enterococcal and nonenterococcal species will be referred to collectively as group D streptococci, except where specific species are discussed.

Microbial Physiology and Structure

The group D streptococci are gram-positive cocci typically arranged in pairs and short chains. Frequently, the microscopic morphology of these isolates cannot be differentiated from *S. pneumoniae* when grown in broth culture. The enterococci grow readily on blood agar media, producing large white colonies after 24 hours of incubation. The nonenterococcal streptococci produce smaller colonies on agar media.

The group specificity of these streptococci is determined by the glycerol teichoic acid associated with the cytoplasmic membrane. Because the antigen is not readily released when the cell wall is hydrolyzed, serologic grouping of these organisms can pose a problem for the clinical laboratory.

Pathogenesis

The enterococci are uniquely suited for survival. They are able to grow in the presence of a high concentration of bile and sodium chloride, which is necessary for survival in the bowel and gall bladder. Additionally, these organisms are resistant to the bactericidal activity of most antibiotics. Thus the introduction of new broad-spectrum antibiotics has permitted these organisms to proliferate and cause serious superinfections.

Epidemiology

E. faecalis is found in small numbers in the upper respiratory tract and small intestines and in large numbers (e.g., 10^7 org/gm of feces) in the large intestine. *E. faecium* has a similar distribution, although it is found less frequently as part of the normal human microbial flora. *S. bovis* is infrequently isolated from healthy individuals. However, it is significantly associated with underlying gastrointestinal disease (e.g., carcinoma of the colon).

Clinical Syndromes

The enterococci have been implicated as a cause of urinary tract infections, intraabdominal abscesses, wound infections, and endocarditis. Enterococci are reported to be a common cause of urinary tract infections in hospitalized patients, particularly those patients with an indwelling catheter and receiving broad-spectrum antibiotics with limited activity against these organisms. The etiologic role of enterococci in intraabdominal abscesses and wound infections is obscure, because the infections are generally polymicrobic and enterococcal infections in an experimental animal model have not been established. Both enterococci and nonenterococcal streptococci are able to cause bacterial endocarditis. This is in part due to the production of dextran, which facilitates adherence to the heart tissue.

Treatment

The nonenterococcal streptococci are inhibited and killed by most commonly used antibiotics, including the penicillins and cephalosporins. In contrast, the enterococci are inhibited by only high concentrations of these antibiotics and are generally not killed by any single antibiotic. Although urinary tract infections may be successfully treated with ampicillin alone, treatment for serious systemic infections requires the combination of a penicillin and an aminoglycoside. Aminoglycosides have poor activity against the enterococci when used alone, but killing is achieved when combined with ampicillin. Recently, isolates of enterococci have been discovered that are resistant to this synergistic killing because the aminoglycosides at even very high concentrations are ineffective. These strains are particularly troublesome because this resistance is plasmid mediated and can be conjugatively transferred to other bacteria. At present, no combination of antibiotics has proved bactericidal activity against these organisms.

VIRIDANS STREPTOCOCCI

The viridans group of streptococci are a heterogeneous collection of α- and nonhemolytic streptococci. Taxonomic nomenclature for these species is confusing because a consensus between European and American microbiologists has not been reached. Thus different species names are often used interchangeably in the literature. Although most isolates of viridans streptococci do not possess a group-specific carbohydrate, cross reactions with groups, A, C, E, F, H, K, M, and O have been reported.

Like most other streptococci, these species are nutritionally fastidious, requiring complex media supplemented with blood products and an incubation atmosphere frequently augmented with 5% to 10% carbon dioxide. Some strains are called "nutritionally deficient streptococci" because they can grow only in the presence of exogenously supplied pyridoxal, the active form of vitamin B_6. These organisms will usually grow initially in blood cultures but will fail to grow when subcultured unless pyridoxal-supplemented media are used.

The viridans streptococci are the most common group of organisms in the oropharynx and can also be isolated from the gastrointestinal and urogenital tracts. Although these organisms can cause a variety of infections, they are most commonly associated with dental caries, subacute endocarditis, and suppurative intraabdominal infections. Adherence to tooth enamel or previously damaged heart valves is believed to be due to the production of insoluble dextran from glucose. This is most commonly observed with *S. mutans* and *S. sanguis. S. milleri,* which includes the species *S. MG-intermedius* and *S. anginosus-constellatus,* is the species most commonly associated with pyogenic infections. The pathogenesis of this abscess formation has not been defined.

Most strains of viridans streptococci are highly susceptible to penicillin with MICs ≤ 0.1 μg/ml. Moderately resistant streptococci (penicillin MIC 0.2 to 0.5 μg/ml) have been observed in as many as 10% of some species. Infections with these isolates can generally be treated with a combination of penicillin and an aminoglycoside. High-level resistance, due to an alteration of the penicillin-binding proteins, is rare. Tolerance to the killing activity of penicillin has also been reported, but the clinical significance is controversial.

BIBLIOGRAPHY

Beachey EH and Courtney HS: Bacterial adherence: the attachment of group A streptococci to mucosal surfaces, Rev Infect Dis 9(Suppl): 475-481, 1987.

Breese BB: A simple scorecard for the tentative diagnosis of streptococcal pharyngitis, Am J Dis Child 131:514-517, 1977.

Christensen KK, Christensen P, and Ferrieri P: Neonatal group B streptococcal infections, vol 35, Basel, Switzerland, 1985, Karger.

Crowe CC, Sanders WE, and Longley S: Bacterial interference. II. Role of the normal throat flora in prevention of colonization by group A streptococcus, J Infect Dis 128:527-532, 1973.

Handwerger S and Tomasz A: Alterations in penicillin-binding proteins of clinical and laboratory isolates of pathogenic *Streptococcus pneumoniae* with low levels of penicillin resistance, J Infect Dis 153:83-89, 1986.

Habif TP: Clinical dermatology: a color guide to diagnosis and therapy, St Louis, 1985, The CV Mosby Co.

Hendley JO, Sande MA, Stewart PM, and Gwaltney JM, Jr: Spread of *Streptococcus pneumoniae* in families. I. Carriage rates and distribution of types. J Infect Dis 132:55-61, 1975.

Lancefield, RC: A serological differentiation of human and other groups of hemolytic streptococci, J Exp Med 57:571-595, 1933.

Mandell GL, Douglas RG, Jr, and Bennett JE: Principles and practice of infectious diseases, ed 2, New York, 1985, John Wiley & Sons.

Peter G and Smith AL: Group A streptococcal infections of the skin and pharynx, N Engl J Med 297:311-317 and 365-370, 1977.

Rodnan GP and Schumacher HR, editors: Primer on the rheumatic diseases, ed 8, Atlanta, 1983, The Arthritis Foundation.

Skinner FA and Quesnel LB: Streptococci, New York, 1978, Academic Press.

Sneath PHA, Mair NS, Sharpe ME, and Holt JG: Bergey's manual of systematic bacteriology, vol 2, Baltimore, 1986. Williams & Wilkins.

Quie PG, Giebink GS, and Winkelstein JA: Symposium: the pneumococcus, Rev Infect Dis, 3:183-395, 1981.

Wannamaker LW: Differences between streptococcal infections of the throat and of the skin, N Engl J Med 282:23, 1970.

Wannamaker LW and Matsen JM: Streptococci and streptococcal diseases: recognition, understanding, and management, New York, 1972, Academic Press.

CHAPTER 7

Neisseriaceae

The family Neisseriaceae includes the genera *Neisseria, Branhamella, Moraxella, Kingella,* and *Acinetobacter.* These organisms are either cocci, frequently arranged as diplococci with flattened adjacent sides, or short bacilli. Members of this family are easily differentiated by morphological and biochemical parameters (Table 7-1). The best known species are *Neisseria meningitidis* and *Neisseria gonorrhoeae. N. meningitidis* can either colonize the upper respiratory tract or cause significant human disease. In contrast, *N. gonorrhoeae* is always considered pathogenic—even in individuals with asymptomatic colonization. The other *Neisseria* species, and the other members of the family Neisseriaceae, can normally colonize mucous membranes and the skin surface and are rare causes of disease.

NEISSERIA MENINGITIDIS

The meningococci are encapsulated gram-negative diplococci that can asymptomatically colonize the nasopharynx of healthy individuals or cause fulminant meningitis, pneumonia, or overwhelming sepsis (meningococcemia) (Box 7-1).

Microbial Physiology and Structure

The meningococci form transparent, nonpigmented, nonhemolytic colonies on chocolate blood agar with enhanced growth in a moist atmosphere with 5% carbon dioxide. Isolates with large capsules will appear as mucoid colonies. Meningococci are oxidase positive and are differentiated from other neisseria by acid production from glucose and maltose but not sucrose or lactose (Table 7-2).

N. meningitidis is subdivided into serogroups and immunotypes (Table 7-3). Serogroups A, B, C, Y, and W135 are most commonly associated with meningococcal disease. All group A meningococci have the same outer membrane proteins and belong to a single serotype, whereas multiple serotypes have been described for groups B and C. Membrane proteins and serotype classification are shared between the

Table 7-1 Differentiation of Members of the Family Neisseriaceae

Characteristics	*Neisseria*	*Branhamella*	*Moraxella*	*Kingella*	*Acinetobacter*
Cell morphology					
Cocci	+	+	−	−	−
Coccobacilli	−	−	+	+	+
Oxidase	+	+	+	+	−
Catalase	+	+	+	−	+
Acid from glucose	+	−	−	+	±
Nitrite reduction	+	±	−	+	−
Mol% G + C of DNA	46-54	40-48	40-48	47-55	38-47

Modified from Bergey's Manual of Systematic Bacteriology, vol 1, Baltimore, 1984, Williams & Wilkins.

Symbols: +, positive for the majority of strains; ±, positive and negative reactions reported equally; −, all strains negative.

Box 7-1   Infections Associated With *Neisseria meningitidis* and *Neisseria gonorrhoeae*

N. meningitidis	Meningitis
	Septicemia
	Pneumonia
	Arthritis
N. gonorrhoeae	Urethritis
	Cervicitis
	Salpingitis
	Proctitis
	Septicemia
	Arthritis
	Conjunctivitis
	Pharyngitis
	Pelvic inflammatory disease

two groups. Serotype 2 is the most common isolate from groups B and C.

Pathogenesis

Three major factors are responsible for meningococcal disease: ability of *N. meningitidis* to colonize the nasopharynx (mediated by pili), systemic spread without antibody-mediated phagocytosis (protection afforded by polysaccharide capsule), and expression of toxic effects (mediated by the lipopolysaccharide endotoxin). Experiments with nasopharyngeal tissue organ cultures have demonstrated that meningococci attach selectively to specific receptors for meningococcal pili on nonciliated columnar cells of the nasopharynx (Figure 7-1). Meningococci without pili have decreased binding to these cells. The organisms are

Table 7-2   Differential Characteristics of *Neisseria* and *Branhamella*

Characteristics	*Neisseria gonorrhoeae*	*Neisseria meningitidis*	*Neisseria lactamica*	*Neisseria sicca*	*Neisseria mucosa*	*Branhamella catarrhalis*
GROWTH ON:						
Blood agar	−	−	+	+	+	+
MTM, ML*	+	+	+	−	−	−
ACID FROM:						
Glucose	+	+	+	+	+	−
Maltose	−	+	+	+	+	−
Lactose	−	−	+	−	−	−
Sucrose	−	−	−	+	+	−
Nitrate reduction	−	−	−	−	+	−

*MTM, Modified Thayer-Martin agar; ML, Martin-Lewis agar.

Table 7-3   Antigenic Determinants of *Neisseria meningitidis*

Epidemiologic Classification	Antigenic Determinant	Number Described
Serogroup	Polysaccharide capsule	13
Serotype	Outer membrane protein	>20
Immunotype	Lipopolysaccharide	8

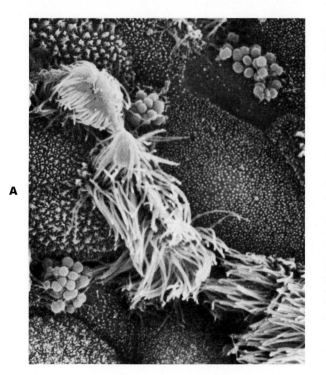
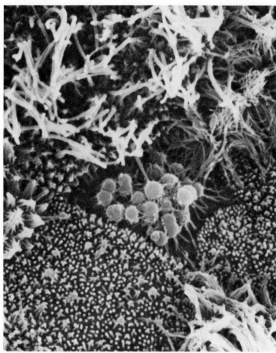

FIGURE 7-1   Scanning electron micrographs showing interaction of *Neisseria meningitidis* with human nasopharyngeal mucosa. The meningococci attach by pili to the microvilli of nonciliated cells but not to ciliated cells **(A)**. Attachment stimulates folding of the epithelial cell membrane around the bacteria and subsequent internalization **(B)**. (From Stephens DS, Hoffman LH, and McGee ZA: J Infect Dis 148:369, 1983.)

then internalized in phagocytic vacuoles, and after 18 to 24 hours the meningococci are found in the subepithelial space. The antiphagocytic properties of the polysaccharide capsule protects *N. meningitidis* from phagocytic destruction.

Bactericidal activity also requires complement activity. Individuals with deficiencies in C5, C6, C7, or C8 in the complement system are at increased risk for meningococcal disease. The diffuse vascular damage associated with meningococcal infections (e. g., endothelial damage, inflammation of vessel walls, thrombosis, disseminated intravascular coagulation [DIC]) is in large part attributed to the action of the lipopolysaccharide endotoxin. The endotoxin is present in the outer membrane. *N. meningitidis* produces excess membrane fragments that are released into the extracellular space (Figure 7-2). This continuous hyperproduction of endotoxin is most likely responsible for the severe endotoxic reaction associated with meningococcal disease.

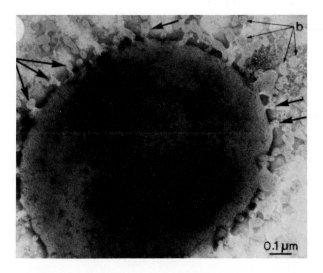

FIGURE 7-2   Negative-stained electron micrograph of *Neisseria meningitidis*. Excess outer membrane fragments containing endotoxin are released into the extracellular space in actively growing cells. (Reproduced from The Journal of Experimental Medicine, 1973, 138:1156, by copyright permission of The Rockefeller University Press.)

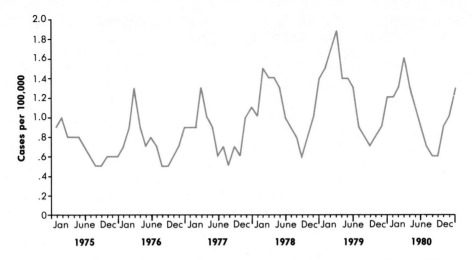

FIGURE 7-3   Attack rate of meningococcal disease in the United States from 1975 to 1980. The peak incidence of disease (1 to 2 cases per 100,000 population) is during the spring months. (From Band JD et al: J Infect Dis 148:754, 1983.)

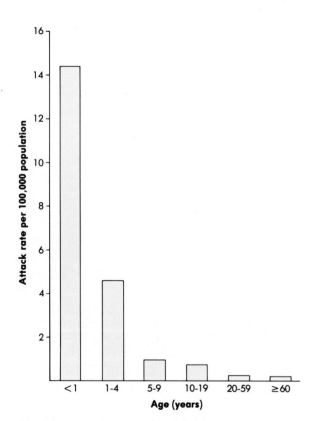

FIGURE 7-4   Attack rate of meningococcal disease in the United States from 1975 to 1980. The highest incidence of disease is in children less than 1 year of age. (From Band JD et al.: J Infect Dis 148:754, 1983.)

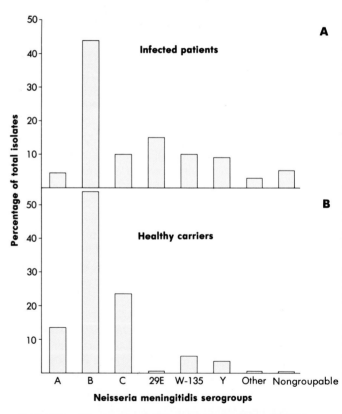

FIGURE 7-5   Serogroups of *Neisseria meningitidis* collected from 1963 to 1982. Serogroups A, B, C, W135, and Y are isolated from 99.2% of the infected patients **(A)** but only 76.8% of the healthy carriers **(B)**. (From Feldman HA: Rev Infect Dis 8:288, 1986.)

Epidemiology

Meningococcal disease can occur worldwide, in epidemics that last for 3 to 6 years, or as endemic disease. Serogroup A *N. meningitidis* was responsible for most epidemics before 1970 but has now been replaced by serogroup B. Endemic disease can be associated with any serogroup, although groups B and C are most commonly isolated. The seasonal incidence of endemic disease is diagramed in Figure 7-3. Transmission of *N. meningitidis* is by respiratory droplets among persons who have prolonged, close contacts, such as family members living in the same household, and within crowded communities such as the military. Classmates and hospital employees are not considered close contacts and are not at significantly increased risk unless they are in direct contact with respiratory secretions.

Endemic disease is most common in children less than 5 years of age (Figure 7-4). During epidemics with meningococci, older individuals are infected, particularly in closed populations. During the first 6 months of life, maternal bactericidal antibodies are protective. However, as passive immunity wanes and before acquired immunity develops, children are susceptible to infection. Acquired immunity develops in asymptomatically colonized persons, with bactericidal antibodies detectable within 2 weeks of colonization. Cross-reacting antibodies providing immunity to *N. meningitidis* can occur with antigenically related strains of meningococci or with bacteria of other genera (e. g., *Escherichia coli* serotype K1 cross-reacts with group B *N. meningitidis*). The majority of meningococcal strains acquired during carriage are typeable (Figure 7-5). Although carriage occurs in 10% to 30% of healthy adults and in a higher proportion during epidemics, disease is rare and unrelated to the prevalence of meningococci in the community.

Clinical Syndromes

Meningitis Meningococcal meningitis usually begins abruptly with headache, meningeal signs, and fever. However, very young children may have only such nonspecific signs as fever and vomiting. Mortality approaches 100% in untreated patients but is less than 15% when appropriate antibiotics are promptly instituted. The incidence of neurologic sequelae is low, with hearing deficits and arthritis most commonly reported.

Meningococcemia Septicemia (meningococcemia) with or without meningitis is a life-threatening disease with a mortality rate of 25%, even in patients who are promptly treated. Thrombosis of small blood vessels and multiorgan involvement is characteristic. Small petechial skin lesions on the trunk and lower extremities are common and may coalesce to form larger hemorrhagic lesions. The disease may progress to overwhelming disseminated intravascular coagulation with shock and includes bilateral destruction of the adrenal glands **(Waterhouse-Frederichsen syndrome)**.

A milder, chronic septicemia has also been described. Bacteremia can persist for days or weeks in these patients, with the only signs of infection being low-grade fever, arthritis, and petechial skin lesions. Response to antibiotic therapy is generally excellent.

Other Syndromes Additional infections associated with *N. meningitidis* include pneumonia, arthritis, and urethritis. Meningococcal pneumonia is usually preceded by a previous respiratory infection, symptoms include cough, chest pain, rales, fever, and chills. Evidence of pharyngitis is observed in the majority of patients. The prognosis for this infection is good.

Laboratory Diagnosis

A definitive identification of *N. meningitidis* is important for the initiation of specific therapy and prophylaxis for contacts when indicated. The most useful specimens for detection of meningococci are blood and cerebrospinal fluid (CSF). Although most patients with systemic disease will have positive blood cultures, additives present in the blood culture broths can be toxic for *Neisseria* and inhibit or delay bacterial growth. If the laboratory is notified that meningococcal disease is suspected, alternative blood-culturing methods can be selected. Growth of the organism from CSF is relatively easy in untreated patients because more than 10^7 org/ml of fluid is normally found. However, detection of viable organisms can be adversely affected in patients previously treated with antibiotics.

Because the bacterial count in CSF is high, the gram-negative diplococci are readily seen within polymorphonuclear leukocytes in Gram-stained specimens. Soluble polysaccharide antigen can also be detected in CSF by counterimmunoelectrophoresis or agglutination of latex particles coated with specific antibodies. These tests are particularly useful for patients with partially treated meningitis.

Treatment

Antibiotic therapy and supportive management of the complications of meningococcal disease have significantly reduced the associated mortality. Although sulfonamides were the basis for the initial therapeutic successes, widespread resistance has negated their effectiveness. The antibiotic of choice now is penicillin, for which the mean MIC is virtually unchanged since 1960; resistance has rarely been observed. Alternative antibiotics include chloramphenicol and the broad-spectrum cephalosporins, which show dramatic activity in vitro.

Prevention and Control

Eradication of the pool of healthy carriers is unlikely. Efforts have instead concentrated on the prophylactic treatment of persons who have significant exposure to diseased patients and the enhancement of immunity to serogroups most commonly associated with disease. Although sulfonamides were used for prophylaxis, this is no longer considered to be reliable because of the increased resistance. In addition, penicillin is ineffective in eliminating the carrier state. Minocycline and rifampin have both been used effectively for antibiotic-mediated chemoprophylaxis; however, toxic effects have been associated with minocycline, and rifampin-resistant *N. meningitidis* can arise during rifampin treatment. At present, prophylaxis with a sulfonamide is recommended for persons exposed to sulfonamide-susceptible strains and rifampin for sulfonamide-resistant strains.

Vaccines directed against the group-specific capsular polysaccharides have been developed for antibody-mediated immunoprophylaxis. An effective polyvalent vaccine, which can be administered to children older than 2 years of age, has been developed against groups A, C, Y, and W135. However, the vaccine cannot be administered to high-risk children in younger age-group. Moreover, the group B polysaccharide is poorly immunogenic and cannot induce a protective antibody response. Thus immunity to *N. meningitidis* group B, the most common cause of significant meningococcal disease, must develop naturally after exposure to cross-reacting antigens.

NEISSERIA GONORRHOEAE

Infection with *Neisseria gonorrhoeae* is the most common sexually transmitted disease in the United States. Clinical manifestations include urethritis, cervicitis, arthritis, conjunctivitis, and a number of local and systemic complications.

Microbial Physiology and Structure

Gonococci, like meningococci, are small, gram-negative diplococci that grow readily on chocolate agar and specialized media such as modified Thayer-Martin agar or Martin-Lewis agar. The organisms are inhibited, however, by the fatty acids and trace metals present in the peptone hydrolysates and agar in other common laboratory media. The gonococci are susceptible to desiccation, and generally require an atmosphere of 5% carbon dioxide and incubation temperature of 35° to 37° C for initial growth in culture. Five morphologically distinct colony types (T1 through T5) have been described, based on such features as color, size, and opacity, with virulence associated with T1 and T2 (Table 7-4). In vitro transfer of

Table 7-4 Characteristics of *Neisseria gonorrhoeae* Colony Types

Type	Size (mm)	Color	Elevation	Edge	Opacity	Consistency
T1	0.5	Dark gold	Convex	Entire	Translucent	Viscid
T2	0.5	Dark gold	Convex	Crenated	Translucent	Friable
T3	1.0	Light brown	Low convex	Entire	Translucent	Viscid
T4	1.0	Colorless	Low convex	Entire	Transparent	Viscid
T5	1.0	Dark gold	Low convex	Rough	Translucent	Viscid

isolates results in a phase transition from T1 and T2 to T3 through T5, with reversion accomplished by in vitro inoculation onto human tissue culture cells. Identification of isolates is based on typical morphology, the presence of cytochrome oxidase, and strict oxidative metabolism of glucose but not other carbohydrates (see Table 7-2).

The structure of the gonococcus is similar to *N. meningitidis* (Figure 7-6). The outer surface is covered with a loosely associated capsule of unknown composition. Protruding through the surface of the bacteria are filamentous, protein pili. These are present only in the virulent T1 and T2 colony types. The major protein in the outer membrane is protein I, which is arranged in trimers forming surface pores. Sixteen serotypes of protein I have been described and are useful for the epidemiologic classification of isolates. Protein II is a minor membrane protein found in avirulent, opaque colonies. The presence of this protein is associated with intercell adhesiveness, as well as increased adherence of gonococci to cultured eukaryotic cells. The endotoxin lipopolysaccharides of *N. gonorrhoeae* resemble those found in *N. meningitidis*. The endotoxins contain lipid A and a core polysaccharide, although the strain-specific O side chains that are present in many gram-negative bacilli are absent. The lipopolysaccharide is present on the outermost portion of the cell membrane and is released in an active form into the extracellular space much like with *N. meningitidis*. Other proteins associated with the gonococci are a protease capable of cleaving immunoglobulin A and β-lactamase that hydrolytically destroys penicillin. The β-lactamase was initially isolated in Southeast Asia and now has worldwide distribution. It is plasmid-mediated and is identical to the enzyme found in other gram-negative bacteria (e. g., *Haemophilus influenzae,* Enterobacteriaceae).

Pathogenesis

Similar to infections with *N. meningitidis,* gonococci attach to mucosal cells, pentrate into the cells and multiply, and then pass through the cells into the subepithelial space where infection is established. The presence of pili is important for the initial attachment (Box 7-2). Nonpiliated cells such as those present in colony types T3 to T5 are avirulent. In the absence of specific opsonic antibodies and complement, the capsule protects against phagocytosis by polymorphonuclear leukocytes. Piliated strains are also more resistant to phagocytosis than are strains without pili. The gonococcal endotoxin is responsible for tissue destruction in cell culture and is believed to be the major virulence factor in vivo. In persistent infection, chronic inflammation and fibrosis occur

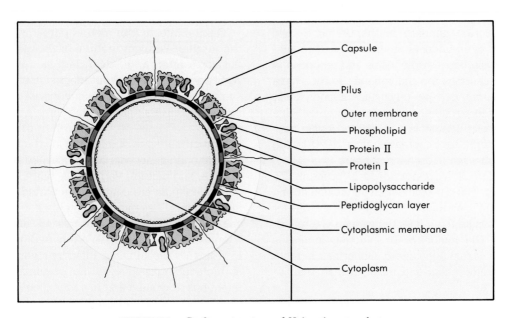

FIGURE 7-6   Surface structure of *Neisseria gonorrhoeae.*

Box 7-2 *Neisseria gonorrhoeae* Virulence
Factors

Capsule	Antiphagocytic
Pili	Attachment to human cells, antiphagocytic, facilitates genetic transformation, binds iron and other cations
Protein I	Major surface antigen, forms surface pores, specific serotypes associated with virulence
Protein II	Presence associated with avirulent, opaque colonies; responsible for intercellular adherence
Lipopolysaccharide	Endotoxin activity
IgA protease	Destroys immunoglobulin IgA1
β-lactamase	Hydrolyses β-lactam ring in penicillin

Box 7-3 Characteristics of Disseminated
Neisseria gonorrhoeae Infection

Patient	Primarily women
Initial infection	Usually asymptomatic genital infection
Site of dissemination	Blood, skin, joints
Strain characteristics	Resistant to serum killing; very susceptible to penicillin*; auxotrophic growth requirements for arginine, hypoxanthine, and uracil*

---

*Penicillin susceptibility and growth requirements may vary depending on the prevalent strain in the community.

that can lead to sterility, arthritis with joint destruction, or blindness.

Epidemiology

Gonorrhea is a disease found only in humans; it has no other known reservoir. More than 1 million cases are reported annually in the United States. However, this is an underestimation of the true incidence of disease because diagnosis and reporting of gonococcal infections are incomplete. Studies have documented that private physicians report fewer than 20% of the patients treated for gonorrhea. The peak incidence of disease is in the 20 to 24-year age-group, with significant increases since 1970 in the incidence of disease in teenage patients.

Transmission of *N. gonorrhoeae* is primarily by sexual contact. The risk of infection for women after a single exposure to an infected man is 50%; the risk for men after exposure to an infected woman is approximately 20%. The incidence of infection increases with multiple sexual encounters. Gonorrhea is more common in homosexual and bisexual men than in heterosexual men.

The major reservoir for the gonococcus is the asymptomatically infected individual. Although it is

difficult to determine a true incidence of asymptomatic infection, it is more common in women than in men. As many as one half of all infected women will have mild or asymptomatic infections, whereas most men will be initially symptomatic. In untreated disease, however, the symptoms will generally clear within a few weeks and asymptomatic carriage can be established. These carriers can transmit disease. Carriage is also influenced by the site of infection; rectal and pharyngeal infections are more commonly asymptomatic than are genital infections.

Clinical Syndromes

Genital infection in men is primarily restricted to the urethra. Purulent urethral discharge and dysuria develops after a 2- to 7-day incubation period. Approximately 95% of all infected men have acute symptoms. Although complications are rare, epididymitis, prostatitis, and periurethral abscesses can occur.

The primary site of infection in women is the cervix, although gonococci can be isolated in the vagina, urethra, and rectum. Vaginal discharge, dysuria, and abdominal pain are commonly reported in symptomatic patients. Ascending genital infection, including salpingitis, tubo-ovarian abscesses, and pelvic inflammatory disease (PID) are reported in 10% to 20% of women. Disseminated infections with septicemia and infection of skin and joints occurs in 1% to 3% of infections in women and in a much lower percentage of infected men. The increased proportion of disseminated infections in women is caused by the large number of untreated asymptomatic infections in this population (Box 7-3). Clinical manifestation of disseminated disease include fever, migratory arthralgias, suppurative arthritis in the wrists, knees, and ankles, and a pustular rash on an erythematous base over the extremities but sparing the head and trunk. *N. gonorrhoeae* is a leading cause of purulent arthritis in adults.

Other diseases associated with *N. gonorrhoeae* include perihepatitis **(Fitz-Hugh-Curtis syndrome),** purulent conjunctivitis—particularly in newborns infected during vaginal delivery **(opthalmia neonatorum),** anorectal gonorrhea in homosexual males, and pharyngitis.

Laboratory Diagnosis

Microscopy The Gram stain is a very sensitive (>90%) and specific (98%) test for the detection of gonococcal infection in men with purulent urethritis (Figure 7-7). However, the test sensitivity is 60% or less for asymptomatic men. The test is also relatively insensitive for detection of gonococcal cervicitis in both symptomatic and asymptomatic women, although a positive result is reliable when an experienced microscopist sees gram-negative diplococci within polymorphonuclear leukocytes. Thus the Gram stain can be reliably used to diagnose infections in men with purulent urethritis, but all negative results in women and asymptomatic men must be confirmed by culture. The Gram stain is useful early in the course of purulent arthritis but is insensitive for detection of *N. gonorrhoeae* in skin lesions, anorectal infections, and pharyngitis. The presence of commensal *Neisseria* in the oropharynx and morphologically similar bacteria in the gastrointestinal tract compromises the specificity of the stain for these specimens.

Culture *N. gonorrhoeae* can be readily isolated from clinical specimens if appropriate precautions are followed (Figure 7-8). Because other commensal organisms normally colonize mucosal surfaces, cultures of genital, rectal, and pharyngeal specimens must be inoculated onto both selective (e. g., modified Thayer-Martin medium) and nonselective (e. g., chocolate blood agar) media. The use of selective media will suppress the growth of contaminating organisms. However, a nonselective medium should also be inoculated because some gonococcal strains are inhibited by vancomycin present in most selec-

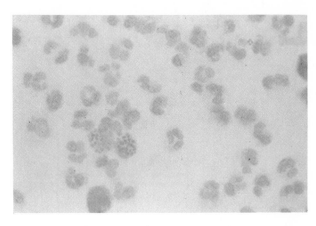

FIGURE 7-7   Gram stain of *Neisseria gonorrhoeae* associated with polymorphonuclear leukocytes in urethral discharge.

tive media. *N. gonorrhoeae* is susceptible to drying and cold temperatures, so these conditions should be avoided by inoculating the specimen directly onto prewarmed media at the time of collection. The endocervix must be properly exposed to ensure that an adequate specimen is collected. Although the endocervix is the most common site of infection in women, the rectum may be the only positive specimen in women who have asymptomatic infections, as well as homosexual and bisexual men. In patients with disseminated disease, blood cultures are generally positive only during the first week. In addition, special handling of these specimens is required to ensure adequate recovery, because supplements present in the blood culture media can be toxic to *Neisseria* organisms. Cultures of infected joints are frequently positive, but skin cultures are generally unrewarding.

Serology Although serologic tests have been developed to detect both gonococcal antigens and antibodies directed against the organism, these tests are neither sensitive nor specific and cannot be recommended.

Treatment

Penicillin has been the treatment of choice, and at present most isolates of *N. gonorrhoeae* are still susceptible. However, three changes have been observed. First, the concentration of penicillin required to inhibit growth of *N. gonorrhoeae* has steadily increased, necessitating significantly higher doses for clinical cures. The recommended therapeutic dose

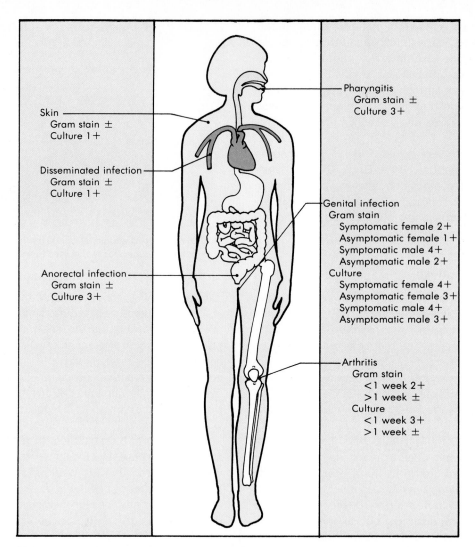

Skin
  Gram stain ±
  Culture 1+

Disseminated infection
  Gram stain ±
  Culture 1+

Anorectal infection
  Gram stain ±
  Culture 3+

Pharyngitis
  Gram stain ±
  Culture 3+

Genital infection
  Gram stain
    Symptomatic female 2+
    Asymptomatic female 1+
    Symptomatic male 4+
    Asymptomatic male 2+
  Culture
    Symptomatic female 4+
    Asymptomatic female 3+
    Symptomatic male 4+
    Asymptomatic male 3+

Arthritis
  Gram stain
    <1 week 2+
    >1 week ±
  Culture
    <1 week 3+
    >1 week ±

FIGURE 7-8   Laboratory detection of *Neisseria gonorrhoeae.* The reliability of the Gram stain is compromised when specimens are collected from sites such as the cervix, pharynx, or rectum that are contaminated with commensal *Neisseria* and morphologically similar organisms. Rapidly growing commensal organisms can also overgrow and obscure *N. gonorrhoeae,* unless selective media is used.

of penicillin G for uncomplicated gonorrhea has risen from 200,000 units in 1945 to 4.8 million units in 1988. Second, penicillin resistance mediated by enzymatic hydrolysis of the β-lactam ring was initially reported in Southeast Asia and now has worldwide distribution. The first β-lactamase producing *N. gonorrhoeae* in the United States was reported in March 1976. During the next 4 years relatively few new cases were detected, and most were related to imported cases.

However, from 1980 the number of resistant strains has risen rapidly, and the increase is not related to foreign travel (Figure 7-9). The genetic information for this resistance is encoded on a transmissible plasmid. Thus this increase in resistance should continue. Finally, strains of penicillin-resistant *N. gonorrhoeae* have been isolated that do not produce β-lactamase. This chromosomally mediated resistance is not limited to the penicillin antibiotics but also

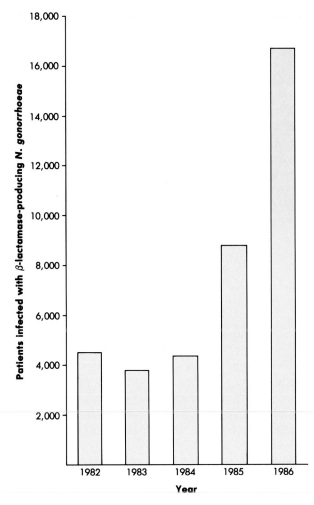

FIGURE 7-9   Cases of β-lactamase producing *Neisseria gonorrhoeae* isolated in the United States from 1982 to 1986.

includes resistance to tetracyclines, erythromycin, and aminoglycosides. This resistance appears to be the result of changes on the cell surface that prevent antibiotic penetration into the gonococcal cell.

Uncomplicated genital infections caused by penicillin-susceptible *N. gonorrhoeae* can be treated with penicillin G, although it is normally combined with tetracycline for the management of dual infections with penicillin-resistant bacteria (e. g., *Chlamydia*). Infections with gonococci-producing β-lactamase can be treated with spectinomycin or a β-lactamase–resistant cephalosporin such as ceftriaxone. Treatment of chromosomally mediated resistant gonococci is dictated by in vitro susceptibility test results.

Strains of gonococci responsibility for disseminated disease are generally highly susceptible to penicillin and can be easily treated. Penicillin or tetracycline can be used to treat pharyngeal and rectal disease; however, spectinomycin is ineffective.

Prevention and Control

Immunity to infections with *N. gonorrhoeae* is poorly understood. Antibodies can be detected to pili antigens, as well as to protein I and the lipopolysaccharide. However, multiple infections in sexually promiscuous individuals is common. This lack of protective immunity is explained in part by the antigenic diversity of gonococcal strains. The variable region at the carboxy terminus of the pili protein is the immunodominant portion of the molecule. Antibodies developed against this region protect against reinfection with a homologous strain, but cross-protection against heterologous strains is incomplete. This also explains the ineffectiveness of vaccines developed against pili proteins.

Chemoprophylaxis is also ineffective, except in protection against gonococcal eye infections in newborns (ophthalmia neonatorum) where 1% silver nitrate, 1% tetracycline, or 0.5% erythromycin eye ointments are routinely used. Prophylactic use of penicillin to prevent genital disease has been demonstrated to be ineffective and may select for resistant strains.

The major efforts to stem the epidemic of gonorrhea encompass education, aggressive documentation of disease by culture, and follow-up screening of sexual contacts. It is important to realize that gonorrhea is not an insignificant disease. Chronic infections can lead to sterility, and asymptomatic infections perpetuate the reservoir of disease, as well as lead to a higher incidence of disseminated infections.

NEISSERIA SPECIES AND BRANHAMELLA

Neisseria species, such as *N. sicca* and *N. mucosa,* and *Branhamella (Neisseria) catarrhalis* are commensal organisms in the oropharynx and genital tract. Until the 1970s these organisms were associated with

> **Box 7-4** Infections Associated With Other
> *Neisseria* Species, *Branhamella*, *Kingella*, and
> *Acinetobacter*

Other *Neisseria* species	Otitis media
	Sinusitis
	Bronchopulmonary
	infections
Moraxella	Conjunctivitis
	Endophthalmitis
Kingella	Endocarditis
Acintobacter	Pneumonia
	Bacteremia
	Wound infection

only isolated reports of meningitis, osteomyelitis, and endocarditis. However, these organisms are now recognized as a cause of bronchopulmonary infections, acute otitis media, and acute sinusitis (Box 7-4). The true incidence of respiratory infections caused by these organisms is not known because most specimens are contaminated with oral secretions. However, serologic studies have documented the important role of *Branhamella* in exacerbation of lower respiratory disease in patients with chronic bronchitis and obstructive pulmonary disease, and preliminary evidence supports a similar role for *N. sicca* and *N. mucosa*. Colonization, as well as disease, with this group of organisms is highest during the cold months of the year. All isolates of *N. sicca* and *N. mucosa* have been susceptible to penicillin. However, 75% or more of the *Branhamella* isolates produce β-lactamase and are penicillin resistant. These isolates are uniformly susceptible to erythromycin, tetracycline, trimethoprim-sulfamethoxazole, and the combination of ampicillin with a β-lactamase inhibitor (e. g., clavulanic acid).

MORAXELLA

Six species comprise the genus *Moraxella*. Members of the genus are normal inhabitants of the upper respiratory tract and urogenital tract. *Moraxella* species are rare causes of disease with conjunctivitis or endophthalmitis most commonly reported. Care must be used to differentiate these organisms from pathogenic *Neisseria* species isolated in genital specimens. Although most *Moraxella* are susceptible to penicillin, β-lactamase–mediated resistance has been reported.

KINGELLA

Kingella, formerly *Moraxella*, consists of three species, with *Kingella kingae* the most commonly isolated. Members of the genus are commensals of the human respiratory tract, with nasopharyngeal carriage of *K. kingae* documented in about 1% of healthy individuals. The most common infection associated with *Kingella* is endocarditis, primarily in children and young adults with underlying valvular heart disease. Isolates are highly susceptible to a variety of antibiotics, and most infections are successfully treated with penicillin and surgical intervention when indicated.

ACINETOBACTER

The taxonomic classification of the genus *Acinetobacter* is currently unsettled. Although a proposal has been made to subdivide this heterogeneous genus into six species, the use of a single species (*Acinetobacter calcoaceticus*) with two biochemically distinct varieties (var. *anitratus* and var. *lwoffi*) as described in *Bergey's Manual of Systematic Bacteriology* will be used here. Most clinically significant infections are associated with var. *anitratus*. Acinetobacters are commensal organisms of the upper respiratory tract, skin, and genitourinary tract. Although the majority of isolates represent simple colonization, clinically significant disease is well documented. Hospital-acquired pneumonia in patients with tracheostomies or endotracheal tubes is the most common infection caused by *Acinetobacter* organisms. Community-acquired pneumonia in patients predisposed to aspiration of oral secretions has also been reported. Less common infections associated with this group of organisms include wound infections (related to the high incidence of carriage on the skin surface), urinary tract infections, and bacteremia with disseminated disease. As with *Moraxella*, these organisms morphologically resemble *Neisseria*. A negative oxidase test with *Acinetobacter* will readily differentiate these organisms from *Neisseria*. *Acinetobacter* tend to be highly resistant to antibiotics. Successful therapy commonly requires a combination of carbenicillin with an aminoglycoside (e. g., gentamicin).

BIBLIOGRAPHY

Band JD, Chamberland ME, Platt T, Weaver RE, Thornsberry C, and Fraser DW: Trends in meningococcal disease in the United States, 1975-1980, J Infect Dis 148:754, 1983.

Bouvet PJM and Grimont PAD: Taxonomy of the genus *Acinetobacter* with the recognition of *Acinetobacter baumannii* sp. nov., *Acinetobacter haemolyticus* sp. nov., *Acinetobacter johnsonii* sp. nov., and *Acinetobacter junii* sp. nov. and emended descriptions of *Acinetobacter calcoaceticus* and *Acinetobacter lwoffii,* Internatl J System Bacteriol 36:228, 1986.

Buxton AE, Anderson RL, Werdegar D, and Atlas E: Nosocomial respiratory tract infection and colonization with *Acinetobacter calcoaceticus:* epidemiologic characteristics, Am J Med 65:507, 1978.

DeVoe IW: The meningococcus and mechanisms of pathogenicity, Microbiol Rev 46:162, 1982.

Ellison RT, Kohler PF, Curd JG, Judson FN, and Reller LB: Prevalence of congenital or acquired complement deficiency in patients with sporadic meningococcal disease, N Engl J Med 308:913, 1983.

Feldman HA: The meningococcus: a twenty year perspective, Rev Infect Dis 8:288, 1986.

Hagar H, Verghese A, Alvarez S, and Berk SL: *Branhamella catarrhalis* respiratory infections, Rev Infect Dis 9:1140, 1987.

Hook EW and Holmes KK: Gonococcal infections, Ann Intern Med 102:229, 1985.

Krieg NR and Holt JG: Bergey's manual of systematic bacteriology, vol 1, Baltimore, 1984, Williams & Wilkins.

Leinonen M, Luotonen J, Herva E, Valkonen K, and Makela PH: Preliminary serologic evidence for a pathogenic role of *Branhamella catarrhalis,* J Infect Dis 144:570, 1981.

McGee ZA, Stephens DS, Hoffman LH, Schlech WF III, and Horn RG: Mechanisms of mucosal invasion by pathogenic *Neisseria,* Rev Infect Dis 5 (suppl):S708, 1983.

Odum L, Jensen KT, and Slotsbjerg TD: Endocarditis due to *Kingella kingae,* Eur J Clin Microbiol 3:263, 1984.

Petola H: Meningococcal disease: still with us, Rev Infect Dis 5:71, 1983.

Retailliau HR, Hightower AW, Dixon RE, and Allen JR: *Acinetobacter calcoaceticus:* a nonsocomial pathogen with an unusual seasonal pattern, J Infect Dis 139:371, 1979.

Roberts RB: The gonococcus, New York, 1977, Wiley Medical Publications.

Stephens DS, Hoffman LH, and McGee ZA: Interaction of *Neisseria meningitidis* with human nasopharyngeal mucosa: attachment and entry into columnar epithelial cells, J Infect Dis 148:369, 1983.

Van Hare GF, Shurin PA, Marchant CD, Cartelli NA, Johnson CE, Fulton D, Carlin S, and Kim CH: Acute otitis media caused by *Branhamella catarrhalis:* biology and therapy, Rev Infect Dis 9:16, 1987.

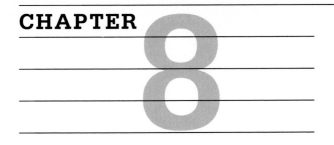

Miscellaneous Cocci

Although the most common cocci associated with human disease are members of the genera *Staphylococcus, Streptococcus,* or *Neisseria,* other cocci commonly colonize the skin and mucous membranes of humans or are present in the environment. These organisms include isolates of *Aerococcus, Peptococcus, Peptostreptococcus,* and *Veillonella,* and are opportunistic pathogens that can cause human disease (Figure 8-1).

AEROCOCCUS

Aerococcus viridans, the only species in the genus, is a microaerophilic, gram-positive coccus that forms tetrads when grown in broth cultures. The organism resembles the viridans group of streptococci when grown on blood-containing media, with the formation of slow-growing, α-hemolytic (green) colonies. The biochemical identification of the *Aerococcus* resembles the group D *Enterococcus,* although *Aerococcus* is weakly catalase positive and lacks the group-specific antigen.

The genus name originates from the observation that these organisms are isolated in the air ("air coccus"), as well as from dust, vegetables, and inanimate objects in hospitals. Despite the ubiquitous presence of *Aerococcus,* these organisms are opportunistic pathogens and rarely cause infections except those secondary to air-borne contamination (e.g., postsurgical wound infections). Most isolates in clinical specimens are clinically insignificant and represent contaminants. *Aerococcus viridans* is susceptible to most antibiotics, including penicillin, streptomycin, erythromycin, and tetracycline.

PEPTOCOCCUS AND PEPTOSTREPTOCOCCUS

The taxonomic classification of anaerobic gram-positive cocci was extensively changed in 1983 (Table 8-1). *Peptococcus niger* is currently the only member of the genus, with other peptococci transferred to the genus *Peptostreptococcus.* Separation of these two genera cannot be easily accomplished by biochemical or morphological criteria. However, the spectrum of diseases and antimicrobial therapy is similar for both groups.

More than 25% of the anaerobes isolated in clinical specimens are members of the genera *Peptococcus* and *Peptostreptococcus.* These gram-positive cocci normally colonize the oral cavity, gastrointestinal tract, genitourinary tract, and skin, and cause infections by spread from these sites to adjacent sterile areas. Infections associated with the anaerobic gram-positive cocci include pleuropulmonary infections following aspiration of oral secretions, sinusitis and brain abscesses after spread of the organisms from the oropharynx or hematogenously from the lungs, intraabdominal sepsis with abscess formation after spread from the intestines, pelvic infections (endometritis, pelvic abscess, puerperal sepsis, salpingitis), soft tissue infections (e.g., Meleney's gangrene, synergistic necrotizing cellulitis), endocarditis, and osteomyelitis.

Most infections are polymicrobial mixtures of aerobic and anaerobic bacteria. Only about 1% of all anaerobic bacteremias are caused by gram-positive cocci, with the majority of the bacteremias caused by *Peptococcus* and *Peptostreptococcus* originating from the genital tract in women. This is because

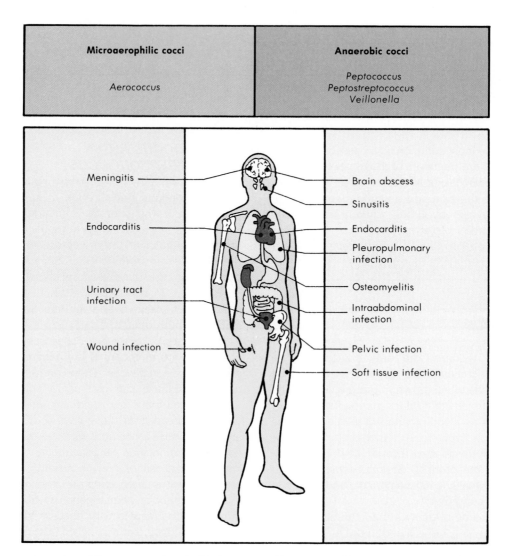

FIGURE 8-1 Diseases associated with *Aerococcus, Peptococcus, Peptostreptococcus,* and *Veil-lonella.*

Table 8-1 Classification of anaerobic gram-positive cocci

Current Classification*	Previous Classification
Peptococcus niger (type species for genus)	
Peptostreptococcus anaerobius (type species for genus)	
Peptostreptococcus asaccharolyticus	Peptococcus asaccharolyticus
Peptostreptococcus prevotii	Peptococcus prevotii
Peptostreptococcus indolicus	Peptococcus indolicus
Peptostreptococcus magnus	Peptococcus magnus
Peptostreptococcus tetradius	Gaffkya anaerobia

*In 1983 the taxonomy of the anaerobic gram-positive cocci was changed, with the transfer of four of the five species of *Peptococcus* and the single *Gaffkya* species into the genus *Peptostreptococcus*.

anaerobic gram-positive cocci are among the predominant organisms in the vagina, in contrast with other body sites colonized with anaerobes where gram-negative bacilli predominate. Bone and joint infections caused by anaerobes, particularly *Peptostreptococcus magnus,* are usually associated with surgical procedures where bacteria residing on the skin contaminate prosthetic material and establish a chronic, indolent infection.

Members of the genera *Peptococcus* and *Peptostreptococcus* are usually susceptible to penicillin, the cephalosporins, clindamycin, and chloramphenicol; they are usually intermediate to erythromycin and the tetracyclines and resistant to the aminoglycosides. Specific therapy is generally indicated when isolated in a monomicrobic infection but may not be necessary in polymicrobic infections.

VEILLONELLA

Veillonella parvula, the major anaerobic gram-negative coccus of clinical significance, is part of the normal bacterial flora of the mouth, gastrointestinal tract, and vagina. Infections with this organism usually originate by spread of a polymicrobial mixture of bacteria from a mucosal surface to a normally sterile adjacent site. *Veillonella* have a relatively low degree of virulence and represent only 1% of all anaerobes isolated in clinical specimens. This organism, either alone or in a mixture, has been associated with a variety of infections. Isolates of *Veillonella* are usually susceptible to penicillin cephalosporins, clindamycin, choloramphenicol, and metronidazole. However, specific therapy against this organism is generally unnecessary unless it is present as a single isolate.

BIBLIOGRAPHY

Bourgault AM, Rosenblatt JE, and Fitzgerald RH: *Peptococcus magnus:* significant human pathogen, Ann Intern Med 93:244, 1980.

Evans JB and Kerbaugh MS: Recognition of *Aerococcus viridans* by the clinical microbiologist, Health Lab Sci 7:76, 1970.

Ezaki T, Yamamoto N, Ninomiya K, Suzuki S, and Yabuuchi E: Transfer of *Peptococcus indolicus, Peptococcus asaccharolyticus, Peptococcus prevotii,* and *Peptococcus magnus* to the genus *Peptostreptococcus* and proposal of *Peptostreptococcus tetradius* sp. nov., Internatl J System Bacteriol 33:683, 1983.

Pien FD, Wilson WR, Kunz K, and Washington JA II: *Aerococcus viridans* endocarditis, Mayo Clin Proc 59:47, 1984.

Topiel MS and Simon GL: Peptococcaceae bacteremia, Diagn Microbiol Infect Dis 4:109, 1986.

CHAPTER 9

Enterobacteriaceae

The family Enterobacteriaceae is the largest, most heterogeneous collection of medically important gram-negative bacilli. Currently, at least 27 genera and 7 enteric groups, with more than 110 species, have been described (Table 9-1). These genera have been classified on the basis of DNA homology, biochemical properties, serologic reactions, susceptibility to genus- and species-specific bacteriophages, and antibiotic susceptibility patterns. The genetic relationship among the more common genera is illustrated in Figure 9-1. Despite the complexity of this family, more than 95% of medically important isolates belong to 10 genera.

Table 9-1 Family Enterobacteriaceae	
Tribe	**Genus (Number of Species)**
Escherichieae	*Escherichia* (5)
	Shigella (4)
Edwardsielleae	*Edwardsiella* (3)
Salmonelleae	*Salmonella* (5)
Citrobactereae	*Citrobacter* (3)
Klebsielleae	*Klebsiella* (7)
	Enterobacter (10)
	Hafnia (3)
	Serratia (9)
Proteeae	*Proteus* (4)
	Morganella (1)
	Providencia (4)
Yersinieae	*Yersinia* (8)
Erwinieae	*Erwinia* (2)

Other genera included in the family Enterobacteriaceae are *Budvicia, Buttiauxella, Cedecea, Kluyvera, Koserella, Leminorella, Moellerella, Obesumbacterium, Rhanella, Tatumella, Xenorhabdus,* and *Yokenella,* as well as a number of unnamed enteric groups.

Enterobacteriaceae are ubiquitous organisms that are found worldwide in soil, water, vegetation and are part of the normal microbial flora of virtually all animals, including humans. Some members of the family (e.g., *Shigella, Salmonella, Yersinia pestis*) are always associated with disease when isolated from man, whereas others (e.g., *Escherichia coli, Klebsiella pneumoniae, Proteus mirabilis*) are members of the normal commensal flora that can cause opportunistic infections. Infections caused by the Enterobacteriaceae can originate from an animal reservoir (e.g., most *Salmonella* infections), from a human carrier (e.g., *Shigella* and *Salmonella typhi*), or by endogenous spread of organisms in a susceptible patient (e.g., *Escherichia*), and the infections can involve virtually all body sites (Figure 9-2). More than 5% of hospitalized patients develop nosocomial infections, with Enterobacteriaceae responsible for the majority of these infections.

MICROBIAL PHYSIOLOGY AND STRUCTURE

Members of this family are moderate-sized (0.3-1.0 × 1.0-6.0 μm) gram-negative bacilli, either motile with peritrichous flagella (except for the infrequently isolated *Tatumella*) or nonmotile, and do not form spores. All members grow aerobically and anaerobically (facultative anaerobes), with growth observed generally after 18 to 24 hours of incubation on a variety of nonselective and selective media. The Enterobacteriaceae have simple nutritional requirements, ferment glucose, reduce nitrates to nitrites, and are oxidase negative. The absence of cytochrome oxi-

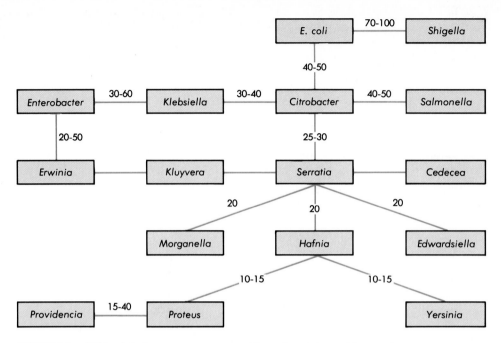

FIGURE 9-1   DNA relatedness among common Enterobacteriaeae. The numbers represent the approximate percentage of relatedness. (From Krieg NR, and Holt JG, editors: Bergey's manual of systematic bacteriology, vol 1, Baltimore, 1984, Williams & Wilkins.)

dase activity is an important characteristic because it can distinguish the Enterobacteriaceae from many other fermentative and nonfermentative gram-negative bacilli.

Morphologic characteristics have been used as a rapid method to identify members of the family Enterobacteriaceae. The ability to ferment lactose has been exploited as a differential characteristic for separating most strains of *Escherichia, Klebsiella,* and Enterobacter, which ferment lactose, from most other common Enterobacteriaceae that do not ferment lactose. The red-colored colonies of lactose-fermenting organisms are readily differentiated on MacConkey agar (a selective medium commonly used for isolation of gram-negative bacilli) from the colorless non-lactose fermenting colonies. Resistance to bile salts present in some selective media has also been used to separate the enteric pathogens *Shigella* and *Salmonella* from commensal Enterobacteriaceae present in the gastrointestinal tract. Some members of the family have prominent capsules *(Klebsiella),* whereas most other strains are surrounded by a loose-fitting, diffusible slime layer.

The serologic classification of Enterobacteriaceae is based on three major groups of antigens: somatic O lipopolysaccharides, capsular K antigens, and the H protein antigens present on the bacterial flagella (Figure 9-3). The heat-stable O lipopolysaccharide is the major cell wall antigen. The antigenically variable O polysaccharide, together with a core polysaccharide common to all Enterobacteriaceae (common antigen) and lipid A, form the lipopolysaccharide (LPS). LPS, also called **endotoxin,** is common to all gram-negative bacteria. Specific O antigens are associated with each genus, although cross-reactions between closely related genera are common (e.g., *Salmonella* and *Citrobacter, Escherichia* and *Shigella).* The antigens are detected by agglutination with specific antisera. The capsular K antigens are either protein or polysaccharides. The heat-labile K antigens may interfere with detection of the O antigens, necessitating removing the capsular antigen by boiling the suspension of organisms. The capsular antigen of *Salmonella typhi* is referred to as the Vi antigen. K antigen are shared by different genera both within and outside the family Enterobacteriaceae (*Escherichia coli* K1

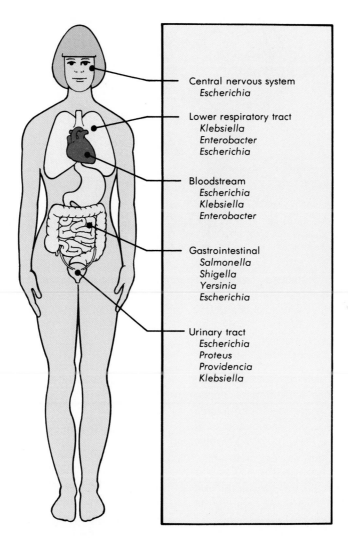

FIGURE 9-2   Sites of infections with members of the Enterobacteriaceae.

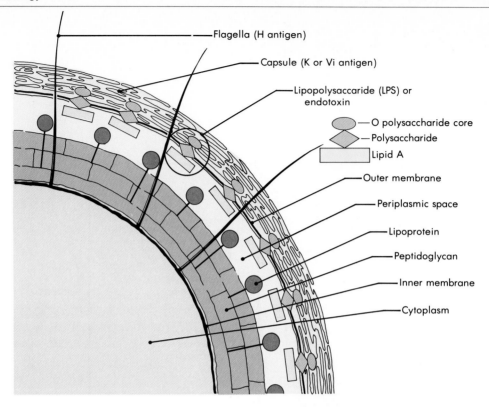

Flagella (H antigen)

Capsule (K or Vi antigen)

Lipopolysaccharide (LPS) or
endotoxin

O polysaccharide core
Polysaccharide
Lipid A

Outer membrane

Periplasmic space

Lipoprotein

Peptidoglycan

Inner membrane

Cytoplasm

FIGURE 9-3   Antigenic structure of Enterobacteriaceae.

cross-reacts with *Neisseria meningitidis* and *Haemophilus meningitidis; Klebsiella pneumoniae* cross-reacts with *Streptococcus pneumoniae*). Organisms with specific K antigens have been associated with increased virulence (e.g., *Escherichia coli* K1 with neonatal meningitis). The H antigens are heat-labile, flagellar proteins. These can be absent from a cell or undergo antigenic variation and be present in two phases. As with the capsular antigens, specific H antigens have been associated with disease.

PATHOGENESIS

Endotoxin

Many of the toxic manifestations of infections with gram-negative bacilli are due to endotoxin, the LPS associated with the outer membrane, that is released upon cell lysis (Box 9-1 and Figure 9-4). Specifically,

toxicity is associated with the lipid A component of LPS.

O, K, and H Antigens

Specific antigens have been associated with meningitis, gastroenteritis, and urinary tract infections. However, the role these somatic, capsular, and flagellar antigens play in the pathogenesis of the infection has not been clearly defined. Some capsular antigens are poor antigens and protect against antibody-mediated phagocytosis. Flagellar antigens probably play a role in adherence of nephrogenic strains to the uroepithelium before the establishment of infection.

Pili

Pili, or fimbriae, are hairlike projections on the surface of the bacilli that mediate attachment to host cells.

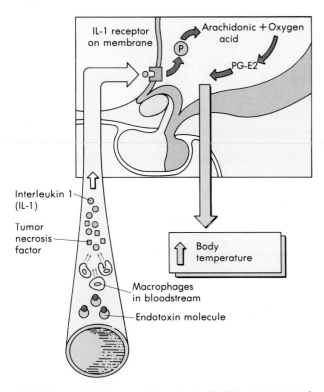

FIGURE 9-4   Production of fever by the lipid A component of lipopolysaccharide. Endotoxin, specifically lipid A, interacts with macrophages and stimulates the initial release of tumor necrosis factor, followed by the production and release of interleukin-1 (IL-1). IL-1 then interacts with the thermoregulatory center of the brain, the hypothalamus, to increase body temperature and decrease peripheral blood flow. The activity of IL-1 on the hypothalamus is mediated by the release of arachidonic acid, with subsequent catalytic conversion to prostaglandin E2 (PG-E2). PG-E2 modulates the neural pathways of the hypothalamus.

ESCHERICHIA COLI

Epidemiology

The genus *Escherichia* consists of at least five species, with *Escherichia coli* the most frequently isolated. *E. coli* is present in the gastrointestinal tract in large numbers and is the Enterobacteriaceae most frequently associated with bacterial sepsis, neonatal meningitis, infections of the urinary tract, and gastroenteritis in travelers to countries with poor hygiene. Most infections (with the exception of gastroenteritis) are endogenous; that is, the individual's normal microbial flora is able to establish infection under conditions where the host defenses are compromised.

The antigenic composition of *E. coli* is complex, with more than 170 O antigens, 56 H antigens, and numerous K antigens described. The serologic classification of *E. coli* isolates is useful for epidemiologic purposes, and specific serotypes are associated with increased virulence.

Clinical Syndromes

Septicemia *Escherichia coli* is the most common gram-negative bacillus isolated from septic patients (Figure 9-5). The focus of infection is commonly either an infection of the urinary tract or spread of organisms from the gastrointestinal tract. The mortality associated with *E. coli* septicemia is influenced by the source of infection and the underlying disease of the patient, with a significantly higher incidence of

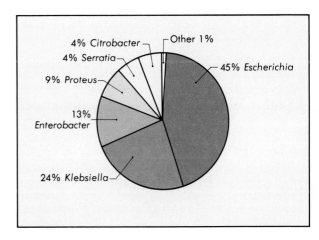

FIGURE 9-5   Enterobacteriaceae associated with bacteremia (Courtesy Barnes Hospital, St. Louis).

death in immunocompromised patients or with infections originating from intestinal perforation.

Urinary Tract Infections *Escherichia coli* is responsible for more than 80% of all community-acquired urinary tract infections and the majority of hospital-acquired infections. Infecting strains originate from the gastrointestinal tract, with disease associated with specific serotypes. Ten major 0 serogroups (01, 02, 04, 06, 07, 09, 015, 016, 018, and 075) cause 80% of the infections. The ability of the bacteria to resist killing in serum, produce hemolysins, and bind to uroepithelial cells is associated with increased virulence. In vitro adherence of *E. coli* to uroepithelial cells has been associated with P pili, with maximum binding demonstrated in patients with pyelonephritis and a lower binding affinity in patients with cystitis and asymptomatic bacteriuria. However, expression of the pili-mediated binding has not been conclusively demonstrated *in vivo*. Other evidence has demonstrated that uropathogens such as *E. coli* can produce a glycocalyx or slime layer that mediates cell adhesion. Regardless of the mechanism, it appears that the ability of the bacteria to adhere to the uroepithelial cells is important for the development of infection.

Meningitis *E. coli,* together with group B streptococci, are the most common causes of neonatal meningitis; 75% of these *Escherichia* strains possess the K1 capsular antigen. Although colonization of infants with *E. coli* at the time of delivery is common, disease is relatively infrequent.

Gastroenteritis Strains of *E. coli* that cause gastroenteritis are subdivided into at least four groups: enterotoxigenic, enteroinvasive, enteropathogenic, and enterohemorrhagic (Table 9-2).

Gastroenteritis produced by **enterotoxigenic *E. coli* (ETEC)** is mediated by heat-labile and heat-stable exotoxins. The action of the heat-labile toxin is similar to the well-characterized *Vibrio cholera* toxin (Chapter 11), with activation of adenylate cyclase and subsequent hypersecretion of fluids and electrolytes into the small intestine. The heat-stable toxin activates guanylate cyclase and stimulates fluid secretion. The production of both toxins is plasmid mediated, and maximum virulence is associated with a specific adhesive pilus: K88 in piglets, K99 in calves, and colonization factors CFA I and CFA II in humans. Disease caused by ETEC follows a 1- to 2-day incubation period and persists for an average of 3 to 4 days. Symptoms are characteristically mild, with cramps, nausea, and vomiting most commonly experienced. Disease mediated by either toxin is indistinguishable. Toxin production is not associated with specific serotypes, so detection of toxigenic strains requires tissue culture or animal model assays for toxin activity. Nucleic acid probes have also been used to detect the toxin genes.

Enteroinvasive *E. coli* (EIEC) are able to invade and destroy the colonic epithelium, producing a disease characterized by fever and cramps, with blood and leukocytes in stool specimens. Disease has been associated with specific O serotypes of *E. coli;* how-

Table 9-2   Gastroenteritis Caused by *Escherichia coli*

Organism	Site of Action	Disease	Pathogenesis
Enterotoxigenic *E. coli* (ETEC)	Small intestine	Traveler's diarrhea; infant diarrhea in underdeveloped countries; watery diarrhea, cramps, nausea, low-grade fever	Heat-stable and/or heat-labile enterotoxins; stimulate guanylate or adenylate cyclase activity with fluid and electrolyte loss
Enteroinvasive *E. coli* (EIEC)	Large intestine	Fever, cramping, watery diarrhea followed by development of dysentery with scant, bloody stools	Plasmid-mediated invasion and destruction of epithelial cells lining colon
Enteropathogenic *E. coli* (EPEC)	Small intestine	Infant diarrhea with fever, nausea, vomiting, nonbloody stools	Plasmid-mediated adherence and destruction of epithelial cells
Enterohemorrhagic *E. coli* (EHEC)	Large intestine	Hemorrhagic colitis with severe abdominal cramps, watery diarrhea initially followed by grossly bloody diarrhea, little or no fever	Mediated by cytotoxic "verotoxin"

ever, serologic classification of isolates cannot reliably identify invasive strains. Invasiveness must be confirmed by the Sereny test, in which EIEC inoculated into the eye of a guinea pig causes keratoconjunctivitis.

The **enteropathogenic _E. coli_ (EPEC)** are historically important agents of childhood diarrhea, particularly in impoverished countries. Although specific O serotypes have been associated with nursery outbreaks of EPEC diarrhea, serotyping _E. coli_ isolated in random or endemic disease is discouraged except in epidemiologic investigations. Disease is caused by the ability of the organism to adhere to the enterocyte plasma membrane and cause destruction of the adjacent microvilli. Adhesiveness is mediated by plasmid-coded pili.

Enterohemorrhagic _E. coli_ (EHEC) produce cytotoxin called verotoxin, which was so named because the toxin causes a cytopathic effect in the Vero cell line of tissue culture cells. EHEC characteristically is responsible for hemorrhagic colitis, with severe abdominal pain, bloody diarrhea, and little or no fever. Serologic classification of isolates has proved to be useful because approximately 80% of EHEC strains are serotype 0157:H7. Disease is most prevalent in the warm months of the year, with the greatest incidence in children under 5 years of age.

SALMONELLA

The taxonomic classification of the genus _Salmonella_ has been fraught with problems. More than 2,000 unique serotypes of _Salmonella_ have been described, which at one time were classified as separate species. It has also been recommended that the genus should be restricted to three species _(Salmonella cholerae-suis, Salmonella typhi,_ and _Salmonella enteritidis)_ based upon biochemical and serologic reactions, with subtype classifications assigned based upon serologic testing. Most recently it has been proposed that the genus consists of a single species _(Salmonella enterica)_ which can be subdivided into six separate but related subspecies (Table 9-3). Unfortunately, the confusing nomenclature associated with this genus will not be eliminated because many of the serotype designations (e.g., _Salmonella typhi, Salmonella cholerae-suis, Salmonella enteritidis)_ are well-established in the medical literature and are likely to remain in common usage.

Table 9-3 Taxonomic Classification of _Salmonella_

Salmonella Subgroup	Source and Significance
Subgroup 1	Obligate parasites and pathogens of man and warm-blooded animals; subgroup most often implicated in human disease (e.g., _S. cholerae-suis, S. enteritidis, S. typhi, S. paratyphi, S. typhimurium)_
Subgroup 2	Majority of strains found in cold-blooded animals or environment
Subgroup 3a	Majority of strains found in cold-blooded animals or environment; occasionally implicated in human disease (_S. arizonae)_
Subgroup 3b	Majority of strains found in cold-blooded animals or environment; occasionally implicated in human disease
Subgroup 4	Usually isolated from cold-blooded animals or environment; rarely implicated in human disease
Subgroup 5	Usually isolated from cold-blooded animals or environment; questionably pathogenic for humans

Epidemiology

Salmonella are widely distributed in the animal kingdom, with isolates recovered from poultry, reptiles, livestock, rodents, domestic animals, birds, and man. Serotypes such as _Salmonella typhi_ and _Salmonella paratyphi_ are highly adapted to man and do not cause disease in nonhuman hosts. Other _Salmonella_ strains are adapted to animals and, when they infect humans, can cause severe human disease (e.g., _Salmonella cholerae-suis)_. Finally, many strains have no host specificity and cause disease in both human and nonhuman hosts.

The source of most infections is ingestion of contaminated water or food-products or direct fecal-oral spread in children. The peak incidence of disease is in young children infected during the warm months of the year at a time when consumption of contaminated food such as egg salad can occur at outdoor social gatherings. The most common sources of human infections are poultry, eggs, and dairy products. An animal reservoir is maintained by animal-to-animal spread and the use of _Salmonella_-contaminated animal feeds. It is estimated that as many as

3,000,000 new cases of *Salmonella* infections occur annually. *Salmonella typhi* is spread by ingestion of food or water contaminated by infected food handlers and, in contrast with other *Salmonella* infections, fewer than 500 cases of typhoid fever occur annually in the United States, with the majority of these associated with foreign travel.

Although exposure to *Salmonella* is frequent, a large inoculum (10^{6-8} bacteria) is required for the development of symptomatic disease. Disease occurs when the organism has an opportunity to multiply to a high density, such as in improperly refrigerated contaminated food products, or in individuals at increased risk for disease because of age, immunosuppression or underlying disease (leukemia, lymphoma, sickle cell disease), or reduced gastric acidity, which effectively reduces the infectious dose. The highest incidence of salmonellosis is in children, particularly those less than 1 year of age, and infections are most severe in the very young and the elderly.

Clinical Syndromes

Salmonella infections occur in one of four forms: gastroenteritis, bacteremia, enteric fever, and asymptomatic colonization.

Gastroenteritis This is the most common form of salmonellosis (Figure 9-6). Symptoms generally appear 6 to 48 hours after consumption of the contaminated food or water, with the initial presentation of

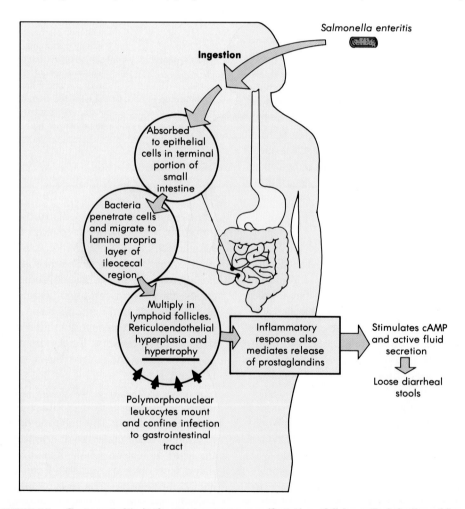

FIGURE 9-6 Gastroenteritis is the most common manifestation of *Salmonella* infection. After passage through the stomach, the bacteria absorb to the brush border of the epithelial cells lining the terminal small intestine and the colon. The bacteria migrate to the lamina propria layer, where they multiply in the lymphoid follicles, stimulating a leukocytic response. Stimulation of prostaglandin-mediated production of cyclic AMP and active fluid secretion also occurs.

nausea, vomiting, and nonbloody diarrhea. Elevated temperature, abdominal cramps, myalgias, and headache are also common. Symptoms can persist from 2 days to 1 week before spontaneous resolution.

Septicemia All *Salmonella* can cause bacteremia, although this is most common with infections caused by *S. cholerae-suis, S. paratyphi, S. typhi,* and *S. dublin. Salmonella* bacteremia is also more common in pediatric and geriatric patients. The clinical presentation of *Salmonella* bacteremia is not unlike other gram-negative bacteremias; however, localized suppurative infections such as osteomyelitis, endocarditis, or arthritis can occur in as many as 10% of the patients.

Enteric Fever *S. typhi* and *S. paratyphi* produce a febrile illness referred to as typhoid fever or paratyphoid fever, respectively. The clinical presentations of both diseases is similar, although paratyphoid fever is generally milder. After a 10- to 14-day incubation period following ingestion of the bacilli, the patient will have a gradually increasing remittent fever with nonspecific complaints of headache, myalgias, malaise, and anorexia. These symptoms persist for 1 or more weeks and are followed by gastrointestinal symptoms. This cycle corresponds to an initial bacteremic phase followed by reinfection of the intestines (Figure 9-7).

Asymptomatic Carriage *S. typhi* infection is

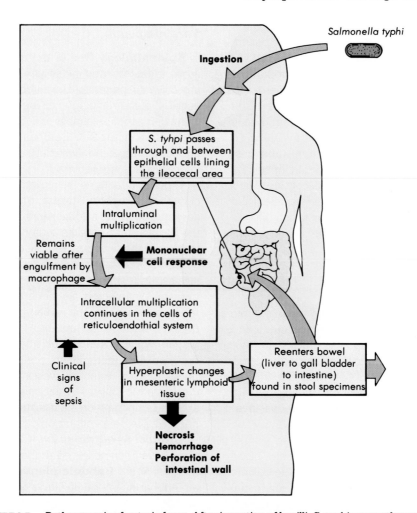

FIGURE 9-7   Pathogenesis of enteric fever. After ingestion of bacilli, *S. typhi* passes through the epithelial cells lining the terminal portion of the small intestine and the colon. The bacilli are engulfed by macrophages and then are carried to the cells of the reticuloendothelial cells, where multiplication continues in the liver, spleen, and bone marrow. Signs of sepsis are seen after a 10-14-day incubation period. The bacilli will spread from the liver through the gall bladder and into the intestines. This stimulates gastrointestinal symptoms.

maintained by human carriage. Chronic carriage for more than 1 year after symptomatic disease will develop in 1% to 5% patients, with the gall bladder the reservoir in most patients. Chronic carriage with other *Salmonella* occurs in less than 1% of patients and does not represent a significant source of infection.

SHIGELLA

Epidemiology

Unlike the genus *Salmonella,* the taxonomic classification of *Shigella* is quite simple. Four species and approximately 32 serotypes have been described: *Shigella dysenteriae, Shigella flexneri, Shigella boydii,* and *Shigella sonnei. Shigella sonnei* is the most common cause of shigellosis in the industrial world, and *Shigella flexneri* is the most common in underdeveloped countries. Shigellosis is primarily a pediatric disease, with most isolates in children from 1 to 4 years of age; epidemic outbreaks of disease are commonly associated with day-care centers, nurseries, and custodial institutions. Infection is transmitted by the fecal-oral route, primarily by way of contaminated hands and less commonly in food or water. Because as few as 200 bacilli can establish disease, shigellosis spreads rapidly in communities where sanitary standards and levels of personal hygiene are low.

Clinical Syndrome

Shigellosis is characterized by abdominal cramps, diarrhea, fever, and bloody stools. Clinical signs and symptoms of shigellosis appear 1 to 3 days after the bacilli are ingested. The bacilli colonize the small intestine and begin to multiply within the first 12 hours. The initial sign of infection, profuse watery diarrhea without histologic evidence of mucosal invasion, is mediated by an enterotoxin. However, the cardinal feature of shigellosis is lower abdominal cramps and tenesmus, with abundant pus and blood in the stool. This results from invasion of the colonic mucosa by the bacilli, destruction of the superficial mucosal layer, and production of mucosal ulcerations. Abundant neutrophils, erythrocytes, and mucus is found in the stool. The bacilli rarely penetrate beyond the mucosal layer, and bacteremia is uncommon. Infection is generally self-limited, although antibiotic

treatment is recommended to reduce the risk of secondary spread to family members and other contacts. Asymptomatic carriage develops in a small number of patients and can represent a persistent reservoir for infection in a community.

YERSINIA

Although the genus *Yersinia* is composed of at least seven species, *Yersinia pestis* and *Yersinia enterocolitica* are the two best-known members. Because the clinical presentations of these infections are dissimilar, they will be considered separately.

Yersinia pestis

Epidemiology One of the most devastating diseases in history was caused by *Yersinia pestis.* During a 5-year period in the middle of the Fourteenth century, epidemic plague or the Black Death claimed 25 million people—almost one fourth of the European population (Figure 9-8). Epidemics continued through the beginning of the Twentieth century, and sporadic infections are still reported.

Y. pestis infections are maintained in two epidemiologic forms: **urban plague,** the disease that was so devastating in the Middle Ages, and **sylvati plague,** the disease that persists today in many countries—including the western region of the United States (Figure 9-9). Urban plague is maintained in rat populations and spread between rats or from rats to humans by infected fleas. With effective control of rats and better hygiene, urban plague has been eliminated from most communities. In contrast, sylvatic plague will be difficult or impossible to eliminate because the mammalian reservoirs (prairie dogs, mice, rabbits, rats) are widespread. Although the organism is highly infectious, human-to-human spread is uncommon unless the patient has pulmonary involvement (pneumonic plague).

Clinical Syndromes Two forms of *Y. pestis* infections have been observed: bubonic plague and pneumonic plague. **Bubonic plague** is characterized by an incubation period of 7 days or less after a bite from an infected flea. Patients will have a high fever and a painful bubo (inflammatory swelling of lymph node) in the groin or axilla. In the absence of treatment patients will rapidly progress to bacteremia, and as many as 75% will die. This was the form of plague that

FIGURE 9-8   During a plague epidemic in Marseilles, France, Charles Delorme, personal physician to King Louis XIII, became the first to advocate special protective clothing to safeguard doctors against infection when treating plague patients. The beaklike mask was supplied with aromatic substances that acted like a filter against the odors of the dying victims. In his hand the doctor holds a stick that had to be carried by everyone coming into contact with plague patients. (From Schreiber W and Mathys FK: Infectio, Basel, Switzerland, 1987, Editiones Roche.)

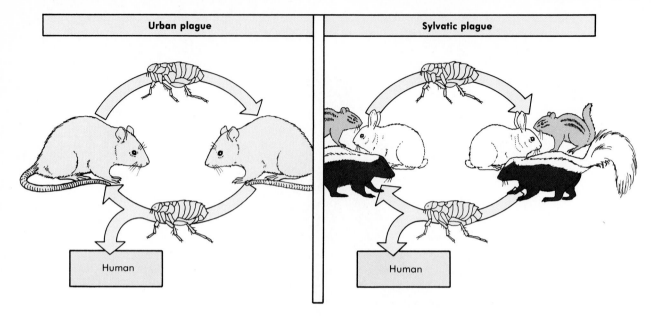

FIGURE 9-9   Epidemiology of *Y. pestis* infections.

was so common during the pandemic of the Middle Ages. Patients with the second form of *Y. pestis* infection, **pneumonic plague,** experience a shorter incubation period (2 to 3 days), initially have fever and malaise, and then develop pulmonary signs within 1 day. The fatality rate with pneumonic plague is in excess of 90% in untreated patients.

Yersinia enterocolitica

Epidemiology   *Yersinia enterocolitica* is a common cause of enterocolitis in Scandinavian and other European countries, as well as in the colder areas of North America. Although most studies indicate that infections are more common during the cold months of the year, not all investigators have documented this observation. The speculation that *Y. enterocolitica* is clinically more active in cold climates is attractive because this parallels the increased metabolic activity of the organisms at 22° to 25°C. Virulence with these organisms has also been associated with specific serotypes: 0:3 and 0:9 in Europe, Africa, Japan, and Canada, and 0:8 in the United States. *Y. enterocolitica* has been found in a variety of sources, including water, milk, and wild and domestic animals. Although an animal reservoir is generally considered to be important, the source of sporadic infections is rarely identified. Epidemic outbreaks have

been associated with contaminated meat or milk.

Clinical Syndromes   Approximately two thirds of all *Y. enterocolitica* infections are enterocolitis, as the name would imply. The gastroenteritis is characterized by diarrhea, fever, and abdominal pain lasting for as long as 1 to 2 weeks, although a chronic form of the disease can develop and persist for months to more than 1 year. Disease involves the terminal ileum and, with enlargement of the mesenteric lymph nodes, can mimic acute appendicitis, necessitating exploratory laparotomy. *Yersinia* infections are most common in children, with pseudoappendicitis particularly troublesome in this age-group. Other manifestations reported in adults include septicemia, arthritis, intraabdominal abscess, hepatitis, and osteomyelitis.

OTHER ENTEROBACTERIACEAE

Klebsiella

Members of this genus have a prominent capsule that is responsible for the mucoid appearance of isolated colonies and enhanced virulence of the organisms *in vivo*. The most commonly isolated member of this genus is *Klebsiella pneumoniae,* which—as the name implies—is associated with community-acquired primary lobar pneumonia. Alcoholics and individuals with compromised pulmonary function are at increased risk for pneumonia, particularly with *Klebsiella,* because of their inability to clear aspirated oral secretions from the lower respiratory tract. Pneumonia caused by *Klebsiella* is frequently associated with necrotic destruction of alveolar spaces, with cavity formation and the production of blood-tinged sputum. *Klebsiella* are also associated with wound, soft tissue, and urinary tract infections.

Proteus

Infections of the urinary tract caused by *Proteus mirabilis* are the most common diseases produced by this genus. *Proteus* strains produce large quantities of urease which splits urea into CO_2 and NH_3. This raises the urine pH and facilitates the formation of renal stones. The increased alkalinity of the urine is also toxic for the uroepithelium. Despite the serological diversity of these organisms, infection has not been associated with any specific serotype. Furthermore, in contrast with *Escherichia coli,* the presence of pili may actually decrease virulence of *Proteus* by enhancing phagocytosis of the bacilli.

Enterobacter, Citrobacter, Serratia, Providencia

Primary infections in immunocompetent patients are rarely caused by *Enterobacter, Citrobacter, Serratia,* or *Providencia* organisms. They are more commonly associated with hospital-acquired infections in patients with a compromised immune system. Antibiotic therapy can be complicated, because resistance to multiple antibiotics is frequently seen.

LABORATORY DIAGNOSIS

Culture

Members of the family Enterobacteriaceae grow readily when cultured in vitro. Specimens collected from normally sterile sources such as spinal fluid or tissue collected at surgery can be inoculated onto nonselective blood agar media; a selective medium (e.g., MacConkey agar, Eosin Methylene Blue [EMB] agar) is used for specimens normally contaminated with other organisms (e.g., sputum, feces). The use of these selective differential agars has the advantage of differentiating empiric lactose-fermenting Enterobacteriaceae from nonfermentative strains, information that can be used to guide empiric antimicrobial therapy. Highly selective or organism-specific media are useful for the recovery of organisms such as *Salmonella* in stool specimens, where large numbers of normal flora could obscure the presence of significant pathogens. The recovery of *Yersinia enterocolitica* is complicated because this organism grows slowly at traditional incubation temperatures and prefers cooler temperatures where it is metabolically more active. This property has been exploited, however, in the clinical laboratory by mixing the fecal specimen with saline and then storing the specimen at 4° C for 2 weeks or more before it is subcultured to agar media. This **"cold enrichment"** permits the growth of *Yersinia* while inhibiting or killing other organisms present in the specimen. Although use of cold enrichment does not aid in the initial management of a patient with *Yersinia* gastroenteritis, it has helped elucidate the role of this organism in both acute and chronic intestinal disease.

Biochemical Identification

Because a large number of diverse species are present in the family Enterobacteriaceae, a review of their biochemical properties is beyond the scope of this book. Table 9-4 summarizes the properties used to separate the most commonly identified species. With increasing sophistication of biochemical test systems, virtually all members of the family can be accurately identified in 4 to 24 hours with one of a number of commercially available identification systems. Furthermore, the most common organisms can be identified with 1-minute spot tests (Figures 9-10 and 9-11).

Serologic Classification

Serology is very useful for determining the clinical significance of an isolate (e.g., serotyping specific pathogenic strains such as *E. coli* Kl, *E. coli* 0157:H7, *Y. enterocolitica* 0:8) or for classifying isolates for epidemiologic purposes (e.g., characterizing isolates in a suspected outbreak of salmonellosis). However, the usefulness of this procedure is limited by cross-reactions with antigenically related Enterobacteriaceae and other organisms.

Table 9-4 Biochemical Characteristics of Clinically Significant Genera of the Family Enterobacteriaceae

Characteristic	Citrobacter	Enterobacter†	Escherichia	Klebsiella	Morganella	Proteus	Providencia	Salmonella	Serratia	Shigella	Yersinia
Primary test battery											
Adonitol	V	V	V	+	−	−	V	−	V	−	−
Arginine	V	V	V	−	−	−	−	V	−	V	−
Citrate	+	+	V	V	−	V	+	V	+	−	−
DNase	−	−	−	−	−	V	−	−	+	−	−
Gas	+	+	+	V	V	V	V	V	V	−	V
H$_2$S	V	−	−	−	−	V	−	V	−	−	−
Indole	V	−	+	V	+	V	+	−	V	V	V
Lysine	−	V	+	V	−	−	−	V	V	−	−
Motility	+	+	V	−	+	+	+	+	+	−	−
Ornithine	V	+	V	−	+	V	−	V	V	V	V
Phenylalanine	−	−	−	−	+	+	+	−	−	−	−
Sucrose	V	+	V	V	−	V	V	−	V	−	V
Urease	V	V	−	V	+	+	V	−	−	−	V
VP	−	+	−	V	−	V	−	−	+	−	−
Secondary test battery											
Arabinose	+	+	+	+	−	−	−	V	V	V	V
Inositol	−	V	−	V	−	−	V	V	V	−	V
KCN	V	V	−	+	+	+	+	−	V	−	−
Lactose	V	V	V	V	−	−	−	−	V	−	V
Malonate	V	V	V	V	−	−	−	−	V	−	−
Mannitol	+	+	+	+	−	−	V	+	+	V	+
Melibiose	V	+	V	+	−	−	−	V	V	V	V
ONPG	+	+	+	V	−	−	−	−	+	V	V
Raffinose	V	+	V	+	−	−	−	−	V	V	V
Rhamnose	+	+	V	V	−	−	V	V	V	V	V
Salicin	V	V	V	+	−	V	V	−	V	−	V
Sorbitol	+	V	V	+	−	−	−	+	V	V	V
Trehalose	+	+	+	+	V	V	V	V	+	V	+
Xylose	+	+	+	+	−	+	+	V	V	−	V

From Kelly MT, Brenner DJ, and Farmer JJ: Enterobacteriaceae. In Lennette EH, Balows A, Hausler, WJ, and Shadomy HJ, editors: Manual of clinical microbiology, ed 4, Washington DC, 1985, American Society for Microbiology.
*Symbols: −, < 10% of strains positive; V, 10% to 89% of strains positive; +, ≥ 90% of strains positive. Only clinically significant species are included in the percentage values for each genus.
†Not including *E. agglomerans*.

TREATMENT

Antibiotic therapy for infections with Enterobacteriaceae must be guided by in vitro susceptibility test results and clinical experience. Whereas some organisms such as *Escherichia coli* and *Proteus mirabilis* are susceptible to many antibiotics, others can be highly resistant. Furthermore, susceptible organisms can rapidly develop resistance when exposed to sub-therapeutic concentrations of antibiotics in a hospital setting. In general, antibiotic resistance is more common in infections acquired in the hospital compared with community-acquired infections. Antibiotic therapy is not recommended for some infections. For example, symptomatic relief but not antibiotic treatment is usually recommended for *Salmonella* gastroenteritis because antibiotics can prolong fecal carriage of this organism.

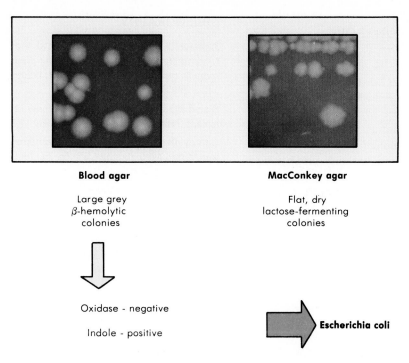

Blood agar

Large grey
β-hemolytic
colonies

MacConkey agar

Flat, dry
lactose-fermenting
colonies

Oxidase - negative

Indole - positive

Escherichia coli

FIGURE 9-10   Flat, dry, lactose-fermenting colonies on MacConkey agar and large, grey, β-hemolytic on blood agar, that are oxidase negative and indole positive, can be identified as *E. coli.*

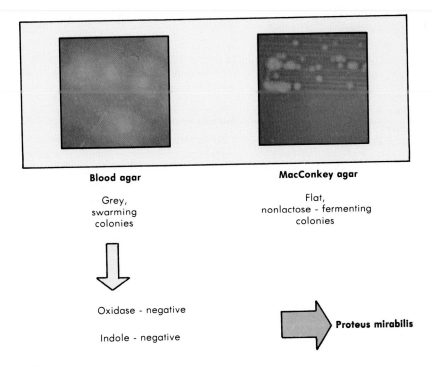

Blood agar

Grey,
swarming
colonies

MacConkey agar

Flat,
nonlactose - fermenting
colonies

Oxidase - negative

Indole - negative

Proteus mirabilis

FIGURE 9-11   Flat, nonlactose-fermenting colonies on MacConkey agar and grey, swarming colonies on blood agar, that are oxidase negative and indole negative, can be identified as *P. mirabilis.*

PREVENTION AND CONTROL

Prevention of infections is difficult because the Enterobacteriaceae are a major part of the endogenous microbial population. However, some risk factors for infections should be avoided, including the unrestricted use of antibiotics that can select for resistant bacteria, performance of procedures that traumatize mucosal barriers without prophylactic antibiotic coverage, and use of urinary catheters. Unfortunately, many of these factors are present in patients at greatest risk for infection (e.g., immunocompromised individuals confined to the hospital for extensive periods).

Exogenous infection with Enterobacteriaceae is theoretically easier to control. The source of infections with organisms like *Salmonella* is well-defined. However, these bacteria are ubiquitous in poultry and eggs. Unless care is used for the preparation and refrigeration of foods, little can be done to control *Salmonella* infections. Transmission of *Shigella* is predominantly in young children, and interruption of fecal-mouth transmission in this population is difficult. Control can only be effective with education and introduction of appropriate infection control procedures (e.g., hand washing, proper disposal of soiled linens) in a setting with an identified outbreak.

Vaccination with formalin-killed *Yersinia pestis* has proved to be effective for those at high risk. Chemoprophylaxis with tetracycline has also proved useful for individuals in close contact with a patient with pneumonic plague. Live, attenuated *Salmonella typhi* vaccines have reduced typhoid fever by 50% in populations with a high endemic rate of disease. However, the duration of this protection is short lived. Inactivated whole-cell vaccines, as well as vaccination with purified Vi antigen, the polysaccharide capsular antigen of *Salmonella typhi* associated with virulence, is also protective.

BIBLIOGRAPHY

Acharya IL, Lowe CU, Thapa R, et al: Prevention of typhoid fever in Nepal with the Vi capsular polysaccharide of *Salmonella typhi:* a preliminary report, N Engl J Med 317:1101-1104, 1987.

Blaser MJ and Newman LS: A review of human salmonellosis. I. Infective dose, Rev Infect Dis 4:1096-1106, 1982.

Cantey JR: Infectious diarrhea: pathogenesis and risk factors, Am J Med 78(suppl 6B):65-75, 1985.

Cornelis G, Laroche Y, Balligand G, et al: *Yersinia enterocolitica,* a primary model for bacterial invasiveness, Rev Infect Dis 9:64-87, 1987.

Dinarello CA and Wolff SM: Molecular basis of fever in humans, Am J Med 72:799-819, 1982.

Edelman R and Levine MM: Summary of an international workshop on typhoid fever, Rev Infect Dis 8:329-349, 1986.

Ewing WH: Edwards and Ewing's identification of Enterobacteriaceae, ed 4, New York, 1986, Elsevier Science Publishing Co.

Harber MJ: Bacterial adherence, Eur J Clin Microbiol 4:257-261, 1985.

Levine MM: *Escherichia coli* that cause diarrhea: enterotoxigenic, enteropathogenic, enteroinvasive, enterohemorrhagic and enteroadherent, J Infect Dis 155:377-389, 1987.

Pajic JK and Davey RB: The genus *Yersinia:* epidemiology, molecular biology, and pathogenesis. In Contributions to microbiology and immunology, vol 9, New York, 1987, Karger Publishers.

Swaminathan B, Harmon MC, and Mehlman IJ: A review: *Yersinia enterocolitica,* J Appl Bacteriol 52:151-183, 1982.

Uhshen MH and Rollo JL: Pathogenesis of *Escherichia coli* gastroenteritis in man—another mechanism, N Engl J Med 302:99-101, 1980.

Vantrappen G, Geboes K, and Ponette E: *Yersinia* enteritis, Med Clin North Am 66:639-653, 1982.

CHAPTER 10

Pseudomonadaceae

The family Pseudomonadaceae is a complex mixture of opportunistic plant, animal, and human pathogens. *Bergey's Manual of Systematic Bacteriology* includes four genera in the family: *Pseudomonas, Xanthomonas, Frateuria,* and *Zoogloea* (Table 10-1). Because the majority of infections in the family are caused by *Pseudomonas,* this chapter will focus on this genus with brief comments about *Xanthomonas,* and will have no further discussion of *Frateuria* and *Zoogloea.*

Pseudomonads are ubiquitous organisms, found in soil, decaying organic matter, vegetation, and water. They are also, unfortunately, found throughout the hospital environment on moist reservoirs such as food, cut flowers, sinks, toilets, floor mops, respiratory therapy equipment, and even disinfectant solutions. Persistent carriage as part of the normal microbial flora in humans is uncommon unless the individual is hospitalized or is an ambulatory, immunocompromised host (Figure 10-1). The broad environmental distribution of *Pseudomonas* is afforded by their simple growth requirements. More than 30 organic compounds can be used as a source of carbon and nitrogen, and some strains can even grow in distilled water by utilizing trace nutrients. *Pseudomonas* also possess a number of structural factors and toxins that enhance the virulence potential of the organism, as well as render them resistant to most commonly used antibiotics. Indeed, it is surprising that these organisms are not more common pathogens, with their ubiquitous presence, ability to grow in virtually any environment, virulence properties, and broad-based antimicrobial resistance. Instead, *Pseudomonas* infections are primarily opportunistic (i.e., restricted to patients with compromised host defenses).

MICROBIAL PHYSIOLOGY AND STRUCTURE

Pseudomonads are straight or slightly curved gram-negative bacilli (0.5-1.0 × 1.5-5.0 μm), motile by means of polar flagella; the organisms utilize relatively few carbohydrates (e.g., glucose, ribose, gluconate) by oxidative metabolism but not by fermentative metabolism. Oxygen is the terminal electron acceptor, and the presence of cytochrome oxidase in *Pseudomonas* is used to differentiate this group from the Enterobacteriaceae. Anaerobic growth can occur, with nitrate used as an alternate electron acceptor. Some strains appear mucoid because of the abundance of a polysaccharide capsule, and some produce diffusible pigments (e.g., pyocyanin [blue], fluorescein [yellow], or pyorubin [red-brown]). The genus consists of a number of different species subdivided by biochemical and genetic differences. The species isolated most frequently in the clinical laboratory are summarized in Table 10-2. By far, *Pseudomonas aeruginosa* is the most common clinically significant

Table 10-1	Pseudomonadaceae
Organism	**Pathogenic Status**
Pseudomonas	Pathogenic for humans, animals, or plants
Xanthomonas	Pathogenic for humans or plants
Frateuria	Pathogenic for plants
Zoogloea	Nonpathogenic

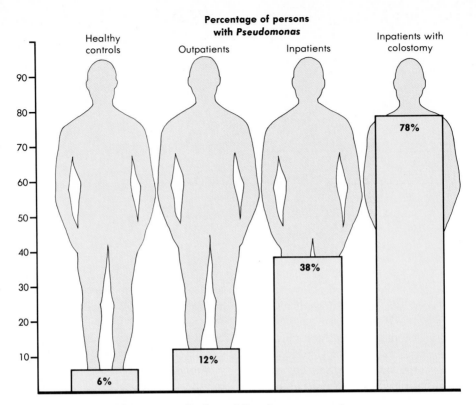

FIGURE 10-1   Carriage of *Pseudomonas aeruginosa.*

Table 10-2 Clinically Important Pseudomonadaceae Species		
Affiliation	**Group**	**Genus and Species**
RNA group I	Fluroescent group	*P. aeruginosa*
		P. fluorescens
		P. putida
	Stutzeri group	*P. stutzeri*
		P. medocina
	Alcaligenes group	*P. alcaligenes*
		P. pseudoalcaligenes
RNA group II	Pseudomallei group	*P. mallei*
		P. pseudomallei
		P. cepacia
RNA group III	Acidovoran group	*P. acidovorans*
		P. testosteroni
RNA group IV	Diminuta group	*P. diminuta*
		P. vesicularis
Species of uncertain affiliation		*P. mesophilica*
		P. paucimobilis
		P. putrefaciens
RNA group V	Maltophilia group	*Xanthomonas (Pseudomonas) maltophilia*

pseudomonad and the best-characterized member of the genus.

PATHOGENESIS

Pseudomonas aeruginosa has a number of virulence factors, including structural components, toxins, and enzymes (Table 10-3). In fact, it is difficult to define the role each factor plays in disease caused by this organism, and most experts in this field believe *Pseudomonas* virulence is multifactorial.

Pili or Fimbriae

These hairlike structures mediate adherence of the bacterium to the respiratory epithelium. The expression of other pseudomonad virulence factors is facilitated by this prolonged colonization on mucosal surfaces.

Polysaccharide Capsule

The surface of *P. aeruginosa* is covered with a polysaccharide layer that protects the organism from phagocytosis. This layer can also anchor the bacteria to cell surfaces, particularly in patients with cystic fibrosis or other chronic respiratory diseases who are predisposed to colonization with mucoid strains of *P. aeruginosa*.

Endotoxin

As is true with other gram-negative bacilli, pseudomonads possess a lipopolysaccharide endotoxin as a major cell wall antigen. The lipid A component of endotoxin mediates the various biological effects described in Chapter 9 and summarized in Table 10-3.

Exotoxin A

One of the most important virulence factors produced by pathogenic strains of *P. aeruginosa* is exotoxin A (Figure 10-2). This toxin blocks eukaryotic cell protein synthesis in a manner similar to that described for diphtheria toxin (Chapter 19). However, these toxins are structurally and immunologically different, and exotoxin A is less potent compared with diphtheria toxin.

Table 10-3 Virulence Factors Associated With *Pseudomonas*

Virulence Factor	Biologic Effect
Pili	Adherence to respiratory epithelium
Polysaccharide capsule	Adherence to tracheal epithelium, antiphagocytic
Endotoxin	Fever, shock, oliguria; leukopenia or leukocytosis; disseminated intravascular coagulation; metabolic abnormalities
Exotoxin A	Inhibition of protein synthesis
Exoenzyme S	Inhibition of protein synthesis
Elastase	Vascular tissue damage, inhibition of neutrophil function
Alkaline protease	Tissue damage, anticomplementary, inactivation of IgG, inhibition of neutrophil function
Phosopholipase C	Tissue damage
Leukocidin	Inhibition of neutrophil and lymphocyte function

Exoenzyme S

This extracellular toxin is produced by one third of the clinical isolates of *P. aeruginosa* and can inhibit protein synthesis. Both exotoxin A and exoenzyme S are ADP-ribosyltransferases, but they are distinguished by the heat stability of exoenzyme S.

Elastase

This enzyme can catalyze the destruction of the elastic fiber in blood vessel walls, resulting in hemorrhagic lesions **(ecthyma gangrenosum)** associated with disseminated *Pseudomonas* infections.

Other Proteases

Other proteases have been described in pseudomonads that mediate tissue destruction, inactivation of antibodies, and inhibition of neutrophils.

Phospholipase C

Phospholipase C breaks down lipids and lecithin, facilitating tissue destruction. The exact role of this

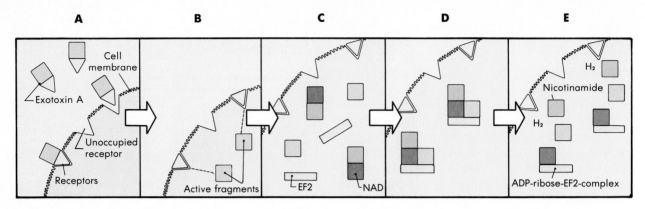

FIGURE 10-2   Mode of action of exotoxin A. **A,** Exotoxin A, composed of fragments A and B, inhibits eukaryotic cell protein synthesis by binding to specific receptors in the cell membrane. **B,** After fragment B binds to a cell receptor, fragment A enters the cell. **C,** Fragment A catalyzes the binding of nicotinamide adenine dinucleotide (NAD) to Elongation Factor 2 (EF2), which is required for translocation of nascent polypeptide chains on eukaryotic ribosomes. **D** and **E,** The reaction terminates in the irreversible formation of an adenosine diphosphate ribose. EF2 diphosphate—EF2 complex with the release of nicotinamide and hydrogen.

enzyme in infections of the respiratory and urinary tracts is unclear, although there is a significant association between hemolysin production and disease at these sites.

EPIDEMIOLOGY

Pseudomonads are opportunistic pathogens present in a variety of environmental habitats. The ability to isolate these organisms from moist surfaces may be limited only by one's interest in searching for the organsism. Pseudomonads have minimal nutritional requirements, can tolerate a wide range of temperatures (4° to 42° C), and are resistant to many antibiotics and disinfectants. Indeed, the simple recovery of *Pseudomonas* from an environmental source (e.g., hospital sink or floor) means very little without epidemiologic evidence that the contaminated site is a reservoir for infection. Furthermore, isolation of *Pseudomonas* from a hospitalized patient is worrisome but does not normally justify therapeutic intervention without evidence of disease. It is important to note that recovery of *Pseudomonas,* particularly species other than *P. aeruginosa,* from a clinical specimen may represent contamination of the specimen during collection or laboratory processing. Because their organisms are opportunistic pathogens, the significance of an isolate must be measured by assessing the clinical presentation of the patient.

CLINICAL SYNDROMES

Pseudomonas aeruginosa

Many of the serious infections caused by *P. aeruginosa* are listed in Box 10-1. A discussion follows.

Bacteremia and Endocarditis   Bacteremia caused by *P. aeruginosa* is clinically indistinguishable from other gram-negative infections, although the mortality rate is higher. This is due in part to the predilection of this organism for immunocompromised patients and in part to the inherent virulence of *Pseudomonas. P. aeruginosa* bacteremia is particularly common in patients with neutropenia, diabetes mellitus, extensive burns, and hematologic malignancies (Figure 10-3). Most *Pseudomonas* bacteremias originate from infections of the lower respiratory tract, urinary tract, and skin and soft tissue (particularly burn wound infections). Although seen in a minority of patients, characteristic skin lesions (ecthyma gangrenosum) may develop. The lesions are seen initially as erythematous vesicles that progress to hemorrhage, necrosis, and ulceration. Microscopic examination of the lesion shows abundant organisms with vascular destruction, which explains the hemorrhagic nature of the lesions, and the absence of neutrophils as would be expected in neutropenic patients.

Pseudomonas endocarditis is most commonly observed in intravenous drug abusers; the source of infection is drug paraphernalia contaminated with

Box 10-1 *Pseudomonas Aeruginosa* Infections

Bacteremia
Endocarditis
Pulmonary infections
   Tracheobronchitis
   Necrotizing bronchopneumonia
Ear infections
   Chronic external otitis
   Malignant external otitis
   Chronic otitis media
Burn wound infections
Urinary tract infections
Gastroenteritis
Eye infections
Musculoskeletal infections

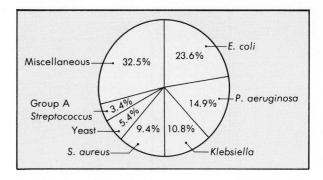

FIGURE 10-3 Septicemia in patients with leukemia or lymphoma. *P. aeruginosa* is the second most common cause of septicemia in this population (Redrawn from Singer C, Kaplan MH, and Armstrong D: Am J Med 62:731-742, 1977.)

water-borne organisms. The tricuspid valve is most often involved and is associated with a chronic course and a more favorable prognosis compared with infections of the aortic or mitral valve, which generally have an acute and frequently fatal course.

Pulmonary Infections *P. aeruginosa* infections of the lower respiratory tract can range from colonization or benign tracheobronchitis in severe necrotizing bronchopneumonia. Colonization is seen in patients with cystic fibrosis, other chronic lung diseases, and neutropenia. *Pseudomonas* pulmonary infection in patients with cystic fibrosis has been associated with exacerbation of the underlying disease, as well as invasive disease in pulmonary parenchyma. Neutropenic and other immunocompromised patients are frequently exposed to *Pseudomonas* following use of contaminated respiratory therapy equipment. Invasive disease in this population is characterized by a diffuse, typically bilateral bronchopneumonia with microabscess formation and tissue necrosis. Bacteremia, with an associated high mortality rate, can be observed in severe infections.

Ear Infections External otitis is most frequently due to *P. aeruginosa,* with swimming **("swimmer's ear")** a significant risk factor (Figure 10-4). Although this localized infection can be managed with topical antibiotics and drying agents, a more virulent form of disease **(malignant external otitis)** can invade the underlying tissues and be life threatening. Aggressive antimicrobial and surgical intervention is

required. *P. aeruginosa* is also commonly associated with chronic otitis media.

Burn Infections *P. aeruginosa* colonization of a burn wound, followed by localized vascular damage and tissue necrosis, and ultimately bacteremia, is not uncommon in patients who have sustained severe burns (Figure 10-5). The moist surface of the burn and absence of neutrophilic response to tissue invasion predispose patients to *Pseudomonas* infections. Use of topical creams and wound management has controlled *Pseudomonas* colonization with only limited success.

Other Infections *P. aeruginosa* is associated with a variety of other infections, including those localized in the gastrointestinal and urinary tracts, eye, central nervous system, and musculoskeletal system. The underlying conditions required for most *Pseudomonas* infections are the presence of the organism in a moist reservoir and the circumvention or absence of host defenses (e.g., cutaneous trauma, elimination of normal microbial flora by injudicious use of antibiotics, neutropenia.)

Pseudomonas pseudomallei

Melioidosis *P. pseudomallei* is a saprophyte found in soil, water, and vegetation in Southeast Asia, India, the Philippines, Guam, Indonesia, New Guinea, and Australia. Although the organism is rarely isolated in the western hemisphere, latent disease occurs in persons who have been in Southeast Asia. The disease, melioidosis, has protean manifestations.

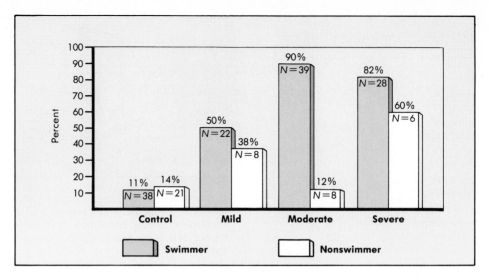

FIGURE 10-4   Recovery of *P. aeruginosa* from patients with external otitis.

The incubation period varies from a few days to as long as 26 years. Most individuals exposed to *P. pseudomallei* remain asymptomatic. Some with cutaneous exposure develop a localized suppurative infection with regional lymphadenitis, fever, and malaise. This form of disease can resolve without incidence or rapidly progress to overwhelming sepsis. The third form of infection is pulmonary disease, which may range from mild bronchitis to necrotizing pneumonia. Cavitation can develop in the absence of appropriate antimicrobial therapy. Isolation of *P. pseudomallei* for diagnostic purposes should be approached carefully because the organism is considered highly infectious.

Other Pseudomonads

The organisms listed in Table 10-2, particularly *Pseudomonas cepacia* and *Xanthomonas (Pseudomonas) maltophilia,* are all capable of opportunistic infections in immunocompromised patients and are frequently isolated in the clinical laboratory. The clinical significance of an isolate is often difficult to assess because specific signs and symptoms of disease may be absent in this population of patients and the organism may be an insignificant water-borne contaminant. The majority of true infections with these organisms have been localized to the respiratory tract in patients with underlying pulmonary disease or to the urinary tract following instrumentation or catheterization.

LABORATORY DIAGNOSIS

Culture

Because pseudomonads have simple nutritional requirements, they grow easily on such common isolation media as blood agar or MacConkey agar. Aerobic incubation is required (unless nitrate is available), so growth in broth is generally confined to the broth-air interface.

Identification

The colonial morphology (colony size, hemolytic activity, pigmentation, odor) combined with selected rapid biochemical tests (e.g., positive oxidase reaction) are used for the preliminary identification of isolates. For example, *P. aeruginosa* grows rapidly and has flat colonies with a spreading border, green pigmentation (caused by production of blue pyocyanin and yellow fluorescein), and a characteristic sweet, grapelike odor. Definitive identification of pseudomonads requires an extensive battery of physiologic tests. Specific classification of isolates for epidemiologic purposes is accomplished by determination of biochemical profiles, antibiotic susceptibility patterns, susceptibility to bacteriophages, production of pyocins, or serologic typing. Although classification by biochemical or susceptibility patterns is used most often in clinical laboratories, these are the least discriminatory methods. Phage typing, pyocin typing,

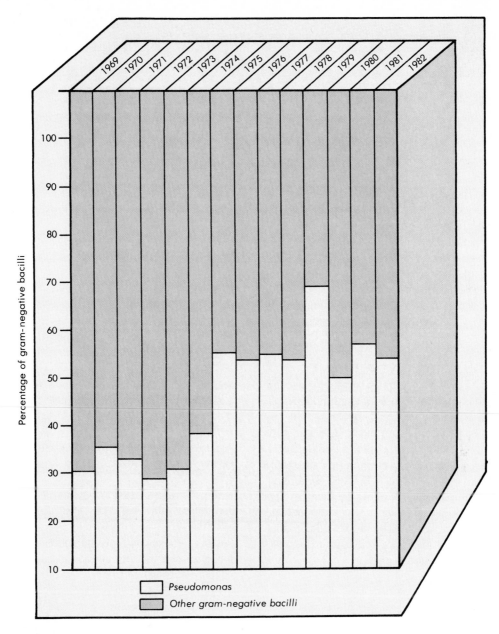

FIGURE 10-5   Gram-negative bacilli recovered in burn wound cultures collected over a 14-year period at the Fort Sam Houston Burn Center. (Data from Pruitt BA Jr and McManus AT: Am J Med 76[suppl]:146-154, 1984.)

and serotyping have all been used successfully but are performed only in reference laboratories.

TREATMENT

Antimicrobial therapy for *Psuedomonas* infections is frustrating because the infected patient with com-

promised host defenses is unable to augment the antibiotic activity, and pseudomonads are typically resistant to most antibiotics (Table 10-4). Even in susceptible organisms resistance can develop during therapy by the induction of antibiotic inactivating enzymes (e.g., β-lactamases) or the transfer of plasmid-mediated resistance from a resistant organism to a susceptible one. Furthermore, some groups of anti-

Table 10-4 Mechanisms of Antibiotic Resistance in *Pseudomonas*

Antibiotic	Resistance Mechanisms
Penicillins and cephalosporins	β-lactamase hydrolysis, altered binding proteins, decreased permeability
Aminoglycosides	Enzymatic hydrolysis by acetylation, adenylation, or phosphorylation; decreased permeability; altered ribosomal target
Chloramphenicol	Enzymatic hydrolysis by acetyltransferase; decreased permeability
Quinolones	Altered target (DNA gyrase); decreased permeability

biotics such as the aminoglycosides are ineffective at the site of infection (poor activity in the acidic environment of an abscess). Successful therapy for serious infections generally requires the combination use of an aminoglycoside and β-lactam antibiotic, which have documented activity against the isolate. Augmentation of compromised immune function with hyperimmune globulin and granulocyte transfusions have beneficial effects with selected patients.

PREVENTION AND CONTROL

Attempts to eliminate *Pseudomonas* from the hospital environment are practically useless with the ubiquitous presence of the organism in water supplies. Effective infection-control practices should concentrate on prevention of contamination of sterile equipment such as respiratory therapy machines, and cross-contamination of patients by medical personnel. The inappropriate use of broad-spectrum antibiotics should be avoided, because this can suppress the normal microbial flora and permit the overgrowth of resistant pseudomonads.

BIBLIOGRAPHY

Bisbe J, Gatell JM, Puig J, et al: *Pseudomonas aeruginosa* bacteremia: univariate and multivariate analyses of factors influencing the prognosis in 133 episodes, Rev Infect Dis 10:629-635, 1988.

Bodey GP, Jadeja L, and Elting L: *Pseudomonas* bacteremia: retrospective analysis of 410 episodes, Arch Intern Med 145:1621-1629, 1985.

Brook I and Finegold SM: Bacteriology of chronic otitis media, J Am Med Assoc 241:487-488, 1979.

Doggett RG: *Pseudomonas aeruginosa:* clinical manifestations of infection and current therapy, New York, 1979, Academic Press.

Doroghazi RM, Nadol JB, Jr, Hyslop NE, Jr, et al: Invasive external otitis: report of 21 cases and review of the literature, Am J Med 71:603-614, 1981.

Gustafson TL, Band JD, Hutcheson RH, Jr, and Schaffner W: *Pseudomonas* folliculitis: an outbreak and review, Rev Infect Dis 5:1-8, 1983.

Hoadley AW and Knight DE: External otitis among swimmers and nonswimmers, Arch Environ Health 30:445-448, 1975.

Jacoby GA, Jr: Properties, problems, and therapeutic progress: *Pseudomonas aeruginosa,* Roche Seminars on Bacteria 5:1-22, 1987.

Pollack M and Young LS: Protective activity of antibodies to exotoxin A and lipopolysaccharide at the onset of *Pseudomonas aeruginosa* septicemia in man, J Clin Invest 63:276-286, 1979.

Pruitt BA, Jr and McManus AT: Opportunistic infections in severely burned patients, Am J Med 76(suppl):146-154, 1984.

Sadoff JC and Sanford JP: Symposium on *Pseudomonas aeruginosa* infections, Rev Infect Dis 5(suppl):833-1004, 1983.

Singer C, Kaplan MH, and Armstrong D: Bacteremia and fungemia complicating neoplastic disease: a study of 364 cases, Am J Med 62:731-742, 1977.

Young LS: The role of exotoxins in the pathogenesis of *Pseudomonas aeruginosa* infections, J Infect Dis 142:626-630, 1980.

Young LS: *Pseudomonas aeruginosa*—biology, immunology, and therapy: a cefsulodin symposium, Rev Infect Dis (suppl):603-776, 1984.

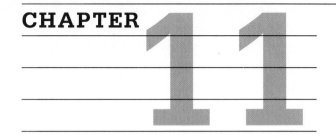

Vibrionaceae and Campylobacter

Members of the family Vibrionaceae are curved or straight bacilli, capable of aerobic or anaerobic growth, oxidase positive, and nonspore formers. They are primarily found in water and are well known for their ability to produce gastrointestinal disease. The family includes three genera associated with human disease: *Vibrio, Aeromonas,* and *Plesiomonas. Campylobacter* organisms, although not a member of the family Vibrionaceae, were formerly classified as *Vibrio.* They are a cause of diarrheal disease and septicemia and are associated with gastritis and pyloric ulcers.

VIBRIO

The genus *Vibrio* is composed of gram-negative, curved bacilli that differ from Enterobacteriaceae by their positive oxidase reaction, polar flagella, and growth on alkaline but not acidic media. The most commonly isolated pathogenic species are listed in Box 11-1.

Box 11-1 Common Pathogenic *Vibrio* Species

V. cholerae O1	Gastroenteritis
V. cholerae non-O1	Gastroenteritis
V. parahaemolyticus	Gastroenteritis
V. vulnificus	Septicemia, wound infection
V. alginolyticus	Wound infection, external otitis

Microbial Physiology and Structure

Vibrio species can grow aerobically or anaerobically on a variety of simple media, with a broad temperature range (from 18° to 37° C) for optimal growth. *Vibrio cholerae*, the best-known member of the genus, can be serologically subdivided into six groups based on somatic O antigens, with most pathogens in the O1 group. However, toxigenic *V. cholerae* non-O1 isolates can also cause human disease. *V. cholerae* O1 can be subdivided into two biotypes (**el tor** and **cholerae**), as well as three serologic subgroups (**ogawa, inaba,** and **hikojima**). These groups are important for the epidemiologic classification of isolates. Isolates of *V. parahaemolyticus* can also be subdivided by differences in the somatic O antigens.

Pathogenesis

The mechanism by which *V. cholerae* causes cholera is well established. The exotoxin produced by the organism is a complex molecule that is able to bind to specific receptors in the small intestine, enter into the mucosal cells, and effect a series of reactions that result in the rapid secretion of sodium, potassium, and bicarbonate into the intestinal lumen (Figure 11-1). Severely infected patients can lose as much as 1 liter of fluid per hour during the height of disease. The tremendous loss of fluid would normally flush the organism out of the gastrointestinal tract; however, *V. cholerae* is able to penetrate through the mucus covering the surface of the intestine and adhere to the mucosal cell layer. Nonadherent strains are unable to establish infection. Similarly, atypical or toxin-negative strains of *V. cholerae* O1 are avirulent.

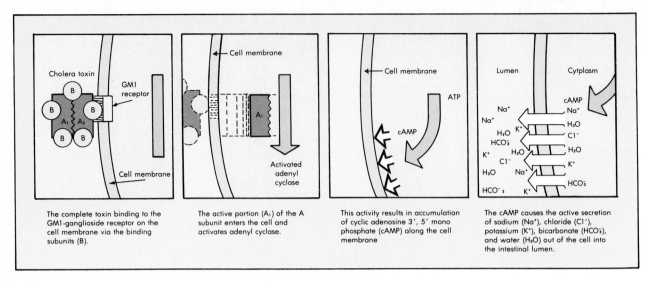

The complete toxin binding to the GM1-ganglioside receptor on the cell membrane via the binding subunits (B).

The active portion (A₁) of the A subunit enters the cell and activates adenyl cyclase.

This activity results in accumulation of cyclic adenosine 3′, 5′ mono phosphate (cAMP) along the cell membrane

The cAMP causes the active secretion of sodium (Na⁺), chloride (Cl⁻), potassium (K⁺), bicarbonate (HCO₃), and water (H₂O) out of the cell into the intestinal lumen.

FIGURE 11-1   Mechanism of cholera toxin action. The toxin molecule is composed of the A subunit, which determines biologic activity, and the B subunit, which binds the toxin to the GM1 ganglioside receptor on the membrane of intestinal cells. The A subunit consists of two peptides: A_1, with toxin activity, and A_2, which is a linking molecule to the B subunit. The B subunit consist of five identical peptides. After the toxin molecule binds to the cell receptor, the A subunit is transferred into the cell and the A_1 peptide is activated. Through a series of steps, adenyl cyclase activity is increased with a corresponding increase in cyclic adenosine 3′,5′-monophosphate (cAMP). The increased cAMP concentration mediates the active secretion of electrolytes and water into the lumen of the intestine.

The means by which other *Vibrio* species cause disease is not as clearly characterized. Most virulent strains of *V. parahaemolyticus* produce a heat-stable hemolysin (**Kanagawa-positive** strains) that is cytotoxic and cardiotoxic in experimental animals. Virulent strains are also able to adhere to and invade the intestinal tissue (in contrast with *V. cholerae*, which is considered to be a noninvasive pathogen). The role toxin production and tissue invasion play in the development of gastroenteritis has not been delineated. *V. alginolyticus* and *V. vulnificus* are associated with systemic infections and, although toxins have been identified in these organisms, the role of toxins in disease is undefined.

Epidemiology

Vibrio cholerae is found in freshwater ponds and estuaries in Asia, the Middle East, Africa, parts of Europe, and along the mid-Atlantic and southeastern U.S. seacoasts. Although the major reservoir is believed to be human carriage, some evidence indicates that infected crustaceans may also be a significant source of infection. Disease is spread by contaminated water and food, most commonly during the warm months of the year. Person-to-person spread is unusual because a high inoculum (e.g., 10^8 to 10^{10} organisms) is required to establish infection in an individual with normal gastric acidity. Achlorhydria or hypochlorhydria can reduce the infectious dose to 10^3 to 10^5 organisms. For this reason cholera is usually seen in communities with poor sanitation. Once the reservoir for this organism is established, elimination is particularly difficult. For that reason sporadic disease has occurred for centuries, and seven major pandemics have been observed since 1817.

In contrast with *V. cholerae*, *V. parahaemolyticus*, *V. vulnificus*, and *V. alginolyticus* are halophilic marine vibrios that require salt for growth. These species are free-living vibrios that inhabit estuaries and coastal waters worldwide (Figure 11-2). Because they are also rapidly killed by gastric acids, the infectious dose is generally high. Gastroenteritis with *V. parahaemolyticus* and septicemia caused by *V. vulnificus* commonly follow ingestion of raw or improperly handled seafood such as oysters. *V. parahaemolyticus* is the major cause of diarrheal disease in Japan, where consumption of raw fish is high. Wound infections caused by *Vibrio* species are usually associated with exposure to seawater or laceration with a seashell.

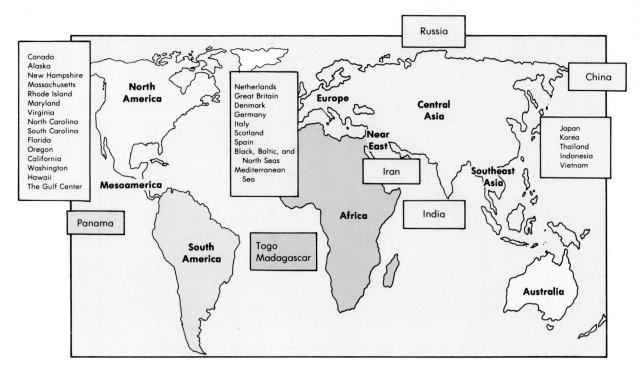

FIGURE 11-2   Worldwide Distribution of *Vibrio parahaemolyticus.*

Clinical Syndromes

Vibrio Cholera Cholera will initially occur an average of 2 to 3 days after ingestion of the bacilli, with the abrupt onset of vomiting and severe watery diarrhea. The stool specimens are colorless and odorless, free of protein, and speckled with mucus flecks (rice-water stools). The severe fluid loss can lead to dehydration, metabolic acidosis (bicarbonate loss), hypokalemia (potassium loss), and hypovolemic shock with cardiovascular collapse. The mortality is 60% in untreated patients but less than 1% in patients promptly treated to replace lost fluids and electrolytes. Cholera will spontaneously resolve after a few days of symptoms. Disease caused by *V. cholera* biotype cholerae is more severe than with biotype el tor. *Vibrio cholerae* non-O1 causes a gastrointestinal disease similar to *V. cholerae* 01, although it is generally less severe.

Vibrio Parahaemolyticus Disease Gastroenteritis caused by *V. parahaemolyticus* can range from self-limiting diarrhea to a cholera-like illness. In general, the disease will present after a 5-hour to 92-hour incubation period (mean—24 hours), with an explosive, watery diarrhea and no gross blood or mucus in the stool specimens except in very severe cases.

Headache, abdominal cramps, nausea, vomiting, and low-grade fever may persist for 72 hours or more. Recovery is generally uneventful.

Vibrio Vulnificus Disorders *V. vulnificus*, initially classified as a "lactose-positive" vibrio, is a particularly virulent *Vibrio* species responsible for rapidly progressive wound infections after exposure to contaminated seawater and for septicemia after consumption of raw oysters (Figure 11-3). The wound infections are characterized by initial swelling, erythema, and pain, followed by the development of vesicles or bullae and eventual tissue necrosis. Systemic signs of fever and chills are usually seen. Although death is uncommon, surgical intervention and antibiotic therapy are required. Septicemia caused by *V. vulnificus* has a mortality rate of 50%. Many patients with septicemia have a history of previous hepatic disease, although normal hosts can also be infected.

Vibrio Alginolyticus Infection *V. alginolyticus*, originally classified as *V. parahaemolyticus* biotype 2, can cause infection in superficial wounds contaminated with seawater. Infections of the ear, eye, and gastrointestinal tract have also been rarely reported.

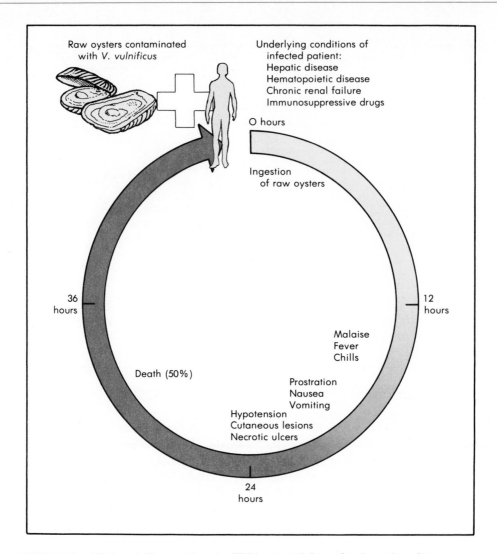

Raw oysters contaminated
with *V. vulnificus*

Underlying conditions of
infected patient:
Hepatic disease
Hematopoietic disease
Chronic renal failure
Immunosuppressive drugs

O hours

Ingestion
of raw oysters

36
hours

12
hours

Malaise
Fever
Chills

Death (50%)

Prostration
Nausea
Vomiting

Hypotension
Cutaneous lesions
Necrotic ulcers

24
hours

FIGURE 11-3   *Vibrio vulnificus* septicemia. Within several days after ingestion of raw oysters contaminated with *V. vulnificus,* the patient will develop malaise followed rapidly by fever, chills, and prostration. Unless antimicrobial therapy is initiated rapidly, progression to hypotension and death will ensue. In advanced disease, cutaneous bullae and necrotic ulcers will develop.

Laboratory Diagnosis

Microscopy *Vibrio* species are small (0.5 × 1.5-3 µm), curved, gram-negative bacilli. The organisms are rarely observed in Gram-stained stool or wound specimens; however, the detection of characteristic motile bacilli by an experienced observer using dark-field microscopy may be useful.

Culture *Vibrio* survive poorly in an acidic or dry environment. Specimens must be collected early in the course of disease and cultured promptly onto appropriate media. If culturing is delayed, the speci-

men should be mixed in Cary-Blair transport medium. Vibrios survive poorly in buffered glycerol-saline, the transport medium used for most enteric pathogens.

Vibrios grow on most media used in clinical laboratories for stool cultures, including blood agar, MacConkey agar, and xylose-lysine-deoxycholate (XLD) agar. Special selective agar for vibrios (e.g., **TCBS,** or **thiosulfate-citrate-bile-sucrose agar**) can also be used (Table 11-1), as well as an enrichment broth (e.g., alkaline peptone broth, pH 8.6). Identification of isolates is performed with selected biochemical tests.

Table 11-1 Isolation of *Vibrio* Species on TCBS Agar

Vibrio Species	Sucrose Fermentation	Colony
V. cholerae O1	Positive	Yellow
V. cholerae non-O1	Positive	Yellow
V. alginolyticus	Positive	Yellow
V. parahaemolyticus	Negative	Dark blue-green
V. vulnificus	Negative	Dark blue-green

Tests with halophilic vibrios will require supplementation of the media with 1% Sodium Chloride.

Treatment

Cholera must be promptly treated with fluid and electrolyte replacement before massive fluid loss results in hypovolemic shock. Antibiotic therapy, although of secondary value, can reduce exotoxin production and eliminate the organism. Tetracycline is the drug of choice, but vibrios are also susceptible to ampicillin, chloramphenicol, and trimethoprim-sulfamethoxazole. *V. parahaemolyticus* gastroenteritis is usually a self-limited disease, although antibiotic therapy can be used to supplement fluid and electrolyte therapy in severe infections. *V. vulnificus* wound infections and septicemia must be promptly treated with antibiotic therapy. Tetracycline is the most effective drug in vivo, although some success has been reported with aminoglycosides.

Prevention and Control

Because vibrios are free-living in freshwater and marine reservoirs and human carriage of *V. cholerae* can range from 1% to 20% in previously infected patients, it is unlikely that the reservoir for this organism will be eradicated. Disease can be controlled effectively only by improved hygiene. This involves adequate sewage management and water purification systems to eliminate contamination of the water supply and appropriate steps to prevent contamination of food.

Although a killed cholera vaccine is available, the protection is short lived and useful only for individuals who will be in an endemic area for less than 6 months. Tetracycline prophylaxis has also been used to reduce the risk of infection in endemic areas. However, because the infectious dose of *V. cholerae* is 10^8 organisms or more, antibiotic prophylaxis is generally unnecessary if appropriate hygiene is used.

AEROMONAS

Aeromonas is a gram-negative aerobic and facultative anaerobic bacillus that morphologically resembles members of the Enterobacteriaceae. However, *Aeromonas* can be readily differentiated by positive oxidase reactivity and the presence of polar flagellum. As the name implies, the most commonly isolated species, *Aeromonas hydrophila*, is found in fresh and brackish water. Human infections follow exposure to untreated water, but disease has not been associated with consumption of shellfish or freshwater fish. Asymptomatic gastrointestinal carriage of this organism is less than 3%, with a peak in both carriage and disease during the warm months of the year.

The organism causes opportunistic systemic disease in immunocompromised patients—particularly those with hepatobiliary disease or an underlying malignancy—and diarrheal disease in otherwise healthy individuals (Table 11-2). Gastrointestinal disease in children is usually an acute, severe illness, whereas adults tend to have chronic diarrhea. The acute diarrheal disease is self-limited, and only supportive care is indicated. Chronic diarrheal disease or systemic infection requires antimicrobial therapy. *Aeromonas* is resistant to penicillins, cephalosporins, and erythromycin, with only gentamicin, tetracycline, trimethoprim-sulfamethoxazole, and chloramphenicol consistently active.

PLESIOMONAS

Plesiomonas is a facultative anaerobic gram-negative bacillus that is oxidase-positive, has multiple

Table 11-2 Characteristics of *Aeromonas* and *Plesiomonas* Gastroenteritis

Epidemiologic and Clinical Features	*Aeromonas*	*Plesiomonas*
Natural habitat	Fresh or brackish water	Fresh or brackish water
Source of infection	Contaminated water	Uncooked shellfish
Clinical presentation		
Diarrhea	Present	Present
Vomiting	Present	Present
Abdominal cramps	Present	Present
Fever	Absent	Absent
Blood and PMNs in stool	Absent	Present
Pathogenesis	Enterotoxin (?)	Invasive

polar flagella, and is differentiated from *Aeromonas* by selected biochemical reactions. *Plesiomonas shigelloides*, the species responsible for human disease, is serologically related to *Shigella sonnei*. The organism is found in brackish water and is acquired by consumption of uncooked shellfish, particularly oysters and shrimp. Disease has also been associated with foreign travel, usually to Mexico or the Caribbean. The primary disease caused by *P. shigelloides* is self-limiting gastroenteritis, with the onset of disease 48 hours after exposure to the organism. When antibiotic therapy is indicated, the organism has been found to be susceptible to gentamicin, chloramphenicol, cephalothin, and trimethoprim-sulfamethoxazole but not ampicillin, carbenicillin, or erythromycin.

CAMPYLOBACTER

The genus *Campylobacter*, from the Greek word "campylo" for curved, consists of comma-shaped, gram-negative bacilli that are oxidase- and catalase-positive, motile by means of a polar flagella, and require a microaerophilic atmosphere for growth. Five species and five subspecies have been classified as campylobacters, as well as five provisional species (Box 11-2).

Campylobacter jejuni is the most common cause of bacterial gastroenteritis in the United States. *Campylobacter coli* and *Campylobacter laridis* have also been associated with diarrheal disease in humans. *Campylobacter cinaedi* and *Campylobacter fennelliae* have been isolated from homosexual men with proctitis, proctocolitis, or enteritis. These five species are primarily restricted to the gastrointestinal tract, with

Box 11-2 *Campylobacter* Species and Their Associated Diseases

C. jejuni	Diarrhea in humans, enteritis in cattle, abortion in sheep
C. coli	Diarrhea in humans
C. laridis	Diarrhea in humans
C. fetus	
Subspecies fetus	Septicemia in humans, abortion in sheep and cattle
Subspecies venerealis	Abortion and infertility in cattle
C. sputorum	
Subspecies sputorum	No known disease
Subspecies bubulus	No known disease
Subspecies mucosalis	Intestinal tumors in swine
*C. fecalis**	Diarrhea in cattle
*C. hyointestinalis**	Ileitis in swine
*C. cinaedi**	Diarrhea in homosexual men
*C. fennelliae**	Diarrhea in homosexual men
*C. pylori**	Gastritis and pyloric ulcers in humans

*Provisional names.

bacteremia observed in fewer than 1% of the infections. In contrast, *Campylobacter fetus* ssp. fetus is most commonly associated with systemic infections such as bacteremia, septic thrombophlebitis, arthritis, septic abortion, and meningitis. *Campylobacter pylori*, a recently recognized organism, is associated with gastritis and pyloric ulcers.

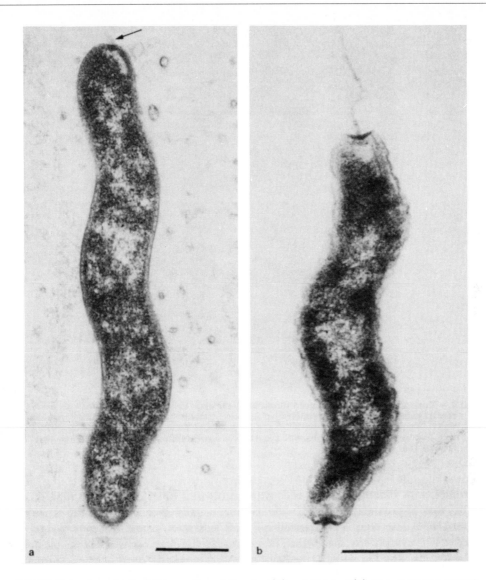

FIGURE 11-4   Longitudinal thin-section of *C. pylori* (**A**) and *C. jejuni* (**B**). *C. pylori* has a smooth surface, bluntly rounded ends, and a single sheathed polar flagellum (*arrow*). *C. jejuni* has a typical rough cell wall, polar pits, and unsheathed bipolar flagella. (Bar = 500 nm). (From Goodwin CS et al: J Med Microbiol 19:257, 1985.)

Microbial Physiology and Structure

Campylobacter has a typical gram-negative cell wall structure. The major antigen of the genus is the lipopolysaccharide of the outer membrane. Serologic heterogeneity of *C. jejuni* isolates is common, with more than 60 different somatic O polysaccharide antigens recognized and 50 capsular and flagellar antigens. Based on morphological differences and analysis of cellular fatty acids, *C. pylori* is either an atypical member of the genus *Campylobacter* or a representative of a new genus (Figure 11-4).

Pathogenesis

The pathogenesis of *C. jejuni* gastrointestinal disease is not understood. Disease is characterized by destruction of the mucosal surfaces of the jejunum (as implied by the name), ileum, and colon. Grossly, the

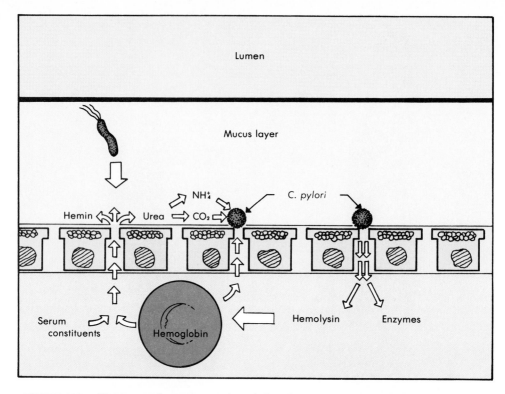

FIGURE 11-5   Diagrammatic representation of *C. pylori* colonization of gastric mucosa. The motile organism moves rapidly through the viscous mucus toward chemotactic growth factors, urea and hemin, present in the gastric pits. The stomach acidity is neutralized by the urease activity, and hemin stimulates growth of *C. pylori*. Infiltration of inflammatory cells and release of hydrolytic enzymes in response to the proliferating organism leads to gastritis. (Redrawn from Hazell SL et al: J Infect Dis 153:658, 1986.)

mucosal surface appears edematous and bloody. Histologic examination reveals ulceration of the mucosal surface, crypt abscesses in the epithelial glands, and infiltration into the lamina propria, with neutrophils, mononuclear cells, and eosinophils. This inflammatory process is consistent with invasion of the organisms into the intestinal tissue. Cytopathic toxins have also been isolated in *C. jejuni*. The roles of microbial invasion and toxin production in the pathogenesis of campylobacter gastroenteritis have not been delineated.

Campylobacter fetus has a propensity to spread from the gastrointestinal tract to the bloodstream and distal foci. This is particularly common in such debilitated and immunocompromised patients as those with liver disease, diabetes mellitus, chronic alcoholism, or a malignancy. In vitro studies have demonstrated that *C. fetus* is resistant to complement- and antibody-mediated serum killing, whereas *C. jejuni* is rapidly killed. Thus, although *C. jejuni* is a more common pathogen, resistance to serum bactericidal activity can account for the observation that *C. fetus* is isolated more frequently in blood.

Campylobacter pylori is associated with histologically documented gastritis and ulcerative disease but is rarely found in normal stomach mucosa (Figure 11-5).

Epidemiology

Campylobacters are commensals of cattle, sheep, dogs, cats, rodents, and fowl. Lifelong asymptomatic carriage in animal reservoirs is common. Human infections result from consumption of contaminated food (particularly poultry), milk, or water. Fecal-oral transmission from person to person may also occur, but transmission from food handlers is uncommon. The source of *C. pylori* has not been identified.

It has been estimated that more than 2 million *C. jejuni* infections occur annually, it is thus more com-

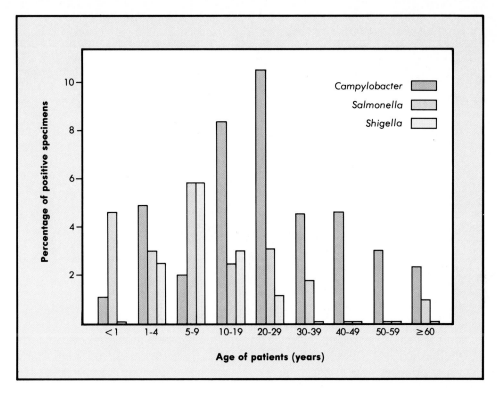

FIGURE 11-6   Age distribution for diarrheal disease caused by *Campylobacter, Salmonella,* and *Shigella* organisms.

mon than *Salmonella* and *Shigella* infections combined (Figure 11-6). *C. fetus* infections are relatively uncommon, with fewer than 250 cases reported. In contrast with *C. jejuni, C. fetus* infects immunocompromised, elderly individuals. *C. cinaedi* and *C. fenelliae* are reportedly isolated from 16% of homosexual men with intestinal symptoms. The incidence of carriage of these organisms in asymptomatic individuals is less than 3%. *C. pylori* is identified in 60% to 100% of patients with gastritis, gastric ulcers, and duodenal ulcers but is infrequently isolated from patients without histologic evidence of gastritis.

Clinical Syndromes

C. jejuni infections are seen most commonly as acute enteritis with diarrhea, malaise, fever, and abdominal pain. Ten or more bowel movements per day can occur during the peak of disease, and grossly bloody stools may be seen. The disease is generally self-limiting, although symptoms may last for 1 week or longer. The range of clinical manifestations can also include colitis, such as that seen in ulcerative

colitis or Crohn's disease, acute abdominal pain, and bacteremia. *C. fetus* most commonly is seen as septicemia with dissemination to multiple organs, although the initial presentation may be referrable to the gastrointestinal tract or abdomen. Endovascular localization is reported.

Laboratory Diagnosis

Microscopy *C. jejuni* is thin (0.3 mm diameter) and cannot be easily seen in Gram-stained specimens. When observed, the organisms appear as small curved bacilli. Pairs of bacteria will resemble the wings of a seagull. The organism, with its characteristic darting motility, can be reliably detected in freshly collected stool specimens when examined by darkfield or phase-contrast microscopy. *C. pylori* is detected by histologic examination of gastric biopsies. Although the organism can be seen in specimens with hemotoxylin and eosin stain or Gram stain, the Warthin-Starry silver stain is the most sensitive.

Culture *C. jejuni* was unrecognized for many years because isolation requires growth in a microae-

Table 11-3 Identification Tests for *Campylobacter*

Test Reactions	*C. jejuni*	*C. coli*	*C. laridis*	*C. fetus* ssp. *fetus*	*C. cinaedi*	*C. fennelliae*	*C. pylori*
Growth at:							
25° C	−	−	−	+	−	−	−
37° C	+	+	+	+	+	+	+
42° C	+	+	+	±	−	−	−
Oxidase	+	+	+	+	+	+	+
Catalase	+	+	+	+	+	+	+
Urease	−	−	−	−	−	−	+
Nitrate reduction	+	+	+	+	+	−	−
Hippurate hydrolysis	+	−	−	−	−	−	−
Susceptibility to:							
Nalidixic acid	S	S	R	R	S	S	R
Cephalothin	R	R	R	S	S	S	S

Modified from Lennette EH et al.: *Manual of Clinical Microbiology*, 1985.

rophilic atmosphere (e.g., 5% to 7% oxygen, 5% to 7% carbon dioxide, and the balance nitrogen), elevated incubation temperature (e.g., 42° C), and selective media. The organism is slow-growing, usually requiring incubation for 48 to 72 hours or longer. *C. fetus* and *C. pylori* are not thermophilic and cannot grow at 42° C. However, a microaerophilic atmosphere is still required for isolation.

Identification Preliminary identification of isolates is based on growth under selective conditions, typical microscopic morphology, and rapid detection of oxidase and catalase activity. The rapid detection of urease activity can be used to classify *C. pylori*. Definitive identification of all isolates is guided by the reactions summarized in Table 11-3.

Treatment

Campylobacters are susceptible to a wide variety of antibiotics, including erythromycin, the tetracyclines, aminoglycosides, chloramphenicol, and clindamycin. Most isolates are resistant to penicillins, cephalosporins, and sulfonamide antibiotics. Erythromycin is antimicrobial and used when indicated to treat enteritis; an aminoglycoside is used for systemic infections. Antibiotics alone are generally ineffective in eradicating *C. pylori*. However, when antibiotics are combined with bismuth, successful elimination of the organism has been reported in preliminary studies.

Prevention and Control

Campylobacter gastroenteritis is prevented by the proper preparation of food, particularly poultry, consumption of pasteurized milk, and safeguards to prevent contamination of water supplies. Elimination of campylobacter carriage in animal reservoirs is unlikely.

BIBLIOGRAPHY

Blake PA: Diseases of humans (other than cholera) caused by vibrios, Ann Rev Microbiol 34:341, 1980.

Blaser MJ, Berkowitz ID, LaForce FM, Cravens J, Reller LB, and Wang WL: *Campylobacter* enteritis: clinical and epidemiologic features, Ann Intern Med 91:179, 1979.

Blaser MJ and Reller LB: *Campylobacter* enteritis, N Engl J Med 305:1443, 1981.

Blaser MJ, Wells JG, Feldman RA, Pollard RA, and Allen JR: *Campylobacter* enteritis in the United States: a multicenter study, Ann Intern Med 98:360, 1983.

Craig JP, Hardegree MC, Pierce NF, and Richardson SH: The structure and functions of enterotoxins. A workshop held at the National Institutes of Health, J Infect Dis 133(suppl):5-156, 1976.

Goodwin CS, Armstrong JA, and Marshall BJ: *Campylobacter pyloridis*, gastritis, and peptic ulceration, J Clin Pathol 39:353, 1986.

Guerrant RL, Lahita RG, Winn WC, Jr, and Roberts RB: Campylobacteriosis in man: pathogenic mechanisms and review of 91 bloodstream infections, Am J Med 65:584, 1978.

Holmberg SD and Farmer JJ III: *Aeromonas hydrophila* and *Plesiomonas shigelloides* as causes of intestinal infections, Rev Infect Dis 6:633, 1984.

Holmberg SD, Schell WL, Fanning GR, et al: *Aeromonas* intestinal infections in the United States, Ann Intern Med 105:683, 1986.

Holmberg SD, Wachsmuth IK, Hickman-Brenner FW, et al: *Plesiomonas* enteric infections in the United States, Ann Intern Med 105:690, 1986.

Joseph SW, Colwell RR, and Kaper JB: *Vibrio parahaemolyticus* and related halophilic vibrios, CRC Crit Rev Microbiol 10:77, 1982.

Marshall BJ: *Campylobacter pyloridis* and gastritis, J Infect Dis 153:650, 1986.

Morris JG and Black RE: Cholera and other vibrios in the United States, N Engl J Med 312:343, 1983.

Tison DL and Kelly MT: *Vibrio* species of medical importance, Diagn Microbiol Infect Dis 2:263, 1984.

Totten PA, Fennell CL, Tenover FC, Wezenberg JM, Perine PL, Stamm WE, and Holmes KK: *Campylobacter cinaedi* and *Campylobacter fennelliae:* two new *Campylobacter* species associated with enteric disease in homosexual men, J Infect Dis 151:131, 1985.

CHAPTER 12

Pasteurellaceae

The family Pasteurellaceae consists of three genera: *Haemophilus, Actinobacillus,* and *Pasteurella,* with *Haemophilus* the most common human pathogen (Table 12-1). Members of the family are small (0.2 × 0.3-2.0 μm), gram-negative bacilli, non-sporeforming, nonmotile, and aerobic or facultative anaerobes. Most have fastidious growth habits, requiring enriched media for isolation, and have limited biochemical reactivity.

HAEMOPHILUS

Haemophilus are small, sometimes pleomorphic, gram-negative bacilli that are obligate parasites on the mucous membranes of humans and animal species. *Haemophilus influenzae,* an obligate human pathogen, is the *Haemophilus* species most commonly associated with disease. Other pathogenic *Haemophilus* include *H. parainfluenzae, H. aphrophilus,* and *H. ducreyi.*

Microbial Physiology and Structure

Most species of *Haemophilus* (from the Greek words for "blood-loving") require supplementation of media with growth-stimulating factors, specifically X factor (hematin) and/or V factor (nicotinamide adenine dinucleotide, NAD). Although both factors are present in blood-enriched media, the blood must be gently heated to release the factors and destroy inhibitors of V factor. For this reason heated blood agar (i.e., chocolate agar) is used for the in vitro isolation of *Haemophilus.*

The cell wall structure of *Haemophilus* is typical of other gram-negative bacilli. Lipopolysaccharide endotoxic activity is present, and the surface of the cell is covered with strain- and species-specific protein antigens (Figure 12-1). The surfaces of many but not all strains of *H. influenzae* are covered with a polysaccharide capsule, with six antigenic serotypes (a through f) recognized. Serotype b is associated with invasive disease in greater than 95% of all *Haemophilus* infections. Other species of *Haemophilus* do not have capsules.

Table 12-1 Most Common Pasteurellaceae Associated With Human Disease

Organism	Primary Diseases
H. influenzae	Meningitis, epiglottis, cellulitis, otitis, sinusitis, pneumonia, conjunctivitis, bacteremia
H. parainfluenzae	Bacteremia, endocarditis
H. aphrophilus	Bacteremia, endocarditis
H. ducreyi	Chancroid
P. multocida	Soft tissue infection, respiratory infection, bacteremia
A. actinomycetemcomitans	Juvenile periodontitis, endocarditis

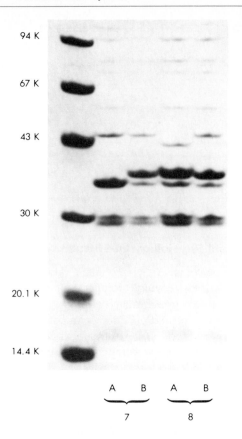
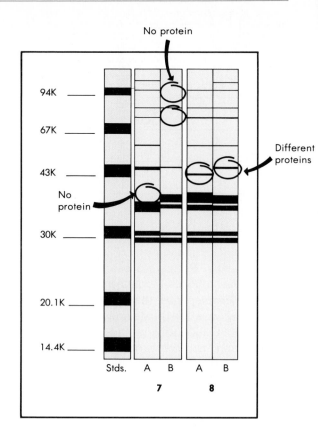

FIGURE 12-1   Migration patterns of *H. influenzae* type b outer membrane protein preparations on sodium dodecyl sulfate-polyacrylamide gels can be used to classify isolates for epidemiologic investigations. The outer membrane protein profiles of four isolates from two day care centers (*7* and *8*) are demonstrated here. The isolates from each center have distinct profiles and therefore are not epidemiologically related. The first lane has six molecular weight standards. (From Barenkamp SJ et al, J Infect Dis 144:210, 1981, published by the University of Chicago.)

Pathogenesis

Haemophilus species, particularly *H. parainfluenzae* and nonencapsulated *H. influenzae,* colonize the upper respiratory tract in virtually all individuals within the first few months of life. These organisms can spread locally and cause disease in the ears (otitis media), sinuses (sinusitis), and lower respiratory tract (bronchitis, pneumonia), but disseminated disease is relatively uncommon. In contrast, encapsulated *H. influenzae* (particularly serogroup b) is infrequently present in the upper respiratory tract but is the most common cause of epiglottitis and pediatric meningitis. The organism is able to penetrate through the nasopharyngeal submucosa and enter the bloodstream. In the absence of specific opsonic antibodies directed against the polysaccharide capsule, high-grade bacteremia can develop, with dissemination to the meninges or other distal foci.

The major virulence factor in *H. influenzae* type b is the antiphagocytic polysaccharide capsule, which contains ribose, ribitol, and phosphate. Phagocytosis and killing of bacteria are greatly stimulated by antibodies directed against the capsule, which develops as the result of natural infection, passive transfer of antibodies, or following vaccination with the purified capsule. The lipopolysaccharide lipid A component induces meningeal inflammation in an animal model and may be responsible for initiating this response in human disease. The roles of other membrane antigens and pili are poorly defined but are not believed to be significant virulence factors. IgA1-specific proteases are produced by *H. influenzae* (both encapsulated and nonencapsulated strains) and may facilitate colonization of mucosal surfaces.

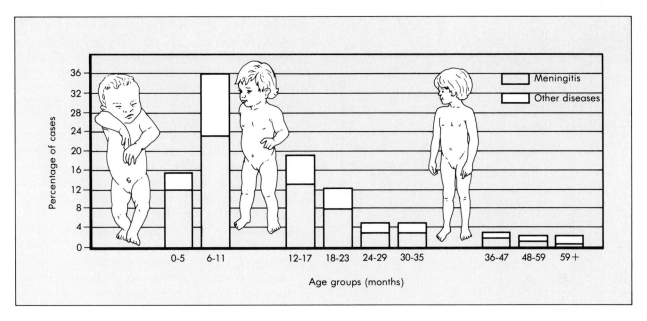

FIGURE 12-2 Age distribution of 600 children with invasive *H. influenzae* type b infections. Two thirds of the infections are meningitis (*black bars*); the other invasive diseases constitute less than one third of the infections. More than 80% of the infections occurred in children less than 24 months of age and 95% in children less than 5 years of age.

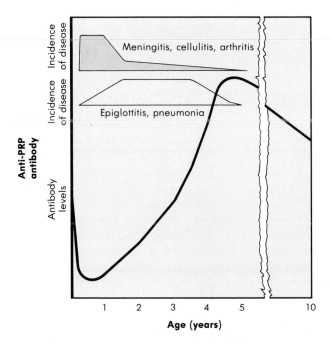

FIGURE 12-3 Antibodies directed against the polyribophosphate capsule of *H. influenzae* type b protect against invasive disease. During the first 3 months of life, the newborn is protected by passively acquired maternal antibodies. By 5 years of age, natural antibodies have developed after exposure to *H. influenzae* type b or cross-reacting antigens. However, from age 3 months to 5 years low antibodies levels leave the child susceptible to *Haemophilus* disease.

Epidemiology

The incidence of invasive *Haemophilus* disease has increased gradually over the last 50 years, with an estimated 15,000 cases reported annually. Two thirds of the infections are meningitis, and the majority are observed in children less than 2 years of age (Figure 12-2). It is well established that the primary risk factor for invasive *Haemophilus* disease is the absence of protective antibodies directed against the polysaccharide capsule (Figure 12-3).

Patient factors that decrease the immunologic clearance of the *H. influenzae* and increase the risk for disease include deficiencies in immunoglobulin synthesis, sickle cell disease, and splenectomy. The prevalence of disease is also higher in Eskimos, American Indians, Hispanics, and blacks. Additionally, the risk of disease for young children in crowded family units and day care centers is increased. Although this disease has historically been considered an endemic disease, focal outbreaks in day care centers and in families with young susceptible children have been well documented (Figure 12-4).

A biphasic seasonal trend for *H. influenzae* type b disease is reported with a peak in October and November and a second peak in the spring months.

The epidemiology of disease caused by nonencapsulated *H. influenzae* and other *Haemophilus* species

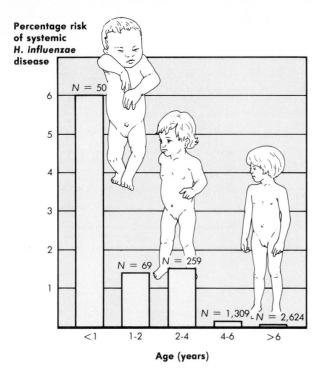

Percentage risk of systemic *H. influenzae* disease

N = 50

N = 69   N = 259

N = 1,309   N = 2,624

<1   1-2   2-4   4-6   >6

Age (years)

FIGURE 12-4   Ward et al. (N. Engl. J. Med. 301:122, 1979) studied the risk of invasive *Haemophilus influenzae* type b disease among household contacts of patients with *H. influenzae* meningitis. Children less than 12 months of age were at greatest risk, with three of the 50 children developing disease. Approximately 1.5% of contact children from 1 to 4 years of age developed disease, while only one child in the 4- to 6-year age-group and one of the 2,624 individuals older than 6 years developed invasive disease. Thus the risk of invasive disease in secondary contacts parallels the risk for primary *H. influenzae* type b disease.

is distinct. Whereas increased exposure to *H. influenzae* type b is associated with increased risk for invasive disease, disease caused by nonencapsulated *Haemophilus* strains is not increased in crowded communities because of ubiquitous presence of the organisms and their decreased virulence. Invasive disease is less common, and infections are not restricted to young children. Ear and sinus infections due to these organisms are primarily pediatric diseases but can be observed in adults, and pulmonary disease is most commonly observed in the elderly, particularly those with a history of underlying obstructive pulmonary disease or conditions predisposing to aspiration (e.g., alcoholism, altered mental state).

Clinical Syndromes

The clinical syndromes that accompany infections

of *H. influenzae* are illustrated in Figure 12-5. A discussion follows.

Meningitis *H. influenzae* type b is the most common cause of pediatric meningitis. Disease follows bacteremic spread from the nasopharynx and cannot be differentiated clinically from other causes of bacterial meningitis. The initial presentation is generally preceded by a 1- to 3-day history of mild upper respiratory disease. With prompt therapeutic intervention mortality is less than 5%, but the incidence of permanent sequelae is still 50% or greater. These defects include hearing loss, delayed language development, mental retardation, and seizures.

Epiglottitis This disease, characterized by cellulitis and swelling of the supraglottic tissues, represents a life-threatening emergency. Although epiglottitis is a pediatric disease, the peak incidence of disease is primarily in children from 2 to 4 years of age, in contrast with meningitis, which is in children from 3 to 18 months of age. The child will have pharyngitis, fever, and breathing difficulties, which can rapidly progress to complete obstruction of the airway and death.

Cellulitis This disease is seen in very young children. Patients have fever and cellulitis characterized by a reddish-blue color on the cheek or periorbital area. The etiologic diagnosis is strongly suggested by the typical clinical presentation, proximity to the oral mucosa, and age of the child.

Arthritis Infection of single large joints, secondary to bacteremic spread of *H. influenzae* type b, is the most common form of arthritis seen in children less than 2 years of age. Disease in older children and adults is reported, but it is very uncommon and generally occurs in joints with preexisting damage or in immunocompromised patients.

Conjunctivitis and Brazilian Purpuric Fever Epidemic, as well as endemic, conjunctivitis can be caused by *H. influenzae,* biotype *aegypticus* (formerly called Haemophilus aegypticus). A specific strain of this organism has also been associated with Brazilian purpuric fever, a fulminant pediatric disease characterized by an initial conjunctivitis, followed a few days later by an acute onset of fever, vomiting, and abdominal pain, and then the rapid development of petechiae, purpura, shock, and death. The pathogenesis of Brazilian purpuric fever and the specific virulence characteristics of the etiologic agent are poorly understood.

Otitis, Sinusitis, and Lower Respiratory Tract Disease Other infections of the upper and low-

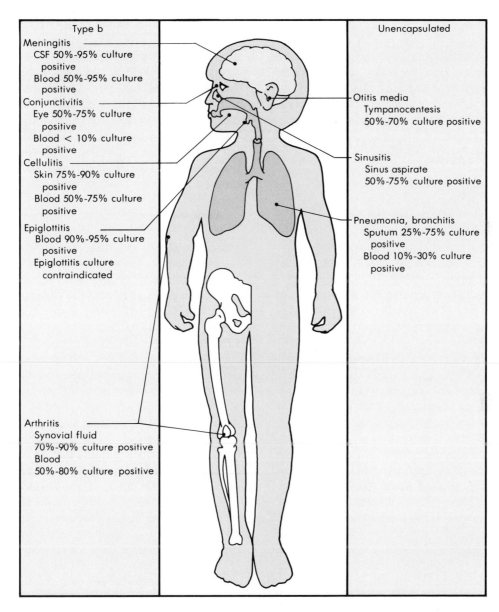

FIGURE 12-5 Infections caused by *H. influenzae* type b are usually invasive, and meningitis, epiglottitis, cellulitis, arthritis, and conjunctivitis are reported most commonly. Cultures of cerebrospinal fluid and blood in patients with meningitis are almost always positive unless the patient has been pretreated with antibiotics. Blood cultures are also usually positive for patients with epiglottitis, arthritis, and cellulitis. In contrast, unencapsulated strains of *H. influenzae* usually cause localized diseases that are diagnostic problems, because blood cultures are usually negative and appropriate respiratory specimens or aspirates from the middle ear or sinuses are difficult to collect.

er respiratory tract are usually caused by nonencapulated strains of *H. influenzae*. Most studies indicate that *H. influenzae* and *Streptococcus pneumoniae* are the two most common causes of both acute and chronic otitis and sinusitis. Primary pneumonia is probably uncommon in children and adults who have normal pulmonary function. However, these organisms commonly colonize patients who have chronic pulmonary disease and frequently are associated with exacerbation of bronchitis, as well as frank pneumonia.

Laboratory Diagnosis

See Figure 12-5 for an illustration of *H. influenzae* diagnosis and the discussion that follows.

Specimen Collection and Transport The specific diagnosis of *Haemophilus* meningitis requires collection of cerebrospinal fluid. In untreated meningitis, the concentration of organisms is approximately 10^7 bacteria/ml fluid; therefore, 1 to 2 ml of fluid is generally adequate for microscopy, culture, and antigen detection tests. Blood should also be cultured for the diagnosis of epiglottitis, cellulitis, arthritis, and pneumonia. These cultures are less useful for patients with localized upper respiratory diseases. Specimens should not be collected from the posterior pharynx for the specific diagnosis of epiglottitis because this may stimulate complete obstruction of the airway.

Microscopy The direct microscopic detection of *Haemophilus* in clinical specimens is a sensitive and specific test when performed carefully. Small gramnegative coccobacilli can be detected in greater than 80% of cerebrospinal fluid specimens from patients with untreated *Haemophilus* meningitis. However, these bacilli may resemble *Streptococcus pneumoniae* if the specimen is not adequately decolorized. Gram stained specimens are also useful for the rapid diagnosis of arthritis and lower respiratory disease. The specific diagnosis of *H. influenzae* type b disease can be made by the quellung test, in which anticapsular antibodies are mixed with the clinical specimen and then examined microscopically for capsular swelling (increased refractility due to coating of the capsule with type-specific antibodies).

Culture Isolation of *H. influenzae* from clinical specimens is relatively easy if media supplemented with growth factors are inoculated. Chocolate agar is used in most laboratories, the bacteria appear as 1- to 2-mm smooth opaque colonies after 24 hours of incubation. Growth on unheated blood agar plates can also be detected surrounding colonies of *Staphylococ-*

cus aureus **(satellite phenomenon).** The staphylococci provide required growth factors by lysing the erythrocytes in the medium and secreting V factor. The size of the colonies of *H. influenzae* is much smaller than on chocolate agar because the V factor inhibitors are not inactivated.

As stated previously, blood cultures should be performed on all patients who have meningitis, epiglottitis, cellulitis, arthritis, and lower respiratory tract diseae. Although positive cultures will be detected for most patients (with the possible exception of patients with pneumonia), growth is generally delayed because most commercially prepared blood culture broths are not supplemented with optimal concentrations of X and V factors.

Antigen Detection The immunologic detection of *H. influenzae* antigen, specifically the **polyribophospate (PRP) capsule,** is a rapid and sensitive method for diagnosing *Haemophilus* disease. PRP can be detected by counterimmunoelectrophoresis or particle agglutination, with the particle agglutination procedure five to 10 times more sensitive (able to detect <1 ng/ml PRP compared with 5 to 10 ng/ml by CIE). Antibody-coated latex particles are mixed with the clinical specimen, and—in the presence of PRP—agglutination occurs. Antigen can be detected in both cerebospinal fluid and urine (where the antigen is eliminated in intact form). This test has a sensitivity comparable to the Gram stain.

Identification *H. influenzae* is readily identified by demonstrating a requirement for both X and V factors and the specific biochemical properties summarized in Table 12-2. *H. influenzae* can be subgrouped for epidemiologic purposes into seven biotypes, based on indole production, urease activity, and ornithine decarboxylase activity. *H. influenzae* biotype *aegypticus* is closely related to or identical to *H. influenzae* biotype III. Further subgrouping of *H. influenzae* can be made by electrophoretic characterization of membrane protein antigens (see Figure 12-1).

Treatment

Unless prompt antimicrobial therapy is initiated, mortality with *H. influenzae* meningitis and epiglottitis approaches 100%. Historically, all isolates were susceptible to either ampicillin or chloramphenicol; however, resistance to these antibiotics is now widely recognized. In 1974 the first strains resistant to ampicillin were reported. This resistance was mediated by a TEM-1 **β-lactamase,** which is present on a

Table 12-2 Differential Characteristics of Common Members of the Family Pasteurellaceae

Organism	Growth factor requirement X	Growth factor requirement V	Enhanced growth with CO_2	Catalase	Fermentation of: Glucose	Fermentation of: Sucrose	Fermentation of: Lactose
HAEMOPHILUS							
H. influenzae	+	+	−	+	+	−	−
H. parainfluenzae	−	+	−	V	+	+	−
H. aphrophilus	−	−	+	−	+	+	+
H. ducreyi	+	−	−	−	−	−	−
ACTINOBACILLUS							
A. actinomycetemcomitans	−	−	+	+	+	−	−
PASTEURELLA							
P. multocida	−	−	−	+	+	+	−

transmissible plasmid and commonly found in other gram-negative bacilli such as *Escherichia coli.* Unfortunately, this β-lactamase is now ubiquitous (found in more than 30% of type b strains and 20% of non-encapsulated strains of *H. influenzae*). A second β-lactamase (designated ROB-1) has also been found in rare strains of *H. influenzae.* Finally, less than 1% of strains have non-β-lactamase-mediated ampicillin resistance. The mechanism of this resistance is most likely alterations of the penicillin-binding proteins (the target of antibiotic activity). As a result of these changes, ampicillin cannot be reliably used as empiric therapy for severe *H. influenzae* infections. Chloramphenicol resistance, mediated by strains producing **chloramphenicol acetyltransferase,** has also been recognized but is relatively uncommon (less than 1% of all *H. influenzae*). Thus infections can be treated either with this antibiotic or a β-lactamase-resistant cephalosporin with good penetration into the cerebrospinal fluid (e. g., cefotaxime). Less severe infections such as sinusitis or otitis can be treated with ampicillin (if susceptible), trimethoprim-sulfamethoxazole, or cefaclor. Because localized infections with *H. influenzae* type b (e. g., cellulitis) can spread via the bloodstream into the central nervous system, all of these infections should be treated with active antibiotics that penetrate into these sites.

Prevention and Control

Active Immunization The major approach to prevention of *Haemophilus* disease is immunization with purified capsular polyribophosphate (PRP). As discussed previously, the anti-PRP activity in serum corresponds to immunity to severe disease. Immunization has attempted to boost the level of protective antibodies during the period of greatest susceptibility to infection. However, response to the purified PRP is disappointing in the population at greatest risk for severe disease, children less than 24 months of age. This problem may be circumvented with a combined vaccine with *H. influenzae* type b PRP and diptheria toxoid. Preliminary trials demonstrate enhanced antibody response to this conjugated vaccine in children less than 18 months of age (Figure 12-6).

Passive Immunization Limited use of hyperimmune globulins active against *H. influenzae* type b has been successful in protecting high-risk children after exposure to the organism. Further trials are necessary to define when this should be administered.

Chemoprophylaxis Antibiotic chemoprophylaxis is used to eliminate carriage of *H. influenzae* type b in children at high risk for disease (e,g., children less than 2 years of age in a family or day care center where systemic disease is documented). Rifampin prophylaxis has been used in these settings.

Other *Haemophilus* Species

Haemophilus parainfluenzae Although *H. parainfluenzae* is the most prevalent *Haemophilus* species in the oropharynx, it is a rare cause of serious infections. The most common disease attributed to this organism is subacute endocarditis. Other infections such as meningitis, arthritis, epiglottitis, and pneumonia have been reported, but because the

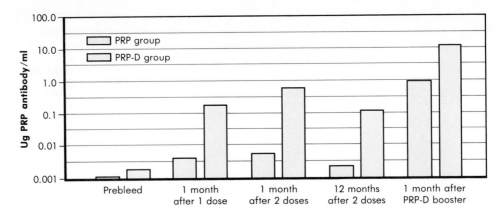

FIGURE 12-6   The antibody response to polyribophosphate (PRP) capsular material of *H. influenzae* type b is poor in children less than 2 years of age. However, when the purified PRP is coupled with a carrier such as diphtheria toxoid, the response is better in young children. In this study reported by Lepow (Pediatr. Infect. Dis. J. 6:804, 1987) children from 9 months to 14 months of age were immunized with either PRP or PRP combined with diphtheria toxoid (PRP-D). The antibody response with PRP-D is significantly higher than with PRP alone. Antibody response in both populations was substantial 1 month after the children were administered a booster injection with PRP-D.

identification of this organism may be confused with *H. influenzae*, the accuracy of these reports is unknown. *H. parainfluenzae* can be readily grown in the laboratory and identified by the growth and biochemical properties outlined in Table 12-2. The antimicrobial susceptibility patterns are similar to *H. influenzae*, although ampicillin resistance is more common with *H. parainfluenzae*.

Haemophilus aphrophilus This organism, also a common member of the normal oropharyngeal flora, is almost exclusively associated with subacute endocarditis. *H. aphrophilus* characteristically grows slowly in vitro and produces adherent colonies, properties that are consistent with the chronic presentation of disease and predilection to adherence to damaged or artificial heart valves. The taxonomic classification of this organism as a member of the genus *Haemophilus* is uncertain because it does not require V factor and the requirement for X factor is rapidly lost with *in vitro* culture.

Haemophilus ducreyi The etiologic agent for **chancroid,** or soft chancre, was first described by Ducrey. This sexually transmitted disease is distinguished from syphilis by the presence of a painful, nonindurated genital ulcer with well-defined margins. The disease has worldwide distribution, including outbreaks reported in the United States, and is generally present in poor socioeconomic areas. Laboratory diagnosis is difficult because the organism sometimes fails to grow in vitro and, when it does

grow, incubation for 3 or more days is required. A Gram stain of clinical material aspirated from the ulcer margin or enlarged lymph node may reveal abundant gram-negative coccobacilli. However, other bacteria can also mimic this appearance. In general, diagnosis is based on the clinical presentation of the patient and a history of chancroid in the community. Oral erythromycin or trimethoprim-sulfamethoxazole are effective antibiotics for treating chancroid, whereas resistance to ampicillin and penicillin is common.

ACTINOBACILLUS

Actinobacillus species are small facultative anaerobic, gram-negative bacilli that grow slowly (generally requiring 2 to 3 days of incubation) and need carbon dioxide for growth in vitro on chocolate or blood agar media. Five species of *Actinobacillus* have been described, with *A. actinomycetemcomitans* the only significant human pathogen. The cumbersome name is derived from the fact this organism is frequently associated with (comitans is latin for "accompanying") *Actinomyces,* the casual agent of actinomycosis (see Chapter 21). This organism is part of the normal oropharyngeal population, detectable in approximately 20% of healthy individuals. *A. actinomycetemcomitans* is associated with juvenile and adult periodontitis and subacute endocarditis. Underlying cardiac disease and spread of the organism from the oral cavity following dental procedures or in patients with periodontitis is reported for virtually all patients with *Actinobacillus* endocarditis. Most serious infections are treated with ampicillin alone or in combination with an aminoglycoside.

PASTEURELLA

Six species of *Pasteurella* are currently recognized; *Pasteurella multocida* is the most common human pathogen. The natural reservoir of *P. multocida* is the upper respiratory tract of such domestic animals as dogs and cats, with human infection related to animal exposure (e.g., animal bite, cat scratch). Three general forms of disease are reported: (1) localized cellulitis and lymphadenitis following an animal bite or scratch, (2) exacerbation of chronic respiratory disease in patients with underlying pulmonary dysfunction (presumably due to colonization of the patients' oropharynx followed by aspiration of oral secretions), and (3) systemic infection in immunocompromised patients, particularly with underlying hepatic disease. The organism is susceptible to a variety of antibiotics; penicillin is the antibiotic of choice, and tetracycline or a cephalosporin are acceptable alternatives. Isolation of *P. multocida* can be readily accomplished on either blood or chocolate agar but not Mac-Conkey agar. Large buttery colonies will grow after overnight incubation and have a characteristic musty odor due to indole production. Identification can be readily accomplished, as indicated in Table 12-2.

BIBLIOGRAPHY

Albritton WL: Infections due to *Haemophilus* species other than *H. influenzae,* Ann Rev Microbiol 36:199-216, 1982.

Barenkamp, SJ et al: Outer-membrane protein subtypes of *Haemophilus influenzae* type b and spread of disease in day-care centers, J Infect Dis 144:210, 1981.

Brenner DJ, Mayer LW, Carlone GM, et al: Biochemical, genetic, and epidemiologic characterization of *Haemophilus influenzae* biogroup *aegyptius* (Haemophilus aegyptius) strains associated with Brazilian purpuric fever, J Clin Microbiol 26:1524-1534, 1988.

Broome CV: Epidemiology of *Haemophilus influenzae* type b infections in the United States, Pediatr Infect Dis J 6:779-782, 1987.

Campos J, Garcia-Tornel S, Roca J, and Iriondo M: Rifampin for eradicating carriage of multiply resistant *Haemophilus influenzae* b, Pediatr Infect Dis J 6:719-721, 1987.

Coonrod JD, Kunz LJ, and Ferraro MJ: The direct detection of microorganisms in clinical samples, New York, 1983, Academic Press.

Doern GV, Jorgensen JH, Thornsberry C, et al: National collaborative study of the prevalence of antimicrobial resistance among clinical isolates of *Haemophilus influenzae,* Antimicrob Agents Chemother 32:180-185, 1988.

Hansen EJ, McCracken GH, and Syrogiannopoulos G: *Haemophilus influenzae* type b lipooligosaccharide induces meningeal inflammation, Pediatr Infect Dis J 6:1150, 1987.

Lepow M: Clinical trials of the *Haemophilus influenzae* type b capsular polysaccharide-diptheria toxoid conjugate vaccine, Pediatr Infect Dis J 6:804-807, 1987.

Mortimer EA: Efficacy of *Haemophilus* b polysaccharide vaccine: an enigma, JAMA 260:1454-1455, 1988.

Murphy TF and Apicella MA: Nontypable *Haemophilus influenzae:* a review of clinical aspects, surface antigens, and the human immune response to infection, Rev Infect Dis 9:1-15, 1987.

Parr TR and Bryan LE: Mechanism of resistance of an ampicillin-resistant, β-lactamase-negative clinical isolate of *Haemophilus influenzae* type b to β-lactam antibiotics, Antimicrob Agents Chemother 25:747-753, 1984.

Peter G: Treatment and prevention of *Haemophilus influenzae* type b meningitis, Pediatr Infect Dis J 6:787-790, 1987.

Santosham M, Reid R, Ambrosino DM, et al: Prevention of *Haemophilus influenzae* type b infections in high risk infants treated with bacterial polysaccharide immune globulin, N Engl J Med 317:923-929, 1987.

Sell SH: *Haemophilus influenzae* type b meningitis: manifestations and long term sequelae, Pediatr Infect Dis J 6:775-778, 1987.

Wallace RJ, Musher DM, Septimus EJ, et al: *Haemophilus influenzae* infections in adults: characterization of strains by serotypes, biotypes, and β-lactamase production, J Infect Dis 144:101-106, 1981.

Ward J: Newer *Haemophilus influenzae* type b vaccine and passive prophylaxis, Pediatr Infect Dis J 6:799-803, 1987.

Williams JD and Moosdeen F: Antibiotic resistance in *Haemophilus influenzae:* epidemiology, mechanisms, and therapeutic possibilities, Rev Infect Dis 8(suppl 5):555-561, 1986.

13

Bordetella, Francisella, and Brucella

A number of gram-negative bacilli are classified together as genera of uncertain affiliation. Three clinically significant members are *Bordetella, Francisella,* and *Brucella* (Table 13-1).

BORDETELLA

Bordetella are extremely small (0.2 to 0.5 × 1 μm), strictly aerobic gram-negative bacilli. The genus consists of three species: *B. pertussis* (Latin for "severe cough"; the agent responsible for **pertussis,** or **whooping cough**), *B. parapertussis* ("like pertussis"; responsible for a milder form of pertussis), and *B. bronchiseptica* (responsible for respiratory disease in dogs, swine, laboratory animals, and occasionally humans).

Microbial Physiology and Structure

Differentiation of the species is based on growth

Table 13-1 Human Pathogens

Genus	Species	Disease
Bordetella	pertussis	Pertussis
	parapertussis	Pertussis
	bronchiseptica	Bronchopulmonary disease
Francisella	tularensis	Tularemia
Brucella	melitensis	Brucellosis
	abortus	Brucellosis
	suis	Brucellosis
	canis	Brucellosis

characteristics, biochemical reactivity, and antigenic properties. *B. pertussis* is unable to grow on common laboratory media (requiring instead an enriched basal medium supplemented with charcoal, starch, blood, or albumin to adsorb toxic substances present in agar), is nonmotile, does not ferment carbohydrates, and is generally metabolically inert. The other *Bordetella* species are less fastidious, can grow on blood and MacConkey agars, and can be identified by their biochemical reactivity.

Pathogenesis

Infection with *B. pertussis* and the development of whooping cough require exposure to the organism, bacterial attachment to the ciliated epithelial cells of the bronchial tree, proliferation of the bacteria, and the production of localized and systemic tissue damage. Although fimbria are present on the bacterial surface, they are not involved in attachment to the target cell, unlike with other bacteria. Instead, filamentous hemagglutinin and pertussis toxin are responsible for bacterial attachment to ciliated epithelial cells. Other toxins, such as those listed in Table 13-2 and discussed here, are responsible for the clinical manifestions of whooping cough.

Pertussis Toxin Consistent with the multiple biologic properties associated with this toxin, pertussis toxin has also been called histamine sensitizing factor, lymphocytosis promoting factor, islet cell activating protein, and pertussigen. The heat-labile pertussis toxin consists of five unique protein subunits: subunit 1 has ADP-ribosylating activity for membrane surface proteins; subunits 2 to 5 are responsible for toxin binding and internalization. The ADP-ribosylat-

Table 13-2   Virulence Factors Associated With *Bordetella pertussis*

Virulence Factors	Biologic Effects
Pertussis toxin	ADP-ribosylation of guanine nucleotide-binding proteins, lymphocytosis, hypoglycemia, mediates attachment to respiratory epithelium
Adenylate cyclase toxin	Impairment of leukocyte chemotaxis and killing, local edema
Tracheal cytotoxin	Ciliastasis and then extrusion of ciliated epithelial cells
Dermonecrotic toxin	Tissue necrosis and shock in animal model
Filamentous hemagglutinin	Mediates attachment to ciliated epithelial cells, agglutinates erythrocytes
Lipopolysaccharide	Endotoxin, antiviral activity

ing activity is responsible for interference with the transfer of signals from cell-surface receptors (guanine nucleotide–binding proteins) to intracellular mediator systems (homeostatic inhibitory regulation of adenylate cyclase activity). Pertussis toxin can also block immune effector cells, including neutrophils, monocytes, macrophages, and natural killer cells. Finally, the toxin is believed to be important for bacterial binding to ciliated epithelial cells, because bacterial attachment is reduced when the toxin is not produced. Antibodies to pertussis toxin confer immunity.

Adenylate Cyclase Toxin   This toxin, released during exponential cell growth, catalyzes the conversion of endogenous ATP to cAMP in eukayrotic cells. Adenylate cyclase toxin interferes with immune effector cell function at the site of bacterial activity with inhibition of leukocyte chemotaxis, phagocytosis, and killing; oxidative activity of alveolar macrophages; and cell lysis by natural killer cells.

Tracheal Cytotoxin   The tracheal cytotoxin has a specific affinity for ciliated epithelial cells. Most likely, low concentrations of the cytotoxin cause ciliostasis (inhibition of cilia movement), whereas higher concentrations of the toxin are produced later in the infection and are responsible for extrusion of the ciliated cells. Nonciliated cells are spared in infections with *B. pertussis.*

Dermonecrotic Toxin   This heat-labile toxin can cause vasoconstriction of peripheral blood vessels with localized ischemia, movement of leukocytes to extravascular spaces, and hemorrhage. The toxin probably is responsible for localized tissue destruction in pertussis.

Filamentous Hemagglutinin   The hemagglutinin can agglutinate a variety of animal erythrocytes and is believed to be important in attachment of *B. pertussis* to ciliated cells. Antibodies directed against the filamentous hemagglutinin interfere with bacterial attachment and are protective.

Lipopolysaccharide   The lipopolysaccharide consists of two lipids (lipid A and lipid X) and two distinct polysaccharides. Endotoxin activity is associated with lipid X, whereas antiviral activity is associated with lipid A.

Epidemiology

Pertussis, a disease with worldwide endemicity, has been recognized for centuries. The incidence of disease, with its associated morbidity and mortality, has been reduced in recent years with the widespread availability of effective vaccines. Most infections now are associated with inadequately immunized children (Figure 13-1). *B. pertussis* has no animal or environmental reservoir, so it must be spread from patients with clinically apparent pertussis or with mild, unrecognized disease. Theoretically, use of the vaccine could eliminate *B. pertussis* in the same manner smallpox virus was eradicated. Unfortunately, widespread acceptance of the vaccine has not been achieved because of fears of vaccine-related toxicity and reports of decreased efficacy. The limited use of the vaccine in Great Britan, Japan, and areas in the United States has increased the incidence of pertussis and its complications.

Clinical Syndromes

Infection is initiated by inhalation of infectious aerosol droplets and attachment and proliferation of the bacteria on ciliated epithelial cells. After a 7- to 10-day incubation period the patient will experience the first of three stages (Figure 13-2). The **catarrhal stage** resembles a common cold, with serous rhinorrhea, sneezing, malaise, anorexia, and a low grade fever. Because the peak production of bacteria is during this stage and the cause of the disease is unrecognized, patients in the catarrhal stage pose the highest risk to their contacts. After 1 to 2 weeks the **paroxysmal stage** begins, with the classic whooping cough paroxysms. The paroxysms are characterized by a series of repetitive coughs followed by an inspiratory whoop. Mucus production in the respiratory tract is common and is partially responsible for the restricted airway. The paroxysms are frequently terminated with vomiting and exhaustion. A marked lymphocytosis is also prominent during this stage. As many as 40 to 50 daily paroxysms may occur during the height of the illness. After 2 to 4 weeks the disease enters the **convalescent stage** when the paroxysms diminish in number and severity, but secondary complications can occur.

Laboratory Diagnosis

Specimen Collection and Transport *B. pertussis* is extremely sensitive to drying and does not survive unless care is used for collection and transport to the laboratory. The optimal diagnostic specimen is a nasopharyngeal aspirate. Pernasal or oropharyngeal swabs have a lower yield and require use of synthetic fiber swabs. Cotton swabs are toxic to *B. pertussis* and must be avoided. The specimen must be either directly inoculated at the patient's bedside onto freshly prepared isolation media (e.g., **Regan-Lowe** medium, **Bordet-Gengou** medium) or placed in a suitable transport medium (e.g., Regan-Lowe transport medium). The organism does not survive transport in traditional transport media.

Culture When received in the laboratory, the inoculated media are incubated in a humidified chamber for up to 7 days. Generally, tiny colonies are observed after 3 or more days of incubation.

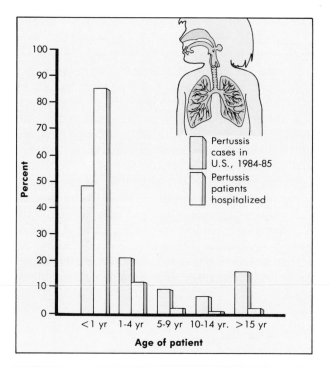

FIGURE 13-1   Age distribution and severity of illness reported for pertussis infection during 1984 and 1985.

	Incubation	Catarrhal	Paroxysmal	Convalescent
Duration	7-10 days	1-2 weeks	2-4 weeks	3-4 weeks (or longer)
Symptoms	None	Rhinorrhea, malaise, fever sneezing, anorexia	Repetitive cough with whoops, vomiting, leukocytosis	Diminished paroxysmal cough, development of secondary complications (pneumonia, seizures, encephalopathy)
Bacterial culture				

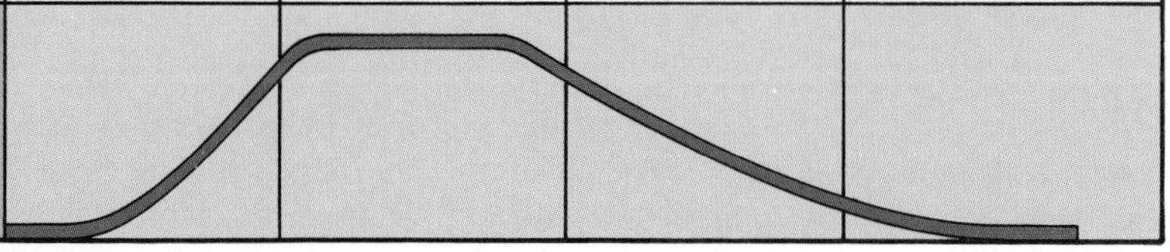

FIGURE 13-2   Clinical presentation of *Bordetella pertussis* disease.

Microscopy Specimens can also be examined by a direct fluorescent antibody (DFA) procedure, in which the aspirated specimen is smeared onto a microscopic slide, air-dried and heat-fixed, and then stained with fluorescein-labeled antibodies directed against *B. pertussis*. Antibodies against *B. parapertussis* should also be used to detect mild forms of pertussis caused by this organism.

Identification *B. pertussis* is identified by its characteristic microscopic and colonial morphology on selective media and its reactivity with specific antiserum (either in an agglutination reaction or with the reagents used in the DFA test). *B. pertussis* can be differentiated from *B. parapertussis* by the ability of *B. parapertussis* to grow on common laboratory media, hydrolyze urea, and utilize citrate as a carbon source (all negative reactions with *B. pertussis*).

Treatment

Treatment for pertussis is primarily supportive, with close nursing supervision during the paroxysmal and convalescent stages of illness. Antibiotics do not ameliorate the clinical course, although erythromycin is effective in eradicating the organisms and can reduce the stage of infectivity. However, this has limited value because the illness is usually unrecognized during the height of contagiousness.

Prevention and Control

Use of the currently available whole cell, inactivated vaccine has been demonstrated to be effective in eliminating symptomatic disease. However, concern about the associated complications has limited the acceptance of the current vaccine. Efforts to develop vaccines directed against specific toxins (e.g., pertussis toxin) are ongoing.

Because pertussis is highly contagious in a susceptible population, and unrecognized infections in family members of a symptomatic patient can maintain disease in a community, erythromycin has been used for prophylaxis.

Other Bordetella

B. parapertussis is responsible for 10% to 20% of the cases of pertussis seen annually in the United States. *B. bronchiseptica* causes respiratory disease primarily in animals but has been associated with human respiratory colonization and bronchopulmonary disease. Both organisms can be readily isolated on conventional laboratory media, and—in contrast with *B. pertussis*—both have easily recognizable metabolic properties.

FRANCISELLA TULARENSIS

Francisella tularensis is the causative agent of **tularemia** (also called glandular fever, rabbit fever, tick fever, deerfly fever) in animals and humans. The taxonomic classification of the organism is derived from Edward Francis, who demonstrated the association between rodent disease in Tulare County, California, and human disease. The colloquial terms for the disease refer to the most common clinical presentation, reservoir, or vectors of tularemia. *F. tularensis* is a very small (0.2 × 0.2 to 0.7 μm), faintly staining gram-negative coccobacillus. The organism has fastidious growth requirement, and isolation from clinical specimens generally requires prolonged incubation on specially enriched media. *F. tularensis* is an intracellular parasite that can survive for prolonged periods in the macrophages of the reticuloendothelial system.

Pathogenesis

Little is known about specific virulence factors associated with *F. tularensis*. An antiphagocytic capsule is present in pathogenic strains, and loss of the capsule is associated with decreased virulence. Additionally, this organism has endotoxin activity like all gram-negative bacilli. Because *F. tularensis* is an intracellular parasite, the organism is protected from humoral antibodies.

Epidemiology

F. tularensis has worldwide distribution, although the prevalence of the two biochemical varieties (biovars tularensis and palaearctica) is somewhat restricted (Table 13-3). The organism is found in a large number of wild mammals, domestic animals, birds, fish, and blood-sucking arthropods, as well as contaminated water. The most common reservoirs of *F. tularensis* in the United States are rabbits and ticks. The organism can be maintained in the tick population by transovarian passage, so eradication of this pathogen

Table 13-3 Differentiation Between *Francisella tularensis,* Biovar Tularensis, and Biovar Palaearctica

Characteristic	Biovar Tularensis	Biovar Palaearctica
Geographic distribution	Jellison A; nearctica North America	Jellison B Europe, Asia, the Americas
Biochemical properties		
Glycerol fermentation	Positive	Negative
Citrulline-ureidase	Positive	Negative
Source of infection	Ticks, rabbits	Contaminated water or rodents
Human disease	Severe	Mild

Table 13-4 Manifestations of Tularemia

Type of Disease	Infections (%)	Characteristics
Ulceroglandular	75-85	Ulcers at the site of exposure and adenopathy of the draining lymph nodes
Glandular	5-10	Adenopathy but no ulcers
Typhoidal	5-15	Systemic signs but no adenopathy or ulcers
Oculoglandular	1-2	Eye involvement
Oropharyngeal	<1	Oropharyngeal involvement

is virtually impossible. Tularemia can be acquired after the bite of an infected arthropod, contact with an infected animal or domestic pet that has caught an infected animal (e.g., rabbit), consumption of contaminated meat or water, or inhalation of an infectious aerosol (most commonly in a laboratory or during dressing an infected animal). Infection with *F. tularensis* requires as few as 10 organisms when exposed to an arthropod bite or contamination of unbroken skin, 50 organisms when inhaled, and 10^8 organisms when ingested.

The reported incidence of disease is low, with fewer than 200 cases each year. The largest number of infections occur during the summer when exposure to infected ticks is highest and during the winter when hunters are exposed to infected rabbits. However, the actual number of infections is likely to be much higher because tularemia is frequently unsuspected or difficult to confirm by laboratory tests. Persons at greatest risk for infection include hunters, those exposed to ticks, and laboratory personnel. In endemic areas, if a rabbit is moving so slowly that it can be shot by a hunter or caught by a pet animal, then the rabbit could be infected. Disease is reported most commonly in Arkansas, Missouri, Oklahoma, Texas, and Utah.

Clinical Syndromes

After a 3- to 5-day incubation period, tularemia symptons include an abrupt onset of fever, chills, malaise, and fatigue. The clinical classification of *F. tularensis* disease is based on the site of infection and the presence of skin ulcers and lymphadenopathy. Thus tularemia can be subdivided into several manifestations (Table 13-4). In the ulceroglandular type of tularemia, axillary adenopathy is generally present after exposure to infected rabbits (presumably resulting from infection of the hands), and inguinal adenopathy is common with tick-borne disease (due to tick bites on the lower extremities). *F. tularensis* pneumonia is present in the majority of patients with typhoidal disease. Focal necrosis in the liver and spleen can also be found.

Laboratory Diagnosis

Microscopy The detection of *F. tularensis* in Gram-stained aspirates from infected nodes or ulcers is almost always unsuccessful because the organism is extremely small and stains faintly. A more sensitive and specific approach is direct staining of the clinical specimen with fluorescein-labeled antibodies directed against the organism.

Culture It is stated that *F. tularensis* cannot be isolated on common laboratory media because the organism requires sulfhydryl-containing substances (e.g., cysteine) for growth. However, chocolate blood agar plates that are used in most laboratories are supplemented with cysteine, and *F. tularensis* will grow on this medium. Specialized media such as cysteine blood agar or glucose cysteine agar are not required for most isolates. If infection with *F. tularensis* is suspected, the laboratory should be notified. *F. tularensis* grows slowly and will be overlooked unless the cultures are incubated for a prolonged period. In addition, the organism is highly infectious and can penetrate through unbroken skin. All cultures must be handled by gloved personnel, with processing of the specimen performed in a biohazard hood. Blood cultures are generally negative, whereas cultures of sputum and aspirates of lymph nodes or draining sinuses are usually positive.

Identification *F. tularensis* is obligately aerobic, weakly catalase positive and oxidase negative. Biochemical characterization of the organism is of little value. Identification is confirmed if there is evidence of reactivity with specific antisera in fluorescent antibody or agglutination tests. Only one serotype of *F. tularensis* has been identified.

Serology Most cases of tularemia are diagnosed by a fourfold or greater increase in antibodies during the course of the illness (Figure 13-3). An agglutinating antibody titer of 1:160 or greater is consistent with exposure to *F. tularensis*, although the persistence of antibodies (including IgG, IgM, and IgA) for many years prevents differentiation between past and current disease. Antibodies directed against *Brucella* can also cross-react with *Francisella*.

Treatment

Streptomycin is the antibiotic of choice for all forms of tularemia. Gentamicin is an acceptable alternative, although less clinical experience exists with this drug. Tetracycline and chloramphenicol have been used to treat infections; however, these are associated with an unacceptably high rate of relapses. The mortality rate is less than 1% with appropriate antibiotic therapy.

Prevention and Control

Prevention is accomplished by avoiding the reservoirs and vectors of infection (e.g., rabbits, ticks), although this is frequently difficult. At a minimum,

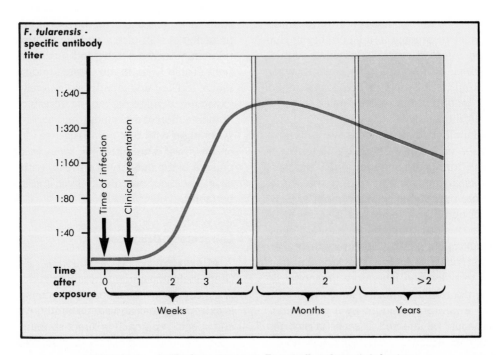

FIGURE 13-3   Antibody response to *Francisella tularensis* infections.

ill-appearing rabbits should not be handled and gloves should be worn when animals are skinned and eviscerated. Ticks should be promptly removed. Because the organism is present in the arthropod's feces and not saliva, the tick must feed for a prolonged time before the infection is transmitted.

Use of live, attenuated vaccines is not completely effective but does reduce the severity of disease. Inactivated vaccines are not protective.

BRUCELLA

The genus Brucella consists of six species, four of which are associated with human **brucellosis:** B. abortus, B. melitensis, B. suis, and B. canis. Brucellosis has been referred by a number of names, based on the original microbiologists who isolated and described the organisms (Sir David Bruce and Bernhard Bang; Bang's disease), the clinical presentation (undulant fever), or the site of recognized outbreaks (Malta fever, Mediterranean remittant fever, rock fever of Gibraltar, county fever of Constantinople, fever of Crete). The individual species are distinguished by their reservoir host, growth properties, biochemical reactivities, and cell wall fatty acid composition. Brucella are small (0.5×0.6 to $1.5 \ \mu$m), nonmotile, nonencapsulated gram-negative coccobacilli. The organism grows slowly on culture, is aerobic with some strains requiring supplemental carbon dioxide for growth, and does not ferment carbohydrates. B. abortus, B. melitensis, and B. suis share two surface antigens; the highest concentration of A antigen is found in B. abortus and the M antigen is in B. meli-

tensis. B. canis is antigenically distinct. Some common characteristics of the four species that cause human disease are summarized in Table 13-5.

Pathogenesis

Like Francisella, Brucella is an intracellular parasite of the reticuloendothelial system and is sheltered from humoral immunity. A capsule is not present, although smooth and rough forms are recognized.

Epidemiology

Brucella infections have a worldwide distribution; more than 500,000 documented cases are reported annually. However, the incidence of disease in the United States is much lower, with approximately one reported case per 1 million individuals (an incidence similar to tularemia). The reason for the paucity of cases in the United States is related to control of disease in the animal reservoir.

Brucella causes mild or asymptomatic disease in the natural host: B abortus in cattle, B. melitensis in goats and sheep, B. suis in swine, and B. canis in dogs, foxes, and coyotes. The organism has a predilection for organs rich in erythritol, which is metabolized by many Brucella strains in preference to glucose. Animal (but not human) tissues rich in erythritol include breast, uterus, placenta, and epididymis. Thus Brucella localizes in these tissues in the nonhuman reservoirs and can cause sterility, abortions, or asymptomatic carriage. Human disease occurs in persons in direct contact with infected animals (veterinarians, slaughterhouse workers, farmers) or their

Table 13-5 Characteristics of Human Brucellosis

| Species | Animal Reservoir | CO$_2$ Required | Growth on Dyes | | Clinical Disease |
			Basic Fuchsin	Thionin	
B. melitensis	Goats, sheep	−	+	+	Severe, acute disease with complications common
B. abortus	Cattle	+	+	−	Mild disease with suppurative complications uncommon
B. suis	Swine	−	±	+	Suppurative, destructive disease with chronic manifestations
B. canis	Dogs	−	−	+	Mild disease with suppurative complications uncommon

products (consumption of unpasteurized milk or cheese). Laboratory personnel are also at significant risk for infection through direct contact or inhalation of the organism. The areas in the United States with the highest number of reported cases are Texas, California, Florida, Virginia, and Iowa.

Clinical Syndromes

The disease spectrum of brucellosis is influenced by the infecting organism. *B. abortus* and *B. canis* tend to produce mild disease with rare suppurative complications. In contrast, *B. suis* is associated with destructive lesions and a prolonged course. *B. melitensis,* the most common cause of brucellosis, also causes severe disease with a high incidence of serious complications.

After exposure to the organism, a localized abscess can develop at the site of inoculation, which is followed by bacteremia. Organisms are phagocytized by macrophages and monocytes and then localized in tissues of the reticuloendothelial system (e.g., spleen, liver, bone marrow, lymph nodes, and kidneys). A *Brucella* infection may remain subclinical, with an increase in *Brucella*-specific antibodies the only evidence of disease, or may be seen clinically in a subacute or acute form. Severe toxic disease is particularly common with *B. melitensis.* The subacute and chronic manifestations of brucellosis can initially include malaise, chills, sweats, fatigue, weakness, myalgias, weight loss, arthralgias, and nonproductive cough. Fever is found in almost all patients and can be intermittent (hence the name undulant fever). The subacute and chronic forms of brucellosis make diagnosis difficult because specific localizing signs are frequently absent. Granuloma formation in the liver, spleen, and bone marrow, as well as destructive changes in many other organs, have been observed.

Laboratory Diagnosis

Microscopy Although *Brucella* organisms stain readily by conventional techniques, their small size makes them difficult to detect in clinical specimens. The direct fluorescent antibody technique has proved useful, but the test reagents are not readily available.

Culture *Brucella* are slow-growing, fastidious organisms on primary isolation. Isolates will grow on most enriched blood agar and occasionally on Mac-Conkey agar; however, growth is slow, requiring incubation for 3 or more days. The isolation from blood cultures can take as long as 4 weeks. Incubation in a carbon-dioxide enriched atmosphere is required for *B. abortus.*

Identification An isolate can be preliminarily identified as *Brucella* by its microscopic and colonial morphology, positive oxidase reaction, and reactivity with antibodies directed against *Brucella.* Species identification is determined by biochemical reactivity, growth in the presence of the dyes basic fuchsin and thionin, and agglutination in specific antisera. However, most laboratories in the United States are unable to identify specific species because the organism is now infrequently isolated.

Serology Subclinical brucellosis and many cases of acute and chronic diseases are identified by a specific antibody response in the infected individual. Antibodies are detected in virtually all patients, with an initial IgM response followed by the production of both IgG and IgA antibodies (Figure 13-4). Antibodies can persist for many months or years, so a significant increase in antibody titer is required for definitive serologic evidence of current disease. A presumptive diagnosis can be made with a ≥1:160 titer of agglutinating antibodies. The antigen used in the *Brucella* agglutination test is from *B. abortus.* Antibodies directed against *B. melitensis* or *B. suis* will cross-react with this antigen; however, there is no cross-reactivity with *B. canis.* The specific use of *B. canis* antigen is required for the diagnosis of this infection. Cross-reactivity with *B. abortus* antigen is also reported with *Yersinia enterocolitica, Francisella tularensis,* and specific serotypes of *Salmonella, Vibrio cholerae,* and *Escherichia coli.*

Treatment

Tetracycline is generally active against most strains of *Brucella;* however, this is a bacteriostatic drug, and relapse after an initial successful response is common. The combination of tetracycline with either streptomycin or gentamicin has proved efficacy with a low incidence of relapse. Long-term therapy with high doses of trimethoprim-sulfamethoxazole is an acceptable alternative regimen, and the addition of rifampin is useful for CNS disease.

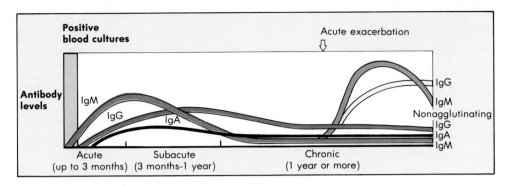

FIGURE 13-4   Antibody responses in untreated brucellosis.

Prevention and Control

Control of human brucellosis is accomplished by controlling disease in livestock, as has been demonstrated in the United States. This requires the systematic identificaiton and elimination of infected herds and animal vaccination. Further efforts to prevent brucellosis include protective clothing for abattoir workers and the avoidance of unpasteurized dairy products. Vaccination of high-risk individuals has limited utility.

BIBLIOGRAPHY

Biellik RJ, Patriarca RA, Mullen JR, et al: Risk factors for community and household acquired pertussis during a large scale outbreak in central Wisconsin, J Infect Dis 157:1134-1141, 1988.

Evans ME: *Francisella tularensis,* Infect Control 6:381-383, 1985.

Fine PE and Clarkson JA: Reflections on the efficacy of pertussis vaccines, Rev Infect Dis 9:866-883, 1987.

Francis E. Tularemia, JAMA 84:1243-1250, 1925.

Goodnow RA: Biology of *Bordetella bronchiseptica,* Microbiol Rev 44:722-738, 1980.

Linnemann CC and Perry EB: *Bordetella parapertussis:* recent experience and a review of the literature, Am J Dis Child 131:560-563, 1977.

Morbidity and mortality report. Pertussis surveillance—United States 1984 and 1985, 36:2013-2014, 1987.

Tuomanen EI and Hendley JO: Adherence of *Bordetella pertussis* to human respiratory epithelial cells, J Infect Dis 148:125-130, 1983.

Weiss AA and Hewlett EL: Virulence factors of Bordetella pertussis, Ann rev Microbiol 40:661-686, 1986.

Young EJ: Human brucellosis, Rev. Infect Dis 5:821-842, 1983.

CHAPTER 14

Legionellaceae

In the summer of 1976 public attention was focused on an outbreak of severe pneumonia that had a high mortality rate for members of the American Legion convention in Philadelphia. After months of intensive investigations the etiologic agent was isolated—a previously unknown gram-negative bacillus. In subsequent studies this organism, named *Legionella pneumophila,* has been associated with at least four epidemics before 1976 and multiple epidemic and sporadic infections after 1976. The existence of this organism was previously unappreciated because it neither stains with conventional dyes nor grows on common laboratory media. Despite the initial problems with the isolation of *Legionella,* it is now recognized to be an ubiquitous aquatic saprophyte and a common cause of human respiratory disease.

The family Legionellaceae consists of one genus, *Legionella;* 25 species and 42 serotypes are currently described. The number of species and serotypes will undoubtedly expand in the future. Despite the diversity observed with this family, the number of species responsible for human disease is limited to approximately 10 (Figure 14-1) with the others found only in environmental sources. *Legionella pneumophila* is responsible for almost 85% of all infections, with serotype 1 isolated most commonly. Differentiation of the species is based on determination of DNA homology, analysis of branched cell wall fatty acids, and biochemical testing. Immunologic serotyping further subdivides individual isolates.

MICROBIAL PHYSIOLOGY AND STRUCTURE

Members of the genus are gram-negative bacilli, 0.3 to 0.9 μm wide and 2 to 5 μm long. The cell wall structure is typical of gram-negative bacilli (Figure 14-2). The organisms characteristically appear as short coccobacilli in tissue but are very pleomorphic when cultured on artificial media (Figure 14-3). *Legionella* in clinical specimens do not stain with common reagents but can be seen in tissues stained with the Dieterle silver stain. One species, *Legionella micdadei,* can also be stained with weak acid-fast stains, but this property is lost when the organism is isolated in culture and is not observed with other *Legionella* species. Growth of *Legionella* in vitro requires supplementation of the media with l-cysteine and is enhanced with iron salts. The organisms are nonfermentative and derive energy from metabo-

PATHOGENESIS

Respiratory disease caused by *Legionella* follows inhalation of infectious aerosols by susceptible individuals. *Legionella* are intracellular parasites, capable of multiplication in alveolar macrophages and monocytes (Figure 14-4). After phagocytosis of viable organisms, the bacteria are protected from intracellular killing by inhibition of lysosomal fusion and acidification. Strains that cause disease in guinea pigs are resistant to serum killing, whereas avirulent strains are rapidly killed in serum and do not survive phagocytosis by macrophages. The physiologic differences between virulent and avirulent strains is poorly understood, although the tissue destruction associated with *Legionella* infections is likely due to production of proteolytic enzymes, phosphatase, lipase, and nuclease. Despite this activity, all strains of *Legionella* have a relatively low potential for causing disease. Most infections are in individuals with compromised

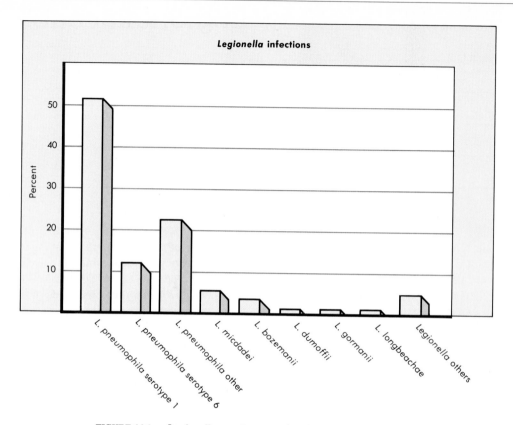

FIGURE 14-1   *Legionella* species associated with human disease.

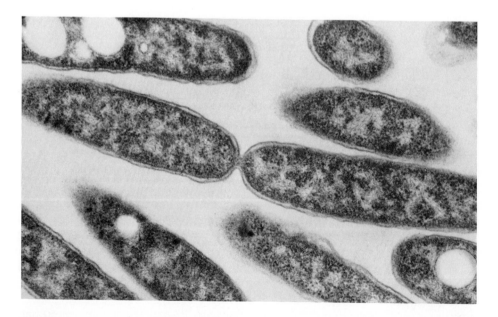

FIGURE 14-2   Electron micrograph of *Legionella pneumophila* grown in broth. Note the periplasmic space separating the inner and outer membranes. (From Bergey's manual of systemic bacteriology, vol 1, New York, 1984, Williams & Wilkins.)

immune status or pulmonary function. For example, persons at greatest risk for infection include renal or cardiac transplant recipients, patients with T-cell dysfunction or chronic pulmonary disease, and heavy smokers.

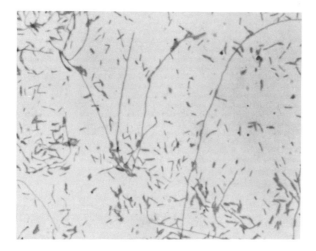

FIGURE 14-3   Stained *Legionella pneumophila* grown on charcoal-yeast extract agar. Note the pleomorphic bacilli with short coccobacilli and very long filaments. (From Bergey's manual of systematic bacteriology, vol 1, New York, 1984, Williams & Wilkins.)

EPIDEMIOLOGY

Sporadic and epidemic legionellosis has a worldwide distribution. The bacteria are commonly present in natural bodies of water such as lakes and streams, air conditioning cooling towers and condensers, and potable water systems (e.g., showers). The organisms are capable of survival in moist environments for prolonged periods of time and at relatively high temperatures.

The incidence of infections caused by *Legionella* is poorly understood because documentation of disease is difficult. Prospective studies have reported the proportion of pneumonias due to *Legionella* ranges from less than 1% to greater than 30%, and it has been estimated that between 25,000 and 50,000 cases of pulmonary legionellosis occur annually. Serologic studies indicate a significant proportion of the population has developed immunity to this group of organisms. Based on these studies and the knowledge that *Legionella* are ubiquitous aquatic saprophytes, it is reasonable to conclude that contact with the organism and immunity following asymptomatic infection is common, and the incidence of true disease is probably underestimated.

Although sporadic disease outbreaks are reported throughout the year, most epidemic infections occur

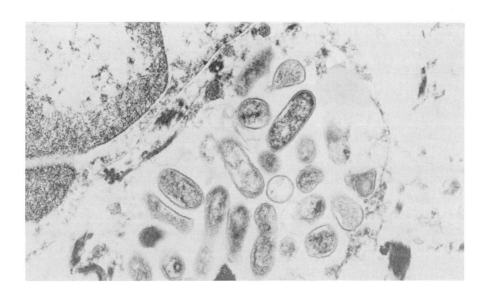

FIGURE 14-4   Lung specimen showing numerous intracytoplasmic *Legionella* within alveolar macrophages. The nucleus is in the upper left corner. (From Weisenburger, DD: Arch Pathol Lab Med 105:130, copyright 1981, American Medical Association.)

Table 14-1   Comparison of Diseases Caused by *Legionella*

	Pontiac Fever	Legionnaires' Disease
EPIDEMIOLOGY		
Presentation	Epidemic	Epidemic, sporadic
Attack rate	>90%	<5%
Person-to-person spread	No	No
Underlying pulmonary disease	No	Yes
Time of onset	Throughout year	Late summer, fall
CLINICAL MANIFESTATIONS		
Incubation period	1-2 days	2-10 days
Pneumonia	No	Yes
Course	Self-limited	Antibiotic therapy
Mortality	<1%	15%-20%

in the late summer and fall months, presumably because of the proliferation of the organisms in water reservoirs during the warm months of the year. The elderly, with decreased cellular immunity or pulmonary function, are at greatest risk for disease. Person-to-person spread has not been demonstrated.

CLINICAL SYNDROMES

Although asymptomatic *Legionella* infections may be detected only by an immune reaction to the organism, symptomatic infections primarily affect the lungs and are seen in one of two forms (Table 14-1): as a flu-like illness (referred to as **Pontiac fever**) or as a severe form of pneumonia (e.g., **Legionnaires' disease**).

Pontiac Fever

Legionella pneumophila was responsible for an epidemic influenza-like illness, without evidence of pneumonia, in the Pontiac, Michigan, Public Health Department in 1968. The disease was characterized by fever, chills, myalgia, malaise, and headache that developed over a 12-hour period, persisted for 2 to 5 days, and then resolved spontaneously with minimal morbidity and no mortalities. Additional epidemics of Pontiac fever, with a high incidence of disease in exposed individuals, have been documented.

Legionnaires' Disease

Legionnaires' disease is characteristically more severe, with significant morbidity and mortality unless prompt therapy is initiated. After an incubation period of 2 to 10 days, patients experience an abrupt onset of fever, chills, dry nonproductive cough, headache, and systemic signs of an acute illness. Multiorgan disease, with involvement of the gastrointestinal tract, central nervous system, and abnormality of liver and renal function, is commonly observed. The primary manifestation of disease is pneumonia, with multilobe involvement and inflammation and microabscesses in lung tissue observed on histopathology. Untreated disease in susceptible patients is characterized by progressive deterioration of pulmonary function. The overall mortality is 15% to 20% but can be much higher in patients who have severe depression of their cell-mediated immunity (e.g., renal or cardiac transplant recipients).

LABORATORY DIAGNOSIS

Table 14-2 lists the sensitivity and specificity of diagnostic tests for *Legionella pneumophila*.

Microscopy

Legionella in clinical specimens stain poorly with the Gram stain (Table 14-3). Nonspecific staining

Table 14-2   Summary of the Sensitivity and Specificity of Diagnostic Tests for *Legionella pneumophila* *

Diagnostic Tests	Sensitivity (%)	Specificity (%)
Microscopy		
Direct fluorescent antibody (DFA)	60-75	>99
Antigen detection (RIA, ELISA, agglutination)	90	>99
respiratory specimens		
urine	80	>99
Nucleic acid probes	75	>99
Culture	75	100
Serology		
Indirect fluorescent antibody (IFA)	75	96

Legionella pneumophila is responsible for approximately 85% of diseases caused by *Legionella* spp. With the exception of culture and DNA probes, the diagnostic tests listed here are specific for *L. pneumophila* or specific serotypes of this species. Thus the other diagnostic tests would not reliably detect the small proportion of legionelloses caused by other species. For optimal detection of *Legionella* infections a combination of microscopy, culture, and serology should be used. The clinical specimen should be initially examined by the DFA test or, alternatively, nucleic acid hybridization. The specimen should also be cultured on media that would support the growth of *Legionella* and inhibit contaminating organisms that might overgrow and obscure the pathogens. Finally, serum for antibody measurements should be collected early in the course of disease and then again at 3 to 4 weeks and, if necessary, at 6 to 8 weeks.

Table 14-3   Staining Reactions of Common *Legionella* Species in Smears and Tissue Sections

Legionella	Smears			Tissue Sections		
	Gram	Gimenez	Modified Acid-Fast	Brown-Brenn	Kinyoun	Dieterle
L. pneumophilia	Negative	Positive	Negative	Negative	Negative	Positive
L. micdadei	Negative	Positive	Acid-fast*	Negative	Acid-fast*	Positive
L. bozemanii	Negative	Positive	Negative	Negative	Negative	Positive
L. dumoffii	Negative	Positive	Negative	Negative	Negative	Positive

*10% to 50% of bacilli will appear acid-fast.

methods such as the Dieterle silver stain or Gimenez stain can be used to visualize the organisms but are of little value with specimens contaminated with normal oral bacteria. The most sensitive procedure for the microscopic detection of *Legionella* in clinical specimens is the **direct fluorescent antibody (DFA) test,** in which fluorescein-labeled monoclonal or polyclonal antibodies directed against *Legionella* species are used. The test is very specific; false-positive reactions are observed only rarely with *Pseudomonas, Bacteroides,* and other organisms. These cross-reactions appear to be limited to the polyclonal reagents. However, the sensitivity is low because the antibody preparations are serotype- or species-specific, and abundant organisms must be present in the specimen for detection. The latter problem is due to the

relative small size and predominantly intracellular location of the bacteria. Positive tests will revert to negative after about 4 days of treatment.

Culture

Despite the fact that *Legionella* were initially difficult to grow in vitro, this can be readily accomplished now. *Legionella* require l-cysteine, and growth is enhanced with iron (supplied in hemoglobin or ferric pyrophosphate). The medium must be carefully buffered at pH 6.9. The medium most commonly used for isolation of *Legionella* is **buffered charcoal yeast extract (BCYE)** agar, although other supplemented media have also been used. Antibiotics can be added to the medium to suppress the growth of other bacteria-contaminating respiratory specimens such as sputum. Alternatively, the specimen can be treated with potassium chloride-sodium chloride for the selective elimination of contaminating organisms. *Legionella* will grow in air or 3% to 5% carbon dioxide at 35° C after 3 to 5 days, appearing as small (1 to 3 mm) colonies with a ground-glass appearance.

Antigen Detection

Detection of *Legionella* in respiratory specimens or urine has been performed by enzyme-linked immuno-assays, radioimmunoassays, agglutination of anti-body-coated latex particles, and nucleic acid hybrid-ization studies. Although the immunologic assays are relatively sensitive, the serotype-specific reagents limit the clinical utility of the test. Additionally, anti-gen excretion in urine may persist for as long as 1 year. The commercially available nucleic acid tests are very specific but relatively insensitive and should be used only as an adjunct to culture.

Serology

The diagnosis of legionellosis is commonly made by measuring a serologic response to infection with the **indirect fluorescent antibody test.** A fourfold or greater increase in antibody titer (to ≥1:128) is considered to be diagnostic. However, the response may be delayed, with a significant serologic increase observed for only 20% to 40% of patients after 1 week of illness and 75% after 8 weeks. This limits the test sensitivity and clinical utility for patient manage-ment. Antibody titers ≥1:256 are presumptive evi-dence of previous exposure to the bacteria, although high titers can persist for prolong periods.

Identification

Identification of an isolate as *Legionella* can be easily accomplished by demonstrating typical mor-phology and specific growth requirements. *Legionella* will appear as weakly staining, pleomorphic, thin gram-negative bacilli. Growth on BCYE, but not on BCYE without cysteine or on enriched blood agar media, is presumptive evidence the isolate is *Legion-ella.* The identification can be confirmed by specific staining with fluorescein-labeled antibodies. In con-trast with the identification of the genus, species

classification is problematic and generally relegated to reference laboratories. Although biochemical tests and the ability to fluoresce under long-wave ultravio-let light are useful differential tests, definitive species classification is accomplished by analysis of the major branched-chain fatty acids in the cell wall and DNA homology.

TREATMENT

In vitro susceptibility tests with *Legionella* are not routinely performed because growth is poor on media commonly used for these tests. However, the limited in vitro data and substantial clinical experience indi-cate erythromycin is the antibiotic of choice. β-lac-tam antibiotics are ineffective because most isolates produce β-lactamases. Erythromycin can substantial-ly reduce the morbidity and mortality of Legionaires' disease. Specific therapy for Pontiac fever is generally unnecessary because this is a self-limiting disease.

PREVENTION AND CONTROL

Prevention of legionellosis requires identification of the environmental source of the organism and reduction of the microbial burden. Hyperchlorination of the water supply and maintenance of elevated water temperatures have proved to be moderately successful. However, complete elimination of *Legion-ella* from a water supply is often difficult or impossi-ble. Because the organism has a low potential for causing disease, reduction of the number of organ-isms in the water supply is frequently adequate for controlling infections.

BIBLIOGRAPHY

Blackmon JA, Chandler FW, Cherry WB, et al: Review arti-cle: legionellosis, Am J Pathol 103:429-465, 1981.

Edelstein PH: Laboratory diagnosis of infections caused by Legionellae, Eur J Clin Microbiol 6:4-10, 1987.

Edelstein PH, Meyer RD, and Finegold SM: Laboratory diagnosis of Legionnaires' disease, Am Rev Resp Dis 121:317-327, 1980.

England AC and Fraser DW: Sporadic and epidemic noso-comial legionellosis in the United States: epidemiologic fea-tures, Am J Med 70:707-711, 1981.

England AC, Fraser DW, Plikaytis BD, et al: Sporadic legionellosis in the United States: the first thousand cases, Ann Intern Med 94:164-170, 1981.

Fallon RJ: The Legionellaceae, Med Lab Sci 43:64-71, 1986.

Fraser DW, Tsai TR, Orenstein W, et al: Legionnaires' disease: description of an epidemic of pneumonia, N Engl J Med 297:1189-1197, 1977.

Garbe PL, Davis BJ, Weisfeld JS, et al: Nosocomial Legionnaires' disease: epidemiologic demonstration of cooling towers as a source, JAMA 254:521-524, 1985.

McDade JE, Shepard CC, Fraser DW, et al: Legionnaires' disease: isolation of a bacterium and demonstration of its role in other respiratory disease, N Engl J Med 297:1197-1203, 1977.

Meyer RD: Legionella infections: a review of five years of research, Rev Infect Dis 5:258-278, 1983.

Meyer RD: Legionnaires' disease: aspects of nosocomial infection, Am J Med 76:657-663, 1984.

Reingold AL: Role of Legionellae in acute infections of the lower respiratory tract, Rev Infect Dis 10:1018-1028, 1988.

Sathapatayavongs B, Kohler RB, Wheat LJ, et al: Rapid diagnosis of Legionnaires' disease by urinary antigen detection: comparison of ELISA and radioimmunoassay, 72:576-582, 1982.

Weisenburger DD, Helms CM, and Renner ED: Sporadic Legionnaires' disease: a pathologic study of 23 fatal cases, Arch Pathol Lab Med 105:130-137, 1981.

Winn WC and Myerowitz RL: The pathology of the Legionella pneumonias: a review of 74 cases and the literature, Human Pathol 5:401-422, 1981.

CHAPTER 15

Bacteroidaceae

The family Bacteroidaceae consists of 13 genera of obligate anaerobic, nonsporeforming, gram-negative bacilli. Members of the family are an important part of the microbial flora that colonizes the human oropharynx, gastrointestinal tract, and genital tract. Infections with these bacteria arise from their endogenous spread from colonized mucosal surfaces to normally sterile body sites. The two genera most commonly associated with human disease are *Bacteroides* and *Fusobacterium*. Although each genus contains many species, only a few are regularly isolated in clinical specimens (Box 15-1). *Bacteroides fragilis* is considered the prototypic anaerobic pathogen.

MICROBIAL PHYSIOLOGY AND STRUCTURE

Members of the family Bacteroidaceae range in size from small coccobacilli (*Bacteroides melaninogenicus* group) to thin, fusiform-shaped bacilli *(Fusobacterium nucleatum)*. *Bacteroides fragilis* is pleomorphic in size and shape, with a casually examined Gram stain appearing to contain a number of different organisms. Most members of the family stain weakly with the Gram reaction, so stained specimens must be carefully examined. Some members grow rapidly in culture (*B. fragilis* group), whereas others are difficult to isolate (*B. melaninogenicus* group and some *Fusobacterium* species). The *B. fragilis* group is a collection of biochemically related, antibiotic-resistant species. At one time these organisms were considered to be members of the same species, but DNA homology studies have conclusively demonstrated the distinctive nature of each species. Likewise,

---

Box 15-1 Common Clinically Significant Isolates of the Family Bacteroidaceae

Bacteroides fragilis group
   B. fragilis
   B. thetaiotaomicron
   B. distasonis
   B. vulgatus
   B. ovatus
Bacteroids melaninogenicus group
   B. melaninogenicus
   B. intermedius
   B. asaccharolyticus
   B. gingivalis
Bacteroides bivius
Bacteroides disiens
Fusobacterium nucleatum
Fusobacterium necrophorum
Fusobacterium mortiferum
Fusobacterium varium

---

black-pigmented *Bacteroides* were formerly considered to be one species, *B. melaninogenicus,* with three subspecies. These bacteria have now been reclassified into eight individual species but are conveniently referred to as the *B. melaninogenicus* group.

Both *Bacteroides* and *Fusobacterium* have a typical gram-negative cell wall structure (Figure 15-1). A major component of the cell wall is a surface lipopolysaccharide (LPS). LPS in *Fusobacterium* is biochemically and biologically similar to the endotoxin present in other gram-negative bacilli. In contrast, LPS in

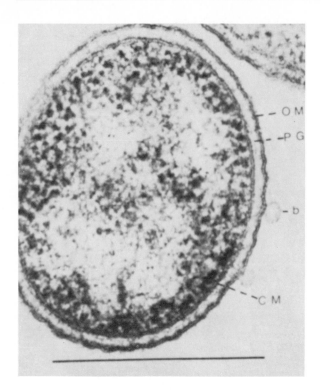

FIGURE 15-1   Electron micrograph of *Bacteroides fragilis*. Note the outer membrane *(OM)* and cytoplasmic membrane *(CM)* is separated by a periplasmic space and peptidoglycan layer *(PG)*. Blebs *(B)* of outer membrane are also seen. The bar is 1 μm. (From Kasper, DL, et al, Rev Infect Dis 1:278, 1979.)

Bacteroides has little or no endotoxin activity and has been found to lack the carbohydrates heptose and 2-keto-3-deoxyoctonate, which are characteristically present in other LPS molecules.

PATHOGENESIS

Despite the variety of anaerobic species that colonize the human body, relatively few are responsible for disease. For example, *B. distasonis* and *B. thetaiotaomicron* are the predominant *Bacteroides* found in the gastrointestinal tract; however, more than 80% of intraabdominal infections are associated with *B. fragilis,* an organism that comprises less than 1% of the gastrointestinal flora. The enhanced virulence of this and other anaerobic bacteria has been carefully studied.

Capsule

A polysaccharide capsule, detectable by staining and immunologic techniques, is present in most *B. fragilis* isolates, as well as some other members of the *B. fragilis* and *B. melaninogenicus* groups (Figure 15-2). The capsule is antiphagocytic and promotes abscess formation (Figure 15-3).

Lipopolysaccharide

As stated previously, the *Fusobacterium* LPS has classic endotoxic properties, whereas this biologic activity is absent in *Bacteroides* LPS. However, both *Fusobacterium* and *Bacteroides* LPS can stimulate leukocyte migration and chemotaxis. This is mediated by activation of the alternate complement pathway, with the subsequent cleavage of the complement component, C5a, a strong inducer of leukocyte migration and chemotaxis.

Agglutinins

Adherence to erythrocytes and epithelial surfaces by the *B. melaninogenicus* group and *F. nucleatum* has been associated with fimbriae-like agglutinins on the cell surface. Encapsulated *B. fragilis* strains can also adhere to peritoneal surfaces more effectively than nonencapsulated strains. The role adherence plays in virulence is not known, although presumably this interferes with macrophage mediated bacterial clearance.

Enzymes

A variety of enzymes have been associated with virulent *Bacteroides* and *Fusobacterium* species (Table 15-1). Although many of these enzymes are found in both virulent and avirulent isolates, the ability potentially to cause tissue destruction and inactivate immunoglobulins most likely plays an important role in anaerobic infections.

Oxygen Tolerance

Anaerobes capable of causing disease are generally able to tolerate exposure to oxygen. Catalase and superoxide dismutase, which inactivate hydrogen peroxide and the superoxide free radicals (O_2^-) respec-

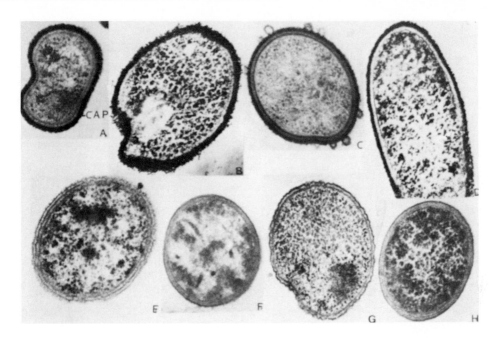

FIGURE 15-2   Electron micrograph of various *bacteroides* species stained with ruthenium red for detection of capsules. A prominent capsule *(CAP)* is observed for four strains of *Bacteroides fragilis* (**A-D**). No capsule is observed for the other strains of the *Bacteroides fragilis* group: *Bacteroides vulgatus* (**E**), *Bacteroides ovatus* (**F**), *Bacteroides distasonis* (**G**), and *Bacteroides thetaiotaomicron* (**H**). Subsequent studies have demonstrated capsules can be present rarely in these other *B. fragilis* group organisms, as well as in members of the *Bacteroides melaninogenicus* group. (From Kasper, DL, et al: J Infect Dis 136:75, 1977, published by the University of Chicago.)

Table 15-1   Distribution of Enzymes Associated With Pathogenic Bacteroidaceae

Enzymes	B. fragilis Group	B. melaninogenicus Group	B. bivius	Bacteroides, Other Sp.	Fusobacterium Sp.
Hyaluronidase	X				
Collagenase	X	X		X	
Condroitin sulfatase	X				
Fibrinolysin	X	X			
Neurominidase	X	X	X	X	
Heparinase	X				
DNase	X	X		X	X
IgM protease		X			
IgG protease		X			
IgA protease		X			
Superoxide dismutase	X	X			X
β-lactamase	X	X	X	X	X

A variety of extracellular and cell-bound enzymes have been detected in many *Bacteroides* and *Fusobacterium* species. Although pathogenicity cannot be attributed to any one enzyme, it is important to note that the most virulent *Bacteroides* contain many enzymes that can facilitate tissue destruction and afford protection against humoral immunity, oxygen toxicity, and antibiotic therapy.

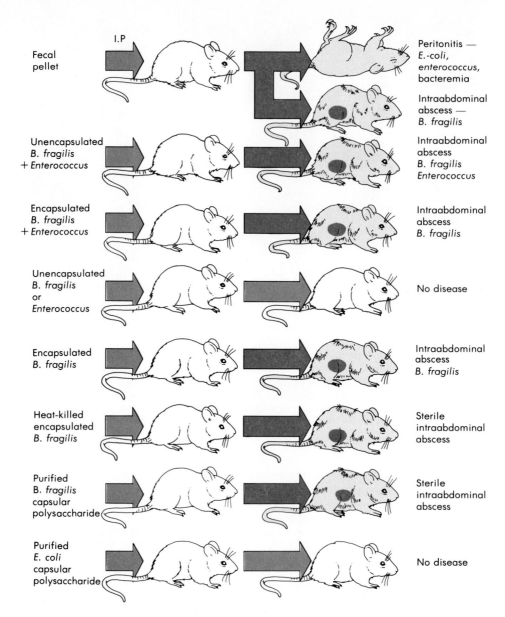

FIGURE 15-3   Demonstration of the role of encapsulation in *Bacteroides fragilis* virulence. In a series of experiments Kasper and associates were able to demonstrate that the capsule present in *Bacteroides fragilis* is responsible for abscess formation, which is characteristic of these infections. In the presence of multiple other bacterial species, abscess formation does not depend on encapsulated strains. However, in the absence of other bacteria (e.g., *Enterococcus* as depicted in this figure), abscess formation is rare unless the *B. fragilis* strain possesses a capsule. Additional experiments demonstrated that sterile abscesses will develop if nonviable encapsulated *B. fragilis* strains or purified capsular material is injected intraperitoneally into mice or Wistar rats. This is not observed with rats inoculated with *Escherichia coli* capsular material or encapsulated *Streptococcus pneumoniae*.

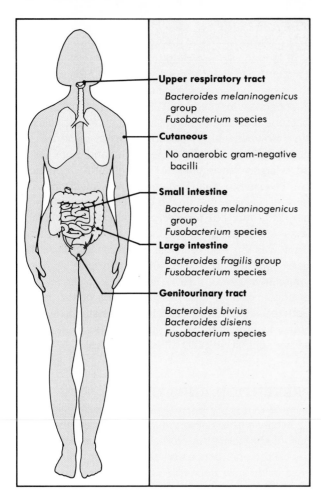

Upper respiratory tract

Bacteroides melaninogenicus
group
Fusobacterium species

Cutaneous

No anaerobic gram-negative
bacilli

Small intestine

Bacteroides melaninogenicus
group
Fusobacterium species

Large intestine

Bacteroides fragilis group
Fusobacterium species

Genitourinary tract

Bacteroides bivius
Bacteroides disiens
Fusobacterium species

FIGURE 15-4   Bacteroidaceae colonization of human body sites.

Table 15-2   Frequency With Which Anaerobes Are Associated With Human Disease

Infection	Anaerobes Involved (%)*
Bacteremia	10-20
Central nervous system	
Brain abscess	89
ENT-dental	52
Thoracic	
Aspiration pneumonia	93
Lung abscess	93
Empyema (nonsurgical)	76
Intraabdominal	
General infection	86
Liver abscess	50-100
Appendicitis	96
Obstetric-gynecologic	
Vulvovaginal abscess	74
Pelvic abscess	92

Summarized by Finegold, SM: Anaerobic Bacteria in Human Disease, New York, 1977, Academic Press.
*Anaerobic cocci and bacilli are commonly associated with a variety of endogenous human disease. Most of these infections are polymicrobic, involving a mixture of aerobic and anaerobic bacteria.

tively, is present in pathogenic strains of *Bacteroides* and *Fusobacterium.*

EPIDEMIOLOGY

The anaerobic gram-negative bacilli form a major component of the human microbial flora. Anaerobes outnumber aerobic bacteria by roughly a factor of 10 fold in the oropharynx, a factor of 100 in the urogenital tract, and a factor of 1,000 in the gastrointestinal tract; members of the family Bacteroidaceae are well-represented at each site (Figure 15-4). Although *Bacteroides fragilis* is commonly associated with pleuropulmonary, intraabdominal, and genital infections, the organism comprises less than 1% of the colonic flora and is rarely isolated from the oropharynx or gen-

ital tract of healthy persons unless highly selective techniques are used. The prominent role this organism plays in human disease is attributed to its enhanced virulence.

CLINICAL SYNDROMES

The cardinal features of infections with the Bacteroidaceae are: (1) endogenous infections, (2) polymicrobic mixture of organisms, and (3) abscess formation. As the infectious process develops, the endogenous bacterial population is able to spread by trauma or disease from normally colonized mucosal surfaces to sterile tissues or fluids. Although many organisms may be initially involved in the infection, the oxygen-susceptible and avirulent organisms are rapidly cleared by the host's defense mechanisms, with a resultant proliferation of virulent organisms and destruction of host tissue. Encapsulated bacteria such as *B. fragilis* are prominently featured in these infections and, as would be expected, are usually associated with abscess formation. Thus the Bacteroidaceae most commonly cause purulent infections at body sites proximal to colonized mucosal surfaces (Table 15-2). *B. fragilis* is isolated in 15% to 20% of pleuropulmonary infections, two thirds or more of suppurative pelvic infections, and virtually all intraabdominal infections. More than 75% of all Bacteroidaceae bactermias are due to *B. fragilis*. *B. melaninogenicus* group and *Fusobacterium* species isolates are commonly isolated in pleuropulmonary or central nervous system infections. *Bacteroides bivius* and *B. disiens* are frequently isolated in genital infections.

LABORATORY DIAGNOSIS

Microscopy

Microscopic examination of specimens from suspected anaerobic infections can be useful. Although the bacteria may stain faintly, the presence of multiple bacterial species and their characteristic morphology (e. g., pleomorphic, fusiform bacilli) can be useful preliminary information. Fluorescein-labeled antibodies directed against the *B. fragilis* and *B. melaninogenicus* organism groups are available for more specific microscopic identification. Cross-reactions with other bacteria are uncommon with these reagents.

Culture

The in vitro recovery of most significant anaerobes is not difficult because the organisms are relatively aerotolerant. However, specimens should be collected and transported to the laboratory in an oxygen-free system, promptly inoculated onto specific media for the recovery of anaerobes, and incubated in an anaerobic environment. Most Bacteroidaceae grow within 2 days, although some (e. g., *B. melaninogenicus* group, *Fusobacterium*) are more fastidious and can require prolonged incubation.

Biochemical Identification

Identification of most Bacteroidaceae is relatively simple, but definitive results may require 2 days or more. Detection of metabolic by-products (organic acids) by gas chromatography has proved to be a useful, simple technique to supplement biochemical testing or for the preliminary analysis of clinical specimens. However, the detection of organic acids by this technique in a specimen from a polymicrobic infection has limited utility.

TREATMENT

B-lactamases are produced by virtually all members of the *B. fragilis* group, as well as 50% to 75% of the other *Bacteroides* species and a small number of *Fusobacterium* species. These enzymes render the bacteria resistant to penicillin and many cephalosporins. However, high concentrations of some penicillins (e. g., carbenicillin, piperacillin) and other selected B-lactam antibiotics (e. g., cefoxitin, imipenem) can be used to treat Bacteroidaceae infections. Tetracycline was at one time highly active against the *B. fragilis* group, but resistance is now common. Likewise, clindamycin resistance in this group of organisms was previously never observed but now has been reported in 5% to 40% of the isolates (mean approximately 7% to 10%). Resistance to clindamycin is uncommon in the other members of the family. Both tetracycline and clindamycin resistance is mediated by transferrable plasmids. Metronidazole is active against virtually all Bacteroidaceae.

PREVENTION AND CONTROL

Because the Bacteroidaceae form an important part of the normal microbial flora and infections are due to the endogenous spread of the organisms, disease is virtually impossible to control. However, it is important to recognize that diagnostic or surgical procedures that disrupt the natural barriers surrounding the mucosal surfaces can introduce these organisms into normally sterile sites. If these barriers are invaded, prophylactic treatment with antibiotics may be indicated.

BIBLIOGRAPHY

Carlsson J, Wrethen J, and Beckman G: Superoxide dismutase in *Bacteroides fragilis* and related *Bacteroides* species, J Clin Microbiol 6:280-284, 1977.

Cuchural GJ, Tally, FP, Jacobus NV, et al: Susceptibility of the *Bacteroides fragilis* group in the United States: analysis by site of isolation, Antimicrob Agents Chemother 32:717-722, 1988.

Finegold SM: Anaerobic bacteria in human disease, New York, 1988, Academic Press.

Hofstad T: Pathogenicity of anaerobic gram-negative rods: possible mechanisms, Rev Infect Dis 6:189-199, 1984.

Hofstad T and Sveen K: The chemotactic effect of *Bacteroides fragilis* lipopolysaccharide, Rev Infect Dis 1:342-346, 1979.

Holdeman LV, Good IJ, and Moore WEC: Human fecal flora: variation in bacterial composition within individuals and a possible effect of emotional stress, Appl Environ Microbiol 31:359-375, 1976.

Kasper DL, Hayes ME, Reinap BG, et al: Isolation and identification of encapsulated strains of *Bacteroides fragilis*, J Infect Dis 136:75-81, 1977.

Kasper DL, Onderdonk AB, Pold BF, and Bartlett JG: Surface antigens as virulence factors in infection with *Bacteroides fragilis*, Rev Infect Dis 1:278-288, 1979.

Mansheim BJ, Onderdonk AB, and Kasper DL: Immunochemical characterization of surface antigens of *Bacteroides melaninogenicus*, Rev Infect Dis 1:26302275, 1979.

Onderdonk AB, Kasper DL, Cisneros RL, and Bartlett JG: The capsular polysaccharide of *Bacteroides fragilis* as a virulence factor: comparison of the pathogenic potential of encapsulated and unencapsulated strains, J Infect Dis 136:82-89, 1977.

16

Miscellaneous Gram-Negative Bacilli

A number of gram-negative bacilli can cause uncommon but medically important diseases (Table 16-1). The following is a discussion of seven such organisms.

CARDIOBACTERIUM

Microbial Physiology and Structure

Carbiobacterium hominis, named for this bacterium's predilection to cause endocarditis in humans, is the only member of the genus. *C. hominis* isolates are nonmotile, characteristically small (1×1 to $2 \mu m$) but sometimes pleomorphic, gram-negative bacilli.

Epidemiology

Endocarditis caused by *C. hominis* is uncommon, with less than 50 cases reported in the literature, although many cases are unreported or undiagnosed. This is due to the low virulence of this organism and to the difficulties associated with its in vitro isolation. *C. hominis* is present in the upper respiratory tract of almost 70% of healthy individuals.

Clinical Syndrome

Endocarditis is the only human disease caused by *C. hominis.* Most patients with *C. hominis* endocarditis have a history of dental disease or dental procedures before clinical symptoms developed, as well as preexisting heart disease. The organism is able to enter the bloodstream from the oropharynx, adhere to the damaged heart tissue, and then slowly multiply.

The course of disease is insidious and subacute, with a history of symptoms (fatigue, malaise, low-grade temperature) for months before the patient seeks medical care. Complications are rare, and complete recovery following appropriate antibiotic therapy is common.

Laboratory Diagnosis

Isolation of *C. hominis* from blood cultures confirms the diagnosis of endocarditis. The organism grows slowly in culture and requires 1 to 2 weeks before growth is detected, which is why infections with these organisms have not been confirmed in some patients. *C. hominis* appears in broth cultures as discrete clumps that can be easily overlooked when cultures are examined. The organism requires enhanced carbon dioxide and humidity for growth on agar media, with pinpoint (1 mm) colonies seen on blood or

Table 16-1 Miscellaneous Gram-Negative Bacilli and Their Diseases

Organism	Disease
Cardiobacterium hominis	Endocarditis
Eikenella corrodens	Opportunistic infections
Flavobacterium meningo-septicum	Opportunistic infections
Calymmatobacterium granulomatis	Granuloma inguinale
Streptobacillus moniliformis	Rat-bite fever
Spirillum minor	Rat-bite fever
Cat scratch bacillus	Cat scratch disease

chocolate agar plates after 3 days of incubation. The organism does not grow on MacConkey agar or other selective media commonly used for gram-negative bacilli. *C. hominis* can be readily identified by its growth properties, microscopic morphology, and reactivity in biochemical tests.

Treatment

C. hominis is susceptible to multiple antibiotics, and most infections are successfully treated with either penicillin or ampicillin for 2 to 6 weeks.

Prevention and Control

C. hominis endocarditis in individuals with preexisting heart disease is prevented by maintenance of good oral hygiene and use of antibiotic prophylaxis at the time of dental manipulations. A long-acting penicillin is effective prophylaxis, but erythromycin should not be used because resistance is common.

EIKENELLA

Microbial Physiology and Structure

Eikenella corrodens is a moderate-sized (0.2 × 2 μm), nonmotile, nonsporeforming, facultatively anaerobic, gram-negative bacillus. The organism is named after Eikin, who characterized the bacterium, and the ability of the organism to pit or ''corrode'' agar.

Epidemiology

E. corrodens is a normal inhabitant of the upper respiratory tract, although its fastidious growth characteristics make it difficult to detect without specific selective culture media. It is an opportunistic pathogen, causing infections in immunocompromised patients and in patients with diseases or trauma of the oral cavity.

Clinical Syndromes

E. corrodens has been associated with human bites or fist-fight injuries (frequently complicated with septic arthritis or osteomyelitis if initially treated improperly), sinusitis, meningitis, brain abscesses, pneumonia, lung abscesses, and endocarditis. Because most infections originate from the oropharynx, polymicrobial mixtures of aerobic and anaerobic bacteria are frequently present. The etiologic role of *Eikenella* in these infections is established by the failure to cure the infection unless therapy specific for *Eikenella* is administered.

Laboratory Diagnosis

E. corrodens is a slow-growing, fastidious organism that requires 5% to 10% carbon dioxide for isolation. Small (0.5 to 1 mm) colonies are observed on common laboratory media after 48 hours of incubation. Pitting in agar is a useful differential characteristic but is seen in less than half of all isolates.

Treatment

E. corrodens is susceptible to penicillin and ampicillin, but resistant to oxacillin, first generation cephalosporins, clindamycin, and the aminoglycosides. Unfortunately, this means *E. corrodens* is resistant to many antibiotics that are empirically selected to treat bite wound infections.

FLAVOBACTERIUM

Microbial Physiology and Structure

The genus *Flavobacterium* consists of seven species that can cause opportunistic human diseases, of which *Flavobacterium meningosepticum* is the best known. The organisms are small (0.5 × 1 to 2 μm), nonmotile, aerobic, gram-negative bacilli that grow on common laboratory media after 24 hours of incubation.

Epidemiology

Flavobacteria are widely disseminated in nature, found in water, wet soil, and moist reservoirs in the hospital (e.g., sinks, faucets, humidifiers, respiratory therapy equipment). The organism is not part of the normal human microbial flora.

Clinical Syndromes

Flavobacterium meningosepticum and other *Flavobacterium* species are opportunistic pathogens in immunocompromised patients. As the name implies, *F. meningosepticum* has been associated with meningitis and septicemia, particularly in neonates. Other infections have developed following exposure to the organism in contaminated solutions.

Laboratory Diagnosis

Flavobacterium spieces are relatively easy to isolate in culture and can be differentiated from related organisms by reactivity in biochemical tests.

Treatment, Prevention, and Control

Flavobacterium species are resistant to many antibiotics, including aminoglycosides, penicillins, and cephalosporins. Effective therapy requires selection of specific agents that have proven in vitro activity against *Flavobacterium* and can reach therapeutic concentrations at the site of infection. Infection can be prevented through established infection control practices that reduce or eliminate exposure to contaminated water supplies and solutions.

CALYMMATOBACTERIUM

Microbial Physiology and Structure

Calymmatobacterium (Donovania) *granulomatis* is the etiologic agent of **granuloma inguinale,** a gran-

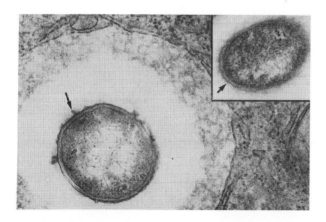

FIGURE 16-1   Electron micrograph of *Calymmatobacterium granulomatis* within a mononuclear cell phagosome. Note the prominent capsule surrounding the cell, surface projections (*arrow*), and typical gram-negative trilaminar membrane structure (*insert*). (From Kuberski T et al: J Infect Dis 142:744-749, 1980, published by the University of Chicago.)

ulomatous disease of the genitalia and inguinal area. The organism was originally discovered by Donovan and subsequently renamed to reflect its appearance in tissues as encapsulated bacilli (**calymma** is Greek for sheathed; ''sheathed bacterium''). *C. granulomatis* is difficult to culture in vitro and is usually detected by staining infected tissues with a Wright or Giemsa stain. The organisms appear as small (0.5 to 1.0 × 1.5 μm) bacilli within phagosomes of mononuclear phagocytes. From 1 to 25 bacteria per phagocytic cell can be seen; a prominent capsule surrounds the organisms (Figure 16-1).

Epidemiology

Granuloma inguinale is a rare disease in the United States but is common in tropical areas such as the Caribbean and New Guinea. Transmission after repeated exposure can occur by sexual intercourse or nonsexual trauma to the genitalia.

Clinical Syndrome

After a prolonged incubation period of weeks to months subcutaneous nodules appear on the genitalia or in the inguinal area. The nodules subsequently break down, revealing one or more painless granulomatous lesions that can extend and coalesce.

Diagnosis

Laboratory confirmation of granuloma inguinale is made by the histopathologic examination of infected tissue. Pathognomonic ''Donovan bodies'' are observed within mononuclear phagocytes.

Treatment, Prevention, and Control

Tetracycline, ampicillin, or trimethoprim-sulfamethoxazole have been used successfully for treatment. Antibiotic prophylaxis for prevention and control has not been proven effective.

STREPTOBACILLUS AND SPIRILLUM

Microbial Physiology and Structure

Streptobacillus moniliforms and *Spirillum minor* are the two etiologic agents of distinct diseases

Table 16-2	Comparison of *Streptobacillus* and *Spirillum* Infections	
	Streptobacillus moniliformis	***Spirillum minor***
Distribution	Worldwide, including United States	Worldwide, primarily Asia
Reservoir	Rats and other small rodents	Rat and other small rodents
Transmission	Rat bite or contact; consumption of contaminated liquids or food	Rat bite or contact
Disease	Rat-bite fever	Rat-bite fever
Incubation period	1-22 days	7-28 days
Clinical presentation and course	Abrupt; high fever; chills, headache, myalgias; rash, arthritis/arthralgia; recurrent fevers if untreated	Abrupt onset; fever, chills, rash, lymphangitis, lymphadenopathy; recurrent fevers if untreated
Mortality	10% if untreated	6% if untreated
Treatment	Penicillin	Penicillin
Diagnosis	Culture, serology, false-positive syphilis serology	Darkfield, animal inoculation, false-positive syphilis serology

referred to collectively as **rat-bite fever** (Table 16-2). *S. moniliformis* is a long (0.1 to 0.5 × 1 to 5 μm), thin, pleomorphic, gram-negative bacillus that may align in long chains similar to streptococci. *S. minor* is a coiled gram-negative bacillus with polar flagella that moves by darting motions.

Epidemiology

Both *Streptobacillus* and *Spirillum* are found in the oropharynx of rats and other small rodents, as well as transiently in animals that feed on rodents (e.g., dogs, cats). Turkeys exposed to rats and mice have also been implicated in *Streptobacillus* infections.

Clinical Syndrome

Although the epidemiology and clinical syndrome of recurrent fevers are similar for both *Streptobacillus* and *Spirillum* infections, distinct differences have been observed. *Streptobacillus* infections have a shorter incubation period, and the clinical course that includes myalgias, arthralgias, and frank arthritis is usually absent in *Spirillum* infections. Ulceration at the bite site, with lymphadenopathy and lymphangitis, is observed in *Spirillum* infections. Recurrent febrile episodes are observed with both organisms in untreated disease.

Laboratory Diagnosis

For patients seen in the United States with suspected rat-bite fever, *Streptobacillus moniliformis* can be cultured in standard broth media or on agar media supplemented with horse or rabbit serum. The organism is slow-growing and 3 or more days of incubation are required for isolation. Serologic tests for the detection of antibodies against *Streptobacillus* antigens are also available in reference laboratories. *Spirillum minor*, which is more common in countries outside the United States, has not been cultured in vitro. Detection of *S. minor* infections is by darkfield examination of blood or infected tissues or by animal inoculations. *S. minor* can be detected in the blood of rodents 1 to 3 weeks after intraperitoneal inoculation with the clinical specimen.

Treatment

Penicillin is the antibiotic of choice for treating rat-bite fever. Tetracycline can be used in penicillin-allergic patients.

CAT SCRATCH BACILLUS

Cat scratch disease is a benign infection characterized by chronic regional adenopathy of the lymph nodes draining the site of a cat scratch or bite. After repeated efforts to isolate the etiologic agent, recent investigations have observed small coccobacilli in the infected tissue and have isolated slow-growing gram-negative bacilli.

Epidemiology

Cat scratch disease is primarily a pediatric infection associated with a history of contact with a cat. Presumably the organism is part of the normal oropharyngeal microbial flora of cats, although this remains to be demonstrated. Most infections occur in the fall and winter, although this seasonality is less obvious in warm climates. Person-to-person transmission has not been seen.

Clinical Syndrome

After a 1-week incubation period a papule or pustule develops at the bite or scratch site. This is followed by the development of regional lymphadenopathy, which can be very prominent and persist for 3 months or more. Systemic symptoms can occur but are rare. The hyperplastic lymph nodes are characterized by granuloma formation followed by abscess development.

Laboratory Diagnosis

The diagnosis is suggested by a history of contact with a cat, clinical presentation, and the absence of common bacteria in cultures of aspirated lymph nodes. The diagnosis is confirmed by observation of the bacterium in histopathologic specimens (using the Warthin-Starry silver stain) or in vitro isolation. Culture of the organism is difficult and has been accomplished in only a few laboratories.

Treatment

Antibiotic therapy has generally not affected the course of disease. This may reflect the bacterium's resistance to commonly used antibiotics (e.g., penicillin, ampicillin, cephalothin, tetracycline, clindamycin). However, the organism is susceptible to cefoxitin, cefotaxime, and the aminoglycosides, and these antibiotics may prove to be effective therapeutic agents.

BIBLIOGRAPHY

Cabrera HA and Davis GH: Epidemic meningitis of the newborn caused by flavobacteria. I. Epidemiology and bacteriology, Am J Dis Child 101:289-295, 1961.

English CK, Wear DJ, Margileth AM, et al: Cat-scratch disease: isolation and culture of the bacterial agent, JAMA 259:1347-1352, 1988.

Kuberski T, Papadimitriou JM, and Phillips P: Ultrastructure of *Calymmatobacterium granulomatis* in lesions of granuloma inguinale, J Infect Dis 142:744-749, 1980.

Stamm WE, Cotella JJ, Anderson RL, and Dixon RE. Indwelling arterial catheters as a source of nosocomial bacteremia: an outbreak caused by *Flavobacterium* species, N Engl J Med 292:1099-1102, 1975.

Stoloff AL and Gillies ML: Infections with *Eikenella corrodens* in a general hospital: a report of 33 cases, Rev Infect Dis 8:50-53, 1986.

Suwanagool S, Rothkopf MM, Smith SM, et al: Pathogenicity of *Eikenella corrodens* in humans, Arch Intern Med 143:2265-2268, 1983.

Wormser GP and Bottone EJ: *Cardiobacterium hominis:* review of microbiologic and clinical features, Rev Infect Dis 5:680-691, 1983.

CHAPTER 17

Bacillus

The medically important, aerobic and facultatively anaerobic, gram-positive bacilli that form spores are classified in the genus *Bacillus*. *Bergey's Manual of Systematic Bacteriology* lists 34 *Bacillus* species; however, only *Bacillus anthracis* and *Bacillus cereus* are significant human pathogens (Table 17-1). Other *Bacillus* species can cause infections, but only in severely immunocompromised patients or when introduced traumatically into normally sterile tissues. Isolation of *B. anthracis*, the causative agent of **anthrax,** is always considered clinically significant, whereas *B. cereus,* which can cause food poisoning and panophthalmitis, is commonly found in the environment and is frequently isolated as a contaminant.

BACILLUS ANTHRACIS

Microbial Physiology and Structure

Bacillus anthracis is a large (1 × 3 to 5 μm) bacillus that occurs as single or paired bacilli in clinical specimens and in long serpentine chains and clumps in culture (Figure 17-1). Although spores are readily observed in 2- or 3-day-old cultures, they are not seen in clinical specimens. A prominent polypeptide capsule (consisting of glutamic acid) is observed in clinical specimens, but it is not produced in vitro unless special growth conditions are used. In addition to the capsular antigen, a polysaccharide cell wall somatic antigen and a toxin are associated with *B. anthracis* (Box 17-1). **Anthrax toxin** consists of three antigenically distinct, heat-labile components: protective antigen, lethal factor, and edema factor. Although

Table 17-1 *Bacillus* Species and Their Diseases	
Organism	**Disease**
Bacillus anthracis	Anthrax
Bacillus cereus	Gastroenteritis
	Emetic form
	Diarrheal form
	Panophthalmitis
	Opportunistic infections
Other *Bacillus* species	Opportunistic infections

none of the components are active alone, the combination of protective antigen with either of the other two components has toxic properties.

Pathogenesis

The two major factors responsible for *B. anthracis* virulence are presence of the capsule and toxin production. The capsule is antiphagoctyic, and antibod-

Box 17-1 *Bacillus anthracis* Major Antigens
Glutamic acid capsule
Polysaccharide somatic antigen
Anthrax toxin
Protective antigen
Lethal factor
Edema factor

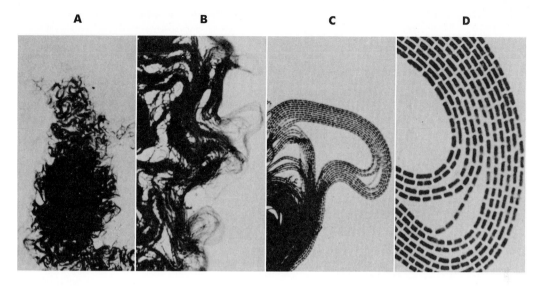

FIGURE 17-1   The surface of *Bacillus anthracis* colonies (**A** and **B**) resembles a tangled web of hair and has been referred to a "medusa head" appearance. On microscopic examination the individual hairs are composed of parallel chains of bacilli (**C** and **D**). This alignment is observed in aspirates from clinical specimens where individual bacilli are most commonly seen. (From Stein E, Ann NY Acad Sci 48:507, 1947.)

ies directed against the capsule are not protective. Only one capsular type has been identified, presumably because the capsule is composed of only glutamic acid. The anthrax toxin protective antigen produces edema in experimental animals when combined with the edema factor and death when combined with the lethal factor. This toxin can be detected in edematous fluid collected from patients with anthrax.

Epidemiology

Bacillus anthracis is an organism found in soil and on vegetation (Figure 17-2). The ability to form spores allows the organism to survive for years under adverse conditions without the need to replicate. Anthrax is a Disease primarily of herbivores; humans may be accidentally infected by exposure to contaminated animals or animal products. Anthrax is rarely seen in the United States, and most infections are reported in Iran, Turkey, Pakistan, and Sudan.

Human disease is acquired by one of three routes: inoculation, inhalation, or ingestion. Approximately 95% of anthrax infections are due to the inoculation of *Bacillus* spores through exposed skin surfaces, either from contaminated soil or infected animal products such as hides, goat hair, or wool. Inhalation anthrax, also called **Woolsorter's disease,** results from inhalation of *B. anthracis* spores during the processing of goat hair. Ingestion anthrax is very rare in humans but is a common route of infection in herbivores. Person-to-person transmission does not occur.

Clinical Syndromes

Cutaneous anthrax is characterized by the development of a painless papule at the site of inoculation that rapidly progresses to an ulcer surrounded by vesicles and then to a necrotic eschar (Figure 17-3). Massive edema due to the anthrax toxin and systemic signs can develop. Mortality in patients with untreated cutaneous anthrax is 20%.

Inhalation anthrax is initially seen as a viral respiratory illness and then rapidly progresses to severe pulmonary disease. Mortality is high even in appropriately treated patients, because the disease is usually not suspected until the course is irreversible.

Gastrointestinal anthrax is a very rare human disease with varied clinical presentations. Mesenteric adenopathy, hemorrhage, and ascites production are all reported and associated with high mortality.

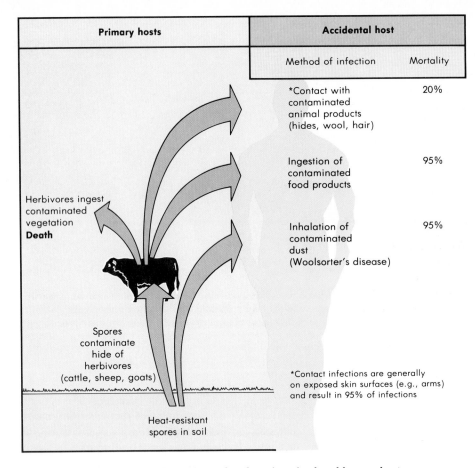

Primary hosts	Accidental host	
	Method of infection	Mortality
	*Contact with contaminated animal products (hides, wool, hair)	20%
	Ingestion of contaminated food products	95%
	Inhalation of contaminated dust ("Woolsorter's disease)	95%

Herbivores ingest contaminated vegetation
Death

Spores contaminate hide of herbivores (cattle, sheep, goats)

Heat-resistant spores in soil

*Contact infections are generally on exposed skin surfaces (e.g., arms) and result in 95% of infections

FIGURE 17-2 Epidemiology of anthrax in animal and human hosts.

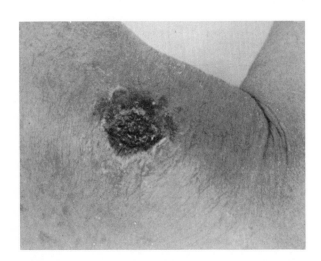

FIGURE 17-3 Large necrotic eschar on forearm seen in cutaneous anthrax after 14 days. (From Laforce FM, *Bacillus anthracis* (anthrax). In Mandell GL, Douglas RG, and Bennett JE: Principles and practices of infectious diseases, ed 2, Churchill Livingstone, New York, 1985.)

Laboratory Diagnosis

B. anthracis can be readily detected by the microscopic examination and culture of cutaneous papules or ulcers. Large gram-positive bacilli, without spores, are seen in the tissue. Staining with specific fluorescein-labeled antibodies can confirm the clinical suspicion, although these reagents are not available in most clinical laboratories. *B. anthracis* will grow on nonselective laboratory media, with the appearance of nonhemolytic, rapidly growing, adherent colonies. The absence of hemolysis, sticky consistency of the colonies, and the microscopic appearance of serpentine chains of bacilli ("medusa head") are useful characteristics for distinguishing *B. anthracis* from other *Bacillus* species. Definitive identification of *B. anthracis* requires selected biochemical tests, including demonstration that *B. anthracis* is nonmotile.

<table>
<tr><td>

Box 17-2   *Bacillus cereus*
Major Virulence Antigens

Heat-labile enterotoxin
Heat-stable enterotoxin
Cerelysin
Phospholipase C

</td></tr>
</table>

Treatment, Prevention, and Control

Penicillin is the antibiotic of choice for treating anthrax, with tetracycline or chloramphenicol used as alternative agents for penicillin-allergic patients. Control of human disease requires control of animal anthrax. This involves vaccination of animal herds in endemic regions and burning or burial of animals who die of anthrax. Vaccination is an effective control measure in animal herds and has been used to protect humans in areas with endemic disease. Complete eradication of anthrax is unlikely because the spores of the organism can exist for many years in soil.

BACILLUS CEREUS AND OTHER *BACILLUS* SPECIES

Pathogenesis

Gastroenteritis caused by *Bacillus cereus* is mediated by one of two enterotoxins (Box 17-2). The heat-stable enterotoxin is responsible for disease characterized by vomiting **(emetic form),** and the heat-labile enterotoxin causes the **diarrheal form** of *B. cereus* disease. The heat-labile enterotoxin is similar to the enterotoxin produced by *Escherichia coli* and *Vibrio cholera;* each stimulates the adenylate cyclase–cyclic AMP system in intestinal epithelial cells and can be assayed by measuring fluid accumulation in rabbit ileal loops inoculated with the toxin. The mechanism of action of the heat-stable enterotoxin is unknown.

The pathogenesis of *B. cereus* panophthalmitis is also incompletely defined. At least three toxins have been implicated: **necrotic toxin** (a heat-labile enterotoxin), **cerelysin** (a potent hemolysin named after the species), and **phospholipase C** (a potent lecithinase). It is likely that the rapid destruction of the eye that is characteristic of *B. cereus* infections is the result of the interaction of these toxins and other unidentified factors.

Epidemiology

Bacillus cereus and other *Bacillus* species are ubiquitous organisms, present in virtually all environmental sites. The isolation of the bacteria from clinical specimens, in the absence of characteristic disease, usually represents insignificant contamination.

Clinical Syndromes

As mentioned previously, *Bacillus cereus* is responsible for two forms of food poisoning (Table 17-2): vomiting disease (emetic form) and diarrheal disease (diarrheal form). The emetic form is associated with consumption of contaminated rice. During the initial cooking of the rice, the vegetative bacilli are killed but the heat-resistant spores survive. If the rice is not refrigerated, the spores germinate and the bacilli can multiply rapidly, releasing the heat-stable enterotoxin that is not destroyed when the rice is reheated. After ingestion of the enterotoxin and a short incubation period, the patient develops a disease of short duration characterized by vomiting, nausea, and abdominal cramps. Fever and diarrhea are generally absent. In contrast, the diarrheal form of

Table 17-2   *Bacillus cereus* Food Poisoning

	Emetic Form	Diarrheal Form
Implicated food	Rice	Meat, vegetables
Incubation period	<6 hours (mean, 2 hours)	>6 h (mean, 9 hours)
Symptoms	Vomiting, nausea, abdominal cramps	Diarrhea, nausea, abdominal cramps
Duration	8-10 hours (mean, 9 hours)	20-36 hours (mean, 24 hours)
Enterotoxin	Heat stable	Heat labile

B. cereus food poisoning is associated with consumption of contaminated meat or vegetables. There is a longer incubation period during which the organism multiplies *in situ* and produces a heat-labile toxin, and then the diarrhea, nausea, and abdominal cramps develop. This form of disease generally lasts for 1 day or longer.

Bacillus cereus is one of the most common causes of posttraumatic eye infections. The source of the organisms can be either soil contamination of the object penetrating the eye or inoculation of organisms colonizing the eye surface into the orbit. *Bacillus* panophthalmitis is a rapidly progressive disease that almost universally ends in complete loss of light perception within 48 hours of the injury. Massive destruction of the vitreal and retinal tissue is observed.

Other infections seen with *B. cereus* and other *Bacillus* species include endocarditis (most commonly in drug abusers), as well as pneumonitis, bacteremia, and meningitis in severely immunosuppressed patients.

Laboratory Diagnosis

As with *Bacillus anthracis,* the other *Bacillus* species can be readily grown in the laboratory. For confirmation of food-borne disease, the implicated food (e.g., rice, meat, vegetable) should be cultured. Isolation of the organism from the patient should not be attempted because fecal colonization is commonplace.

Treatment, Prevention, and Control

Because of the short and uncomplicated course of *Bacillus cereus* gastroenteritis, symptomatic treatment is adequate. Treatment of other *Bacillus* infections is complicated by the rapid, progressive course of the infections and the high incidence of multidrug resistance that is observed with these organisms (most isolates are resistant to penicillins and cephalosporins, as well as other antibiotics). Clindamycin, vancomycin, and the aminoglycosides can be used, although susceptibility must be confirmed by in vitro testing. Food poisoning can be prevented by the proper refrigeration of food products after cooking and before serving.

BIBLIOGRAPHY

Davey RT, Jr and Tauber WB: Posttraumatic endophthalmitis: the emerging role of *Bacillus cereus* infection, Rev. Infect Dis 9:110-123, 1987.

Ihde DC and Armstrong D: Clinical spectrum of infection due to *Bacillus* species, Am J Med 55:839-845, 1973.

Shamsuddin D, Tuazon CU, Levy C, and Curtin J: *Bacillus cereus* panophthalmitis: source of the organisms, Rev Infect Dis 4:97-103, 1982.

Terranova W and Blake PA: *Bacillus cereus* food poisoning, N Engl J Med 298:143-144, 1978.

Tuazon CU, Murray HW, Levy C, et al: Serious infections from *Bacillus* sp, JAMA 241:1137-1140, 1979.

Van Ness GB: Ecology of anthrax, Science 172:103-109, 1971.

CHAPTER 18

Clostridium

Most medically important, anaerobic, gram-positive, spore-forming bacilli are classified in the genus *Clostridium* (Table 18-1). Unfortunately, the identification of many of these organisms poses a problem. The genus consists of more than 80 species that have diverse biochemical properties and limited genetic relatedness. Although most members of the genus are strict anaerobes, some are aerotolerant (e. g., *C. tertium, C. histolyticum*) and can grow on agar media exposed to air. Some clostridia can appear gram-negative (e. g., *C. ramosum*), and spores are normally not observed microscopically in some species (*C. per-*

fringens, C. ramosum). Thus classification of an isolate into the genus *Clostridium* must be accomplished by a combination of diagnostic tests, including demonstration of spores, optimal growth anaerobically, and a complex pattern of biochemical reactivity.

The organisms are ubiquitous, present in soil, water, and as part of the normal microbial flora in the gastrointestinal tract of animals and humans. Clostridia are well-recognized human pathogens, with a clearly documented history of causing such diseases as tetanus *(C. tetani)*, botulism *(C. botulinum)*, and gas gangrene (*C. perfringens, C. novyi, C. septicum, C. bifermentans,* and others). Despite the notoriety of these diseases, we now know that clostridia are more commonly associated with skin and soft tissue infections, food poisoning, and pseudomembranous colitis. Their remarkable capacity for causing diseases is attributed to their ability to survive adverse environmental conditions by spore formation, rapid rate of growth in a nutritionally enriched, oxygen-deprived environment, and production of numerous histiolytic toxins, enterotoxins, and neurotoxins.

Table 18-1 Common *Clostridium* Species Associated With Human Disease

Organism	Disease
C. perfringens	Bacteremia; myonecrosis (gas gangrene); soft tissue infections (e. g., cellulitis, fasciitis); food poisoning; enteritis necroticans
C. tetani	Tetanus
C. botulinum	Food-borne botulism, infant botulism, wound botulism
C. difficile	Antibiotic-associated diarrhea, antibiotic-associated pseudomembranous colitis
Other *Clostridium* species (e. g., *C. septicum, C. ramosum, C. novyi, C. bifermentans*)	Bacteremia, myonecrosis, soft tissue infections

CLOSTRIDIUM PERFRINGENS

Microbial Physiology and Structure

Clostridium perfringens, the clostridial species most frequently isolated from clinical specimens, can either be associated with simple colonization or can cause severe, life-threatening disease. *C. perfringens* is a large, rectangular, gram-positive bacillus, with spores rarely observed in vivo or following in vitro cul-

FIGURE 18-1   *C. perfringens* on sheep blood agar plates. Note the flat. rapidly spreading colonies and the hemolytic activity of the organisms. A preliminary identification of *C. perfringens* can be made by recognition of the zone of complete hemolysis and a wider zone of partial hemolysis, combined with the characteristic microscopic morphology of these bacilli.

tivation. The organism is one of the few nonmotile clostridia, but rapid spreading growth on laboratory media (resembling growth of motile organisms) is characteristic (Figure 18-1). *C. perfringens* is hemolytic and metabolically active, which facilitates the rapid identification of these organisms in the laboratory.

Pathogenesis

Clostridium perfringens can cause a spectrum of diseases, from self-limited gastroenteritis to overwhelming destruction of tissue (e. g., clostridial myonecrosis) with very high mortality despite early appropriate medical intervention. This pathogenic potential is attributed to the numerous toxins and enzymes produced by this organism (Table 18-2). Five toxins lethal for experimental animals have been described: alpha, beta, epsilon, iota, and delta (Figure 18.2). **Alpha toxin,** the most important *C. perfringens* toxin, is a lecthinase (phospholipase C) that lyses erythrocytes, platelets, leukocytes, and endothelial cells (Figure 18-3). Massive hemolysis, bleeding, tissue destruction, and hepatic toxicity is associated with this toxin. The largest quantities of alpha toxin are produced by *C. perfringens* type A. Beta, epsilon, iota, and theta toxins are associated with increased capillary permeability. Beta toxin production is also implicated in **necrotizing enterocolitis** (enteritis necroticans, pig-bel). Theta toxin is an oxygen-sensitive hemolysin that is immunologically related to streptolysin O, pneumolysin, and tetanolysin (*Clos-*

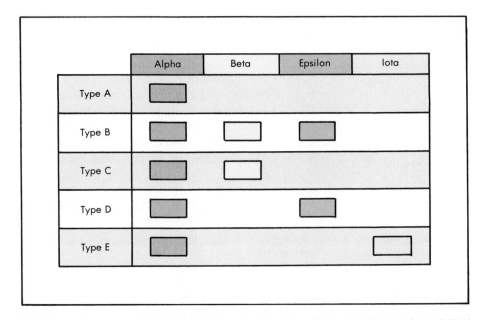

FIGURE 18-2   Four major lethal toxins are described for *C. perfringens* and are used to subdivide isolates into five types (A through E). Type A is further subdivided into many epidemiologic serotypes.

Table 18-2   Virulence Factors Associated With *Clostridium perfringens*

Virulence Factor	Activity
Alpha toxin	Phospholipase C, hemolytic
Beta toxin	Increases capillary permeability
Epsilon toxin	Increases capillary permeability
Iota toxin	Increases capillary permeability
Theta toxin	Increases capillary permeability
Delta toxin	Hemolytic
Kappa toxin	Collagenase
Mu toxin	Hyaluronidase
Nu toxin	Deoxyribonuclease
Lambda toxin	Protease
Neuraminidase	Hydrolyzes serum glycoproteins
Enterotoxin	Reverses water, sodium, and chloride transport in intestine
Gamma toxin	?
Eta toxin	?

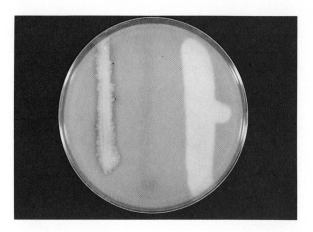

FIGURE 18-3   Growth of *C. perfringens* on egg yolk agar. The alpha toxin produced by *C. perfringens* hydrolyzes phospholipids in serum and egg yolk, producing an increased turbidity (right). This precipitate is not observed when the organism is grown in the presence of antitoxin (left). This reaction (named the Nagler reaction after the microbiologist who first described it) can be used for the presumptive identification of *C. perfringens*.

tridium tetani hemolysin). Intravenous injection of theta toxin causes intravascular hemolysis, pulmonary edema, and heart block. Delta toxin, produced by *C. perfringens* types B and C, is a hemolysin for sheep erythrocytes (commonly used in laboratory media) and is lethal for experimental animals. Tissue destruction and spread of the organism is facilitated by kappa, mu, nu, and lambda toxins, as well as neuraminidase. Enterotoxin is a heat-labile protein produced in the colon and released during spore formation. It is produced primarily by type A strains but also by a few type C and D strains and disrupts ion transport in the intestinal lumen in much the same way *Vibrio cholerae* does. A large number of vegetative cells must be ingested with the contaminated food to produce the enterotoxic effects. Antibodies to enterotoxin, indicating previous exposure, are commonly found in adults, although these are not protective.

Epidemiology

Clostridium perfringens is a common inhabitant of the intestinal tract of man and animals and is widely distributed in nature, particularly in soil and water contaminated with feces. These organisms form spores under adverse environmental conditions and can survive for prolonged periods. Approximately 80% of the environmental isolates are type A, which,

together with types C and D, are responsible for human disease. Gas gangrene and food poisoning are primarily caused by *C. perfringens* type A; enteritis necroticans by type C.

Clinical Syndromes

Bacteremia The isolation of *C. perfringens* or other clostridial species in blood cultures can be alarming. However, more than half of the isolates are clinically insignificant, representing transient bacteremia or, more likely, contamination of the culture with clostridia colonizing the skin surface. The significance of an isolate must be viewed in the light of other clinical findings.

Myonecrosis (Gas Gangrene) This life-threatening disease illustrates the full virulence potential of clostridia. The onset of disease, characterized by intense pain, generally begins within 1 week after clostridia are introduced into tissue by trauma or surgery. The progression from the time of onset through extensive muscle necrosis, shock, renal failure, and death is rapid, frequently occurring in less than 2 days. Macroscopic examination of muscle reveals devitalized necrotic tissue, with the presence of gas due to the metabolic activity of the rapidly dividing bacteria (hence, the name gas gangrene). Microscopically, abundant box car–shaped gram-positive bacilli are seen in the absence of cellular material. Extensive hemolysis and bleeding due to clostridial toxins are characteristic. Most clostridial myonecrosis is caused by *C. perfringens*, although other species can also produce this disease (e. g., *C. septicum*, *C. bifermentans*, and *C. novyi*).

Cellulitis, Fasciitis, and Other Soft Tissue Infections Clostridial species can colonize wounds and the skin surface with no clinical consequences. Indeed, most isolates of *C. perfringens* and other clostridial species from wound cultures are insignificant. However, these organisms can also initiate cellulitis (Figure 18-4) or a rapidly progressive, destructive process (fasciitis) whereby the organisms spread through fascial planes, causing suppuration and gas formation. Fasciitis is distinguished from myonecrosis by the absence of muscle involvement, but it shares a dismal outcome. Surgical intervention is generally unsuccessful because of the rapidity with which the organisms spread. Most infections are

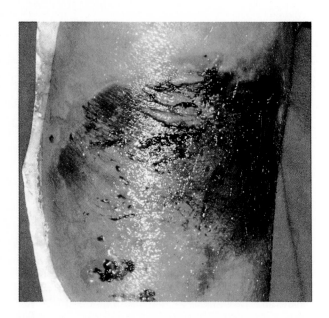

FIGURE 18-4   Clostridial cellulitis. Clostridia can be introduced into tissue following surgery or a traumatic injury. This patient suffered a compound fracture of the tibia. Five days after the injury, skin discoloration with bullae and necrosis, and the presence of serosanguinous revealed subcutaneous gas but no evidence of muscle necrosis. The patient made an uneventful recovery. (From Lambert HP and Farrar WE: Infectious diseases illustrated, London, 1982, Gower Medical Publishing.)

caused by *C. perfringens*, *C. septicum*, or *C. ramosum*.

Food Poisoning Clostridial food poisoning is characterized by a short incubation period (8 to 24 hours), clinical presentation of abdominal cramps and watery diarrhea in the absence of fever or vomiting, and a clinical course of less than 24 hours. Disease is due to ingestion of meat products contaminated with large numbers (10^{8-9} organisms) of enterotoxin-producing *C. perfringens*.

Enteritis Necroticans This rare disease is an acute necrotizing process in the small intestine that is characterized by abdominal pain, bloody diarrhea, shock, and peritonitis. The incidence of death approaches 50%. Beta toxin–producing type C *C. perfringens* is responsible for this disease.

Laboratory Diagnosis

The laboratory performs a confirming role in the diagnosis of clostridial disease, because therapy must be initiated immediately. The microscopic detection of gram-positive bacilli in clinical specimens, usually in the absence of leukocytes, can be very useful because these organisms have a characteristic morphology. Culture of the anaerobes is also relatively simple, with detection of *C perfringens* accomplished on simple media after incubation for 1 day or less. Under appropriate conditions *C. perfringens* can divide every 8 minutes, so growth on agar media or in blood culture broths can be detected after incubation for only 4 to 6 hours. Recovery of other clostridia such as *C. septicum* can require much longer incubation. Documentation of the role of *C. perfringens* in food poisoning is demonstrated by recovery greater than 10^6 organisms per gram of food. Isolation of the organism from fecal specimens may represent normal colonization.

Treatment, Prevention, and Control

Treatment for systemic *C. perfringens* infections such as fasciitis and myonecrosis requires aggressive surgical debridement and high-dose penicillin therapy. The use of antiserum directed against alpha toxin (antitoxin) and treatment in a hyperbaric oxygen chamber (presumably to inhibit growth of the anaerobe) have poorly defined benefits. Despite all therapeutic efforts, the prognosis with these diseases is poor, with mortality reported from 40% to almost

100%. Less serious, localized clostridial diseases can be successfully treated with penicillin, with resistance only rarely reported for species other than *C. perfringens*. Antibiotic therapy for clostridial food poisoning is unnecessary.

Prevention and control of *C. perfringens* infections is difficult because of the ubiquitous distribution of the organisms. Disease requires introduction of the organism into devitalized tissues and maintenance of an anaerobic environment favorable for bacterial growth. Thus most infections can be prevented by proper wound care and judicious use of prophylactic antibiotics.

CLOSTRIDIUM TETANI

Microbial Physiology and Structure

Clostridium tetani is a small, motile, spore-forming bacillus that frequently stains gram-negative. Round, terminal spores are produced that give the organisms a drumstick appearance (Figure 18-5). In contrast with *C. perfringens*, *C. tetani* is difficult to grow in vitro (because of its sensitivity to oxygen toxicity) and is relatively inactive metabolically.

Pathogenesis

Tetanus is caused by a potent, heat-labile, neurotoxin (tetanospasmin) that is produced during the stationary phase of growth and released when cell lysis occurs. Tetanospasmin is synthesized as a single 150,000 d peptide that is cleaved into a light and heavy chain by an endogenous protease when the neurotoxin is released from the cell. The two chains are held together by a disulfide bond and noncovalent forces. The carboxy-terminal portion of the heavy (100,000 d) chain binds to gangliosides on neuronal membranes. The toxin is then internalized and moves from the peripheral nerve terminals to the central nervous system by retrograde axonal transport. It is released from the postsynaptic dendrites, crosses the synaptic cleft, and is localized within vesicles in the presynaptic nerve terminals (Figure 18-6). Tetanospasmin acts by blocking the release of neurotransmitters (e. g., gamma-aminobutyric acid or GABA, glycine) for inhibitory synapses, thus permitting unregulated excitatory synaptic activity (spastic paralysis).

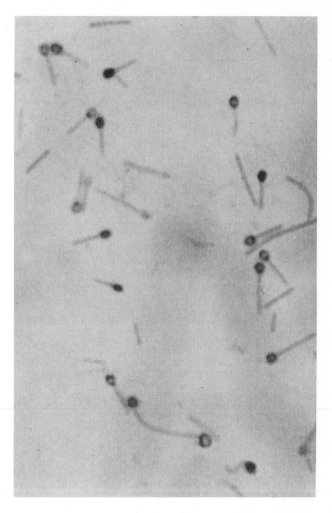

FIGURE 18-5   *Clostridium tetani*. Note the "drumstick" or "tennis racket" shape due to the terminal spore.

Epidemiology

Clostridium tetani is ubiquitous, found in fertile soil and colonizing the gastrointestinal tract of many animals, including humans (Figure 18-7). The vegetative forms of *C. tetani* are extremely susceptible to oxygen toxicity, but the organisms sporulates readily and can survive in nature for prolonged periods. Tetanus is relatively rare in the United States due to the high incidence of immunity following vaccination; however, it is still responsible for significant mortality in underdeveloped areas where vaccination is unavailable or medical practices are lax. Tetanus in neonates and the unprotected elderly is associated

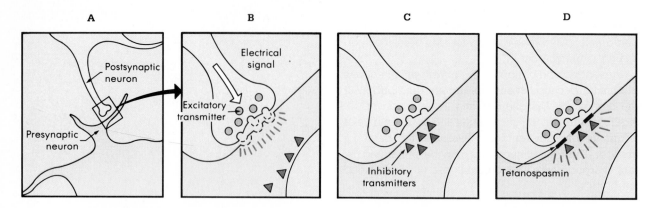

FIGURE 18-6   Mechanism of tetanospasmin activity. **A,** Neurotransmission is controlled by the balance between excitatory and inhibitory neurotransmitters. **B,** The inhibitory neurotransmitters (e.g., GABA, glycine) prevent depolarization of the postsynaptic membrane and conduction of the electrical signal. **C,** Tetanospasmin does not interfere with production or storage of GABA or glycine, but rather their release (presynaptic activity). **D,** In the absence of inhibitory neurotransmitters, excitation of the neuroaxon is unrestrained.

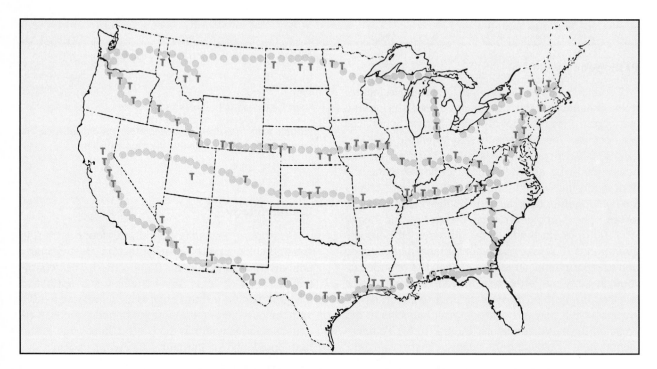

FIGURE 18-7   Distribution of *Clostridium tetani* in the soil of the United States. Soil samples were collected every 50 miles on four east-west transects across the United States. *C. tetani (T)* was present in 30% of the soil samples. (Modified from Smith LDS: Health Lab Science 15:74-80, 1978.)

Table 18-3   Clinical Manifestations of Tetanus

Disease	Clinical Manifestations
Generalized	Involvement of bulbar and paraspinal muscles (trismus or lockjaw, risus sardonicus, difficulty in swallowing, irritability, opisthotonos); involvement in the autonomic nervous system (sweating, hyperthermia, cardiac arrhythmias, fluctuations in blood pressure); disease in newborns called tetanus neonatorum; prognosis related to patient age, immune status, and site of primary infection
Cephalic	Primary infection in head, particularly ear; isolated or combined involvement of cranial nerves, particularly seventh cranial nerve; very poor prognosis
Localized	Involves muscles in area of primary injury; may precede generalized disease; favorable prognosis

with a high mortality rate. Fewer than 100 cases of tetanus are reported annually in the United States; virtually all patients are inadequately immunized. Drug abusers who inject drugs subcutaneously (''skin poppers'') are susceptible to tetanus.

Clinical Syndromes

See Table 18-3 for a summary of clinical symptoms. The incubation period for tetanus is variable, ranging from a few days to weeks. The length of the incubation period is directly related to the distance of the primary wound infection from the central nervous system.

Generalized tetanus is the most common form seen. Involvement of the masseter muscles (**trismus, or lockjaw**) is the presenting sign in the majority of

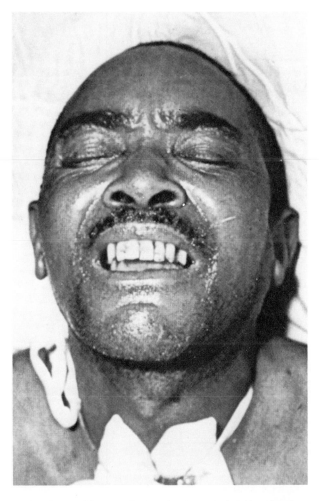

FIGURE 18-8   Risus sardonicus in tetanus, due to spasms in the masseter muscles. (From Lambert HP and Farrar WE: *Infectious diseases illustrated*, London, 1982, Gower Medical Publishing.)

FIGURE 18-9   An infant with tetanus and opisthotonus resulting from persistent spasms in the back muscles. (From Lambert HP and Farrar WE: *Infectious diseases illustrated*, London, 1982, Gower Medical Publishing.)

patients. The sardonic smile characteristic of sustained trismus is known as **"risus sardonicus"** (Figure 18-8). Other early signs include drooling, sweating, irritability, and persistent back spasms (**opisthotonos;** Figure 18-9). More severe disease is seen with involvement of the autonomic nervous system, with cardiac arrhythmias, fluctuations in blood pressure, profound sweating, and dehydration.

Another form of *C tetani* disease is **localized tetanus,** where the disease remains confined to the musculature at the site of primary infection. A variant of localized tetanus is **cephalic tetanus,** where the primary site of infection is the head. In contrast with localized tetanus, the prognosis for patients with cephalic tetanus is very poor.

Laboratory Diagnosis

The diagnosis of tetanus, like most other clostridial diseases, is made on the basis of clinical presentation. Microscopic detection or in vitro isolation of *C. tetani* is useful but frequently unsuccessful. Only about 30% of patients with tetanus have positive cultures.

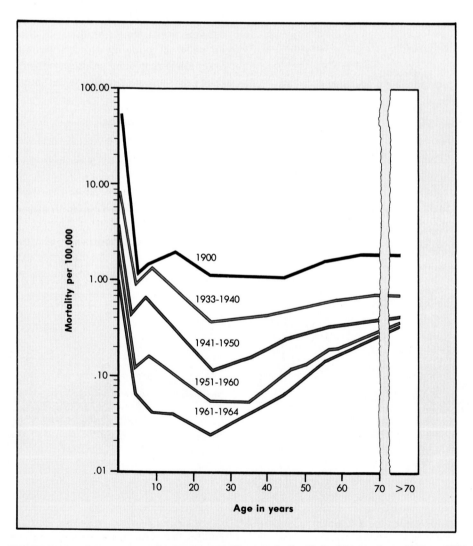

FIGURE 18-10   Average annual tetanus mortality rate in the United States from 1900 to 1964. (Redrawn from LaForce FM et al: Reprinted by permission of the New England Journal of Medicine 280:569, 1969.)

Treatment, Prevention, and Control

Mortality associated with tetanus has steadily decreased during the last century (Figure 18-10), due in large part to the decreased incidence of tetanus in the United States. The highest incidence of mortality is in newborns and in patients who experience an incubation period of less than 1 week before the onset of disease (Figure 18-11).

Treatment of tetanus requires debridement of the primary wound (which may appear innocuous), administration of penicillin, and passive immunization with human tetanus immunoglobulin. Wound care and penicillin therapy eliminates vegetative bacteria—producing toxin, while the antitoxin antibodies bind free tetanospasmin molecules. The toxin bound to nerve endings are protected from antibodies. Thus the toxic effects must be controlled symptomatically until normal regulation of synaptic transmission is restored.

The level of toxin that reaches the bloodstream is inadequate to stimulate an immune reaction; therefore all surviving patients should complete a course of active immunizations with tetanus toxoid (inactive tetanus toxin). Furthermore, the single most effective way to prevent tetanus is to complete tetanus immunizations at an early age with the combined tetanus-diphtheria-pertussis vaccine, and then maintain protective immunity with booster vaccinations every 10 years with tetanus toxoid. Passive protection of non-immune individuals in a population of immunized individuals (herd immunity) is not effective, because tetanus is caused by exposure to *C. tetani* in contaminated soil, not by contact with infected persons.

CLOSTRIDIUM BOTULINUM

Microbial Physiology and Structure

Clostridium botulinum, the etiologic agent of botulism, is a fastidious, spore-forming, anaerobic bacillus. Botulism is caused by a potent neurotoxin, the production of which is regulated by a specific bacteriophage.

Pathogenesis

Eight antigenically distinct **botulinum toxins** (A, B, C alpha, C beta, D, E, F, and G) have been described, with human disease associated with types A, B, and E. Only one toxin is produced by most individual isolates. Like tetanus toxin, *Clostridium botulinum* toxin is a 150,000 d protein that is cleaved into two chains after synthesis. Also similar to tetanus toxin, the 100,000 d heavy chain is responsible for binding to ganglioside receptors. Neither chain by

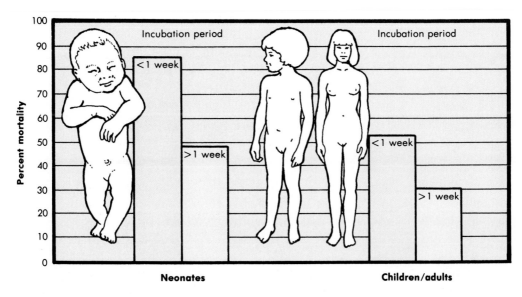

FIGURE 18-11   Mortality in patients with tetanus: correlation with length of incubation period and age of patient.

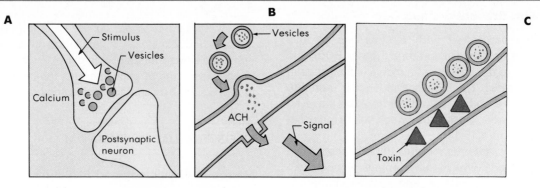

FIGURE 18-12   Mechanism of botulinum toxin activity. Synaptic activity at cholinergic synapses is mediated by the neurotransmitter acetylcholine (ACH). **A,** As a nerve stimulus enters the presynaptic nerve ending, an influx of calcium is stimulated. **B,** Calcium is required for the fusion of synaptic vesicles with the presynaptic membrane and release of ACH into the synaptic space. ACH crosses to the postsynaptic membrane and acts on specific receptors. ACH is rapidly hydrolzyed by acetylcholinesterase. The combined activity of many ACH-mediated stimuli produces an electrical potential change and transmission of the signal. **C,** Botulinum toxin interrupts this transmission by interfering with the release of ACH from the synaptic vesicles. The synthesis of ACH and its packaging in synaptic vesicles is not affected.

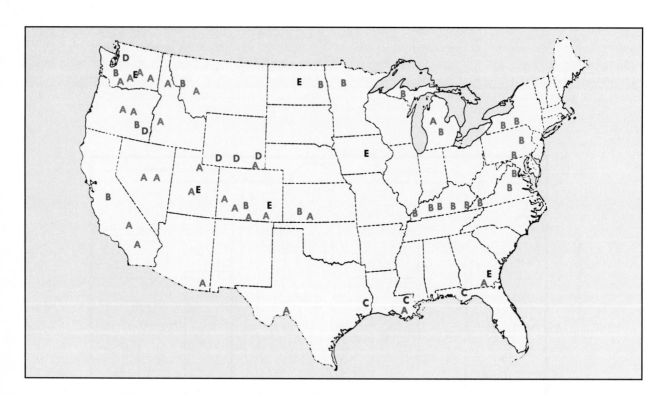

FIGURE 18-13   Distribution of *Clostridium botulinum* in soil samples collected in the United States. **A,** Type A strains were found mainly in neutral or alkaline soil in the western part of the United States. **B,** Type B strains were found primarily in the eastern part of the United States in rich, organic soil. **C,** Types C, D, and E were recovered less frequently, with type E mostly restricted to wet soil. (From Smith LDS: Health Lab Sciences 15:74-80, 1978.)

itself is toxic. Botulinum is very specific for cholinergic nerves. The toxin blocks neurotransmission at peripheral cholinergic synapses by preventing release of the neurotransmitter acetylcholine (Figure 18-12).

Epidemiology

Clostridium botulinum is distributed worldwide, with spores commonly isolated in soil samples. Generally, type A toxin–producing strains predominate west of the Mississippi River and type B strains east (Figure 18-13). Food-borne botulism is usually associated with home-canned foods and occasionally preserved fish (type E toxin). The food may or may not appear spoiled, but even a small taste can cause full-blown clinical disease. Botulism in infants has been associated with the consumption of honey contaminated with botulinal spores.

Clinical Syndromes

Three forms of botulism have been identified: classical or food-borne, infant, and wound botulism (Table 18-4).

Food-Borne Botulism After consumption of contaminated food and a 1- to 2-day incubation period, the patient develops weakness and dizziness. The initial signs of botulism include blurred vision, painful oropharyngeal dryness, constipation, and abdominal pain. Bilateral, descending weakness of the peripheral muscles develops in progressive disease **(flaccid paralysis),** with death most commonly attributed to respiratory paralysis. Despite aggressive management of the patient, the disease may continue to progress because of the neurotoxin's irreversible binding and long-term inhibitory activity on the release of excitatory neurotransmitters. Complete recovery in patients who survive this initial period frequently requires many months to years. Although

mortality once approached 70%, this has been reduced to 10% with the use of better supportive care, particularly in the management of respiratory complications.

Infant Botulism Although this disease was first recognized in 1976, it is now believed to be the most common form of botulism seen in the United States, with an estimated 250 cases each year. In contrast with food-borne botulism, this disease is caused by the in vivo production of neurotoxin by *C. botulinum* colonizing the gastrointestinal tract of young infants. Disease is reported in infants less than 9 months of age (most between 1 to 6 months) with the initial symptoms nonspecific (e. g., constipation or "failure to thrive"). Progressive disease with extensive flaccid paralysis and respiratory arrest can develop, although the mortality rate in documented infant botulism is very low (1% to 2%). However, some infant deaths attributed to other conditions (e. g., sudden infant death syndrome) may be due to botulism.

Wound Botulism This is the rarest form of botulism in the United States. As the name implies, wound botulism develops from in vivo toxin production by *C. botulinum* in contaminated wounds. The symptoms of disease are identical to those of food-borne disease; however, the incubation period is generally longer (4 days or more) and gastrointestinal symptoms are less prominent.

Laboratory Diagnosis

The clinical diagnosis of botulism is confirmed by isolating the organism or demonstrating toxin activity. In food-borne disease an attempt should be made to culture *C. botulinum* from the patient's feces and the implicated food if it is available (Figure 18-14). Isolation of *C. botulinum* from specimens contaminated with other organisms can be improved by first heating the specimen for 10 minutes at 80° C to kill all vegetative cells. Culture of the heated specimen on nutritionally enriched anaerobic media permits the germination of the heat-resistant *C. botulinum* spores. The food, stool specimen, and patient's serum should also be tested for toxin activity by a mouse bioassay. The specimen is divided, and one portion is mixed with antitoxin. Both portions are then inoculated intraperitoneally into mice. If antitoxin treatment protects the mice, then toxin activity is confirmed.

The diagnosis of infant botulism is supported by

Table 18-4	*Clostridium botulinum* Disease
Disease	**Source of Toxin**
Classical botulism	Ingestion of contaminated food
Infant botulism	In vivo production in colonized infant
Wound botulism	In vivo production in infected wound

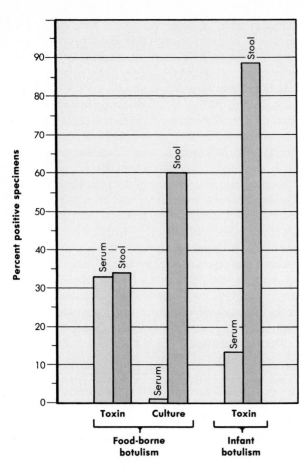

FIGURE 18-14   Detection of *Clostridium botulinum* and botulinal toxin in serum and stool specimens collected from patients with food-borne botulism and infants with botulism. Confirmation of food-borne disease requires testing the implicated food and the patient's serum and stool specimen for botulinal toxin activity, as well as culturing the food and stool specimen for the organism. No single test for food-borne botulism has a sensitivity greater than 60%. Serum and stool specimens from infants with botulism should be tested for toxin activity and the stool specimen cultured. (Data from Dowell VR et al: JAMA 238:1829, 1977; and Hatheway CL and McCroskey LM: J Clin Microbiol 25:2334, 1987.)

the isolation of *C. botulinum* from feces or detection of toxin activity in feces or serum. The organism can be isolated from stool cultures in virtually all patients, because carriage of the organism may persist for many months, even after the baby has recovered. Wound botulism is confirmed by isolation of the organism in the wound or detection of toxin activity in wound exudate or serum.

Treatment, Prevention, and Control

Treatment of botulism requires adequate ventilatory support, elimination of the organism from the gastrointestinal tract by the judicious use of gastric lavage and penicillin therapy, and the use of antitoxin to bind toxin circulating in the bloodstream. Ventilatory support has had an unquestioned benefit in significantly reducing mortality. The value of the other treatment modalities has been less clearly defined, although the use of polyvalent antitoxin (mixture of antibodies against types A, B, and E toxins) is recommended.

Prevention of disease involves destruction of the spores in food products (virtually impossible for practical reasons), prevention of spore germination (by maintaining the food in an acid pH or storage at 4° C or colder), or destruction of the preformed toxin (by heating the food for 20 minutes at 80° C). Infant botulism is strongly associated by consumption of honey contaminated with *C. botulinum* spores, so children less than 1 year of age should not eat honey.

CLOSTRIDIUM DIFFICILE

Until the mid-1970s the clinical importance of *Clostridium difficile* was not appreciated. This organism was infrequently isolated in fecal cultures and rarely associated with human disease. However, systematic studies have now clearly demonstrated that toxin-producing *C. difficile* is responsible for antibiotic-associated gastrointestinal diseases ranging from relatively benign, self-limiting diarrhea to severe, life-threatening pseudomembranous colitis (Figures 18-15 and 18-16). Consistent with the etiologic role of this organism in gastrointestinal disease, a similar disease can be reproduced in hamsters exposed to *C. difficile*.

C. difficile produces two toxins: an **enterotoxin** (toxin A) and a **cytotoxin** (toxin B). These toxins are immunologically distinct, although physical separation has proved to be difficult. The enterotoxin can elicit a hemorrhagic hypersecretion of fluid when injected into rabbit intestines, whereas the cytotoxin caused destructive changes in monolayer cells (cytopathic effect). The precise role each toxin plays in disease is still unclear.

C. difficile is part of the normal intestinal flora in a small proportion of healthy persons. When exposed to antibiotics, the normal enteric flora is altered, permit-

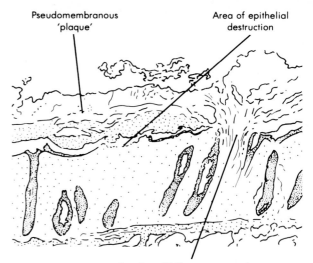

Pseudomembranous 'plaque'   Area of epithelial destruction

Eruption of inflammatory exudate through epithelial surface

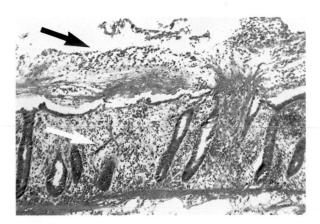

FIGURE 18-15   Antibiotic-associated colitis mediated by *Clostridium difficile*. The lumen of the colon is covered with white plaques that overlay the relatively intact epithelial surface. The plaque or pseudomembrane consists of mucus, leukocytes, dead epithelial cells, and bacteria.

FIGURE 18-16   Antibiotic-associate colitis: histological section of colon showing intense inflammatory response with the characteristic "plaque" (*black arrow*) overlying the intact intestinal mucosa (white arrow). Hematoxylin and eosin stain. (Top from Lambert HP and Farrar WE: Infectious Diseases Illustrated, London, 1982, Gower Medical Publishing.)

ting the overgrowth of these relatively resistant organisms. Proliferation of the organisms with localized production of their toxins leads to disease—from self-limiting diarrhea to pseudomembranous colitis.

Specific diagnosis of *C. difficile* involvement is accomplished by culture of feces on highly selective media, detection of the cytotoxin by an in vitro cytotoxicity assay with tissue culture cells, or detection of a cellular protein by a latex agglutination procedure (Figure 18-17). The single most specific test for *C. difficile* disease is the **cytotoxicity assay.** However,

maximal sensitivity is accomplished by using a combination of diagnostic tests.

Discontinuation of the implicated antibiotic (e. g., ampicillin, clindamycin) is generally sufficient to alleviate mild disease. However, specific therapy with either metronidazole or vancomycin is required for serious disease. Relapses after completion of therapy may occur in as many as 20% to 30% of patients, because spores of *C. difficile* are resistant to antibiotic treatment. Retreatment with the same antibiotic is frequently successful. Prevention is difficult because

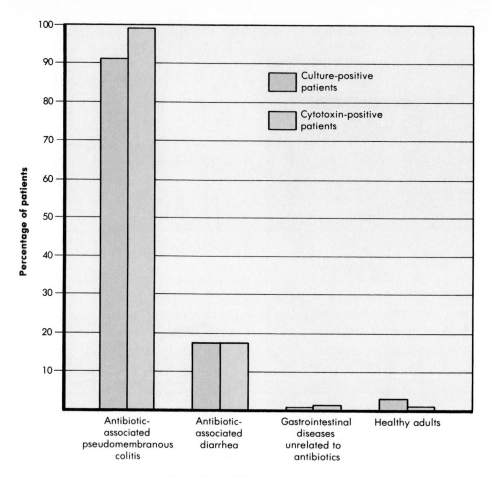

FIGURE 18-17   Association of *Clostridium difficile* with pseudomembranous colitis and diarrhea.

the organism is commonly isolated in hospital environments, particularly in areas adjacent to infected patients. The spores of *C. difficile* are difficult to destroy, thus the organism can contaminate an environment for many months and be the source of nosocomial hospital outbreaks of *C. difficile* disease.

BIBLIOGRAPHY

Bartlett JG: Antibiotic-associated pseudomembranous colitis, Rev Infect Dis 1:530-539, 1979.

Bizzini B: Tetanus toxin, Microbiol Rev 43:224-240, 1979.

Dowell VR, McCroskey LM, Hatheway CL, et al: Coproexamination for botulinal toxin and *Clostridium botulinum*. A new procedure for laboratory diagnosis of botulism, JAMA 238:1829-1832, 1977.

Eidels L, Proia RL, and Hart DA: Membrane receptors for bacterial toxins, Microbiol Rev 47:596-620, 1983.

Hatheway CL and McCroskey LM: Examination of feces and serum for diagnosis of infant botulism in 336 patients, J Clin Microbiol 25:2334-2338, 1987.

Lyerly DM, Lockwood DE, Richardson SH, and Wilkins TD: Biological activities of toxins A and B of *Clostridium difficile*, Infect Immun 35:1147-1150, 1982.

Mellanby J and Green J: How does tetanus toxin act? Neuroscience 6:281-300, 1981.

Middlebrook JL and Dorland RB: Bacterial toxins: cellular mechanisms of action, Microbiol Rev 48:199-221, 1984.

Mills DC and Arnon SS: The large intestine as the site of *Clostridium botulinum* colonization in human infant botulism, J Infect Dis 156:997-998, 1987.

Smith LDS: The occurrence of *Clostridium botulinum* and *Clostridium tetani* in the soil of the United States, Health Lab Sci 15:74-80, 1978.

Smith LDS: Virulence factors of *Clostridium perfringens*, Rev Infect Dis 1:254-260, 1979.

Sugiyama H: *Clostridium botulinum* neurotoxin, Microbiol Rev 44:419-447, 1980.

CHAPTER 19

Corynebacterium and Propionibacterium

The nonspore-forming gram-positive bacilli comprise a heterogeneous collection of organisms, many of which are poorly characterized. The most commonly isolated organisms belong to the genera *Corynebacterium* and *Propionibacterium,* organisms also referred to respectively as aerobic and anaerobic diphtheroids. The best-known member of this group of organisms is *Corynebacterium diphtheriae,* the etiologic agent of **diphtheria.** Other *Corynebacterium* and *Propionibacterium* species can cause human disease (Table 19-1), although this is uncommon. It has been proposed that some *Corynebacterium* species (e.g., *C. haemolyticum*) should be transferred to other genera. However, because they have been conventionally assigned to this genus and their precise tax-

onomic classification has not been resolved, they will be referred to as *Corynebacterium* in this chapter.

MICROBIAL PHYSIOLOGY AND STRUCTURE

The *Corynebacterium* and *Propionibacterium* are small, usually pleomorphic, gram-positive bacilli. The organisms may appear in short chains, V or Y configurations, or in clumps resembling "Chinese characters." Metachromatic granules within the cells may be seen with special stains. The organisms are nonmotile, catalase-positive, and ferment carbohydrates, producing lactic acid (*Corynebacterium*) or propionic

Table 19-1 Diseases Associated With *Corynebacterium* and *Propionibacterium* Species

Organism	Animal Reservoir and Disease	Human Disease
C. diphtheriae	—	Diphtheria
C. ulcerans	Cattle, mastitis	Pharyngitis (mild to diphtheria-like)
C. pseudotuberculosis	Sheep, horses, cattle, goats, deer; suppurative lymphadenitis, caseous bronchopneumonia	Chronic lymphadenitis
C. haemolyticum	—	Pharyngitis, cutaneous infection
C. pyogenes	Cattle, sheep, swine; suppurative infection	Miscellaneous opportunistic infections
C. pseudodiphtheriticum	—	Endocarditis
C. equi	Horses, cattle, swine; suppurative pneumonia	Suppurative pneumonia
C. bovis	Cattle; mastitis	Miscellaneous opportunistic infections
C. xerosis		Miscellaneous opportunistic infections
Group JK		Miscellaneous opportunistic infections
P. acnes		Acne, miscellaneous opportunistic infections

Table 19-2   Virulence Factors in *Corynebacterium* Species

Virulence Factor	Activity	Organism
Diphtheria exotoxin	Blockage of protein synthesis by inhibition of peptide translocation on ribosome	*C. diphtheriae* *C. ulcerans* *C. pseudotuberculosis*
Dermonecrotic toxin	Sphingomyelinase that increases vascular permeability	*C. ulcerans* *C. pseudotuberculosis*
Hemolysin	Undefined activity	*C. pyogenes*

acid (*Propionibacterium*). Corynebacteria are aerobic or facultatively anaerobic, and propionibacteria are anaerobic, although some are aerotolerant. The organisms generally grow slowly on an enriched medium. Corynebacteria and propionibacteria are found in plants and animals, and they normally colonize the skin, upper respiratory tract, gastrointestinal tract, and urogenital tract of humans.

PATHOGENESIS

Although most of these organisms are opportunistic pathogens, specific virulence factors have been isolated from the more pathogenic species (Table 19-2).

Diphtheria Exotoxin

The mechanism of action of diphtheria exotoxin is well known. The **"tox" gene** that codes for the exotoxin is introduced into *Corynebacterium diphtheriae* by a lysogenic bacteriophage **(B-corynephage).** Strains of *C. diphtheriae* that do not have this bacteriophage are unable to produce toxin. Diphtheria exotoxin is a 63,000 d protein that consists of two fragments. The B fragment mediates binding to the cell surface, permitting the enzymatically active A fragment to enter the cell. In the presence of limiting amounts of iron, the A fragment blocks protein synthesis in a manner similar to *Pseudomonas aeruginosa* exotoxin A (see Chapter 10). The A fragment catalyzes the irreversible inactivation of elongation factor 2 (EF-2), which is required for movement of nascent peptide chains on ribosomes. Because the turnover of EF-2 is very slow and only about 1 molecule per ribosome is present in a cell, it has been estimated that one exotoxin molecule can inactivate the entire content of EF-2 in a cell. Diphtheria exotoxin is also found in some strains of *C. ulcerans* and *C. pseudotuberculosis* that have the "tox" B-corynephage. However, the concentration of toxin in these bacteria is generally low.

Dermonecrotic Toxin

Dermonecrotic toxin is produced by strains of *C. pseudotuberculosis, C. ulcerans,* and some strains of *C. haemolyticum*. The activity of this glycoprotein toxin is concentrated on the sphingomyelin of vascular endothelial cells; it may be responsible for the increased vascular permeability seen in diphtheria infections.

Hemolysin

A soluble hemolysin has been described in *C. pyogenes* that may facilitate the dermonecrotic activity associated with this organism.

EPIDEMIOLOGY

Diphtheria is a disease with worldwide distribution among urban poor, where crowding exists and a protective level of vaccine-induced immunity is low. *Corynebacterium diphtheriae* is maintained in the population by asymptomatic carriage in the oropharynx or on the skin and is transmitted person-to-person by respiratory droplets or skin contact. Immune individuals (either after exposure to *C. diphtheriae* or immunization) can carry the organism for years.

An active immunization program has made diphtheria uncommon in the United States; fewer than

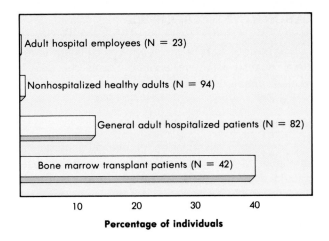

FIGURE 19-1   *Corynebacterium* group JK colonization in four adult populations. Although *Corynebacterium* group JK does not commonly colonized healthy individuals or hospital employees, there is an increased incidence of colonization in hospitalized patients, particularly highly immunocompromised patients that have received broad-spectrum antibiotics during a prolonged hospitalization. (Redrawn from Stamm WE et al: Ann Intern Med 91:167, 1979.)

200 cases are confirmed annually. Although primarily a pediatric disease, the incidence of disease has shifted toward older age-groups in populations where active immunization programs for children exist. Skin infections with toxigenic *C. diphtheriae* (cutaneous or wound diphtheria) is seen more commonly in tropical areas of the world.

Human infection with some strains of corynebacteria (e. g., *C. ulcerans, C. pseudotuberculosis, C. pyogenes, C. equi,* and *C. bovis*) requires exposure to an animal reservoir (see Table 19.1), whereas other corynebacteria and propionibacteria are part of the indigenous human oropharyngeal or skin flora (e. g., *C. haemolyticum, C. pseudodiphtheriticum, C. xerosis,* group JK, and *P. acnes*). With only a few exceptions the indicence of disease caused by these organisms is relatively rare. The incidence of respiratory infections caused by *C. haemolyticum* may be underestimated because the organism is relatively difficult to recognize in culture and infections, which mimic streptococcal pharyngitis, respond to penicillin treatment. *Corynebacterium* group JK is a recently recognized opportunistic pathogen in immunocompromised patients, particularly those with hematologic disorders. Although carriage of this organism is uncommon in healthy persons, approximately 40% or more of specific hospital populations can be colonized (Figure

19-1). Predisposing conditions for disease include prolonged hospitalization, granulocytopenia, prior or concurrent antimicrobial therapy or chemotherapy, and a mucocutaneous portal of entry. The highest incidence of disease is seen in elderly men.

Propionibacterium acnes is carried by most individuals. Disease is observed in two populations: acne in teenagers and young adults and opportunistic infections in patients with prosthetic devices or intravascular lines (e. g., artificial valves, joints, and shunts or indwelling catheters). Local production of propionic acid by *P. acnes* with the associated inflammatory response and accumulation of pus is the basis of acne development. *P. acnes* can also adhere to synthetic materials such as prosthetic valves and joints and cause chronic infections.

CLINICAL SYNDROMES

Diphtheria

The clinical presentation of diphtheria is determined by the site of infection, immune status of the patient, and virulence of the organism. Exposure to *Corynebacterium diphtheriae* can result in asymptomatic colonization in fully immune individuals, mild respiratory disease in partially immune patients, or a fulminant, sometimes fatal disease in nonimmune patients. Patients with diphtheria involving the respiratory tract will develop symptoms after a 2- to 6-day incubation period. Organisms will multiply locally, on epithelial cells in the pharynx or adjacent surfaces, and initially produce localized damage by exotoxin activity. The onset is sudden, with malaise, sore throat, exudative pharyngitis, and a low-grade temperature. The exudate evolves into a thick **pseudomembrane,** composed of bacteria, lymphocytes, plasma cells, fibrin, and dead cells, that can cover the tonsils, uvula, and palate, and extend up into the nasopharynx or down into the larynx. The pseudomembrane is firmly adherent to the respiratory tissue and is difficult to dislodge without causing bleeding of the underlying tissue. As the patient recovers after the approximately 1-week course of the disease, the membrane dislodges and is expectorated. Complications in patients with severe disease include breathing obstruction and myocarditis.

Cutaneous diphtheria is acquired by skin contact

with other infected persons. The organism, which colonizes the skin surface, gains entry into the subcutaneous tissue through breaks in the skin (e. g., following an insect bite). A papule will develop and evolve into a chronic nonhealing ulcer, sometimes covered with a greyish membrane. Systemic signs of disease due to the exotoxin can be seen. The ulcer may also be superinfected with *Staphylococcus aureus* or group A *Streptococcus.*

Other Clinical Syndromes

A variety of diseases have been associated with other species of *Corynebacterium* and *Propionibacterium* (see Table 19-1). The most important include mild to severe pharyngitis or a diphtheria-like illness produced by toxigenic *C. ulcerans;* pharyngitis (frequently with a diffuse, erythematous, macular skin rash on the extremities and trunk) or cutaneous infections by *C. haemolyticum;* opportunistic infections by *Corynebacterium* group JK; and skin or opportunistic infections by *P. acnes.* The development of bacteremia, pneumonitis, wound infections, and other diseases by *Corynebacterium* group JK is particularly troublesome, because the organism is characteristically resistant to antibiotics and severely immunocompromised patients are most susceptible to infections.

LABORATORY DIAGNOSIS

Diagnosis of diphtheria depends on clinical parameters because definitive laboratory tests can take 1 week or more.

Microscopy

Microscopic examination of clinical material is unreliable. Metachromatic granules in *C. diphtheriae* stained with methylene blue have been described. However, this is not specific for this organism, and interpretation of the smear requires technical expertise.

Culture

Corynebacteria and propionibacteria grow on common laboratory media, although some species may require prolonged incubation. It is recommended that specimens for the recovery of *C. diphtheriae* should be collected from both the nasopharynx and throat and then inoculated onto nonselective media, as well as media developed specifically for this organism (e. g., cysteine-tellurite agar, serum tellurite agar, Loeffler medium). *C. diphtheriae* has a characteristic grey-to-black color on tellurite agar, and the microscopic morphology is best seen on Loeffler medium.

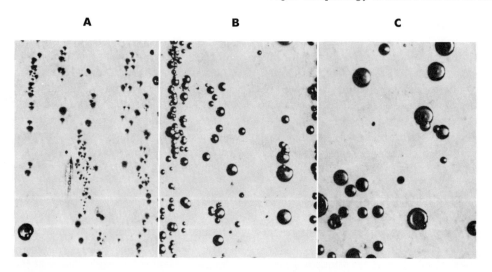

FIGURE 19-2   Colonial morphologies of *Corynebacterium diphtheriae.* Three distinct morphologies have been described for *C. diphtheriae.* The mitis type **(A)** are small, round, convex, black colonies; the intermedius type **(B)** are small, flat, gray colonies; and the gravis type **(C)** are large irregular, gray colonies. These differences are useful for epidemiologic classification of isolates.

Three colonial morphologies of *C. diphtheriae* have been described on cysteine-tellurite agar: **gravis, intermedius,** and **mitis** (Figure 19-2). Although these were initially related to severity of the illness, the distinctions are not valid now. However, the colonial morphologies are useful for epidemiologic classification of isolates. Precise identification of *C. diphtheriae* and the other corynebacteria is based on specific biochemical tests. Resistance to all antibiotics except vancomycin is a helpful differential property for the identification of *Corynebacterium* group JK.

Toxigenicity Testing

All isolates of *C. diphtheriae* should be tested for production of exotoxin. This can be done by either an in vitro immunodiffusion assay **(Elek test)** or in vivo by animal inoculation (Figure 19-3).

TREATMENT

The most important factor for treatment of diphtheria is early use of diphtheria antitoxin for the specific neutralization of exotoxin. Antibiotic therapy with penicillin or erythromycin has proved to be effective in eliminating *Corynebacterium diphtheriae* from patients who have the disease and also those who are asymptomatic carriers. Bed rest, isolation to prevent secondary spread, and maintenance of an open airway in patients with respiratory diphtheria are all appropriate.

PREVENTION AND CONTROL

Symptomatic diphtheria can be prevented by active immunization with diphtheria toxoid during childhood and booster immunizations every 10 years throughout life. The nontoxic, immunogenic toxoid is prepared by formalin treatment of toxin. Immunization with this preparation, in conjunction with pertussis and tetanus **(DPT vaccine),** is initially performed in three monthly injections followed by regular booster injections (normally with tetanus only).

Immunity in a population can be determined with the **Schick test** (Figure 19-4). Nonimmune persons who are exposed to a patient with diphtheria should be immunized with a booster dose of the vaccine to stimulate protective antibodies.

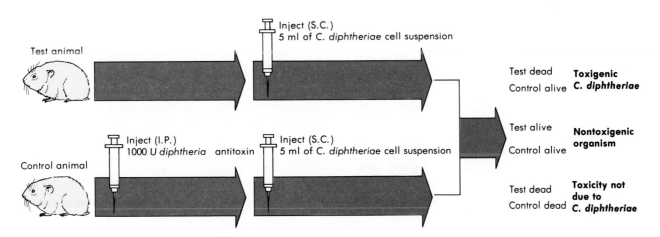

FIGURE 19-3 In vivo detection of *Corynebacterium diphtheriae* exotoxin. The clinical presentation of diphtheria is caused by production of exotoxin by "tox" positive strains of *C. diphtheriae*. The most sensitive way to detect toxin production is by injecting a guinea pig subcutaneously with a suspension of *C. diphtheriae* grown in broth culture. After 1 to 4 days the test animal will show signs of localized tissue necrosis and systemic signs of disease, including death. To ensure this is due to toxin, a control guinea pig is preimmunized with an intraperitoneal injection of diphtheria antitoxin. When this control is injected with the toxic cell suspension, the antitoxin will neutralize the exotoxin and protect the animal.

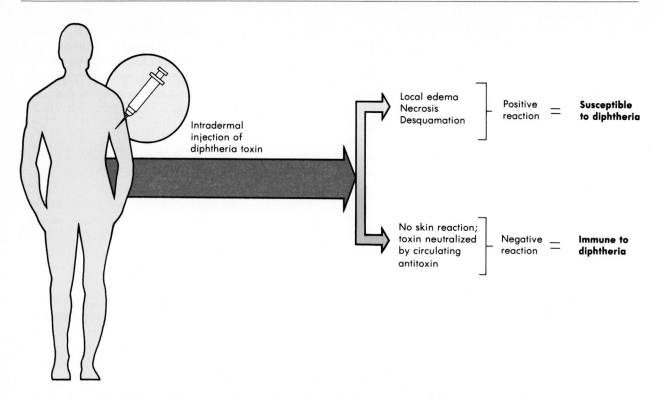

FIGURE 19-4   Schick test. The Schick test determines the immune status of a patient by measuring for the presence of neutralizing antibodies (antitoxin) in the individual's circulatory system.

BIBLIOGRAPHY

Hodes HL: Diphtheria, Pediatr Clin North Am 26:445-459, 1979.

Lipsky BA, Goldberger AC, Tompkins LS, and Plorde JJ: Infections caused by nondiphtheria corynebacteria, Rev Infect Dis 4:1220-1235, 1982.

Miller RA, Brancato F, and Holmes KK: *Corynebacterium haemolyticum* as a cause of pharyngitis and scarlatiniform rash in young adults, Ann Intern Med 105:867-872, 1986.

Papenheimer AM, Jr and Gill DM: Diphtheria, Science 182:353-358, 1973.

Pearson TA, Braine HG, and Rathbun HK: *Corynebacterium* sepsis in oncology patients—predisposing factors, diagnosis, and treatment, JAMA 238:1737-1740, 1977.

Stamm WE, Tompkins LS, Wagner KF, et al: Infection due to *Corynebacterium* species in marrow transplant patients, Ann Intern Med 91:167-173, 1979.

20

Miscellaneous Gram-Positive Bacilli

Gram-positive bacilli, belonging to a number of unrelated genera, have been described (Table 20-1). Some of these organisms are infrequent human isolates but are always clinically significant *(Listeria, Erysipelothrix)*, whereas others are a major component of the normal human microbial flora and can be commonly isolated in clinical specimens, although their role in disease is either incompletely defined *(Gardnerella)* or unimportant *(Lactobacillus, Eubacterium, Bifidobacterium).*

LISTERIA

Listeria monocytogenes, the only member of this genus that is a human pathogen, is a gram-positive, nonspore-forming, facultatively anaerobic bacillus capable of causing meningoencephalitis, bacteremia, and endocarditis, as well as a variety of other diseases. These small coccobacilli can microscopically resemble corynebacteria or gram-positive diplococci (e. g., *Streptococcus pneumoniae, Enterococcus*). The organisms are motile, with a characteristic tumbling motility in liquids incubated at room temperature, which is a useful differential characteristic for their preliminary identification. Although listeria have a widespread distribution in nature, human disease in uncommon and restricted to several well-defined populations—neonates, pregnant women, and immunocompromised patients, particularly those with a malignancy or following renal transplantation.

Pathogenesis

Listeria monocytogenes produces no identifiable toxins or other major virulence factors. All virulent strains produce a β-hemolysin, although the role this plays in disease is unknown. The major property associated with virulence is the ability to survive and multiply in mononuclear phagocytes. In this protected environment the organism is able to spread throughout the patient until specific cellular immunity can activate the phagocytic cell and induce intracellular bacterial killing. In the presence of depressed cell-mediated immunity, progressive disease will occur.

Epidemiology

L. monocytogenes is isolated from soil, water, vegetation, and a variety of mammals, birds, fish, insects, and other animals (Figure 20-1). Asymptomatic carriage, as well as disease, is well-documented in both mammals and humans. Although the incidence of human carriage is unknown, fecal carriage in healthy

Table 20-1 Human Diseases Caused by *Listeria, Erysipelothrix,* and *Gardnerella*

Pathogen	Disease
Listeria monocytogenes	Granulomatosis infantiseptica
	Meningitis
	Septicemia
	Endocarditis
Erysipelothrix rhusiopathiae	Erysipelas
	Septicemia
	Endocarditis
Gardnerella vaginalis	Bacterial vaginosis

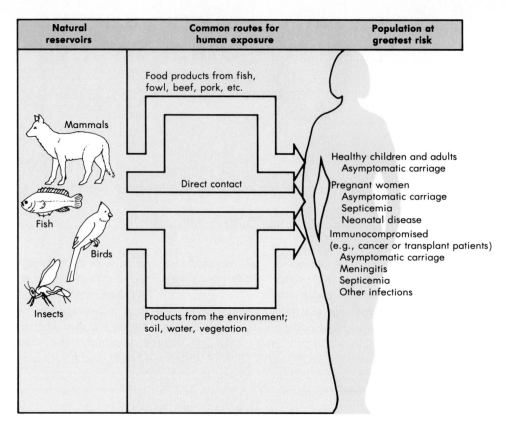

Natural reservoirs	Common routes for human exposure	Population at greatest risk

Food products from fish, fowl, beef, pork, etc.

Mammals

Direct contact

Fish

Birds

Insects

Products from the environment; soil, water, vegetation

Healthy children and adults
  Asymptomatic carriage

Pregnant women
  Asymptomatic carriage
  Septicemia
  Neonatal disease

Immunocompromised
(e.g., cancer or transplant patients)
  Asymptomatic carriage
  Meningitis
  Septicemia
  Other infections

FIGURE 20-1   Epidemiology of *Listeria* infections.

individuals is estimated to be 1% to 5%. The incidence of human disease is also unknown but is apparently increasing in frequency. Listeria are the most common cause of meningitis in renal transplant patients and in adult patients with cancer. These organisms are the fifth most common cause of meningitis overall. Human listeriosis is a sporadic disease seen throughout the year, with a peak in the warmer months. This is in contrast to bovine listeriosis, which peaks during the cold months. The source of most human infections is not known; however, focal epidemics have been associated with consumption of contaminated milk, meat, and cabbage. Listeria are able to grow in cold temperatures—a fact that has been exploited for the selective isolation of this organism **("cold enrichment").** Thus refrigeration of contaminated food products permits the slow multiplication of the organism to an infectious dose.

Clinical Syndromes

Listeria infections found in adults are illustrated in Figure 20-2. A discussion follows.

Neonatal Disease Two forms of neonatal disease have been described: **early-onset disease** acquired transplacentally in utero and **late-onset disease** acquired at or soon after birth. Early-onset disease, also called **granulomatosis infantiseptica,** is a devastating disease with a high mortality unless promptly treated. The disease is characterized by disseminated abscesses and granulomas in multiple organs. Late-onset disease occurs 2 to 3 weeks after birth as meningitis or meningoencephalitis with septicemia. The clinical signs and symptoms are not unique; thus other causes of neonatal central nervous system disease must be excluded.

Meningitis in Adults Meningitis is the most common form of listeria infections in adults, with most infections in patients with depressed cell-mediated immunity. Although the presentation is not specific for this organism, listeria should always be suspected in organ transplant patients or patients with a malignancy who develop meningitis.

Primary Bacteremia Bacteremic patients may have an unremarkable history of chills and fever (commonly observed in pregnant women) or have a more

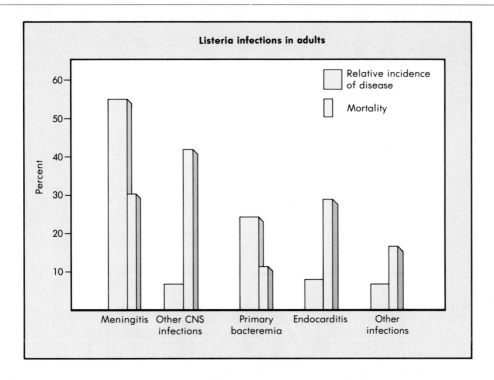

FIGURE 20-2   *Listeria* infections in adults. The most common manifestation of *Listeria* disease in adults is meningitis or meningoencephalitis. Less commonly seen diseases include septicemia, endocarditis, endophthalmitis, peritonitis, amnionitis, osteomyelitis, lymphadenitis, and cholecystitis. (Data from Nieman RE and Lorber B: Rev Infect Dis 2:207-227, 1980.)

acute presentation with high grade fever and hypotension. Mortality appears to be restricted to severely immunocompromised patients and the infants of septic pregnant women.

Endocarditis and Other Infections One complication of listeria bacteremia is endocarditis. A variety of other inflammatory and purulent infections have also been reported.

Laboratory Diagnosis

Microscopy Gram stain preparations of cerebrospinal fluid are usually negative because listeria are generally present in concentrations 100 to 1,000 times less than observed with other causes of bacterial meningitis. In a positive stain, intracellular and extracellular gram-positive coccobacilli are seen. Care must be used in interpreting the Gram stain because the organism can resemble corynebacteria, *Streptococcus pneumoniae,* or (if the preparation is overdecolorized) *Haemophilus.*

Culture Listeria grow on most conventional laboratory media; small, round colonies are observed on

agar media after incubation for 1 to 2 days. The use of selective media and cold enrichment may be required to detect listeria in specimens contaminated with more rapidly growing bacteria. Useful differential tests for the preliminary identification of listeria are listed in Table 20-2. β-hemolysis on sheep blood agar media can separate listeria from morphologically similar bacteria; however, the hemolysis is generally weak and may not be observed initially. Motility in liquid medium or semisolid agar is also helpful for the preliminary identification of listeria (Figure 20-3). All gram-positive bacilli isolated from blood and cerebrospinal fluid should be identified to distinguish between *Corynebacterium* (presumably contaminants) and *Listeria.*

Identification Definitive identification is achieved with selected biochemical and serologic tests. At least 11 serotypes have been described, with 1a, 1b, and 4b the most common.

Treatment, Prevention, and Control

Penicillin or ampicillin, either alone or in combina-

Table 20-2 Differential Features of *Listeria monocytogenes, Erysipelothrix rhusiopathiae,* and *Corynebacterium* Species

Test Reactions	Listeria	Erysipelothrix	Corynebacterium
Hemolysis on blood agar	β	α	Variable
Motility	+	−	−
Catalase	+	−	+
H₂S on TSI*	−	+	−

*Hydrogen sulfide production on triple sugar iron agar.

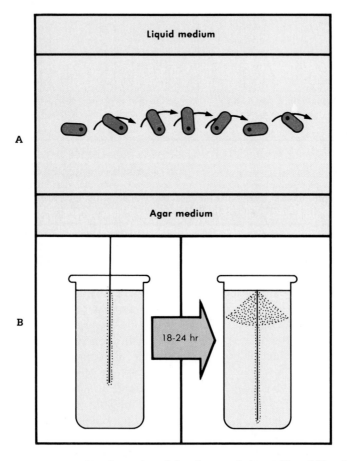

FIGURE 20-3   *Listeria* motility. Detection of the characteristic motility of *Listeria* in liquid and agar media incubated at 25° C can be used to differentiate these organisms from morphologically similar bacteria. In liquid medium the bacilli will tumble end-over-end **(A)**, whereas in agar medium an ''umbrella'' motility is observed **(B)**. Organisms are inoculated into the agar medium by stabbing the semisolid agar with an inoculating loop. After overnight incubation at room temperature, an umbrella of growth is observed, indicating motility is greatest under decreased oxygen tension.

tion with an aminoglycoside, is the treatment of choice for infections with *Listeria monocytogenes*. Because listeria are widespread and most infections are sporadic, prevention and control are difficult. A vaccine is not available, and prophylactic antibiotic therapy for high-risk patients has not been evaluated.

ERYSIPELOTHRIX

Microbial Physiology and Structure

Erysipelothrix rhusiopathiae, the sole member of the genus, is a gram-positive, nonspore-forming, facultative anaerobic bacillus that has worldwide distribution in wild and domestic animals. The bacilli are small, slender (0.2 to 0.4 × 0.5 to 2.5 μm), and sometimes pleomorphic with a tendency to form long filaments. The organisms are microaerophilic, preferring growth in a reduced oxygen atmosphere, with small, grayish, α-hemolytic colonies observed after 2 to 3 days of incubation. Animal disease—particularly in swine—is widely recognized, but human disease is uncommon.

Pathogenesis

Little is known about specific virulence factors in *Erysipelothrix*. Disease in swine has been associated with neuraminidase production, but, because this organism is an uncommon human pathogen, similar studies in humans have not been performed.

Epidemiology

Erysipelothrix is an ubiquitous organism with worldwide distribution in many wild and domestic animals, including mammals, birds, and fish. The first report of disease with this organism was swine erysipelas, and swine remain the most important animal associated with human disease. **Erysipelas** is an occupational disease in humans, with greatest risk for butchers, meat processors, farmers, poultry workers, fish handlers, and veterinarians. Infections follow subcutaneous inoculation of the organism through an abrasion or puncture wound while handling contaminated animal products or soil. Most infections are seen from July to October, although this is a rare human pathogen.

Clinical Syndromes

Human infection with *Erysipelothrix rhusiopathiae* is seen as an inflammatory skin lesion, erysipelas, that develops at the site of trauma after a 1- to 4-day incubation period. The lesion, most commonly present on the fingers or hands, is characterized by an erythematous, raised edge that slowly spreads peripherally as the discoloration in the central area fades. The lesion is pruritic with a burning or throbbing sensation; suppuration is uncommon. Spontaneous resolution can occur but can be hastened with appropriate antibiotic therapy.

Septicemia with *Erysipelothrix* is uncommon, but when present it is frequently associated with endocarditis.

Erysipelothrix endocarditis may have an acute onset but is usually subacute. Involvement of previously undamaged heart valves, particularly the aortic valve, is common, and valvular destruction and metastatic pyogenic complications are reported with this rare disease.

Laboratory Diagnosis

Microscopic examination of skin lesions is not useful. Likewise, culture of material collected from skin lesions with swabs or by aspiration is frequently negative. The most reliable diagnostic specimen is a full-thickness skin biopsy at the leading edge of the erysipeloid lesion.

Erysipelothrix rhusiopathiae will grow on most conventional laboratory media. Initial classification of the organism can be performed as detailed in Table 20-2, with definitive identification performed by biochemical testing.

Treatment, Prevention, and Control

Erysipelothrix is highly susceptible to penicillin, the antibiotic treatment of choice. Cephalosporins, erythromycin, and clindamycin are also active in vitro, but the organism is resistant to the sulfonamides, aminoglycosides, and vancomycin. Whereas erysipelas is generally a nonfatal disease, approximately one third of the patients with endocarditis die despite appropriate antibiotic therapy.

Infections in individuals with an increased occupational risk are prevented by use of gloves and appropriate covering on exposed skin surfaces. Vaccination is not available.

GARDNERELLA

Gardnerella vaginalis, an organism formerly classified in the genera *Haemophilus* and *Corynebacterium,* morphologically resembles a gram-negative bacilli but has the cell wall structure of a gram-positive organism. *G. vaginalis* is pleomorphic, averaging 0.5 × 1.5 μm, nonmotile, and does not form spores or possess a capsule. The organism is facultatively anaerobic and grows slowly in a carbon dioxide enriched atmosphere.

Gardnerella vaginalis is part of the normal vaginal flora in 20% to 40% of healthy women. In women with bacterial vaginosis (nonspecific vaginitis), the number of *Gardnerella* and various obligate anaerobes (primarily *Bacteroides* species and *Peptostreptococcus* species) significantly increases. However, the pathogenic role of *G. vaginalis* in this disease is incompletely defined. This organism has also been associated with postpartum bacteremia, endometritis, and vaginal abscesses.

The simple isolation of *Gardnerella* from vaginal secretions has little diagnostic significance because the organism is frequently recovered in healthy women. The preferred diagnostic test is microscopic examination of vaginal secretions. *G. vaginalis* is suspected when **"clue cells"** are observed (Figure 20-4).

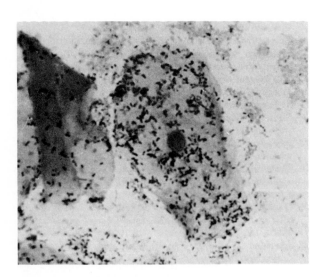

FIGURE 20-4 Vaginal epithelial cells covered with *G. vaginalis* ("clue" cells). Although *G. vaginalis* is gram-positive, the bacteria will typically appear as small gram-negative or gram-variable bacilli. (From Holmes K et al: Sexually transmitted diseases, New York, 1984, McGraw-Hill.)

The abundance of small gram-negative and gram-variable bacilli and the absence of larger gram-positive bacilli (lactobacilli) is consistent with bacterial vaginosis.

Gardnerella vaginalis is inhibited by ampicillin and metronidazole, both of which have been used to treat bacterial vaginosis.

LACTOBACILLUS, EUBACTERIUM, AND BIFIDOBACTERIUM

Other gram-positive bacilli that are part of the normal human bacterial flora include *Lactobacillus, Eubacterium,* and *Bifidobacterium.* These are obligate anaerobes, although some are facultative anaerobes, and are present in the gastrointestinal tract and urogenital tract in large numbers. Each of these organisms can be frequently isolated in clinical specimens, but they have a very low virulence potential and usually represent insignificant contaminants. To demonstrate their etiologic role in an infection, they should be isolated repeatedly in large numbers and in the absence of other pathogenic organisms.

BIBLIOGRAPHY

Amsel R, Totten PA, Spiegel CA, et al: Nonspecific vaginitis: diagnostic criteria and microbial and epidemiologic associations, Am J Med 74:14-22, 1983.

Barza M: Listeriosis and milk (Editorial), N Engl J Med 312:438-440, 1985.

Gorby GL and Peacock JE, Jr: *Erysipelothrix rhusiopathiae* endocarditis: microbiologic, epidemiologic, and clinical features of an occupational disease, Rev Infect Dis 10:317-325, 1988.

Grieco MH and Sheldon C: *Erysipelothrix rhusiopathiae,* Ann NY Acad Sci 174:523-532, 1970.

Nieman RE and Lorber B: Listeriosis in adults: a changing pattern. Report of eight cases and review of the literature, 1968-1978, Rev Infect Dis 2:207-227, 1980.

Stamm AM, Dismukes WE, Simmons BP, et al: Listeriosis in renal transplant recipients: report of an outbreak and review of 102 cases, Rev Infect Dis 4:665-682, 1982.

Actinomyces and Nocardia

The phylogenetic identity of the genera *Actinomyces* and *Nocardia* has historically been obscured by their morphologic similarity to fungi. These organisms typically form delicate filamentous forms, or hyphae, similar to fungi when detected in clinical specimens or isolated in culture. In fact, the name *Actinomyces* is Greek for "ray fungus." However, these organisms are true bacteria, can be readily differentiated from fungi and from each other (Table 21-1), and are responsible for significant human diseases.

ACTINOMYCES

Microbial Physiology and Structure

The genus *Actinomyces* consists of facultative anaerobic or strict anaerobic, gram-positive bacilli that typically are arranged in hyphae that can fragment into short bacilli (Figure 21-1). *Actinomyces* is not acid-fast, grows slowly in culture, and tends to produce chronic, slowly-developing infections. The species responsible for most human infections is *Actinomyces israelii* (named after Israel, who first described human actinomycosis), although disease is also associated with other species—including *A. naeslundii, A. odontolyticus,* and *A. visosus.*

Pathogenesis

Actinomycosis is a chronic infection produced by opportunistic organisms that normally colonize the upper respiratory tract, gastrointestinal tract, and female genital tract. The organisms have a low viru-

lence potential and cause disease only when the normal mucosal barriers are disrupted. When introduced into normally sterile tissues, *Actinomyces* establishes a chronic, suppurative infection that can spread unchecked through normal anatomic barriers, producing a multiorgan disease. Actinomycosis is characterized by multiple abscesses connected by sinus tracts. In the affected tissues and the sinus tracts macroscopic colonies of *Actinomyces* organisms can be seen. These colonies, called **sulfur granules** because they frequently appear yellow or orange, are masses of filamentous organisms bound together by calcium phosphate. The areas of suppuration are

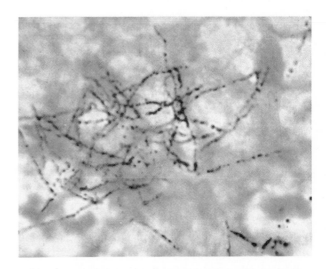

FIGURE 21-1 Gram stain of *Actinomyces* in pus, demonstrating irregularly stained, beaded, branching filaments. (Reprinted with permission from HP Lambert and WE Farrar. Ed. Infectious Diseases Illustrated, London, Gower Medical Publishing 1982.)

Table 21-1 Differential Characteristics of *Actinomyces, Nocardia,* and Fungi

Characterisitcs	*Actinomyces*	*Nocardia*	Fungi
Structural properties			
Nuclear membrane	Absent	Absent	Present
Mitochondria	Absent	Absent	Present
Chitin in cell wall	Absent	Absent	Present
Reproduction			
Bacterial fission	Yes	Yes	No
Budding	No	No	Yes
Spore formation	No	No	Yes
Inhibition by antibiotics			
Penicillin	Yes	No	No
Sulfonamides	No	Yes	No
Antifungal	No	No	Yes
Staining characteristics			
Gram stain	Positive	Positive	Positive
Acid-fast stain	Negative	Positive	Negative
Atmosphere of growth			
Aerobic	No	Yes	Yes
Anaerobic	Yes	No	No
Epidemiology			
Endogenous infections	Yes	No	Yes
Exogenous infections	No	Yes	Yes

commonly surrounded by fibrosing granulation tissue, which gives the surface overlying the involved tissues a hard or woody consistency.

Epidemiology

Actinomycosis is an endogenous infection. There is no evidence of person-to-person spread of disease originating from an external source such as soil or water. All age-groups can be affected, and there is no seasonal or occupational predilection. Disease is classified by the organ systems involved (Box 21-1). **Cervicofacial** infections are seen in patients with poor oral hygiene, following a dental procedure (e. g., tooth extraction), or after trauma to the oral cavity. Patients with **thoracic** infections generally have a history of aspiration, with establishment of disease in the lungs and then spread to adjoining tissues. **Abdominal** infections are most commonly preceded by surgery or trauma to the bowel. **Pelvic** infections can be a secondary manifestation of abdominal actinomycosis or may be a primary infection in women with intrauterine contraceptive devices (Figure 21-2).

Central nervous system infections usually represent secondary spread from another focus.

Clinical Syndromes

The majority of all cases of actinomycosis are cervicofacial. The disease may be seen either as an acute pyogenic infection or more commonly as a slowly evolving, relatively painless process. Tissue swelling with fibrosis and scarring, and open draining sinus tracts along the angle of the jaw and neck, should alert the physician to the possibility of actinomycosis. Symptoms of thoracic actinomycosis are nonspecific. Abscess formation in the lung tissue may be observed in early disease, with subsequent spread into adjoining tissues as the disease progresses. Abdominal actinomycosis can spread throughout the abdomen, involving virtually every organ system. Pelvic actinomycosis can occur as a relatively benign form of vaginitis or with more extensive tissue destruction, including tuboovarian abscesses or urethral obstruction. The most common manifestation of central nervous system actinomyco-

Box 21-1	*Actinomyces* Infections

Cervicofacial actinomycosis
Thoracic actinomycosis
Abdominal actinomycosis
Pelvic actinomycosis
Central nervous system actinomycosis

sis is a solitary brain abscess, but meningitis, subdural empyema, or epidural abscess can also be seen.

Laboratory Diagnosis

Laboratory confirmation of actinomycosis is frequently difficult because the organisms are concentrated in sulfur granules and sparsely distributed in the involved tissues. If sulfur granules can be detected in a sinus tract or in tissue, then the granule should be crushed, Gram stained, and examined microscopically. Thin, gram-positive branching bacilli along the periphery of the granules will be seen (Figure 21-3). The diagnosis can be confirmed by culturing the granules, as well as collecting tissue and purulent material from the site of infection. A large amount of tissue or pus should be collected because relatively few organisms will be present despite

extensive tissue destruction. This is not uncommon in chronic infections caused by slow-growing organisms. Furthermore, *Actinomyces* is fastidious and grows slowly under anaerobic conditions; isolation requires incubation for 2 weeks or more. Facultatively anaerobic isolates grow best in 10% carbon dioxide. Colonies appear white, with a domed surface that can become irregular with incubation for a week or more and resemble a "molar" tooth (Figure 21-4).

It is important to recognize that *Actinomyces* is commonly found on mucosal surfaces. Care must therefore be exercised in the collection of the specimen to avoid contamination with *Actinomyces* organisms that are members of the normal mucosal microbial flora. The significance of isolating *Actinomyces* from a poorly collected specimen cannot be determined.

Identification of *Actinomyces* is accomplished by the selective use of differential biochemical tests. Although the organisms are biochemically active, their slow growth frequently delays definitive test results. Serologic tests for organism identification are generally restricted to reference laboratories.

Treatment, Prevention, and Control

Treatment for actinomycosis involves the combination of surgical debridement of the involved tissues and long-term administration of antibiotics. *Actino-*

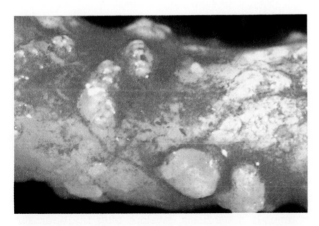

FIGURE 21-2   Pelvic actinomycosis. The female genital tract is normally colonized with *Actinomyces*. Although they rarely cause significant disease, the organisms can colonize the surface of an intrauterine contraceptive device (IUD) and establish an infection. Colonies of *Actinomyces israelii* cover the surface of this IUD.

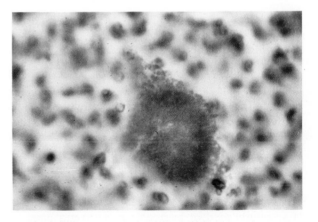

FIGURE 21-3   Sulfur granule collected from sinus tract in patient with actinomycosis. Note the delicate filamentous bacilli at the periphery of the crushed granule.

FIGURE 21-4   Molar tooth appearance of *A. israelii* after incubation for 1 week.

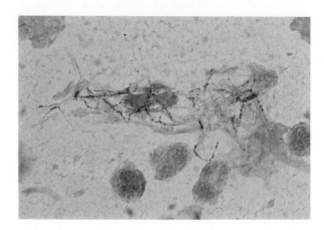

FIGURE 21-5   Gram stain of *N. asteroides* in expectorated sputum. Note that the delicate beaded filaments cannot be distinguished from *Actinomyces*.

myces is uniformly susceptible to penicillin (considered the antibiotic of choice), as well as tetracycline, erythromycin, and clindamycin. For infections that do not appear to respond to prolonged therapy (e. g., 4 to 12 months), an undrained focus should be suspected. Despite extensive tissue destruction by *Actinomyces*, clinical response is generally good. Prevention of these endogenous infections is difficult. However, good oral hygiene should be maintained and appropriate antibiotic prophylaxis should be used when the mouth or gastrointestinal tract is penetrated.

NOCARDIA

Microbial Physiology and Structure

The genus *Nocardia* consists of strict aerobic bacilli that form branched hyphae in both tissues and culture. The organisms are gram-positive, although many stain poorly and appear to be gram-negative with intracellular gram-positive beads (Figure 21-5). *Nocardia* organisms have a cell wall structure similar to mycobacteria with mycolic acids present, and, like the mycobacteria, are acid-fast. However, this property is demonstrated only when weak mineral acids (1% sulfuric acid or hydrochloric acid) are used to decolorize the bacteria. This partial acid-fastness is a helpful differential characteristic separating *Nocardia* from morphologically similar organisms such as *Actinomyces*. *Nocardia* species are catalase positive, utilize carbohydrates oxidatively, and can grow on most nonselective laboratory media; however, isolation can require incubation for 1 week or more. Colonies vary in morphology from dry to waxy and from orange to white.

The genus was named after Nocard, who first described bovine disease characterized by pyogenic pulmonary and cutaneous lesions. *Nocardia asteroides*, *N. brasiliensis*, and *N. caviae* are the most common human pathogens, capable of causing acute or chronic suppurative infections (Table 21-2). *N. asteroides*, the organism isolated most frequently in human infections, causes primarily bronchopulmonary disease, with a high predilection for hematogenous spread to the central nervous system or skin. *N. brasiliensis* and *N. caviae* are most commonly responsible for primary cutaneous infections with infrequent hematogenous dissemination.

Pathogenesis

Members of the genus *Nocarida* are found worldwide in soil rich with organic matter and are rarely isolated as commensal organisms in humans. Bronchopulmonary infections most likely develop after the initial colonization of the oropharynx and then aspiration of oral secretions into the lower airways. Primary cutaneous nocardiosis develops after the traumatic introduction of organisms into subcutaneous tissues. Disease is characterized by necrosis and abscess formation, similar to that caused by other pyogenic bacteria. Chronic infections with sinus tract formation can be observed, particularly with some cutaneous

Table 21-2 *Nocardia* Organisms Associated With Human Disease

Organism	Disease
N. asteroides	Bronchopulmonary disease with secondary dissemination to skin or central nervous system
N. brasiliensis	Cutaneous disease with rare secondary dissemination
N. caviae	Same as *N. brasiliensis*

infections. Although sulfur granules are observed with *Actinomyces*, these are uncommon with *Nocardia* and seen only with cutaneous involvement.

Epidemiology

In contrast with actinomycosis, *Nocardia* infections are exogenous; that is, caused by organisms that are not part of the normal human flora. *N. asteroides* is relatively avirulent, causing infections primarily in patients with defective cell–mediated immunity produced either by disease (e. g., leukemia) or immunosuppressive therapy (e. g., patients receiving corticosteroids). Renal and cardiac transplant recipients are particularly at risk for *Nocardia* infections. Patients infected with *N. brasiliensis* are usually immunocompetent. The organism is introduced into subcutaneous tissues by trauma and rarely causes infections at secondary sites. When systemic disease is observed, it is usually in immunocompromised patients. *N. caviae* is a rare cause of human disease.

Clinical Syndromes

Bronchopulmonary infections caused by *Nocardia* cannot be distinquished from infections caused by other pyogenic organisms. Signs such as cough, dyspnea, and fever are usually present but are not diagnostic. Cavitation and spread into the pleura are common. Although the clinical picture is not specific for *Nocardia*, these organisms should be considered when immunocompromised patients develop pneumonia with cavitation, particularly when evidence of dissemination to the central nervous system or skin exists. **Cutaneous infections** may be seen as cellulitis, pustules, pyoderma, subcutaneous ab-

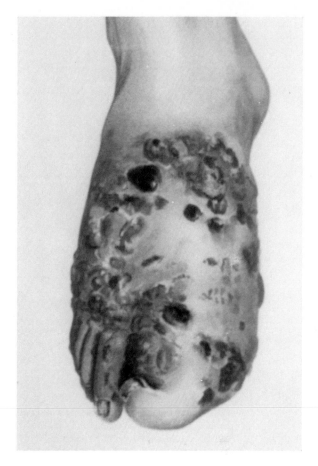

FIGURE 21-6   Myetoma caused by *N. brasiliensis*. The foot is grossly enlarged and covered with multiple draining sinus tracts. (From Binford CH and Connor DH, editors: Pathology of tropical and extraordinary diseases, vol 2, 1976, Washington, DC, Armed Forces Institute of Pathology.)

scesses, lymphocutaneous syndrome, or mycetoma. **Mycetoma** is a chronic, granulomatous disease most frequently present on the lower extremities where the etiologic organism is traumatically introduced (Figure 21-6). The underlying connective tissues, muscle, and bone can be involved, and draining sinus tracts usually open on the skin surface. A variety of organisms can cause mycetoma, although *N. brasiliensis* is the most common cause in North, Central, and South America.

Central nervous system infections, most commonly single or multiple brain abscesses, can be observed in as many as one third of all patients (Figure 21-7).

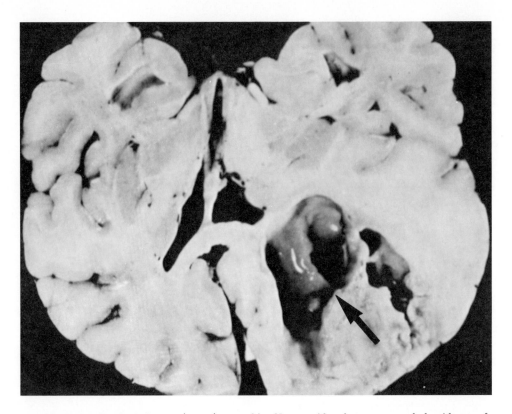

FIGURE 21-7   Cerebral abscess (*arrow*) caused by *N. asteroides*. Autopsy revealed evidence of brochopulmonary and cutaneous disease. (From Binford CH and Connor DH, editors: Pathology of tropical and extraordinary diseases, vol 2, 1976, Washington, DC, Armed Forces Institute of Pathology.)

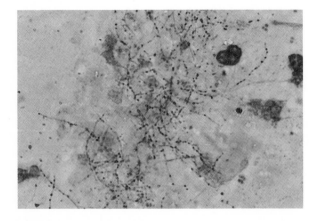

FIGURE 21-8   Acid-fast stain of *N. asteroides* in expectorated sputum.

Laboratory Diagnosis

In contrast with *Actinomyces, Nocardia* organisms are usually distributed throughout the clinical specimen, making microscopic detection and in vitro isolation relatively easy. The delicate hyphae of *Nocardia* in tissues resemble *Actinomyces*; however, *Nocardia* Gram stain poorly, and most isolates are partially acid-fast (Figure 21-8). Although the organism grows on most laboratory media incubated in an atmosphere of 5% to 10% carbon dioxide, the presence of this slow-growing organism may be obscured by more rapidly growing commensal bacteria. Attempts have been made to isolate this organism selectively by decontamination of respiratory secretions or use of selective media. However, these procedures have had limited success. For this reason care should be used in collecting respiratory specimens, although similar problems are generally not encountered for other specimens. It is important to

Table 21-3 Identification of *Nocardia* Responsible for Human Diseases

	N. asteroides	*N. brasiliensis*	*N. caviae*
Decomposition of:			
Casein	−	+	−
Tyrosine	−	+	−
Xanthine	−	−	+
Urea	+	+	+
Acid from:			
Lactose	−	−	−
Xylose	−	−	−

The test reactions must be incubated for 2 weeks at ambient temperature before the reactions are interpreted.

notify the laboratory that nocardiosis is suspected, because most laboratories do not routinely incubate clinical specimens for more than 1 to 3 days, a period of time inadequate for the isolation of *Nocardia*, particularly from contaminated specimens.

Identification of *Nocardia* is uncomplicated. A preliminary classification can be made after demonstration of acid-fast hyphal elements. However, definitive identification is frequently delayed because *N. asteroides*, the species most commonly isolated, is nonreactive with most differential tests used for biochemical classification (Table 21-3).

Treatment, Prevention, and Control

Nocardia infections are treated with antibiotics combined with appropriate surgical intervention. Sulfonamides are the antibiotics of choice for treating nocardiosis. Tobramycin, amikacin, and some of the newer β-lactams also have good in vitro activity but unproven in vivo effectiveness. Antibiotic therapy should be extended for 6 weeks or more. Whereas clinical response is favorable in localized infections, the prognosis is poor for patients with disseminated disease.

Because *Nocardia* are ubiquitous, exposure to the organisms cannot be avoided. However, bronchopulmonary disease is uncommon in immunocompetent persons, and primary cutaneous infections can be prevented by proper wound care. The complications associated with disseminated disease can be minimized by consideration of nocardiosis in the differential diagnosis for immunocompromised patients with cavitary pulmonary disease.

BIBLIOGRAPHY

Arroyo JC, Nichols S, and Carroll GF: Disseminated *Nocardia caviae* infection, Am J Med 62:409-412, 1977.

Beaman BL, Burnside J, Edwards B, and Causey W: Nocardial infections in the United States, 1972-1974, J Infect Dis 134:286-289, 1976.

Brown JR: Human actinomycosis: a study of 181 subjects, Hum Pathol 4:319-330, 1973.

Frazier AR, Rosenow EC, and Roberts GD: Nocardiosis. A review of 25 cases occurring during 24 months, Mayo Clin Proc 50:657-663, 1975.

Smega RA: Actinomycosis of the central nervous system, Rev Infect Dis 9:855-865, 1987.

Smego RA and Gallis HA: The clinical spectrum of *Nocardia brasiliensis* infection in the United States, Rev Infect Dis 6:164-180, 1984.

Weese WC and Smith IM: A study of 57 cases of actinomycosis over a 36 year period. A diagnostic failure with "good" prognosis after treatment, Arch Intern Med 135:1562-1568, 1975.

Mycobacterium

The genus *Mycobacterium* consists of nonmotile, nonspore-forming aerobic bacilli, 0.2-0.6 × 1-10 μm in size. The bacilli occasionally form branched filaments, but these can be readily disrupted. The cell wall is rich in lipids, making the surface hydrophobic and the mycobacteria resistant to many disinfectants, as well as to such common laboratory stains as the Gram and Giemsa. Once stained, the bacilli are also refractory to decolorization with acid solutions, hence the name **acid-fast bacilli.** Because the mycobacterial cell wall is complex and this group of organisms is fastidious, most mycobacteria grow slowly, dividing every 12 to 24 hours. Isolation of the "rapidly growing" mycobacteria requires incubation for 3 days or more; the slow growing organisms (e.g., *Mycobacterium tuberculosis, M. avium-intracellulare)*

require 3 to 8 weeks. *M. leprae,* the etiologic agent of leprosy, cannot be grown in cell-free cultures.

Human infections caused by mycobacteria are described in the Bible (e.g., leprosy) and in writings from ancient Greece (e.g., wasting diseases resembling tuberculosis). With our ability now to treat most mycobacterial diseases effectively, it is difficult to remember that tuberculosis was a leading cause of death less than 250 years ago. It is still a significant cause of morbidity and mortality in countries with limited medical resources, and *M. tuberculosis,* as well as other mycobacteria (e.g., *M. avium-intracellulare),* can cause overwhelming, disseminated disease in immunocompromised patients. The mycobacteria that are associated with human disease are listed in Table 22-1. *M. tuberculosis* and *M. bovis* (as well as

Table 22-1 Mycobacteria Associated With Human Disease*

Mycobacterium	Environmental contaminant	Reservoir
M. tuberculosis	No	Humans
M. bovis	No	Humans, cattle, other mammals
M. leprae	No	Humans, armadillos
M. kansasii	Rarely	Water, cattle, swine (rarely)
M. marinum	Rarely	Fish, water
M. simiae	No	Primates, possibly water
M. scrofulaceum	Possibly	Soil, water, foodstuffs
M. szulgai	No	Unknown
M. avium-intracellulare	Possibly	Soil, water, swine, cattle, birds
M. xenopi	Possibly	Water
M. ulcerans	No	Unknown
M. fortuitum	Yes	Soil, water, animals, marine life
M. chelonae	Yes	Soil, water, animals, marine life

*Human disease has also been rarely associated with other mycobacterial species.

M. africanum and *M. microti*) can be differentiated by physiologic properties, despite the fact that these organisms share almost complete nucleic acid homology. It is likely that these mycobacteria will be combined into a single genus in the future; however, they will be considered separately in this discussion. Although at least 37 species of mycobacteria have been described, more than 95% of all human infections are caused by six species: *M. tuberculosis, M. avium-intracellulare, M. kansasii, M. fortuitum, M. chelonae,* and *M. leprae.*

MICROBIAL PHYSIOLOGY AND STRUCTURE

Mycobacteria possess a complex cell wall, with the structural foundation formed by the peptidoglycan skeleton with covalently linked arabinogalactan-mycolate molecules and overlayed with free lipids and polypeptides (Figure 22-1).

Approximately 25% of the dry weight is free lipids located on the outer layers of the cell. The lipids include waxes, species-specific mycosides (complex glycolipids and peptidoglycolipids), and **cord factor** (6,6''-dimycolate of trehalose). Cord factor is associated with the parallel alignment of rows of bacilli ("cord" formation), a characteristic of virulent strains of *M. tuberculosis.*

The peptide chains in the outer layer comprise 15% of the cell wall weight and are biologically important antigens, stimulating the patient's cellular immune response to infection. Extracted and partially purified preparations of these protein derivatives are used as a skin testing reagent to measure exposure to *M. tuberculosis.* Similar preparations from other mycobacteria have been used as species-specific skin testing reagents.

The peptidoglycan skeleton is relatively uniform in all mycobacterial species and forms the major component of the cell wall (Figure 22-1). As with other bacteria, intrapeptide bridges between the peptidoglycan chains bind the skeleton into a rigid structure. Attached to these chains by covalent bonds are **mycolic acids** linked to *D*-arabinose and *D*-galactose. The mycolic acids are the major lipids in the mycobacterial cell wall.

Growth properties and colonial morphology are used for the preliminary identification of mycobacteria. *M. tuberculosis* and closely related species are slow-growing bacteria, and the colonies are nonpigmented (buff-colored; Figure 22-2). Runyon classified the other mycobacteria into four groups based on their rate of growth and their ability to produce pig-

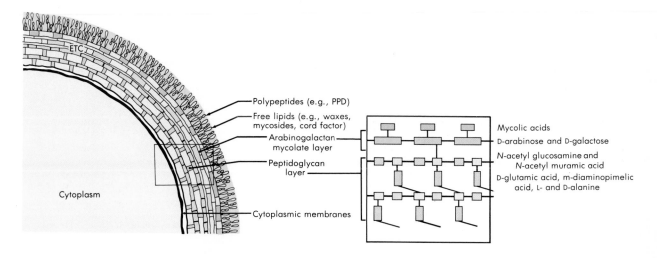

FIGURE 22-1   Structural arrangement of mycobacterial cell wall layers. Mycolic acid is esterified to the arabinose-galactose layer at the arabinose side chain. Phosphodiester linkage binds the arabinogalactan layer to the underlying peptidoglycan layer at the muramic acid subunit. This linkage occurs at every tenth repeating arabinogalactan unit. (Redrawn from Kubica GP and Wayne LG, editors: The mycobacteria: a sourcebook, New York, 1984, Marcel Dekker, Inc.)

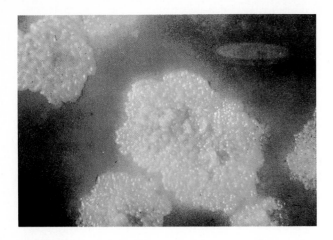

FIGURE 22-2   *M. tuberculosis* colonies on Lowenstein-Jensen agar after 8 weeks of incubation. (From Finegold SM and Baron EJ: Bailey and Scott's diagnostic microbiology, ed, St Louis, 1986, The CV Mosby Co.)

FIGURE 22-3   *M. kansasii* colonies on Middlebrook agar after exposure to light. (From Finegold SM and Baron EJ: Bailey and Scott's diagnostic microbiology, ed, St Louis, 1986, The CV Mosby Co.)

Table 22-2   Classification of Important Mycobacteria

Mycobacteria	Clinical Significance	Pigmentation*	Rate of Growth
UNCLASSIFIED MYCOBACTERIA			
M. leprae	Strict pathogen	†	†
M. tuberculosis	Strict pathogen	No	Slow
M. bovis	Strict pathogen	No	Slow
M. ulcerans	Strict pathogen	No	Slow
RUNYON GROUP I			
M. kansasii	Usually pathogenic	Photo	Slow
M. marinum	Usually pathogenic	Photo	Moderate
M. simiae	Usually pathogenic	Photo	Slow
RUNYON GROUP II			
M. scrofulaceum	Rarely pathogenic	Scoto	Slow
M. szulgai	Strict pathogen	Scoto	Slow
M. xenopi	Rarely pathogenic	Scoto	Slow
M. gordonae	Nonpathogenic	Scoto	Slow
M. flavescens	Nonpathogenic	Scoto	Moderate
RUNYON GROUP III			
M. avium-intracellulare	Usually pathogenic	No	Slow
RUNYON GROUP IV			
M. fortuitum	Rarely pathogenic	No	Rapid
M. chelonae	Rarely pathogenic	No	Rapid

Modified from the Manual of clinical microbiology, ed 4, Washington, DC, American Society for Microbiology, 1985.

*Pigmentation: nonpigmented, no; photochromogen, photo; scotochromogen, scoto.

†*M. leprae* does not grow in vitro.

ments in the presence or absence of light (Table 22-2). The pigmented mycobacteria produce intense yellow pigmented carotenoids (Figure 22-3). Photochromogens are organisms that produce these pigments only after exposure to light, whereas the scotochromogenic organisms can produce the pigments in the dark as well as the light. The **Runyon classification** can be used to differentiate the most common mycobacterial isolates, although definitive identification requires biochemical testing.

PATHOGENESIS

Mycobacterium tuberculosis

Tuberculosis is the classic human mycobacterial disease. The route of infection is inhalation of infectious aerosols which are able to reach the terminal airways. Following engulfment by alveolar macrophages, the bacilli are able to replicate freely, with eventual destruction of the phagocytic cells. This leads to further cycles of phagocytosis by macrophages and lymphocytes that migrate to the site of infection, followed by cellular destruction. Infected macrophages also spread during the initial phase of disease to the local lymph nodes, as well as into the bloodstream and other tissues (e.g., bone marrow, spleen, kidneys, bone, central nervous system).

This process of slow growth, tissue destruction, and dissemination continues until the patient develops a cellular immune reaction to the organisms. This is heralded by the development of positive skin test reactivity to mycobacterial antigens. Although unrelenting growth and tissue destruction can continue in progressive active tuberculosis, in most patients the cellular immune response arrests the disease at this stage, with the only signs of infection a lifelong positive skin test and radiographic evidence of calcification of the initial active foci in the lungs or other organs.

Histologic response to infection and growth of mycobacteria is characterized by granulomatous inflammation. When the tissue is initially infected, macrophages and polymorphonuclear leukocytes accumulate. After several weeks, and before an effective immune reaction is initiated, the macrophages become predominant, some of which form multinuclear giant cells or **"Langhans cells."** In extensive, active disease, caseous (cheesy) necrosis and cavitation are seen. This tissue destruction is mediated by the patient's response to growth of the bacilli; no mycobacterial toxins have been described.

Reactivation of quiescent bacilli can develop years later when the patient's immunologic responsiveness wanes, either due to old age or to immunosuppressive disease or therapy.

Other Mycobacteria

The histologic hallmark of tuberculosis (granulomatous inflammation) is seen with other mycobacterial infections. Disseminated infections with *M. kansasii* and *M. avium-intracellulare* are observed in immunocompromised patients. However, infections with *M. marinum* and *M. ulcerans* are generally restricted to the skin surfaces, because these organisms grow preferentially at cool temperatures. *M. fortuitum* and *M. chelonae* rarely cause disseminated infections, because these mycobacteria are relatively avirulent.

EPIDEMIOLOGY

Mycobacterium tuberculosis

Tuberculosis is spread by close person-to-person contact through infectious aerosols. On rare occasions, disease can also be acquired by ingestion or skin trauma. Although disease can be established in primates and laboratory animals such as guinea pigs, humans are the only natural reservoir.

As the incidence of tuberculosis has decreased in the United States, the opportunity for exposure to this organism has likewise diminished. Thus primary exposure is generally restricted to selected populations such as the urban poor, patients with compromised immune systems (e.g., AIDS patients), and immigrants from areas with a high incidence of disease (e.g., Southeast Asia). Medical care workers are also at risk for infection, because many patients with tuberculosis are admitted to hospitals without the disease initially suspected.

Mycobacterium bovis

M. bovis can infect a variety of animals, with cattle the most significant source of human exposure. Bovine tuberculosis is rarely seen in the United

States, because disease in animal herds is controlled by vaccination of healthy animals, quarantine and destruction of infected herds, and pasteurization of milk. In contrast with *M. tuberculosis,* most infectious with *M. bovis* follow consumption of contaminated milk.

Mycobacterium avium and M. intracellulare

These species are generally considered together because differentiation by physiologic parameters is difficult and their diseases in humans are identical. The organisms are isolated in soil and water, as well as infected poultry, swine, and other animals. Although human exposure is common, significant human disease is rare except in immunocompromised patients. Patients with AIDS are at particular risk for disseminated disease with these mycobacteria, although the reasons for this are not entirely clear. Person-to-person spread does not occur.

Mycobacterium leprae

Although leprosy is a relatively rare mycobacterial disease in the United States (approximately 250 cases are reported annually), more than 12 million cases are recognized worldwide. Less than 10% of the infections recognized in the United States are in native-born Americans, with most infections in immigrants from Mexico, Asia, the Pacific Islands, and Africa. The largest number of cases in the United States are reported in California, Hawaii, and Texas, and leprosy is endemic in these states, as well as Louisiana. Endemic disease has also been demonstrated in armadillos.

Disease is spread by person-to-person contact. Although the most important route of infection is unknown, it is believed that *M. leprae* can be spread by either inhalation of infectious aerosols or skin contact with respiratory secretions, wound exudates, or arthropod vectors. Large numbers of *M. leprae* are observed in nasal secretions in patients with lepromatous leprosy.

Other Mycobacterial Species

Most other mycobacteria associated with human disease are frequently isolated from environmental sources (see Table 22-1). Person-to-person spread is not observed.

CLINICAL SYNDROMES

Mycobacterium tuberculosis

Tuberculosis can involve any organ, although most infections are restricted to the lungs. Infection is initiated after inhalation of contaminated aerosol droplets. The initial pulmonary focus is the middle or lower lung fields, where the tubercle bacilli can multiply freely. Within 3 to 6 weeks the patient's cellular immunity is activated and replication ceases for most patients. However, approximately 5% of patients exposed to *M. tuberculosis* will develop active disease within 2 years, and another 5 to 10% will develop disease sometime later in life.

The clinical signs and symptoms of tuberculosis reflect the site of infection, with primary disease usually restricted to the lower respiratory tract. The onset is generally insidious, with nonspecific complaints of malaise, weight loss, cough, and night sweats. Sputum production may be scant or bloody and purulent. Sputum production and hemoptysis is usually associated with cavitary disease. The clinical diagnosis is supported by radiographic evidence of pulmonary disease, positive skin test reactivity, and the laboratory detection of mycobacteria either by microscopy or isolation of organisms in culture. When active disease develops with pneumonitis or with abscess formation and cavitation, one or both upper lobes of the lungs are usually involved.

Extrapulmonary tuberculosis resulting from hematogenous spread of the bacilli during the initial phase of multiplication can also be seen. The most common sites of infection include lymph nodes, pleura, and the genitourinary tract. With disseminated or miliary tuberculosis there is frequently no evidence of pulmonary disease.

Mycobacterium avium-intracellulare Complex

This group of organisms was historically an uncommon human pathogen. Isolation of *M. avium-intracellulare* frequently represented transient colonization in asymptomatic patients. When disease was observed, it was generally restricted to patients with compromised immune functions and was clinically identical to tuberculosis, with pulmonary disease seen most frequently. More recently these organisms have assumed renewed importance, representing the most common mycobacterial isolate in many clinical

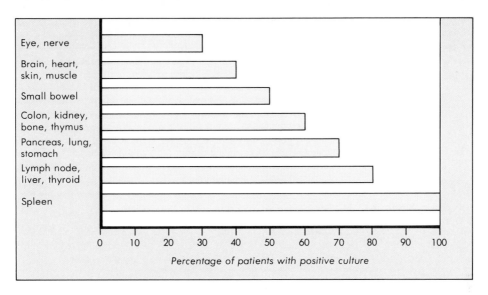

FIGURE 22-4 Distribution of *M. avium-intracellulare* in AIDS patients. (Redrawn from Hawkins CC et al: Ann Inter Med 105:184-188, 1986.

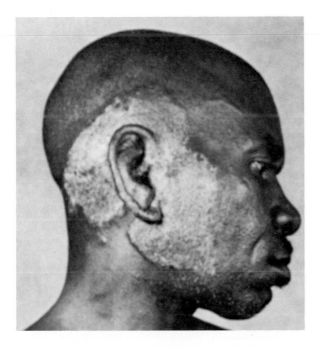

FIGURE 22-5 Tuberculoid leprosy. An anesthetic plaque with well-defined edges and scaly surface extends over the cheek and into the scalp in a man from Zaire. (From Binford CH and Connor DH, editors: Pathology of tropical and extraordinary diseases: an atlas, Washington, DC, 1976, Armed Forces Institute of Pathology.

FIGURE 22-6 Lepromatous Leprosy. Advanced lepromatous lesions over most of the body and gynecomastia in this adult Zairian man. Reprinted with permission (From Binford Ch and Connor DH, editors: Pathology of tropical and extraordinary diseases: an atlas, Washington, DC, 1976, Armed Forces Institute of Pathology.

Table 22-3   Clinical and Immunologic Manifestations of Leprosy

	Tuberculoid	Lepromatous
Skin lesions	Few erythematous or hypopigmented plaques with flat centers and raised, demarcated borders; peripheral nerve damaged with complete sensory loss; nerves visibly enlarged	Many erythematous macules, papules, or nodules; extensive tissue destruction (e.g., nasal cartilage and bone, ears); diffuse nerve involvement with patchy sensory loss; nerves not enlarged
Histopathology	Infiltration of lymphocytes around center of epithelial cells; Langhan cells present; few or no acid-fast bacilli present	Predominantly "foamy" macrophages with few lymphocytes; Langhan cells absent; acid-fast bacilli abundant in skin lesions and internal organs
Infectivity	Low	High
Immune response		
Delayed hypersensitivity	Reactivity to lepromin	Nonreactive to lepromin
Immunoglobulin levels	Normal	Polyclonal hypergamma-globulinemia
Erythema nodosum leprosum	Absent	Usually present

laboratories, because they are the major mycobacterial pathogen in AIDS patients seen in the United States. In contrast with disease in other groups of patients, *M. avium-intracellulare* infection in AIDS patients is typically disseminated, with virtually no organ spared (Figure 22-4). The tissues of some patients are literally filled with the mycobacteria and hundreds to thousands of bacilli per milliliter of blood have been isolated. The magnitude of these infections is remarkable, because overwhelming sepsis is initiated by fewer than 1 organism per milliliter of blood in most other bacteremic patients (e.g., infected with *Escherichia coli* or *Staphylococcus aureus*). Thus these mycobacteria must be relatively avirulent to persist in such high concentrations for a prolonged period of time without clinical evidence of disease.

Mycobacterium leprae

The clinical presentation of leprosy (also called **Hansen's disease**) is subdivided into **tuberculoid leprosy** and **lepromatous leprosy,** with intermediate forms also recognized. The intermediate forms can evolve into either tuberculoid or lepromatous leprosy. Each form has characteristic clinical and immunologic manifestations (Table 22-3; Figures 22-5 and 22-6). Lepromatous leprosy is the form characteristically associated with disfiguring skin lesions and large numbers of acid-fast bacilli in the infected tissues. It is also the most infectious form of leprosy.

Other Mycobacteria

M. kansasii is the most common photochromogen responsible for human disease. Isolation from respiratory specimens can represent transient colonization or significant pulmonary disease. Chronic pulmonary disease is the most common presentation of *M. kansasii* infection, although disseminated disease in immunocompromised patients is also seen.

M. marinum and *M. ulcerans* are organisms that grow preferentially at cooler temperatures. Thus they usually cause disease restricted to the skin. *M. marinum* infections are acquired by contact with contaminated fresh water or salt water. For this reason these infections are termed **"swimming pool granulomas."** Cutaneous infection may be seen either as a nodular lesion that progresses to ulceration or as a series of nodular lesions along the lymphatics draining the area of primary ulceration. This latter appearance resembles cutaneous disease produced by the fungus *Sporothrix schneckii*.

M. ulcerans infections are not seen in the United States but are restricted primarily to Africa, Mexico, and Australia. The infection initially is seen as a subcutaneous nodule that ulcerates. The indolent infection can progress over a period of months, with

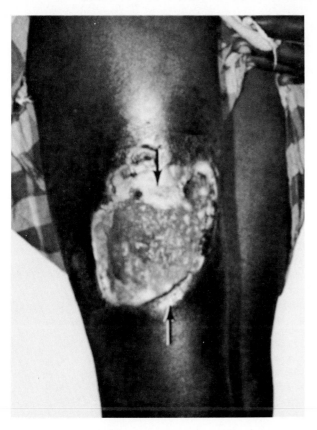

FIGURE 22-7 *Mycobacterium ulcerans* infection. Infection begins as a painless nodule. As the nodule expands, the overlying skin can break down and ulcerate. Despite extensive lesions the patient generally feels well, with no fever, regional lymphadenopathy, malaise, or leukocytosis. Secondary infections can produce systemic signs and symptoms. Note the proximal healing *(upper arrow),* but the ulcer is extending distally *(lower arrow).* (From Binford CH and Connor DH, editors: Pathology of tropical and extraordinary diseases: an atlas, Washington, DC, 1976, Armed Forces Institute of Pathology.

involvement of the subcutaneous tissues and extensive ulceration (Figure 22-7).

The rapidly growing mycobacteria (e.g., *M. fortuitum* and *M. chelonae*) are most commonly associated with disease following the introduction of bacteria into the deep subcutaneous tissues by trauma or iatrogenically (e.g., infections introduced with a intravenous catheter, contaminated wound dressing, or prosthetic device such as a heart valve). Unfortunately, infections with these organisms are increasing as more invasive procedures are performed on hospital-ized patients and advanced medical care is able to extend the survival of immunocompromised patients.

LABORATORY DIAGNOSIS

Skin Test

Reactivity to intradermal injection of mycobacterial antigens can differentiate between infected and noninfected individuals. Tests with antigens extracted from *M. tuberculosis* have been used most commonly and are best standardized, although skin tests with species-specific antigens from other mycobacteria have also been developed.

The methods of antigen preparation and skin inoculation have undergone a number of changes since the tests were first developed. Two preparations of tuberculin antigens are currently used: "old tuberculin" and "purified protein derivative" **(PPD).** The standardized PPD is recommended. Tuberculin can be inoculated into the intradermal layer of skin by either intradermal injection of a specific amount of antigen or by placing a drop of the antigen on the skin surface and then injecting the antigen with the use of a multiprong inoculator **(Tine test).** The latter test is not recommended because the amount of antigen injected into the intradermal layer cannot be accurately controlled.

The most common dose of antigen is 5 tuberculin units (equivalent to 0.1 µg of PPD). Skin test reactivity is measured 48 hours after intradermal injection, with 10 mm or more induration considered positive for exposure to *M. tuberculosis*. A positive PPD reaction usually develops within 3 to 4 weeks after exposure. Some infected patients may have less than 10 mm induration, but this level of reactivity generally represents exposure to other mycobacteria (Figure 22-8). Patients infected with *M. tuberculosis* may not respond to tuberculin skin testing if they are anergic; thus control antigens should always be used with tuberculin tests.

Microscopy

Detection of acid-fast bacilli in clinical specimens is extremely valuable for rapid laboratory confirmation of mycobacterial disease. The clinical specimen is

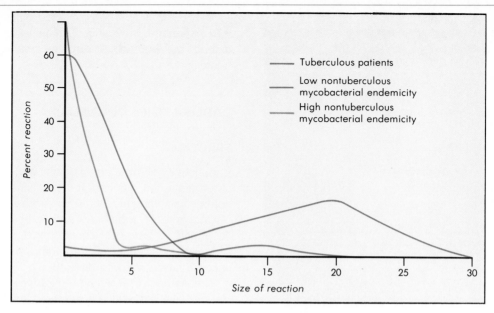

FIGURE 22-8   Hypothetical reactivity to *Mycobacterium tuberculosis* skin test antigens (PPD). The majority of patients exposed to *M. tuberculosis* have a 14- to 20-mm skin test reaction (induration) when tested with 5 tuberculin units of PPD. In communities where there is a low incidence of tuberculosis, as well as an exposure to other mycobacteria, the majority of patients either fail to react to PPD or have less than 10 mm induration (representing exposure to related mycobacteria and immunologic cross-reactivity). In communities where exposure to mycobacteria other than *M. tuberculosis* is high the majority of reactions to PPD are less than 10 mm.

stained with either carbolfuchsin (Ziehl-Neelsen and Kinyoun stains) or fluorochrome dyes (e.g., Truant auramine-rhodamine), decolorized with acid-alcohol solution, and then counterstained. The specimens are then examined with either a light microscope or, if fluorochrome dyes are used, a fluorescent microscope (Figures 22-9 and 22-10). The fluorochrome stain is more sensitive because the specimen can be rapidly scanned for fluorescence with low magnification and then confirmed using higher magnification.

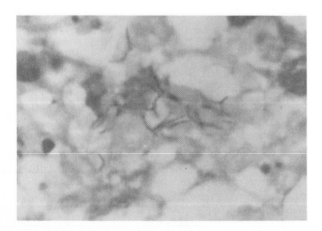

FIGURE 22-9   Acid-fast stain of *M. tuberculosis* stained by the Kinyoun method with carbolfuchsin.

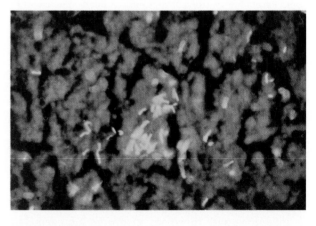

FIGURE 22-10   Acid-fast stain of *M. tuberculosis* stained by the Truant fluorochrome method with the fluorescent dyes auramine and rhodamine.

Approximately one third to one half of all culture-positive specimens will have a positive acid-fast stain. The test sensitivity is improved with respiratory specimens, particularly from patients with radiographic evidence of cavitation, and specimens with large numbers of mycobacteria isolated in culture. The test specificity is greater than 95% when carefully performed.

Culture

Patients with pulmonary tuberculosis, particularly cavitary disease, can have large numbers of organisms in their respiratory secretions (e.g., 10^8 bacilli or more). A positive culture will be obtained in virtually all patients with active pulmonary disease if an early morning sputum is collected for 3 consecutive days. Recovery of *M. tuberculosis* from other sites in disseminated disease (e.g., genitourinary tract, cerebrospinal fluid) is sometimes more difficult and requires collection of additional cultures or processing a large volume of fluid.

Isolation of mycobacteria from clinical specimens is complicated by the fact that most isolates grow slowly and can be obscured by the rapidly growing bacteria normally present in clinical specimens. Thus specimens such as sputum are initially treated with a decontaminating reagent (e.g., 2% sodium hydroxide). Because mycobacteria are tolerant to brief alkali treatment, this process kills the rapidly growing bacteria and permits the selective isolation of mycobacteria. Extended decontamination of the specimen will kill mycobacteria, so the procedure is not performed with normally sterile specimens or when small numbers of mycobacteria are expected.

Traditionally, specimens are inoculated onto both egg-based media (e.g., Lowenstein-Jensen) and agar-based media (e.g., Middlebrook). Detection of *M. tuberculosis, M. avium-intracellulare,* and other important slow-growing mycobacteria generally requires incubation for 3 weeks or more. This detection time has recently been shortened by the use of specially formulated broth cultures where the metabolism of ^{14}C-labeled palmitic acid is measured in an ion chamber system (BACTEC; Table 22-4).

Nucleic Acid Probes

Probes have been developed for the detection of nucleic acid sequences specific for mycobacteria. This approach has been very successful for the identification of *M. tuberculosis* and *M. avium-intracellulare* isolated in culture. Although these probes have had limited success with the direct detection of mycobacteria in clinical specimens, it is anticipated that gene amplification techniques should remedy this problem.

Preliminary Identification

Growth properties and colonial morphology can be used for the preliminary identification of *M. tuberculosis* and mycobacteria other than *M. tuberculosis* (commonly referred to as MOTT; see Table 22-2). It is important that the clinical laboratory provide this preliminary classification because person-to-person transmission of MOTT strains generally does not occur; thus isolation precautions and prophylactic antibiotics are not required. Additionally, antimicrobial therapy must be selected for the specific isolates.

Definitive Identification

The definitive identification of most mycobacteria requires the use of selected biochemical tests (Table

Table 22-4 Recovery of Mycobacteria

Organism (Number of Isolates)	Mean Time of Detection (Days)		
	Egg Medium	Agar Medium	BACTEC Broth
M. tuberculosis (52)	30.2	28.2	11.7
M. avium-intracellulare (86)	33.1	35.8	11.4
M. fortuitum-chelonei (33)	16.1	16.5	8.4

Data collected at Barnes Hospital during a 9-month period.

Table 22-5   Selected Biochemical Reactions for Five Commonly Isolated Mycobacteria

Organism	Niacin	Nitrate Reductase	Heat-Stable Catalase	Tween-80 Hydrolysis	Iron Uptake	Arylsulfatase	Urease
M. tuberculosis	+	+	−	−		−	+
M. kansasii	−	+	+	+		−	+
M. avium-intracellulare	−	−	±	−		−	−
M. fortuitum	−	+	+	V	+	+	+
M. chelonei	V*	−	V	V	−	+	+

Modified from Manual of clinical microbiology, ed 4, Lennette EH, Balows A, Hausler WJ Jr, and Shadomy HJ: Washington DC, 1985, American Society for Microbiology.
*V, variable.

22-5). Key tests for the identification of *M. tuberculosis* include production of niacin and nitrate reductase. Unfortunately, biochemical identification of mycobacteria requires 3 weeks or more before results are available. Inoculation of guinea pigs and demonstration of disease has also been used for identification of *M. tuberculosis,* although this test is now rarely used.

Because *M. leprae* cannot be grown in cell-free cultures, laboratory confirmation of leprosy requires histopathology consistent with clinical disease and skin test reactivity to lepromin or the presence of acid-fast bacilli in the lesions.

Serology

A number of tests (e.g., RIA, ELISA, latex agglutination) have been developed for the serologic diagnosis of active mycobacterial disease. Unfortunately, tests for the detection of mycobacterial antigens or specific antibody are insensitive and nonspecific. Somewhat better success has been achieved with the use of highly purified antigens in the diagnosis of tuberculous meningitis. However, these tests are only useful in chronic, extensive disease. Further work is necessary before mycobacterial serology can be recommended as a diagnostic tool.

TREATMENT

Most mycobacteria are resistant to antibiotics used to treat other bacterial infections. Effective therapy for infection with *M. tuberculosis* requires use of at least two antimycobacterial agents in order to avoid the selection of resistant organisms during treatment. Isoniazid (INH) combined with rifampin is most commonly used, although therapy should be guided by the results of in vitro susceptibility tests. Because mycobacteria multiply slowly, treatment has traditionally been for 18 to 24 months. Recent studies indicate that therapy for 9 months or less can be effective for most patients with tuberculosis.

M. avium-intracellulare is resistant to most common antimycobacterial agents. Effective therapy is prolonged, with a combination of primary or secondary drugs demonstrated to be active in vitro. Unfortunately, the toxic effects of many of these therapeutic combinations are significant.

In contrast with related species, the rapidly growing mycobacteria associated with human disease (e.g., *M. fortuitum* and *M. chelonae*) are resistant to most commonly used antimycobacterial agents but susceptible to some aminoglycosides, cephalosporins, tetracyclines, and quinolones. However, the specific activity of these agents must be determined by in vitro tests. Because infections with these mycobacteria are generally confined to the skin or associated with prosthetic devices, surgical intervention is also necessary.

Treatment of *M. leprae* infections is based on clinical experience because in vitro testing is not possible and in vivo testing with animal models (e.g., mouse foot pad inoculations) is not practical. Drug resistance develops rapidly if a single agent is used. Therefore it is recommended that primary therapy consist of dapsone, rifampin, and either clofazimine or ethionamide. If this disease is diagnosed early in its course

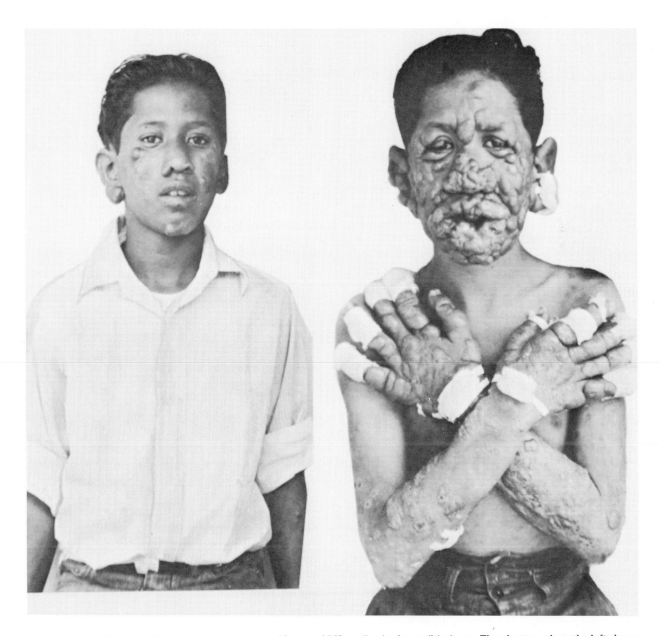

FIGURE 22-11   Advance of lepromatous leprosy in a 13-year-old Hawaiian in the antibiotic era. The photograph on the left shows the patient on admission in 1931. The photograph to the right was taken in 1933. Had antibiotics been available on admission, the disease would not have progressed but would have regressed and probably become bacteriologically negative after several years of treatment. (From Binford CH and Connor DH: Pathology of tropical and extraordinary diseases: an atlas, Washington, DC, 1976, Armed Forces Institute of Pathology.)

and if treatment is initiated promptly, clinical response is satisfactory. However, delays in treatment can be devasting (Figure 22-11). Therapy is administered for a minimum of 2 years and lifelong in some patients.

PREVENTION AND CONTROL

Chemoprophylaxis

Prophylaxis with INH for 1 year is advocated for individuals with significant exposure to *M. tuberculosis* or recent conversion of skin test reactivity. In reality, this approach is aimed at treating subclinical infection rather than prophylaxis. Hepatotoxicity, particularly in older persons, is associated with INH prophylaxis. Chemoprophylaxis is unnecessary for persons exposed to patients with MOTT infections.

Immunoprophylaxis

Vaccination with attenuated *M. bovis* (bacille Calmette-Guerin or **BCG**) is commonly used in countries where tuberculosis is endemic and responsible for significant morbidity and mortality. This vaccination significantly reduces the incidence of tuberculosis, similar to INH prophylaxis. One problem with BCG immunization is that all patients develop positive skin test reactivity; thus skin testing cannot be used to detect previous exposure to *M. tuberculosis*. For this reason BCG immunizations are not used in the United States and other countries where the incidence of tuberculosis is low.

BIBLIOGRAPHY

Binford CH and Connor DH: editors: Pathology of tropical and extraordinary diseases: an atlas, Washington, DC, 1976, Armed Forces Institute of Pathology.

David HL: Bacteriology of the mycobacterioses, Pub No CDC 76-8316, 1976, Washington, DC, US Department of Health, Education, and Welfare.

Davidson PT: Tuberculosis: new views of an old disease, N Engl J Med 312:1514-1515, 1985.

Glassroth J, Robins AG, and Snyder DE Jr: Tuberculosis in the 1980s, N Engl J Med 302:1441-1450, 1980.

Good RC: Serologic methods for diagnosing tuberculosis, Ann Intern Med 110:97-98, 1989.

Hastings RC, Gillis TP, Krahenbuhl JL, and Franzblau SG: Leprosy, Clin Microbiol Rev 1:330-348, 1988.

Hawkins CC, Gold JWM, Whimbey E, et al: *Mycobacterium avium* complex infections in patients with the acquired immunodeficiency syndrome, Ann Intern Med 105:184-188, 1986.

Kubica GP and Wayne LG: The mycobacteria: a sourcebook, New York, 1984, Marcel Dekker, Inc.

Lipsky BA, Gates J, Tenover FC, and Plorde JJ: Factors affecting the clinical value of microscopy for acid-fast bacilli, Rev. Infect Dis 6:214-222, 1984.

MacGregor RR: A year's experience with tuberculosis in a private urban teaching hospital in the postsanatorium era, Am J Med 58:221-228, 1975.

Neill MA, Hightower AW, and Broome CV: Leprosy in the United States, 1971-1981, J Infect Dis 152:1064-1069, 1985.

Wayne LG: The atypical mycobacteria: recognition and disease association, CRC Critical Rev. Microbiol 12:185-222, 1986.

Weir MR and Thornton GF: Extrapulmonary tuberculosis: experience of a community hospital and review of the literature, Am J Med 79:467-478, 1985.

Wolinsky E: Nontuberculous mycobacteria and associated diseases, Am Rev Resp Dis 119:107-159, 1979.

Woods GL and Washington JA II: Mycobacteria other than *Mycobacterium tuberculosis:* review of microbiologic and clinical aspects, Rev. Infect Dis 9:275-294, 1987.

Young LS: *Mycobacterium avium* complex infection, J Infect Dis 157:863-867, 1988.

CHAPTER 23

Treponema, Borrelia, and Leptospira

The collection of bacteria in the order Spirochaetales has been grouped together based on their common morphological properties. These "spirochetes" are thin, helical (0.1-0.5 × 5-20 μm), gram-negative bacteria (Figures 23-1 and 23-2). The physiologic properties of this group of organisms are of only minor clinical importance, because relatively few species are cultured in vitro. The specific detection and identification of most pathogens are based on morphological or serologic tests.

The order Spirochaetales is subdivided into two families and five genera, of which three *(Treponema, Borrelia,* and *Leptospira)* are responsible for human disease (Table 23-1).

TREPONEMA

The treponemal species responsible for human disease are *T. pallidum* and *T. carateum. T. pallidum* is subdivided into subspecies *pallidum* (referred to as *T. pallidum* in this chapter), the etiologic agent of **syphilis,** and subspecies *pertenue* (referred to as *T. pertenue*), the agent responsible for **yaws.** The two organisms are nearly 100% genetically homologous and are most likely the same species, differing only in their clinical manifestations. *T. carateum* is responsible for **pinta.** Syphilis will be discussed initially in this chapter, with yaws and pinta presented at the end of the section.

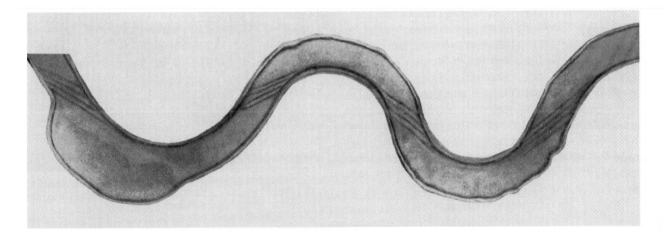

FIGURE 23-1   Helical arrangement of *Treponema pallidum.* The name "spirochete" is derived from the Greek words for "coiled hair." This is an appropriate term for bacteria longer than most other associated with human disease and that are frequently too thin to be seen by light microscopy. Note the periplasmic flagella that run along the length of the spirochete. (From Binford CH and Connor DH: Pathology of tropical and extraordinary diseases, vol 1, 1976, Washington DC, 1976, Armed Forces Institute of Pathology.)

231

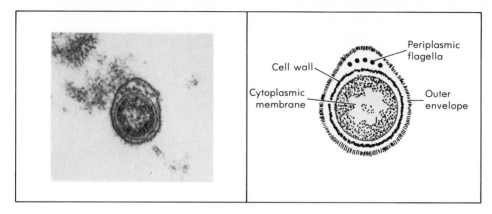

FIGURE 23-2   Electron micrograph of cross section through *Borrelia burgdorferi,* the agent responsible for Lyme disease. The protoplasmic core of the bacterium is enclosed in a cytoplasmic membrane and conventional cell wall. This in turn is surrounded by an outer envelope, or sheath. Between the protoplasmic core and outer sheath are periplasmic flagella (also called axial fibrils), which are anchored at either end of the bacterium and wrap around the protoplasmic core. From two to more than 100 flagella are present, a number characteristic of individual species. These flagella are essential for motility. (From Steere AC, et al: N Engl J Med 308:733-740, 1983.)

Microbial Physiology and Structure

Syphilis is a sexually transmitted disease that has plagued mankind for centuries. The spirochete responsible for this disease is a strict human pathogen. Natural syphilis is not found in any other species, and experimental syphilis has been established only in rabbits. *T. pallidum* is a thin, coiled spirochete (0.1 × 5-15 μm) that cannot be grown in cell-free cultures. Limited growth has been achieved in cultured rabbit epithelial cells, but replication is slow (doubling time, 30 hours) and can be maintained only for a few generations. The spirochetes cannot be seen with light microscopy in specimens stained with Gram or Giemsa stains, but motile forms can be visu-

Table 23-1   Order Spirochaetales

	Human Diseases	**Etiologic Agent**
FAMILY SPIROCHAETALES		
Genus *Spirochaeta*	None	
Genus *Cristispira*	None	
Genus *Treponema*	Syphilis	*T. pallidum,* spp. *pallidum*
	Yaws	*T. pallidum,* spp. *pertenue*
	Pinta	*T. carateum*
Genus *Borrelia*	Epidemic relapsing fever	*B. recurrentis*
	Endemic relapsing fever	Many *Borrelia* species
FAMILY LEPTOSPIRACEAE		
Genus *Leptospira*	Leptospirosis (subclinical infections, influenza-like syndrome, Weil's disease)	*L. interrogans*

alized by darkfield illumination or by staining with specific antitreponemal antibodies labeled with fluorescent dyes. *T. pallidum* is extremely labile, unable to survive exposure to drying or disinfectants. Thus inanimate objects such as toilet seats cannot contribute to the spread of syphilis. Direct person-to-person contact is required for transmission.

Pathogenesis

The clinical course of syphilis evolves through three phases. The initial, or primary, phase is characterized by one or more skin lesions **(chancres)** at the site of spirochete penetration. Although dissemination of spirochetes in the bloodstream occurs soon after infection, the chancre represents the primary site of initial replication. Histologic examination of the lesion reveals endarteritis and periarteritis, characteristic of syphilitic lesions at all stages, and infiltration of the ulcer with polymorphonuclear leukocytes and macrophages. Ingestion of spirochetes by the phagocytic cells is seen, but the organisms frequently survive. The secondary phase of syphilis heralds clinical signs of disseminated disease, with prominent skin lesions dispersed over the entire body

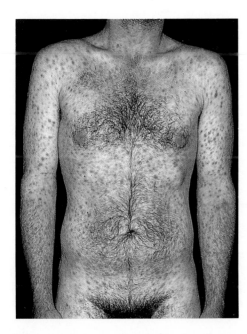

FIGURE 23-3   Disseminated rash in secondary syphilis. (From Habif TP, editor: Clinical dermatology, St Louis 1985, The CV Mosby Co.)

surface (Figure 23-3). Spontaneous remission may occur after either the primary or secondary stages, or the patient may develop late manifestations of disease in which virtually all tissues may be involved. Each phase represents localized multiplication of the spirochete and tissue destruction. Although replication is slow, large numbers of organisms are present in the initial chancre, as well as in the secondary lesions following dissemination of the spirochetes in the bloodstream.

The inability to grow *T. pallidum* to high concentrations in vitro has limited detection of specific virulence factors in this organism. However, several investigators have now successfully cloned *T. pallidum* genes in *E. coli* and isolated the protein products. Several gene products have been specifically associated with virulent strains, although their role in pathogenesis remains to be delineated. The outer membrane proteins are associated with adherence to the surface of host cells, and virulent spirochetes produce hyaluronidase, which may facilitate perivascular infiltration. Virulent spirochetes are also coated with host cell fibronectin, which can protect against phagocytosis.

Epidemiology

Syphilis is found worldwide, and is the third most common sexually transmitted disease in the United States (after *Neisseria gonorrhoeae* and *Chlamydia* infections). More than 35,000 cases of primary and secondary syphilis were reported in the United States in 1987, for an incidence of 14.6 cases per 100,000 persons between the ages of 15 to 64 years. However, the large number of unreported infections contributes to a gross underestimation of the true incidence of this disease. The highest incidence of disease is in the black and Hispanic population (Figure 23-4) and in urban areas.

Since natural syphilis is exclusive to humans, and has no other known natural hosts, the most common method of spread is by direct sexual contact. However, this disease can also be acquired congenitally or rarely by transfusion with contaminated blood. The risk of disease following a single sexual contact is estimated to be 30%, but this is influenced by the stage of disease in the infectious individual. As mentioned previously, the spirochetes are unable to survive on dry skin surfaces. Thus *T. pallidum* is transferred primarily during the early stages of disease,

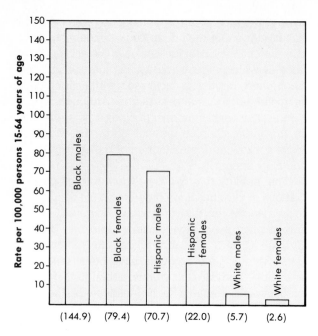

FIGURE 23-4   Incidence of syphilis in the United States during 1987.

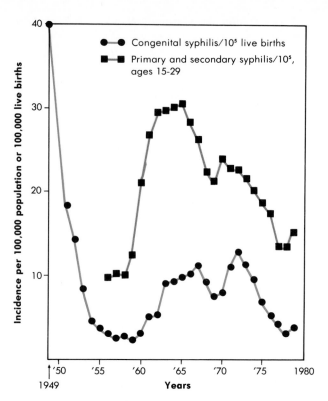

FIGURE 23-5   Annual incidence of congenital syphilis in the United States compared with that of primary and secondary forms of the disease. The trend of infant congenital disease follows disease in young adult women by 12 to 18 months. (Modified from Holmes KK, et al: Sexually transmitted diseases, New York, 1984, McGraw Hill.)

when large numbers of organisms are present in moist cutaneous or mucosal lesions. Congenital transmission to the fetus can take place soon after the mother is infected, because bacteremia characteristically occurs early during the course of the disease. A woman with untreated disease can have spontaneous bacteremia for as long as 8 years, transmitting the spirochetes to fetal tissues if she becomes pregnant during this period. After 8 years the disease can remain active, but bacteremia does not occur.

With the advent of effective antimicrobial therapy, the incidence of late (tertiary) syphilis has markedly decreased. Although antibiotic therapy has decreased the length of infectivity in infected individuals, the incidence of primary and secondary syphilis has remained high because of sexual practices. The incidence of congenital syphilis corresponds to the pattern of syphilis in women of childbearing age (Figure 23-5).

Clinical Syndromes

Figure 23-6 illustrates the natural history of untreated syphilis.

Primary Syphilis The initial syphilitic chancre develops at the site of inoculation. The lesion starts as a papule but then erodes to form a painless ulcer with raised borders. In most patients painless regional lymphadenopathy develops 1 to 2 weeks after the appearance of the chancre, which represents a local focus for proliferation of spirochetes. Abundant spirochetes, which are able to disseminate throughout the patient via the lymphatics and bloodstream, are present in the chancre. The fact that this ulcer heals spontaneously within 2 months gives the patient a false sense of relief.

Secondary Syphilis Clinical evidence of disseminated disease marks the second stage of syphilis. This stage is characterized by a flu-like syndrome, with sore throat, headache, fever, myalgias, and anorexia, generalized lymphadenopathy, and a generalized mucocutaneous rash. The flu-like syndrome and lymphadenopathy generally appear first and then are followed a few days later by the disseminated skin

Course of disease and blood tests

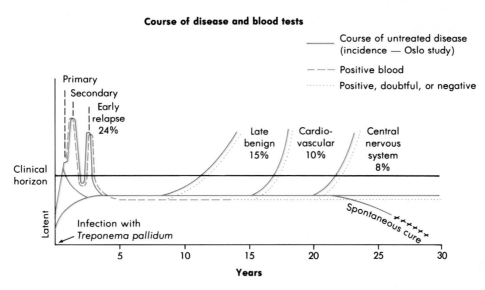

FIGURE 23-6   The natural history of untreated acquired syphilis was carefully chronicled at the University of Oslo, where almost 2,200 untreated patients were studied. The incubation period from the time of infection to onset of primary disease varies from 10 to 90 days (average 21 days). Without treatment, the chancre will heal in 3 to 6 weeks. Asymptomatic dissemination occurs during this period, and secondary lesions develop from 2 weeks to 6 months (average, 6 weeks) after the chancre initially appeared. These lesions will last for 2 to 10 weeks. At the end of the secondary stage of syphilis, patients enter a latent phase from which they can undergo spontaneous cure or relapse into the secondary stage manifestations (observed in 24% of the patients). Tertiary syphilis can occur years later with the development of systemic granulomas (called gummas) in soft tissues (15% of the patients), cardiovascular disease (10%), or central nervous system lesions (8%). (Reprinted by permission from the Southern Medical Journal, 26:18, 1933; incidence from Clark EG, Danbolt N: J Chron Dis 2:311, 1955.)

lesions. The rash can be quite variable (macular, papular, pustular), cover the entire skin surface, including palms and soles, and can resolve slowly over a period of weeks to months (see Figure 23-3). As with the primary chancre, the rash in secondary syphilis is highly infectious. Gradually, the rash and symptoms resolve spontaneously and the patient enters the latent or clinically inactive stage of disease.

Late Syphilis   A small proportion of patients can progress to the tertiary stage of syphilis. Progressive disease can cause devastating destruction of virtually any organ or tissue. The nomenclature of late syphilis reflects the organs of primary involvement (e. g., neurosyphilis or cardiovascular syphilis). An increased incidence of neurosyphilis despite adequate therapy for early syphilis has been documented in patients with acquired immunodeficiency syndrome (AIDS).

Congenital Syphilis   In utero infections can lead to significant fetal disease, with death, multiorgan malformations, or latent infections. Most infected infants are born without clinical evidence of disease but then develop rhinitis followed by a widespread desquamating maculopapular rash. Late bony destruction and cardiovascular syphilis are common in untreated infants who survive the initial course of disease.

Laboratory Diagnosis

Microscopy   The diagnosis of syphilis (primary, secondary, congenital stages) can be made rapidly by darkfield examination of the exudate from skin lesions (Table 23-2). The test is reliable only when the clinical material with actively motile spirochetes is examined immediately—the spirochetes will not survive transport to the laboratory. Material collected from oral lesions should not be examined because nonpathogenic, oral spirochetes can contaminate the specimen. Specific identification of *T. pallidum* can be made by use of fluorescein-labeled antitreponemal

Table 23-2 Diagnostic Tests for Syphilis

Diagnostic Test	Method or Examination
Microscopy	Darkfield
	Direct fluorescent antibody staining
Culture	Not used
Serology	Nontreponemal tests
	VDRL
	RPR
	Treponemal tests
	FTA-ABS
	MHA-TP

antibodies. Histologic staining of tissue lesions may be beneficial. Silver stains have been used most commonly.

Culture Efforts to culture *T. pallidum* have been generally unsuccessful and should not be attempted.

Serology The diagnosis of syphilis in most patients is made by serologic testing. The two general types of tests used are biologically nonspecific (nontreponemal) tests and the specific treponemal tests.

Nontreponemal tests measure IgG and IgM antibodies (also called **reagin** antibodies) developed against lipids released from damaged cells during the early stage of disease and also present on the cell surface of treponemes. The antigen used for the nontreponemal tests is cardiolipin, derived from beef heart. The two tests used most commonly are the Venereal Disease Research Laboratory **(VDRL)** test and the rapid plasma reagin **(RPR)** test. Both tests measure flocculation of cardiolipin antigen by the patient's serum. Both tests are rapid, although complement in serum must be inactivated for 30 minutes before the VDRL test can be performed. Only the VDRL test can be used to test cerebrospinal fluid from patients with suspected neurosyphilis.

Treponemal tests are specific antibody tests used to confirm positive reactions with the VDRL or RPR tests. The treponemal tests can also be positive before the nontreponemal tests become positive in early syphilis, or remain positive when the nonspecific tests revert to negative in some patients who have late syphilis. The tests most commonly used are the Fluorescent Treponemal Antibody Absorption **(FTA-ABS)** test and the Microhemagglutination Test for *T. pallidum* **(MHA-TP).** The MHA-TP test is technically easier to perform and interpret than the FTA-ABS test.

Because positive reactions with the nontreponemal tests develop late during the first phase of disease, many patients who initially have chancres will have negative serologic findings (Table 23-3). However, within 3 months all patients will develop positive serologic results that remain positive in untreated patients with secondary syphilis. Patients who have late syphilis can develop negative serologic findings (Figure 23-7). Serologic results revert to negative following successful treatment, so the nonspecific tests can be used to monitor therapy. The specific treponemal tests, particularly FTA-ABS, become positive earlier in the course of disease and remain positive for life. Thus specific treponemal tests can be used to confirm the diagnosis of syphilis but cannot be used to follow therapeutic response.

The specificity of the nontreponemal tests is 98% or greater. Transient false-positive reactions are seen in patients with acute febrile diseases, following immunizations, and in pregnant women. Chronic false-positive reactions occur most often in patients with chronic autoimmune diseases or infections. The specificity of the treponemal tests is also 98% to 99%,

Table 23-3 Sensitivity of Serologic Tests in Untreated Syphilis

	Stage of Disease			
	Primary	Secondary	Latent	Late
VDRL slide test	59-87	100	73-91	39-94
FTA-ABS test	86-100	99-100	96-99	96-100
MHA-TP	64-87	99-100	96-100	94-100

From Holmes KK, Mardh PA, Sparling PF, and Wiesner PJ: Sexually transmitted diseases, New York, 1984, McGraw-Hill Inc.

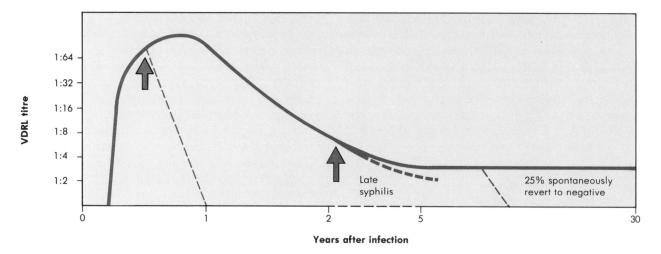

FIGURE 23-7   Variation of VDRL titer in syphilis. In untreated patients the antibody titer peaks during the first year and then slowly decreases during late syphilis. Approximately 25% of patients who have untreated latent syphilis will be serologically negative. Patients treated during the early stage of syphilis generally have negative results within the first year. The antibody titers decrease more slowly for patients treated in the late stage of syphilis. (Reproduced, with permission, from: Hart G: Ann Intern Med, 1986, 104:368.)

with most false-positive reactions observed in patients with connective tissue diseases (Box 23-1).

Positive serologic tests in infants of infected mothers can represent passive transfer of antibodies or a specific immunologic response to infection. These two possibilities are distinguished by collecting sera over a 6-month period. In noninfected infants the antibody titers will decrease to undetectable levels within 3 months of birth. The antibody titers will remain elevated in infants who have congenital syphilis.

Treatment, Prevention, and Control

Penicillin is the drug of choice for treating infections with *T. pallidum;* long-acting benzathine penicillin is used for the early stages of syphilis, and penicillin G is recommended for congenital and late syphilis. Tetracycline, erythromycin, and chloramphenicol can be used as alternative antibiotics for patients allergic to penicillin. Only penicillin or chloramphenicol can be used for patients with neurosyphilis.

Because protective vaccines are not available, control of syphilis requires the practice of safe sex techniques and adequate contact and treatment of sex partners of patients who have documented infections.

Other Treponemes

T. pertenue and *T. carateum* are the other treponemes responsible for human disease. These organ-

---

Box 23-1   Medical Conditions Associated With False-Positive Treponemal and Nontreponemal Serology Tests

NONTREPONEMAL TESTS

Viral infection	Recent immunization
Collagen-vascular disease	Heroin addiction
Acute or chronic illness	Leprosy
Pregnancy	Malaria

TREPONEMAL TESTS (FTA-ABS)

Pyodermas	Psoriasis
Skin neoplasms	Systemic lupus erythematosus
Acne vulgaris	Pregnancy
Mycoses	Drug addiction
Crural ulcerations	Herpes genitalis
Rheumatoid arthritis	

isms, which have not been isolated in vitro, are morphologically identical to *T. pallidum*. *T. pertenue* is the etiologic agent of yaws, a granulomatous disease with early skin lesions (Figure 23-8) and then late destructive lesions of the skin, lymph nodes, and bones. The disease is present in primitive, tropical areas in parts of South America, Central Africa, and Southeast Asia (Figure 23-9) and is spread by direct contact with infected skin lesions.

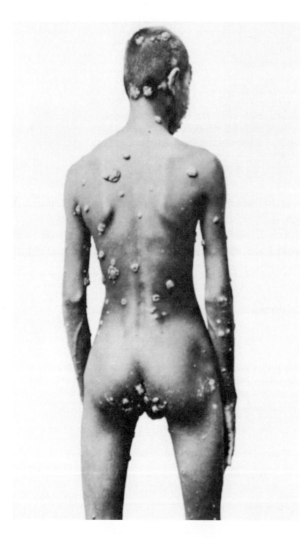

FIGURE 23-8   Yaws. The elevated papillomatous nodules characteristic of early yaws are widely distributed and painless. They contain numerous spirochetes easily demonstrable by darkfield examination. (From Binsford CH and Connor DH: Pathology of tropical and extraordinary diseases, vol 1, Washington, DC, 1976, Armed Forces Institute of Pathology.)

T. carateum is responsible for Pinta, a disease primarily restricted to the skin. After a 1- to 3-week incubation period small pruritic papules develop on the skin surface. These lesions enlarge and persist for months to years before resolution (Figure 23-10). Disseminated, recurrent, hypopigmented lesions can develop over years, resulting in scarring and disfigurement (Figure 23-11). Pinta is present in Mexico and Central and South America, and is also spread by direct contact with infected lesions.

Both yaws and pinta are diagnosed by their clinical presentation in an endemic area. The diagnoses are confirmed by the detection of spirochetes in skin lesions by darkfield microscopy. Serologic tests for syphilis are also positive but may develop only late in the disease course.

Penicillin, tetracycline, and chloramphenicol have been used to treat both diseases. Control of the disease is managed by treating infected individuals and eliminating person-to-person spread.

BORRELIA

Members of the genus *Borrelia* are responsible for two important human diseases: **relapsing fever** and **Lyme disease.** Relapsing fever is a febrile illness characterized by recurrent episodes of fever and septicemia, separated by afebrile periods. Two forms of the disease are recognized. *Borrelia recurrentis* is the etiologic agent of epidemic or louse-borne relapsing fever and is spread person-to-person by the human body louse *(Pediculus humanus).* Endemic relapsing fever is caused by many species of borreliae and is spread by infected ticks.

Borrelia burgdorferi is the etiologic agent of Lyme disease, another tick-borne disease characterized by the initial presentation of an erythematous skin lesion (erythema chronicum migrans) followed by chronic neurologic, cardiac, and rheumatic manifestations. The disease is named after the Connecticut town where the first cases in the United States were described.

Microbial Physiology and Structure

Members of the genus *Borrelia* are weakly staining gram-negative bacilli that resemble other spirochetes, although they tend to be larger (0.2-0.5 × 3-30 μm), stain well with Giemsa stain, and can be easily

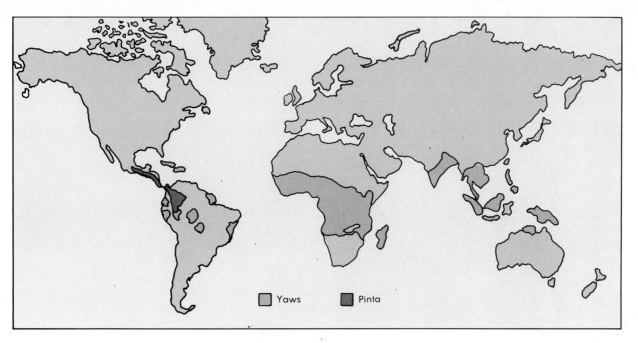

FIGURE 23-9 Geographic distribution of yaws and pinta. (Redrawn from Mandell GL, Douglas RG, Jr, and Bennett JE: Principles and practices of infectious disease, New York, John Wiley & Sons, 1985.)

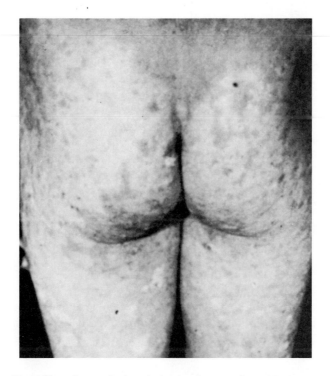

FIGURE 23-10 Pinta. Note the marked variation in pigmentation with the presence of several achromic areas which cannot be distinguished from vitiligo. Depigmentation is commonly seen as a late sequel of any of the treponemal diseases. (From Binford CH and Connor DH: Pathology of tropical and extraordinary diseases, vol 1, Washington DC, 1976, Armed Forces Institute of Pathology.)

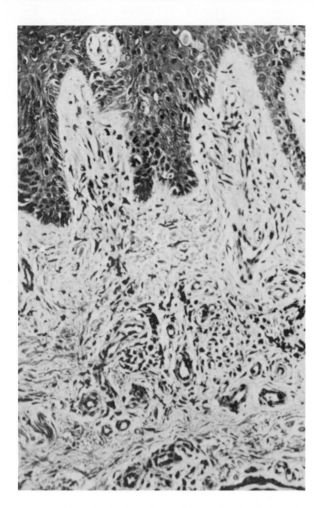

FIGURE 23-11   Pinta. The dyschromic changes of pinta are related to hyperemia and to migration of melanophores into both the epidermis and the upper dermis. The basal layer gradually becomes devoid of pigment. (From Binsford CH and Connor DH: Pathology of tropical and extraordinary diseases, vol 1, Washington DC, 1976, Armed Forces Institute of Pathology.)

seen in smears of peripheral blood collected from patients with relapsing fever (Figure 23-12). From 7 to 20 periplasmic flagella (depending on the individual species) are present between the periplasmic cylinder and the outer envelope and are responsible for the organism's twisting motility. Borreliae are microaerophilic and have complex nutritional requirements, making recovery in culture difficult. The few species that have been successfully cultured have generation times of 18 hours or longer. Because culture is generally unsuccessful, diagnosis of diseases caused by

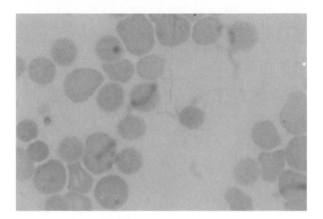

FIGURE 23-12   *Borrelia* present in the blood of a patient with endemic relapsing fever.

borreliae is by microscopy (relapsing fever) or serology (Lyme disease).

Pathogenesis

After exposure to infected arthropods, borreliae are able to spread in the bloodstream to multiple organs. Members of the genus do not produce recognized toxins and are rapidly removed when a specific antibody response is mounted. The periodic febrile and afebrile cycles of relapsing fever are due to the ability of the borreliae to undergo antigenic variation. When specific antibodies are formed, the borreliae are rapidly cleared from the bloodstream. However, organisms residing in internal tissues are able to alter their outer envelope antigens and reemerge as antigenically novel organisms. The cycles of bacteremia and immunologic clearance are terminated when bactericidal antibodies are produced.

B. burgdorferi (named after the microbiologist who originally isolated the organism) is present in low numbers in the skin tissues when erythema chronicum migrans develops. It is not known whether viable organisms are responsible for the late manifestations of disease or if these represent immunologic cross-reactivity to borrelia antigens.

Epidemiology

The vectors for relapsing fever are soft-shelled **ticks** (*Ornithodoros* species) and the human **body louse** (Figure 23-13). Humans are the only reservoir for

Infection	Reservoir	Vector
Relapsing fever Epidemic (louse-borne)	Humans	Body louse 
Replapsing fever Endemic (tick-borne)	Rodents, soft-shelled ticks	Soft-shelled tick 
Lyme disease	Rodents, deer domestic pets, hard-shelled ticks	 Hard-shelled tick

FIGURE 23-13   Epidemiology of *Borrelia* infections.

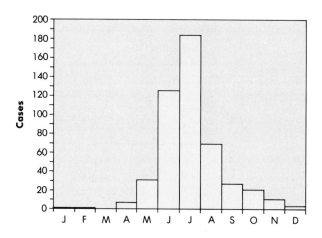

FIGURE 23-14   Cases of Lyme disease reported by month of onset of illness in the United States during 1982. (From Schmid, et al: J Infect Dis 151:1144, 1985, published by the University of Chicago.)

B. recurrentis, the etiologic agent of louse-borne epidemic relapsing fever. Infected lice do not survive for more than a few months, so maintenance of disease requires crowded, unsanitary conditions (wars, natural disasters) that permit frequent contact with infected lice. In contrast, tick-borne relapsing fever is a zoonotic disease, with rodents and ticks acting as the main reservoirs. Despite the fact the borreliae produce a disseminated infection in ticks, the arthropods are able to survive and maintain an endemic reservoir of infection by transovarian transmission. Furthermore, ticks can survive for months to years between feedings.

Louse-borne relapsing fever is endemic in both tropical and temperate regions of the world, particularly Central and Eastern Africa and South America. Tick-borne disease is widespread.

Hard-shelled ticks (*Ixodes* species) are the vectors of Lyme disease. Despite the relative recent recognition of Lyme disease in the United States, retrospective studies have demonstrated that the disease was present for many years in other countries. Currently, Lyme disease has been described on 6 continents, in at least 20 countries, and in 35 states of the United States. The three principal foci of infection in the United States are the Northeast (Massachusetts to Maryland), upper Midwest (Minnesota and Wisconsin), and Pacific West (California and Oregon). The tick vector in the Northeast and Midwest is *Ixodes*

dammini, whereas *I. pacificus* has been found on the West Coast. *I. ricinus* is the tick vector in Europe.

Only 30% of patients with Lyme disease recall a specific tick bite. The reason is that larvae and nymph stages of the tick, as well as adult ticks, can transmit disease. The small nymph forms are responsible for the majority of infections. Most Lyme disease infections are reported from spring to fall, corresponding to activity of infected ticks (Figure 23-14). Person-to-person spread has not been reported.

Clinical Syndromes

Relapsing Fever   The clinical presentation of epidemic louse-borne and endemic tick-borne relapsing fever is essentially the same (Figure 23-15). After a 1-week incubation period, the abrupt onset of disease is heralded with shaking chills, fever, muscle aches, and headache. Splenomegaly and hepatomegaly are commonly present. The symptoms correspond to the bacteremic phase of the disease and are relieved after 3 to 7 days, when the borreliae are cleared from the blood. After a 1-week afebrile period bacteremia and fever return. The clinical symptoms are generally milder and shorter during this and subsequent febrile episodes. Two or three relapses are common, although the number of relapses can be as many as 13. The clinical course and outcome of epidemic relapsing fever tend to be more severe than with endemic disease, but this may be related to the patient's underlying poor state of health. Mortality

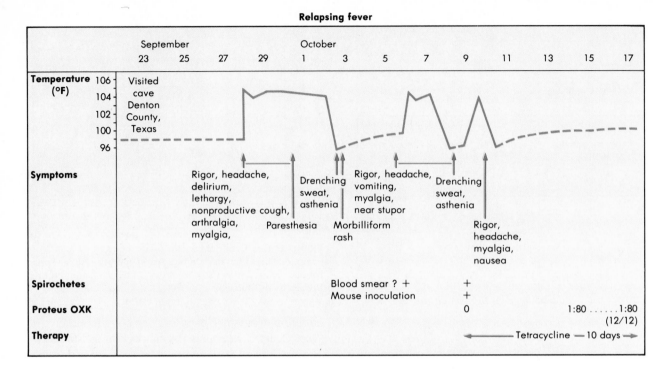

FIGURE 23-15   The clinical evolution of tick-borne relapsing fever in a 14-year-old boy. The initial febrile episode is typically the most severe. Subsequent episodes tend to be shorter and less intense. Although nonspecific reactivity with Proteus OXK antigens are frequently observed, the specific diagnosis is made by observing borreliae in peripheral blood smears. These are positive only during the febrile periods. (Modified from PM Southern and JP Sanford: Medicine 48:129-149, © by Williams & Wilkins, 1969.)

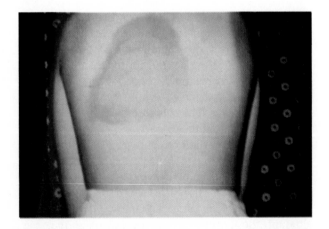

FIGURE 23-16   Erythema chronicum migrans in a patient with Lyme disease. (From Steere, AC, et al: Ann Intern Med 86:685, 1977.)

with endemic disease is less than 5% but can be from 4% to 40% in epidemic disease.

Lyme Disease After an incubation period of from 3 to 30 days one or more skin lesions develops at the site of the tick bite. The lesion **(erythema chronicum migrans)** begins as a small macule or papule and then enlarges over the next few weeks, covering an area ranging from 3 to 68 mm in diameter (Figure 23-16). As the lesion develops, it will typically appear with a red, flat border and central clearing; however, erythema, vesicle formation, and central necrosis can also be seen. Within weeks the lesion will fade and disappear, although new, transient lesions may subsequently appear. These skin lesions appear in 85% of the patients with documented Lyme disease. Other early signs and symptoms of Lyme disease include malaise, severe fatigue, headache, fever, chills, musculoskeletal pains, myalgias, and lymphadenopathy. These will last for an average of 4 weeks. Late manifestations develop in almost 80% of untreated

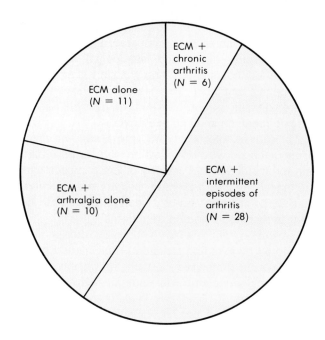

FIGURE 23-17   Frequency of the types of joint involvement in patients with Lyme arthritis. The spectrum ranges from arthralgias to intermittent episodes of arthritis to chronic arthritis. (From Steere, AC, et al: Ann Intern Med 107:725-731, 1987.)

Table 23-4   Sensitivity of Diagnostic Tests for Borrelia Infections

Diagnostic Test	Test Sensitivity	
	Relapsing Fever	Lyme Disease
Microscopy	Good	Poor
Culture	Poor	Poor
Serology	Not available	Good

patients with Lyme disease; these occur within a week of the onset of disease to more than 2 years later. Two phases can be seen. The first involves neurologic symptoms (meningitis, encephalitis, peripheral nerve neuropathy) and cardiac dysfunction (heart block, myopericarditis, congestive heart failure). These symptoms are seen in 10% to 15% of patients and can last for days to months. The second phase of late disease is characterized by arthralgias and arthritis (Figure 23-17). These complications can persist for months to years, during which spirochetes are rarely visualized in the involved tissue or isolated in culture. These reactions are most likely the result of immunologic cross-reactions with host tissues.

Laboratory Diagnosis

The sensitivity for the following diagnostic tests is outlined in Table 23-4.

Microscopy Because of their relatively large size, the borreliae responsible for relapsing fever can be seen during the febrile period by preparing a Giemsa or Wright stain of the blood. Examination of Giemsa-stained blood smears is the most sensitive method for diagnosing relapsing fever, more than 70% of the patients may have positive smears. The test sensitivity can be improved by inoculating a mouse with blood from infected patients and then examining the mouse after 1 to 10 days for the presence of borreliae in the bloodstream.

In contrast, the microscopic examination of blood or tissues collected from patients with Lyme disease cannot be recommended because *B. burgdorferi* is rarely seen in clinical specimens.

Culture Some borreliae, including *B. recurrentis* and *B. hermsi* (a common cause of endemic relapsing fever in the United States), can be cultured in vitro on specialized media. However, the cultures are rarely performed in most clinical laboratories because the media are not readily available and replication is slow. The in vitro isolation of *B. burgdorferi* has had limited success, with positive skin biopsy cultures from the early lesions reported for 6% to 45% of the patients and only rarely from the affected joints of patients who have late manifestations of disease.

Serology Because the borreliae responsible for relapsing fever undergo antigenic phase variation, serologic tests are not useful.

In contrast, serologic testing is an important confirmatory test for patients who clinically diagnosed as having Lyme disease. The tests most commonly used are the immunofluorescence assay and the enzyme-linked immunosorbent assay (ELISA); ELISA is preferred because of its greater sensitivity and specificity for all stages of Lyme disease. Unfortunately, all serologic tests are relatively insensitive during the early acute stage of disease; only about half of the patients who have erythema chronicum migrans and the other early symptoms give positive test results. In untreated patients IgM antibodies appear first, peak 3

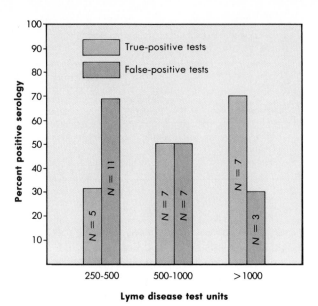

FIGURE 23-18   Predictive value of positive serologic tests for Lyme disease. A positive ELISA test unit of 250 is defined as positive for the test used in this study. Although false-positive tests are relatively uncommon, these can have a significant impact on the interpretation of a serologic test for a disease with a low prevalence. The majority of false-positive tests are from patients with other spirochetal infections, infectious mononucleosis, and autoimmune diseases. These patients primarily have low test values with screening tests, so the predictive value of a positive test result improves at the higher test values. (Modified from Mayo Clin Proc 63:1116-21, 1988.)

to 6 weeks after exposure to the infected tick, and then gradually fade. The IgG antibodies appear later and persist during the late manifestations of the disease. Thus all patients who have late complications of Lyme disease have detectable antibodies to *B. burgdorferi,* although the antibody level may be low in patients treated with antibiotics. Although cross-reactions are uncommon, positive serologic results must be interpreted carefully, particularly when the titers are low (Figure 23-18). The serologic tests should not be performed in the absence of clinical symptoms of Lyme disease.

Treatment, Prevention, and Control

Relapsing fever has been treated most effectively with tetracycline or chloramphenicol. Tetracycline is the drug of choice except for pregnant women and young children, for whom the drug is contraindi-

cated. A Jarisch-Herxheimer reaction (shocklike picture with an increase in temperature, decrease in blood pressure, rigors, and leukopenia) can occur within a few hours after therapy is started and must be carefully managed. This reaction corresponds to the rapid killing of borreliae and the possible release of toxic products such as endotoxin.

The early manifestations of Lyme disease are effectively managed with either tetracycline or penicillin, and the incidence and severity of late complications are ameliorated with therapy. Despite this intervention, Lyme arthritis and other complications can be seen in a small proportion of patients.

Prevention of tick-borne borrelia diseases is attempted by avoiding ticks and their natural habitats, wearing protective clothing, and using insect repellants. Rodent control for endemic relapsing fever is also important. Epidemic louse-borne disease is controlled by the use of delousing sprays and improved hygienic conditions. Vaccines are not available for either relapsing fever or Lyme disease.

LEPTOSPIRA

Microbial Physiology and Structure

The genus *Leptospira* consists of two species: *Leptospira interrogans* (subdivided further into 19 serogroups and 172 serotypes) and *Leptospira biflexa* (38 serogroups and 65 serotypes). The species names are derived from the fact that *Leptospira* are thin, coiled bacilli (0.1 × 6-12 μm) with a hook at one or both ends (interrogans meaning shaped like a question mark; biflexa for twice bent; Figure 23-19). *L. interrogans* is pathogenic for many wild and domestic animals, as well as humans. *L. biflexa* is a free-living saprophyte found in moist environmental sites and is not associated with disease. The three serotypes of *L. interrogans* responsible for human disease in the United States are icterohemorrhagiae, canicola, and pomona. Occasionally, citations in the literature will incorrectly refer to these serotypes as species of *Leptospira.*

The pathogenic leptospires are obligatively aerobic and motile by means of two periplasmic flagella, each anchored at opposite ends of the bacterium. They utilize fatty acids or alcohols as sources of carbon and energy. In contrast with most other spirochetes, the leptospires can be grown on specially for-

FIGURE 23-19 Leptospira interrogans serotype canicola. These spirochetes have a single hooked end. (From Bergey's manual of systematic bacteriology, vol 1, Baltimore, © 1986, the Williams & Wilkins Co., Baltimore.)

mulated media enriched with rabbit serum or bovine serum albumin. However, the generation time is from 6 to 16 hours and incubation for up to 2 weeks may be required.

Pathogenesis

L. interrogans can cause subclinical infection, a mild flu-like febrile illness, or severe systemic disease **(Weil's disease),** with renal and hepatic failure, extensive vasculitis, myocarditis, and death. The severity of disease is influenced by the number of infecting organisms, the host's immunologic defenses, and the virulence of the infecting strain.

Because leptospires are thin and highly motile, they can penetrate intact mucous membranes or skin surfaces through small cuts or abrasions. They then are able to spread in the bloodstream into all tissues, including the central nervous system. Multiplication of *L. interrogans* proceeds and damages the endothelium of small blood vessels, which is responsible for the major clinical manifestations of disease (e. g., meningitis, hepatic and renal dysfunction, and hemorrhage). Organisms can be readily demonstrated in blood and cerebrospinal fluid early in the course of disease and in urine during the later stages. Clearance of leptospires occurs when humoral immunity develops. However, some clinical manifestations may be related to immunologic reactions with the organisms. For example, meningitis develops after the organisms have been removed from the cerebrospinal fluid.

Epidemiology

Leptospirosis has worldwide distribution. Fewer than 100 human infections are normally reported in the United States each year, but the incidence of disease is significantly underestimated because most infections are mild and misdiagnosed as a "viral syndrome."

Many wild and domestic animals are colonized with leptospires, and dogs, cattle, rodents, and wild animals are the most common sources for human disease in the United States. Chronic carriage in humans has not been demonstrated.

Streams, rivers, standing water, and moist soil can be contaminated with urine from infected animals and serve as a source for human infection. A moist, alkaline environment is required for survival of leptospires. Most human infections are due to either recreational exposure to contaminated water or occupational exposure to infected animals (farmers, slaughterhouse workers, veterinarians). Most human infections are reported during the warm months of the year when recreational exposure is greatest. Person-to-person spread is very rare.

Clinical Syndromes

The majority of infections with *L. interrogans* are clinically inapparent and detected only by demonstra-

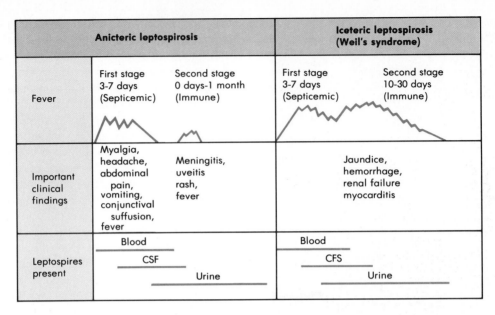

Anicteric leptospirosis		Icteric leptospirosis (Weil's syndrome)	
Fever	First stage 3-7 days (Septicemic) Second stage 0 days-1 month (Immune)	First stage 3-7 days (Septicemic) Second stage 10-30 days (Immune)	
Important clinical findings	Myalgia, headache, abdominal pain, vomiting, conjunctival suffusion, fever Meningitis, uveitis rash, fever	Jaundice, hemorrhage, renal failure myocarditis	
Leptospires present	Blood CSF Urine	Blood CFS Urine	

FIGURE 23-20   Stages of icteric and anicteric leptospirosis. (From Feigin RD and Anderson DC: CRC Crit Rev Clin Lab Sci 5:413, 1975.)

tion of specific antibodies. Symptomatic infections develop after a 1- to 2-week incubation period (Figure 23-20). The initial presentation is similar to a flu-like illness, with fever and myalgias. During this phase the patient is bacteremic with the leptospires, and the organisms can frequently be isolated in cerebrospinal fluid even though no meningeal symptoms are present. The fever and myalgias may remit after 1 week with no further difficulties, or the patient may progress to more advanced disease—including aseptic meningitis—or to a generalized illness, with headache, rash, vascular collapse, thrombocytopenia, hemorrhage, and hepatic and renal dysfunction.

Leptospirosis confined to the central nervous system can be mistaken for viral meningitis, because the course of disease is generally uncomplicated with a very low mortality. Culture of the cerebrospinal fluid is usually negative at this stage. In contrast, the icteric form of generalized disease (approximately 10% of all symptomatic infections) is more severe and has a mortality approaching 10%. Although hepatic involvement with jaundice is striking in severe leptospirosis, hepatic necrosis is not seen and surviving patients do not suffer permanent hepatic damage. Similarly, most patients recover full renal function.

Congenital leptospirosis can also occur. The dis-

ease is characterized by a sudden onset with headache, fever, myalgias, and a diffuse rash.

Laboratory Diagnosis

The sensitivity for the following diagnostic tests is outlined in Table 23-5.

Microscopy Because leptospires are thin and at the limits of the resolving power of a light microscope, they cannot be seen by conventional light microscopy. Darkfield microscopy is also relatively insensitive and can be nonspecific. Although leptospires can be seen in blood specimens during the early course of disease, protein filaments from erythrocytes can be easily mistaken for organisms. Fluorescein-labeled antibody preparations have been used to stain leptospires but are not available in most clinical laboratories.

Culture Leptospires can be readily cultured in vitro if specially formulated media are available. *L. interrogans* can be recovered in blood during the first week of infection and in urine thereafter for 3 months or more. The concentration of organisms in urine may be low, so multiple specimens should be collected if leptospirosis is considered. Organisms are also present in cerebrospinal fluid, but this is usually

Table 23-5   Diagnostic Tests for Leptospirosis

Diagnostic Test	Method	Sensitivity
Microscopy	Gram stain	Organisms cannot be seen
	Darkfield	Insensitive, nonspecific
	Direct fluorescent antibody	Insensitive, generally unavailable
Culture	Blood	Positive during first week
	CSF	Positive during first or second week
	Urine	Positive after first week
Serology	Microscopic agglutination	Sensitive, specific; positive after 1 week; peak at 1 month

before the patient develops meningeal symptoms. Leptospires grow slowly, requiring incubation for a minimum of 1 to 2 weeks and sometimes as long as 6 weeks.

Serology Because most clinical laboratories do not prepare media for the isolation of leptospires, the majority of infections are diagnosed by serologic techniques, which are both sensitive and specific. The microscopic agglutination test (MAT) is used most often to detect specific antibodies directed against leptospires. Because the test is directed against specific serotypes, the use of pools of leptospiral antigens is necessary. If an infection is caused by a serotype of leptospire not contained in the antigen pool, then it would be undetected. Serial dilutions of the patient's serum are mixed with the test antigens and then examined microscopically for agglutination. Infected patients have a titer of at least 1:100 and may be 1:25,000 or higher. Agglutinating antibodies appear 6 to 12 days after the onset of infection and reach a maximum titer in the third or fourth week of illness. Patients treated with antibiotics may have a diminished antibody response and nondiagnostic titers. The agglutinating antibodies are detectable in low titer for many years after the acute illness, so their presence in a treated patient may represent either a blunted antibody response in acute disease or residual antibodies from a distant, unrecognized infection with leptospires.

Treatment, Prevention, and Control

Leptospirosis is usually not fatal, particularly in the absence of icteric disease. Treatment with either penicillin or tetracycline can shorten the clinical symptoms and complications of leptospirosis if started early. Doxycycline has also been used both to treat infections and to prevent disease in individuals exposed to infected animals or water contaminated with urine. The total eradication of leptospirosis is difficult because the disease is widespread in wild and domestic animals. However, vaccination of livestock and pets has proved successful in reducing disease in these populations and therefore subsequent human exposure. Rodent control is also effective in eliminating leptospirosis in communities.

BIBLIOGRAPHY

Barbour AG: Laboratory aspects of Lyme borreliosis, Clin Microbiol Rev 1:399-414, 1988.

Barbour AG and Hayes SF: Biology of *Borrelia* species, Microbiol Rev 50:381-400, 1986.

Butler T, Hazen P, Wallace CK, et al: Infection with Borrelia recurrentis: pathogenesis of fever and petechiae, J Infect Dis 140:665-672, 1979.

Canale-Parola E: Physiology and evolution of spirochetes, Microbiol Rev 41:181-204, 1977.

Duffy J, Mertz LE, Wobig GH, and Katzmann JA: Diagnosing Lyme disease: the contribution of serologic testing, Mayo Clin Proc 63:1116-1121, 1988.

Feigen RD and Anderson DC: Human leptospirosis, CRC Crit Rev Clin Lab Sci 5:413-467, 1975.

Hart G: Syphilis tests in diagnostic and therapeutic decision making, Ann intern Med 104:368-376, 1986.

Heath CW, Alexander AD, and Galton MM: Leptospirosis in the United States, analysis of 483 cases in man 1949-1961, N Engl J Med 273:857-864, 915-922, 1965.

Holmes KK, Mardh PA, Sparling PF, and Wiesner PJ: Sexually transmitted diseases, New York, 1984, McGraw-Hill.

Johnson RC, Hyde FW, and Rumpel CM: Taxonomy of the Lyme disease spirochetes, Yale J Biol Med 57:529-537, 1984.

Meyerhoff J: Lyme disease, Am J Med 75:663-670, 1983.

Schmid GP, Horsley R, Steere AC, et al: Surveillance of Lyme disease in the United States, 1982, J Infect Dis 151:1144-1149, 1985.

Southern PM and Sanford JP: Relapsing fever—a clinical and microbiological review, Medicine 48:129-149, 1969.

Steere AC, Malawista SE, Bartenhagen NH, et al: The clinical spectrum and treatment of Lyme disease, Yale J Biol Med 57:453-461, 1984.

Steere AC, Schoen RT, and Taylor E: The clinical evolution of Lyme arthritis, Ann Intern Med 107:725-731, 1987.

CHAPTER 24

Mycoplasma and *Ureaplasma*

The family Mycoplasmataceae consists of two genera: *Mycoplasma*, with 69 recognized species, and *Ureaplasma*, with two species. Both genera will be referred to collectively as mycoplasmas in this text. Despite the ubiquity of the bacteria in plants and animals, only three species are definitively recognized as human pathogens (Table 24-1). *Mycoplasma pneumoniae* (also called **Eaton's agent** after the investigator who originally isolated it) is responsible for respiratory disease, and *M. hominis* and *Ureaplasma urealyticum* cause genitourinary tract diseases. These and other mycoplasmas that colonize humans have been associated with a variety of other maladies (e. g., infertility, spontaneous abortion, vaginitis, cervicitis, epididymitis, prostatitis), but their etiologic role in these diseases has not been demonstrated conclusively.

MICROBIAL PHYSIOLOGY AND STRUCTURE

Mycoplasma and *Ureaplasma* are the smallest free-living bacteria (Table 24-2). They are pleomorphic, with an average diameter of 0.2 to 0.8 μm; filamentous forms also exist. Many of these bacteria are able to pass through 0.45 μm filters that are used to remove bacteria from solutions. In addition, mycoplasmas do not have a cell wall; the cytoplasmic contents are enclosed only by a plasma membrane. The absence of the cell wall renders the organisms resistant to penicillins, cephalosporins, and other antibiotics that interfere with the integrity of the cell wall. For these reasons the mycoplasmas were thought originally to be viruses. However, the organisms divide by binary fission (typical of all bacteria) and are gram-negative. Most mycoplasmas are facultatively anaerobic (*M. pneumoniae* is a strict aerobe), grow on arti-

Table 24-1 Diseases Caused by *Mycoplasma* and *Ureaplasma*

Organism	Disease
Mycoplasma pneumoniae	Pneumonia
	Tracheobronchitis
	Pharyngitis
Mycoplasma hominis	Pyelonephritis
	Pelvic inflammatory disease
	Postabortal fever, postpartum fever
Ureaplasma urealyticum	Nongonococcal urethritis

Table 24-2 Properties of *Mycoplasma* and *Ureaplasma*

Properties	Characteristics
Size	0.2 to 0.8 μm
Cell wall	Absent
Growth requirements	Sterols
Atmosphere requirements	Facultatively anaerobic*
Replication	Binary fission
Generation time	1 to 6 hours
Antibiotic susceptibility:	
Penicillins	Resistant
Cephalosporins	Resistant
Tetracycline	Susceptible
Erythromycin	Susceptible†

M. pneumoniae is aerobic.
†*M. hominis* is resistant.

249

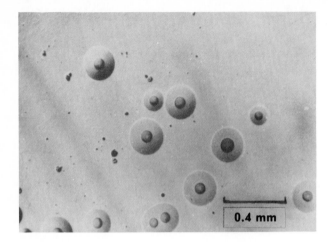

Figure 24-1   ''Fried egg'' colonies of mycoplasmas. This morphology is typically seen with all mycoplasmas except *M. pneumoniae*. *M. pneumoniae* has a strict aerobic atmosphere requirement, is one of the slowest growing mycoplasmas, and appears as homogeneous granular colonies after incubation for 1 week or more. The other mycoplasmas generally grow within 1 to 4 days. (From Razin S and Oliver O: J General Microbiol 24:225-237, 1961.)

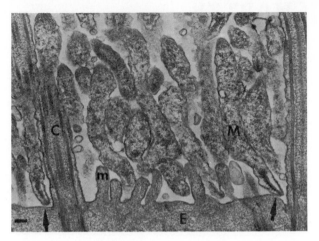

Figure 24-2   *M. pneumoniae* infection of a tracheal ring organ culture after 72 hours. Note the heavy parasitization of the mycoplasmas (*M*) with specialized tips (*arrows*) interacting with the epithelial cell (*E*) and in close apposition to the bases of the cilia (*C*) and microvilli (*m*). The bar marker represents 0.1 μm. (From Wilson MH and Collier AM: J Bacteriol 125:332-339, 1976.)

ficial cell-free media, and require exogenous sterols supplied by the addition of animal serum to the growth medium. The mycoplasmas grow slowly with a generation time from of 1 to 6 hours, and form small colonies that have a ''fried egg'' appearance (Figure 24-1). Colonies of *Ureaplasma* (also called **T-strains,** for tiny strains) are extremely small, 10 to 50 μm in diameter. The three human pathogens can be differentiated by their ability to metabolize glucose (*M. pneumoniae*), arginine (*M. hominis*), or urea (*U. urealyticum*).

Because these organisms do not have a cell wall, the major antigenic determinants are membrane proteins and glycolipids. Cross-reactivity of these antigens with human tissues and other bacteria is observed.

PATHOGENESIS

M. pneumoniae is an extracellular pathogen that adheres to the respiratory epithelium by a specialized terminal protein attachment factor (Figure 24-2). This factor, called **P1,** interacts specifically with a glycoprotein receptor on the epithelial cell surface. Ciliostasis occurs following attachment and then destruc-

tion of the superficial layer of epithelial cells. The mechanism of this cytopathic effect is unknown. Because exposure to *M. pneumoniae* is common in childhood and disease is more severe in older individuals, the observed pathology may be due to the host's immune response. Dissemination to extrapulmonary sites is rare.

EPIDEMIOLOGY

Pneumonia caused by *M. pneumoniae* occurs worldwide throughout the year, with no consistent increased seasonal activity (Figure 24-3). However, because pneumonia caused by other infectious agents (e.g., *Streptococcus pneumoniae*, viruses) is frequently more common during the cold months of the year, *M. pneumoniae* disease is proportionally more common during the summer and fall. Epidemic disease is reported every 4 to 8 years. Disease is most common in school-age children and young adults (Figure 24-4), with disease uncommon in children less than 5 years of age or adults more than 20 years of age. *M. pneumoniae* is the most common cause of pneumonia in children from 5 to 15 years of age. Infections in adults are more severe than in children;

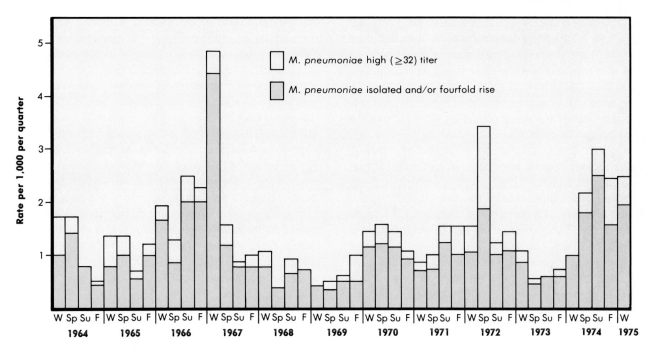

Figure 24-3   *Mycoplasma pneumoniae* infections in children less than 15 years of age are seen throughout this 12-year study in Seattle. Although not shown here, the proportion of pneumonias due to mycoplasmas is generally highest in the warm months of the year when other respiratory pathogens are less active. (From Foy HM et al: J Infect Dis 139:681-687, 1979, published by the University of Chicago.)

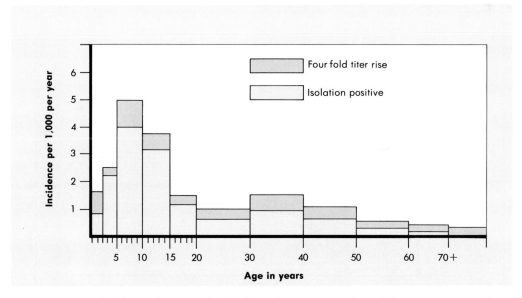

Figure 24-4   Incidence of pneumonia with *Mycoplasma pneumoniae* in different age-groups. The greatest proportion of infections is in patients from 5 to 15 years of age. (From Foy HM et al: J Infect Dis 139:681-687, 1979, published by the University of Chicago.)

however, hospitalization is normally not required. Infection is spread by nasal secretions, requires close contact, and usually occurs among classmates or within a family. The attack rate is higher among children than adults (overall average approximately 60%). The incubation period and time of infectivity are prolonged. Thus disease can persist for weeks to months in a classroom or family.

Colonization of infants, particularly girls, with *M. hominis* and *Ureaplasma* occurs at birth, with *Ureaplasma* isolated more frequently. Carriage of these mycoplasmas usually does not persist, although a small proportion of prepubertal children will remain colonized. The incidence of genital mycoplasmas increases after puberty, corresponding to sexual activity. Approximately 15% of sexually active men and women are colonized with *M. hominis,* and 45% to 75% are colonized with with *Ureaplasma. Ureaplasma* is more common in women than in men. The incidence of carriage in adults who are sexually inactive is no greater than in prepubertal children.

CLINICAL SYNDROMES

Three diseases associated with *M. pneumoniae* are pneumonia, tracheobronchitis, and pharyngitis.

Pneumonia caused by this organism has been referred to as "primary atypical pneumonia" and "walking pneumonia"—indicative of the comparatively mild, although protracted, course of the disease (Figure 24-5). The onset of clinical pneumonia is insidious following a long incubation period. The initial symptoms include malaise, low-grade fever, and headache. These symptoms increase in severity, and after 2 to 4 days a nonproductive cough develops. Auscultation of the chest reveals rhonchi and rales, and patchy bronchopneumonia is seen on chest x-ray films. Myalgias are common, and a maculopapular rash may develop. Secondary complications can be seen that include otitis media, erythema multiforme, hemolytic anemia, myocarditis, pericarditis, and neurologic abnormalities. Resolution of the disease is slow. Secondary infections can occur because immunity is incomplete.

Tracheobronchitis is a common form of mycoplasma disease, with an onset and symptoms similar to pneumonia. Disease primarily involves inflammation of the bronchials with peribronchial infiltration of lymphocytes and plasma cells. Pharyngitis can either be a complication of pneumonia or tracheobronchitis, or it can be the predominant manifestation. Fever, headache, sore throat, pharyngeal exudates, and cervical lymphadenopathy are common. *M. pneumoniae*

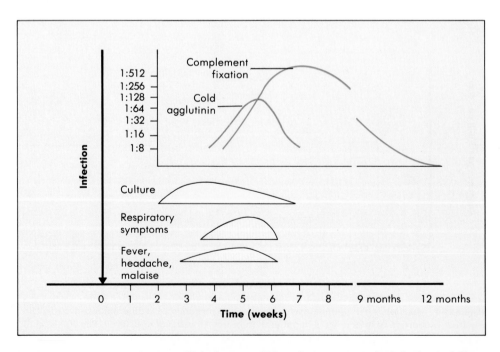

Figure 24-5   Correlation between clinical course of *Mycoplasma pneumoniae* infection and diagnostic tests.

pharyngitis is indistinguishable from pharyngitis due to group A *Streptococcus,* Epstein-Barr virus, or other upper respiratory tract viruses.

M. hominis is casually associated with pyelonephritis, pelvic inflammatory disease, and postabortal and postpartum fevers. Although *M. hominis* commonly colonizes asymptomatic, sexually active individuals, the etiologic role of this mycoplasma in disease is supported by isolation of the organism from the site of inflammation, as well as blood, and serologic response to infection. Similar evidence has supported the role of *Ureaplasma* in urethritis, where this organism is implicated in as much as one half of all urethral infections not caused by *Neisseria gonorrhoeae* or *Chlamydia.*

LABORATORY DIAGNOSIS

An assessment of the diagnostic tests for *M. pneumoniae* infections is summarized in Table 24-3.

Microscopy

Although mycoplasmas are taxonomically classified as gram-negative bacteria, they stain poorly because the cell wall is absent.

Culture

M. pneumoniae, unlike other mycoplasmas, is a strict aerobe. This mycoplasma can be isolated from throat washings or expectorated sputum, although most patients produce scant amounts of sputum. The specimen should be inoculated into special media supplemented with serum (provides cholesterol), yeast extract (for nucleic acid precursors), glucose, a pH indicator, and penicillin (to inhibit other bacteria). Growth in culture is slow, with a generation time of 6 hours. Thus the cultures must be incubated for up to 3 weeks. Metabolism of glucose with a corresponding pH change indicates of growth. Colonies of *M. pneumoniae* are small and have a homogeneous granular appearance, unlike the "friend egg" morphology of other mycoplasmas. Identification of isolates can be confirmed by inhibition of growth with specific antisera.

M. hominis is a facultative anaerobe, grows within 1 to 4 days, and metabolizes arginine but not glucose. The colonies have a typical large fried-egg appearance. Specific differentiation from other genital mycoplasmas is by inhibition of growth with specific antisera. *Ureaplasma* requires urea for growth but is inhibited by the increased alkalinity associated with metabolism of urea. Thus the growth medium must be both supplemented with urea and also highly buffered. Despite this, these mycoplasmas die rapidly after initial isolation.

Serology

Serologic tests are available only for *M. pneumoniae.* Detection of local production of IgA is generally not useful because the antibody rapidly disappears after onset of infection. Detection of IgG by complement fixation is more useful; the antibody is initially detected soon after onset of infection, peaks within 4 weeks, and persists for 6 to 12 months. This is the standard specific test for *M. pneumoniae* infections. Detection of a significant increase in antibody titers is observed in only 58% to 65% of patients because sera is collected in many patients too late in the course of disease to measure seroconversion. A high antibody titer (e. g., $\geq$ 1:256) can be considered presumptive evidence of infection. Because the antibodies are directed against outer membrane glycolipids that are common to other organisms and tissues, false-positive reactions are reported with other mycoplasmas and some plant antigens, as well as in patients with bacterial meningitis, syphilis, and pancreatitis. Although complement fixation tests are cumbersome,

Table 24-3 Diagnostic Tests for *Mycoplasma pneumoniae* Infections

Test	Assessment
Microscopic	Negative, due to absence of cell wall
Culture	Slow (1 to 3 weeks for positives); not available in most labs
Serologic	
Complement fixation	Peaks at 4 weeks, persists for 6-12 months; diagnostic titer $\geq$1:256 or fourfold increase; seroconversion in 58% to 65%; some false-positives
Cold agglutinins	Diagnostic titer $\geq$1:128 or fourfold increase; seroconversion in 34% to 68%; nonspecific
Nucleic acid probes	Preliminary studies report sensitivity and specificity to be >95%

the use of ELISA or immunofluorescent assays has not significantly improved test reliability.

Nonspecific reactions to the outer membrane glycolipids can also be measured. The most useful is production of cold agglutinins (e.g., IgM antibodies that bind the I antigen on the surface of human erythrocytes at 4° C). A positive cold agglutinin assay is observed in approximately 65% of patients with *M. pneumoniae* infections, particularly in symptomatic patients. Because this test is not specific for *M. pneumoniae,* cross-reactions occur in respiratory diseases caused by other organisms (e. g., infectious mononucleosis, adenovirus). Strongly reactive cold agglutinin titers (>1:32) or a significant increase in titer can provide strong presumptive evidence of mycoplasma disease.

Nucleic Acid Probes

Genetic probes directed against *M. pneumoniae* have been commercially developed and appear to be very promising for the rapid direct detection of *M. pneumoniae* in clinical specimens. Additional experience with these probes will be necessary to confirm these preliminary findings.

TREATMENT

Both erythromycin and tetracycline are equally effective in treating *M. pneumoniae* infections, although tetracycline is reserved for treating adult infections. Tetracycline—which has the advantage of also being active against *Chlamydia,* a common cause of nongonococcal urethritis—is also active against *M. hominis* and *Ureaplasma.* For strains of *Ureaplasma* resistant to tetracycline, erythromycin or spectinomycin are active antibiotics. In contrast with the other mycoplasmas, *M. hominis* is resistant to erythromycin.

PREVENTION AND CONTROL

Prevention of mycoplasma disease is problematic. *M. pneumoniae* infections are spread by close contact: thus isolation of infected individuals could theoretically reduce the risk of infection. However, the prolonged infectivity of the patient, even while receiving appropriate antibiotics, makes this approach impractical. Inactivated vaccines, as well as attenuated live vaccines, have also been disappointing. Protective immunity has been low, and concern exists that disease in immunized persons may be more severe if the immune response participates in the pathogenesis of disease. Infections with *M. hominis* and *Ureaplasma* are transmitted by sexual contact, the disease can be prevented by avoidance of sexual activity or use of proper barrier precautions.

BIBLIOGRAPHY

Barile MF, Razin S, Smith PF, and Tully JG: Current topics in mycoplasmology, Rev Infect Dis 4:1-277, 1982.

Cassell GH and Cole BC: *Mycoplasmas* as agents of human disease, N Engl J Med 304:80-89, 1981.

Foy HM, Kenny GE, Cooney MK, and Allen ID: Long-term epidemiology of infections with *Mycoplasma pneumoniae,* J Infect Dis 139:681-687, 1979.

Murray HW, Masur H, Senterfit LB, and Roberts RB: The protean manifestations of *Mycoplasma pneumoniae* infection in adults, Am J Med 58:229-238, 1975.

Tilton RC, Dias F, Kidd H, and Ryan RW: DNA probe versus culture for detection of *Mycoplasma pneumoniae* in clinical specimens, Diagn Microbiol Infect Dis 10:109-112, 1988.

Wilson HM and Collier AM: Ultrastructural study of *Mycoplasma pneumoniae* in organ culture, J Bacteriol 125:332-

25

Rickettsia

The family Rickettsiaceae consists of aerobic, gram-negative bacilli that, with one exception *(Rochalimaea),* are obligate intracellular parasites (Box 25-1). The Rickettsiaceae (referred to collectively in this chapter as rickettsia) were originally thought to be viruses because they are small (0.3 × 1-2 μm), Gram stain poorly, and grow only in the cytoplasm of eukaryotic cells. However, rickettsia are now known to be structurally similar to gram-negative bacilli, to contain DNA and RNA, enzymes for the Krebs cycle, and ribosomes for protein synthesis, to multiply by binary fission, and are inhibited by antibiotics (tetracycline, chloramphenicol).

Four genera *(Rickettsia, Coxiella, Rochalimaea,* and *Ehrlichia)* are associated with human disease. The pathogenic species are maintained in animal reservoirs and transmitted by arthropod vectors (e.g., ticks, mites, lice, fleas). Humans are usually accidental hosts. The pathogenic rickettsia can be subdivided into six groups: spotted fever, typhus, scrub typhus, Q (for "query") fever, trench fever, and ehr-

lichiosis, with each group associated with one or more diseases (Table 25-1). Rocky Mountain spotted fever, Q fever, and ehrlichiosis are the most common rickettsial diseases in the United States; diseases caused by other rickettsia are relatively rare.

MICROBIAL PHYSIOLOGY AND STRUCTURE

The cell wall structure of rickettsia, with a peptidoglycan layer and lipopolysaccharide, is typical of gram-negative bacilli. The bacteria do not have flagella or attachment proteins but are surrounded by a loosely adherent slime layer. All rickettsia are seen best with the Giemsa or Gimenez stains but react poorly with the Gram stain. With the exception of *Rochalimaea,* all rickettsia are strict intracellular parasites. The intracellular location of the rickettsia vary: *Rickettsia* species are generally found free in the cytoplasm, whereas *Coxiella* and *Ehrlichia* multiply in cytoplasmic vacuoles. *Rickettsia rickettsii,* the bacterium responsible for Rocky Mountain spotted fever, can also grow to high concentrations in the nucleus (Figure 25-1). *Rochalimaea* multiplies readily on the surface of eukaryotic cells but rapidly dies following phagocytosis. Rickettsia can also grow to high concentrations in the yolk sac of embryonated eggs and in animal models.

The rickettsia enter eukaryotic cells by "induced phagocytosis," or rickettsial-stimulated phagocytosis by nonprofessional phagocytes. After phagocytosis occurs, rickettsia must be released into the cytoplasm or the organism fails to survive. The phagosome membrane is degraded by production of phospholi-

---

Box 25-1 Medically Important Rickettsiaceae

ORDER Rickettsiales

Family Rickettsiaceae

  Tribe I Rickettsieae

    Genus 1 *Rickettsia*

    Genus 2 *Coxiella*

    Genus 3 *Rochalimaea*

  Tribe II Ehrlichieae

    Genus 1 *Ehrlichia*

Table 25-1   Geographic Distribution of Rickettsiaceae Associated With Human Disease

Group	Organism	Disease	Distribution
Spotted fever	*Rickettsia rickettsii*	Rocky Mountain spotted fever	Western hemisphere
	R. akari	Rickettsialpox	United States, Soviet Union, Africa, Korea
Typhus	*R. prowazekii*	Epidemic typhus	Africa, Asia, South America
		Brill-Zinsser disease	Worldwide
	R. typhi	Murine typhus	Worldwide
Scrub typhus	*R. tsutsugamushi*	Scrub typhus	Asia, Australia, South Pacific
Q fever	*Coxiella burnetii*	Q fever	Worldwide
Trench fever	*Rochalimaea quintana*	Trench fever	Mexico, Europe, Middle East, North Africa (rare)
Ehrlichiosis	*Ehrlichia canis*	Ehrlichiosis	United States

pase A. Multiplication by binary fission then proceeds slowly compared with other bacteria (generation time, 6 to 10 hours) until destruction of the host cell occurs, freeing the rickettsia to infect new cells. In contrast with the other rickettsia, multiplication of *Coxiella* and *Ehrlichia* proceeds within phagolysosomes rather in the cytoplasm, and host cell lysis is low.

Why rickettsia must grow inside eukaryotic cells is not completely understood. The bacteria are capable of protein synthesis and can produce ATP via the tri-carboxylic acid cycle. They also have a transport system for ADP and ATP and are capable of intracellular concentration of adenine nucleotides. It would appear that rickettsia are energy parasites that utilize the host cell ATP as long as it is available. The rickettsia also use the host cell coenzyme A and nicotinamide adenine dinucleotide.

Once released from the host cell, most rickettsia are unstable and die quickly. The exception is *Coxiella,* which is highly resistant to desiccation and remains viable in the environment for months to years. This characteristic is extremely important in the epidemiology of *Coxiella* infections.

PATHOGENESIS

The hallmark of rickettsial infections is vasculitis (Figure 25-2). After human exposure via an arthropod vector, most rickettsia initially proliferate at the site of inoculation and then spread to the endothelial cells lining the small blood vessels. Active destruction of the endothelial lining leads to focal hyperplasia, inflammation, and formation of microthrombi with fibrin deposits. The clinical manifestion of this process is localized infarction of multiple organs and tissues (e.g., skin, heart, adrenals, kidneys, brain). In rickettsialpox (caused by *R. akari*) and scrub typhus (caused by *R. tsutsugamushi*) the initial multiplication of rickettsia at the site of infection leads to the formation of a papule, which in turn ulcerates and then forms an eschar.

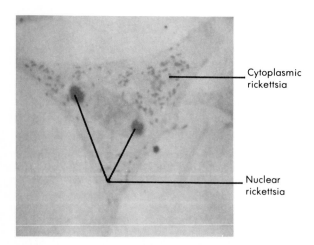

Cytoplasmic rickettsia

Nuclear rickettsia

FIGURE 25-1   Cultured chick embryo fibroblast cells infected with *Rickettsia rickettsii.* The bacilli are sparsely distributed in the cytoplasm but concentrated in two distinct masses within the nucleus. (From Bergey's Manual of Systematic Bacteriology, vol 1, Baltimore, 1984, Williams & Wilkins.)

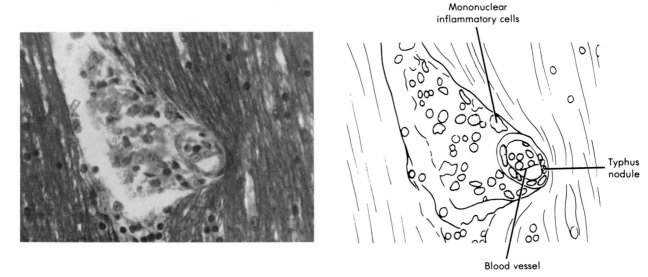

FIGURE 25-2   Vasculitis in a small blood vessel is characteristic in Rocky Mountain spotted fever. Note the prominent perivascular clustering of mononuclear inflammatory cells forming a "typhus nodule." (From Lambert HP and Farror WE, editors: Infectious diseases illustrated: an integrated text and color atlas, New York, 1982, Gower Medical Publishing Ltd.)

Disease	Organism	Vector	Reservoir
Rocky Mountain spotted fever	R. rickettsia	Tick-borne 	Wild rodents, dogs
Ehrlichiosis	E. canis		Unknown
Rickettsialpox	R. akari	Mite-borne 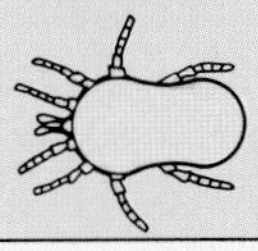	Wild rodents
Scrub typhus	R. tsutsugamushi		Wild rodents
Epidemic typhus	R. prowazekii	Louse-borne 	Humans, flying squirrels
Trench fever	R. quintana		Humans
Murine typhus	R. typhi	Flea-borne 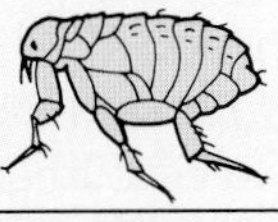	Wild rodents
Q fever	C. burnetti	None*	Cattle, sheep, goats

*Tick vectors may be responsible for animal-to-animal transmission.

FIGURE 25-3   Epidemiology of common rickettsial infections.

Bacteria in the genus *Ehrlichia* (organisms named after Paul Ehrlich, one of the fathers of microbiology) are leukocytic rickettsiae, which parasitize lymphocytes, neutrophils, and monocytes. They also infect endothelial cells of small blood vessels. Erythrocytes are spared.

Understanding of the pathogenesis of Q fever, caused by *Coxiella burnetii,* is limited, because acute disease is rarely fatal. However, available studies indicate the pathogenesis of Q fever differs substantially from disease caused by the other rickettsia. Human infection most commonly follows inhalation of airborne particles from a contaminated environmental source rather than from the bite of an arthropod vector. *Coxiella* then proliferate locally in the respiratory tract, with subsequent dissemination to other organs. In severe, acute infections, necrotizing hemorrhagic pneumonia involving primarily the alveolar cells is observed. Hepatic destruction with granuloma formation is also observed in both acute and chronic infections. The most common clinical manifestation of chronic Q fever is endocarditis.

EPIDEMIOLOGY

The epidemiology of selected rickettsial infections of humans is shown in Figure 25-3. As noted, the reservoir for ehrlichiosis is currently unknown, although because *E. canis* or a closely related species is responsible, dogs may serve as a reservoir for infection. However, there is no direct evidence for canine-to-human spread of disease. Transovarian transmission of rickettsia has been demonstrated in Rocky Mountain spotted fever (tick), rickettsialpox (mites), and scrub typhus (mites).

Rocky Mountain Spotted Fever

R. rickettsii is responsible for Rocky Mountain spotted fever, which is the most common rickettsial disease in the United States with 600 to 800 documented cases reported annually. Although the disease was first described in Montana by Howard Ricketts, most cases are now reported in the Southeast Atlantic and South Central states (Figures 25-4 and 25-5). The reason for this shift is unknown, but it could result from either inefficient transovarian transmission in the Rocky Mountain tick population or from the presence of avirulent rickettsia that compete with virulent rickettsia for the tick vector. The vectors

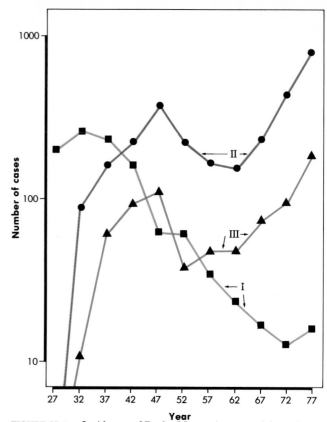

FIGURE 25-4   Incidence of Rocky Mountain spotted fever in the United States from 1925 to 1979. Each symbol is a mean of 5 years: Mountain and Pacific States (blue); South Atlantic and South Central States (green); New England, mid-Atlantic, and North Central States (red). (From Weiss E: Ann Rev Microbiol 36:348, 1982.)

are the wood tick *(Dermacentor andersoni)* in Rocky Mountain states, the dog tick *(Dermacentor variabilis)* in Southeastern states, and the Lone Star tick *(Amblyomma americanum)* in South Central states.

More than 90% of all infections occur from April through September, corresponding to the period of greatest tick activity. There are two peaks (biphasic distribution): the first is in the spring when adult ticks become active and the second is during the summer months corresponding with the maturation of larval ticks, infected by transovarian transmission. History of a tick bite is elicited from most but not all patients.

The animal reservoirs for Rocky Mountain spotted fever are wild rodents and dogs. Studies with infected dogs have demonstrated that high-grade rickettsemia (rickettsia in the bloodstream) can persist for more than 2 weeks, favoring transmission to the tick vec-

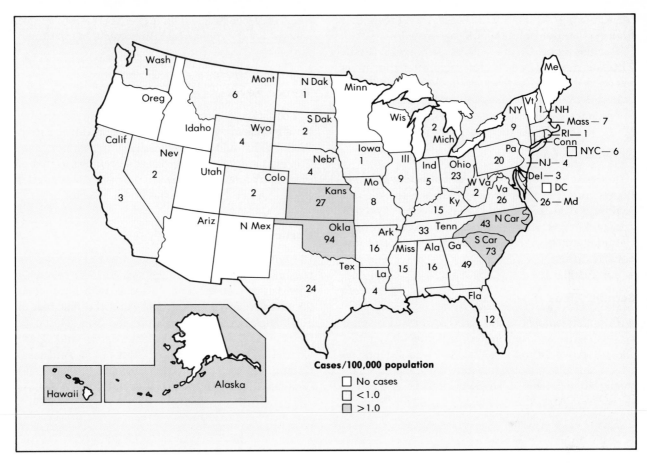

FIGURE 25-5   Distribution of Rocky Mountain spotted fever in the United States, 1985. (From the Morbidity and Mortality Weekly Report 35, 1986.)

tors. Following exposure to *R. rickettsii,* dogs are symptomatic and will not survive infection when large numbers of organisms are present.

Human infection with *R. rickettsii* occurs when the tick defecates into the wound after feeding or inoculates rickettsia directly through the skin during the feeding process. Transmission of disease requires prolonged contact with the infected ticks and can usually be avoided if the tick is promptly removed. The incidence of Rocky Mountain spotted fever is highest in children and teenagers, but mortality is greatest in those over 40 years of age.

Rickettsialpox

Infections with *R. akari* are maintained in the rodent population by the bite of mouse ectoparasites (e.g., mites) and in mites by transovarian transmis-

sion. Humans are accidental hosts when bitten by infected mites. Rickettsialpox is relatively uncommon in the United States; infections have been reported in Russia, Korea, and Africa.

Epidemic Typhus

R. prowazekii is the etiologic agent of epidemic typhus, also called louse-borne typhus, and the principal vector is the human body louse, *Pediculus humanus.* In contrast with most other rickettsial diseases, humans are the primary resevoir, although flying squirrels have also been implicated recently. Epidemic typhus is associated with crowded, unsanitary conditions that favor the spread of body lice, such as during wars, famines, and natural disasters. The disease is reported in South America, Africa, Asia, and less commonly in the United States. Lice die from their infection within 2 to 3 weeks, thus preventing transovarian transmission of *R. prowazekii.*

Recrudescent disease with *R. prowazekii* **(Brill-Zinsser disease)** can occur years after the initial infection. Infected individuals in the United States are primarily Eastern European immigrants exposed to epidemic typhus during World War II.

Endemic Typhus

Endemic or murine typhus is caused by *R. typhi.* Rodents are the primary reservoir, and the rat flea *Xenopsylla cheopis)* is the principal vector. Approximately 40 to 60 cases are reported annually in the United States from the Southeast and Gulf states (especially Texas).

Scrub Typhus

R. tsutsugamushi is the etiologic agent for scrub typhus, a disease transmitted to humans by mites. The reservoirs for this rickettsia are wild rodents, as well as the mite population where it is transmitted by transovarian means. Disease is present in Eastern Asia, Australia, and Japan and other Western Pacific islands. Scrub typhus is also imported into the United States.

Trench Fever

Rochalimaea quintana is the rickettsia responsible for trench fever, a disease that was prevalent during World Wars I and II but is relatively uncommon now. The epidemiology of this disease is similar to that of epidemic typhus. Humans are the principal reservoir; the human body louse is the primary vector.

Ehrlichiosis

Ehrlichia canis or a closely related species is responsible for human ehrlichiosis. This is a relatively newly appreciated human disease, with fewer than 50 cases reported during 1987 and the first half of 1988. Most infections are reported from March to October and are from the Southeastern and South Central portion of the United States (e.g., Arkansas, Georgia, Mississippi, South Carolina, Texas, and Oklahoma). Ticks are the primary vector.

Table 25-2   Clinical Course of Common Rickettsial Diseases

Disease	Average Incubation Period (Days)	Clinical Presentation	Rash	Eschar	Mortality
Rocky Mountain spotted fever	7	Abrupt onset; fever, chills, headache, myalgia	>90%; macular; centripetal spread	No	3%
Rickettsialpox	9-14	Abrupt onset; fever, headache, chills, myalgia, photophobia	100%; papulovesicular; generalized	Yes	<1%
Epidemic typhus	8-12	Abrupt onset; fever, headache, chills, myalgia	>80%; macular; centrifugal spread	No	Variable
Endemic typhus	7-14	Gradual onset; fever, headache, myalgia, cough	>55%; maculopapular rash on trunk	No	<1%
Scrub typhus	10-12	Abrupt onset; fever, headache, myalgia spread	<50%; maculopapular rash; centrifugal	No	<1%
Ehrlichiosis	12	Abrupt onset; fever, headache, myalgia, malaise, leukopenic, thrombocytopenia	20%; nonspecific	No	<1%
Q fever (acute)	20	Abrupt onset; fever, headache, chills, myalgia; granulomatous hepatitis	No	Yes	1%
Q fever (chronic)	Months to years	Chronic disease with subacute onset; endocarditis, hepatic dysfunction	No	No	High

Q Fever

The epidemiology of Q fever is completely different from other rickettsial infections. The etiologic agent, *Coxiella burnetii,* is extremely stable in harsh environmental conditions and can survive in soil for months to years. A large number of wild mammals and birds are infected with this rickettsia, as are sheep, cattle, and goats, the primary reservoirs associated with human disease. A number of different genera of ticks are also infected. The rickettsia can reach high concentrations in the placenta of infected livestock. Dried placenta following parturition, feces, and urine, as well as tick feces, can contaminate soil, which in turn can serve as a focus for infection when airborne and then inhaled *C. burnetii* is also excreted in urine, feces, and milk. Consumption of contaminated unpasteurized milk can cause human infection.

Q fever has a worldwide distribution. Although only 20 to 30 cases are reported annually in the United States, this is certainly an underestimation of the prevalence of this disease. Infection is common in the livestock in the United States, although actual disease is very rare. Human exposure, particularly for ranchers, veterinarians, and food handlers, is frequent, and experimental studies have demonstrated that the infectious dose of rickettsia is small. Thus most human infections are asymptomatic. Additionally, *C. burnetii* is frequently not considered when patients have symptomatic disease. At the present time the diagnosis can made only by specific serologic testing, which is not readily available.

CLINICAL SYNDROMES

Rickettsial infection should be suspected if fever, headache, myalgias, rash, and a history of exposure to arthropods (e.g., ticks, mites, lice, fleas) are present. The specific diagnosis is confirmed with a careful medical history and appropriate diagnostic tests. Table 25-2 summarizes the clinical findings in many of these diseases.

Rocky Mountain Spotted Fever

Infections with *R. rickettsii* can range from asymptomatic to fulminant disease. Evidence of asymptomatic disease is demonstrated by serologic surveys in endemic areas. Clinically symptomatic disease develops after an incubation period of from 1 to 14 days (average 7 days). The onset is abrupt, with fever, chills, headache, and myalgias. From 3 to 5 days later a macular rash develops, initially on the hands, wrists, feet, and ankles (including the palms and soles), and then spreads centripetally to the trunk and face (Figure 25-6). In severe infections the rash may

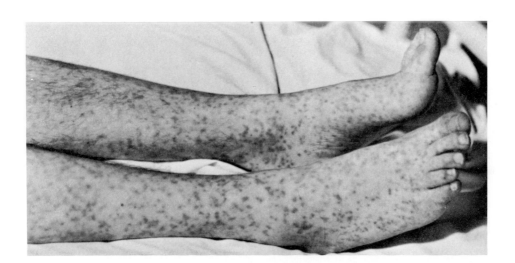

FIGURE 25-6   The rash of Rocky Mountain spotted fever consists of generally distributed, sharply defined purpuric macules initially involving the extremities, including the palms and soles and then spreading to the trunk. (From Binford CH and Connor DH, editors: Pathology of tropical and extraordinary diseases: an atlas, vol 1, Washington, DC, 1976, Armed Forces Institute of Pathology.)

progress to become petechial and then purpuric. More than 90% of patients will develop the characteristic rash, although for some this is late in the course of illness. Complications of Rocky Mountain spotted fever can include splenomegaly, neurologic disturbances, thrombocytopenia, disseminated intravascular coagulation, and heart failure. Complications increase and the prognosis worsens when the characteristic rash fails to develop or develops late in the course of disease and the diagnosis is delayed.

Rickettsialpox

Clinical infection with *R. akari* is biphasic. Initially, a papule develops at the site of contact with the infected mite. This occurs approximately 1 week after the bite and quickly progresses to ulceration and then eschar formation. During this period the rickettsia spread systemically. After an incubation period of 7 to 24 days (average, 9 to 14 days), the second phase of disease develops abruptly with high fever, severe headache, chills and sweats, myalgias, and photophobia. Within 2 to 3 days a generalized papulovesicular rash forms. A ''poxlike'' progression of the rash is seen, with vesicle formation and then crusting. Despite the appearance of the disseminated rash, the course of rickettsialpox is usually mild and uncomplicated, with complete healing in untreated patients within 2 to 3 weeks. Specific antibiotic therapy speeds this process.

Epidemic Typhus

Clinical disease with *R. prowazekii* develops after a 5- to 15-day incubation period (average, 8 to 12 days). As is seen with other rickettsial diseases, the onset is abrupt, with fever, chills, myalgias, and a severe headache. A maculopapular rash develops on day 5, first on the upper trunk and then spreading centrifugally to cover the body, sparing only the face, palms, and soles. Complications can include myocarditis and central nervous system dysfunction; a mortality as high as 66% has been reported in some epidemics. The high mortality is undoubtedly due to the lack of antibiotic therapy and proper supportive medical care. In uncomplicated disease the temperature returns to normal within 2 weeks, but complete convalescence may require 3 months or more.

Reactivation or recrudescent epidemic typhus (Brill-Zinsser disease) can occur years after the initial disease. The course is generally milder than with epidemic typhus, and convalescence is shorter.

Endemic Typhus

The incubation period for *R. typhi* disease is 7 to 14 days. The onset is more gradual than that seen with epidemic typhus, but the symptoms include headache, myalgia, and fever. A nonproductive cough is also frequently seen. After 3 to 5 days, from 55% to 80% of the patients develop a maculopapular rash primarily on the chest and abdomen. In a small proportion of patients the rash is more widespread. The course of disease is generally uncomplicated, lasting less than 3 weeks even in untreated patients.

Scrub Typhus

R. tsutsugamushi disease develops after a 6- to 18-day incubation period (average, 10 to 12 days), appearing suddenly with severe headache, fever (increasing to 104° F within a few days), and myalgias. A macular to papular rash develops on the trunk in fewer than half of the patients and spreads centrifugally to the extremities. Generalized lymphadenopathy, splenomegaly, central nervous system complications, and heart failure can occur. Fever in untreated patients will disappear after 2 to 3 weeks, whereas patients who receive appropriate treatment will respond promptly.

Ehrlichiosis

Ehrlichia infections in humans are poorly understood because they have been only recently described. The course of disease is similar to Rocky Mountain spotted fever. Approximately 12 days after the tick bite (range, from 1 to 3 weeks) the patient develops a high fever, headache, malaise, and myalgia. Leukopenia, caused by infection and destruction of leukocytes by the rickettsia, and thrombocytopenia are observed. In contrast with other rickettsial infections, a rash is observed only in 20% of the diseased patients. The absence of a rash has contributed to the difficulty in diagnosing this disease. However, the prognosis is generally good, with complete recovery normally seen.

Q Fever

Acute and chronic presentations of *Coxiella burnetii* infections are recognized. Acute disease is characterized by a long incubation period (average 20 days) followed by sudden onset with severe headache, high-grade fever, chills, and myalgias. Respiratory symptoms are generally mild but can be severe. Hepatosplenomegaly is present in approximately half of the patients. Histologically diffuse granulomas are seen in the liver of most patients who have either acute or chronic Q fever. The most common presentation of chronic Q fever is subacute endocarditis, generally on a prosthetic or previously damaged heart valve. The incubation period for chronic Q fever can be months to years and the presentation insidious. Unfortunately, the progression of chronic disease is frequently unrelenting, and the prognosis poor.

LABORATORY DIAGNOSIS

Microscopy

As stated earlier, direct visualization of rickettsial intracellular inclusions is best seen in specimens stained with Giemsa or Gimenez stains. The bacteria stain poorly with Gram stain. Rickettsia can also be detected in the perivascular endothelial cells by direct staining with fluorescein-labeled antibodies (Figure 25-7). This test has proved useful for the rapid laboratory diagnosis of Rocky Mountain spotted fever, although it is insensitive if antibiotic therapy is initiated before the skin biopsy is collected. Specimens must be collected when the rash develops.

Culture

Although rickettsia can be cultured in embryonated eggs or in cell cultures, the organisms are highly infectious and considered dangerous to handle except in the most experienced laboratories. For that reason the diagnosis for most infections is made by serologic testing.

Serology

A variety of serologic tests have been used, with the most common test the **Weil-Felix agglutination reaction.** This test is based on the observation

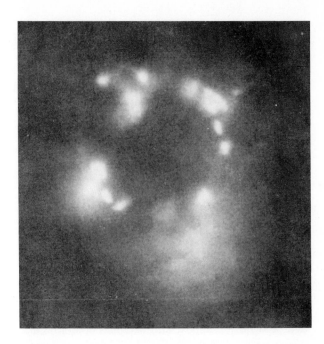

FIGURE 25-7   Rickettsia infect and multiply in endothelial cells that line small blood vessels. When stained with specific antibodies that are labeled with fluorescein, the infected cells can be readily visualized. (From Fleisher E: J Pediatr July 1979.)

by Weil and Felix that serum from patients with some rickettsial diseases can agglutinate certain strains of *Proteus vulgaris*—specifically strains OX-2, OX-19, and OX-K. The expected agglutination patterns are summarized in Table 25-3. Positive agglutination reactions are found in patients with Rocky Mountain spotted fever, epidemic typhus, murine typhus, and scrub typhus. Agglutination is initially detected 7 to 14 days after the onset of clinical disease, although this may be delayed by 4 weeks or more in treated patients. The positive reaction, indicated by a fourfold increase in titer or a single titer ≥1:320, also falls rapidly during convalescence. Because this test is not specific for rickettsia, positive reactions must be supported by an appropriate epidemiologic history and clinical evidence of disease.

A more appropriate test for the diagnosis of rickettsial disease is the **complement fixation test.** Positive reactions generally develop later in the course of disease (e.g., at the end of the second week for untreated patients, after 6 to 8 weeks for treated

Table 25-3   Weil-Felix Agglutination Patterns for Common Rickettsial Diseases

Disease	Weil-Felix Serology		
	OX-19	OX-2	OX-K
Rocky Mountain spotted fever	+	+	−
Rickettsialpox	−	−	−
Epidemic typhus	+	±	−
Brill-Zinsser disease	±	−	−
Murine typhus	+	±	−
Scrub typhus	−	−	+
Q fever	−	−	−
Ehrlichiosis	−	−	−

patients) and persist for a longer period. Cross-reactivity is observed between Rocky Mountain spotted fever and rickettsialpox, and between epidemic and murine typhus. However, this should not pose a problem when the clinical history and presentation are considered.

The **indirect fluorescent antibody test** is also useful for the detection of IgM and IgG antibodies against rickettsia. In fact, this is the diagnostic test of choice for ehrlichiosis. Unfortunately, reagents for this test are generally available only in state public health laboratories.

Serologic tests for Q fever must be considered separately from those for other rickettsia. *Coxiella burnetii* differs by its ability to undergo **phase variation,** giving phase I and phase II antigens. The phase I antigens are only weakly antigenic. Thus in acute Q fever, IgM and IgG antibodies develop primarily against phase II antigens. Diagnosis of acute Q fever disease is demonstrated by a fourfold increase in antibody titers, an IgM titer $\geq$1:64, or an IgG titer $\geq$1:256. The IgM complement fixation titer first appears positive after 2 weeks in untreated patients and reverts to negative after 12 weeks. The IgG titer is positive after 12 weeks and persists for more than 1 year in 90% of patients. Diagnosis of chronic Q fever disease is confirmed by demonstrating antibodies against both phase I and phase II antigens, with higher titers to the phase I antigen (ratio of phase I/phase II > 1). An indirect fluorescent antibody test for Q fever infections is more sensitive but not readily available in clinical laboratories.

TREATMENT

Rickettsia are susceptible to tetracycline and chloramphenicol. Prompt diagnosis and initiation of appropriate therapy usually result in a satisfactory prognosis. Unfortunately, because serologic diagnosis generally cannot be made until 2 to 4 weeks after the onset of disease, diagnosis must be based on clinical presentation. This can be delayed if the characteristic rash is not observed. Mortality associated with untreated Rocky Mountain spotted fever and scrub typhus was reported to approach 30%, and to be 60% or greater with untreated epidemic typhus (see Table 25-2). However, with the use of effective antibiotics and supportive care, mortality resulting from Rocky Mountain spotted fever in the United States is 3%. For most other rickettsial infections, mortality is substantially less, except for chronic Q fever, for which the prognosis is poor, despite antibiotic therapy and surgical intervention in heart disease.

PREVENTION AND CONTROL

The prevention of disease by immunoprophylaxis with vaccines has had only limited success. Vaccines are available only for epidemic typhus and Q fever. The inactivated *R. prowazekii* vaccine is available for high-risk groups. Both inactivated and attenuated vaccines for Q fever have been developed and shown to be protective for humans, although the duration of protection is unknown. However, neither vaccine is commercially available. There is minimal incentive for eradication of infection in livestock herds. The infected animals are asymptomatic, and it is unknown what the impact of an animal immunization program would have on the incidence of human disease. Therefore vaccination of these herds has also been poorly accepted.

Control of rickettsial diseases is difficult because the range of reservoir hosts and arthropod vectors is extensive. Furthermore, many infected hosts and vectors can survive for extended periods (e.g., ticks infected with *R. rickettsii* can survive for 4 years without feeding). Therefore attempts to control disease by eradicating carriage in the reservoir hosts or vectors has proved ineffective.

The risk for exposure to tick-borne diseases can be reduced by (1) avoiding tick-infested areas, (2) wearing protective clothing, (3) using insect repellants, and (4) regularly inspecting for and removing attached ticks. Diseases transmitted by the human body louse can be controlled by maintaining appropriate sanitary hygiene and using delousing sprays when indicated. Rodent control in urban areas will help reduce the risk for associated rickettsial diseases.

BIBLIOGRAPHY

Human ehrlichiosis—United States. Morbidity and Mortality Weekly Report 37:270-277, 1988, Atlanta, Centers for Disease Control.

Maeda K, Markowitz N, Hawley RC, et al: Human infection with *Ehrlichia canis,* a leukocytic rickettsia, N Engl J Med 316:853-856, 1987.

McDade JE and Fishbein DB: 1988. Rickettsiaceae: the Rickettsiae. In Balows A, Hausler W Jr., Ohashi M, and Turano A, editors: Laboratory diagnosis of infectious diseases principles and practice, vol 2, Viral, rickettsial, and chlamydial diseases, 1988, Springer-Verlag New York Inc.

Philip RN, Casper EA, MacCormack JN, et al: A comparison of serologic methods for diagnosis of Rocky Mountain spotted fever, Am J Epidemiol 105:56-67, 1977.

Sawyer LA, Fishbein DB, and McDade JE: Q fever: current concepts, Rev Infect Dis 9:935-946, 1987.

Taylor JP, Betz TG, Fishbein DB, et al: Serological evidence of possible human infection with *Ehrlichia* in Texas, J Infect Dis 158:217-220, 1988.

Taylor JP, Betz TG, and Rawlings JA: Epidemiology of murine typhus in Texas: 1980-1984, JAMA 255:2173-2176, 1986.

Tobin MJ, Cahill N, Gearty G, et al: Q fever endocarditis, Am J Med 72:396-400, 1982.

Walker DH: Biology of rickettsial diseases, Boca Raton, Florida. 1988, CRC Press.

Weiss E: The biology of rickettsiae, Ann Rev Microbiol 36:345-370, 1982.

Walker DH, Cain BB, and Olmstead PM: Laboratory diagnosis of Rocky Mountain spotted fever by immunofluorescent demonstration of *Rickettsia rickettsii* in cutaneous lesions, Am J Clin Pathol 69:619-624, 1978.

Chlamydiae

Members of the family Chlamydiaceae are obligate intracellular bacteria that were once regarded as viruses. Chlamydiae possess inner and outer membranes similar to those of gram-negative bacteria, contain both DNA and RNA, pcssess procaryotic ribosomes, synthesize their own proteins, nucleic acids, and lipids, and are susceptible to numerous antibiotics. Unlike other gram-negative bacteria, however, chlamydiae lack a peptidoglycan layer between the inner and outer membranes and undergo a unique growth cycle. *Chlamydiae* possess a small genome of approximately 6.6×10^8 d.

Chlamydiae are divided into two distinct species, *C. trachomatis* and *C. psittaci.* A third group of chlamydiae has recently been described, the TWAR strain, which shares the *Chlamydia* genus-specific antigen but is otherwise serologically distinct from *C. trachomatis* and *C. psittaci.* The species can also be separated on the basis of susceptibility to sulfonamides, the morphology of the elementary bodies, the formation of inclusions containing glycogen, and the natural host range (Table 26-1).

Microbial Physiology and Structure

Chlamydiae exist in two morphologically distinct forms: the small (300 to 400 nm) extracellular, infectious **elementary body** (EB) and the larger (800 to 1000 nm) intracellular, noninfectious **reticulate body** (RB) (Figure 26-1). The small size of the EB places it among the smallest of the procaryotes. The EB is rigid and resistant to disruption as a result of disulfide linkages among its outer membrane pro-

Table 26-1 Differentiation of *Chlamydia* Species

Property	Species		
	C. trachomatis	*C. psittaci*	**TWAR**
Host range	Humans, mice	Birds, humans, lower mammals	Humans
Elementary body morphology	Round	Round	Pear-shaped
Inclusion morphology	Round, vacuolar	Variable, dense	Round, dense
Glycogen in inclusions	Yes	No	No
Susceptibility to sulfonamides	Yes	No	No
DNA homology	10%	10%	10%
Plasmid DNA	Yes	Yes	No

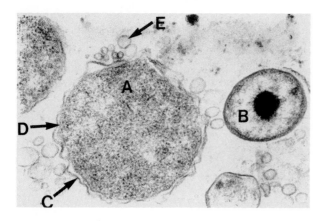

FIGURE 26-1   A transmission electron photomicrograph of *C. trachomatis* serovar F, showing a reticulate body *(A)* and a condensing form *(B)* that has nearly completed its transition to an elementary body. Note the electron-dense nucleic acid core in the condensing form. Separation between the outer membrane *(C)* and inner membrane *(D)* can be seen most clearly in the reticulate body. The smaller forms *(E)* are membrane blebs. (Courtesy BA Collett and WJ Newhall, Indiana University. From Batteiger BE and Jones RB: Chlamydial infections, Infect Dis Clin North Am 1:55-81, 1987.)

teins. The RB, the metabolically active form, is osmotically fragile by comparison. It is relatively deficient in the major outer membrane proteins, which are crosslinked to a lesser extent than in EB.

A group-specific antigen has been detected that is heat-stable and periodate-sensitive. It can be extracted from infected tissue with ether or with detergents. This antigen is used in the complement fixation test to diagnose the chlamydial infections psittacosis and lymphogranuloma venereum.

As with other gram-negative bacteria, chlamydiae possess a lipopolysaccharide (LPS). The chlamydial LPS contains determinants that cross-react with LPS of gram-negative bacteria, and a determinant that is genus specific.

Chlamydiae are nonmotile and nonpiliated. However, surface projections have been identified that presumably allow uptake of nutrients from the host cytoplasm.

Chlamydiae replicate via a unique growth cycle that occurs within susceptible host cells (Figure 26-2). The cycle is initiated by attachment of the infectious EB to microvilli of susceptible cells. In vivo, *C. trachomatis* infects only nonciliated, columnar, or

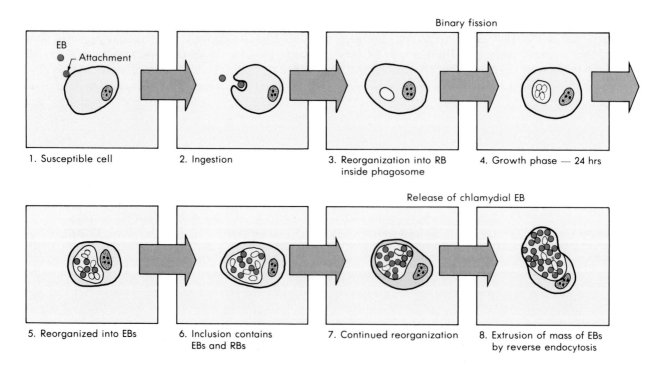

FIGURE 26-2   Schematic depiction of the growth cycle of *C. trachomatis*. (Redrawn from Batteiger BE and Jones RB: Chlamydial infections, Infect Dis Clin North Am 1:55-81, 1987.)

cuboidal epithelial cells found on various mucosal surfaces such as that of the conjunctivae, endocervix, urethra, rectum, endometrium, and fallopian tubes. The chlamydial strains responsible for lymphogranuloma venereum appear to replicate in macrophages. The EB actively penetrates the host cell; heat-killed or antibody-treated chlamydiae are not ingested. Once internalized, the chlamydiae remain within phagosomes where the replicative cycle proceeds. Fusion of cellular lysosomes with the EB-containing phagosome, which ordinarily would result in release of bactericidal substances into the phagosome, is specifically inhibited. Prevention of phago-lysosomal fusion may be mediated by components of the chlamydial cell envelope.

Within 6 to 8 hours after entering the cell, the EB, which is metabolically inactive, reorganizes into the metabolically active, dividing RB. RBs are able to synthesize their own DNA, RNA, and protein but lack the necessary metabolic pathways to produce their own high-energy phosphate compounds. They have been termed **"energy parasites"** because of this defect. The RBs repeatedly divide by binary fission approximately 8 to 24 hours after infection. As replication proceeds, the phagosome is termed an **inclusion.** Approximately 18 to 24 hours after infection, the RBs reorganize into the smaller EBs. Between 48 to 72 hours the cell ruptures and infective EBs are released.

C. TRACHOMATIS

To understand *Chlamydia trachomatis* infections it is important to recognize the species has been sub-divided into 3 biovars: LGV (lymphogranuloma venereum), trachoma, and a third biovar that contains the mouse pneumonitis agent (Table 26-2). The human biovars have been further divided into 15 serotypes (commonly called serovars) based on antigenic differences among the strains. There are three serovars of the LGV biovar (L_1, L_2, L_3) and 12 serovars of the trachoma biovar: serotypes A, B, Ba, and C are associated with blinding trachoma; serotypes D through K are associated with oculogenital disease.

Pathogenesis

The pathogenesis of *C. trachomatis* infections is poorly understood. Whereas LGV strains cause a systemic infection involving lymphoid tissues, the trachoma strains are restricted in vivo to replication in squamocolumnar-columnar epithelial cells. The clinical manifestations of chlamydial infection are thought to result from the direct destruction of cells during replication, as well as from the host inflammatory response.

Chlamydiae gain access through minute abrasions or lacerations. In LGV the lesions form in the lymph nodes, draining the site of primary infection (Figure 26-3). Abscesses are composed of aggregates of mononuclear cells surrounded by endothelial cells. The lesions may become necrotic, thereby attracting polymorphonuclear leukocytes and causing spread of the inflammatory process of surrounding tissue. Subsequent rupture of the lymph node causes formation of abscesses, fissures, sinus tracts, or fistulas. Non-LGV serotypes of *C. trachomatis* infect columnar or transitional epithelial cells of mucosal surfaces. Infection stimulates formation of a severe inflammation,

Table 26-2 Human Diseases Caused by *Chlamydia trachomatis*

Biovar	Serovar	Clinical Disease or Syndrome	Geographic Distribution
Trachoma	A, B, Ba, C	Trachoma	Primarily endemic in Asia and Africa
	D-K	Inclusion conjunctivitis, pneumonia in infants, nongonococcal urethritis, other genital infections	Worldwide
Lymphogranuloma venereum	LGV 1, 2, 3	LGV	Worldwide

From Joklik WK, Willett HP, Amos DB, and Wilfert CM, editors: Zinsser microbiology, ed 19, Norwalk, Conn, 1988, Appleton & Lange.

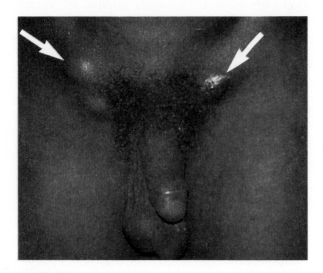

FIGURE 26-3   Lymphogranuloma venereum *(LGV)* showing bilateral inguinal buboes with adenopathy above and below the inguinal ligament on one side (the groove sign) and thinning of the skin over the adenopathy on the opposite side where the node is about to rupture. As in most patients with inguinal LGV, there is no primary lesion present. (From Holmes, KK et al: Sexually transmitted diseases, New York, 1984, McGraw-Hill Book Co.)

which consists of neutrophils, lymphocytes, and plasma cells. Eventually, true lymphoid follicles with germinal centers are induced.

Clinical Syndromes

C. trachomatis is responsible for a wide range of clinical diseases (Table 26-3).

Trachoma   Trachoma, which is characterized by the pebbled appearance of the infected conjunctiva, is a chronic keratoconjunctivitis caused by serotypes A, B, Ba, and C. Trachoma is the leading cause of preventable blindness in developing countries, affecting an estimated 500 million individuals. This disease is seen initially as a follicular conjunctivitis with a diffuse inflammation involving the entire conjunctiva. Progression of the disease leads to conjunctival scarring, producing inturned eyelids. Subsequently, inturned eyelashes cause constant abrasion of the cornea, which eventually results in corneal ulceration, scarring, pannus formation (invasion of vessels into the cornea), and loss of vision.

Trachoma is endemic in the Middle East, North Africa, and India. Infections occur predominantly in children; the incidence declines during late childhood and adolescence. In endemic areas children are the chief reservoir of the agent. However, the incidence of blindness continues to rise during adulthood

Table 26-3   Clinical Spectrum of *Chlamydia trachomatis* Infections

Serovars	Host	Infection	Complications
A, B, Ba, C	Females, males, children	Trachoma	Blindness
B, D-K	Females	Cervicitis, urethritis, proctitis, conjunctivitis	Salpingitis, endometritis, perihepatitis, ectopic pregnancy, infertility, postpartum endometritis
B, D-K	Males	Urethritis, postgonococcal urethritis, proctitis, conjunctivitis	Epididymitis, Reiter's syndrome
B, D-K	Infants	Conjunctivitis, pneumonia, asymptomatic pharyngeal and gastrointestinal tract carriage; otitis media(?)	
L_1, L_2, L_3	Females, males	Lymphogranuloma venereum	Rectal strictures, draining sinuses, lymphatic obstruction

?, Relationship has not been conclusively established. Modified from Bell TA, Grayston JT: Centers for Disease Control guidelines for prevention and control of *Chlamydia trachomatis* infections: summary and commentary, Ann Intern Med 104:524, 1986; Centers for Disease Control: *Chlamydia trachomatis* infections: policy guidelines for prevention and control, MMWR 34(suppl 3S):53S, 1985; Thompson SE, Washington AE: Epidemiology of sexually transmitted *Chlamydia trachomatis* infections, Epidemiol Rev 5:96, 1983; and Handsfield HH, editor: Sexually transmitted diseases, Infect Dis Clin North Am 1(1):62, 1987.

as the disease progresses. Trachoma is transmitted eye-to-eye by droplet, hands, contaminated clothing, and by eye-seeking flies, which transmit ocular discharges to eyes of other children. Because a high percentage of children in endemic areas harbor *C. trachomatis* in the respiratory and gastrointestinal tracts, transmission may also occur by respiratory droplet or by fecal contamination. Communities where trachoma is endemic generally are characterized by crowded living conditions, poor sanitation, and poor personal hygiene—risk factors that promote transmission of infections.

Trachoma has been classified into four stages based on the degree of inflammation of the conjunctival lining of the upper lid (Box 26-1). Recurrences are common after apparent healing. Inapparent or subclinical infection has been documented in children in endemic areas and in individuals who acquired trachoma in childhood and then emigrated to the United States.

Adult inclusion conjunctivitis An acute follicular conjunctivitis caused by the *C. trachomatis* strains associated with genital infections has been documented in sexually active adults. The infection is characterized by mucopurulent discharge, keratitis, corneal infiltrates, and occasionally some corneal vascularization (Figure 26-4, *B*). In chronic cases corneal scarring has been reported. Most cases occur in adults between the ages of 18 and 30, with probable genital infection before eye involvement. Autoinocu-

Box 26-1	The Four Stages of Trachoma
Stage I	Asymptomatic with little conjunctival exudate
Stage II	Follicular and papillary hypertrophy
Stage III	Scarring of the conjunctiva and pannus
Stage IV	Healed trachoma with no evidence of inflammation

lation or oral-genital contact are believed to be the routes of transmission.

Neonatal conjunctivitis Inclusion conjunctivitis in the newborn is acquired by passage through an infected maternal birth canal. *C. trachomatis* conjunctivitis has been documented in 2% to 6% of neonates. It usually occurs 2 to 30 days after birth and is generally caused by the serotypes implicated in genital infections (serotypes D through K). After an incubation period of 5 to 12 days, swelling of the lids, hyperemia, and copious purulent discharge appear (Figure 26-4, *A*). Untreated infections may run a course of up to 12 months, accompanied by conjunctival scarring and corneal vascularization. Without therapy or with topical therapy only, infants are at risk for development of *C. trachomatis* pneumonia.

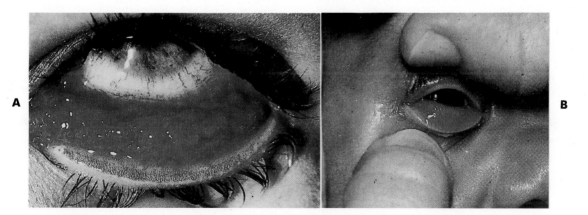

FIGURE 26-4   A, Acute recurrent chlamydial conjunctivitis in a 2-month-old infant. The follicular hyperplasia of the lower palpebral conjunctiva, seen in this recurrent infection, is usually not seen in the first episode of neonatal chlamydial conjunctivitis. **B,** Paratrachoma in an adult, caused by a genital immunotype of *C. trachomatis.* Note the marked follicular appearance. (From Holmes, KK et al: Sexually transmitted diseases, New York, 1984, McGraw-Hill Book Co.)

Table 26-4 Clinical Urogenital Infections caused by *C. trachomatis*

Site of Infection	Clinical Syndrome
Men	
Urethra	Nongonococcal urethritis, postgonococcal urethritis
Epididymis	Epididymitis
Rectum	Proctitis
Conjunctiva	Conjunctivitis
Systemic	Reiter's syndrome
Women	
Urethra	Acute urethral syndrome
Bartholin's gland	Bartholinitis
Cervix	Cervicitis, cervical dysplasia (?)
Fallopian tube	Salpingitis
Conjunctiva	Conjunctivitis
Liver capsule	Perihepatitis
Systemic	Arthritis, dermatitis

From Holmes KK, Mardh PA, Sparling PF, and Wiesner PJ: Sexually transmitted diseases, New York, 1984, McGraw-Hill.

Ocular lymphogranuloma venereum The LGV serotypes of *C. trachomatis* have been implicated as a cause of Parinaud's oculoglandular conjunctivitis, which is a conjunctival inflammation associated with preauricular, submandibular, and cervical lymphadenopathy.

Urogenital Infections *C. trachomatis* is the single most frequent cause of sexually transmitted disease in the United States, causing an estimated 4 million cases per year. *C. trachomatis* has been implicated as a cause of various clinical syndromes (Table 26-4). Genital tract infections are caused by serotypes D through K, and LGV is caused by serotypes L_1, L_2, and L_3.

Most genital tract infections in women are asymptomatic but can nevertheless spread to cause symptomatic disease, including cervicitis, endometritis, urethritis, salpingitis, bartholinitis, and perihepatitis.

Because of the high incidence of asymptomatic infections, chlamydial infection may be overlooked. Women at high risk for chlamydial infection include female sex partners of men with **nongonoccal urethritis (NGU)** caused by chlamydia. Symptomatic infections produce mucopurulent discharge and hypertrophic ectopy and generally yield greater numbers of organisms on culture than do asymptomatic infections. Untreated infections may spontaneously resolve or may persist for months. Complications of untreated infections include pelvic inflammatory dis-

ease, ectopic pregnancy, perihepatitis, and possibly cervical dysplasia.

Urethritis caused by *C. trachomatis* may occur with or without concurrent cervical infection. Chlamydial infection of Bartholin's ducts may occur alone or with gonococcal infection. Spread of lower tract infection with chlamydiae can produce endometritis, salpingitis, and perihepatitis.

Unlike genital infections in women, the majority of genital infections in men caused by *C. trachomatis* are symptomatic. However, it has recently become apparent that up to 25% of chlamydial infections in men may be asymptomatic (Figure 26-5). Approximately 35% to 50% of cases of nongonococcal urethritis are caused by *C. trachomatis*. Postgonococcal urethritis results from coinfection with both *N. gonorrhoeae* and *C. trachomatis*, and symptomatic illness develops after successful treatment of the gonorrhea because of the longer incubation period of chlamydiae. Epididymitis may occur in association with urethritis due to chlamydiae or coliform bacteria. Up to 15% of proctitis in homosexual men may be caused by *C. trachomatis*.

Reiter's syndrome (urethritis, conjunctivitis, polyarthritis, and mucocutaneous lesions) is believed to be initiated by genital infection with *C. trachomatis*. The disease usually occurs in young white males. Approximately 50% to 65% of patient's with Reiter's syndrome have a chlamydial genital infection at the onset of arthritis. Serologic studies suggest that over

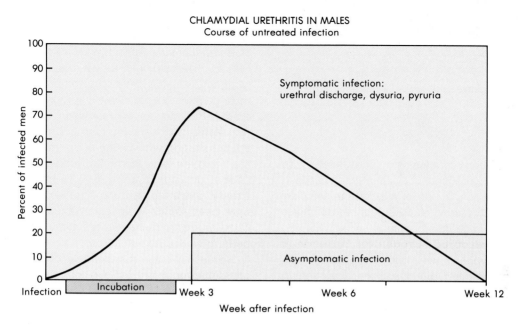

CHLAMYDIAL URETHRITIS IN MALES
Course of untreated infection

Symptomatic infection:
urethral discharge, dysuria, pyruria

Asymptomatic infection

FIGURE 26-5   Time course of chlamydial urethritis in men.

80% of men with Reiter's syndrome have evidence of a preceding or concurrent infection with *C. trachomatis*.

Lymphogranuloma venereum (LGV) is a chronic sexually transmitted disease caused by *C. trachomatis* serotypes L_1, L_2, and L_3. LGV occurs sporadically in North America, Australia, and Europe, but it is highly prevalent in Africa, Asia, and South America. In the United States male homosexuals are the major reservoir of disease. Acute LGV is reported more frequently in men primarily because symptomatic infection is less common in women.

After an incubation period of 1 to 4 weeks, a primary lesion may appear at the site of infection (e.g., penis, urethra, glans, scrotum, vaginal wall, cervix, vulva). Small, painless, and inconspicuous, the lesion is often overlooked. Fever, headache, and myalgia may accompany the lesion, which heals rapidly.

The second stage of infection is marked by inflammation and swelling of the lymph nodes draining the site of initial infection (Table 26-5). Most commonly, the inguinal nodes are involved, producing painful, fluctuant "buboes" that gradually enlarge. Systemic manifestations may include fever, pain, hepatitis, pneumonitis, and meningoencephalitis. In women, proctitis is common because of lymphatic spread from the cervix or vagina. In men, proctitis develops after anal intercourse or by lymphatic spread from the urethra. Untreated LGV may resolve at this stage or progress to the development of genital ulcers, fistulas, strictures, or genital elephantiasis.

Laboratory Diagnosis

Ocular infections The diagnosis of acute phase ocular infections can be made by culture of the organism. In neonatal conjunctivitis and early trachoma,

Table 26-5 Site of Primary Lymphogranuloma Venereum Infection Determining Subsequent Lymphatic Involvement

Site of Primary Infection	Affected Lymph Nodes
Penis, anterior urethra	Superficial and deep inguinal
Posterior urethra	Deep iliac, perirectal
Vulva	Inguinal
Vagina, cervix	Deep iliac, perirectal, retrocrural, lumbosacral
Anus	Inguinal
Rectum	Perirectal, deep iliac

From Holmes KK, Mardh PA, Sparling PF, and Wiesner PJ: Sexually transmitted diseases, New York, 1984, McGraw-Hill.

Box 26–2 Diagnostic Tests for Chlamydiae

Cytology	Culture
Direct antigen detection	Serology

direct immunofluorescence of conjunctival cells with fluorescein-conjugated monoclonal antibody is sensitive and specific. Detection of antibody in tears by microimmunofluorescence is useful for the diagnosis of chronic eye infections in adults.

Non-LGV serotype urogenital infections Diagnosis of *C. trachomatis* can be achieved by cytologic samples, culture, direct antigen detection in clinical specimens, and serologic testing (Box 26-2). The sensitivity of each method depends on the patient population examined, the sex of the patient, and the specimen site. For example, as stated earlier, symptomatic infections generally yield a higher number of inclusion-forming units than do asymptomatic infections. Infected epithelial cells may be recovered from the urethra, cervix, conjunctiva, nasopharynx, rectum, and aspirates from fallopian tubes and epididymis.

Cytologic examination of cell scrapings for the presence of inclusions was the first method used for the diagnosis of *C. trachomatis*. However, this method is insensitive when compared to culture or direct immunofluorescence.

Isolation of *C. trachomatis* in cell culture remains the most sensitive and specific method of diagnosis of *C. trachomatis* infections. Cycloheximide-treated McCoy cells have been shown to produce consistently higher inclusion counts than other cell lines and are most commonly used. Centrifugation of the specimen onto the cell monolayer greatly enhances the isolation rate. After incubation for 48 to 72 hours the monolayers are stained with iodine or an immunofluorescent stain and are examined microscopically for the presence of inclusions (Figure 26-6). Despite the high sensitivity of culture, a number of technical problems have been shown to adversely affect the recovery of *C. trachomatis*. Moreover, relatively few laboratories perform cell culture isolation.

Because of these shortcomings of cell culture, antigen detection methods have been developed and are now commercially available. Direct immunofluorescence (DFA) staining employs fluorescein-isothiocyanate-conjugated monoclonal antibodies to *C. trachomatis* for the detection of elementary bodies in smears from clinical samples (Figure 26-7). The Microtrak DFA (Syva, Palo Alto, California) is the most widely evaluated DFA reagent. This DFA achieved ≥90% sensitivity as compared to culture when evaluated in symptomatic men and high-risk women, i.e., sex partners of chlamydia positive males, and women attending sexually transmitted disease clinics. In populations with asymptomatic infections, the sensitivity of DFA decreases, presumably because fewer inclusion-forming units are present in asymptomatic infections.

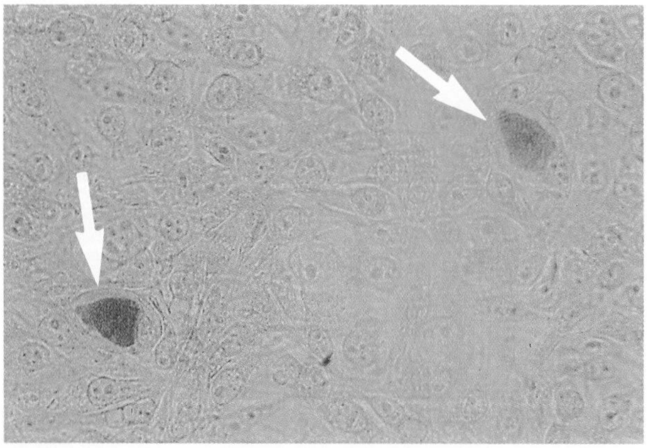

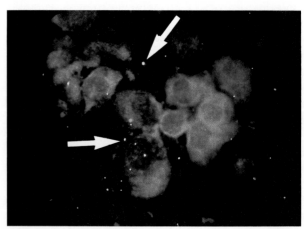

FIGURE 26-6 Iodine-stained *C. trachomatis* inclusion bodies *(arrows)*.

FIGURE 26-7 Fluorescent-stained elementary bodies *(arrows)* in a clinical sample.

Chlamydial antigen has also been detected in clinical specimens by enzyme-linked immunoassay (ELISA). A number of commercially available kits are marketed, but to date only the Chlamydiazyme assay (Abbott Laboratories, Abbott Park, Illinois) has been extensively evaluated. Although the sensitivity of this test is >90% for symptomatic women at high risk for disease, the sensitivity is unacceptably low for infected men and asymptomatic women. In a population with a low prevalence of disease (≤5%), culture remains the test of choice.

Antibodies to a genus-specific antigen can be detected by complement fixation. For *C. trachomatis* type-specific antibodies, the microimmunofluorescence assay, a very tedious and difficult test, is used.

Serologic testing has limited value in the diagnosis of urogenital infections in adults because the test cannot differentiate between current and past infection. Most adults with chlamydial infection have had a previous exposure to *C. trachomatis* and thus are seropositive. Although IgM may not always be produced, the presence of a high titer of IgM (≥1:128) suggests a recent infection. A negative serologic test may prove to reliably exclude infection. Detection of IgM to *C. trachomatis* is useful in the diagnosis of neonatal infection with *C. trachomatis*.

LGV Diagnosis of LGV is established by the isolation of an LGV serotype from a bubo or other infected site. However, recovery rates of only 24% to 30% have been reported. A skin test using intradermal injections of LGV antigen, the **Frei test,** suffers from lack of sensitivity in early LGV and lack of specificity, since the Frei antigen is a genus-specific antigen. Moreover, the Frei test can remain positive for many years, thus limiting its usefulness. Antibody detection by complement fixation is also nonspecific, again because the antigen used is genus-specific. Other serologic tests such as microimmunofluorescence and neutralization are performed only in specialized laboratories.

Treatment

Ocular infections Systemic therapy with a tetracycline is recommended for treatment of ocular infections in older children and adults. In areas with endemic trachoma, therapy without improved sanitation is of limited value. Surgical correction of lid deformities reduces trachoma-induced blindness. For neo-natal conjunctivitis or disease in young children systemic erythromycin therapy is used.

Urogenital infections Systemic therapy with tetracycline or erythromycin in recommended for treatment of urogenital infections. Treatment of the sexual partner is also recommended. LGV infections require aspiration of fluctuant lymph nodes in addition to antibiotic therapy. Complications of LGV, such as strictures, fistulas, and elephantiasis, require surgery.

Prevention and Control

Prevention of *C. trachomatis* infections is difficult because the populations with endemic disease frequently have limited access to medical care. The blindness associated with advanced stages of this disease can only be prevented by prompt treatment of early disease and prevention of subsequent reexposure. Although treatment can be initiated, the eradication of disease within a population, and thus prevention of reinfections, is difficult.

Chlamydia conjunctivitis and genital infections are prevented by the use of safe sexual practices and the prompt treatment of both symptomatic patients and their contacts.

C. PSITTACI

C. psittaci is the cause of **psittacosis (parrot fever),** which can be transmitted to humans. The disease was first described in parrots, thus the name psittacosis (Psittakos, Greek for parrot). The natural reservoir of *C. psittaci* is birds; virtually any species of bird can become infected and thus serve as a source of human infection. The organism is present in the blood, tissues, feces, and feathers of infected birds.

Epidemiology

Transmission from birds to humans is usually by inhalation of dried bird excrement. Person-to-person transmission is rare. Transmission has also been documented after handling of infected tissue or plumage. Thus veterinarians, zoo keepers, pet shop workers, and poultry processing plant employees are at risk for this infection. Despite the high potential for transmission, only approximately 100 cases of psittacosis were reported in the United States in 1988. However, this is likely an underestimation of the true incidence of disease.

Pathogenesis

Infection occurs via the respiratory tract, with subsequent spread to the reticuloendothelial cells of the liver and spleen. Multiplication of the organisms occurs in these sites, producing focal necrosis. Subsequently, the lung and other organs are seeded by hematogenous spread, causing a predominantly lymphocytic inflammatory response in both the alveolar and interstitial spaces. Edema, thickening of the alveolar wall, infiltration of macrophages, necrosis, and occasionally hemorrhage occur at these sites. Mucus plugs develop in the bronchioles, causing cyanosis and anoxia.

Clinical Syndromes

After an incubation period of 7 to 15 days, onset of illness is usually manifested by headache, high fever,

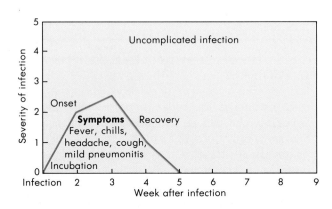

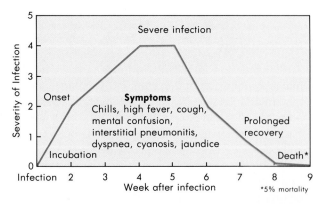

FIGURE 26-8   Time course of *Chlamydia psittaci* infection.

and chills (Figure 26-8). Other early manifestations include malaise, anorexia, myalgia, arthralgia, and occasionally a pale macular rash. Pulmonary signs include a nonproductive cough, rales, and consolidation. Central nervous system involvement is common, usually consisting of headache but in severe cases encephalitis, convulsions, coma, and death may occur. Gastrointestinal symptoms such as nausea, vomiting, and diarrhea may be present. Other systemic symptoms include carditis, hepatomegaly, splenomegaly, and follicular keratoconjunctivitis.

Laboratory Diagnosis

Psittacosis is usually diagnosed by serologic testing. A fourfold rise in titer by complement fixation (CF) of paired acute and convalescent phase sera is diagnostic. Titer rises are usually observed by the end of the second week of illness. However, antibiotic therapy may delay the appearance of antibody. A single CF titer of ≥1:32 is evidence of recent infection.

Isolation of the organism in cell culture is readily achieved. However, this is attempted only in specialized laboratories because of the possibility of laboratory-acquired infections.

Treatment

Tetracycline is the drug of choice for psittacosis. Erythromycin has also been successful.

Prevention

Antibiotic-supplemented feed has reduced the incidence of psittacosis among bird handlers. Medication of parrots before importation into the United States has also reduced transmission of *C. psittaci*. No vaccine currently exists for this bacterium.

TWAR STRAIN

The TWAR strain of *Chlamydia* was first isolated from the conjunctiva of a child in Taiwan. It was initially considered a psittacosis strain because of the morphology of the inclusions produced in cell culture. Subsequently, it was shown that the Taiwan isolate (TW-183) was serologically related to a pharyngeal isolate designated AR-39. The strain was designated TWAR, an acronym for TW and AR (acute respiratory). DNA homology studies have demonstrated that the TWAR strain is distinct from both *C. trachomatis* and *C. psittaci,* and a new species name, *C. pneumoniae* has been proposed. At present, only a single serotype has been identified.

Seroepidemiology studies have demonstrated the worldwide occurrence of TWAR infections. Up to 50% of individuals in a given population can be antibody positive. Most infections are seen between the ages of 7 and 30. The infection appears to be transmitted by human contact; no animal reservoir has been identified.

Clinically, TWAR has been associated with pneumonia, bronchitis, pharyngitis, sinusitis, and a flu-like illness. In young adults the infection is usually mild to moderate in severity. In elderly patients and those with chronic respiratory diseases severe pneumonia is not uncommon. Treatment with tetracycline or erythromycin has been successful.

Diagnosis of TWAR infection can be achieved by cell culture isolation and serologic tests. A cell culture procedure similar to that employed for isolation of *C. trachomatis* has been used successfully. Serologic diagnosis can be made by complement fixation or by microimmunofluorescence. The complement fixation test uses a genus-specific antigen and thus is not specific for TWAR infection; the microimmunofluorescence test uses TWAR elementary bodies as antigen. A fourfold titer rise in either IgM or IgG is diagnostic, whereas an IgM titer of $\geq 1:16$ or an IgG titer of $\geq 1:512$ are suggestive of recent infection. However, the microimmunofluorescence test is only available in a few specialized laboratories.

BIBLIOGRAPHY

Batteiger BE, and Jones RB: Chlamydial infections, Infect Dis Clin North Am 1:55-81, 1987.

Bowie WR, and Holmes KK: *Chlamydia trachomatis* (trachoma, inclusion conjunctivitis, lymphogranuloma venereum, and nongonococcal urethritis). In Mandell GL, Douglas RG Jr, and Bennett JE, editors: Principles and pactices of infectious diseases, ed 2, John Wiley & Sons, New York, 1985.

Campbell LA, Kuo CC, and Grayston JT: Characterization of the new *Chlamydia* agent, TWAR, as a unique organism by restriction endonuclease analysis and DNA-DNA hybridization, J Clin Microbiol, 25:1911-1916, 1987.

Grayston JT, Kuo CC, Wang SP, and Altman J: A new *Chlamydia psittaci* strain, TWAR, isolated in acute respiratory tract infections, N Engl J Med 315:161-168, 1986.

Holmes KK, Mardh PA, Sparling PF, and Wiesner PJ: Sexually transmitted diseases, New York, 1984, McGraw-Hill.

Mahony JB, Chernesky MA, Bromberg K, and Schacter J: Accuracy of immunoglobulin M immunoassay for diagnosis of chlamydial infections in infants and adults, J Clin Microbiol 24:731-735, 1986.

Oriel D, Ridgway G, Schacter J, Taylor-Robinson D, and Ward M, editors: Chlamydial infections: proceedings of the sixth international symposium on human chlamydial infections, Cambridge, England, 1986, Cambridge University Press.

Schacter J: Biology of *Chlamydia trachomatis*. In Holmes KK, Mardh PM, Sparling PF, and Wiesner PJ, editors: Sexually transmitted diseases, New York, 1984, McGraw-Hill.

Schaffner, W: Chlamydia psittaci (psittacosis). In Mandell GL, Douglas RG Jr, and Bennett JE, editors: Principles and practices of infectious diseases, ed 2, John Wiley & Sons, New York, 1985.

Smith TF, Brown SD, and Weed LA: Diagnosis of *Chlamydia trachomatis* infections by cell cultures and serology, Lab Medicine 13:92-100, 1982.

Stamm WE: Diagnosis of Chlamydia trachomatis genitourinary infections, Ann Intern Med 108:710-717, 1988.

Stamm WE, and Holmes KK: *Chlamydia trachomatis* infections of the adult. In Holmes KK, Mardh PM, Sparling PF, and Wiesner PJ, editors: Sexually transmitted diseases, New York, 1984, McGraw-Hill.

Wyrick PB, Gutman LT, and Hodinka RL: Chlamydiae. In Joklik WK, Willett HP, Amos DB, and Wilfert CM, editors: Zinsser Microbiology, ed 19, Norwalk, Conn, 1988, Appleton & Lange.

CHAPTER 27

Oral Microbiology

Most diseases of the oral cavity result directly from bacterial colonization and infections. However, other factors may also influence the course of oral diseases. The predominant etiologic agents are colonized microorganisms and their products; collectively, this adherent mass on the teeth is known as dental plaque.

The principal bacterial-associated diseases of the oral cavity are dental caries and the periodontal diseases, both of which are considered among the most prevalent infections of humans.

Before delving into oral microbiology and related pathogenetic factors in dental diseases, the possible econiches of the oral cavity in which these bacteria reside will be reviewed. The oral cavity may be broadly subdivided into three major components: the teeth, the supporting structures of the teeth (periodontium), and other intraoral structures, including the lips, tongue, floor of the mouth, buccal mucosa, palate, temporomandibular joint, fauces, and the tonsils. This chapter focuses on the teeth and periodontium.

TEETH

Development

Humans experience two dentitions: deciduous and permanent (adult).

Deciduous Dentition The deciduous dentition is composed of 20 teeth (Figure 27-1): 10 teeth for each dental arch (maxillary arch and mandibular arch). Eruption of the deciduous teeth usually begins between 6 and 7½ months of age with the lower central incisors. The second molars are the last deciduous teeth to erupt, occurring between 20 and 24 months of age. Along with eruption of the deciduous teeth is an ever-increasing complex of bacterial microflora.

Permanent Dentition The permanent or adult dentition consists of 32 teeth (Figure 27-2). The integrity of the teeth in each dental arch is contingent on the arrangement of the teeth, with each tooth in close contact with neighboring teeth. Malpositioning of the teeth permits greater bacterial colonization and growth because bacterial plaque-retentive areas develop. These plaque-retentive areas harbor and foster the growth and development of greater numbers and varieties of bacteria.

The permanent teeth begin erupting between 6 and 7 years of age; the final teeth, third molars or "wisdom teeth," erupt between 17 and 21 years of age.

Structure and Morphology

Teeth can be subdivided into four basic substructures: enamel, dentin, cementum, and pulp (Figure 27-3).

Dental **enamel** is the hardest material in the body and consists nearly entirely of inorganic salts (97% to 98%). This material covers the crown of the tooth and is usually resistant to abrasive wear. However, depending on the enamel's morphology, bacterial colonies can begin to act on basic enamel structures. Teeth with deep grooves, pits, and fissures provide bacteria with areas sheltered from normal oral physiologic cleansing mechanisms, including cheek and tongue movements. Dental caries begins when the

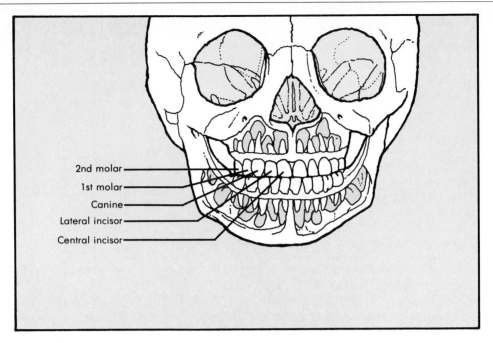

FIGURE 27-1   Dentition of 4-year-old child, anterior view.

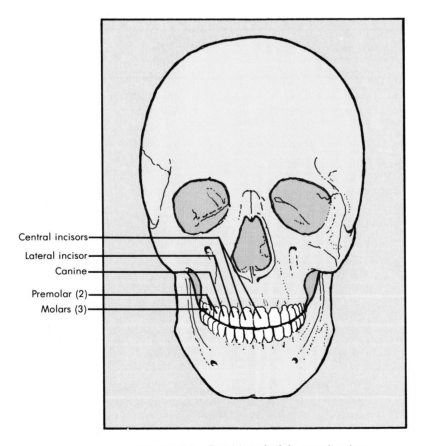

FIGURE 27-2   Dentition of adult, anterior view.

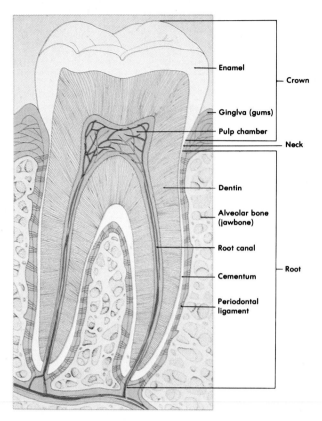

FIGURE 27-3   Cross-section of the tooth and its supporting structures.

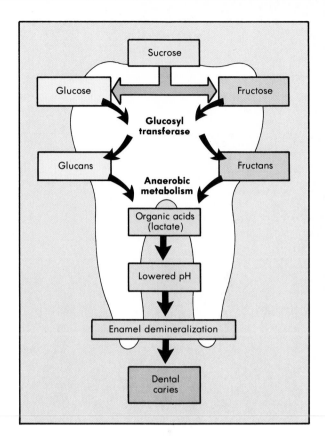

FIGURE 27-4   Dental plaque metabolism of sucrose.

demineralization of enamel occurs as a direct result of acid production through the plaque constituents' metabolism of sugars (Figure 27-4).

Dentin (see Figure 27-3), which is approximately 65% to 70% inorganic salts, is the calcified tissue that forms the main structure of the tooth. Dentin is the second structure subjected to dental caries. Pain is generally not perceived by patients unless dentin has been destroyed by the caries process.

The **cementum** is thin and covers the roots of the teeth (see Figure 27-3). One function of cementum is to incorporate collagenous periodontal ligament fibers and thus mediate the connection between teeth and alveolar bone. The cementum may also be colonized by bacteria, and dental caries may result (root caries). This form of dental caries is predominant in the geriatric population, among whom root structures are frequently exposed to the oral cavity (gingival recession).

Dental **pulp** occupies the central area of the crown

of the tooth and the roots (see Figure 27-3). It is soft tissue, primarily a collagen mass with a network of blood vascular and nervous systems, that remains from the formative organ of dentin. If dental caries continues through the enamel, cementum, and dentin, the pulp may be invaded.

PERIODONTIUM

The supporting structures of the teeth include collagen fibers known as the **periodontal ligament,** as well as alveolar bone (Figure 27-5, *B*). These supporting structures are primarily responsible for maintaining the teeth in proper alignment for form and function within the maxilla and mandible. The specialized portion of the jaw bone surrounding the roots of the teeth is called the alveolar bone (see Figure 27-5). Overlying the alveolar bone and periodontal ligament structures are the soft tissues known as the gingiva (see Figure 27-5, *A*).

The **gingiva** (Figure 27-6) is attached to the tooth at the cementum of the roots and alveolar bone. It is salmon pink and stippled (see Figure 27-5, *A*), and is essential for the protection of the underlying structures—alveolar bone and periodontal ligament. An inadequate width of gingiva may initiate or accelerate gingival recession.

The area of the gingiva between the teeth circumscribes a triangular space known as the **papilla** or interproximal zone (Figure 27-6). This is a frequent site for the initiation of periodontal disease and usually exhibits the earliest clinical signs of inflammation of the gingiva.

The nonkeratinized periodontal tissues adjacent to the gingiva are referred to as alveolar mucosa (see Figure 27-6). Alveolar mucosa is thin and transparent, covering the inside lining of the cheeks, vestibule areas of the dentition, and floor of the mouth. It is usually red and glistening, reflecting the paucity of keratin that alveolar mucosa structures have.

Bacterial colonies can adhere to any of the structures of the soft tissues of the periodontium, and, when not removed on a regular basis, can initiate inflammatory processes consistent with the periodontal diseases.

TOOTH-ASSOCIATED MATERIALS

The discussion of dental plaque should be put into proper perspective along with all tooth-associated materials. The tooth-associated materials collectively include food debris, acquired pellicle, materia alba, bacterial plaque, and calculus.

Food Debris

Food debris is retained in the mouth after mastication of food. This debris, unless wedged between the teeth or inside the periodontal-tooth interface, is usually removed by saliva or oral musculature action.

Pellicle

Acquired pellicle is a very thin (0.1 to 0.8 µm), primarily protein film that forms on surfaces of the teeth. The pellicle is derived from salivary components and is acquired very soon after the teeth are

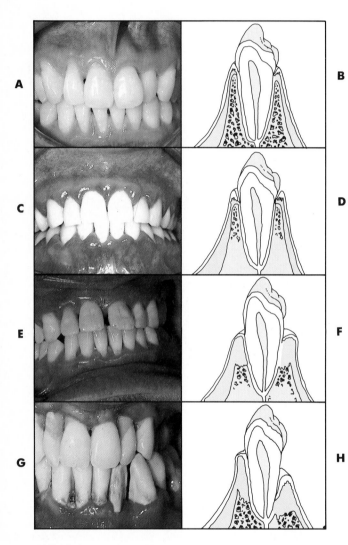

A
B
C
D
E
F
G
H

FIGURE 27-5 Progress of periodontal disease. **A** and **B,** Normal healthy gingiva (gums). Healthy gums and bone anchor teeth firmly in place. **C** and **D,** Gingivitis. Plaque and its byproducts irritate the gums, making them tender, inflamed, and likely to bleed. **E** and **F,** Periodontitis. Unremoved, plaque hardens into calculus (tartar). As plaque and calculus continue to build up, the gums begin to recede (pull away) from the teeth, and pockets form between the teeth and gums. **G** and **H,** Advanced periodontitis. The gums recede farther, destroying more bone and the peridontal ligament. Teeth—even healthy teeth—may become loose and need to be extracted.

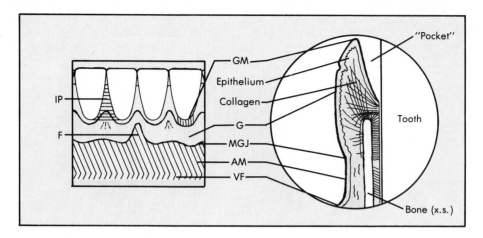

FIGURE 27-6   Diagrammatic illustration of surface characteristics of the clinically normal gingiva showing: IP, interdental papilla; F, frenum; GM, gingival margin; G, gingiva; MGJ, mucogingival junction; AM, alveolar mucosa; VF, vestibular fornix.

cleaned. It is basically acellular and bacteria free. The pellicle consists of high molecular–weight glycoproteins that are derived from the saliva and are selectively absorbed to the surfaces of the teeth. The pellicle is reformed within minutes of being removed from the surfaces of the teeth via brushing or flossing. It is not removed by even forceful rinsing. Acquired pellicle is considered the first stage in the formation of dental plaque, since bacterial adherence followed by colonization quickly ensues.

Materia Alba

Materia alba, literally "white matter," is a mixture of salivary proteins, bacteria, desquamated epithelial cells, and dying leukocytes. Clinically, materia alba appears as a soft white structure that is loosely adherent to the surfaces of the teeth. Materia alba can usually be flushed off with rinsing or forceful sprays of fluid.

Bacterial Plaque

Bacterial plaque is a mat of densely packed, colonized microorganisms, growing and tenaciously attached to the teeth. It is embedded in a matrix of pellicle adhering to the surfaces of the teeth. Total bacterial counts are estimated to be approximately 10^8 to 10^{11} bacteria per gram of plaque. Bacterial

plaque is not removed by rinsing or forceful spraying with fluids but is easily removed by mechanical means.

Subpopulations of the dental plaque develop further with varying compositions and metabolisms. An understanding of plaque development, and especially how bacteria inhabit the various econiches of the teeth and periodontium, will enhance prevention of dental disease and improve modalities of treatment.

Factors enhancing the rate of bacterial colonization in the oral cavity include oral hygiene, diet, malocclusion, malposed teeth, oral immune factors, saliva flow, and roughness of teeth surfaces, among others. Without proper brushing of the teeth, selected, extensive colonies of bacteria form on the surfaces of the teeth within 1 to 3 days. (Figure 27-7). The colonies are localized within the pits and fissures of the coronal (crown) portion of the tooth (see Figure 27-3) and are also found along the gingiva (see Figure 27-5, *D, F,* and *H*). Within a few days these bacterial colonies begin to coalesce and fuse to form a continuous "matlike" deposit.

Within approximately 7 to 10 days dental plaque increases in thickness on the surfaces of the teeth and in the region between the teeth and gingiva (sulcus) (see Figure 27-5, *D*). Streptococci and gram-positive bacilli form the predominant structures of early plaque development, which is located primarily supragingivally (at or above the gingiva). As dental

A   **B**   **C**

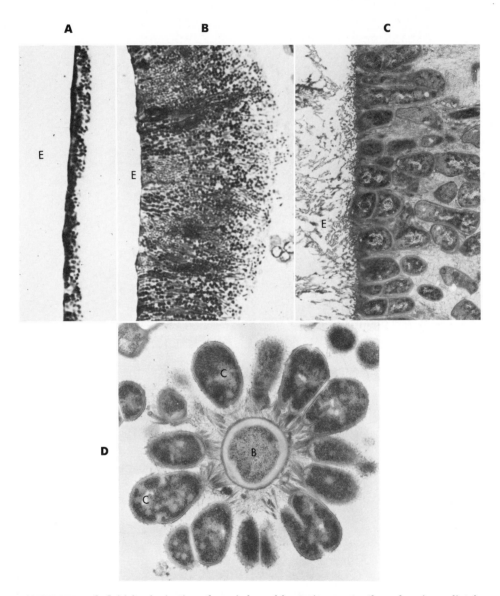

D

FIGURE 27-7   A, Initial colonization of cocci-shaped bacteria on a tooth surface immediately adjacent to healthy gingiva. *E,* enamel (×2000). **B,** Maturation of dental plaque colonies depicting thick layers of cocci-shaped bacteria on a tooth surface adjacent to healthy gingiva (×2000). **C,** Higher magnification using electron microscopy of **A.** Cocci-shaped bacteria are in contact with the pellicle (glycoprotein layer) of the enamel (×11,000). **D,** Formations of cocci-shaped bacteria adherent to a central filament-shaped bacteria. These are often referred to as "corncob" structures (×23,000). (**A** and **C** from Grant DA, Stern IB, and Listgarten MA: Periodontics, St. Louis, 1988, The CV Mosby Co. (**B** from Listgarten MA: J Periodontol 47:1, 1976. **D** from Listgarten MA: J Periodontol 46:10, 1975.)

FIGURE 27-8   Maturation of plaque continuing for 4 days with scanning electron microscopy. Demonstrated are large colonies of filament-shaped bacteria at the tooth surface (×6000). (From Grant DA, Stern IB, and Listgarten MA: Periodontics, St. Louis, 1988, The CV Mosby Co.)

plaque matures on the surfaces of the teeth, the bacterial mat becomes increasingly complex (see Figure 27-7, D). Bacteria proliferate within the environment to result in a complex, gram-negative motile, assaccharolytic (protein-utilizing) mass of bacteria.

Bacterial plaque, initially supragingival, continues to proliferate and extends underneath the gingiva (subgingival) (Figure 27-5, F). As bacteria continue to proliferate subgingivally, relatively low oxygen tension environments are created (Figure 27-5, H). In this econiche below the gingiva, facultative and eventually strict anaerobic bacteria are favored for growth and development. Ultrastructurally, the subgingival plaque consists primarily of bacilli and filamentous bacteria (Figure 27-8).

The mixed, complex colonies contain more than 200 different species of bacteria. The types of bacteria found in dental plaque vary considerably among individuals with the age of the dental plaque mass itself and the area of the oral cavity from which it was collected.

Calculus

Dental calculus is a result of extremely mature plaque that has completed mineralization; it occurs as relatively hard, firmly adhering materials on the crowns and root surfaces of the teeth. Dental calculus is composed of permeable and toxic products from bacterial plaque associated with it, and these products are released into the local environment.

Calculus may also be classified by its location on the surfaces of the teeth. "Supragingival calculus" is usually quite abundant at the orifice of the major salivary glands, and in particular on the lingual (tongue) surface of the lower anterior teeth. It may be enhanced by special intraoral situations such as malposed teeth, malocclusion, rough surfaces, and improperly restored teeth.

Subgingival calculus is found underneath the gingival margin on the root surfaces of the teeth. The extent of subgingival plaque frequently correlates with the extent of the space created between the gingiva and tooth as a result of the disease process (periodontal "pocket"). The radius of infectivity of dental plaque and calculus is believed to be approximately 1 to 2 mm (i.e., the pathologic processes in the periodontium are related to the location of the developing plaque and calculus).

PATHOGENESIS

The pathogenesis of dental caries is initiated at selected sites in the dentition and usually includes the pits and fissures of the occlusal (biting) surfaces of the teeth, as well as the interproximal areas (between the teeth) of the dentition. These localized sites are protected from natural cleansing mechanisms of the oral cavity, such as the tongue, oral mucosa and musculature, and saliva.

The demineralization and decalcification of tooth structure by acids results in a "leather-like" consistency of the tooth. A pH of 4.6 or less (see Figure 27-4) can result in hydrolysis of enamel phosphoprotein by phosphoprotein phosphatase. Solubilization and demineralization of tooth enamel is accelerated by organic acids. As enamel is destroyed, oral bacteria, including the streptococci, penetrate into the enamel

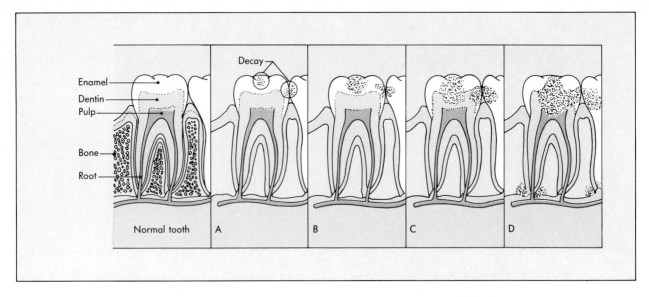

FIGURE 27-9   Dental caries. **A,** Decay often begins in hard-to-clean areas. **B,** Left untreated, a small cavity becomes larger. **C,** Decay spreads beneath the enamel, to the dentin, destroying more tooth structure. **D,** Once decay enters the pulp, an abscess may occur. The tooth will need endodontic treatment or may need to be extracted.

matrix. The acids formed during the metabolism of carbohydrates by oral bacteria in dental plaque are capable of dissolving the enamel structures of the teeth. The more frequently that acids form on the surfaces of the teeth and the longer the time intervals in which acids remain on the surfaces of the teeth, the greater the likelihood of enamel demineralization.

Streptococci are present in the oral cavity in very large numbers and play a dominant role in the formation and pathogenesis of dental caries. The most frequently isolated bacteria associated with dental caries is *Streptococcus mutans*. Their numbers increase with initiation of dental caries and decrease when caries are treated, exhibiting a strong correlation with this disease process.

Dental caries extends through the enamel and into the dentin where the carious lesion must be restored. (Figure 27-9 *A* to *C*) Without appropriate treatment at this time, the carious lesion may further extend through the dentin and into the dental pulp region. (Figure 27-9 *D*) Sensitivity to heat and cold or sensitivity to percussion are possible symptoms related to extension of the caries to the pulp.

After bacterial involvement of the pulpal regions during caries pathogenesis, a periapical lesion (granuloma or abscess) may develop at the tip of the root.

Root canal therapy (endodontic therapy) is even more critical at this time and includes mechanical removal and debridement of the pulpal contents of the tooth. When this is performed, the periapical lesion frequently resolves without the use of antibiotic therapy.

Periodontitis is an inflammatory process that extends beyond the soft tissues and into the alveolar bone, periodontal ligament, and cementum (see Figure 27-5 *E* to *H*). Generally, the connective tissue fibers surrounding the teeth and connecting the teeth to alveolar bone are disrupted and in the process of dissolution.

The junctional epithelium, which connects the soft tissues to the root surfaces of the teeth, migrates downward, resulting in a relatively deep space (pocket) between the tooth and the gingiva. A pocket is the space around the teeth that is pathologically deepened by periodontal disease. This results directly from extension of the spread of inflammation caused by bacterial products (see Figure 27-5 *B, D, F, H*). Inflammatory cells can spread along the course of blood vascular channels because of the loose connective tissue surrounding the neurovascular bundles. This area offers less resistance than the dense connective tissue fibers of the periodontal ligament and

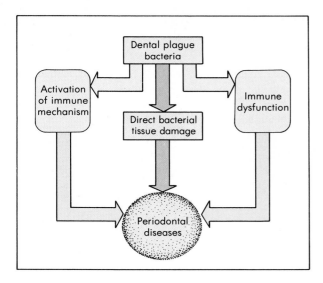

FIGURE 27-10   Pathogenic mechanisms of periodontal diseases.

connective tissue. The location of the inflammation surrounding the alveolar bone depends on the course of inflammatory cells along the vascular channels. Frequently, the blood vessels extending on the alveolar crest between the teeth are the site of direct spread of inflammation into the periodontal ligament space. This inflammatory process is further driven by bacterial plaque as it develops in the pocket.

The inflammatory process in the hard tissues around the teeth may result in marked osteoclastic activity, and the progressive extension results in alveolar bone destruction. Periodontitis is the loss of alveolar bone and clinical gingival attachment. Radiographically, this loss of bone can be observed (see Figures 27-13 and 27-14). The conversion clinically from gingivitis to periodontitis reflects the progression, histopathologically, from the extension of inflammation from the soft tissues into the hard tissues. The reasons for this progression remain unclear but may represent colonization and infection by highly pathogenic plaque bacteria or reflect aberrations of host cell responsiveness to bacterial plaque or both.

Three plausible hypotheses can be proposed for periodontal destruction (Figure 27-10): (1) direct destruction caused by bacterial plaque and metabolic products; (2) immune hyper-responsiveness precipitated by immune complexes, lymphocyte blastogenesis, or activation of complement pathways; or (3)

immune deficiencies involving neutrophil function (chemotaxis, phagocytosis, superoxide radical formation), neutropenia, or the autologous mixed lymphocyte response (AMLR).

EPIDEMIOLOGY

Recent surveys (1987) completed by the Department of Health and Human Services indicate that 91.9% of adults and 91.2% of seniors have dental caries (coronal). This represents a substantial reduction from previous surveys. Factors responsible for this trend include fluoridation of drinking water, school-based preventive programs, knowledge of dietary factors, and increased patient awareness of health.

Gingival bleeding was found in 43.6% of adults and 46.9% of seniors (65 years of age or older.). Gingival bleeding is elicited by the manipulation of gingival tissues with a periodontal probe.

Loss of attachment ($\geq$2 mm) was observed in 76.6% of adults and 95.1% of seniors. This index is a diagnostic sign of periodontitis (any form) and is measured with a periodontal probe. Gingivitis, a reversible dental disease, does not have gingival or periodontal attachment loss.

Periodontitis, an irreversible dental disease, has loss of attachment around the teeth. Early periodontal destruction would have 2 to 4 mm of loss of attachment. Moderate periodontal destruction would have 4 to 6 mm loss of attachment, and advanced destruction would have 7 mm or greater loss of attachment.

CLINICAL SYNDROMES

Dental Caries

Dental caries is a bacterial disease that can destroy the teeth. The normally hard tooth structure is converted to a soft, leather-like consistency. This is detected clinically with a sharp dental instrument (explorer) or with dental radiographs.

Periodontal Diseases

Periodontal infections cause the most common diseases of the periodontium and are among the most common of all diseases in humans. The periodontal

Table 27-1   Classification of Gingivitis

Form	Primary Etiology	Secondary Etiology
1. Plaque-associated	Dental plaque	—
2. ANUG	Dental plaque	Stress, fatigue
3. Steroid hormone–influenced	Dental plaque	Steroids
4. Medication-influenced gingival overgrowth	Dental plaque	Medications (e. g., phenytoin, cyclosporins)
5. Miscellaneous	Dental plaque	Nutrition, other (?)

From Suzuki JB: Dent Clin North Am 32(2):195-216, 1988.

diseases can be broadly subdivided into two main categories: gingivitis and periodontitis.

Gingivitis Gingivitis is an inflammatory process precipitated by the accumulations of bacteria along the gingiva of the teeth. Gingivitis is primarily a reversible disease, and the removal of the bacteria from the teeth at the margin of the gingiva usually results in the resolution of inflammation. The inflammatory process in gingivitis is confined primarily to the gingiva (soft tissues) and is seen clinically as redness around the necks of the teeth. Bleeding of the gingiva (either spontaneous or induced) is a frequent accompanying sign.

The primary etiologic agent of gingivitis is bacterial plaque. This forms the basis for a classification of gingivitis, with "plaque-associated" gingivitis being the most common form (Table 27-1). The primary bacteria involved in gingivitis are gram-positive cocci mixed with bacilli and filamentous forms.

Other modifying secondary etiologic factors permit classification of other forms of gingivitis (see Table 27-1). Acute necrotizing ulcerative gingival stomatitis (ANUG), commonly known as "**trench-mouth**," "Vincent's infection," has several secondary etiologic factors and constitutes another form of gingivitis. Stress and anxiety are perhaps the most significant contributing factors. Other factors include fatigue, lowered resistance, nutritional impairment, smoking, mouth-breathing, calculus, and gross neglect of oral hygiene. However, the primary etiologic agent for ANUG is bacterial plaque.

The clinical features of ANUG are frequently seen by the practicing physician as periodontal tissues with gingival craters covered by a grayish-white pseudomembrane (Figure 27-11). In these areas of gingival necrosis, acute areas of inflammation contribute to the pain, bleeding, soreness, and sensitivity of these lesions. The extent of ANUG lesions may be isolated in the interproximal regions between the teeth, or they may be generalized throughout the entire dentition. This is one of the most painful and malodorous of the dental diseases. Boggy, edematous, keratinized gingiva is frequently present, with lymphadenopathy, malaise, and pyrexia as frequent accompanying signs. The primary bacteria involved in ANUG lesions are spirochetes, vibrios, and other filamentous forms.

A third form of gingivitis is "steroid hormone–

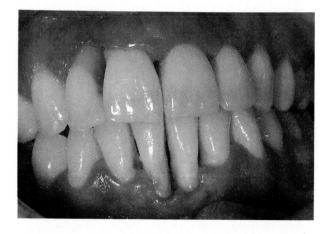

FIGURE 27-11   Clinical presentation of a systemically healthy 22 year old female dental patient with Acute Necrotizing Ulcerative Gingivitis (ANUG) ("Trenchmouth"). The soft tissue of the gingiva around the teeth can be noted, with the diseased areas covered by a "pseudomembrane" of neutrophils, spirochetes and other motile bacteria, and sloughed epithelial cells.

influenced gingivitis." This disease results from the presence of steroid hormones that amplify the clinical inflammatory response of gingivitis. The increased levels of estrogens and progesterones associated with pregnancy, adolescence, or birth-control medication, enhance marginal gingival inflammation. The clinical signs and symptoms of this disease include acute gingival inflammation around one or more teeth, with spontaneous gingival bleeding or bleeding during gentle instrumentation by a dental clinician. Severe cases of steroid hormone–influenced gingivitis may progress to a pyrogenic granuloma, commonly known as a "pregnancy tumor," which is clinically apparent.

The fourth form of gingivitis is "medication-influenced gingival overgrowth." This form of gingivitis is complicated by phenytoin and cyclosporines, among other drugs, and frequently results in a hyperplastic type of tissue growth of the gingiva.

The clinical signs and symptoms include gingival overgrowth in the form of a diffuse swelling of the interdental papilla or multiple tiny nodules on the gingiva of the anterior teeth (Figure 27-12). Other signs include moderate to acute inflammation, soreness, and tenderness of the gingiva.

Table 27-2 Classification and Clinical Features of Periodontitis

Form	Age	Clinical Features
Adult	>35 years	Abundant plaque
Rapidly progressive	18-35 years	Variable plaque
Juvenile	12-26 years	Little plaque
Prepubertal	<12 years	Little plaque

Other forms of gingivitis may be influenced by nutritional deprivation, mouth breathing or HIV infections. The clinical signs and symptoms of nutritionally deficient patients are similar to the other forms of gingivitis.

Periodontitis Clinical observations, coupled with basic microbiology research, have permitted a descriptive classification of the various forms of periodontitis. These forms are generally associated with age of onset but also can be related to the types of bacteria present in the pockets surrounding the teeth (Table 27-2).

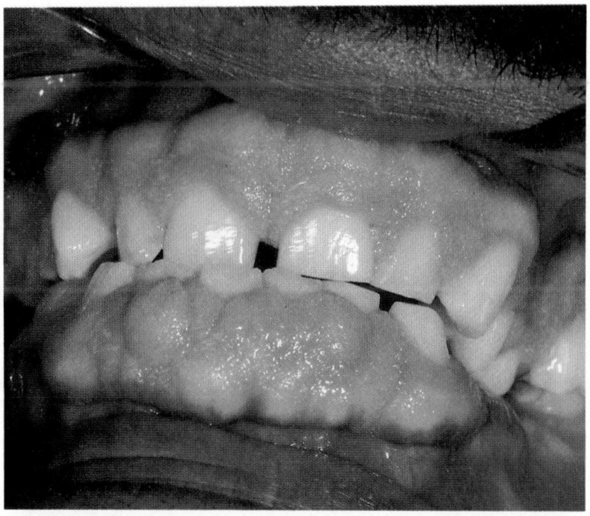

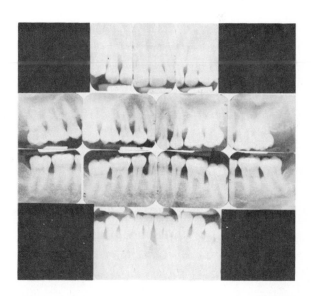

FIGURE 27-12   Clinical presentation of a 16 year old female dental patient with Gingival Hyperplasia (Gingival overgrowth). This condition may result from a combination of dental plaque accumulation with systemic administration of phenytoin (Dilantin), cyclosporins, nifedipine, among other medications.

FIGURE 27-13   Adult periodontitis. Radiographs are from a male patient, age 52. (From Suzuki JB: Dent Clin of North Am 32:195, 1988.)

The most frequent cause of tooth loss in adults is due to adult periodontitis. (Fig. 27-13). Beyond 35 to 40 years of age the majority (70% to 90%) of the adult population of the United States exhibits some sign of periodontitis. The presence of bacterial plaque and calcified deposits (calculus or tartar) is usually related to the amount of destruction of gingiva and alveolar bone tissues.

The disease, adult periodontitis, progresses in a chronic fashion and generally takes years or perhaps decades to develop. Subgingival bacterial plaque formation and maturation cause the inflammatory process to extend to the hard tissues. Favorable eco-niches or environment for the development of subgingival anaerobic forms of bacteria are further established within the pockets surrounding the teeth.

The primary bacteria found in subgingival pockets in adult periodontitis patients are primarily gram-negative, motile, complex, and assaccharolytic forms. The associated microorganisms include *Bacteroides* species, *Eikenella corrodens, Fusobacterium nucleatum, Wolinella recta,* and *Eubacterium* species (Table 27-3).

The clinical dental characteristics associated with adult forms of periodontitis are bacterial and calcified deposits around the teeth commensurate with the disease.

Patients are generally older than 35 years. The destruction to the alveolar bone (seen radiographically in Figure 27-13) and gingiva in adult periodontitis patients, as well as all forms of periodontitis, is generally irreversible in nature.

Another form of periodontitis that occurs in younger adults between 18 to 35 years of age is referred to as rapidly progressive periodontitis. As the name states, a relatively rapid progression of disease occurs that may be related to abnormal host-defense mechanisms. The primary immune cells that may be incompetent include the neutrophils or polymorphonuclear leukocytes.

It has been proposed that deficient host-defense mechanisms permit extremely pathogenic forms of bacteria and dental plaque to develop. These include the *Bacteroides gingivalis, Bacteroides intermedius, Bacteroides forsythus, Actinobacillus actinomycetemcomitans, Eikenella corrodens,* and *Wolinella recta* (see Table 27-3). Frequently, gingival inflammation may be masked because of a deficient neutrophil chemotactic response. In addition, clinically evident deposits of plaque and calculus may not always be associated with rapidly progressive periodontitis. In this disease it is the **quality** of the bacterial plaque rather than the **quantity** of bacterial plaque that may be responsible for the rapid progression of disease in these younger individuals.

Table 27-3   Microbial Profile of Periodontitis

Form	Bacteria
Adult	Bacteroides species
	Eikenella corrodens
	Fusobacterium nucleatum
	Wolinella recta
	Eubacterium species
Rapidly progressive	Bacteroides gingivalis
	Bacteroides intermedius
	Bacteroides forsythus
	Actinobacillus actinomycetemcom-itans
	Eikenella corrodens
	Wolinella recta
Juvenile	Actinobacillus actinomycetemcom-itans
Prepubertal	Actinobacillus actinomycetemcom-itans
	Bacteroides intermedius
	Eikenella corrodens
	Capnocytophaga species

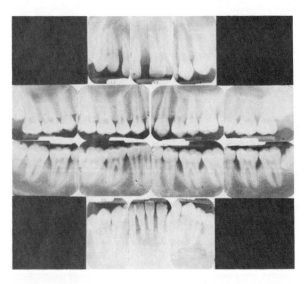

FIGURE 27-14   Juvenile periodontitis. Radiographs are from a female patient, age 13. (From Suzuki JB: Dent Clin of North Am 32:195, 1988.)

The third form of periodontitis is juvenile periodontitis, which affects persons between the ages of 12 and 26 years (Fig. 27-14). Generally, only selected teeth are affected by this disease—specifically, the first molars and the incisor teeth. Ironically, usually no gross clinical dental deposits are observed with juvenile periodontitis. There are, however, genetic and host-defense mechanisms complicating the pathogenesis. The specific bacteria involved is *Actinobacillus actinomycetemcomitans*.

The most recently described form of periodontitis is **prepubertal periodontitis**. As the name implies, this is a form of periodontitis confined to the mixed dentition or occurring during eruption of the primary teeth. Selected host-defense mechanisms (e. g., neutrophil adhesion) may be impaired in prepubertal periodontitis, permitting bacterial plaque to develop at a premature stage. The bacteria responsible for forms of prepubertal periodontitis include *Actinobacillus actinomycetemcomitans*, *Bacteroides intermedius*, *Eikenella corrodens*, and *Capnocytophaga* species.

LABORATORY DIAGNOSIS

Caries is detected by examination of the saliva for numbers of streptococci and lactobacilli; this ascertains which patients are "caries prone." However, these salivary tests have very limited application in clinical practice. Cultures of the caries lesion are not performed, since the clinical diagnosis is fairly accurate.

The diagnosis of gingivitis is made primarily by clinical signs (specifically, gingival inflammation) and symptoms (bleeding or itching gums). Dental radiographs must be used to confirm no loss of alveolar bone structure around the teeth (which would indicate periodontitis). There are no laboratory diagnostic tests specifically for oral bacteria associated with gingivitis.

Clinical and radiographic observations remain the primary diagnostic criteria for the periodontal diseases. Bone loss patterns and severity are related to the patient's age and systemic considerations. Recently, developments using anaerobic culture techniques, immunofluorescence, and DNA probes have permitted more accurate identification of the oral bacteria associated with periodontitis. Dental plaque samples are taken from the space (pocket)

between the teeth and gingiva and submitted to a reference laboratory for analysis.

TREATMENT

Dental Caries

Mechanical removal of the carious lesion is the treatment for most cases. This involves the use of a high-speed rotary instrument and hand instruments. The objective of the mechanical treatment is removal of the damaged enamel and dentin, as well as any bacteria contained within the lesion.

Extension of the carious lesion into the pulp of the tooth results in endodontic involvement (root canals). Mechanical removal and debridement of the pulpal contents is necessary.

Penicillin VK is frequently used as a systemic antibiotic therapy for the management and control of periapical lesions resulting from pulpal involvement by caries. Erythromycin is used as an alternative to penicillin therapy. Warm compresses, both intraoral and extraoral, may improve the clinical problem created by an endodontically affected tooth.

Gingivitis

The treatment of the majority of forms of gingivitis includes improved oral hygiene, including brushing and flossing the teeth, scaling the calcified and soft tissue deposits around the teeth, and knowledge of possible secondary factors influencing the course of various gingivitis states.

Periodontitis

Treatment of adult periodontitis involves measures similar to those taken for the control of gingivitis. However, additional measures may be required to gain access to the deep pocket areas for removal of the deposits around the teeth, and surgical approaches may be necessary. Generally, systemic antibiotics are not required for management of adult periodontitis patients unless periodontal abscesses form. Systemic antibiotics such as penicillin, ampicillin, or the tetracyclines may be necessary in this situation.

Treatment of rapidly progressive periodontitis is similar to treatment of the adult forms, with the addition of systemic antibiotics (usually tetracycline) during the initial stages of clinical management. Young adults who have rapidly progressive periodontitis may be more prone to acute periodontal abscesses and may require emergency dental care, which includes incision and drainage, scaling the affected teeth, irrigation of the pocket area, and a regimen of antibiotics (e. g., penicillin, tetracyclines, or metronidazole).

Palliative therapy for ANUG patients includes gentle irrigation and intraoral rinsing with a mixture of warm water and hydrogen perioxide, diluted 50% each. This can be repeated four to eight times per day as needed. In addition, improved toothbrushing and interproximal cleaning such as flossing ameliorate the situation. Patients who have submandibular or cervical lymphadenopathy or who are febrile may require systemic antibiotics. The drug of choice is penicillin VK, with tetracycline as an alternative. Follow-up by a dental clinician is essential for the purpose of scaling the teeth to mechanically remove the plaque and calcified deposits. In addition, soft tissue management (gingivoplasty or gingivectomy) may be necessary for more extensive regions of ANUG.

The treatment for juvenile periodontitis includes approaches similar to those for rapidly progressive periodontitis. In addition, innovative surgical techniques, including bone grafting (autografts, allografts, or synthetics) and autologous third molar tooth transplantation, may also be considered. Systemic antibiotics, including the tetracyclines, are becoming increasingly important for the management of juvenile periodontitis patients. (Note that tetracyclines cause permanent staining of adult dentition if administered to patients younger than 8 years of age.)

The treatment for prepubertal periodontitis includes improvement of oral hygiene measures, scaling and root planing of the teeth, curettage of the gingival lining of the pockets, and the use of such systemic antibiotics as amoxicillin.

PREVENTION AND CONTROL

Dental research is proceeding in the areas of antibiotic regimens and antiplaque agents because these seem to be the most useful in controlling dental diseases. These measures may be employed in debilitated dental patients, regardless of age, who are not able to perform normal oral hygiene procedures. Dental patients who may benefit from systemic antibiotics either have rapidly progressive, juvenile, or prepubertal forms of periodontitis, are myelosuppressed, or are experiencing exacerbations of the disease (e. g., abscesses, ANUG).

The antiplaque mouth rinses are among the newest applications of technology in the control of microflora associated with dental diseases. Dental patients who may benefit from antiplaque mouth rinses have conditions in the oral cavity favoring dental plaque development, such as the following:

Physical impairment (e. g., arthritis, stroke) affecting the ability to perform oral hygiene

Mental impairment

Malposed teeth

Malocclusion

Extensive bridgework

Diet (high sucrose)

Lack of motivation

Orthodontic therapy

Orthognathic surgical therapy

Intermaxillary splinting (e. g., trauma)

Periodontitis

Antiplaque mouth rinses should not be considered the primary recommendation for treatment and prevention of dental diseases caused by bacteria. They serve as as adjunctive approaches to basic oral hygiene measures such as toothbrushing flossing, and scaling of the teeth during routine dental appointments.

BIBLIOGRAPHY

Department of Health and Human Services, United States Public Health Service, Oral health of United States adults, national findings, NIH publication no. 87-2868, Bethesda, Maryland, August 1987, National Institute of Dental Research.

Genco RJ: Antibiotics in the treatment of periodontal diseases, J Periodontol 52:545, 1981.

Gibbons RJ and Van Houte J: Bacterial adherence in oral microbial ecology, Ann Rev Microbiol 29:19, 1975.

Goldman HM and Cohen DW, editors: Periodontal therapy. ed 6, St Louis, 1980, The CV Mosby Co.

Grant D, Stern I, and Listgarten M, editors: Periodontics. ed 3, St. Louis, 1988, The CV Mosby Co.

Loesche WJ: Bacterial succession in dental plaque: role in dental disease. In Schlessinger D, editors: Microbiology—1985. Washington, DC, 1975, American Society of Microbiologists.

Moore WEC: Variation in periodontal floras, Infect Immun 46:720, 1984.

Newbrun E: Dietary carbohydrates: their role in cariogenicity, Med Clin North Am 63:1069, 1979.

Newbrun E: Sugar and dental caries: a review of human studies, Science 217:418, 1982.

Newbrun E: Cariology, Baltimore, 1983, William & Wilkins.

Nolte W, editor: Oral microbiology with basic microbiology and immunology, St. Louis, 1982, The CV Mosby Co.

Scheinin A and Odont D: Dietary carbohydrates and dental disorders, Am J Clin Nutr 40:19-65, 1982.

Suzuki JB: Diagnosis and classification of the periodontal diseases, Dent Clin North Am 32(2):195-216, 1988.

Suzuki JB et al: Immunologic profile of localized and generalized juvenile periodontitis. II. Neutrophil chemotaxis, phagocytosis, and intracellular spore germination, J Periodontol 19:461, 1984.

Van Houte J: Bacterial specificity in the etiology of dental caries, Int Dent J 30:305, 1980.

MYCOLOGY

PART II

CHAPTER 28

Introduction to the Mycoses

The systematic study of fungi is only about 150 years old, yet the practical manifestations of these organisms have been known since antiquity. In addition to their disease-producing potential in humans, fungi are directly or indirectly harmful to humans in many other ways. They contribute to food spoilage, are the major cause of plant diseases, and destroy timber, textiles, and several synthetic materials. However, as saprobes, they share with bacteria (particularly organisms such as the *Streptomyces*) a role in the decay of complex plant and animal remains in the soil, breaking them down into simpler molecules to be absorbed by future generations of plants. Without this essential decay process the growth of plants, on which life depends, would eventually cease for lack of basic materials.

Fungi are also specifically beneficial to humans. They are used industrially in the production of antibiotics, organic acids, steroids, alcoholic beverages, and other products of fermentation, such as soya sauce. The carbon dioxide produced by fungi makes dough rise in bread-making, and the various ketones, aldehydes, and organic acids that result from the metabolic activities of fungi on milk curd provides the unique cheeses we enjoy. Furthermore, they serve as scientific models to study genetics, biochemical processes, and relationships involving parasite and hosts.

Fungi are a diverse group of organisms and occupy many niches in the environment. In general they are free living and abundant in nature, with only a few in the normal flora of humans. Although tens of thousands of species have been described, fewer than 100 are routinely associated with human diseases. Unlike viruses, protozoan parasites, and some species of

bacteria, fungi do not need to colonize or infect tissues of humans or animals to preserve or perpetuate the species. With only two major exceptions, virtually all fungal infections originate from an exogenous source, either by inhalation or by traumatic implantation.

Fungi are **eukaryotic organisms.** Most important, they possess a nucleus enclosed by a nuclear membrane. Unlike plant cells and most bacteria, they do not contain chlorophyl and cannot synthesize macromolecules from carbon dioxide and energy derived from light rays. Therefore all fungi lead a heterotrophic existence in nature as **saprobes** (organisms that live on dead or decaying organic matter), **symbionts** (organisms that live together where the association is of mutual advantage), **commensals** (two organisms living in a close relationship where one benefits by the relationship and the other neither benefits nor is harmed), or **parasites** (organisms that live on or within a host from which they derive benefits without making any useful contributions in return; in the case of pathogens, the relationship is harmful to the host).

Whereas the ability of fungi to cause disease in humans or animals appears to be accidental, most disease-producing fungi have developed characteristics that enable them to adapt to hostile tissue environments and grow. Fungi that colonize the cutaneous layers of the epidermis or invade hair and nails metabolize *keratin,* the tough, fibrous, insoluble protein that forms the principal matter of these tissues. Other fungi, such as *Histoplasma capsulatum, Blastomyces dermatitidis, Paracoccidioides brasiliensis,* and *Coccidioides immitis,* which cause systemic disease, have developed the capacity to overcome various host cellular defense mechanisms, are able to grow at the higher temperatures of the host (37° C) rather than the temperatures found in natural environments (around 25° C) and can survive in a lowered oxidation-reduction state (a situation found in damaged tissues).

CLASSIFICATION

Organisms are classified for convenience of reference. The scheme of classification should reflect natural relationships between organisms. Biochemical, ultrastructural, molecular biologic, and genetic studies support the concept that at least five kingdoms are necessary to accommodate all living things. These are the Monera (prokaryotes); Protista (protozoa); Fungi; Plantae; and Animalia. The Fungi kingdom is composed of two phyla (Table 28-1), the Zygomycota (organisms that produce a zygote during their sexual cycle) and the Dikaryomycota (organisms that have an extended dikaryotic state as a result of sexual conjugation). The phylum Dikaryomycota is divided into two subphyla, the Ascomycotina and the Basid-

iomycotina, which are defined according to the type of structure that forms and houses the haploid progeny. Those organisms for which a sexual phase has not been observed are classified in the form-class Deuteromycotina or Fungi imperfecti.

All fungi reproduce by asexual processes, and most can reproduce by sexual mechanisms. If a sexual phase is observed for any fungal isolate, the terms **anamorph** (asexual) and **teliomorph** (sexual) are often used to describe the taxonomic status of the organism. This is reflected mostly in the nomenclature used to refer to that organism. For example, the anamorphic designation for the organism causing histoplasmosis is *Histoplasma capsulatum;* if the specific isolate exists in the teleiomorphic state, the name is changed to *Ajellomyces capsulatum.* This convention is used for all fungal organisms in order to maintain taxonomic consistency. However, in clinical situations it is common to refer to the organisms by their asexual designations, since the anamorphic state is isolated from clinical specimens and the sexual phase occurs only under the extremely controlled conditions of a culture.

In considering relationships within and between

Table 28-1 Taxonomic Classification of the Fungi Kingdom

Organism	Sexual Characteristics	Medically Important Genera
Phylum Zygomycota	Sexual reproduction occurs through the fusion of two compatible gametangia to produce a zygote. Asexual reproduction is characterized by production of sporangiospores.	Agents causing zygomycosis
Phylum Dikaryomycota	Organisms whose dikaryotic life cycle includes an extended dikaryotic phase after sexual conjugation (i.e., haploid nuclei do not fuse immediately)	
Subphylum Ascomycotina	Sexual reproduction occurs through the fusion of two compatible nuclei to form a diploid nucleus followed by meiosis to yield haploid progeny. The entire process occurs within a sac called an ascus, and the resultant spores are called ascospores.	Agents causing ringworm, histoplasmosis and blastomycosis
Subphylum Basidiomycotina	Sexual reproduction takes place in a sac called a basidium, where two compatible nuclei fuse to form a diploid nucleus, followed by meiosis to yield haploid progeny. These then mature on the outer surface of the basidium. The haploid progeny are called basidiospores.	Agent causing cryptococcosis
Form-class Deuteromycotina (Fungi imperfecti)	A sexual stage has not been observed in fungi classified in this category.	*Candida, Trichosporon, Torulopsis, Pityrosporum, Epidermophyton, Coccidioides, Paracoccidioides*

certain fungi, conflicts often arise, and no universal consensus may be possible concerning the classification or even the nomenclature of a given organism. Differences of opinion occasionally arise, since one taxonomist may place greater emphasis on certain phenotypic features of an isolate than on criteria deemed more important by other individuals. Another difficulty that contributes to this problem is the conflict between those who group many organisms together and those who prefer to emphasize minor phenotype differences and split them into many groups. The result of these conflicting opinions is the controversy that frequently arises over the proper name of individual fungi.

MORPHOLOGY AND REPRODUCTION

Fungi can be divided into two basic morphologic forms, **yeast** and **hyphae,** and their developmental histories encompass both vegetative and reproductive phases. These phases are frequently present simultaneously in growing cultures and may not be easily separated.

Yeasts are unicellular and reproduce asexually by processes termed **blastoconidia formation** (budding; Figure 28-1, *A*) or **fission** (Figure 28-1, *B*).

Most fungi have branching, threadlike tubular filaments called hyphae (Figure 28-1, *C* to *E*) that elongate at their tips by a process called **apical extension.** These filamentous structures are either **coenocytic** (hollow and multinucleate; Figure 28-1, *D*) or **septate** (divided by partitions; Figure 28-1, *E* and *F*). The collective term for a mass of hyphae is **mycelium** (synonymous with **mold**). Hyphae that grow submerged or on the surface of a culture medium are called **vegetative hyphae,** since they are responsible for absorbing nutrients. Those that project above the surface of the medium are called **aerial hyphae.** Aerial hyphae often produce specialized structures called **conidia** (i.e., asexual reproductive propagules) that are easily airborne and disseminated into the environment. The shape, size, and certain developmental features of conidia are useful to the mycologist in identifying the specific species. Various examples of conidia are illustrated in Figure 28-2.

The morphology of fungi is not fixed, because some (e.g., *H. capsulatum, B. dermatitidis, P. brasili-*

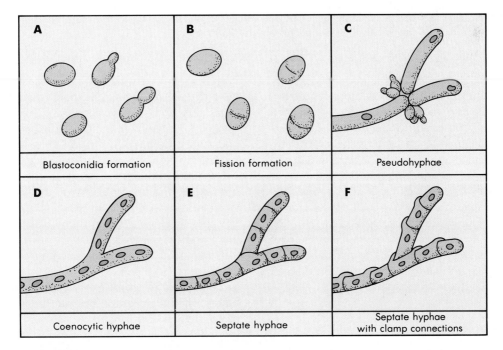

FIGURE 28-1 Fungal cell morphology. **A,** Yeast cells reproducing by blastoconidia formation; **B,** yeast cells dividing by fission; **C,** pseudohyphal development; **D,** coenocytic hyphae; **E,** septate hyphae; and **F,** septate hyphae with clamp connections.

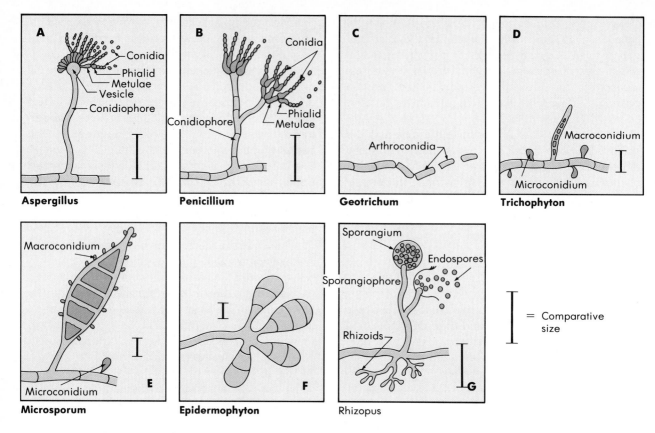

FIGURE 28-2   Conidial development in **A**, *Aspergillus*; **B**, *Penicillium*; **C**, *Geotrichum*; **D**, *Tricho-phyton*; **E**, *Microsporum*; **F**, *Epidermophyton*; and **G**, *Rhizopus*.

ensis) can exist in a mycelial or yeast morphology depending on the environmental conditions of growth (in soil, on decaying vegetation, or in host tissues). The morphology of *Candida albicans,* part of the normal flora of the mouth, gastrointestinal tract, and membranes lining the mucosa of other cavities and tissues, is unique. In addition to being yeastlike or filamentous, this organism can assume a **pseudohyphal** morphology, wherein the cells are elongated and linked together like sausages (Figure 28-1, *C*). Pseudohyphal development is an exaggerated form of budding; the newly formed cells do not take on an oval shape and pinch off from the parent but remain attached and continue to elongate.

CELL STRUCTURE

Fungi have structures typical of eukaryotic cells. Figure 28-3 illustrates a section of a model fungal cell with its component parts. The cell possesses a complex cytosol that contains microvesicles, microtubules, ribosomes, mitochondria, Golgi apparatus, nuclei, a double-membraned endoplasmic reticulum, and other structures. The nuclei of fungi are enclosed by a membrane and contain virtually all of the cellular DNA and have a true nucleolus rich in RNA. An interesting and unique property of this membrane is that during the mitotic cycle it persists throughout metaphase, in contrast to the nuclear membrane of plant and animal cells, which dissolves and then reforms after the chromosomes segregate to their centromeres.

Enclosing the complex cytosol is another membrane called the **plasmalemma** that is composed of glycoproteins, lipids, and ergosterol. The fact that fungi possess ergosterol in contrast to cholesterol, which is the major sterol found in the host tissues of mammals, is important because most antifungal strategies are currently based on its presence in fun-

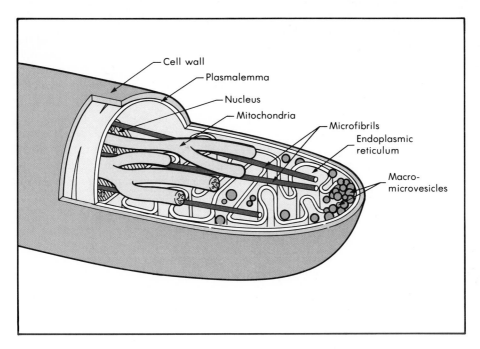

FIGURE 28-3   Diagramatic section of the apical portion of hyphal cell illustrating its component parts.

gal membranes (antifungal agents are discussed later in this chapter).

Unlike mammalian cells, fungi possess a multilayered rigid cell wall immediately exterior to the plasmalemma. The cell wall is structurally and biochemically complex, containing **chitin,** a polymer of *N*-acetyl glucosamine, as its structural foundation. Layered on the chitin are mannans, glucans, and other complex polysaccharides in association with polypeptides. In filamentous fungi the biosynthesis of chitin occurs at the growing tip and is controlled by the activity of chitin synthetase. This enzyme exists in the cytosol as a zymogen in discrete membrane-bound packets called chitosomes. The active form of chitin synthetase is found in the plasmalemma, and polymerization of chitin microfibrils occurs outside this membrane.

In addition to the cell wall, some fungi produce a capsular polysaccharide. This structure isolates the organism from and at the same time serves as a communicator with its surrounding environment; in the case of pathogens it is host tissue. The cell wall and structures such as the capsular material serve to determine virulence and play a role in eliciting host immune responses (see the discussion of *Cryptococcus neoformans*).

Table 28-2 summarizes some of the important features of fungi.

CLINICAL SYNDROMES

The effects of fungi on humans are numerous, but from a medical viewpoint there are only three major categories of importance:

1. Mycotoxicoses
2. Hypersensitivity diseases
3. Colonization of the host and resultant disease processes

Mycotoxicoses

Fungi are metabolically versatile organisms and sources of innumerable alkaloids and other toxic metabolites. The mycotoxicoses produced by fungi are most often the result of the accidental or recreational ingestion of these toxic fungi or their metabolites. Several examples will point out the positive and negative results on human health.

Ergot Alkaloids The pharmacologic properties of ergot alkaloids, which are produced by *Claviceps purpurea* when it infects grain, have been known

Table 28-2 Summary of Important Features of Fungal Cells

Cell Feature	Characteristics
Nucleus	Membrane-bound (eukaryotic); multichromosomal and can be haploid or diploid depending on whether conjugation has taken place; contains RNA-rich area (nucleolus)
Cytosol	Complex; contains several organelles such as nucleus, mitochondria, Golgi apparatus, ribosomes, a well-defined endoplasmic reticulum and other inclusions
Plasmalemma	Composed of glycoproteins, lipids, and ergosterol; differs from mammalian cell membranes, which contain cholesterol
Cell wall	Multilayered; composed of chitin, other polysaccharides such as glucans, mannans, glucomannans, galactomannans and peptides; some fungi produce an extracellular polysaccharide capsule
Shape and size	Yeasts are oval to round; 2 to 10 μm in diameter; molds are filamentous septate or coenocytic and branching; 2 to 10 μm in diameter and can be several hundred micrometers in length
Physiology	Respiration almost exclusively oxidative with limited capacity for anaerobiasis (fermentation); metabolism exclusively heterotrophic; biochemically versatile; produces various primary (e.g., citric acid, ethanol) and secondary (e.g., alpha amanitin, aflatoxin) metabolites; doubling time long (hours) compared to most bacteria (minutes)
Staining properties	Gram-positive; vegetative cells not acid-fast; in histopathologic sections the fungal cell wall can be stained by special procedures (methenamine-silver or periodic Schiff-stain methods)

throughout history. During the Middle Ages disease associated with the consumption of contaminated bread and other bakery products made with rye and other grains occurred in epidemics. These were known as St. Anthony's fire, symptoms consisted of inflammation of the infected tissues (cellular response to injury), followed by necrosis (cell death), and gangrene (death of large masses of tissue). Pharmacologically, we now know that the ergot alkaloids produce **alpha-adrenergic blockade,** which inhibits certain responses to epinephrine and 5-hydroxytryptamine. This creates marked peripheral vasoconstriction, which, if not corrected, restricts the flow of blood and results in necrosis and gangrene.

Another feature of the ergot alkaloids is their extensive activity in directly stimulating smooth muscle. With this in mind, they have been used as oxytocic agents, promoting labor during childbirth by increasing the force and frequency of uterine contractions.

The effect of ergot alkaloids on the central nervous system includes stimulation of the hypothalamus and other sympathetic portions of the midbrain. They have been used to good effect in promoting the medical welfare of humans.

Psychotropic Agents Toxic metabolites produced by fungi have also been used by primitive tribes for religious, magical, and social needs. In recent times problems involving toxins of fungi have been seen with the recreational use of psychotropic agents such as psilocybin and psilocin, as well as the simisynthetic derivative lysergic acid diethylamide (LSD).

Aflatoxins Among the mycotoxicoses that have had a profound economic impact on society is contamination with *Aspergillus flavus*. This fungus produced the outbreak of "Turkey X" disease in the early 1960s in England and almost destroyed the turkey industry. Turkey poults were fed feed that was contaminated with *Aspergillus* and developed lethargy, anorexia, and muscle weakness followed by spasms and death. Postmortem studies on the birds revealed gross hemorrhage and necrosis of the liver. Further histopathologic examination showed parenchymal cell degeneration and extension proliferation of the bile duct epithelial cells, and biochemical and pharmacologic studies showed that the etiologic agents of the disease were toxins produced by *A. flavus* that belonged to a group of compounds called the **bisfuranocoumarin metabolites.** These compounds, the aflatoxins, have become known as potent carcinogens, but have not been shown to play a specific role in human carcinogenesis.

Other Mycotoxicoses Several other mycotoxicoses that have impacted on the well-being of humans have also been described. Among these are yellow rice toxicosis in Japan and alimentary toxic aleukia in the Soviet Union.

Hypersensitivity Diseases

One index used to measure the degree of air pollution is "the fungal spore count" because fungal spores are ubiquitous in nature, supply a good index of environmental contamination, and carry medical relevance. Humans are constantly bombarded by airborne spores and other fungal elements. These elements can be an antigenic stimulus and, depending on the immunologic status of the individual, may induce a state of hypersensitivity resulting from the production of immunoglobulins or sensitized lymphocytes. In **hypersensitivity pneumonitis** the clinical manifestations of these reactions include rhinitis, bronchial asthma, alveolitis, and various forms of atopy. Growth of the fungus in tissues is not required, and the manifestations are seen only in sensitized patients when they are subsequently exposed to the fungus, its metabolites, or other cross-reactive materials.

Colonization and Resultant Diseases

As stated previously, virtually all of the fungal organisms implicated in human disease processes are free living. In general, humans have a high level of innate immunity to fungi, and most of the infections they cause are mild and self-limiting. Intact skin and mucosal surfaces serve as barriers to any infection

---

Box 28-1 Classification of Medically Important Fungi

Superficial mycoses
Cutaneous mycoses
Subcutaneous mycoses
Systemic mycoses
Opportunistic mycoses

---

caused by fungi that primarily colonize the superficial, cutaneous, and subcutaneous layers of skin. Fatty acid content, pH, epithelial turnover, and the normal bacterial flora of the skin appear to contribute to host resistance. Humoral factors such as transferrin have been shown to restrict the growth of several fungi by limiting the amount of available iron, but the role it plays in resistance has not been established.

A handful of fungi are capable of causing significant disease in otherwise healthy humans. Once established, these infections can be classified according to the tissue levels initially colonized, as follows (see Box 28-1):

1. **Superficial mycoses**—infections limited to the outermost layers of the skin and hair
2. **Cutaneous mycoses**—infections that extend deeper into the epidermis, as well as invasive hair and nail diseases
3. **Subcutaneous mycoses**—infections involving the dermis, subcutaneous tissues, muscle, and fascia
4. **Systemic mycoses**—infections that originate primarily in the lung but may spread to many organ systems

In addition to these four categories of fungus infections there is a fifth that includes agents with low pathogenic potential and that produces disease only under unusual circumstances, mostly involving host debilitation. As a result of the many advances in medicine, these organisms, once thought to be innocuous saprobes, have gained prominence as etiologic agents of disease. Among the unusual circumstances leading to infections by these fungi are changes in the normal microbial flora of the gut through the use of broad-spectrum antibacterial drugs, host debilitation produced by the use of therapeutic measures such as cytotoxins, X-irradiation, steroids, and other immunosuppressive drugs, and alteration of the host's immune system through underlying endocrine disorders such as uncontrolled diabetes mellitus or suppression such as that caused by acquired immunodeficiency syndrome (AIDS). The fungal infections that frequently accompany these situations are categorized as **opportunistic mycoses;** if they are not rapidly diagnosed, aggressively managed, and the underlying disorders brought under control, the infections become life-threatening. Among the organisms causing these disorders are *Candida albicans,* a yeast that is found as part of the normal flora of the mouth, buccal mucosa, and other tissues; *Malassezia furfur,* a lipophilic yeast often isolated from areas rich in sebaceous glands; and a few dermatophytes that have become well adapted to humans and are the only reservoir from which they are isolated.

PATHOGENIC FUNGAL CHARACTERISTICS

The fungal pathogens that colonize humans and cause disease possess features that allow categorization into groups according to primarily colonized tissues. The agents causing superficial infections, for example, tend to grow only on the outermost layers of the skin or cuticle of the hair shaft, rarely inducing an immune reaction. The **dermatophytes** are also limited to keratinized tissues of the epidermis, hairs, and nails, but they do have greater invasive properties and, depending on the species involved, may evoke a highly inflammatory reaction from the host. The subcutaneous agents of disease generally have a low degree of infectivity, and infections caused by these organisms are usually associated with some form of traumatic injury. The systemic agents of disease all involve the respiratory tract, and they possess unique morphologic features that appear to contribute to the organism's ability to survive within the host.

ANTIFUNGAL AGENTS

The fact that fungi contain ergosterol in their cell membrane has been exploited for therapeutic purposes. Several different species of *Streptomyces* produce a class of structurally related polyene antibiotics that interact with sterols, change the permeability properties of the membrane of eukaryotic cells, and lead to cell lysis. These polyenes have a macrocyclic lactone ring with a series of conjugated double bonds (Figure 28-4). Over 100 polyene antibiotics have been described, and they show differences in the number of conjugated carbon-to-carbon double bonds in the molecule, the size of the conjugated ring, and the presence or absence of a hexosamine sugar or aromatic moiety in the molecule. Most of the polyenes studied are toxic to human cells and cannot be used to treat fungal infections because they bind preferentially to cholesterol rather than to ergosterol. However, **amphotericin B** (see Figure 28-4) is used to treat life-threatening fungal diseases because it binds more avidly to ergosterol than cholesterol and causes perturbations in membrane function that lead to death of the fungal cell. However, it must be used prudently since it can also bind to cholesterol and cause injury to host cells.

Another group of compounds, collectively called the azole derivatives (e.g., miconazole, ketoconazole, fluconazole, itraconazole; Figure 28-5), have been found useful in treating certain fungal infections. Experimental data indicate that the target for antifungal activity of these compounds is the microsomal kytochrome P450-dependent enzyme systems, which are involved in the 14-α-demethylation of lanosterol, a key step in the biosynthesis of ergosterol (Figure 28-6). The inhibition of ergosterol synthesis and concomitant accumulation of 14-α-methylsterol precursors results in an accumulation of methylated ergosterol precursors that then interfere with fatty acid synthe-

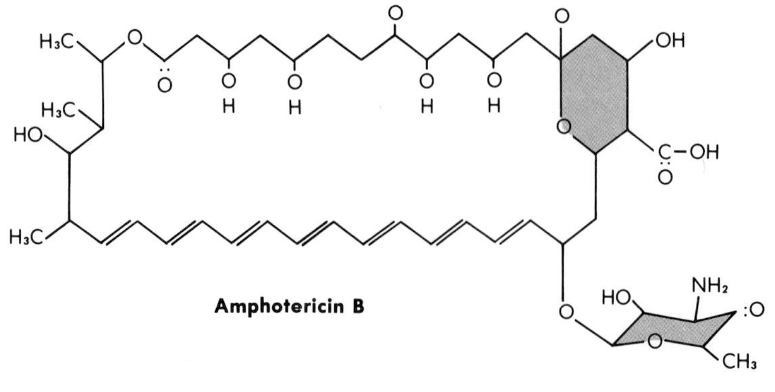

FIGURE 28-4   Chemical structure of the antifungal polyene, amphotericin B.

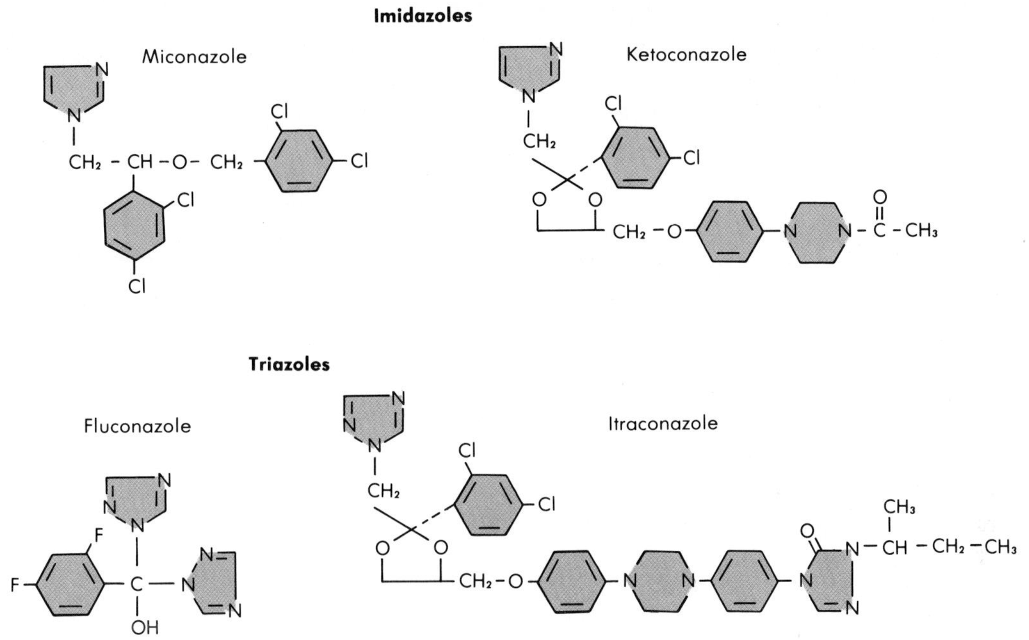

FIGURE 28-5 Chemical structures of representative azole derivatives.

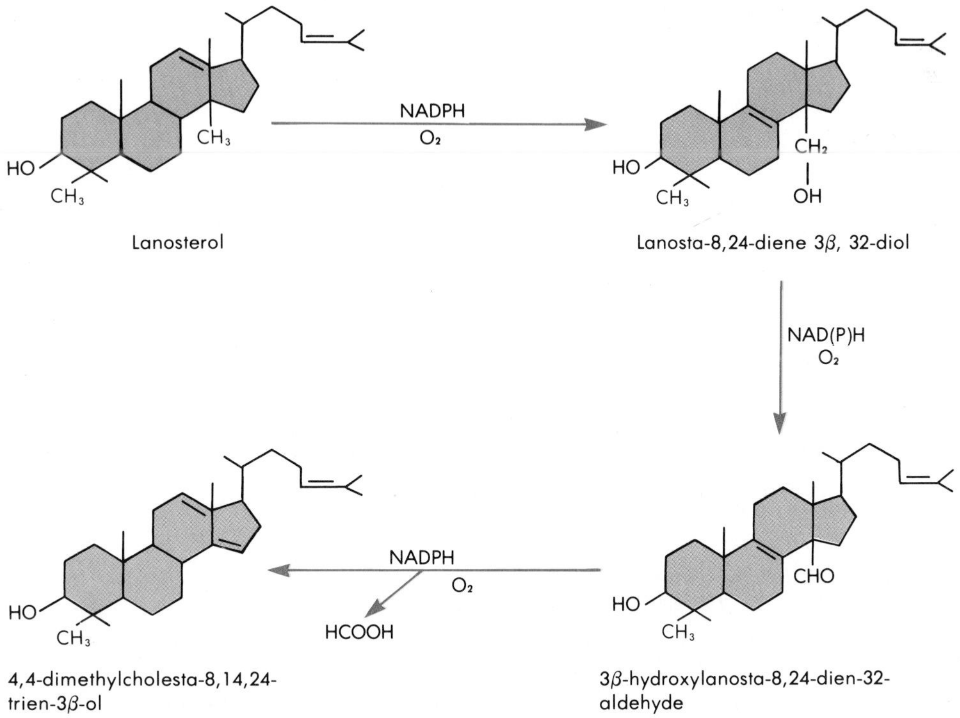

FIGURE 28-6 Metabolic pathway of ergosterol biosynthesis, illustrating sites at which the azole derivatives and amphotericin B exert their antifungal activities.

sis, produce an uncoordinated synthesis of the cell wall, and inhibit certain cytochrome functions and membrane activities.

Since chitin is an important component unique to fungal cells, another approach to developing specific chemotherapy against fungal infections has been to look for compounds that interfere with the synthesis of chitin. Several antibiotics, such as the **polyoxins,** are known to be excellent inhibitors of chitin synthetase. Unfortunately, most tested compounds do not affect medically important fungi because they do not enter the cell where the enzyme resides.

SUMMARY

The study of fungi is in the process of transition. Traditionally, mycology was almost entirely descriptive and taxonomically oriented, a result most probably of its close association with botany. At present, mycology is enjoying a period of rapid growth and interest. Fungi are being employed as models for the elucidation of many genetic and biologic processes common to all living things. They are also being used as models to study cellular differentiation and adaptation, particularly where it concerns host-parasite interactions.

BIBLIOGRAPHY

Alexopoulos CJ and Mims CW: Introductory mycology, New York, 1973, John Wiley & Sons.

Bartnicki-Garcia S: Cell wall construction during spore germination in phycomycetes. In Turian G and Holh HR, editors: The fungal spore: morphogenetic controls, New York, 1981, Academic Press.

Cabib E, Roberts R, and Bowers B: Synthesis of the yeast cell wall and its regulation, Ann Rev Biochem 51:763-793, 1982.

Cole GT: Models of cell differentiation in conidial fungi, Microbiol Rev 50:95-132, 1986.

Medoff G,. Brajtburg J, Kobayashi GS, and Bolard J: Antifungal agents useful in therapy of systemic fungal infections, Ann Rev Pharmacol 23:303-330, 1983.

Howard DH, editor: Fungi pathogenic for humans and animals. Part A, Biology, New York, 1983, Marcel Dekker.

Fromtling RA: Overview of medically important azole derivatives, Clin Microbiol Rev 1:187-217, 1988.

Soll DR: The cell cycle and commitment to alternate cell fates in Candida albicans. In Nurse P and Streibolva E, editors: The microbial cell cycle, Boca Raton, Florida, 1984, CRC Press.

Szaniszlo PJ, editor: Fungal dimorphism: with emphasis on fungi pathogenic for humans, New York, 1985, Plenum Press.

Webster J: Introduction to fungi, ed 2, London, 1980, Cambridge University Press.

Wyllie TD and Morehouse LG, editors: Mycotoxic fungi, mycotoxins, mycotoxicoses: an encyclopedic handbook, vol 3, Mycotoxicoses of man and plants: mycotoxin control and regulatory aspects, New York, 1978, Marcell Dekker.

29

Superficial, Cutaneous, and Subcutaneous Mycoses

Fungal infections can be classified according to the tissue levels that are initially colonized. This chapter focuses infections caused by (1) the **superficial mycoses,** which are infections limited to the outermost layers of the skin and hair; (2) the **cutaneous mycoses,** which include infections that are deeper in the epidermis and invasive hair and nail diseases; and (3) the **subcutaneous mycoses,** which involve the dermis, subcutaneous tissues, muscle, and fascia.

In general, intact skin and mucosal surfaces serve as barriers to infection caused by the superficial, cutaneous, and subcutaneous mycotic agents. Fatty acid content, pH, epithelial turnover of the skin, and the normal bacterial flora appear to contribute to host resistance. The mucosa and ciliary action of cells lining the respiratory tract help to eliminate foreign substances that are accidentally inhaled. Humoral factors such as transferrin have been shown to restrict the growth of several fungi in vitro by limiting the amount of available iron, and they may play a role in restricting the growth of dermatophytes to the outer layers of skin.

SUPERFICIAL MYCOSES

Superficial fungal infections are usually cosmetic problems that are easily diagnosed and treated. Four infections fall into this classification; two involve the skin (pityriasis versicolor and tinea nigra) and two the hair (black piedra and white piedra) (Figure 29-1). Infections of the skin are limited to the outermost layers of the stratum corneum. Those of hair involve only the cuticle. In general these superficial fungal organisms do not elicit a cellular response from the host because they colonize tissues that are not living. Furthermore, these infections cause no physical discomfort to the patient, and the condition is generally brought to the attention of the physician as an incidental finding or for cosmetic reasons. The diseases are easy to diagnose, and specific therapeutic measures usually result in good clinical responses.

Etiology and Clinical Syndromes

Pityriasis Versicolor This fungal infection (Figure 29-1, *A*) is caused by *Malassezia furfur (Pityrosporum orbiculare),* a lipophilic yeastlike organism closely related to *P. ovale.* It is found in areas of the body rich in sebaceous glands and is part of the normal flora of the skin. These organisms require a medium supplemented with saturated or unsaturated fatty acids to support growth. The organism grows as budding yeasts, although hyphal forms are occasionally seen.

The lesions of pityriasis versicolor occur most commonly on the upper torso, arms, and abdomen as discrete hyper- or hypopigmented macular lesions (Figure 29-2). They scale very easily, giving the affected area a dry, chalky appearance. On rare occasions the lesions take on a papular (elevated) appearance, and in some cases where the hair follicle is involved the lesions can cause folliculitis. The clinical diagnosis of pityriasis versicolor is made microscopically by visualizing the characteristic ''spaghetti and meatballs'' appearance of the organism in potassium hydroxide–treated specimens (Figure 29-3).

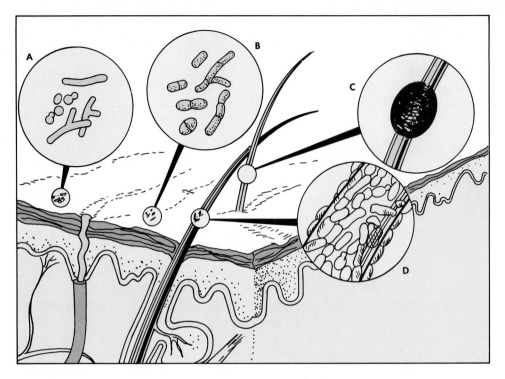

FIGURE 29-1   Schematic illustration of superficial fungal infection and tissue involvement. **A,** Pityraiasis versicolor; **B,** tinea nigra; **C,** black piedra; **D,** white piedra.

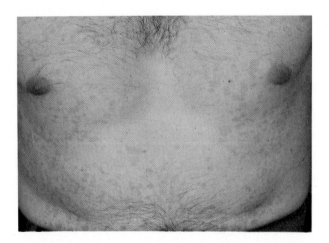

FIGURE 29-2   Clinical presentation of pityriasis versicolor.

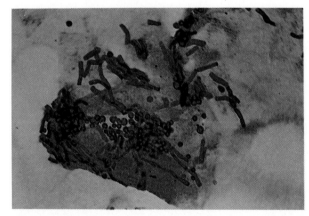

FIGURE 29-3   Skin scrapings stained with periodic-acid Schiff's stain showing typical yeastlike and hyphal fragments of *Malassezia furfur*, the etiologic agent of pityriasis versicolor.

Cultures are not routinely done to confirm the diagnosis, since the organism requires a special medium containing fatty acids.

Tinea Nigra The etiologic agent of tinea nigra (Figure 29-1, *B*), *Exophiala werneckii,* is a dimorphic fungus that produces melanin, imparting a brown to black color to the organism. On primary isolation from clinical material it grows as yeasts with many cells in various stages of cell division, producing characteristic two-celled oval structures (Figure 29-4). As the colony ages, elongate hyphae develop, and in older cultures mycelia and conidia predominate.

The clinical manifestations of tinea nigra are usually asymptomatic and consist of well-demarcated macular lesions (discolored spots on the skin that are not raised above the surface) that enlarge by peripheral extension (Figure 29-5). The brown to black lesions are most often seen on the palms of the hands and soles of the feet but may occur on other areas of the body. Diagnosis is made by visualizing the characteristic darkly pigmented yeastlike cells and hyphal fragments in microscopic examination of potassium hydroxide–treated scrapings taken from the affected areas; it is confirmed by culture.

Black Piedra The etiology of black piedra (Figure 29-1, *C*), a superficial hair infection, is *Piedraia hortai,* an organism that exists in the perfect (teleomorphic) state when it colonizes the hair shaft. Cultures taken from clinical material usually yield only the asexual (anamorphic) state of the fungus, which consists of slow-growing brown to reddish-black mycelia. The teleomorphic state is occasionally found in older cultures and consists of specialized structures within which asci containing spindle-shaped ascospores develop.

The major clinical feature of black piedra are the hard nodules, found along the infected hairshaft (Figure 29-6). The nodules have a hard carbonaceous consistency and house asci. The differential diagnosis includes ruling out nits of pediculosis and abnormal hair growth. The infection is easily diagnosed by microscopic examination of affected hairs.

White Piedra This infection of the hair (Figure 29-1, *D*) is caused by the yeastlike organism *Trichosporon beigelii* (Figure 29-7). The organism grows well on all laboratory media except those containing cycloheximide, an antibiotic used in media for the selective isolation of most pathogenic fungi (e.g., Mycosel agar). Young cultures are white and have a

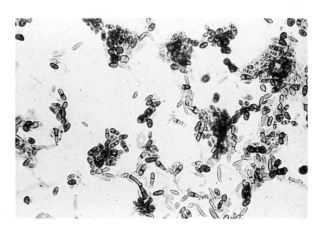

FIGURE 29-4   Yeastlike cells of *Exophiala wernekii,* the causative agent of tinea nigra.

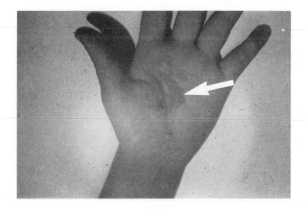

FIGURE 29-5   Clinical presentation of tinea nigra. Note dark pigmentation in the center of the palm.

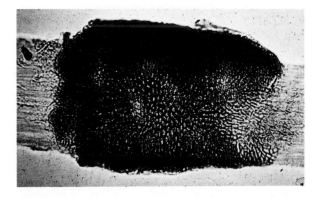

FIGURE 29-6   Hair infected with *Piedraia hortai.* The hard black nodule contains asci and ascospores, the sexual phase of the fungus.

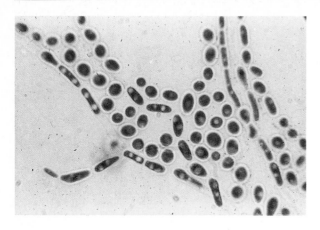

FIGURE 29-7   Hyphae, arthroconidia, and blastospores of *Trichosporon beiglii,* illustrating the dimorphic characteristics of the agent of white piedra.

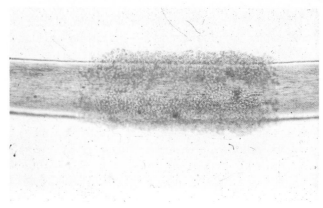

FIGURE 29-8   Clinical presentation of white piedra.

Table 29-1   Important features of the organisms causing the superficial mycoses

Organism	Disease	Tissue	Clinical Features	Diagnostic Procedures
Malassezia furfur	Pityriasis versicolor	Skin	Hyperpigmented or hypopigmented macular lesions that scale readily, giving it a chalky-branny appearance; occurs most frequently on the upper torso of the body	Direct microscopic examination of alkali-stain* treated skin scrapings reveals fungal elements having the classical "spaghetti and meatballs" appearance. Cultures not routinely done to confirm diagnosis. Organism requires fatty acid supplemented medium for growth.
Exophiala werneckii	Tinea nigra	Skin	Grey to black, well-demarcated macular lesions most frequently occurring on the palms of the hand	Direct microscopic examination of skin scrapings treated with alkali-stain.* Culture on Sabouraud's dextrose agar yields pigmented (brown to black) yeasts and hyphae (dimorphic).
Piedraia hortae	Black piedra	Hair	Hard, gritty, brown to black concretions that develop along the hair shaft; structures house the sexual phase (asci and ascospores) of the fungus	Direct microscopic examination of hairs. Culture yield asexual phase of the fungus.
Trichosporon beigelii	White piedra	Hair	Soft, white to creamy yellow granules that form a sleeve-like collarette along the hair shaft	Direct microscopic examination of hairs. Culture on Sabouraud's dextrose agar. Growth is dimorphic—hyphae, arthroconidia, and blastoconidia.

(From Swartz JH and Lamkin B: Arch Dermatol 80:89, copyright 1964, the American Medical Association.)

*A useful alkaline dye containing clearing agent can be made from Parker Super Quick permanent blue-black ink by adding 10 g potassium hydroxide per 100 ml of ink. The solution is centrifuged to sediment the amorphous precipitate that forms. The clear blue supernatant should be decanted and stored in a plastic container to prevent insoluble carbonate precipitates from forming.

have a pasty consistency. As the culture ages, colonies develop deep, radiating furrows and take on a yellowish coloration with a creamy texture. Microscopic examination reveals septate hyphae that fragment rapidly to form arthroconidia (see Figure 29-7). The arthroconidia rapidly round up, and many cells form blastoconidia. The organism is identified by various biochemical tests and its ability to assimilate certain carbohydrates.

White piedra affects hairs of the scalp, mustache, and beard. It is characterized by the development of cream-colored soft pasty growths along infected hair shafts (Figure 29-8). The growth occurs as a sleeve or collarette around the hair shaft and consists of mycelia that rapidly fragment into arthroconidia. The differential diagnosis includes trichomycosis axillaris (caused by *Corynebacterium tenuis*) and the nits of pediculosis. The infection is diagnosed by microscopic examination of infected hairs and confirmed by culture.

Table 29-1 summarizes the important features of the superficial mycoses.

Treatment

These infections in general are cosmetic and easily diagnosed. When proper therapy is instituted, the infections respond well with no consequences. The general approach to treating pityriasis versicolor and tinea nigra is removal of the organism from the skin. This is accomplished by the topical use of keratolytic (chemicals that lyse keratin) agents. Preparations containing selenium disulfide, hyposulfite, thiosulfate, or salicylic acid accomplish this, but recurrences of the disease may occur. Topical preparations containing miconazole nitrate have been used effectively in eradicating the disease.

In hair infections caused by *P. hortae* and *T. beigelii,* effective therapy is achieved by shaving or cropping the infected hairs close to the scalp surface. Recurrence does not occur if proper personal hygiene is practiced.

CUTANEOUS MYCOSES

The cutaneous mycoses involve diseases of the skin, hair, and nails. They are generally restricted to the keratinized layers of the integument and its appendages (Figure 29-9). Unlike the superficial

infections, various cellular immune responses may be evoked, causing pathologic changes in the host that may be expressed in the deeper tissues of the skin. The severity of the reponse appears to be directly related to the immune status of the host and strain or species of fungus involved. The term **dermatophyte** has been used traditionally to describe these agents. However, the suffix "phyte" implies that these organisms are plants, which is misleading because fungi are not phylogenetically related to plants. However, for historical reasons the term dermatophyte will be used in this section in reference to the organisms causing these diseases.

The clinical manifestations of these diseases are also referred to as **"ringworm"** or **"tinea,"** depending on the anatomic site involved (e.g., tinea pedis, feet; tinea capitis, scalp; tinea manus, hands; tinea unguinum, nails; tinea corporis, body). In some cases they are given special names depending on what organism causes the disease (e.g., favus, *Trichophyton schoenleinii;* tokelau, *T. concentrichum*). The term *"tinea"* comes from Latin and means worm or moth. It is used descriptively because of the serpentine (snakelike) and annular (ringlike) lesions that occur on the skin, making it appear that a worm is burrowing at the margin (Figure 29-10).

Etiology

The cutaneous mycoses are caused by a homogeneous group of closely related organisms known as the dermatophytes. Although over 100 species have been described, only about 40 are considered valid and less than half of these are associated with human disease (Table 29-2). In the anamorphic state they are classified in three genera (i.e., *Microsporum, Trichophyton,* and *Epidermophyton*) on the basis of their sporulation patterns (Figure 29-11), certain morphologic features of development, and nutritional requirements. The teleomorphic state of some of the organisms belonging to the genera *Microsporum* and *Trichophyton* are known. They are all Ascomycetes and have been reclassified in the genus *Arthroderma*. As yet, the sexual phase of *Epidermophyton* has not been observed.

Ecology and Epidemiology

The isolation of different species of dermatophytes varies markedly from one ecologic niche to another.

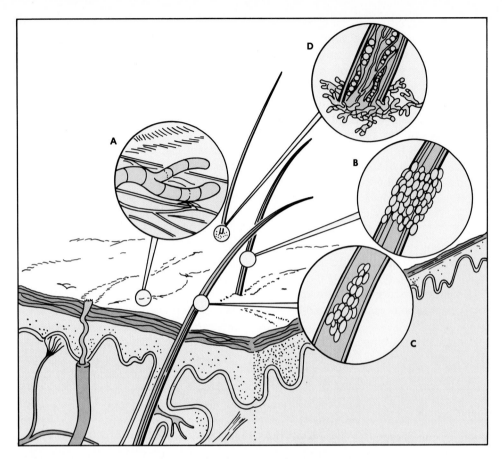

FIGURE 29-9   Schematic of tissues colonized by dermatophytes. **A,** Stratum corneum; **B,** ecto-thrix hair infection; **C,** endothrix hair infection; **D,** favic hair infection.

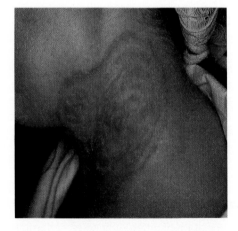

FIGURE 29-10   Clinical presentation of tinea corporis.

Table 29-2   Asexual (Anamorphic) State of Selected Dermatophytes*

Microsporum	Trichophyton	Epidermophyton
M. audouinii	T. concentricum	E. floccosum
M. canis	T. equinum	
M. cookei	T. mentagrophytes	
M. equinum	var. interdigitale	
M. fulvum	T. mentagrophytes	
M. gallinae	var. metagrophytes	
M. gypseum	T. rubrum	
M. nanum	T. tonsurans	
	T. verrucosum	
	T. violaceum	

*At present, 41 species of dermatophytes are recognized as etiologic agents of disease.

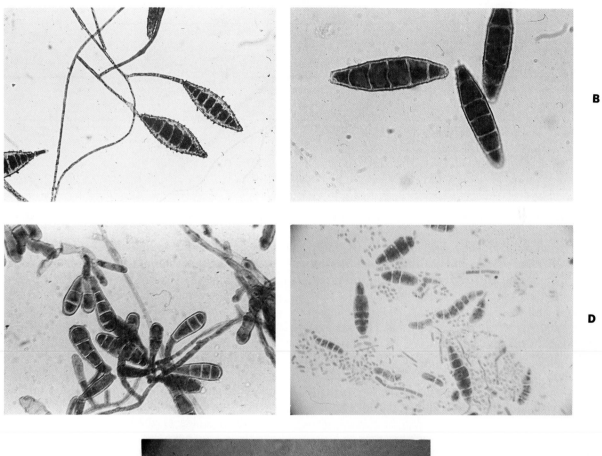

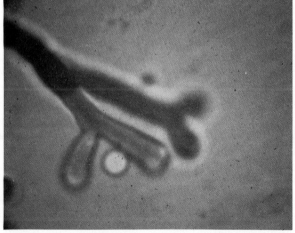

FIGURE 29-11   Sporulation pattern and identifying features of some dermatophytes. **A,** Macroconidia of *Microsporum canis;* **B,** macroconidia of *M. gypseum;* **C,** macroconidia of *Epidermophyton flocosum;* **D,** microconidia and macroconidia of *Trichophyton mentagrophytes;* **E,** favic chandelier of *Trichophyton schoenleini.*

Table 29-3   Classification of Dermatophytes According to Ecologic Niche*

Anthropophilic Dermatophytes	Zoophilic Dermatophytes	Geophilic Dermatophytes
M. audouinii	M. canis	M. cookei
T. mentagrophytes, var. interdigitale	M. equinum	T. gypseum
T. rubrum	M. gallinae	M. fulvum
T. tonsurans	T. equinum	M. nanum
T. violaceum	T. mentagrophytes, var. mentagrophytes	
E. floccosum		

*Includes only those species that are common throughout the world.

Some species are frequently isolated from the soil and have been grouped as **geophylic dermatophytes.** Other species have been found most often in association with domestic and wild animals and birds. These are referred to as **zoophilic dermatophytes.** A third group, the **anthropophilic dermatophytes,** has been found almost exclusively in association with humans and their habitats. Table 29-3 summarizes the groupings of various species that are common throughout the world.

The clinical importance of identifying species of dermatophytes is to determine the possible source of infection. There are also some considerations of prognostic value. The anthropophilic group tends to cause chronic infections and may be difficult to cure. The zoophilic and geophilic dermatophytes tend to cause inflammatory lesions, respond well to therapy, and may occasionally heal spontaneously.

Some species of dermatophytes are endemic in certain parts of the world and have a limited geographic distribution. At the present time, T. yaoundei, T. gourvilli, and T. soudenense are geographically restricted to Central and West Africa; in Japan and its surrounding areas M. ferrugineum predominates; and T. concentricum is confined to islands in the South Pacific and a small area of Central and South America. However, the increasing mobility of the world's population is disrupting several of these patterns. In recent times T. tonsurans has replaced M. audouinii as the principal agent of tinea capitis in the United States, a result of the mass migration of individuals from Mexico and other Latin American countries where T. tonsurans predominates. Less well understood are the epidemics of ringworm that occasionally occur.

The prevalence of dermatophytes and the incidence of disease are difficult to determine, since they are not reportable. Fragmentary surveys from epidemiologic studies and case reports indicate that the cutaneous mycoses are among the most common human diseases. Reports estimate that they are the third most common skin disorder in children under 12 years of age and the second most common in older populations. The occurrence of these diseases varies with age, sex, ethnic group, and cultural and social habits of the population.

The incidence of the cutaneous mycoses and clinical manifestations of the disease among various age groups depend on the anatomic site of involvement and the species of dermatophyte involved. Tinea capitis is a problem of the pediatric population until puberty, when it spontaneously ceases to be a major infectious disorder. On the other hand, tinea pedis, which is rarely a disease in childhood, gradually becomes the predominant infection and remains so into adulthood.

The incidence of tinea capitis in black children in the United States is disproportionately high. In India tinea capitis occurs more often in the native children than in Europeans, whereas the Europeans have a higher incidence of tinea pedis than does the native population. Clinical surveys conducted during the war in Southeast Asia revealed that persons from the United States had a higher incidence of tinea pedis caused by T. mentagrophytes than did the native population. The indigenous population appeared to be highly susceptible to a distinctive strain of T. rubrum. Tinea capitis, tinea pedis, and tinea cruris are more common in men than women, whereas the reverse is true for tinea unguium (infections of the nail plate) of the hand. For nails of the feet it is seen more often in men.

The reasons for these observations are poorly understood. The incidence of cutaneous mycoses is related to customs associated with the type of clothing that is worn and how the feet are shod. Studies on institutionalized populations and family outbreaks of dermatophyte infections indicate that close and crowded living conditions are a factor in the spread of infections. There is evidence that natural resistance to these infections exists in certain individuals. Certain humoral factors are fungistatic, and cell-mediated immunity appears to be important in resistance to dermatophyte infections.

Pathogenesis

A delicate balance appears to operate between host and parasite in dermatophyte infections. Some of these fungi show an evolutionary development toward a parasitic existence. Those that have achieved a high level of coexistence with humans also exhibit a degree of specificity for the tissues that are colonized. These fungi are often referred to as **"keratinophilic fungi"** because they can use keratin as a substrate. Keratinases have been isolated from some of these fungi, but keratin is not an essential metabolite for them. The reason for the high degree of selectivity of tissues containing this protein for growth of the dermatophytes is unknown. As versatile as these fungi might appear, they seem unable to invade organs other than the keratinized layers of skin, hair, and nails.

Laboratory Diagnosis

The diagnosis of disease requires that fungal elements be seen in clinical specimens taken from the lesion and/or confirmed by culture. Skin and nail scrapings or hairs taken from areas suspected to be infected are examined microscopically. The procedure for processing these specimens is similar to the procedures described previously for examining clinical material from superficial fungal infections. The specimen is treated with an alkali solution to clear it of epithelial cells and other debris. Dermatophytes resist the caustic solution and will appear as branching septate hyphal elements in specimens taken from cutaneous lesions or nails (Figure 29-12). Fungus elements in infected hairs examined by this procedure will appear as spores inside (endothrix infection; figure 29-13, *A*) or surrounding (ectothrix infection; Figure 29-13, *B*) the hair shaft. An exception is the hair infection caused by *T. schoenleinii*. Disease caused

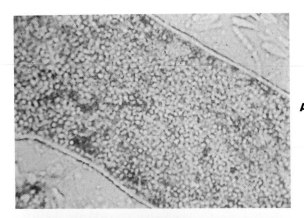

A

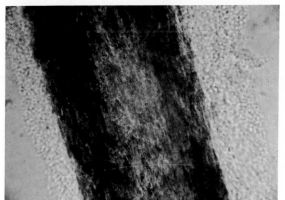

B

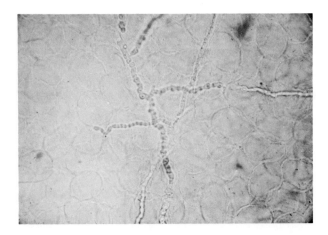

FIGURE 29-12   Skin scraping treated with 10% KOH containing hyphal fragment.

FIGURE 29-13   Fungal infections of the hair. **A,** Endothrix infection; **B,** ectothrix infection.

by this organism is called favus, and the infected hair will have a waxy mass of hyphal elements (scutulum) surrounding the base of the hair follicle at the scalp line (Figure 29-9, *D*). Microscopic examination of the hair reveals degenerated hyphal elements coursing throughout the hair shaft (see Figure 29-9, *D*). Such hairs are called **favic** and are diagnostic of the infection. Direct microscopic examination of clinical material will confirm only the diagnosis of a fungal infection. To identify the specific etiologic agent, cultures must be taken.

Culture and Identification Since the skin surface harbors many bacterial species and saprobic fungi as part of the normal flora or transient colonizers, media such as Sabouraud's dextrose agar are not routinely used for primary culture because these organisms overgrow the culture and inhibit growth of the more slow-growing dermatophytes. For this reason selective media containing antibiotics are recommended when culturing specimens taken from the skin. A medium commonly used for the isolation of dermatophytes from clinical material is one that contains cycloheximide (to suppress saprobic fungal growth) and chloramphenicol (to inhibit growth of bacteria). Clinical material is seeded directly onto the medium and incubated at 25° C. Dermatophytes and most pathogenic fungi grow well on this medium when incubated at 25° C, whereas saprobic fungi and bacteria are inhibited.

Cultures are examined periodically, and all fungi that grow are identified microscopically and by physiologic tests. In general, the mycelia of these fungi are undifferentiated and species identification is based primarily on the conidia that are produced (Table 29-4). The conidia may be large (5-100 μm $\times$ 6-8 μm)

and multicellular (macroconidia), or they may be small (3 μm $\times$ 10 μm) and unicellular (microconidia).

In addition to conidia, some dermatophytes produce spiral hyphae, chlamydospores, nodular bodies, racquet hyphae, and chandeliers. These structures are produced commonly by some species of dermatophytes and infrequently by others; however, they should not be considered distinguishing features of the species. Figure 29-11 illustrates the identifying features of some dermatophytes. All cultures that do not produce growth are routinely held for 4 weeks before being discarded as negative.

Treatment

The clinical nature of the dermatophyte infections frequently poses a challenge to the clinician. Skin infections generally are approached conservatively with topical treatment. The discovery of azole derivatives as effective antifungal agents has provided several new drugs that can be used topically, such as miconazole, clotrimazole, and econazole. All of the azole derivatives appear to work by interfering with the cytochrome P450-dependent enzyme systems at the demethylation step from lanosterol to ergosterol.

For hair infections, griseofulvin, a secondary metabolite of the fungus *Penicilium griseofulvum,* is an effective and safe drug prescribed orally for the management of tinea capitis. This compound is fungistatic and appears to work by affecting the microtubular system of fungi. It interferes with the mitotic spindle and cytoplasmic microtubules. The molecular action of griseofulvin is different from that of other inhibitors, such as cholchicine and the vinca alkaloids, which bind to receptors on tubulin and inactivate the free subunits of microtubules.

SUBCUTANEOUS MYCOSES

The subcutaneous mycoses include a wide spectrum of fungal infections characterized by the development of lesions, usually at sites of trauma where the organism is implanted in the tissue (Figure 29-14). The infections initially involve the deeper layers of the dermis, subcutaneous tissue, or bone. Most infections have a chronic and insidious growth pattern, eventually extending into the epidermis, and are expressed clinically as lesions on the skin surface.

Table 29-4 General Characteristics of Macroconidia and Microconidia of Dermatophytes

Genus	Macroconidia	Microconidia
Microsporum	Numerous, thick-walled, rough*	Rare
Epidermophyton	Numerous, smooth-walled	Absent
Trichophyton	Rare, thin-walled, smooth	Abundant†

M. audouinii is an exception.
†*T. schoenleinii* is an exception.

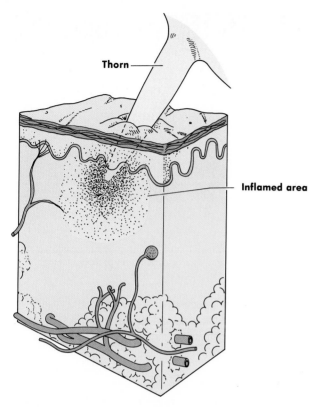

Thorn

Inflamed area

FIGURE 29-14   Schematic of tissue level colonized primarily by agents causing subcutaneous mycoses. The fungus gains access to the deeper layers of skin by traumatic implantation. The organisms implicated in these disease processes are usually common fungi found in the soil.

Several features are common about this group of infections, including the following:

1. The patient can usually associate some form of trauma occurring at the sites of the infection before the lesions developed (e.g., a splinter, a thorn, the implantation of other foreign bodies, or a bite).
2. The infections occur on parts of the body that are most prone to be traumatized (e.g., feet, legs, hands, arms, and buttocks).
3. The etiologic agents are usually organisms commonly found in the soil or on decaying vegetation.
4. Several bacterial infections (e.g., actinomycotic mycetoma, botryomycosis, and atypical acid fast disease) mimic the subcutaneous fungal infections. For this reason it is extremely important that the etiologic agent be established,

since most of the bacerial infections can be managed with antibiotics.

5. With one or two exceptions, the subcutaneous mycoses are difficult to treat and surgical intervention (i.e., excision or amputation) is frequently employed.

Etiologic and Clinical Syndromes

With the exception of lymphocutaneous sporotrichosis, most subcutaneous fungal infections are rare and considered exotic in the United States and other highly developed countries. The diseases that are less frequently or rarely seen are chromoblastomycosis, phaeohyphomycosis, chronic subcutaneous zygomycosis, and eumycotic mycetoma. Two diseases, lobomycosis and rhinosporidiosis, are considered possibly caused by fungi, but confirmation by culture is lacking.

The causative agents are a heterogeneous group of organisms with low pathogenic potential that are commonly isolated from soil or decaying vegetation. The clinical manifestations of these diseases appear to be an interplay between the etiologic agent and host responses. In general, patients who develop disease have no underlying immunologic defect.

Lymphocutaneous Sporotrichosis This chronic infection is characterized by nodular and ulcerative lesions that develop along lymphatics that drain the primary site of inoculation (Figure 29-15). Other infrequently seen forms of sporotrichosis include fixed cutaneous lesions, primary and secondary pulmonary sporotrichosis, and disseminated disease. The causative agent is the dimorphic fungus *sporothrix schenkii.*

Chromoblastomycosis This disease is characterized by the development of verrucous (warty) nodules that appear at sites of inoculation (Figure 29-16). As lesions progress they appear to vegetate and take on a "cauliflower-like" appearance. The organisms responsible for chromoblastomycosis are common soil inhabitants that are collectively called the **dematiaceous fungi.** The term "dematiaceous" is used to describe fungi that have brown to black melanin pigments in their cell wall.

Phaeohyphomycosis Another clinical disease caused by various dematiaceous fungi is **phaeohyphomycosis,** which is a heterogeneous group of cutaneous diseases caused by various dematiaceous fungi. The most common form described is phaeohy-

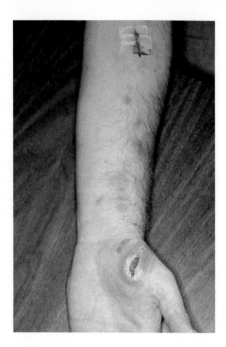

FIGURE 29-15   Clinical case of lymphocutaneous sporotrichosis, illustrating the characteristic pattern of lesions along the lymphatics that drain the site of the original lesion.

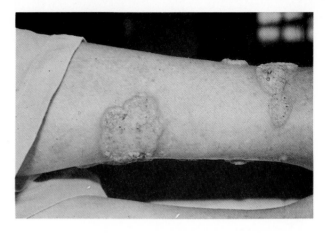

FIGURE 29-16   Clinical case of chromoblastomycosis, illustrating the characteristic verrucous vegetative lesions.

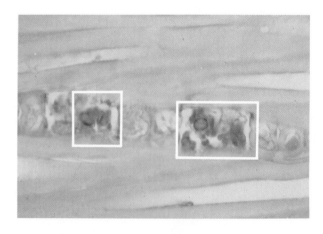

FIGURE 29-17   Tissue section taken from a case of phaehyphomycosis, illustrating the characteristic pigmented hyphal fragments seen in histopathologic examination.

phomycotic cyst. The disease does not exhibit the intense hyperplasia seen in chromoblastomycosis, and when organisms are seen in histopathologic examination of tissues they usually appear as pigmented septate hyphal fragments (Figure 29-17).

The taxonomy of the dematiaceous agents is currently undergoing careful scrutiny, and as yet complete agreement has not been reached. The dematiaceous fungi most often associated with chromoblastomycosis are *Fonsecaea pedrosoi*, *F. compacta*, *Cladosporium carrionii*, and *Phialophora verrucosa*. These organisms are identified according to the pattern and type of sporulation exhibited by the isolate. In many cases the isolate may exhibit more than one pattern of sporulation, and for this reason confusion and conflicts often arise concerning the correct taxonomic placement of the organism.

Similar taxonomic problems exist in the identification of agents implicated in phaeohyphomycosis, for which about 40 different organisms have been imcriminated. Included in this large number of etiologic agents are some fungi that are rare and no doubt reflect the exotic nature of some of these clinical entities.

Eumycotic Mycetoma The term **"mycetoma"** is clinically descriptive and includes a wide spectrum of manifestations involving the skin and deeper tissues of the dermis and subcutaneous tissues. The disease is characterized by indolent, deforming, swollen lesions that contain numerous draining sinus tracts (Figure 29-18).

Other Subcutaneous Mycoses Zygomycosis, lobomycosis, and rhinosporidiosis are rare clinical entities. A detailed description of these diseases is contained in the bibliography to this chapter.

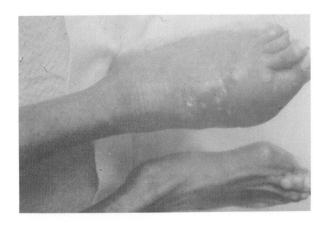

FIGURE 29-18   Clinical case of eumycotic mycetoma, illustrating the degree of induration and draining sinus tracts that characterize this infection. (Courtesy Marilyn Bartlett.)

FIGURE 29-19   Microscopic examination illustrating "rosette pattern" of conidiation in *sporothrix schenckii*.

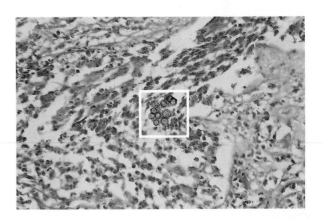

FIGURE 29-20   Tissue section taken from a case of chromoblastomycosis, illustrating the characteristic "sclerotic" cells.

Laboratory Diagnosis

Lymphocutaneous Sporotrichosis In tissue and in cultures incubated at 37° C *Sporothrix schenckii* is a budding yeast cell. Cultures at 25° C develop as delicate radiating colonies that appear within 3 to 5 days on most media. The colonies are moist and white at first; with prolonged incubation they slowly develop a brown to black pigmentation. Microscopic examination reveals delicate branching hyphae, about 2 μm in diameter, with numerous conidia developing in a rosette pattern at the ends of conidiophores (Figure 29-19). Laboratory confirmation is established by converting the mycelial growth to the yeast morphology by subculture at 37° C.

Chromoblastomycosis The diagnosis of chromoblastomycosis is usually made by histopathologic examination of clinical material taken from the lesions. There is a characteristic tissue response termed "pseudoepitheliomatous hyperplasia," which means that the tissue exhibits an epithelial overgrowth caused by an abnormal multiplication in the number of normal cells in normal arrangement in the tissue. In addition to the histopathology, copper-colored spherical cells in various stages of cell division are seen (Figure 29-20). These structures, called **sclerotic** or **Medlar bodies,** are the tissue forms of the fungus.

Eumycotic Mycetoma Examination of the purulent fluid that exudes from the sinus tracts in eumycotic mycetoma often reveals small grains of fungal

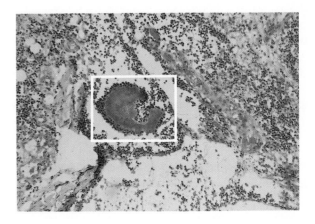

FIGURE 29-21   Histopathologic section of tissue taken from mycetoma, illustrating a microcolony (often referred to as a granule).

tissue. These elements may be white, brown, yellow, or black and can be well demonstrated on histopathologic examination of tissue biopsies of the lesion (Figure 29-21). The etiologic agents causing these diseases consist of various actinomycetes belonging to the genera *Actinomyces, Nocardia, Streptomyces,* and *Actinomadura,* as well as a whole host of fungi, including *Pseudallescheria boydii* and *Madurella grisea.* It is important to establish the cause of the disease by culturing specimens, since the clinical management of the infection will vary depending on the causative organism.

Treatment

Subcutaneous lymphangitic sporotrichosis responds dramatically to a saturated solution of potassium iodide given orally. Adverse side effects include gastrointestinal upset and dermatologic problems that are rapidly reversed by discontinuing therapy. Extracutaneous sporotrichosis invariably requires systemic therapy with amphotericin B.

Cautery and surgical removal of early lesions have been used in the treatment of chromoblastomycosis; however, most patients who seek help have advanced disease. The extensive tissue involvement is often not amenable to surgical intervention and requires chemotherapy. At present, the drug of choice for treating chromoblastomycosis is 5-fluorocytosine. This drug is given orally and acts by inhibiting RNA and DNA synthesis.

As stated previously, the clinical treatment of eumycotic mycetoma varies with the causative agent. In the case of actinomycotic mycetomas, several antibacterial antibiotics can be used. For example, infections caused by *Actinomyces israelli* respond to high doses of penicillin, and those caused by *Nocardia asteroides* respond to sulfa drugs in combination with streptomycin. However, if a fungal organism is the causative agent, the physician will frequently resort to total excision of the lesion or amputation of an affected limb if the disease is extensive, since antifungal therapy in general is unsuccessful.

Tables 29-5 and 29-6 summarize the important features of the subcutaneous mycoses.

Table 29-5 Diseases and Organisms of the Subcutaneous Mycoses

Disease	Organism
Sporotrichosis	*Sporothrix schenkii*
Chromoblastomycosis	*Foncecaea prodrosoi*
	F. compacta
	Wangiella dermatitidis
	Philiaophora verrucosa
	P. richardsiae
	Exophiala jeanselmei
	E. spinifera
	Cladosporium carrionii
	C. bantianum
Phaehyphomycosis	*Alternaria alternata*
	Aureobasidium pullalans
	Chaetomium funiculosum
	Curvularia geniculata
	C. lunata
	Drechslera hawaiensis
	Wangiella dermatitidis
Eumycotic mycetoma	*Pseudallescheria boydii*
	Madurella grisea
	M. mycetomatis
	Acremonium kiliese
	Leptosphaeria senegalensis
	Pyrenochaeta romeroi
Zygomycosis	*Conidiobolus coronatus*
	Basidiobolus ranarum
Lobomycosis	*Loboa-loboi* (not cultured)
Rhinosporidiosis	*Rhinosporidium seeberi* (not cultured)

BIBLIOGRAPHY

Larone DH: Medically important fungi: a guide to identification, ed 2, New York, 1987, Elsevier Science Publishing.

McGinnis MR: Laboratory handbook of medical mycology, New York, 1980, Academic Press.

Rebell G and Taplin D: Dermatophytes: their recognition and identification, Coral Gables, Florida, 1979, University of Miami Press.

Rippon JW: Medical mycology: the pathogenic fungi and the pathogenic actinomycetes, Philadelphia, 1988, WB Saunders Co.

Table 29-6 Summary of the Findings in the Subcutaneous Mycoses

Clinical Disease	Mycologic Findings
Lymphocutaneous sporotrichosis is the disease most commonly associated with *Sporothrix schenkii*. The organism gains access to the deeper layers of skin as a result of traumatic implantation. A small, hard, painless nodule appears at the site of injury. The nodule becomes fixed over a period of time and enlarges to form a fluctuant mass, which eventually breaks down and ulcerates. As the primary lesion enlarges, several other nodules begin to develop along the lymphatics that drain that site. They also become fluctuant and ulcerate. The infection rarely extends beyond regional lymphatics. Other forms of sporotrichosis include pulmonary, fixed cutaneous, mucucutaneous, and extracutaneous disseminated disease.	Dimorphic fungus. At 25° C the organism grows as a mold. White at first, it becomes brown to black with age. It produces conidia in a typical rosette pattern. At 37° C and in tissues the fungus is a yeast.
The most common form of chromoblastomycosis consists of warty, vegetative lesions that resemble a cauliflower.	Histopathologic examination of tissue reveals "sclerotic cells." These are brown to copper spherical cells seen in hematoxylin- and eosin-stained sections. Cultures will grow out a mold that is gray to black. Speciation requires careful examination of conidia development and certain biochemical tests.
Phaehyphomycosis is also known as phaeomycotic cyst. The lesions are subcutaneous or intramuscular, tend to be stationary, and occur singly.	The important difference between phaeohyphyomycosis and chromoblastomycosis is that brown hyphal cells and not "sclerotic cells" are found in tissues.
Eumycotic mycetoma is characterized by localized indolent, deforming, swollen lesions with numerous draining sinus tracts. It involves subcutaneous and cutaneous tissues, fascia, and bone. The lesions are composed of suppurating abscesses, granulomata, and draining sinuses with the presence of "grains," which are microcolonies of the etiologic agent.	Various fungi have been implicated, and the "grains" or microcolonies found in the purulent exudate are useful in establishing cause of disease.
Zygomycosis is a chronic inflammatory or granulomatous disease that is generally restricted to the subcutaneous tissue or nasal submucosa. The disease occurs in horses. When caused by the Basidiobolus ranarum organism, it is characterized by massive palpable, indurated, nonulcerating subcutaneous masses on the limbs, trunk, chest, back, or buttocks. The fungus resides in the gastrointestinal tract of reptiles and amphibians.	Material can be cultured on medium containing streptomycin and penicillin. Conidia are forcibly discharged and reseed other parts of the medium.
Lobomycosis consists of chronic, localized, subcutaneous lesions characterized as keloidal, verrucoid, nodular, and tumorlike.	Budding yeastlike organisms occur in chains. Cultures are negative. There are no confirmed reports on its isolation.
Rhinosporidiosis is marked by large polyps, tumors, or wartlike lesions that are hyperplastic, highly vascularized, friable, and sessile or pedunculate.	The diagnosis rests on observing spherical structures that are a few microns to several microns in diameter. Mature "spherules" are filled with endospores appearing similar to those seen in coccidioidomycosis. Cultures are negative. Reports on its isolation are unconfirmed.

Systemic Mycoses

30

Unlike many other fungi, the organisms classified as systemic mycotic agents are inherently virulent and cause disease in healthy humans. Five fungi are included in this group (Box 30-1): *Histoplasma capsulatum, Blastomyces dermatitidis, Paracoccidioides brasiliensis, Coccidioides immitis,* and *Cryptococcus neoformans.* Each of these fungi exhibits biochemical and morphologic features that enable it to evade host defenses.

Four of these pathogens *(H. capsulatum, B. dermatitidis, P. brasiliensis,* and *C. immitis)* are dimorphic. As saprobes, and in culture at 25° C, they grow as filamentous molds; however, when they infect humans or are cultured at 37° C using special techniques, they transform to a unicellular morphology (Figure 30-1, *A* through *D*). Infections due to *H. capsulatum, B. dermatitidis,* and *P. brasiliensis* are characterized by the presence of budding yeast cells (Figure 30-1, *A* through *C*), whereas *C. immitis* infections are characterized by the presence of spherules (spo-

rangiumlike structures filled with endospores; Figure 31-1, *D*).

Unlike the dimorphic pathogens, *C. neoformans* is monomorphic. The organism grows as a yeast within infected tissue and in culture at 25° C or 37° C (Figure 30-1, *E*). A characteristic feature of the yeasts of *C. neoformans* is that they possess an acidic mucopolysaccharide capsule.

The primary focus of infection for all five systemic mycotic agents is the lung. In the vast majority of cases the respiratory infections are asymptomatic or of very short duration, resolve rapidly without therapy, and are accompanied in the host by a high degree of specific resistance to reinfection. In some cases a secondary spread occurs outside the lungs, with each organism exhibiting a characteristic pattern of secondary organ involvement. Frequently, it is the secondary spread of systemic disease that causes the patient to seek medical attention. The severity of infection depends on the organism and the host. If the immune status of the host is compromised, due to underlying disease or immunosuppressive therapy, the infection can be life threatening if therapy is not rapidly instituted and the underlying disorder corrected. In addition, immunosuppression may cause reactivation of latent infection.

Four of the systemic mycoses (i.e., **histoplasmosis, blastomycosis, paracoccidioidomycosis,** and **coccidioidomycosis**) tend to be restricted to particular geographic regions; however, ease of travel and increases in reactivation disease are starting to blur these distinctions.

Box 30-1 Agents Causing
the Systemic Mycoses

Histoplasma capsulatum
Blastomyces dermatitidis
Paracoccidioides brasiliensis
Coccidioides immitis
Cryptococcus neoformans

Saprobic phase
(25° C)

Parasitic phase
(37° C)

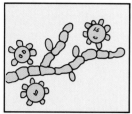

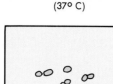

A

A. *Histoplasma capsulatum*

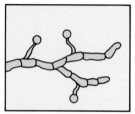

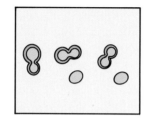

B

B. Blastomyces dermatitidis

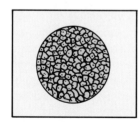

C

C. *Paracoccidioides brasiliensis*

D

D. *Coccidioides immitis*

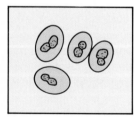

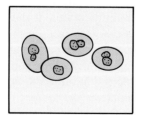

E

E. *Cryptococcus neoformans*

FIGURE 30-1 Schematic illustration of the saprobic (in vitro at 25° C) and parasitic (in vitro at 37° C) phases of systemic pathogenic fungi. **A,** *Histoplasma capsulatum* **B,** *Blastomyces dermatitidis,* and **C,** *Paracoccidioides brasiliensis* exhibit mold-to-yeast transition when infecting susceptible species; **D,** *Coccidioides immitis* exhibits mold-to-spherule transition when it infects susceptible species; **E,** *Cryptococcus neoformans,* which is an encapsulated yeast at 25° C, at 37° C, or in infected tissues.

HISTOPLASMOSIS

Histoplasmosis results from the inhalation of conidia and other infectious elements of *H. capsulatum* (Figure 30-2). It occurs worldwide and is particularly common in the midwestern United States (Figure 30-3). In most cases it is asymptomatic, but in about 1% clinical symptoms of an acute pneumonia occur, followed less often by a progressive disseminated disease. It has also been known as **Darling's disease, reticuloendothelial cytomycosis, cave disease,** and **spelunker's disease** (Table 30-1).

Morphology

The mold phase of *H. capsulatum* is characterized by thin, branching, septate hyphae that produce microconidia and tuberculate macroconidia (Figure 30-4). The parasitic or tissue phase of *H. capsulatum* is a small budding yeast cell 2 to 5 μm in diameter and found almost exclusively within macrophages (Figure 30-5). The sexual state for *H. capsulatum* has been designated *Ajellomyces capsulata* and is classified as an ascomycete.

Epidemiology

The etiologic agent of histoplasmosis, *H. capsulatum* var. *capsulatum,* grows in soil with a high nitrogen content, especially in areas contaminated with the excreta of bats and birds (starlings and chickens in particular). Birds are not infected, whereas natural infection does occur in bats. The fungus has been isolated from soil samples taken from habitats associated with birds and bats (e.g., chicken coops, attics, barns, wood piles, caves, and roosting areas such as city parks and even schoolyards). Numerous well-documented epidemics of respiratory histoplasmosis have occurred when environments harboring the fungus have been disturbed by activities such as exploring caves (spelunking), demolishing old buildings during urban renewal, cleaning chicken coops, and setting up campsites. Many researchers and epidemiologists have accidentally acquired the disease as a result of efforts to document the outbreaks when they explored suspected sites to take soil samples.

Histoplasmosis is widely distributed throughout the temperate, subtropical, and tropical zones of the world. Within these zones are areas of high endemicity, including the Ohio and Mississippi Valley regions of the United States, the southern fringes of the prov-

inces of Ontario and Quebec in Canada, and scattered areas of Central and South America (see Figure 30-3). Surveys of skin test reactivity to histoplasmin indicate that 80% or more of the long-term residents in the Ohio and Mississippi river valley have been infected with *H. capsulatum*. Serial studies of individuals living in an endemic area have shown that skin test reactivity can be lost and reacquired, suggesting that the high incidence of reactivity in these areas results from reinfection.

Cases of histoplasmosis have also been reported in

FIGURE 30-2   Schematic illustration of the natural history of the saprobic and parasitic cycles of *Histoplasma capsulatum.*

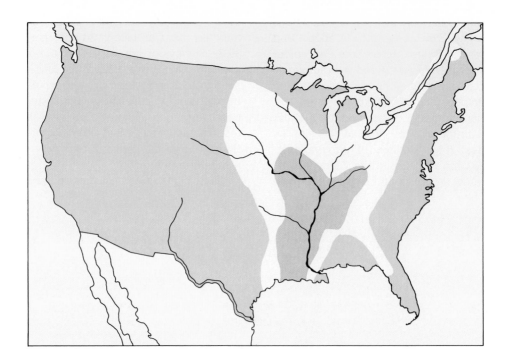

FIGURE 30-3   Endemic areas of histoplasmosis in North America.

Table 30-1   Summary of Histoplasmosis

Etiologic agent	Mycology	Epidemiology and Ecology	Clinical Disease

Asexual phase:
  Histoplasma capsulatum
Sexual phase:
  Ajellomyces capsulatum
	Dimorphic; mycelial at 25° C; typical tuberculate macroconidia 8-14 μ in diameter; microconidia 2-4 μ in diameter. At 37° C and in tissue this organism is a budding yeast 2-3 × 3-4 μ in diameter. Found predominantly in histiocytes.	Occurs throughout temperate, subtropical, and tropical areas of the world. Endemic areas include the Ohio and Mississippi river valleys and parts of Central and South America. The organism has been isolated from numerous soil samples, particularly those contaminated by bat, chicken, and starling droppings. Bats are naturally infected, but birds are not. A unique clinical form occurs in Africa caused by *H. capsulatum* var. *duboisii.*	The clinical symptoms will vary depending on the degree of individual exposure and immunologic state of the patient. About 95% of all primary cases are not referable to specific symptomology. Less than 1% of infections will become progressive and require therapy. In this case there is usually an underlying condition of debilitation or immunosuppression that makes these individual prone to life-threatening disease.

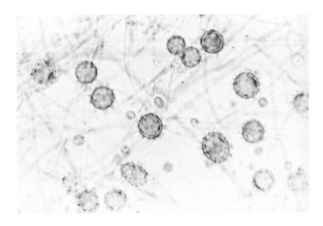

FIGURE 30-4   Tuberculate macroconidia and microconidia of *Histoplasma capsulatum*.

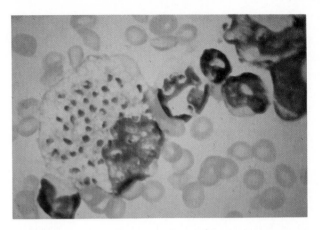

FIGURE 30-5   Yeast cells of *Histoplasma capsulatum* phagocytized by bone marrow mononuclear cells (Giemsa stained section).

Europe and Asia. A variant form of histoplasmosis occurs in Africa; the etiologic agent of this disease has been designated *H. capsulatum* var. *duboisii*.

Clinical Syndromes

The lung is the usual portal of entry for infection. *H. capsulatum* spores or hyphal fragments are inhaled, phagocytized by pulmonary macrophages, and then convert to yeast, which is able to replicate in nonimmune macrophages (see Figure 30-4). In the immunocompetent host, macrophages acquire fungicidal activity and contain the infection. Transient fungemia before the development of immunity accounts for the distribution of calcified granulomas in liver and spleen frequently seen at autopsy of patients from endemic areas. Viable organisms may persist in the host following resolution of uncomplicated histoplasmosis; they are the presumed source of disseminated disease in immunocompromised patients who do not have a history of recent exposure.

The major clinical syndromes associated with *H. capsulatum* infection are summarized in Table 30-2.

An estimated 500,000 persons in the United States are exposed to *H. capsulatum* each year; however, most persons infected with *H. capsulatum* have a high degree of natural resistance to the organism. Few, if any, overt symptoms appear, and the disease resolves rapidly. Approximately 1% of infections result in symptomatic disease, usually an acute self-limited flu-like illness with varying degrees of pulmonary involvement. Symptoms usually resolve without specific antifungal therapy. Occasionally, an overly vigorous host immune response can result in complications such as mediastinal fibrosis (development of hard fibrous tissue in the upper mediastinum, causing compression, distortion, or obliteration of the superior vena cava and sometimes constriction of the bronchi and large pulmonary vessels).

H. capsulatum can also cause progressive and potentially fatal disease when host defenses are impaired. In a small number of cases the initial infec-

Table 30-2   Classification of Histoplasmosis

Type of Infection	Specific Disorder	Comments
Histoplasmosis in normal hosts	Asymptomatic or mild flu-like illness	Occurs with normal exposure
	Acute pulmonary histoplasmosis	Occurs with heavy exposure
	Rare complications	Pericarditis, mediastinal fibrosis, etc.
Opportunistic infection	Disseminated histoplasmosis	Occurs in individuals who have an immune defect
	Chronic pulmonary histoplasmosis	Occurs in individuals who have a structural defect

tion is not cleared and there is progression to disseminated histoplasmosis, which is characterized by continued intracellular replication of *H. capsulatum* yeasts within macrophages, presumably due to a defect in cell-mediated immunity. Clinically, infection ranges from acute, life-threatening, disseminated histoplasmosis to chronic, mild, disseminated histoplasmosis, depending on the extent of parasitization of the mononuclear phagocytic system. Patients frequently complain of fever, night sweats, and weight loss. Mucosal lesions are also common and may be the primary clinical finding in an otherwise healthy-appearing individual.

Severe progressive disseminated histoplasmosis is reported with increasing frequency in adults who have hematologic malignancies, who are receiving immunosuppressive therapy (particularly chronic corticosteroid therapy), or who have acquired immunodeficiency syndrome (AIDS). In these settings disseminated histoplasmosis is best described as an opportunistic infection. Reports from non-endemic areas suggest that infection with the human immunodeficiency virus (HIV-1) may trigger reactivation of dormant *H. capsulatum* in patients previously exposed to histoplasmosis. Central nervous system involvement, an unusual complication of disseminated histoplasmosis, has also been reported in association with HIV infection.

Chronic pulmonary histoplasmosis is a disease most often seen in patients with underlying chronic obstructive pulmonary disease. As a result of structural defects in the lung, *H. capsulatum* can escape normal defense mechanisms and cause progressive, destructive lesions, similar to tuberculosis.

In Africa, a distinct clinical form of histoplasmosis is seen that involves primarily the bone and subcutaneous tissues.

Diagnosis

The diagnosis of histoplasmosis is based on serologic findings, direct histopathologic examination of infected tissue, and culture. Diagnosis of disseminated histoplasmosis requires demonstration of the organism in extrapulmonary sites. This is best accomplished by a combination of culture and histopathologic examination of tissue.

The antigenic reagents used in the serologic tests for histoplasmosis are derived from two sources, the cell-free culture filtrate from the **mycelial phase** of growth **(histoplasmin)** and inactivated whole **yeast phase** cells. Both reagents are used, since neither type of antigen detects antibodies in all cases.

In general, delayed skin test reactivity to histoplasmin develops within 2 weeks after exposure. This test is of little diagnostic or prognostic value, and in certain situations it may be misleading because of the immunogenic properties of histoplasmin. In a significant percentage of the hypersensitive population, serologic titers may become elevates as a result of skin testing with the reagent (anamnestic response). For this reason the skin test has no place in the diagnostic workup of a patient and should not be used.

Two serologic tests are frequently used to diagnose histoplasmosis. The complement fixation test is the standard test and is positive later in disease (6 weeks or longer after symptoms). This test, when properly performed, yields information most relevant to the diagnosis, prognosis, and management of the patient. Complement fixation tests, which measure antibodies directed against *H. capsulatum,* are performed using histoplasmin and intact formalin-treated yeast as the antigens. Serum complement fixation titers $\geq 1 : 16$ or a fourfold rise in titer are suggestive of histoplasmosis; however, false positive reactions can occur due to cross-reactive antibodies associated with other fungal infections and tuberculosis. Complement fixation titers decline following infection in normal hosts; fewer than 5% of individuals have a positive complement fixation in areas of high skin test positivity.

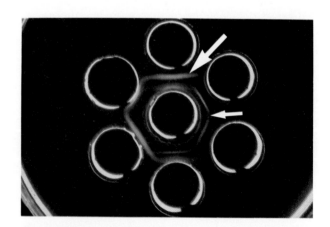

FIGURE 30-6   Immunodiffusion illustrating H (*short arrow*) and M (*long arrow*) precipitin bands that form when histoplasmin is tested against sera containing reactive antibodies.

A single serologic test does not allow a reliable diagnostic or prognostic interpretation and might cause a delay in instituting specific therapeutic measures. However, serologic tests on two or more serum specimens taken at suitable intervals of time during the acute and convalescent phases of infection yield information of great diagnostic and prognostic value.

The immunodiffusion test detects antibodies to **H** and **M antigens** of *H. capsulatum* (Fig. 30-6) and is more specific but less sensitive than complement fixation. Serologic tests can aid in the diagnosis of histoplasmosis but do not distinguish disseminated disease from other forms of histoplasmosis. Furthermore, these tests can be negative in 25% or more of patients who have disseminated histoplasmosis.

In contrast to traditional serologic tests, direct detection of *H. capsulatum* antigens in blood or urine may prove valuable for the rapid diagnosis of disseminated disease.

Intracellular yeast can often be seen by histopathologic examination of infected tissue, especially bone marrow, blood, and lung, using special stains, thus permitting rapid diagnosis. *H. capsulatum* can be cultured from bone marrow or blood in more than 75% of cases of disseminated histoplasmosis. Sputum cultures are useful in the diagnosis of chronic pulmonary histoplasmosis but are usually negative in cases of acute self-limited disease. *H. capsulatum* usually takes 1 to 2 weeks to grow in culture. Preliminary identification of the isolate is based on morphologic features, including delicate septate hyphae with tuberculate macroconidia (see Figure 30-4). In the past, confirmation was based on the ability to convert the culture from the mycelial to the yeast morphology, a process that can require weeks to months. Use of an **exoantigen test** now permits confirmation as soon as sufficient growth occurs. In the exoantigen test, antigens are extracted from the agar medium supporting growth of the fungus and reacted against antihistoplasma antibody in an immunodiffusion test. A positive identification is made when precipitin lines of identity form with control histoplasmin antigens.

Treatment

Amphotericin B remains the mainstay of treatment for disseminated histoplasmosis and other severe forms; however, in AIDS patients relapses following completion of therapy are a significant problem.

BLASTOMYCOSIS

Blastomycosis, also called **Chicago disease, Gilchrist's disease,** and **North American blastomycosis,** is caused by the inhalation of conidia of *B. dermatitidis* (Figure 30-7). The disease occurs mainly in North America and parts of Africa. Primary pulmonary infections are often inapparent and difficult to document even radiologically. The forms of disease

Table 30-3 Summary of Blastomycosis

Etiologic Agent	Mycology	Epidemiology and Ecology	Clinical Disease
Asexual phase: *Blastomyces dermatitidis* Sexual phase: *Ajellomyces dermatitidis*	Dimorphic; mycelia at 25° C; typical pyriform microconidia 2-4 μ in diameter; At 37° C and in tissue this organism is a yeast 8-15 μ in diameter; buds produced singly are attached to parent cell by a broad base.	Geographically delimited to North American continent and parts of Africa. The area of endemicity in the United States overlaps that for histoplasmosis. Blastomycosis is an important veterinary problem, and dogs develop a similar disease. There are a few reports of successful isolation of the organism from soil.	Primary infection in lung is often inapparent, although epidemics of respiratory blastomycosis have been documented. Chronic cutaneous and osseous disease are the most common clinical presentation.

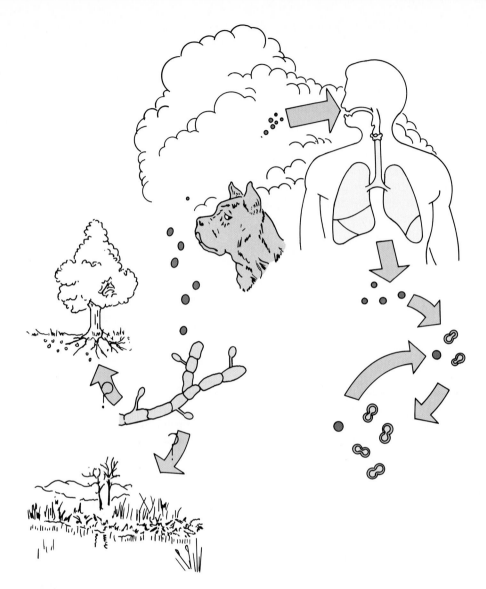

FIGURE 30-7   Schematic illustration of the natural history of the saprobic and parasitic cycle of *Blastomyces dermatitidis.*

most often seen clinically are ulcerative lesions of the skin and lytic bone lesions, both of which represent systemic or disseminated disease (Table 30-3).

Morphology

B. dermatitidis is closely related biochemically and serologically to *H. capsulatum.* The teleomorphic state of *B. dermatitidis* has been discovered, classified as an ascomycete, and designated *Ajellomyces*

dermatitidis, the same genus as the sexual state of *H. capsulatum.*

Epidemiology

The geographic distribution of blastomycosis is limited to the North American continent and parts of Africa. Blastomycosis, like histoplasmosis, is endemic in the Ohio and Mississippi valley region, and to a lesser extent in the Missouri and Arkansas River basins. Additional endemic sites have been found in Minnesota, Southern Manitoba, and Southwest Ontario, including the St. Lawrence River basin. Epidemics have occurred in Wisconsin, Minnesota, Illinois, and in the eastern states of Virginia and North Carolina. In endemic areas natural disease exists among dogs and horses and is an important veterinary problem. The veterinary picture is similar to the clinical and pathologic results seen in human infections. Untreated, blastomycosis has a rapid and fatal course. It has also been reported in a wide geographic area of Africa.

The natural reservoir for the agent of blastomycosis is not known. Unlike *H. capsulatum, B. dermatitidis* cannot be routinely cultured from soil in endemic areas. In a few instances the organism has been cultured from environmental sites that were associated with point source outbreaks. It is believed that the organism is present in the soil but flourishes only in a narrow, as yet undefined, ecologic niche.

Clinical Syndromes

The natural history of *B. dermatitidis* infections is less well documented than are infections caused by *H. capsulatum,* because of the lack of reliable serologic tests and characteristic radiographic findings. Symptomatic blastomycosis is an uncommon disease whose manifestations frequently indicate systemic spread. It is thought that large numbers of asymptomatic infections occur, analogous to histoplasmosis.

Inhalation of *B. dermatitidis* conidia produces a primary pulmonary infection in the nonimmune host. As with *H. capsulatum, B. dermatitidis* conidia convert to yeast and are phagocytized by macrophages, which may carry them to other organs. The initial infection can be symptomatic or asymptomatic; because of the lack of reliable tests the true incidence of asymptomatic infections is unknown. Chest x-ray films may show nonspecific pulmonary infiltrates; however, unlike histoplasmosis, resolution of these lesions is not accompanied by calcifications. Primary pulmonary disease can have three outcomes: resolution without involvement of other organs, resolution of the pulmonary infection followed by systemic disease, or progressive pulmonary disease.

Diagnosis

Serologic and immunologic findings for blastomycosis are unclear. At present, two antigenic prepara-

tions are used in tests to detect the immune response to infection by *B. dermatitidis,* cell-free culture filtrate of the **mycelial phase (blastomycin)** and inactivated whole **yeast phase** cells. The data obtained from skin testing and serologic studies are difficult to interpret because of the poorly defined antigenic preparations that were used. The reagents tend to have a high degree of cross reactivity with other mycoses, particularly histoplasmosis and coccidioidomycosis.

An immunodiffusion test that appears to be specific for blastomycosis has been developed. It is based on the availability of a yeast-phase culture filtrate, shown to possess a high degree of sensitivity and specificity, that contains a specific antigen designated as "A". Suitable control sera containing antibodies that react with "A" antigen must be included in immunodiffusion studies with patient sera.

Diagnosis of blastomycosis requires identification of the organism in infected tissue or isolation in culture. Microscopic examination of potassium hydroxide–treated purulent fluid expressed from abscesses reveals characteristic broad-based budding yeast cells (Figure 30-8). Biopsies of stained skin lesions will reveal characteristic broad-based budding yeast. The organism grows readily in culture. Identification is based on conversion to the yeast or an exoantigen test.

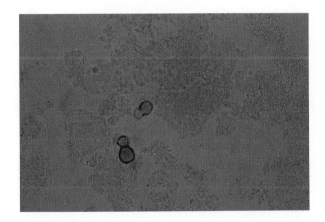

FIGURE 30-8   Broad-based budding yeast cells of *Blastomyces dermatitidis* seen in purulent material expressed from a microabcess.

Treatment

Amphotericin B remains the mainstay of therapy for patients with systemic disease or serious pulmonary disease. Uncomplicated pulmonary disease may respond to ketoconazole.

PARACOCCIDIOIDOMYCOSIS

Paracoccidioidomycosis is a pulmonary disease resulting from the inhalation of infectious conidia of *P. brasiliensis* (Figure 30-9). Pulmonary infections are often asymptomatic; the form most often seen is ulcerative lesions of the oral and nasal cavity. The disease has been called **South American blastomycosis** and **Lutz-Splendore-Almeida's disease** (Table 30-4).

Morphology

P. brasiliensis is dimorphic, growing as a mold in the environment and as budding yeast in infected tissue. The yeast phase is characterized by multiple budding from a single cell (Figure 30-10). The transition from mold to yeast can be induced in vitro by raising the temperature from 25° C to 37° C. Recent studies have shown that as little as 10^{-10} M 17-beta-estradiol significantly inhibits the transformation of mycelia to yeast at the permissive temperature of 37° C. Testosterone, corticosterone, and 17-alpha-estradiol had no inhibitory effect on the mycelia to yeast transition at the elevated temperature. The findings may have clinical significance (see the following discussion on epidemiology). A sexual state for *P. brasiliensis* has not been described.

FIGURE 30-9   Schematic illustration of the natural history of the saprobic and parasitic cycle of *Paracoccidioides brasiliensis.*

Table 30-4 Summary of Paracoccidioidomycosis

Etiologic Agent	Mycology	Epidemiology and Ecology	Clinical Disease
Asexual phase: *Paracocciodes brasiliensis* Sexual phase: Not known	Dimorphic; mycelia at 25° C; no typical pattern of sporulation. At 37° C and in tissues the organism is a yeast with several budding cells attached to the parent cell, some in a "pilot's wheel" arrangement. Yeasts are 2-30 μ in diameter.	This disease is geographically limited to Central and South America. The major focus of the disease is Brazil. Females are as susceptible to infections as males, but the incidence of clinical disease is about 9 times higher in males. The organism has been isolated from the soil on rare occasions.	Primary pulmonary disease is often inapparent. Disseminated disease often causes ulcerative lesions of the buccal, nasal, and occasionally gastrointestinal mucosa.

Epidemiology

This disease is geographically restricted to Central and South America and has a high incidence in Brazil, Venezuela, and Colombia. The endemic areas have been delineated by data taken from extensive skin test surveys and case reports. Unlike *H. capsulatum* and *C. immitis, P. brasiliensis* cannot routinely be cultured from soil in endemic areas. Careful retrospective epidemiologic studies and data taken from a report on the isolation of *P. brasiliensis* from soil samples suggest that the fungus resides in environments that have high humidity and average temperatures around 23° C.

Results of skin test surveys indicate an equal distribution of reactors among males and females. However, when data from clinically significant disease are analyzed, there is a disproportionate number of males affected (approximately 90% males compared with 10% females). This difference has been attributed to factors that place males at higher risk, as well as underlying diseases, malnourishment, and hormonal differences. Inhibition of the mycelia to yeast transition by beta-estradiol may also account for the lower incidence of disseminated paracoccidioidomycosis in adult females.

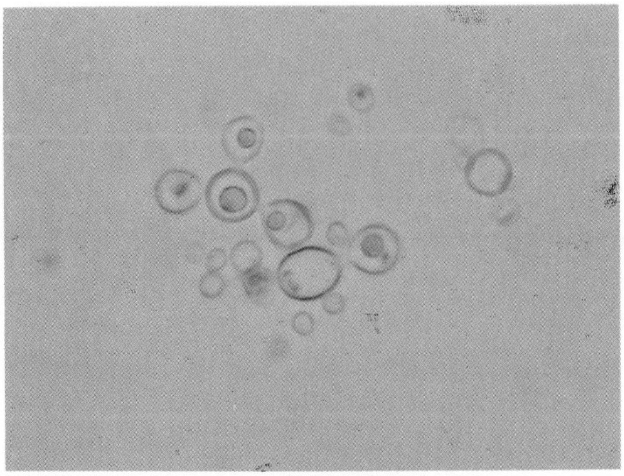

Clinical Syndromes

Because of the prominence of oral and nasal lesions, infection was believed to result from local inoculation; however, it is now known that paracoccidioidomycosis resembles the other systemic mycoses in that the primary infection occurs in the lungs as a result of inhaling conidia. Primary pulmonary paracoccidioidomycosis is frequently asymptomatic but can develop into progressive pulmonary disease or disseminated disease.

Diagnosis

As with histoplasmosis and coccidioidomycosis, diagnosis is based on detection of specific antibod-

FIGURE 30-10 Multipolar budding characteristic of the yeast phase of *Paracoccidioides brasiliensis.*

ies, visualization of the organism in histopathologic material, and isolation of the organism in culture. Two antigenic preparations of *P. brasiliensis,* a cell-free culture filtrate of the **mycelial phase (paracoccidioidin)** and an inactivated whole **yeast phase** preparation, are used for the serologic diagnosis (these serologic reagents are not routinely available in the United States). Specific antibodies are measured by complement fixation and immunodiffusion.

The organism can be seen in KOH preparations of infected material or in silver-stained histologic sections. The presence of multiple small budding cells arranged around a large mature cell ("ship's wheel" pattern) is diagnostic of *P. brasiliensis* (see Figure 30-10); however, in the absence of buds the yeast of *P.*

brasiliensis may be confused with immature spherules of *C. immitis* or the nonbudding yeast of *B. dermatitidis* or *H. capsulatum.* Clinical material cultured on medium and incubated at 25° C will yield a slow-growing white mold after 10 to 14 days of incubation. Microscopic examination of the growth is usually not too revealing, since *P. brasiliensis* does not sporulate readily. Transfer of the culture to incubation at 37° C will yield a yeastlike growth with characteristic multipolar budding cells.

Treatment

Successful treatment of paracoccidioidomycosis generally requires long-term therapy. Amphotericin B

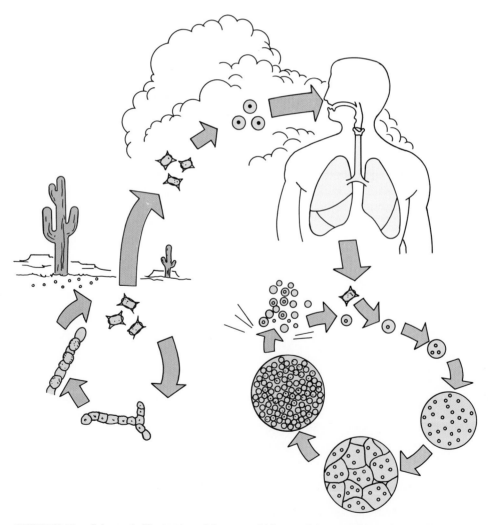

FIGURE 30-11   Schematic illustration of the natural history of the saprobic and parasitic cycle of *Coccidioides immitis.*

is effective against all forms of paracoccidioidomycosis, but because of toxicity and the difficulty of longterm administration it is usually reserved for severe cases. Sulfa drugs have been used for decades to treat paracoccidioidomycosis; however, treatment failures can occur despite therapy that lasts years. Ketoconazole (and the newer azoles) are very active in vitro against *P. brasiliensis* and appear to be clinically effective.

COCCIDIOIDOMYCOSIS

Inhalation of arthroconidia of *C. immitis* causes an acute, self-limited and usually benign respiratory infection that may be asymptomatic or vary in severity from the level of a common cold to disseminated and life-threatening disease (Figure 30-11). The infection is limited to North, Central, and South America. Coccidioidomycosis has also been called **Posada's disease, San Joaquin Valley fever,** and **desert rheumatism** (Table 30-5).

Morphology

C. immitis is a dimorphic fungus that grows as a filamentous mold in the environment. The mycelia of *C. immitis* fragment to produce cylindric arthroconidia (Figure 30-12). The tissue phase of *C. immitis* is the spherule (Figure 30-13), a multinucleated structure that undergoes internal cleavage to produce uninucleate endospores that can then generate new spherules, in contrast to the blastoconidia (produced by budding) seen in *H. capsulatum, B. dermatitidis,* and *P. brasiliensis.* A sexual state has not been observed for *C. immitis.*

Epidemiology

Coccidioidomycosis which can be described as a disease of the "New World," is geographically limited to the North, Central, and South American continents (Figure 30-14). The areas of highest endemicity include the central San Joaquin Valley in California, Maricopa and Pima Counties in Arizona, and several western and southwestern counties in Texas, regions that have a semi-arid climate. The disease is also endemic in the northern states of Mexico and parts of Venezuela, Paraguay, and Argentina. Cases of coccidioidomycosis have also been reported in Central America. The organism can be routinely isolated from soil in areas of high endemicity. Although geographically restricted, the organism has on occasion spread extensively as a result of large dust storms.

Table 30-5 Summary of Coccidioidomycosis

Etiologic Agent	Mycology	Epidemiology and Ecology	Clinical Disease
Asexual phase: *Coccidioides immitis* Sexual phase: Not known	Dimorphic; mycelia at 25° C; as the culture ages, the septate hyphae matures in a manner such that alternate cells develop into arthroconidia being separated by vacuolized cells. The arthroconidia separate readily and have a "barrel-shaped" appearance. In tissue and at 37° C the organism develops into large spherical structures 10-60 μ in diameter, called spherules (sporangia) that are filled with endospores, 2-5 μ in diameter.	This disease is geographically restricted to North, Central, and South America, where there are areas of high endemicity. In North America the disease is highly endemic in the San Joaquin Valley of California, the Southwestern part of the United States and the northern states of Mexico. Natural infection occurs in domestic and wild animals. The organism has been repeatedly isolated from soil samples taken from the endemic area.	Approximately 60% of these infections are asymptomatic. The most common symptoms of primary disease are cough, fever, and chest pain. Night-sweats and joint pain are not unusual. An epidemiologic history should be taken to find out whether the patient has been in an endemic area.

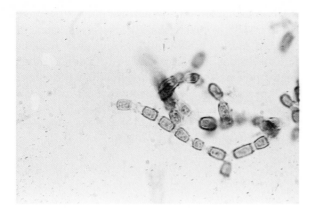

FIGURE 30-12   Hyphae and arthroconidia of *Coccidioides immitis.*

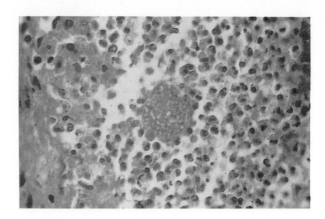

FIGURE 30-13   The spherule phase of *Coccidioides immitis* as seen in tissue section stained.

Clinical Syndromes

The natural history of coccidioidomycosis has been well characterized because large groups of non-immune individuals who have migrated to endemic areas could be studied (e.g., military personnel stationed in the San Joaquin valley during World War II). In contrast to histoplasmosis, exposure to *C. immitis* causes a greater percentage of individuals to undergo a mild febrile to moderately severe respiratory disease. In general, however, a high degree of innate immunity exists in the adult population. Approximately 40% of individuals develop a symptomatic pulmonary infection following exposure to the organism. These primary infections are usually self-limited, but in a small number of cases *C. immitis* cause progressive pulmonary disease or disseminate to produce extrapulmonary disease, mainly involving the meninges and/or skin.

Diagnosis

At present, two sources of antigen, both cell-free culture filtrates, are used in the preparation of serologic reagents: the **mycelial phase** of growth **(coccidiodin)** and the **spherule phase (spherulin).** Skin test reactivity to coccidioidin develops 2 to 4 weeks after symptoms. The tube precipitin and the complement fixation tests are the time-honored serologic procedures used to diagnose coccidioidomycosis. Precipitins appear early, between 2 to 4 weeks after symptoms, followed by the appearance of complement-fixing antibodies.

Other methods of detecting specific antibodies, such as latex particle agglutination and agar immunodiffusion, are now available. These tests have largely replaced the precipitin and complement fixation tests for routine screening because they are more sensitive in detecting infected individuals, are commercially available, and are easily performed. In the immunodiffusion test, two lines of precipitation appear to be significant. One line is associated with the antigen that detects precipitin and agglutinating antibodies, and another line is associated with the complement-fixing antibodies. Once the presence of an infection has been established, complement fixation titers can yield important prognostic information. High titers, or persistent or rising titers, indicate a high probability of disseminated disease. *C. immitis* spherules can be seen on infected tissue stained with hematoxylin and eosin. The organism can be cultured on conventional media; however, it must be handled with caution because *C. immitis* is a leading cause of laboratory-acquired infections. The mold phase of *C. immitis* resembles several saprobic fungi; therefore definitive identification is based on conversion to spherules or a specific exoantigen test.

Treatment

Amphotericin B is the drug of choice for the treatment of serious coccidioidal infections. Meningeal infections are particularly difficult to treat, partly because of the poor penetration of amphotericin B into the cerebrospinal fluid. Ketoconazole is effective in suppressing infections, but relapses occur following cessation of therapy.

CRYPTOCOCCOSIS

Cryptococcosis, also called **Busse-Buschke's disease, torulosis,** or **European blastomycosis,** is a chronic to acute infection caused by *C. neoformans* (Table 30-6). The lung is the primary site of infection; however, the organism has a high predilection for systemic spread to the central nervous system. (Figure 30-15). *C. neoformans* is the leading cause of fungal meningitis and is an important cause of morbidity and mortality in transplant recipients and AIDS patients. *C. neoformans* also produces systemic disease in patients who have no apparent underlying immunologic disorder.

Morphology

Unlike the other systemic mycotic agents, *C. neoformans* is not a dimorphic organism. It grows as a budding yeast in infected tissue and in culture at 25° C and 37° C. The infectious particle has not been identified. The most distinctive feature of *C. neoformans* is the presence of an acidic mucopolysaccharide capsule, which is required for pathogenicity and is important for diagnosis, both in terms of antigen detection and specific histologic staining. The teleomorphic state of *C. neoformans* has been discovered; it is named *Filobasidiella neoformans* and has been classified as a basidiomycete.

FIGURE 30-14   Geographic distribution of coccidioidomycosis in North, Central, and South America (shaded).

Table 30-6   Summary of Cryptococcosis

Etiologic Agent	Mycology	Epidemiology and Ecology	Clinical Disease

Asexual phase:
  Cryptococcus neoformans
Sexual phase:
  Filobasidiella neoformans
	Monomorphic; this organism is a yeast at 25° C, and 37° C. The unique feature of the yeast is the acidic mucopolysaccharide capsule.	This disease is worldwide in distribution. This yeast has been repeatedly isolated from sites inhabited by pigeons, particularly their roosts and droppings. Pigeons are not naturally infected.	Primary pulmonary cryptococcosis is usually inapparent but may be chronic, subacute, or acute. The clinical entity most often seen is cryptococcal meningitis. Osseous and cutaneous disease can be present without apparent neurologic involvement.

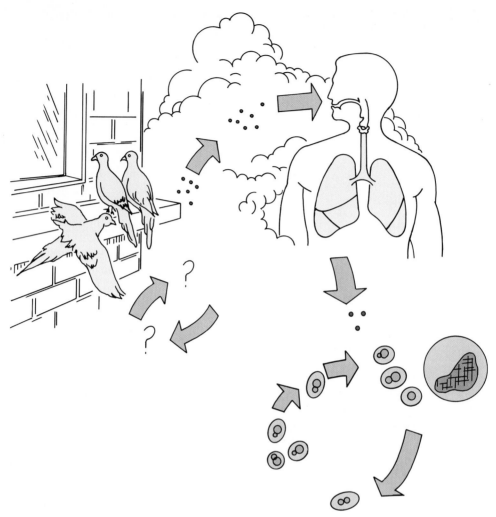

FIGURE 30-15   Schematic illustration of natural history of the saprobic and parasitic cycle of *Cryptococcus neoformans.*

Epidemiology

The etiologic agent of cryptococcosis has been recovered in large numbers from the excreta and debris of pigeon roosts, thus it appears to survive well in a desiccated, alkaline, nitrogen-rich, and hypertonic environment. There is a close relationship to the habitats of pigeons, but the organism does not infect the bird. Cryptococcosis occurs throughout the world. The true prevalence of infection is unknown, due to the lack of a reliable skin test or other serologic screening test, but subclinical infections are believed to be common. Symptomatic cryptococcal disease, mainly meningitis, is frequently seen in individuals who are debilitated, immunosuppressed, or otherwise compromised. However, a large number of patients who develop cryptococcal meningitis have no underlying immune or metabolic defects.

Clinical Syndromes

Primary pulmonary infections are frequently asymptomatic and may be detected as an incidental finding on a routine chest x-ray examination. The most common picture is that of a solitary pulmonary nodule that can mimic a carcinoma; the correct diagnosis is usually made when the mass is resected. *C.*

neoformans can also produce a symptomatic pneumonia characterized by diffuse pulmonary infiltrates.

Cryptococcal meningitis, caused by hematogenous spread of yeast from the lungs to the meninges
surrounding the brain, is the most frequently diagnosed form of cryptococcosis. Symptoms usually
include combinations of headache, mental status
changes, and fever lasting several weeks. Occasionally, cryptococcal disease of the central nervous system may take the form of an expanding intracerebral
mass that causes focal neurologic deficits. Other
common manifestations of disseminated cryptococcosis include skin lesions and osteolytic bone
lesions.

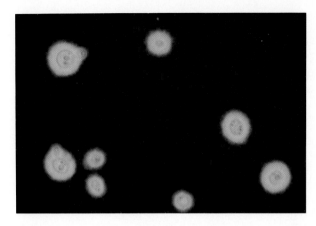

FIGURE 30-16   Encapsulated budding yeast cells of *Cryptococcus neoformans* highlighted with India Ink.

Diagnosis

In contrast to the other systemic mycoses, the
serologic procedures used in the diagnosis of cryptococcosis are based on the detection of antigen, not
antibody. The latex agglutination test for detection of
cryptococcal polysaccharide antigen in cerebrospinal
fluid and serum is now routinely used in clinical laboratories; it is sensitive, specific, and simple to perform. The test involves the use of latex particles that
have been coated with rabbit anticryptococcal antibody. Capsular polysaccharide present in a clinical
specimen binds to the antibodies and agglutinates
the latex particles. Since the latex particles are coated
with antibody, false positive reactions can be caused
by rheumatoid factor (IgM antibodies that bind IgG).
Because of this problem, appropriate control tests
must be performed in which the clinical specimen is
mixed with latex particles coated with nonimmune
rabbit antibodies. In addition, the patient's serum
can be treated with a protease to destroy any proteins
that might cause nonspecific agglutination of the
latex particles. Cross reactions occur with sera from
patients with disseminated infections caused by
Trichosporon beiglii. To evelute the clinical progress
of a patient during therapy, serial samples of cerebrospinal fluid are evaluated for the presence of cryptococcal antigen. A favorable prognosis is indicated by
a decrease in the titer of antigen.

A rapid diagnosis of cryptococcal meningitis can
often be made by examination of an India ink preparation of cerebrospinal fluid. *C. neoformans* appears
as a single cell or budding yeast surrounded by a
white halo due to exclusion of the ink particles by the
polysaccharide capsule (Figure 30-16). Although
diagnosis may be rapid, the India ink preparation is
positive in only half of the cases of cryptococcal meningitis. Culture remains the definitive method for documenting infection. The organism grows well on
standard mycologic media but is inhibited by cycloheximide, which is added to plates used for culturing
dermatophytes. The identification of *C. neoformans* is
based on the presence of a capsule, production of the
enzyme urease, and other specific biochemical reactions.

Treatment

Whereas pulmonary cryptococcosis is frequently a
self-limited infection or can be cured by surgical excision of a solitary nodule, disseminated cryptococcosis
is almost always fatal if untreated. Cryptococcal meningitis remains the model for combination therapy
with antifungal agents. Amphotericin B is active
against *C. neoformans* but exhibits relatively poor
penetration into cerebrospinal fluid; 5-fluorocytosine
has good cerebrospinal fluid penetration, but development of resistant cryptococci is a problem. Controlled clinical trials have shown that combination
therapy with amphotericin B and 5-fluorocytosine for
6 weeks is as effective as amphotericin B alone for 10
weeks. Nonetheless, relapses following either treatment regimen remain a problem. Amphotericin B is
also indicated for the treatment of other forms of disseminated cryptococcosis.

BIBLIOGRAPHY

Drutz DJ, editor: Systemic fungal infections: diagnosis and treatment. I. Med Clin North Am, vol 3, 1988.

Drutz DJ, editor: Systemic fungal infections: diagnosis and treatment. II. Med Clin North Am, vol 4, 1989.

Rippon JW: Medical mycology: the pathogenic fungi and the pathogenic actinomycetes, Philadelphia, 1989, WB Saunders Co.

Sarosi GA and Davies SF, editors: Fungal diseases of the lung, Orlando, Florida, 1986, Grune & Stratton, Inc.

Szaniszlo PJ, editor: Fungal dimorphism: with emphasis on fungi pathogenic for humans, New York, 1985, Plenum Press.

31 Opportunistic Mycoses

In recent years there have been marked increases in the number of serious infections caused by fungi that were traditionally regarded as not being pathogenic. The infections caused by these organisms occur in patients with impaired host defenses secondary to underlying diseases, malignancy, AIDS, or who have had immunosuppressive therapy, breaches of normal barriers, or alterations of the normal flora. Because these fungi take advantage of the host's debilitated condition to become pathogens, they are commonly called **opportunistic mycoses.** If the infection is not diagnosed rapidly and treated aggressively—correcting the underlying immunodeficiency or debilitation as well—it usually becomes fatal.

Although these organisms are "opportunistic pathogens," this term is based on the clinical setting and does not refer to any taxonomic category. The two most common opportunistic pathogens, *Candida albicans* and *Aspergillus fumigatus,* have very different biologic properties and host interactions. In addition, pathogenicity is not an all-or-nothing phenomenon. Individual species within both *Candida* and *Aspergillus* exhibit a range of pathogenicity.

CANDIDIASIS

Of the various species in the genus *Candida* that have been implicated in **candidiasis,** *C. albicans* is the most common. The clinical spectrum of manifestations ranges from superficial infections of the skin to systemic disease. *C. albicans, C. tropicalis, C. glabrata,* and *C. parapsilosis* are often found as part of the normal flora of humans and can be isolated from healthy mucosal surfaces of the oral cavity, the vagi-

na, the gastrointestinal tract, and the rectal area. As many as 80% of normal individuals may show colonization of these sites in the absence of disease. In contrast, *C. albicans* is rarely isolated from the surface of normal human skin and only sporadically from certain intertriginous areas (skin surfaces that appose each other, as in the groin). Under certain circumstances these organisms may gain hematogenous access from the oropharynx or gastrointestinal tract when the mucosal barrier function is breached (eg., inflammation of mucous membranes secondary to chemotherapy) or through contaminated intravenous catheters and syringes. The organs most often involved include the lungs, spleen, kidney, liver, heart, and brain. In the eye, *Candida* may produce an endophthalmitis (inflammation involving the ocular cavities), which indicates that there is seeding of *Candida albicans* to multiple organs. Skin lesions may occur in 10% to 30% of patients who have disseminated infection. The recognition of such lesions may be important in early diagnosis, since antemortem blood cultures are negative in a high percentage of patients with autopsy-proven systemic candidiasis.

Morphology

With one exception, a striking morphologic feature of the yeasts in the genus *Candida* is that they multiply by blastospore formation, producing either pseudohyphae or septate hyphae. *Candida glabrata,* a yeast implicated in urinary tract infections and systemic disease in debilitated individuals, is the exception, producing only yeast cells. *Candida* are identified and speciated according to their ability to assimilate and ferment various carbohydrates, as well as by certain physiologic and morphologic responses they exhibit when grown under controlled nutritional conditions (see discussion of diagnosis).

Clinical Syndromes

The spectrum of infection caused by organisms in the genus *Candida* include the following: localized diseases of the skin and nails; diseases that affect the mucosal surface of the mouth, vagina, esophagus, and bronchial tree; and diseases that disseminate and involve multiple organ systems. The diseases involving skin and nails frequently mimic those seen with the dermatophytes. In all cases that diagnosis must be supported by observing fungi in specimens taken from the lesions and confirmed by culture of the organisms from clinical material.

Chronic mucocutaneous candidiasis (CMC; see discussion of immunity and host factors) is a heterogenous group of clinical syndromes characterized by chronic, treatment-resistant superficial *Candida* infections of the skin, nails, and oropharynx. Despite the extensive cutaneous involvement there is virtually no propensity for disseminated visceral candidiasis to occur. In many cases there are narrow but specific abnormalities in T cell-mediated immunity. In others the defects are more general. Various underlying conditions such as hypoparathyroidism, hypoadrenalism, hypothyroidism, and the presence of circulating autoimmune antibodies have been associated with CMC. In adults CMC is often associated with a thymoma. When cutaneous candidiasis indicates the possibility of an immunologic or endocrine disorder, efforts must be made to uncover and correct the underlying defect so that the fungal infection can be properly treated.

Disseminated candidiasis is usually spread through the bloodstream and therefore involves many organs. The incidence of this form of candidiasis is steadily rising as more patients with serious hematologic malignancies are treated aggressively with potent immunosuppressive drugs and as more patients undergo bone marrow and organ transplants. Disseminated candidiasis continues to be a major problem in immunocompromised hosts, such as in AIDS patients. Other factors leading to a predisposition to disseminated *Candida* infection are shown in Box 31-1.

Immunity and Host Factors

Innate immunity to these organisms in the normal adult human appears to be very good. The immune mechanisms responsible for protection against *Candida* infections include both humoral and cell-mediated processes. Cell-mediated processes are considered more important, in this regard, based on experience with *CMC*, where a defect in cell-mediated

---

Box 31-1 Predisposing Factors for Disseminated *Candida* Infections

Burns
Local occlusion
Obesity
Malnutrition
Extremes of age
Diabetes mellitus
Other endocrinopathies (hypoadrenalism, hypothyroidism, hypoparathyroidism)
Malignancy (especially hematologic, thymoma)
Intrinsic immunodeficiency states (chronic granulomatous disease, chronic mucocutaneous candidiasis, AIDS)
Indwelling catheters
Intravenous drug abuse
X-irradiation
Corticosteroids and other immunosuppressive agents
Antibacterial antibiotics (especially broad-spectrum)

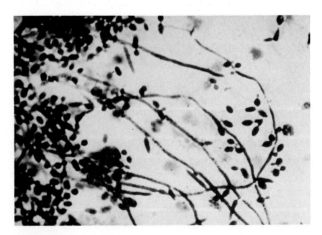

FIGURE 31-1 Budding yeast cells and pseudohyphae of *Candida*.

immunity leads to extensive superficial candidiasis despite normal, or even exaggerated, humoral defenses.

Production of serum antibodies to the principal cell wall glycoprotein antigens of *Candida* occurs in low titers in normal individuals. The protective role of these antibodies, however, is controversial, and their presence may only reflect an immunologic response to colonization of the gastrointestinal tract early in life. Clinically patients with primarily B-cell deficiency states are not at high risk for infections by *Candida*.

In addition to antibodies to *Candida*, several other naturally occurring anti-*Candida* serum factors have been described. The role these factors play in the prevention of or susceptibility to disease is unknown. Probably, various innate nonimmune host factors in conjunction with cell-mediated immunity and complement activation contribute more to host defense against infections caused by *Candida* than does humoral immunity. At present, the serodiagnosis of candidiasis is not routinely performed.

Diagnosis

The organisms responsible for these infections include *C. albicans, C. tropicalis, C. pseudotropicalis, C. krusei,* and *C. parapsilosis*. In histopathologic sections these organisms may produce budding yeasts, pseudohyphae, and septate hyphae (Figure 31-1). On culture the organisms produce yeast and pseudohyphal cells, and the gross appearance of the colony is opaque and cream colored with a pasty consistency. All yeasts cultured from clinical material should be identified as to species. Several procedures are available, most of which combine morphologic, physiologic, and biochemical tests. A rapid and reliable test to identify *C. albicans* is the **germ tube test.** Most isolates of *C. albicans* produce hyphal outgrowths from blastospores when they are suspended in serum and incubated at 37° C (Figure 31-2). For the test to be valid, the incubated suspension must be examined after 2 to 3 hours of incubation, since other species may form outgrowths with longer incubation. a few isolates of *C. albicans* do not form germ tubes under these conditions. Therefore other tests are performed based primarily on the physiologic properties of the organisms, such as their ability to assimilate various sugars or to produce certain morphologic structures, such as chlamydospores, under certain growth conditions (Figure 31-3).

Treatment

Except for diseases of the nails, clinical responses generally result when proper therapy for candidial diseases is instituted. Therapy will vary depending on organ involvement. Topical therapeutic approaches are usually instituted for cutaneous and mucocutaneous disease. For systemic disease, am-

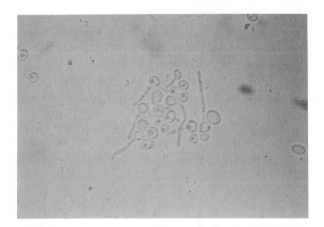

FIGURE 31-2   Development of germ tubes by *Candida albicans* yeast cells after incubation in serum for 2 hours at 37° C.

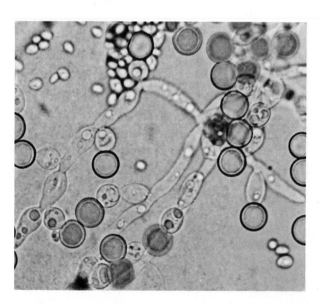

FIGURE 31-3   Formation of chlamydospores by *Candida albicans* when cultured on cornmeal agar at 25° C.

Table 31-1 Diseases That Are Associated With *Aspergillus*

Disease	Etiologic Factors
Mycotoxicoses	Ingestion of contaminated food products
Hypersensitivity pneumonitis	Allergic bronchopulmonary disease
Secondary colonization	Colonization of preexisting cavity (e.g., pulmonary abscess) without invasion into contiguous tissues
Systemic disease	Invasive disease involving multiple organs

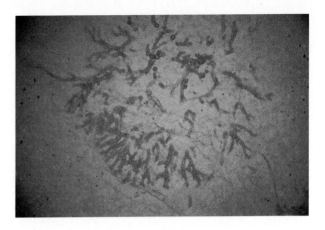

FIGURE 31-4   Dichotomous branching septate hyphae of *Aspergillus fumigatus* in tissue specimen.

photericin B alone or combined with 5-fluorocytosine may be indicated. Sometimes lesions of the oral, vaginal, or perianal areas spread to involve the adjacent skin and become chronic and resistant to therapy. Such situations often indicate an immunologic disorder (e.g., such as occurs in CMC). For complete cures, the immunologic disorder must be corrected.

ASPERGILLOSIS

The spectrum of medical problems caused by various species of *Aspergillus* is broad (Table 31-1).

Morphology

These organisms are identified according to the pattern of conidiophore development, morphologic features, and color of the conidia formed (Figure 31-4). Species of *Aspergillus* are extremely common in the environment, and several have been implicated as etiologic agents. Of the approximately 900 described, *A. fumigatus* and *A. flavus* have been most frequently associated with invasive disease.

Clinical syndromes

The normal healthy individual is not susceptible to aspergillosis. It is purely an opportunistic infection. As with candidiasis, the type of disease evoked depends on the local or general physiologic and immunologic state of the host. Factors that lead to host debilitation are also important in aspergillosis. In contrast to candidiasis, the etiologic agents implicated in aspergillosis are ubiquitous in the environment and not part of the normal flora of humans. These agents are involved in many animal diseases, such as mycotic abortion of sheep and cattle and pulmonary infections of birds, and serve as carcinogenic agents in animals that have ingested contaminated feed. Because of their diverse involvement in both human and animal disease, they pose a great economic problem.

Allergic aspergillosis may be benign early on and severe as the patient grows older. With aging, respiratory distress increases, leading to bronchiectasis (chronic dilation of the air passages), with collapse of a segment of the lung eventually resulting in fibrosis.

In **secondary colonization,** a chronic clinical situation may exist with little distress except occasional bouts of hemoptysis (coughing up blood) and pathologic changes in the lung that lead to the formation of a **"fungus ball"** (Figure 31-5). Histopathologically, the fungus ball is a spherical mass of intertwined septate branching hyphal elements. These structures can also be visualized radiologically as space-occupying spherical structures that move within the cavity as the patient's position is changed.

Systemic aspergillosis is an extremely serious disorder that is usually rapidly fatal unless diagnosed early and treated aggressively. As in disseminated candidiasis, the physiologic and immunologic conditions that contribute to the host's increased susceptibility to the infection must also be reversed for proper management.

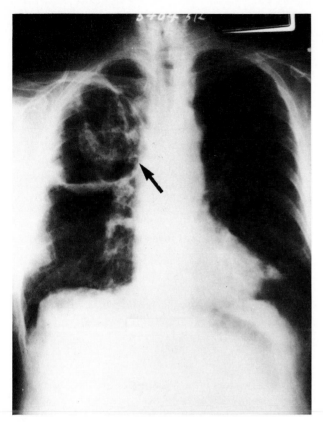

FIGURE 31-5   Chest film illustrating cavity in the upper right lobe with an organizing fungus ball (*arrow*).

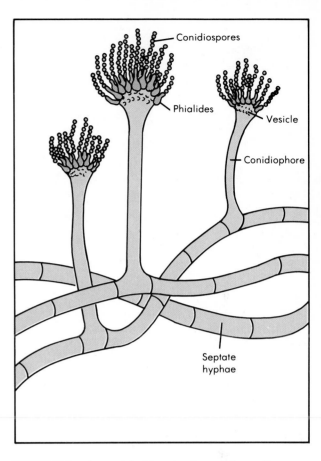

FIGURE 31-6   Asexual fruiting structure of *Aspergillus* species, illustrating septate hyphae, conidiophore, vesicle, phialides, and conidiospores.

Diagnosis

The diagnosis of invasive aspergillosis is considered when septate hyphae that branch at regular intervals and tend to be oriented in the same direction are seen in clinical specimens (see Figure 31-4). Confirmation of invasive aspergillosis is sometimes difficult, since cultures are not always performed or are often negative. Because these organisms exist everywhere in the environment, the clinical and histopathologic diagnosis of invasive aspergillosis can be confirmed only by repeated isolation of the organism in culture.

Species identification of *Aspergillus* rests on certain microscopic morphologic features, mainly the shape of the conidiophore, the vesicle, phialides, and conidiospores (Figure 31-6). Pigmentation of aerial growth of the colony is also used as an identifying feature of the organism.

ZYGOMYCOSIS (MUCORMYCOSIS, PHYCOMYCOSIS)

Morphology

Fungi causing the zygomycosis grow rapidly on all laboratory media not containing cycloheximide. *Rhizopus, Absidia,* and *Mucor* have all been implicated in zygomycosis. They form coenocytic hyphae and reproduce asexually by producing sporangiophores within which develop sporangiospores (Figure 31-7). The organisms are ubiquitous in the environment and are frequently encountered as contaminants. It may be difficult to establish the relevance of the isolate if coenocytic hyphal elements are not seen in histopathologic section. Repeated isolation of the organism from consecutive specimens provides strong evidence that the organism may be relevant, even

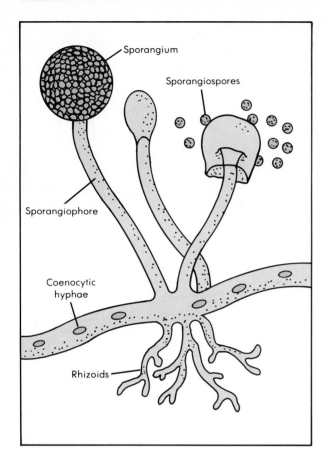

FIGURE 31-7   Asexual fruiting structure of *Rhizopus* species, illustrating sporangium, sporangiophore, sporangiospores, coenocytic hyphae and rhizoids.

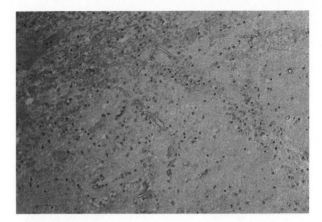

FIGURE 31-8   Histologic section from a case of zygomycosis, illustrating coenocytic hyphae.

struction of blood vessels) and necrosis of adjacent tissues.

Diagnosis

These organisms are filamentous, and their distinct morphology helps identify them in microscopic examination of pathologic material. The hyphal filaments are coenocytic (i.e., not separated by cross walls) and have a ribbonlike appearance in tissue specimens (Figure 31-8). Cultures grown on medium not containing antibiotics are dense and have a "hairy" appearance. Microscopic examination of this growth confirms the coenocytic nature of the hyphae and sporulating species produce characteristic sporangia (asexual fruiting bodies; Figure 31-7) that contain sporangiospores. The most frequently encountered agents of human disease are *Rhizopus arrhizus* and *Absidia corymbifera*. It is often difficult to assess reports on the number and variety of species that may cause disease, since confirmation by culture is frequently lacking, and therefore the diagnosis is often based on histopathologic examination of tissue.

though coenocytic hyphal elements are not seen in histopathologic examination of tissue.

Clinical Syndromes

Various diseases are caused by organisms belonging to the order Mucorales of the phylum Zygomycota. Rhinocerebral zygomycosis is the most common and usually occurs as a terminal event in acidotic patients who have uncontrolled diabetes mellitus. Other forms of zygomycosis seen in immunosuppressed or otherwise debilitated patients involve the lungs, the gastrointestinal tract, and subcutaneous tissues. In severely burned patients these organisms colonize the damaged tissues and tend to become invasive. In disseminated disease the organism shows a marked predilection for invading major blood vessels. The emboli that result cause ischemia (ob-

BIBLIOGRAPHY

Musial CE, Cockerill FR III, and Roberts GD: Fungal infections of the immunocompromised host: clinical and laboratory aspects, Clin Microbial Rev 1:349-364, 1988.

Odds FC: *Candida* and candidosis, London, England, 1988, Bailliere Tindall.

Rinaldi MG: Invasive aspergillosis, Rev Infect Dis, 5:1061-1077, 1983.

PARASITOLOGY

PART

III

32

Introduction to Medical Parasitology

Medical parasitology is the area of microbiology that studies the invertebrate animals capable of producing disease in humans and a variety of other animals. These invertebrate animal parasites are classified within the kingdom Animalia and are separated into two subkingdoms, Protozoa and Metazoa (Table 32-1). The subkingdom Protozoa comprises unicellular animals in which all life functions occur in a single cell. The Metazoa are multicellular animals in which life functions occur in cellular structures organized as tissue and organ systems.

The subkingdom Protozoa comprises seven major groups, or phyla, three of which are the concern of medical parasitology:

1. Phylum Sarcomastigophora consists of the amebae and flagellates, which are unicellular. Locomotion of amebae is accomplished by extrusion of pseudopodia ("false feet"), whereas flagellates move by lashing of the whiplike flagella. Reproduction by both these types involves simple binary fission.

2. Phylum Ciliophora consists of the ciliates, which are also unicellular. Ciliate locomotion involves the coordinated movement of rows of hairlike structures called cilia. Reproduction is accomplished by binary fission or a more complex nuclear exchange called conjugation. A typical organism, *Balantidium coli,* is discussed with the intestinal protozoa.

3. Phylum Apicomplexa organisms are often referred to as sporozoa or coccidia. These unicellular organisms have a system of organelles at their apical end that produce substances that help the organism penetrate host cells to become an intra-

Table 32-1 Classification of Medically Important Parasites (Kingdom Animalia)

Subkingdom	Phylum	Organisms
Protozoa	Sarcomastigophora	Amebae, flagellates
	Ciliophora	Ciliates
	Apicomplexa	Sporozoa, coccidia
Metazoa	Nematoda	Roundworms
	Platyhelminthes	Flatworms
	Trematodes (flukes)	
	Cestodes (tapeworms)	
	Arthropoda	
	Crustacea	Crabs, crayfish, shrimp, copepods
	Arachnida	Mites, ticks, spiders, scorpions
	Insecta	Mosquitoes, flies, lice, fleas, wasps, ants, beetles, moths, roaches, true bugs

Table 32-2 Food-Associated Parasitic Infections

Organism	Contaminated Food/Contaminant	Disease/Infection
INTESTINAL PROTOZOA		
*Entamoeba histolytica**	Water, fresh fruits and vegetables contaminated by cysts from food handlers; primarily from human feces	Diarrhea, dysentery, intestinal invasion and ulceration leading to secondary amebiasis to all parts of body, especially liver and lungs
Entamoeba polecki	As for *E. histolytica;* primarily from pig and monkey feces	Potential pathogen, intestinal irritation
*Giardia lamblia**	As for *E. histolytica;* primarily from human feces; also beavers and muskrats in streams	Diarrhea, intestinal irritation, malabsorption syndrome
*Dientamoeba fragilis**	Probable food and water contamination with pinworm eggs *(Enterobius vermicularis);* thought to be transport mechanism for organism inside eggshell	Diarrhea, intestinal irritation, abdominal discomfort, especially in children
Balantidium coli	Food and water contamination, primarily from pig feces	Diarrhea, intestinal ulceration, rarely to liver as secondary balantidiosis
Isospora belli	Food and water contamination from human feces	Diarrhea, intestinal irritation, abdominal discomfort
Cryptosporidium species*	Food and water; direct human-to-human contamination, as well as from farm animals (sheep, calves, etc.); from oocysts in feces	Watery diarrhea and chronic fluid loss as organisms invade surface of intestinal cells
BLOOD AND TISSUE PROTOZOA		
Toxoplasma gondii	Ingestion of poorly cooked meats or fresh meat juices (beef, pork, lamb); cat feces as contaminant on fingers touching food and water	Lymphadenopathy, myalgia, myocarditis; congenital infections may produce hydrocephaly, microcephaly, retinochoroiditis, and calcification in brain
Sarcocystis hominis	Ingestion of poorly cooked beef; also contamination with dog feces	Diarrhea, intestinal irritation
Sarcocystis lindemanni	Food and water contaminated with dog or cat feces	Invasion of skeletal muscle or myocardium; may become destructive in these tissues
NEMATODES		
Trichinella spiralis	Ingestion of contaminated pork, bear, walrus, ''country sausage'', many carnivores shown to be carriers of cysts in their flesh	Fever, myalgia, periorbital edema, eosinophilia, intestinal irritation; may be lethal in severe infections
Enterobius vermicularis	Food and water contaminated with eggs from food handlers; also direct human-to-human transmission	Rectal discomfort, insomnia, itching; worms migrating in girls may cause chronic salpingitis
Ascaris lumbricoides	Food and water contaminated with infective eggs or human feces (''night soil'') used on gardens	Bronchial damage by migrating larval worms, intestinal obstruction; in severe infections migration of adult worms into bile duct and liver may produce intense tissue damage; perforation of intestine may lead to peritonitis and abscesses

*Also known to be sexually transmitted.

Table 32-2 Food-Associated Parasitic Infections, cont'd

Organism	Contaminated Food/Contaminant	Disease/Infection
Toxocara canis	Food, water, soil contaminated with eggs of dog and cat *Ascaris* worms	Visceral larva migrans as larval worms migrate to all parts of body, especially liver and lungs; may be dangerous in dirt-eating children if migration to eyes occurs
Trichuris trichiura	Food and water contaminated with infective eggs developed in soil	Intestinal irritation, chronic anemia, possible rectal prolapse
Dracunculus medinensis	Ingestion of water from step wells or ponds contaminated with Cyclops-carrying larval worms	Cutaneous blisters; secondary infection, including tetanus, abscesses, pericarditis; fatal cases recorded
TREMATODES		
Fasciola hepatica	Ingestion of aquatic vegetation in salads, especially watercress; larval worms (metacercariae) are encysted on vegetation; primary sources of eggs are sheep, cattle, bison, deer	Worms migrate to liver and bile duct, producing hyperplasia and fibrosis, liver abscesses and necrosis; ingestion of raw, infected livers from sheep and other animals may produce asphyxiation as adult worms lodge in pharyngeal tissue; ingestion of infected liver may lead to false infection as eggs are found in patient's feces
Fasciolopsis buski	As for *Fasciola hepatica;* vegetation, primarily aquatic bulbs (water chestnuts); as bulbs are peeled with teeth, pig and human feces are sources of eggs	Inflammation and ulceration of intestine, abscesses, obstruction, toxic reactions, diarrhea
Paragonimus westermani and related flukes	Ingestion of raw crabs and crayfish infected with encysted metacercariae; human, pig, dog, feline feces are sources of eggs	Adult worms in lungs, abdomen, brain produce cysts, abscesses, fibrotic nodules; in brain type of epilepsy occurs
Opisthorchis sinensis, Heterophyes heterophyes, Metagonimus yokogawai and related flukes	Ingestion of raw fish containing encysted metacercariae, human "night soil" fertilizer, dog, cat, fox and rabbit feces are sources of eggs	*O. sinensis* in liver and bile ducts produces irritation, leading to hyperplasia and malignant transformation; *H. heterophyes* in intestine produces irritation and some necrosis; *M. yokogawai* in intestine produces inflamation and diarrhea
CESTODES		
Diphyllobothrium latum and related tapeworms	Ingestion of uncooked fish containing the larval worm (sparganum); bear and other fish-eating animals are sources of eggs	Large numbers of worms may produce intestinal obstruction, diarrhea, abdominal pain; in mild infections anemia may occur because worm absorbs vitamin B_{12}
Hymenolepis nana	Ingestion of eggs in food or water contaminated by food handlers; also direct person-to-person contamination	Intestinal irritation, diarrhea, abdominal distress; internal autoinfection may occur, leading to severe worm burdens
Hymenolepis diminuta	Ingestion of cereals, flour, baking products containing infected arthropods ("mealworms"); rat and mouse feces are sources of eggs	Usually well tolerated, but in severe infections intestinal irritation, diarrhea, etc., may occur
Taenia solium	Ingestion of poorly cooked pork containing larval worms *(Cysticercus);* human feces primary source of eggs	Adult worms may produce intestinal irritation or obstruction; see also *Cysticercus*

Continued.

Table 32-2   Food-Associated Parasitic Infections, cont'd

Organism	Contaminated Food/Contaminant	Disease/Infection
Taenia saginata	Ingestion of poorly cooked beef containing larval worms *(Cysticercus)*; human feces primary source of eggs	As for *T. solium;* see also *Cysticercus*
Cysticercus species (cysticercosis), *T. solium,* and several related *Taenia* species	Ingestion of eggs in food or water, ''human night soil''; possible autoinfection in persons harboring adult tapeworms	Invasion of all parts of body; subcutaneous nodules, especially dangerous in brain and eyes; cysticercosis produced by *T. saginata* rare but recorded
Echinococcus granulosus, E. multilocularis, and related forms (hydatid cyst, hydatidosis)	Ingestion of eggs in food or water contaminated with feces from dog, fox, wolf, dingo, jackal, hyena, domestic cat and other felines	Invasion of all parts of body, especially liver and lungs and including bones; rupture or puncture of cysts can produce anaphylaxis because of toxicity of fluid or seeding of new sites for cyst development

cellular parasite. The alternation of hosts and of generations typically occurs, such as malaria organisms alternating between humans and mosquitoes and reproducing asexually in humans (schizogony) and sexually in mosquitoes (gametogeny).

The subkingdom Metazoa, which includes all animals that are not Protozoa, has two groups of organisms of major importance in this text, the helminths (''worms'') and the arthropods (crabs, insects, ticks, etc.).

The helminths are separated into two phyla as follows:

1. Phylum Nematoda consists of the roundworms, which have cylindric bodies. The sexes of roundworms are separate, and these organisms have a complete digestive system. The nematodes may be intestinal parasites or may infect the blood and tissue.

2. Phylum Platyhelminthes consists of the flatworms, which have flattened bodies that are leaflike or resemble ribbon segments. Platyhelminthes can be further separated into trematodes and cestodes.

   a. Trematodes, or flukes, have leaf-shaped bodies. Most are hermaphroditic, with male and female sex organs present in a single body. Their digestive systems are incomplete and have only saclike tubes. The life cycles of flukes are complex, with snails serving as first intermediate hosts and with other aquatic animals or plants as second intermediate hosts.

   b. Cestodes, or tapeworms, have bodies composed of ribbons of proglottids, or segments. All are hermaphroditic, and all lack a digestive system, with nutrition being absorbed through the body wall. The life cycles of some cestodes are simple and direct, whereas those of others are complex and require one or more intermediate hosts.

The arthropods are a single phylum with three major classes.

Phylum Arthropoda is the largest group of animals in the Kingdom Animalia. Arthropods have segmented bodies (head, thorax, abdomen), paired jointed appendages, and well-developed digestive and nervous systems. Respiration by aquatic forms is via gills and by terrestrial forms via tubular body structures. All have a hard chitin covering as an exoskeleton. The major classes of arthropods follows.

1. Class Crustacea consists of familiar aquatic forms such as crabs, crayfish, shrimp, and copepods. Several are involved as intermediate hosts in life cycles of various intestinal or blood and tissue helminths.

2. Class Arachnida consists of familiar terrestrial forms such as mites, ticks, spiders, and scorpions. These animals have no wings or antennae, as do insects, and adults have four pairs of legs, as opposed to three pairs for insects. Of medical

importance are those serving as vectors for microbial diseases (mites, ticks) or as venomous animals that bite (spiders) or sting (scorpions).

3. Class Insecta consists of familiar aquatic and terrestrial forms such as mosquitoes, flies, midges, fleas, lice, bugs, wasps, and ants. Wings and antennae are present, and adult forms have three pairs of legs. Of medical importance are the many insects that serve as vectors for various microbial diseases (mosquitoes, fleas, flies, lice, bugs) or as venomous animals that sting (bees, wasps, ants).

This taxonomic review of animal parasites is intended to enhance the reader's comprehension of the interrelationships among parasitic organisms, their epidemiology and transmission of disease, specific disease processes involved, and possibilities for prevention and control of maladies. We have deliberately attempted to simplify this taxonomy by using it to address the major divisions involved in medical parasitology, specifically, intestinal and urogenital protozoa, blood and tissue protozoa, nematodes, trematodes, and cestodes.

Physician awareness of parasitic diseases is undoubtedly more critical now than at any time in the history of medical practice. Rapid global transit and the influx of refugee populations into many communities have changed the approach to what were once considered "exotic tropical diseases." Physicians today must be prepared to answer questions from patients regarding protection from malaria and the risks of drinking water and eating fresh fruits and vegetables in remote areas where they may be traveling. With this knowledge of parasitic diseases, the physician can also evaluate signs, symptoms, and incubation periods in returning travelers and make a diagnosis and begin treatment for a patient with a possible parasitic disease. Proper education regarding parasitic diseases in medical curricula cannot be overemphasized as a requirement for physicians whose practice includes travelers to foreign countries and refugee populations. Many of the important parasites responsible for human diseases are acquired by consumption of contaminated food or water. These are presented in appropriate detail in the following chapters; however, the data in Table 32-2 are provided to help orient the reader.

BIBLIOGRAPHY

Beaver PC, Jung RC, and Cupp EW: Clinical parasitology, ed 9, Philadelphia, 1984, Lea & Febiger.

Brown HW and Neva FA: Basic clinical parasitology, ed 5, Norwalk, Conn, 1983, Appleton-Century-Crofts.

Howard BJ et al: Clinical and pathogenic microbiology, St Louis, 1987, the CV Mosby Co.

Markell EK, Voge M, and John DT: Medical parasitology, ed 6, Philadelphia, 1986, WB Saunders Co.

Schmidt GD and Roberts LS: Foundations of parasitology, ed 4, St Louis, 1989, The CV Mosby Co.

Washington JA II, editor: Laboratory procedures in clinical microbiology, ed 2, New York, 1985, Springer-Verlag New York, Inc.

33

Intestinal and Urogenital Protozoa

Protozoa can be subdivided into four groups: amebae, flagellates, ciliates, and coccidia.

AMEBAE

The amebae are primitive, unicellular microorganisms. For most, their life cycle is relatively simple and divided into two stages: the actively motile feeding stage (**trophozoite**) and the quiescent, resistant, infective stage (**cyst**). Replication is accomplished by binary fission (splitting the trophozoite) or by the development of numerous trophozoites within the mature multinucleated cyst. Motility of amebae is accomplished by extension of a **pseudopod** (''false foot'') with extrusion of the cellular ectoplasm, and then drawing up the rest of the cell in a snail-like movement to meet this pseudopod. The amebic trophozoites remain actively motile as long as the environment is favorable. The cyst form will develop when the environmental temperature or moisture level drops.

Most amebae found in man are commensal organisms (*Entamoeba coli, Entamoeba hartmanni, Entamoeba gingivalis, Endolimax nana, Iodamoeba butschlii*). However, *Entamoeba histolytica* is an important human pathogen. Other amebae, particularly *Entamoeba polecki,* can cause human disease but are rarely isolated. Some free-living amebae *(Naegleria fowleri, Acanthamoeba* species) are present in warm, freshwater ponds or swimming pools and can be opportunistic human pathogens, causing meningoencephalitis or keratitis.

Entamoeba histolytica

Microbial Physiology and Structure Cyst and trophozoite forms of *E. histolytica* are detected in fecal specimens from infected patients (Figure 33-1). Trophozoites can also be found in the crypts of the large intestine. In freshly passed stools actively motile trophozoites can be seen, whereas in formed stools the cysts are usually the only form recognized. To diagnose amebiasis, it is important to distinguish between the *E. histolytica* trophozoites and cysts and those of commensal amebae.

Pathogenesis Following ingestion the cysts pass through the stomach, where exposure to gastric acid stimulates release of the pathogenic trophozoite in the duodenum. The trophozoites divide and produce extensive local necrosis (histolytica for ''tissue lysis'') in the large intestine. The basis for this tissue destruction is incompletely understood, although it is attributed to production of a cytotoxin. Necrosis requires direct contact with the ameba, so lysosomal enzymes (e.g., phospholipase A_2) may be important. Flask-shaped ulcerations of the intestinal mucosa are present with inflammation, hemorrhage, and secondary bacterial infection. Invasion into the deeper mucosa with extension into the peritoneal cavity may occur. This can lead to secondary involvement of other organs, primarily the liver but also the lungs, brain, and heart. Extraintestinal amebiasis is associated with trophozoites. Amebae are found only in environments with a low PO_2 because the protozoa are killed by ambient oxygen concentrations.

Epidemiology *E. histolytica* has a worldwide distribution. Although it is found in cold areas such as

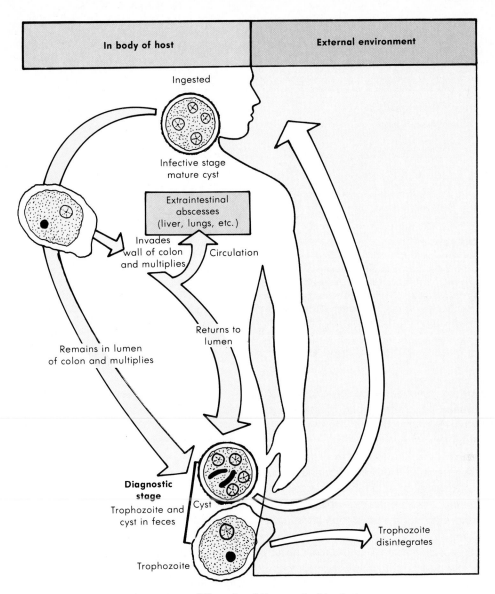

In body of host

External environment

Ingested

Infective stage
mature cyst

Extraintestinal
abscesses
(liver, lungs, etc.)

Invades
wall of colon
and multiplies

Circulation

Returns to
lumen

Remains in lumen
of colon and multiplies

Diagnostic
stage
Trophozoite and
cyst in feces

Cyst

Trophozoite
disintegrates

Trophozoite

FIGURE 33-1   Life cycle of *Entamoeba histolytica*.

Alaska, Canada, and Russia, its incidence is highest in tropical and subtropical regions with poor sanitation and contaminated water. The average prevalence of infection in these areas is 10% to 15%, with as many as 50% of the population infected in some areas. Many of the infected individuals are asymptomatic carriers, which represent a reservoir for spread of *E. histolytica* to others. The prevalence of infection in the United States is 1% to 2%.

Patients infected with *E. histolytica* pass noninfectious trophozoites, as well as the infectious cysts, in their stools. The trophozoites cannot survive in the external environment or in transport through the stomach if ingested. Therefore the main source of water and food contamination is the asymptomatic carrier who passes cysts. This is a particular problem in hospitals for the mentally ill, military and refugee camps, prisons, and crowded day-care centers. Flies and cockroaches can also be a source of *E. histolytica* cysts. Sewage containing cysts can contaminate water systems, wells, springs, and agricultural areas where human waste is used as fertilizer. Finally, cysts

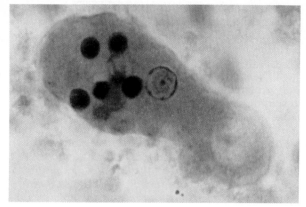
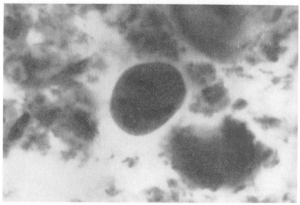

A   B

FIGURE 33-2   *Entamoeba histolytica* trophozoite **(A)** and cyst **(B).** Trophozoites are motile and vary in size from 12 to 60 μm (average 15 to 30). The single nucleus in the cell is round with a central dot (karyosome) and an even distribution of chromatin granules around the nuclear membrane. Ingested erythrocytes may be in the cytoplasm. The cysts are smaller (11 to 20 μm with an average size of 15 to 20 μm) and contain one to four nuclei (usually four). Round chromatoidal bars may be in the cytoplasm. (From Lennette EH et al: Manual of clinical microbiology, ed 4, Washington, DC, 1985, American Society for Microbiology.)

can be transmitted by oral-anal sexual practices, with amebiasis prevalent in homosexual populations. Direct trophozoite transmission in sexual encounters can produce cutaneous amebiasis.

Clinical Syndromes The outcome of infection may result in a carrier state, intestinal amebiasis, or extraintestinal amebiasis. If the strain of *E. histolytica* has a low virulence, if the inoculum is low, or if the patient's immune system is intact, the organisms may reproduce and cysts be passed in stool specimens with no clinical symptoms. Detection of carriers in areas with a low endemicity is important for epidemiologic purposes.

Patients with intestinal amebiasis develop clinical symptoms related to the localized tissue destruction in the large intestine. These include abdominal pain, cramping, and colitis with diarrhea. More severe disease is characterized by numerous bloody stools per day.

Systemic signs of infection (fever, leukocytosis, rigors) are present in patients with extraintestinal amebiasis. The liver is primarily involved because trophozoites in the blood are removed as they pass through this organ. Abscess formation is common. The right lobe is most commonly involved. Pain over the liver with hepatomegaly and elevation of the diaphragm is observed.

Laboratory Diagnosis The identification of *E. histolytica* trophozoites and cysts (Figure 33-2) in stools and trophozoites in tissue is diagnostic of amebic infection. Care must be taken to distinguish between these amebae and commensal amebae, as well as polymorphonuclear leukocytes. Microscopic examination of stool specimens is inherently insensitive because the protozoa are not usually distributed homogeneously in the specimen and the parasites are concentrated in the intestinal ulcers and at the margins of the abscess, not in the stool or the necrotic center of the abscess. For this reason multiple stool specimens should be collected. Extraintestinal amebiasis is sometimes diagnosed using scanning procedures for the liver and other organs. Specific serologic tests, together with microscopic examination of the abscess material, can confirm the diagnosis. Virtually all patients with hepatic amebiasis and most patients (greater than 80%) with intestinal disease have positive serology at the time of clinical presentation. This may be less useful in endemic areas where the prevalence of positive serologic results is higher. Examinations of stool specimens are frequently negative in extraintestinal disease.

Treatment, Prevention, and Control Acute, fulminating amebiasis is treated with metronidazole followed by iodoquinol. Asymptomatic carriage can

Table 33-1 Morphologic Identification of *Entamoeba histolytica* and *Entamoeba coli*

	Entamoeba histolytica	*Entamoeba coli*
Size (μm diameter)		
Trophozoite	12-50	20-30
Cyst	10-20	10-30
Pattern of peripheral nuclear chromatin	Fine, dispersed ring	Coarse, clumped
Karyosome	Central, sharp	Eccentric, coarse
Ingested erythrocytes	+	0
Cyst structure		
Nuclei	1-4	1-8
Chromatoid bars	Rounded ends	Splintered, frayed ends

be eradicated with iodoquinol, diloxanide furoate, or paromomycin. Human infection results from ingestion of food or water contaminated with human feces, or as a result of specific sexual practices. The elimination of the cycle of infection requires introduction of adequate sanitation measures and education about the routes of transmission. Physicians should alert travelers to developing countries of the risks associated with consumption of water (including ice cubes), unpeeled fruits, and raw vegetables. Water should be boiled and fruits and vegetables thoroughly cleaned before consumption.

Other Intestinal Amebae

Other amebae that can parasitize the human gastrointestinal tract include *Entamoeba coli*, *Entamoeba hartmanni*, *Entamoeba polecki*, *Endolimax nana*, and *Iodamoeba buttschlii*. Only *E. polecki*, which is primarily a parasite of pigs and monkeys, can cause human disease, a mild transient diarrhea. The diagnosis of *E. polecki* infection is confirmed by the microscopic detection of cysts in stool specimens. Treatment is the same as for *E. histolytica* infections.

The nonpathogenic intestinal amebae are important because they must be differentiated from *E. histolytica* and *E. polecki*. This is particularly true for *E. coli*, which is frequently detected in stool specimens collected from patients exposed to contaminated food or water. Accurate identification of intestinal amebae requires careful microscopic examination of the cyst and trophozoite forms present in stained and unstained stool specimens (Table 33-1).

Free-Living Amebae

Naegleria, *Acanthamoeba*, and other free-living amebae are found in contaminated lakes, streams, and other water environments. Most human infections with these amebae are acquired during the warm summer months by individuals exposed to the amebae while swimming in contaminated waters. Inhalation of cysts present in dust may account for some infections, whereas ocular infections with *Acanthamoeba* are most frequently associated with contamination of contact lenses with nonsterile cleaning solutions.

Clinical Syndromes *Naegleria* are opportunistic pathogens. Although colonization of the nasal passages is usually **asymptomatic,** these amebae can invade the nasal mucosa and extend into the brain. Destruction of brain tissue is characterized by a fulminant, rapidly fatal **meningoencephalitis.** Symptoms include intense frontal headache, sore throat, fever, blocked nose with altered senses of taste and smell, stiff neck, and Kernig's sign. The cerebrospinal fluid is purulent and may contain many erythrocytes and motile amebae. Clinically, the course of the disease is rapid, with death usually occurring within 4 or 5 days. Postmortem findings show *Naegleria* trophozoites present in the brain but no evidence of cysts. Although all cases were fatal before 1970, survival has now been reported in a few cases in which the disease was rapidly diagnosed and treated.

In contrast to *Naegleria*, *Acanthamoeba* produces **granulomatous amebic encephalitis** primarily in immunocompromised individuals. The course of the disease is slower, with an incubation period of 10 days or more. The resulting disease is a chronic granulomatous encephalitis with edema of the brain tissue.

Eye and skin infection caused by *Acanthamoeba* may also occur. The **keratitis** is usually associated with trauma to the eye preceding contact with contaminated soil, dust, or water. The use of improperly cleaned contact lenses is also associated with this disease. Invasion of *Acanthamoeba* produces corneal ulceration and severe ocular pain. Corneal transplantation, and in some cases enucleation of the eye, may be required.

Laboratory Diagnosis For the diagnosis of *Naegleria* and *Acanthamoeba* infections, nasal discharge, cerebrospinal fluid, and, in the case of eye infections, corneal scrapings should be collected. The specimens should be examined by both a saline wet preparation, as well as iodine-stained smears. *Naegleria* and *Acanthamoeba* are difficult to differentiate except by experienced microscopists. However, the observation of an ameba in a normally sterile tissue is diagnostic (Figure 33-3). The clinical specimens can be cultured on agar plates seeded with live gram-negative enteric bacilli. Amebae present in the specimens are able to use the bacteria as a nutritional source and can be detected within 1 or 2 days by the presence of trails on the agar surface formed as the amebae move.

Treatment, Prevention, and Control The treatment of choice for *Naegleria* infections is amphotericin B combined with miconazole and rifampin. Experimental infections with *Acanthamoeba* appear to respond to sulfadiazine. Amebic keratitis and cutaneous infections may respond to topical miconazole and propamidine isethionate. The wide distribution of these organisms in fresh and brackish waters makes prevention and control of infection difficult. It has been suggested that known sources of infection be off-limits to bathing, diving, and water sports, although this is generally difficult to enforce. Swimming pools with cracks in the walls, allowing soil seepage, should be repaired to avoid creation of a source of infection.

FLAGELLATES

The flagellates of clinical significance include *Giardia lamblia, Dientamoeba fragilis,* and *Trichomonas*

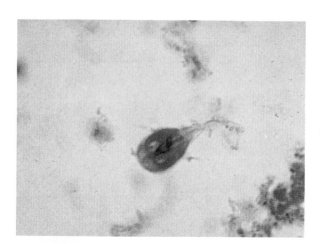

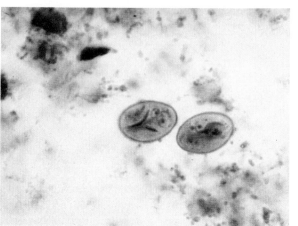

A

B

FIGURE 33-4   *Giardia lamblia* trophozoite **(A)** and cyst **(B)**. Trophozoites are 9 to 12 μm long and 5 to 15 μm wide. Flagella are present, as well as two nuclei with large central karyosomes, a large ventral sucking disk for attachment of the flagellate to the intestinal villi, and two oblong parabasal bodies below the nuclei. The morphology gives the appearance the trophozoites are looking back at the viewer. Cysts are smaller—8 to 12 μm long and 7 to 10 μm wide. Four nuclei and four parabasal bodies are present. (From Feingold SM and Baron EJ: Bailey and Scott's diagnostic microbiology, ed 7, St Louis, 1986, The CV Mosby Co.)

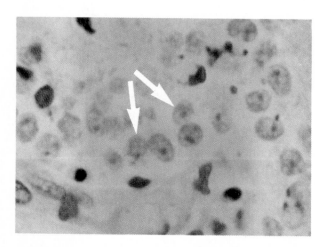

FIGURE 33-3   Numerous *Naegleria* trophozoites present in a section of spinal cord from a patient with amebic meningoencephalitis. (From Rothrock JF and Buchsbaum HW: JAMA 243:2329-30, copyright 1980, American Medical Association.)

vaginalis. Nonpathogenic commensal flagellates such as *Chilomastix mesnili* (enteric) and *Trichomonas tenax* (oral) may also be observed. *Giardia,* like *E. histolytica,* has both cyst and trophozoite stages in its life cycle. In contrast, no cyst stage has been observed for either *Trichomonas* or *Dientamoeba.* Unlike the amebae, most flagellates move by the lashing of flagella that pull the organisms through fluid environments. Diseases produced by flagellates are primarily the result of mechanical irritation and inflammation. For example, *G. lamblia* attaches to the intestinal villi with an adhesive disk, resulting in localized tissue damage. The tissue invasion with extensive tissue destruction, as seen with the *E. histolytica,* is rare with flagellates.

Giardia lamblia

Microbial Physiology and Structure Both cyst and trophozoite forms of *G. lamblia* are detected in fecal specimens from infected patients (Figure 33-4).

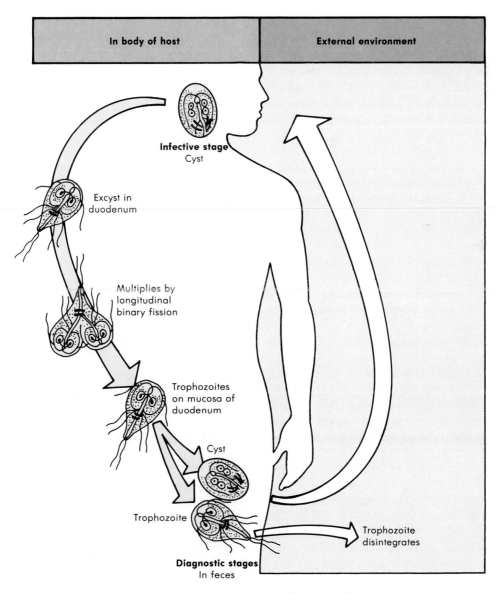

| In body of host | External environment |

Infective stage
Cyst

Excyst in duodenum

Multiplies by longitudinal binary fission

Trophozoites on mucosa of duodenum

Cyst

Trophozoite

Trophozoite disintegrates

Trophozoite

Diagnostic stages
In feces

FIGURE 33-5   Life cycle of *Giardia lamblia.*

Pathogenesis Infection with *G. lamblia* is initiated by ingestion of cysts (Figure 33-5). The minimum infective dose for humans is estimated to be 10 to 25 cysts. Gastric acid stimulates excystation with release of trophozoites in the duodenum and jejunum, where the organisms multiply by binary fission. The trophozoites can attach to the intestinal villi by a prominent ventral sucking disk. Although the tips of the villi may appear flattened and inflammation of the mucosa with hyperplasia of lymphoid follicles may be observed, frank tissue necrosis does not occur. Additionally, metastatic spread of disease beyond the gastrointestinal tract is very rare.

Epidemiology *Giardia* has worldwide distribution, and this flagellate has a sylvatic or "wilderness" distribution in many of our streams, lakes, and mountain resorts. This sylvatic distribution is maintained in reservoir animals such as beavers and muskrats.

Giardiasis is acquired by drinking inadequately treated contaminated water, ingestion of contaminated uncooked vegetables or fruits, or person-to-person spread by the fecal-oral or oral-anal routes. The cyst stage is resistant to chlorine concentrations (1 to 2 parts per million) used in most water treatment facilities. Therefore adequate water treatment should include chemical treatment combined with filtration. Risk factors associated with *Giardia* infections include poor sanitary conditions, travel to known endemic areas, drinking inadequately treated water (e.g., from contaminated mountain streams), day-care centers, and oral-anal sexual practices. Disease is particularly common in children in day-care centers and homosexuals.

Clinical Syndromes *Giardia* infection can result in either asymptomatic carriage (observed in approximately 50% of infected individuals) or symptomatic disease, ranging from mild diarrhea to a severe **malabsorption syndrome.** The incubation period before symptomatic disease develops ranges from 1 to 4 weeks (average 10 days). The onset of disease is sudden, with foul-smelling, watery diarrhea, abdominal cramps, flatulence, and steatorrhea. Blood and pus is rarely present in stool specimens, consistent with the absence of tissue destruction. Spontaneous recovery generally occurs after 10 to 14 days of disease, although a more chronic disease with multiple relapses may develop. This is particularly a problem for patients with IgA deficiency or intestinal diverticuli.

Laboratory Diagnosis With the onset of diarrhea and abdominal discomfort, stool specimens should be examined for cysts and trophozoites. *Giardia* may occur in "showers," with many organisms present in the stool on a given day and few or none detected on the next. For this reason, the physician should never accept a single negative stool specimen as conclusive that the patient is free of intestinal parasites. It is recommended that one stool specimen per day for 3 days should be examined. If stools remain persistently negative in a patient where giardiasis is highly suspected, additional specimens can be collected by duodenal aspiration or biopsy of the upper small intestine.

Treatment, Prevention, and Control It is important to eradicate *Giardia* from both asymptomatic carriers and diseased patients. The drug of choice is quinacrine, with metronidazole an acceptable alternative. Prevention and control of giardiasis involves avoidance of contaminated water and food, especially by the traveler and outdoors person. Protection is afforded by boiling drinking water from streams and lakes or in countries with a high incidence of endemic disease. Maintenance of properly functioning filtration systems in municipal water supplies is also required, because cysts are resistant to standard chlorination procedures. Public health efforts should be made to identify the reservoir of infection to prevent spread of disease. In addition, high-risk sexual behavior should be avoided.

Dientamoeba fragilis

Microbial Physiology and Structure *Dientamoeba fragilis* was initially classified as an ameba; however, the internal structures of the trophozoite are typical of a flagellate. no cyst stage has been described.

Epidemiology *D. fragilis* has a worldwide distribution. The transmission of the delicate trophozoites is not completely understood. Some observers believe the organism can be transported from person-to-person inside the protective shell of worm eggs such as *Enterobius vermicularis,* the pinworm. Transmission by the fecal-oral and oral-anal routes is known to occur.

Clinical Syndromes Most infections with *D. fragilis* are asymptomatic, with colonization of the cecum and upper colon. However, some patients may

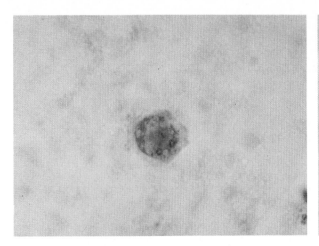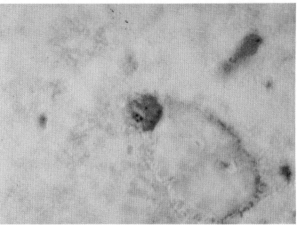

FIGURE 33-6   *Dientamoeba fragilis* trophozoites. The trophozoite is small (5 to 12 μm) with one or two nuclei. The central karyosome consists of four to six discrete granules. (From Feingold SM and Baron EJ: Bailey and Scott's diagnostic microbiology, ed 7, St Louis, 1986, The CV Mosby Co.)

develop symptomatic disease, with abdominal discomfort, flatulence, intermittent diarrhea, anorexia, and weight loss. There is no evidence of tissue invasion with this flagellate, although irritation of the intestinal mucosa occurs.

Laboratory Diagnosis Infection is confirmed by the microscopic examination of stool specimens in which typical trophozoites can be seen (Figure 33-6).

Treatment, Prevention, and Control The therapy of choice for *D. fragilis* infection is iodoquinol, with tetracycline or paramomycin acceptable alternatives. The reservoir for this flagellate and the organism's life cycle are unknown. Thus specific recommendations for prevention and control are difficult. However, infections can be avoided by maintaining adequate sanitary conditions. The eradication of infections with *Enterobius* may also reduce the transmission of *Dientamoeba*.

Trichomonas vaginalis

Microbial Physiology and Structure *Trichomonas vaginalis* is not an intestinal protozoan but rather the cause of urogenital infections. The flagellate possesses four flagella and a short undulating membrane that are responsible for motility. *T. vaginalis* is strictly a parasite of the urogenital system—the vagina in women and the urethra and prostate in men.

Epidemiology This parasite has worldwide distribution with sexual intercourse as the primary mode of transmission (Figure 33-7). Occasionally, infections have been transmitted by fomites (toilet articles, clothing), although this is limited by the lability of the trophozoite form. Infants may be infected by passage through the mother's infected birth canal. The prevalence of this flagellate in developed countries is reported to be from 5% to 20% in women and from 2% to 10% in men.

Clinical Syndromes Most infected women are asymptomatic or have a scant, watery vaginal discharge. Vaginitis may occur with more extensive inflammation and erosion of the epithelial lining, associated with itching, burning, and painful urination. Men are primarily asymptomatic carriers who serve as a reservoir for infections in women. However, occasional men experience urethritis, prostatitis, and other urinary tract problems.

Laboratory Diagnosis The microscopic examination of vaginal or urethral discharge for characteristic trophozoites is the diagnostic method of choice (Figure 33-8). Stained (Giemsa, Papanicolaou) or unstained smears can be examined. The organism can be grown in culture, but this is more time consuming than direct microscopy and does not differen-

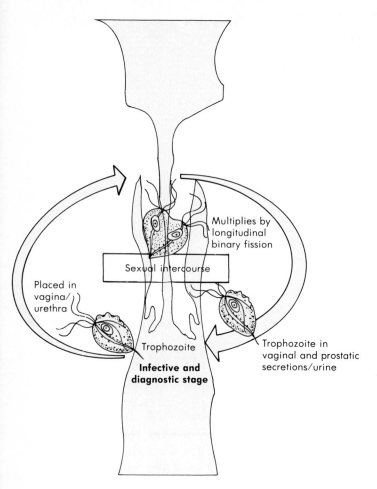

FIGURE 33-7   *Trichomonas vaginalis* life cycle.

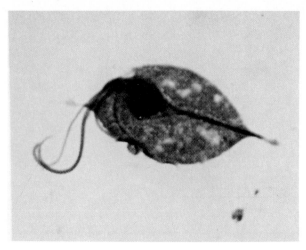

FIGURE 33-8   *Trichomonas vaginalis* trophozoite. The trophozoite is 7 to 23 μm long and 6 to 8 μm wide (average 13 by 7 μm); flagella and a short undulating membrane are present at one side, and an axostyle extends through the center of the parasite. (Reproduced by permission from LR Ash and TC Orihel, Atlas of human parasitology, ed 2, © 1984 by the American Society of Clinical Pathologists Press, Chicago.)

tiate between colonization with small numbers of organisms and disease.

Treatment, Prevention and Control The drug of choice is metronidazole. Both male and female sex partners must be treated to avoid reinfection. Personal hygiene, avoidance of shared toilet articles and clothing, and safe sexual practices are important preventive actions. Elimination of carriage in men is critical for the eradication of disease.

CILIATES

The intestinal protozoan *Balantidium coli* is the only member of the ciliate group that is pathogenic for humans. Disease produced by *B. coli* is similar to amebiasis because the organisms elaborate proteolyt-

ic and cytotoxic substances that mediate tissue invasion and intestinal ulceration.

Balantidium coli

Microbial Physiology and Structure The life cycle is simple, involving ingestion of infectious cysts, excystation, and invasion of trophozoites into the mucosal lining of the large intestine, cecum, and terminal ileum (Figure 33-9). The trophozoite is covered with rows of hairlike cilia that aid in motility. Morphologically more complex than amebae, *B. coli* has a funnel-like primitive mouth called a cytostome, a large and small nucleus involved in reproduction, food vacuoles, and two contractile vacuoles.

Epidemiology *B. coli* is distributed worldwide. Swine and (less commonly) monkeys are the most important reservoirs. Infections are transmitted by the fecal-oral route; outbreaks are associated with contamination of water supplies with pig feces. Person-to-person spread, including food handlers, has been implicated in outbreaks. Risk factors associated with human disease include contact with swine and substandard hygienic conditions.

Clinical Syndromes As with other protozoan parasites, asymptomatic carriage of *B. coli* can exist.

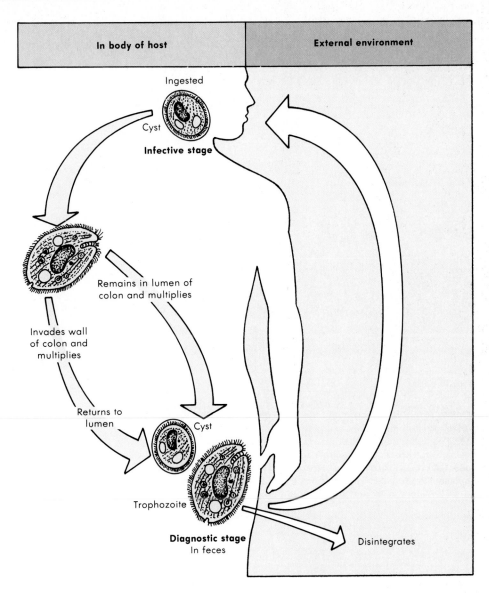

In body of host | External environment

Ingested

Cyst

Infective stage

Remains in lumen of colon and multiplies

Invades wall of colon and multiplies

Returns to lumen

Cyst

Trophozoite

Diagnostic stage
In feces

Disintegrates

FIGURE 33-9 *Balantidium coli* life cycle.

Symptomatic disease is characterized by abdominal pain and tenderness, tenesmus, nausea, anorexia, and watery stools with blood and pus. Ulceration of the intestinal mucosa, as with amebiasis, can be seen, as well as secondary complication due to bacterial invasion into the eroded intestinal mucosa. Extraintestinal invasion of other organs is extremely rare in balantidiasis.

Laboratory Diagnosis Microscopic examination of feces for trophozoites and cysts is performed (Fig-

ure 33-10). *B. coli* is a large organism compared with other intestinal protozoa and is readily detected in fresh, wet microscopic preparations.

Treatment, Prevention, and Control The drug of choice is tetracycline; iodoquinol or metronidazole are alternative antimicrobials. Prevention and control is similar to that used for amebiasis. Appropriate personal hygiene, maintenance of sanitary conditions, and the careful monitoring of pig feces are all important preventive measures.

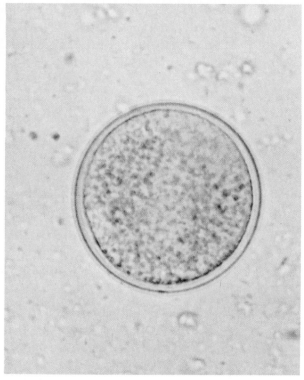

FIGURE 33-10 *Balantidium coli* trophozoite (**A**) and cyst (**B**). The trophozoite is very large, varying in length from 50 to 200 μm and in width from 40 to 70 μm. The surface is covered with cilia, and the prominent internal structure is a macronucleus. A micronucleus is also present. Two pulsating contractile vacuoles are also seen in fresh preparation of the trophozoites. The cyst is smaller (40 to 60 μm in diameter), surrounded by a clear refractile wall and a single nucleus in the cytoplasm. (Reproduced by permission from LR Ash and TC Orihel, Atlas of human parasitology, ed 2, © 1984 by the American Society of Clinical Pathologists Press, Chicago.)

COCCIDIA

Coccidia comprise a very large group called *Apicomplexa,* some members of which are discussed in this section with the intestinal parasites and others with the blood and tissue parasites. All coccidia demonstrate typical characteristics, especially the existence of both asexual (schizogony) and sexual (gametogony) reproduction. Most members of the group also share alternative hosts. A familiar example of this is with malaria, for which mosquitoes harbor the sexual cycle and humans the asexual cycle. The coccidia that are discussed in this chapter are *Isospora, Sarcocystis, Cryptosporidium,* and *Blastocystis.*

Isospora belli

Microbial Physiology and Structure *Isospora belli* is a coccidian parasite of the intestinal epithelium. Both sexual and asexual reproduction in the intestinal epithelium can occur, resulting in tissue damage (Figure 33-11). The end product of gametogenesis is the oocyst, which is the diagnostic stage present in fecal specimens.

Epidemiology *Isospora* is distributed worldwide but has been infrequently detected in stool specimens. Recently, however, this parasite has been reported with increasing frequency in both normal and immunocompromised patients. This is likely due to the increased awareness of disease caused by *Iso-*

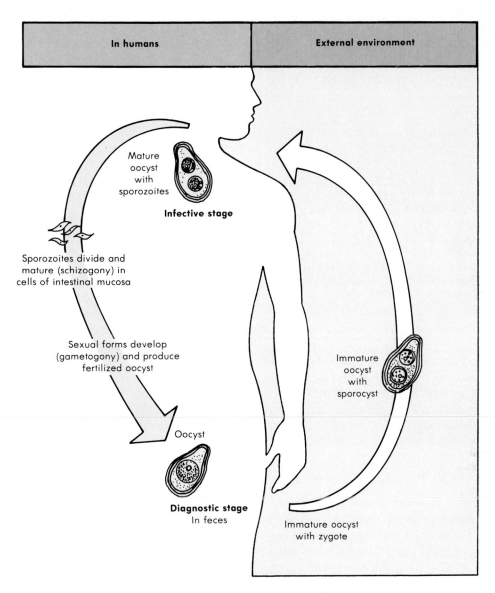

FIGURE 33-11   *Isospora* life cycle.

spora in AIDS patients. Infection with this organism follows ingestion of contaminated food or water, or oral-anal sexual contact.

Clinical Syndromes Infected individuals may be asymptomatic carriers or suffer gastrointestinal disease ranging from mild to severe. Disease most commonly mimics giardiasis, with a malabsorption syndrome characterized by loose, foul-smelling stools. Chronic diarrhea with weight loss, anorexia, malaise,

and fatigue can be seen, although it is difficult to separate this presentation from the patient's underlying disease.

Laboratory Diagnosis Careful examination of concentrated stool sediment and special staining with either iodine or a modified acid-fast procedure will reveal the parasite (Figure 33-12). Small bowel biopsy has been used to establish the diagnosis when stool specimens are negative.

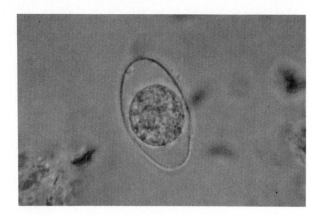

FIGURE 33-12   Immature oocyst of *Isospora*. Oocysts are ovoid (approximately 25 μm long and 15 μm wide) with tapering ends. A developing sporocyst is seen within the cytoplasm. (From Lennette EH et al: Manual of clinical microbiology, ed 4, Washington, DC, 1985, American Society for Microbiology.)

Treatment, Prevention, and Control The drug of choice is trimethoprim-sulfamethoxazole, with the combination of pyrimethamine and sulfadiazine an acceptable alternative. Prevention and control is effected by maintaining personal hygiene, high sanitary conditions, and avoidance of oral-anal sexual contact.

Sarcocystis

Physician awareness of the genus *Sarcocystis* is important only in recognition that it can be detected in stool specimens. *Sarcocystis* species can be isolated from pigs and cattle and are identical in all aspects to *Isospora* with one exception. In contrast to *Isospora*, *Sarcocystis* oocysts rupture before passage in stool specimens and only sporocysts will be present.

Cryptosporidium

Microbial Physiology and Structure The life cycle of *Cryptosporidium* species is typical of coccidians, as is the intestinal disease, but this species differs in the intracellular location in the epithelial cells. In contrast to the deep intracellular invasion observed with *Isospora*, *Cryptosporidium* is found just within the brush border of the intestinal epithelium. The coccidia attach to the surface of the cells and replicate by a process that involves schizogony (Figure 33-13).

Epidemiology *Cryptosporidium* species are dis-tributed worldwide. Infection is reported in a wide variety of animals, including mammals, reptiles, and fish. Zoonotic spread from animal reservoirs to humans, as well as person-to-person spread by fecal-oral and oral-anal routes, are the most common means of infection. Veterinary personnel, animal handlers, and homosexuals are at particularly high risk for infection.

Clinical Syndromes As with other protozoan infections, exposure to *Cryptosporidium* may result in asymptomatic carriage. Disease in previously healthy individuals is usually a mild self-limiting enterocolitis that is characterized by watery diarrhea without blood. Spontaneous remission after an average of 10 days is characteristic. In contrast, disease in immunocompromised patients (e.g., AIDS patients), characterized by 50 or more stools per day and tremendous fluid loss, can be severe and last for months to years. In some AIDS patients disseminated *Cryptosporidium* infections have been reported.

Laboratory Diagnosis *Cryptosporidium* can be demonstrated in stool specimens by staining the oocysts with modified acid-fast stains (Figure 33-14). Antibodies labeled with fluorescein have also been used to visualize the protozoans. Although concentration of the stool specimen has been useful for some patients, this is generally unnecessary for the detection of the organism in immunocompromised patients.

Treatment, Prevention, and Control Unfortunately, no broadly effective therapy has been developed for managing *Cryptosporidium* infections in immunocompromised patients. Spiramycin is the only drug known to have any activity against *Cryptosporidium*, although this treatment is frequently ineffective. Therapy consists primarily of supportive measures to restore the tremendous fluid loss from the watery diarrhea.

Because of the widespread distribution of this organism in humans and other animals, prevention is difficult. The same methods of improved personal hygiene and sanitation used for other intestinal protozoa should be maintained for this disease. In addition, avoidance of high-risk sexual activities is critical.

Blastocystis hominis

Blastocystis hominis, previously regarded as a nonpathogenic yeast, is now the center of considerable controversy concerning its taxonomic position

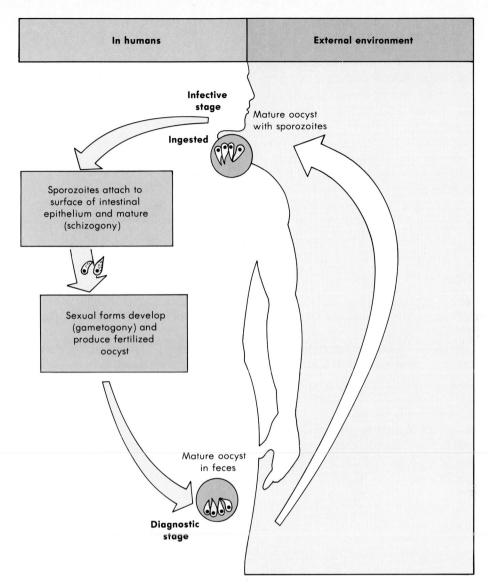

In humans

External environment

Infective stage

Ingested

Mature oocyst with sporozoites

Sporozoites attach to surface of intestinal epithelium and mature (schizogony)

Sexual forms develop (gametogony) and produce fertilized oocyst

Mature oocyst in feces

Diagnostic stage

FIGURE 33-13   *Cryptosporidium* life cycle.

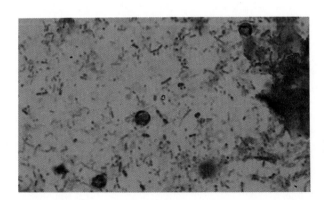

FIGURE 33-14   Acid-fast stained *cryptosporidium* oocysts (approximately 5 to 7 μm in diameter).

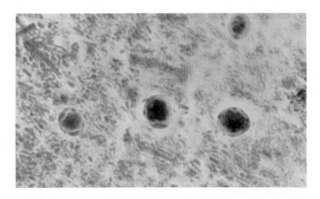

FIGURE 33-15   *Blastocystis hominis* oocysts. Approximately 8 to 10 μm in diameter; the internal morphology is dominated by the large central vacuole, which displaces the nucleus and cytoplasm to the periphery. (From Lennette EH et al: Manual of clinical microbiology, ed 4, Washington, DC, 1985, American Society for Microbiology.)

and its pathgenicity. The organism (Figure 33-15) is found in stool specimens from asymptomatic individuals, as well as from persons with persistent diarrhea. It has been suggested that the presence of large numbers of these parasites (five or more per oil immersion microscopic field) in the absence of other intestinal pathogens is consistent with its role in disease. Treatment with iodoquinol or metronidazole has been successful in eradicating the organisms from the intestine and alleviating symptoms. However, the definitive role of this organism in disease remains to be demonstrated.

BIBLIOGRAPHY

Adams EB and MacLeod IN: Invasive amebiasis. I. Amebic dysentery and its complications, Medicine 56:315-323, 1977a.

Adams EB and MacLeod IN: Invasive amebiasis. II. Amebic liver abscess and its complications, Medicine 56:325-334, 1977b.

Beaver PC, Jung RC, and Cupp EW: Clinical parasitology, ed 9, Philadelphia, 1984, Lea & Febiger.

Brown HW and Neva FA: Basic clinical parasitology, ed 5, Norwalk, Connecticut, 1983, Appleton-Century-Crofts.

Current WL et al: Human cryptosporidiosis in immunocompetent and immunodeficient persons, N engl J Med 308:1252-1257, 1983.

Gay JD, Abell TL, Thompson JH Jr, and Loth V: *Entamoeba polecki* infections in Southeast Asian refugees: multiple cases of a rarely reported parasite, Mayo Clin Proc 60: 523-530, 1985.

Howard BJ et al: Clinical and pathogenic microbiology, St Louis, 1987, The CV Mosby Co.

John DT and De Jonckheere JF: Isolation of *Naegleria australiensis* from an Oklahoma lake, J Protozool 32:571-575, 1985.

Markell EK, Voge M, and John DT: Medical parasitology, ed 6 Philadelphia, 1986, WB Saunders Co.

McLaren LC: Isolation of *Trichomonas vaginalis* from the respiratory tract of infants with respiratory disease, Pediatrics 71:888-890, 1983

Miller RL Wang AL, and Wang CC: Identification of *Giardia lamblia* isolates susceptible and resistant to infection of the double-stranded RNA virus, Exp Parasitol 66:118-123, 1988.

Osterholm MT et al: An outbreak of food-borne giardiasis, N Engl J Med 304:24-28, 1981.

Schmidt DG and Roberts LS: Foundations of parsitology, ed 4, St Louis 1989, The CV Mosby Co.

Washington JA II, editor: Laboratory procedures in clinical microbiology, ed 2, New York, 1985, Springer-Verlag.

Yang J and Scholten T: *Dientamoeba fragilis:* a review with notes on its epidemiology, pathogenicity, mode of transmission, and diagnosis: Am J Trop Med Hyg 26:16-22, 1977.

34

Blood and Tissue Protozoa

The protozoa of blood and tissues are closely related to the intestinal protozoan parasites in practically all aspects except for their sites of infection (Box 34-1). The malaria parasites *(Plasmodium)* infect both blood and tissues and will be discussed first.

PLASMODIUM

Plasmodia are coccidian or sporozoan parasites of blood cells, and—as is seen with other coccidia— they require two hosts: the mosquito for the sexual reproductive stages **(gametogony)**, and humans and other animals for the asexual reproductive stages **(schizogony)**.

The four species of plasmodia that infect humans are *P. falciparum, P. vivax, P. ovale,* and *P. malariae* (Table 34-1). These species share a common life cycle, as illustrated in Figure 34-1. Human infection is initiated by the bite of an *Anopheles* mosquito, which introduces infectious plasmodia sporozoites into the circulatory system. The sporozoites are carried to the parenchymal cells of the liver, where asexual repro-

duction (schizogony) occurs. This phase of growth is termed the **exoerythrocytic cycle** and will last for 8 to 25 days, depending on the plasmodial species. Some plasmodial species (i.e., *P. vivax* and *P. ovale*) can establish a dormant hepatic phase in which the sporozoites (called **hypnozoites** or "sleeping forms") do not divide. The presence of these viable plasmodia can lead to relapse infections months to years after the initial clinical disease (**relapsing malaria**).

The liver hepatocytes eventually rupture, liberating the plasmodia (termed merozoites at this stage), which in turn attach to specific receptors on the surface of erythrocytes and enter the cells, thus initiating the **erythrocytic cycle.** Asexual replication progresses through a series of stages (ring, trophozoite, schizont) that culminate in the rupture of the erythrocyte, releasing up to 24 merozoites, which initiate another cycle of replication by infecting other erythrocytes. Some merozoites also develop within erythrocytes into male and female gametocytes. If a mosquito ingests mature male and female **gametocytes** during a blood meal, the sexual reproductive cycle of malaria can be initiated, with the eventual production of sporozoites infectious for humans. This sexual

Box 34-1 Medically Important Blood and Tissue Protozoa	
Plasmodium	*Pneumocystis*
Babesia	*Leishmania*
Toxoplasma	*Trypanosoma*
Sarcocystis	

Table 34-1   Human Malarial Parasites

Parasite	Disease
Plasmodium vivax	Benign tertian malaria
Plasmodium ovale	Benign tertian or ovale malaria
Plasmodium malariae	Quartan or malarial malaria
Plasmodium falciparum	Malignant tertian malaria

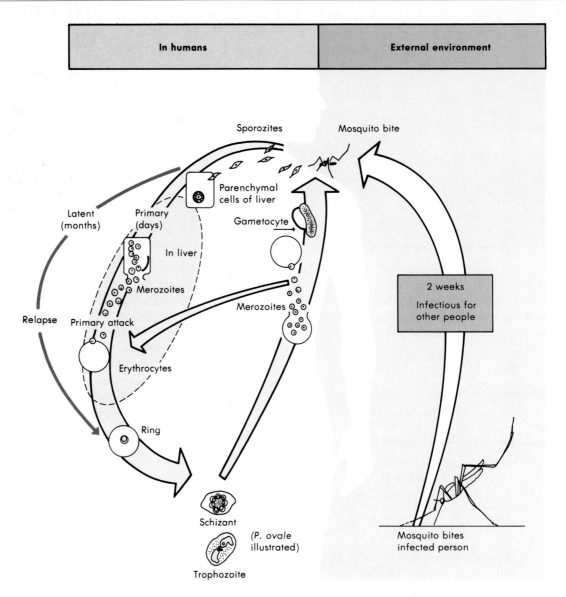

In humans	External environment

Sporozites

Mosquito bite

Parenchymal cells of liver

Latent (months)

Primary (days)

Gametocyte

In liver

Relapse

Merozoites

Primary attack

2 weeks
Infectious for other people

Merozoites

Erythrocytes

Ring

Schizant

(*P. ovale* illustrated)

Mosquito bites infected person

Trophozoite

FIGURE 34-1   Life cycle of *Plasmodium* species.

reproductive stage within the mosquito is necessary for maintenance of malaria within a population.

Most malaria seen in the United States is acquired by visitors or residents from countries with endemic disease (**imported malaria**). However, the appropriate vector *Anopheles* mosquito is found in the eastern and western sections of the United States, and domestic transmission of disease has been observed (**introduced malaria**). In addition to transmission by mosquitos, malaria can also be acquired by blood transfusions from an infected donor (**transfusion**

malaria). This type of transmission can also occur among narcotic addicts who share needles and syringes (**"mainline" malaria**). Congenital acquisition, although rare, is also a possible mode of transmission (**congenital malaria**).

Plasmodium vivax

Physiology and Structure *P. vivax* (Figures 34-2 and 34-3) is selective in that it invades only young, immature erythrocytes. These cells respond to the

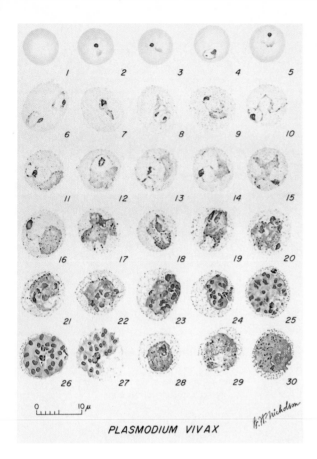

PLASMODIUM VIVAX

FIGURE 34-2 Stages of *Plasmodium vivax* in the erythrocytic cycle. *1,* normal erythrocyte; *2-5,* young trophozoites; *6-16,* growing trophozoites; *17-18,* mature trophozoites; *19-21,* early immature schizonts; *22-23,* older immature schizonts; *24-27,* mature schizonts; *28-29,* macrogametocytes; *30,* microgametocyte. (From Lennette EH, et al, editors: Manual of clinical microbiology, ed 4, Washington DC, 1985, American Society for Microbiology.)

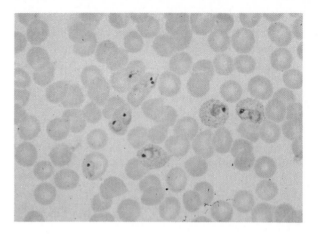

FIGURE 34-3 *Plasmodium vivax* ring forms and young trophozoites. Note the multiple stages of the parasite seen in the peripheral blood smear, enlarged parasitized erythrocytes, and presence of Schüffner's dots with the trophozoite form. These are characteristic of *P. vivax* infections.

parasite by becoming enlarged and distorted. In addition to erythrocyte enlargement and distortion, *P. vivax* produces discrete, red granules called Schüffner's dots in the host cell. These characteristics are helpful in identification of the specific plasmodial species, which is important for the treatment of malaria.

Epidemiology *P. vivax* is the most prevalent of the human plasmodia with the widest geographic distribution, including the tropics, subtropics, and temperate regions.

Clinical Syndromes After an incubation period (usually 10 to 17 days) the patient experiences vague, flu-like symptoms with headache, muscle pains, photophobia, anorexia, nausea, and vomiting.

As the infection progresses, increased numbers of rupturing erythrocytes liberate merozoites, as well as toxic cellular debris and hemoglobin, into the circulation. Together these produce the typical pattern of chills, fever, and malarial rigors. These paroxysms usually reappear periodically (generally every 48 hours) as the cycle of infection, replication, and cell lysis progresses. The paroxysms may remain relatively mild or progress to severe attacks, with hours of sweating, chills, and shaking, persistently high temperatures (103° to 106° F), and exhaustion.

P. vivax causes what is called **"benign tertian malaria,"** referring to the cycle of paroxysms every 48 hours (in untreated patients) and the fact that most patients tolerate the attacks and can survive for years without treatment. If left untreated, however, chronic *P. vivax* infections can lead to brain, kidney, and liver damage as a result of the malarial pigment, cellular debris, and capillary plugging of these organs by masses of adherent erythrocytes.

Laboratory Diagnosis Microscopic examination of both thick and thin films of blood is the method of choice for confirming the clinical diagnosis of malaria and identifying the specific species responsible for disease. The thick films are used to detect the presence of organisms, and the thin film for establishing species identification. Blood films can be taken at any time over the course of the infection, but the best time is midway between paroxysms of chills and fever, when the greatest number of intracellular organisms will be present. It may be necessary to take repeated films at intervals of 4 to 6 hours.

Serologic procedures are available, but they are used primarily for epidemiologic surveys or for screening blood donors. Serologic findings usually remain positive for approximately a year, even after complete treatment of the infection.

Treatment The treatment of *P. vivax* infection involves a combination of supportive measures and chemotherapy. Bed rest, relief of fever and headache, regulation of fluid balance, and in some cases blood transfusion are all supportive therapies.

The chemotherapeutic regimens are as follows:

1. Suppressive—a form of prophylaxis to avoid infection and clinical symptoms
2. Therapeutic—aimed at eradicating the erythrocytic cycle
3. Radical cure—aimed at eradicating the exoerythrocytic cycle in the liver
4. Gametocidal—aimed at destruction of erythrocytic gametocytes to prevent mosquito transmission

Chloroquine is the drug of choice for suppression and therapeutic treatment of *P. vivax*, followed by primaquine for radical cure and elimination of gametocytes. Primaquine is especially effective in preventing relapse from latent forms of *P. vivax* in the liver. Because antimalarial drugs are potentially toxic, it is imperative that physicians carefully review the recommended therapeutic regimens.

Prevention and Control Chemoprophylaxis and prompt eradication of infections are critical in breaking the mosquito-human transmission cycle. Control of mosquito breeding and protection of individuals by screening, netting, protective clothing, and insect repellents are also essential. Immigrants from and travelers to endemic areas must be carefully screened, using blood films or serologic tests to detect possible infection.

The development of vaccines to protect persons living in or traveling to endemic areas is actively under investigation.

Plasmodium ovale

Physiology and Structure *P. ovale* is similar to *P. vivax* in many respects, including its selectivity for

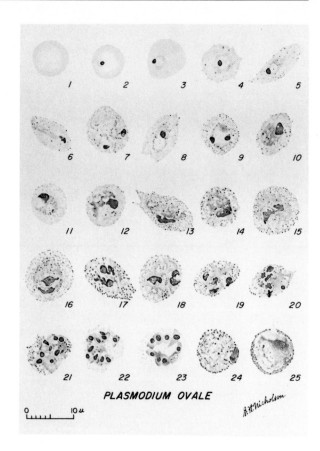

FIGURE 34-4   Stages of *Plasmodium ovale* in the erythrocytic cycle. *2-5*, young trophozoites, *6-12*, growing trophozoites; *13-15*, mature trophozoites; *16-22*, immature schizonts; *23*, mature schizont; *24*, mature macrogametocytes; *25*, mature microgametocyte. (From Lennette EH et al., editors: Manual of clinical microbiology, ed 4, Washington, DC, 1985, American Society for Microbiology.)

young, pliable erythrocytes (Figure 34-4). As a consequence, the host cell becomes enlarged and distorted, usually in an oval form. Schüffner's dots appear as pale pink granules, and the cell border is frequently fimbriated or ragged. The ring forms and gametocytes of *P. ovale*, as for *P. vivax*, are not helpful for identification of the specific species. The schizont of *P. ovale*, when mature, contains about half the number of merozoites seen in *P. vivax*.

Epidemiology *P. ovale* is distributed primarily in tropical Africa, where it is often more prevalent than *P. vivax*. It is also found in Asia and South America.

Clinical Syndromes The clinical picture of tertian attacks for *P. ovale* (**benign tertian** or **ovale malaria**) infection is similar to that for *P. vivax.* Untreated infections last only about a year instead of the several years for *P. vivax.* Both relapse and recrudescence phases are similar to *P. vivax.*

Laboratory Diagnosis As with *P. vivax,* both thick and thin blood films are examined for the typical oval-shaped host cell with Schüffner's dots and a ragged cell wall. Serologic tests reveal cross reaction with *P. vivax* and other plasmodia.

Treatment, Prevention, and Control The treatment regimen, including the use of primaquine to prevent relapse from latent liver forms, is similar to that used for *P. vivax* infections. Preventing *P. ovale* infection involves the same measures as for *P. vivax* and the other plasmodia.

Plasmodium malariae

Physiology and Structure In contrast with *P. vivax* and *P. ovale, P. malariae* is able to infect only mature erythrocytes with firm cell membranes. As a result, the parasite's growth must conform to the size and shape of the red cell (Figure 34-5). This produces no red cell enlargement or distortion, as seen in *P. vivax* and *P. ovale,* but does result in distinctive shapes of the parasite seen in the host cell—"band and bar forms," as well as very compact, dark-staining forms. The schizont of *P. malariae* shows no red cell enlargement or distortion and is usually composed of eight merozoites appearing in a rosette formation. Occasionally, reddish granules called Ziemann's dots appear in the host cell.

As noted for *P. vivax* and *P. ovale,* the ring forms and gametocytes of *P. malariae* are not helpful in establishing species. Unlike *P. vivax* and *P. ovale,* hypnozoites for *P. malariae* are not found in the liver, and relapse does not occur. Recrudesence does occur, and attacks may develop after apparent abatement of symptoms.

Epidemiology *P. malariae* infection occurs primarily in the same subtropical and temperate regions as the other plasmodia but is less prevalent.

Clinical Syndromes The incubation period for *P. malariae* is the longest of the plasmodia, usually lasting 18 to 40 days, but possibly several months to years. The early symptoms are flu-like, with fever patterns of 72 hours (**quartan** or **malarial malaria**) in periodicity. Attacks are moderate to severe and last

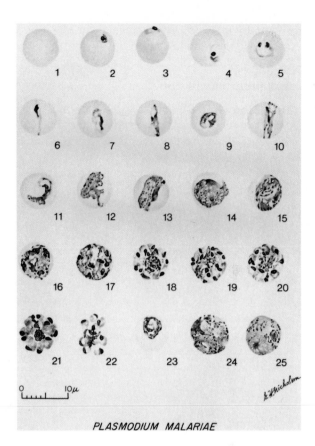

PLASMODIUM MALARIAE

FIGURE 34-5 Stages of *Plasmodium malariae* in the erythrocytic cycle. *2-5,* young trophozoites; *6-11,* growing trophozoites; *12-13,* mature trophozoites; *14-20,* immature schizonts; *21-22,* mature schizonts; *23,* developing gametocyte; *24,* macrogametocyte; *25,* microgametocyte. (From Lennette, EH, et al, editors: Manual of clinical microbiology, ed 4, Washington, DC, 1985, American Society for Microbiology.)

several hours. Untreated infections may last as long as 20 years.

Laboratory Diagnosis Searching thick and thin films of blood for the characteristic bar and band forms, as well as the "rosette" schizont, will establish the diagnosis of *P. malariae* infection. As noted previously, serologic tests cross react with other plasmodia.

Treatment, Prevention, and Control Treatment is similar to that for *P. vivax* and *P. ovale* infections and must be undertaken to prevent recrudescent infections. However, treatment to prevent relapse due to latent liver forms is not required, because these forms do not develop in malarial malaria. Preventive and controlling mechanisms are as discussed for *P. vivax* and *P. ovale*.

Plasmodium falciparum

Physiology and Structure *P. falciparum* demonstrates no selectivity in host erythrocytes and will invade any red blood cell at any stage in its existence (Figure 34-6). Also, multiple merozoites can infect a single erythrocyte. Thus three or even four small rings may be seen in an infected cell. *P. falciparum* is often seen in the host cell at the very edge or periphery of

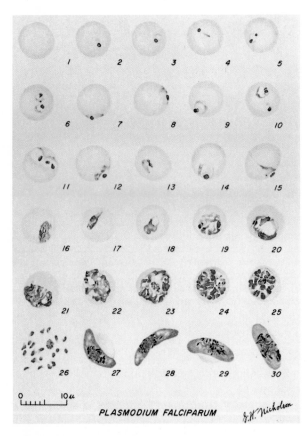

FIGURE 34-6   Stages of *Plasmodium falciparum* in the erythrocytic cycle. *2-11*, young trophozoites; *12-15*, growing trophozoites; *16-18*, mature trophozoites; *19-22*, immature schizonts; *23-26*, mature schizonts; *27-28*, mature macrogametocytes; *29-30*, mature microgametocytes. Only the ring trophozoite forms and gametocytes are found in the peripheral blood. The other forms are sequestered in the capillary beds. (From Lennette EH et al editors: Manual of clinical microbiology, ed 4, Washington, DC, 1985, American Society for Microbiology.)

the cell membrane, appearing almost as if it were stuck on the outside of the cell. This is called the appliqué or accolé position and is distinctive for this species.

Growing trophozoite stages and schizonts of *P. falciparum* are rarely seen in blood films because their forms are sequestered in the liver and spleen. Only in very heavy infections will they be found in the peripheral circulation. Thus examination of peripheral blood smears from patients with *P. falciparum* malaria characteristically contains only young ring forms and occasionally gametocytes. The typical crescentic gametocytes are diagnostic for the species. In contrast to other plasmodia, the ring forms and crescentic gametocytes are helpful in establishing *P. falciparum* infection. Infected red cells do not enlarge and become distorted, as for *P. vivax* and *P. ovale*, and only occasionally will reddish granules known for *P. falciparum* as Maurer's dots be observed.

P. falciparum, like *P. malariae*, does not produce hypnozoites in the liver, and relapses are not known to occur. Recrudescence does occur, and increased attacks may be fatal.

Epidemiology *P. falciparum* occurs almost exclusively in tropical and subtropical regions.

Clinical Syndromes The incubation period of *P. falciparum* is the shortest of the plasmodia ranging from 7 to 10 days, and does not extend over months to years. Following the early flu-like symptoms, *P. falciparum* rapidly produces daily (quotidian) chills and fever, and severe nausea, vomiting, and diarrhea. The periodicity of the attacks then becomes tertian (36 to 48 hours), and fulminating disease develops. The term **"malignant tertian malaria"** is appropriate for this infection. Because it is similar to intestinal infections, the nausea, vomiting, and diarrhea have led to the observation that malaria is "the malignant mimic."

Although any malaria infection may be fatal, *P. falciparum* is the most likely to do so if left untreated. The increased numbers of erythrocytes infected and destroyed result in toxic cellular debris, adherence of red cells to proximal red cells, and the formation of capillary plugging by masses of red cells, platelets, leukocytes, and malarial pigment.

Involvement of the brain (**cerebral malaria**) is most often seen in *P. falciparum* infection. Capillary plugging from an accumulation of malarial pigment and masses of cells can result in coma and death.

Kidney damage is also associated with *P. falcipa-*

rum malaria, resulting in so-called **blackwater fever.** Intravascular hemolysis with rapid destruction of red cells produces a marked hemoglobinuria and can result in acute renal failure, tubular necrosis, nephrotic syndrome, and death. Liver involvement is characterized by abdominal pain, vomiting of bile, severe diarrhea, and rapid dehydration.

Laboratory diagnosis Thick and thin blood films are searched for the characteristic rings of *P. falciparum*, which can frequently occur in multiples within a single cell, as well as in the accolé position (Figure 34-7), and for the distinctive crescentic gametocytes (Figure 34-8).

Laboratory personnel must perform a thorough search of the blood films because mixed infections can occur with any combination of the four species, but most often the combination is *P. falciparum* and *P. vivax*. The detection and proper reporting of a mixed infection bears directly on the treatment modality chosen.

Treatment, Prevention, and Control Treatment of malaria is based on the history regarding travel to endemic areas, prompt clinical review and differential diagnosis, accurate and rapid laboratory work, and correct usage of antimalarial drugs.

Because chloroquine-resistant strains of *P. falciparum* are present in many parts of the world, physicians must review all current protocols for proper treatment of *P. falciparum* infections, noting particularly where chloroquine resistance is known to occur.

If the patient's history indicates the origin is not from a chloroquine-resistant area, the drug of choice is either chloroquine or parenteral quinine. If the patient acquired the *P. falciparum* infection in a known chloroquine-resistant area, the drug of choice is quinine or quinadine. When there is uncertainty whether the *P. falciparum* is chloroquine-resistant, it is advisable to assume the strain is resistant and treat accordingly. If the laboratory reports a mixed infection involving *P. falciparum* and *P. vivax*, the treatment must not only eradicate *P. falciparum* from the erythrocytes but also eradicate the liver stages of *P. vivax* to avoid relapses. Failure on the part of laboratory detection and reporting to establish such a mixed infection can result in inappropriate treatment and unnecessary delay in accomplishing a complete cure.

P. falciparum infection can be prevented and controlled exactly as for *P. vivax* and the other human malarias. The problem with chloroquine resistance complicates the management of these patients, but can be overcome by the physician's awareness of appropriate regimens.

BABESIA

Babesia are intracellular sporozoan parasites that morphologically resemble plasmodia (Figure 34-9). Babesiosis is a zoonosis infecting a variety of animals, such as deer, cattle, and rodents, with humans as

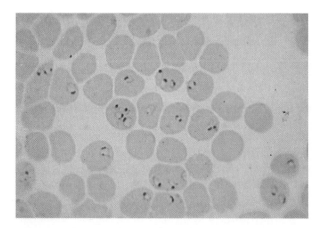

FIGURE 34-7   Ring forms of *Plasmodium falciparum*. Note the multiple ring forms within the individual erythrocytes, characteristic of *P. falciparum*.

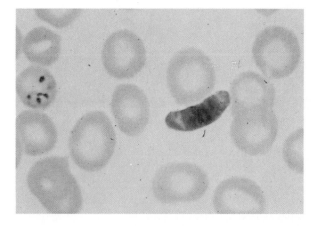

FIGURE 34-8   Mature gametocyte of *Plasmodium falciparum*. The presence of this sausage-shaped form is diagnostic of *P. falciparum* malaria.

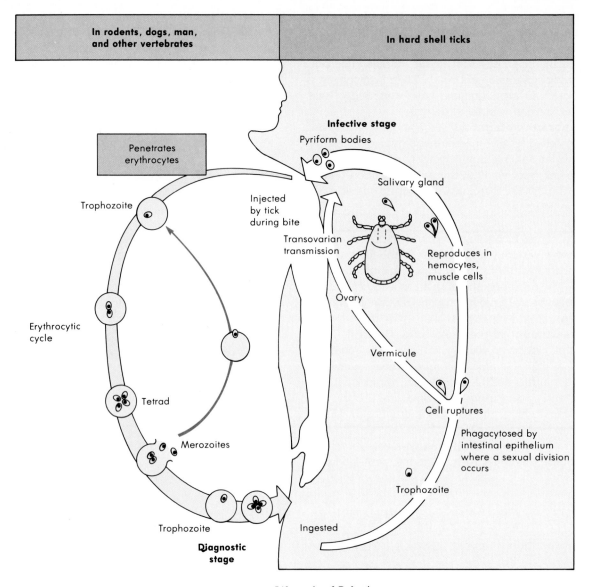

In rodents, dogs, man, and other vertebrates

In hard shell ticks

Penetrates erythrocytes

Trophozoite

Injected by tick during bite

Infective stage
Pyriform bodies

Salivary gland

Transovarian transmission

Reproduces in hemocytes, muscle cells

Erythrocytic cycle

Ovary

Vermicule

Tetrad

Cell ruptures

Merozoites

Phagacytosed by intestinal epithelium where a sexual division occurs

Trophozoite

Trophozoite

Ingested

Diagnostic stage

FIGURE 34-9   Life cycle of *Babesia*.

accidental hosts. Infection is transmitted by the hard shell *(Ixodid)* tick. *Babesia microti* is responsible for human infections.

Physiology and Structure Human infection follows contact with an infected tick (Figure 34-10). The infectious forms are introduced into the bloodstream and infect erythrocytes. The trophozoites multiply by binary fission, forming tetrads, and then lyse the erythrocyte, releasing the merozoites. These can reinfect other cells to maintain the infection. Infected cells can also be ingested by feeding ticks, in which additional replication can take place. Infection in the tick population can also be maintained by transovarian transmission. The infected cells in humans resemble the ring forms of *Plasmodium falciparum*, but malarial pigment or other stages of growth characteristically seen with plasmodial infections are not seen with careful examination of blood smears.

Epidemiology Many species of *Babesia* are found in Africa, Asia, Europe, and North America, with *B. microti* responsible for disease in the northeastern seaboard of the United States (e.g., Nantucket Island,

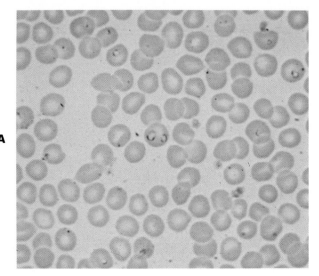

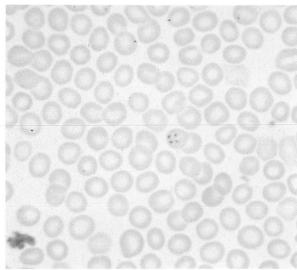

FIGURE 34-10   *Babesia microti* in erythrocytes can resemble malarial parasites. **A**. Erythrocyte with two rings; **B**. Erythrocyte with tetrad.

Martha's Vineyard, Shelter Island). *Ixodes dammini* is the tick vector responsible for transmitting babesiosis in this area, and the natural reservoir hosts are field mice, voles, and other small rodents. Serologic studies in endemic areas have demonstrated a high incidence of past exposure to *Babesia*. Presumably, most infections are asymptomatic or mild. Although most infections follow tick bites, transfusion-related infections have been demonstrated.

Clinical Syndromes After an incubation period of 1 to 4 weeks, symptomatic patients experience general malaise, fever without periodicity, headache, chills, sweating, fatigue, and weakness. As the infection progresses with increased destruction of erythrocytes, hemolytic anemia develops and the patient may experience renal failure. Hepatomegaly and splenomegaly can develop in advanced disease. Low grade parasitemia may persist for weeks. Splenectomy or functional asplenia, immunosuppression, or advanced age increases individual susceptibility to infections, as well as more severe disease.

Laboratory Diagnosis Examination of blood smears is the diagnostic method of choice. Laboratory personnel must be experienced in differentiating *Babesia* and *Plasmodium*. Infected patients may have negative smears due to the low-grade parasitemia. These infections can be diagnosed by inoculating samples of blood into hamsters, who are highly susceptible to infection. Serologic tests are also available for diagnosis.

Treatment, Prevention, and Control The drugs of choice are clindamycin combined with quinine. However, most patients with mild disease recover without specific therapy.

The use of protective clothing and insect repellents can minimize tick exposure in endemic areas, which is critical for prevention of disease. Ticks must feed on humans for several hours before the organisms are transmitted, so prompt removal of ticks can be protective.

TOXOPLASMA GONDII

T. gondii is a typical coccidian parasite related to *Plasmodium, Isospora,* and other members of the phylum Apicomplexa. It is found in a wide variety of animals, including birds and humans. The essential reservoir host of *T. gondii* is the common house cat and other felines.

Physiology and Structure Organisms develop in the intestinal cells of the cat, as well as during an extraintestinal cycle with passage to the tissues via the bloodstream (Figure 34-11). The organisms from the intestinal cycle are passed in cat feces and mature in the external environment within 3 to 4 days into infective oocysts. These oocysts, similar to those of *Isospora belli,* the human intestinal protozoan parasite, can be ingested by mice and other animals (in-

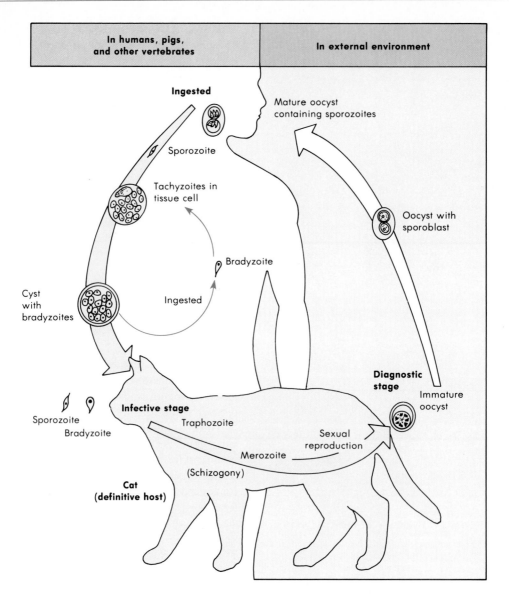

In humans, pigs,
and other vertebrates

In external environment

Ingested

Mature oocyst
containing sporozoites

Sporozoite

Tachyzoites in
tissue cell

Oocyst with
sporoblast

Bradyzoite

Cyst
with
bradyzoites

Ingested

Diagnostic
stage

Immature
oocyst

Sporozoite

Bradyzoite

Infective stage

Traphozoite

Sexual
reproduction

Merozoite

(Schizogony)

Cat
(definitive host)

FIGURE 34-11   Life cycle of *Toxoplasma gondii*.

cluding humans) and produce acute and chronic infection of various tissues, including brain. Infection in cats is established when the tissues of infected rodents is eaten.

Some infective forms or **trophozoites** from the oocyst develop as slender crescentic types called **tachyzoites.** These are rapidly multiplying forms that are responsible for both the initial infection and the tissue damage. Slow-growing, shorter forms called **bradyzoites** also develop and form cysts in chronic infections.

Epidemiology Although symptomatic toxoplasmosis is infrequent, human infection with *T. gondii* is common, occurring worldwide. The wide variety of animals—both carnivores and herbivores, as well as birds—harboring the organism accounts for the widespread transmission.

Humans become infected from two sources: (1) ingestion of improperly cooked meat from animals serving as intermediate hosts and (2) the ingestion of infective oocysts from cat fecal contamination. Transplacental infection from an infected mother can occur with devastating results for the fetus. Serologic studies show an increased prevalence in human populations where the consumption of uncooked meat or meat juices is popular. It is noteworthy that serologic tests of human and rodent populations are negative in the few geographic areas where cats have not existed. Outbreaks of toxoplasmosis in the United States are usually traced to poorly cooked meat (e.g., hamburgers), as well as contact with cat feces.

Transplacental infection can occur in pregnancy either from infection acquired from meat and meat juices or from contact with cat feces. Transfusion infection via contaminated blood can occur but is not common.

Clinical Syndromes Most *T. gondii* infections are benign and asymptomatic, occurring as the parasite moves in the blood to tissues where it becomes an intracellular parasite. When symptomatic disease occurs, the infection is characterized by cell destruction, reproduction of more organisms, and eventual cyst formation.

Symptoms of acute disease include chills, fever, headaches, myalgia, lymphadenitis, and fatigue, occasionally resembling infectious mononucleosis. The chronic form symptoms are lymphadenitis, occasionally a rash, evidence of hepatitis, encephalomyelitis, and myocarditis. In some case chorioretinitis appears and may lead to blindness.

The **congenital toxoplasmosis** results generally from asymptomatic infection in the mother with unrecognized transmission to the fetus. Typically, intracerebral calcification, chorioretinitis, hydrocephaly or microcephaly, and convulsions occur. This is a severe form of the disease and often fatal.

Infection with *T. gondii* thus is asymptomatic in most persons, but in others the disease may be either acute and severe or chronic. An immunosuppressed or compromised patient is especially at risk for severe disease, and reactivation of the cerebral disease is often noted in patients with AIDS.

Laboratory Diagnosis Serologic testing is required, and the diagnosis of acute active infection is established by increasing antibody titers documented in serially collected blood. Because contact with the organism is common attention to increasing titers is essential to differentiate acute, active infection from previous asymptomatic or chronic infection.

Demonstration of these organisms as trophozoites and cysts (Figure 34-12) is seldom attempted because

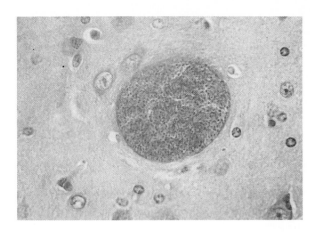

FIGURE 34-12   Cyst of *Toxoplasma gondii* in mouse brain. Hundreds of organisms may be present in the cyst, which may become active and initiate disease with decreased host immunity (e.g., immunosuppression in transplant patients, diseases such as AIDS).

it involves inoculation of laboratory animals or critical examination of tissue specimens, both of which are laborious and time-consuming.

Currently the ELISA test for detecting IgM antibodies appears to be the most reliable procedure because of its simplicity and rapidity in documenting acute infections. The test is not generally satisfactory in AIDS patients with latent or reactivated infections because they fail to produce an IgM response or increasing IgG titer.

Treatment, Prevention, and Control The drug of choice is pyrimethamine, with trisulfapyrimidines or spiramycin as alternatives. Education regarding consumption of raw or poorly cooked meat and meat juices, as well as avoidance of cat feces is essential.

Pregnant women should exercise special caution in handling and cooking meat and in the control of cats. Thorough washing of hands should follow the preparation of meat, including hamburger and frozen meat. The house cat should either be removed from the environment or kept indoors to avoid contact with rodents. Also, the cat should be fed only dry or canned food. The handling of the cat litterbox should be done by another person. If this is not possible, then the pregnant woman should empty the litterbox every day to avoid development of infective oocysts in cat feces. Oocysts require 3 to 4 days to become infective. Thorough washing of hands and use of disposable gloves should be a part of routine handling of cat litter.

Children should be protected from cat feces, especially with regard to open sandboxes, which are often used by cats.

SARCOCYSTIS LINDEMANNI

S. lindemanni is a typical coccidian closely related to the intestinal forms, *S. suihominis, S. bovihominis,* and *Isospora belli,* and the blood and tissue parasite, *Toxoplasma gondii. S. lindemanni* occurs worldwide in various animals, especially sheep, cattle, and pigs. Humans are only accidentally infected as the result of eating meat from these animals.

Most infections are asymptomatic, but occasionally an infection may cause myositis, swelling of muscle, dyspnea, and eosinophilia. Infection of the myocardium has been observed but is extremely rare. There is no specific treatment for the muscle infection.

PNEUMOCYSTIS CARINII

The taxonomic position of this extracellular, opportunistic organism producing interstitial plasma cell pneumonia is not clear, although recent evidence indicates a relationship to fungi. However, because *Pneumocystis* has historically been placed with sporozoans in phylum Apicomplexa, it will be in this section. Pulmonary infections with *P. carinii* occur in a wide variety of domestic and wild animals, as well as humans. The life cycle of *P. carinii* appears to consist of a resistant cyst stage and sporozoites released from the cyst in an asexual reproductive process (sporogony). These become trophozoites and are capable of attaching to the surface of pulmonary epithelial cells. Asexual reproduction of trophozoites on cell surfaces eventually leads to the cyst stage and reproduction, producing additional sporozoites.

Transmission from host-to-host is apparently by droplet inhalation and close contact. This is an opportunistic organism present in many persons who are completely asymptomatic. Symptomatic disease, **pneumocystosis,** develops when some imbalance or debilitating illness (e.g., AIDS) is present, suggesting a long-term carrier state.

Epidemiology This is a cosmopolitan organism in humans and other animals. Rodents especially have been suggested as reservoir hosts.

Pneumocystosis is seen primarily in premature and malnourished children in crowded institutions such as orphanages and hospitals. It is also seen in adults in chronic disease wards and is the most common opportunistic infection seen in patients with AIDS and other immune deficiencies.

Clinical Syndromes Onset of the disease is insidious, but it can be suspected in patients who are malnourished or immunocompromised and who develop fever and pneumonitis not otherwise explainable.

As the disease progresses, the patient experiences weakness, dyspnea, and tachypnea, leading to cyanosis. Radiology of the lungs shows infiltrations spreading from hilar areas, giving the lungs a so-called ''ground glass'' appearance. In these cases the arterial oxygen tension is low and carbon dioxide tension is normal or low. Death is the result of asphyxia.

Laboratory Diagnosis Examination of stained slides of impression smears of lung tissue obtained by brush biopsy, or aspirates of bronchial washings reveals the typical *P. carinii* organisms, as well as material obtained by percutaneous transthoracic needle aspiration (Figure 34-13). The parasites can be stained with silver or Giemsa stain, or with specific fluorescein-labeled antibodies.

Radiology is of value in establishing the typical appearance of the lungs infected with *P. carinii.*

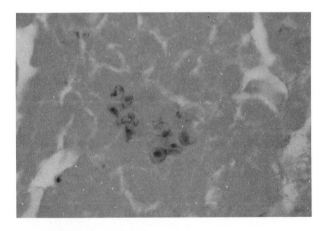

FIGURE 34-13 *Pneumocystis carinii* cysts stained with Gomori methenamine silver.

Serologic tests are not available as a diagnostic procedure.

Treatment, Prevention, and Control The drug of choice is trimethoprim-sulfamethoxazole, with pentamadine as an alternative. Supportive measures such as oxygen and antibiotics may also be indicated.

Education regarding transmission of the organism, avoidance of crowding of infected patients, and eliminating possible contact with droplet transmission are all critical, as are prompt diagnosis and treatment. The organism and its transmission are difficult to control because of the extended carrier state and its presence as an opportunistic pathogen.

LEISHMANIA

The hemoflagellates are flagellated, insect-transmitted protozoa that infect blood and tissues. Three species of *Leishmania,* a protozoan hemoflagellate, produce human disease: *L. donovani, L. tropica,* and *L. braziliensis* (Table 34-2). The diseases are distinguished by the ability of the organism to infect deep tissues **(visceral leishmaniasis)** or replicate only in cooler superficial tissues **(cutaneous** or **mucocutaneous leishmaniasis).** The reservoir hosts and geographic distribution differ for the three species, but transmission by sandflies (belonging to the genus *Phlebotomus* or *Lutzomyia*) is common to all leishmanial species.

Leishmania donavani

Physiology and Structure The life cycles of all leishmania parasites differ in epidemiology, tissues affected, and clinical manifestations (Figure 34-14).

Table 34-2 Leishmaniasis in Humans	
Parasite	**Disease**
L. donovani	Visceral leishmaniasis (Kala-azar, dum dum fever)
L. tropica	Cutaneous leishmaniasis (Oriental sore, Delhi boil)
L. braziliensis	Mucocutaneous leishmaniasis (American leishmaniasis, espundia, Chiclero ulcer)

The **promastigote** stage (long, slender form with a free flagellum) is present in the saliva of infected sandflies. Following a bite, the promastigotes penetrate through the skin, lose their flagella, enter the amastigote stage, and invade reticuloendothelial cells. Reproduction occurs in the amastigote stage and, as cells rupture, destruction of specific tissues develops (e.g., visceral organs such as liver and spleen, cutaneous tissues). The **amastigote** stage (Figure 34-15) is diagnostic for leishmaniasis, as well as the infectious stage for sandflies. Ingested amastigotes transform in the sandfly into the promastigote stage, which multiplies by binary fission in the fly midgut. After development, this stage will migrate to the fly proboscis where new human infection can be introduced during feeding.

Epidemiology *L. donovani* of the classic **kala-azar** or **dum dum fever** type occurs in many parts of Asia, Africa, and Southeast Asia. Except for some rodents in Africa, there are few reservoir hosts. The vector is the *Phlebotomus* sandfly. Variants of *L. donovani* are also recognized. *L. donovani infantum* is present in countries along the Mediterranean basin (European, Near Eastern, and African), and is found in parts of China and the U.S.S.R. Reservoir hosts of this organism include dogs, foxes, jackals, and porcupines. The vector is also the *Phlebotomus* sandfly. *L. donovani chagasi* is found in South and Central America, especially Mexico and the West Indies. Reservoir hosts are dogs, foxes, and cats, and the vector is the *Lutzomyia* sandfly.

Clinical Syndromes The incubation period for **visceral leishmaniasis** may be from several weeks to a year, with gradual onset of fever, diarrhea, and anemia. Chills and sweating that may resemble malaria symptoms are common early in the infection. As organisms proliferate and invade cells of the liver and spleen, marked enlargement of these organs, weight loss, and emaciation occur. Kidney damage may also occur as cells of the glomeruli are invaded. With persistence of the disease, deeply pigmented, granulomatous areas of skin, referred to as "post-kala-azar dermal leishmaniasis," occur. If untreated, visceral leishmaniasis develops into a fulminating, debilitating, and lethal disease in a few weeks, or may persist as a chronic deteriorating disease, leading to death in 1 or 2 years.

Laboratory Diagnosis The amastigote stage can be demonstrated in tissue biopsy, bone marrow examination, lymph node aspiration, and thorough

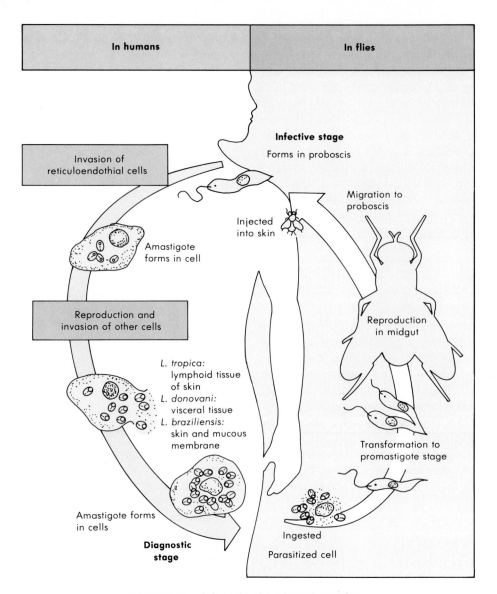

FIGURE 34-14   Life cycle of *Leishmania* species.

examination of properly stained smears. Culture of blood, bone marrow, and other tissues often demonstrates the promastigote stage of the organisms (Figure 34-16). Serologic testing is also available.

Treatment, Prevention, and Control The drug of choice is stibogluconate; pentamidine is the alternative. Prompt treatment of human infections and control of reservoir hosts along with insect control helps eliminate transmission of disease. Protection from sandflies by screening and insect repellents is also essential.

Leishmania tropica

Physiology and Structure The life cycle of *L. tropica* is illustrated in Figure 34-14.

Epidemiology Cutaneous leishmaniasis produced by *L. tropica* is present in many parts of Asia, Africa, Mediterranean Europe, and the southern U.S.S.R. In these regions the reservoir hosts are dogs, foxes, and rodents, and the vector is the sandfly *Phlebotomus*. Two related species are also recognized. *L. aethiopica* is endemic in Ethiopia, Kenya, and Yemen, with dogs and rodents as reservoir hosts and

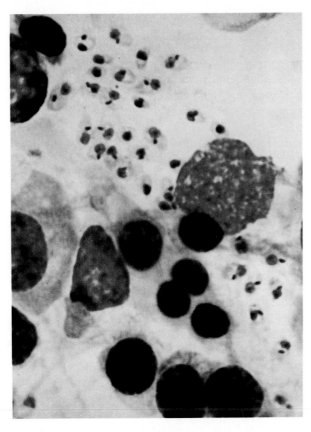

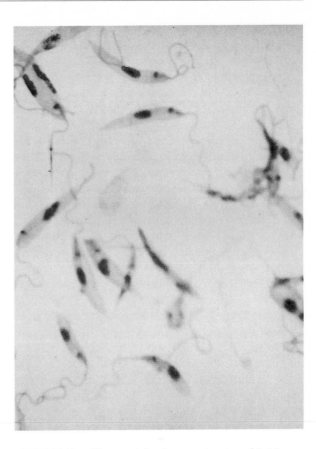

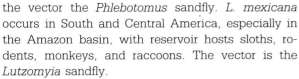

FIGURE 34-15 Giemsa-stained amastigotes (Leishman-Donovan bodies) of *Leishmania donovani* present in bone marrow. A small dark-staining kinetoplast can be seen next to the spherical nucleus in some parasites. (Reproduced by permission from LR Ash and TC Orihel, Atlas of Human parasitology, ed 2, © 1984 by the American Society of Clinical Pathologists Press, Chicago.)

FIGURE 34-16 Giemsa-stained promastogotes of *Leishmania tropica*. (Reproduced by permission from LR Ash and TC Orihel, Atlas of human parasitology, ed 2, © 1984 by the American Society of Clinical Pathologists Press, Chicago.)

the vector the *Phlebotomus* sandfly. *L. mexicana* occurs in South and Central America, especially in the Amazon basin, with reservoir hosts sloths, rodents, monkeys, and raccoons. The vector is the *Lutzomyia* sandfly.

Clinical Syndromes The incubation period after a sandfly bite may be as short as 2 weeks, or 2 months may elapse before the first sign, a red papule, appears at the site of fly's bite and feeding. This lesion becomes irritated, with intense itching, and begins to enlarge and ulcerate. Gradually, the ulcer becomes hard and crusted and exudes a thin, serous material. At this stage secondary bacterial infection may complicate the disease.

The lesion may heal without treatment in a matter of months, but usually leaves a disfiguring scar. A

disseminated nodular type of cutaneous leishmaniasis has been reported from Ethiopia, probably caused by an allergy to *L. aethiopica* antigens.

Laboratory Diagnosis Demonstration of the amastigotes in properly stained smears from touch preparations or ulcer biopsy and culture of ulcer tissue are appropriate laboratory methods to determine the diagnosis. Serologic tests are also available.

Treatment, Prevention, and Control The drug of choice is stibogluconate, with an alternative treatment of applying heat directly to the lesion. Protection from sandfly bites using screening, protective clothing, and repellents is essential. Prompt treatment and eradication of the ulcers to prevent transmission, along with control of sandflies and reservoir hosts, will reduce the incidence of human infection.

Leishmania braziliensis

Physiology and Structure The life cycle of *L. braziliensis* is illustrated in Figure 34-14.

Epidemiology Mucocutaneous leishmaniasis produced by *L. braziliensis* is seen from the Yucatan peninsula into Central and South America, especially in those rain forests where chicle workers are exposed to sandfly bites while harvesting the chicle sap for chewing gum (thus the name **chiclero ulcer**). There are many jungle reservoir hosts, as well as domestic dogs. The vector is the *Lutzomyia* sandfly. The variant *L. braziliensis panamensis* is similar in all respects to *L. braziliensis* except for its more frequent occurrence in Panama and slight difference in growth in cultures. Reservoir hosts and the vector are similar to *L. braziliensis.*

Clinical Syndromes The incubation period and appearance of ulcers for *L. braziliensis* are similar to *L. tropica,* requiring a few weeks to months for the papule to appear. The essential difference in clinical disease is the involvement and destruction of mucous membranes and related tissue structures. This is often combined with edema and secondary bacterial infection to produce severe and disfiguring facial mutilation.

Laboratory Diagnosis The diagnostic tests are similar for all *Leishmania.* Organisms are demonstrated in ulcers or cultured tissue. Serologic testing is also available.

Treatment, Prevention, and Control The drug of choice is stibogluconate with an alternative of amphotericin B. As with all the other leishmania complexes, screening, protective clothing, insect repellents, and prompt treatment are needed to prevent transmission and control disease. The protection of forest and construction workers in endemic areas is most difficult, and disease here may be effectively controlled only by vaccination. Work to develop a vaccine is currently under way.

TRYPANOSOMES

Trypanosoma, another hemoflagellate, causes two distinctly different forms of disease (Table 34-3). One is called **African trypanosomiasis,** or **sleeping sickness,** and is produced by *Trypanosoma brucei gambiense* and *T. brucei rhodesiense.* It is transmitted by tsetse flies. The second infection is called **American trypanosomiasis,** or **Chagas' disease** produced by *T. cruzi.* It is transmitted by true bugs (triatomids, reduviids—so-called kissing bugs).

Trypanosoma brucei gambiense

Physiology and Structure The life cycle of the African forms of trypanosomiasis is illustrated in Figure 34-17. The infective stage of the organism is the **trypomastigote,** present in the salivary glands of transmitting tsetse flies (Figure 34-16). This stage has a free flagellum and an undulating membrane running the full length of the body. The trypomastigotes enter the wound created by the fly bite and find their way into blood and lymph, eventually invading the central nervous system. Reproduction of the trypomastigotes in blood, lymph, and spinal fluid is by binary or longitudinal fission.

These trypomastigotes in blood are then infective for biting tsetse flies, where further reproduction occurs in the midgut. The organisms then migrate to the salivary glands where an **epimastigote** form (having a free flagellum but only a partial undulating membrane) continues reproduction to the infective tryposmastigote stage. Tsetse flies become infective 4 to 6 weeks after feeding on blood from a diseased patient.

Epidemiology *T. brucei gambiense* is limited to tropical West and Central Africa, correlating to the range of the tsetse fly vector. The tsetse flies transmitting *T. brucei gambiense* prefer shaded stream

Table 34-3 Trypansoma Species Responsible for Human Diseases

Parasite	Vector	Disease

T. brucei
  sspp. gambiense
  sspp. rhodesiense
	Tsetse fly	African trypanosomiasis
		  (sleeping sickness)

T. cruzi	Reduviids	American trypanosomiasis (Chagas' disease)

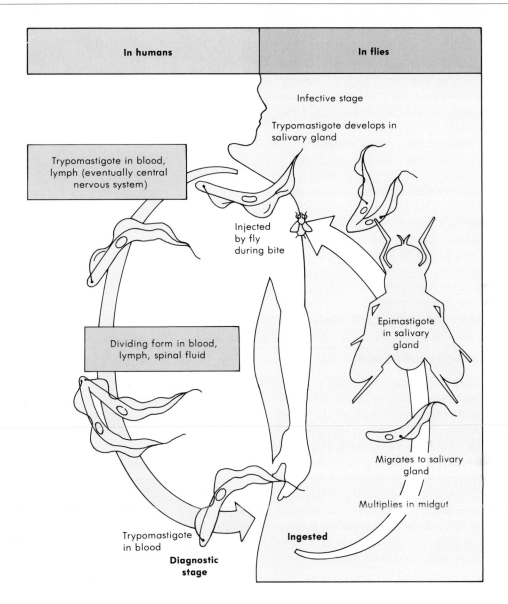

In humans	In flies

Infective stage

Trypomastigote develops in salivary gland

Trypomastigote in blood, lymph (eventually central nervous system)

Injected by fly during bite

Epimastigote in salivary gland

Dividing form in blood, lymph, spinal fluid

Migrates to salivary gland

Multiplies in midgut

Trypomastigote in blood

Diagnostic stage

Ingested

FIGURE 34-17   Life cycle of *Trypanosoma brucei*.

banks for reproduction and proximity to human dwellings. Persons who work in such areas are at greatest risk of infection. There are no known animal reservoir hosts.

Clinical Syndromes The incubation period of **Gambian sleeping sickness** varies from a few days to weeks. One of the earliest signs of disease is an occasional ulcer at the site of the fly bite. As reproduction of organisms continues, the lymph nodes are invaded and fever, myalgia, arthalgia, and lymph

node enlargement result. Patients in this acute phase often exhibit hyperactivity.

Chronic disease progresses to central nervous system involvement with lethargy, tremors, meningoencephalitis, mental retardation, and general deterioration. In the final stages of chronic disease, convulsions, hemiplegia, and incontinence occur, and the patient becomes difficult to arouse or respond, eventually progressing to a comatose state. Death is the result of central nervous system damage, combined with other infections such as malaria or pneumonia.

Laboratory Diagnosis Organisms can be demonstrated in thick and thin blood films, in concentrated anticoagulated blood preparations, and in aspirations from lymph nodes and concentrated spinal fluid. Serologic tests are also available.

Treatment, Prevention, and Control In the acute stages of the disease the drug of choice is suramin, with pentamidine as an alternative. In chronic disease with central nervous system involvement the drug of choice is melarsoprol, with alternatives of tryparsamide combined with suramin.

The most essential elements are control of breeding sites of the tsetse flies by clearing brush, using insecticides, and treating human cases to reduce transmission to flies. Persons going into known endemic areas should wear protective clothing and use screening, bed netting, and insect repellents.

Trypanosoma brucei rhodesiense

Physiology and Structure The life cycle is similar to *T. brucei gambiense* (Figure 34-17), with both trypomastigote and epimastigote stages, and transmission by tsetse flies.

Epidemiology The organism is found primarily in East Africa, especially the cattle-raising countries where tsetse flies breed in the brush rather than stream banks. *T. brucei rhodesiense* also differs from *T. brucei gambiense* in having domestic animal hosts (cattle and sheep) and wild game animals as reservoir hosts. This transmission and vector cycle makes the organism much more difficult to control than *T. brucei gambiense*.

Clinical Syndromes The incubation period for *T. bucei rhodesiense* is shorter than for *T. brucei gambiense*. Acute disease (fever, rigors, and myalgia) occurs more rapidly and progresses to a fulminating, rapidly fatal illness. Infected persons are usually dead within 9 to 12 months if untreated.

This more virulent organism also develops in greater numbers in the blood, without lymphadenopathy, and CNS invasion occurs early in the infection, with lethargy, anorexia, and mental disturbance. The chronic stages described for *T. brucei gambiense* are not often seen because, in addition to rapid CNS disease, the organism produces kidney damage and myocarditis leading to death.

Laboratory Diagnosis Examination of blood and

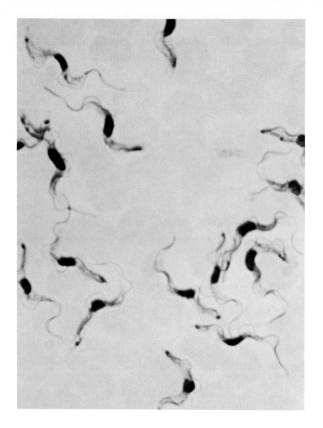

FIGURE 34-18   Trypomastigote stage of *Trypanosoma gambiense* in a blood smear. (Reproduced by permission from LR Ash and TC Orihel, Atlas of human parasitology, ed 2, © 1984 by the American Society of Clinical Pathologists Press, Chicago.)

spinal fluid is carried out as for *T. brucei gambiense* (Figure 34-18). Serologic tests are available.

Treatment, Prevention, and Control The same treatment protocol applies as for *T. brucei gambiense,* with early treatment for the more rapid neurologic manifestations. Similar prevention and control measures are needed: tsetse fly control and use of protective clothing, screens, netting, and insect repellent. In addition, early treatment is essential to control transmission, detect infection, and determine treatment in domestic animals. Control of infection in game animals is difficult, but it can be reduced if tsetse fly control measures are applied, specifically eradication of brush and grassland breeding sites.

Trypanosoma cruzi

Physiology and Structure The life cycle of *T. cruzi* (Figure 34-19) differs from *T. brucei* with the

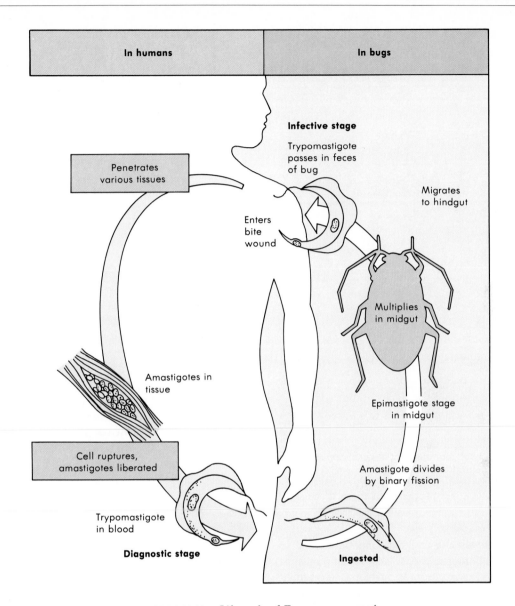

FIGURE 34-19   Life cycle of *Trypanosoma cruzi*.

development of an additional form called an amastigote (Figure 34-20). The amastigote is an intracellular form with no flagellum and no undulating membrane. It is smaller than the trypomastigote, oval, and found in tissues. The infective trypomastigote, present in the feces of a reduviid bug ("kissing bug"), enters the wound created by the biting, feeding bug. The bugs have been called kissing bugs because they frequently bite people around the mouth and in other facial sites. They are notorious for biting, feeding on blood and tissue juices, and then defecating into the wound. The organisms in the feces of the bug enter the wound and are usually aided in penetration by the patient's rubbing or scratching the irritated site.

The trypomastigotes then migrate to other tissues (e.g., cardiac muscle, liver, brain), lose the flagellum and undulating membrane, and become the smaller, oval, intracellular amastigote form. These intracellular amastigotes multiply by binary fission and eventually destroy the host cells. Then they are liberated to enter

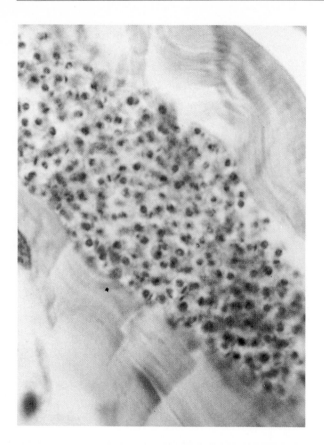

FIGURE 34-20   Amastigote stage of *Trypanosoma cruzi* in skeletal muscle. (Reproduced by permission from LR Ash and TC Orihel, Atlas of human parasitology, ed 2, © 1984 by the American Society of Clinical Pathologists Press, Chicago.)

new host tissue as intracellular amastigotes or to become trypomastigotes infective for feeding reduviid bugs. Ingested trymastigotes develop into epimastigotes in the midgut of the insect and reproduce by longitudinal binary fission. The organisms migrate to the hindgut of the bug, develop into trypomastigotes, and then leave the bug in the feces after biting, feeding, and defecating, initiating a new human infection.

Epidemiology *T. cruzi* occurs widely in both reduviid bugs and a broad spectrum of reservoir animals in North, Central, and South America. Human disease is found most often among children in South and Central America, where there is a direct correlation between infected wild animal reservoir hosts and the presence of infected bugs whose nests are found in human homes. Cases are rare in the United States because the bugs prefer nesting in animal burrows,

and because homes are not as open to nesting as those in South and Central America.

Clinical Syndromes Chagas' disease may be asymptomatic, acute, or chronic. One of the earliest signs is development at the site of the bug bite of an erythematous and indurated area called a **chagoma.** This is often followed by a rash and edema around the eyes and face. Acute infection is also characterized by fever, chills, malaise, myalgia, and fatigue. Death may ensue a few weeks after an acute attack, the patient may recover, or the patient may enter the chronic phase as organisms proliferate and enter the heart, liver, spleen, brain, and lymph nodes.

Chronic Chagas' disease is characterized by hepatosplenomegaly, myocarditis, and enlargement of the esophagus and colon as a result of destruction of tissues controlling the growth of these organs.

Megacardium and electrocardiographic changes are commonly seen in chronic disease. Involvement of the central nervous system may produce granulomas in the brain with cyst formation and a meningoencephalitis. Death from chronic Chagas' disease results from tissue destruction in the many areas invaded by the organisms, and sudden death results from complete heart block and brain damage.

Laboratory Diagnosis *T. cruzi* can be demonstrated in thick and thin blood films or concentrated anticoagulated blood early in the acute stage. As the infection progresses, the organisms leave the bloodstream and become difficult to find. Culture of blood and biopsy of lymph nodes may demonstrate the organisms in the amastigote stage. Serologic tests are also available.

Treatment, Prevention, and Control The drug of choice is nifurtimox. Education regarding the disease, its insect transmission, and the wild animal reservoirs is critical. Bug control, eradication of nests, and construction of homes to prevent nesting of bugs are also essential. The use of DDT in bug-infested homes has demonstrated a drop in both malaria and Chagas' disease transmission.

Development of a vaccine is possible because *T. cruzi* does not have the wide antigenic variation observed with the African trypanosomes.

BIBLIOGRAPHY

Beaver PC, Jung RC, and Cupp EW: Clinical parasitology, ed 9, Philadelphia, 1984, Lea & Febiger.

Benenson MW et al: Oocyst-transmitted toxoplasmosis

associated with ingestion of contaminated water, N Engl J Med 307:666-669, 1982.

Bittencourt AL et al: Esophageal involvement in congenital Chagas' disease: report of a case with megaesophagus, Am J Trop Med Hyg 33:30-33, 1984.

Brown HW and Neva FA: Basic clinical parasitology, ed 5, Norwalk, Connecticut, 1983, Appleton-Century-Crofts.

Bruce-Chwatt LJ: The challenge of malaria vaccine: trials and tribulations, Lancet 1:371-373, 1987.

Corredor A et al: Epidemiology of visceral leishmaniasis in Colombia, Am J Trop Med Hyg 40:480-486, 1989.

Felegie TP et al: Recognition of *Pneumocystis carinii* by Gram stain impression smears of lung tissue: J Clin Microbiol 20:1190-1191, 1984.

Gombert ME et al: Human babesiosis: clinical and therapeutic considerations, JAMA 248:3005-3007, 1982.

Howard BJ et al: Clinical and pathogenic microbiology, St Louis, 1987, The CV Mosby Co.

Kreutzer BD et al: Identification of *Leishmania* sp. by multiple isozyme analysis, Am J Trop Med Hyg 32:703-715, 1983.

Markell EK, Voge M, and John DT; Medical parasitology, ed 6, Philadelphia, 1986, WB Saunders Co.

Mazier D et al: Complete development of hepatic stages of *Plasmodium falciparum* in vitro, Science 227:440-442, 1985.

Ray K et al: Evaluation of serology as a tool for malaria surveillance in East Champaran District of Bihar, India, Ann Trop Med Parasitol 82:225-228, 1988.

Schmidt GD and Roberts LS: Foundations of Parasitology, ed 4, St Louis 1989, The CV Mosby Co.

Sharma GK: Malaria: a critical review, J Commun Dis 19:187-290, 1987.

Spencer HC et al: Imported African trypanosomiasis in the United States, Ann Intern Med 82:633-638, 1975.

Washington JA II, editor: Laboratory procedures in clinical microbiology, ed 2, New York, 1985, Springer-Verlag.

Wong B et al: Central-nervous-system toxoplasmosis in homosexual men and parenteral drug abusers, Ann Intern Med 100:36-42, 1984.

CHAPTER 35

Nematodes

The most common helminths recognized in the United States are primarily intestinal nematodes, although in other countries nematode infections of blood and tissues can cause devastating disease. The nematodes are the most easily recognized form of intestinal parasite because of their large size and cylindric unsegmented bodies; hence the common name, **roundworms.** These parasites live primarily as adult worms in the intestinal tract, and if both male and female worms are present, nematode infections are most commonly confirmed by detecting fertilized eggs in feces. The most common nematodes of medical importance are listed in Table 35-1.

The filariae are long, slender roundworms that are parasites of blood, lymph, subcutaneous, and connec-

tive tissues. All of these nematodes are transmitted by either mosquitoes or biting flies, and most produce larval worms called microfilariae that are demonstrated in blood specimens or in subcutaneous nodules.

ENTEROBIUS VERMICULARIS

Physiology and Structure

Enterobius vermicularis, the pinworm, is a small, white worm that is familiar to parents who find them in the perianal folds or vagina of an infected child. Infection is initiated by ingestion of embryonated eggs. (Figure 35-1). Larvae hatch in the small intes-

Table 35-1 Nematodes of Medical Importance

Parasite	Common Name	Disease
Enterobius vermicularis	Pinworm	Enterobiasis
Ascaris lumbricoides	Roundworm	Ascariasis
Toxocara canis	Dog ascaris	Visceral larva migrans
Toxocara cati	Cat ascaris	Visceral larva migrans
Trichuris trichiura	Whipworm	Trichuriasis
Ancylostoma duodenale	Old World hookworm	Hookworm infection
Necator americanus	New World hookworm	Hookworm infection
Ancylostoma braziliense	Dog or cat hookworm	Cutaneous larva migrans
Strongyloides stercoralis	—	Strongyloidiasis
Trichinella spiralis	Threadworm	Trichinosis
Wuchereria bancrofti	Bancroft's filariasis	Filariasis
Brugia malayi	Malayan filariasis	Filariasis
Loa loa	African eye worm	Loiasis
Mansonella species	Mansonelliasis	Filariasis
Onchocerca volvulus	River blindness	Onchocerciasis
Dracunculus medinensis	Guinea worm	Dracunculosis
Dirofilaria immitis	Dog heartworm	Filariasis

tine and migrate to the large intestine, where they mature into adults in 2 to 6 weeks. Fertilization of the female by the male produces the characteristic asymmetric eggs. These eggs are laid in the perianal folds by the migrating female. The eggs rapidly mature and are infectious within hours.

Epidemiology

Enterobius vermicularis occurs worldwide, but is most common in the temperate regions, where person-to-person spread is greatest in crowded conditions such as in day care centers, schools, and mental institutions. An estimated 500 million cases of pinworm infection are reported worldwide, and this is the most common helminth infection in North America.

Infection occurs when the eggs are ingested and the larval worm is free to develop in the intestinal mucosa. These eggs may be transmitted from hand to mouth by children scratching the perianal folds in response to the irritation caused by the migrating,

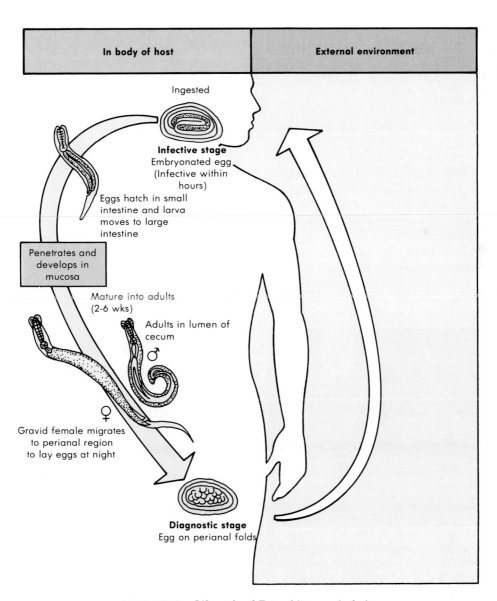

FIGURE 35-1   Life cycle of *Enterobius vermicularis.*

egg-laying female worms, or the eggs may find their way to clothing and play objects in day care centers. They can also survive long periods of time in the dust that accumulates over doors, on windowsills, and under beds in the rooms of infected persons. Egg-laden dust can be inhaled and swallowed to produce infestation. In addition **autoinfection** (''retrofection'') can occur wherein eggs hatch in the perianal folds and the larval worms migrate into the rectum and large intestine. Infected individuals who handle food can also be a source of infection. No animal reservoir for *Enterobius* is known. Physicians should be aware of the related epidemiology of *Dientamoeba fragilis;* this organism correlates well with the presence of *E. vermicularis,* with *D. fragilis* transported in the pinworm eggshell.

Clinical Syndromes

Many children and adults show no symptoms and serve only as carriers. Patients who are allergic to the secretions of the migrating worms experience severe pruritus, loss of sleep, and fatigue. The pruritus may cause repeated scratching of the irritated area and lead to secondary bacterial infection. Worms that migrate into the vagina may produce genitourinary problems and granulomas.

There is evidence that worms attached to the bowel wall may produce inflammation and granuloma for-

mation around the eggs. Penetration through the bowel wall into the peritoneal cavity, liver, and lungs has been recorded infrequently.

Laboratory Diagnosis

Occasionally, the adult worms are seen by laboratory personnel in stool specimens, but the method of choice for diagnosis involves use of an anal swab with a sticky surface that picks up the eggs (Figure 35-2) for microscopic examination. Sampling can be done with clear tape or commercially available swabs. The sample should be collected when the child arises, before bathing or defecation, to pick up eggs laid by migrating worms during the night. Parents can collect the specimen and deliver it to the physician for immediate microscopic examination. Three swabbings, one per day for 3 consecutive days, may be required for detecting the diagnostic eggs. The eggs are rarely seen in fecal specimens.

Treatment, Prevention, and Control.

The drug of choice is pyrantel pamoate; as an alternative drug, mebendazole is used. To avoid reintroduction of the organism and reinfection in the family environment, it is customary to treat the entire family simultaneously.

Personal hygiene, clipping of fingernails, thorough washing of bed clothes, and prompt treatment of infected individuals all contribute to control. When housecleaning is done in the home of an infected family, dusting under beds, on window sills, and over doors should be done with a damp mop to avoid inhalation of infectious eggs.

ASCARIS LUMBRICOIDES

Physiology and Structure

These large, pink worms have a more complex life cycle then *Enterobius vermicularis* (Figure 35-3), but are otherwise typical of an intestinal roundworm.

The ingested infective egg releases a larval worm that penetrates the duodenal wall, enters the bloodstream, is carried to the liver and the heart, and then enters the pulmonary circulation. The larvae break free in the alveoli of the lungs, where they grow and molt. In about 3 weeks the larvae pass from the respi-

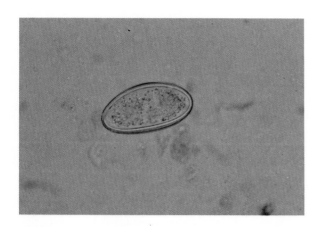

FIGURE 35-2   *Enterobius vermicularis* egg. The thin-walled eggs are 50-60 μm by 20-30 μm, ovoid, and flattened on one side (not because children sit on them—but this is an easy way to correlate the egg morphology with the epidemiology of the disease). A developing larva is seen inside the egg.

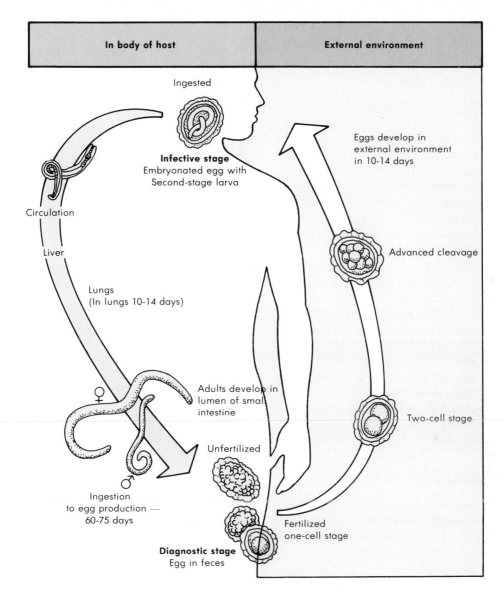

FIGURE 35-3   Life cycle of *Ascaris lumbricoides*.

ratory system to be coughed up, swallowed, and returned to the small intestine.

As the male and female worms mature in the small intestine (primarily jejunum), fertilization of the female by the male initiates egg production, which may amount to 200,000 eggs per day for as long as a year. Female worms can also produce unfertilized eggs in the absence of males. Eggs are found in the feces 60 to 75 days after the initial infection. Fertilized eggs become infectious after approximately 2 weeks.

Epidemiology

Ascaris lumbricoides is prevalent in areas of poor sanitation and where human feces are used as fertilizer. Because both food and water are contaminated with ascaris eggs, this parasite affects more of the world's population than any other. Although no animal reservoir is known for *A. lumbricoides,* an almost identical species from pigs, *Ascaris suum,* can infect humans. This species is seen in swine growers and is associated with the use of pig manure for gardening. Ascaris eggs are quite hardy and can survive extreme temperatures and persist for several months in feces and sewage. Ascariasis is the most common helminth worldwide, with an estimated 1 billion individuals infected.

Clinical Syndromes

Infections that are caused by the ingestion of only a few eggs may produce no symptoms; however, even a single adult ascaris may be dangerous because it can migrate into the bile duct and liver and create tissue dammage. Furthermore, because the worm has a tough, wiry body, it can perforate the intestine, creating a peritonitis with secondary bacterial infection.

Following infection with many larvae, migration of worms to the lungs can produce a pneumonitis resembling an asthmatic attack. Also, a tangled bolus of mature worms in the intestine can result in obstruction, perforation, and occlusion of the appendix. As mentioned previously, migration into the bile duct, gallbladder, and liver can produce severe tissue damage. This migration can occur in response to fever, drugs other than those used to treat ascariasis, and some anesthetics. Patients with many larvae may also experience abdominal tenderness, fever, distention, and vomiting.

Laboratory Diagnosis

Examination of the sediment of concentrated stool reveals the knobby-coated, bile-stained, fertilized and unfertilized eggs (Figure 35-4). Occasionally, adult worms pass with the feces, which can be quite dramatic because of their large size (6 to 12 inches long). Roentgenologists may also visualize the worms in the intestine, and cholangiograms often disclose their presence.

Treatment, Prevention, and Control

The drug of choice is mebendazole; pyrantel pamoate and piperazine are alternatives. Patients with mixed parasitic infections *(Ascaris lumbricoides, Giardia lamblia,* and *Entamoeba histolytica)* in the stool should be treated for ascariasis first to avoid provoking worm migration and possible intestinal perforation. Education, improved sanitation, and avoidance of human feces as fertilizer are critical. A program of mass treatment in highly endemic areas

A   **B**

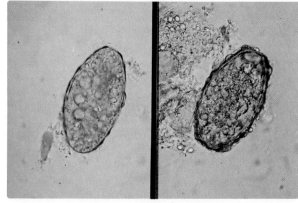

C

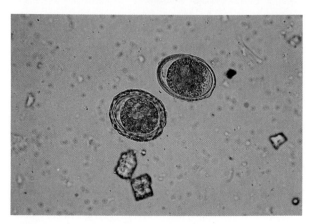

FIGURE 35-4   *Ascaris lumbricoides* eggs. Eggs are oval, measuring 55 to 75 μm long and 35 to 50 μm wide. The thick-walled outer shell can be partially removed (decorticated egg). **A,** Unfertilized egg; **B,** unfertilized decorticated egg; **C,** fertilized eggs with the outer shell removed from the upper egg.

has been suggested, but this may not be economically feasible. Furthermore, eggs can persist in contaminated soil for 3 years or more. Certainly, improved personal hygiene among individuals who handle food is an important aspect of control.

TOXOCARA CANIS AND *T. CATI*

Physiology and Structure

Toxocara canis and *T. cati* are ascaris worms naturally parasitic in the intestines of dogs and cats that accidentally infect humans, producing a disease called **visceral larva migrans** or **toxocariasis.**

When ingested by humans, the eggs of these worms can hatch into larval forms that cannot follow its normal developmental cycle in the natural dog or cat host. They can penetrate the human gut and reach the bloodstream and then migrate as larvae to various human tissues. They do not develop beyond the migrating larval form.

Epidemiology

Wherever infected dogs and cats are present, the eggs are a threat to humans. This is especially true for children who are exposed more readily to contaminated soil and who tend to put objects in their mouth.

Clinical Syndromes

Migration of the larval worm may produce hemorrhage, necrosis, and granulomas. Also, eosinophilia, liver damage, pulmonary inflammation, and ocular problems have been observed. Children often have pronounced eosinophilia, chronic cough and fever, and retinitis when larvae migrate to the eyes.

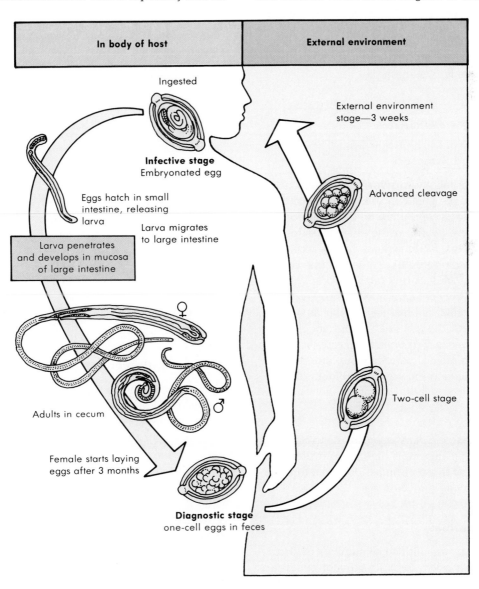

FIGURE 35-5   Life cycle of *Trichuris trichiura*.

Laboratory Diagnosis

Diagnosis of visceral larval migrans is based on clinical findings and confirmed by serologic testing. Examination of feces from infected patients is not useful because egg-laying adults are not present. However, examination of fecal material from infected pets often supports the diagnosis.

Treatment, Prevention, and Control

The drug of choice is diethylcarbamazine or thiabendazole. Mebendazole is an acceptable alternative. This zoonosis can be greatly reduced if pet owners conscientiously eradicate worms from their animals and clean up pet fecal material from yards and school playgrounds. Children's play areas and sandboxes should be carefully monitored.

TRICHURIS TRICHIURA

Physiology and Structure

Commonly called **"whipworm"** because it resembled the handle and lash of a whip, *T. trichiura* has a simple life cycle (Figure 35-5). Ingested eggs hatch into a larval worm in the small intestine and then migrate to the cecum, where they penetrate the mucosa and mature to adults. Three months after the initial infection, the fertilized female worm will start laying eggs and may produce 3,000 to 10,000 eggs per day. Female worms can live for as long as 8 years. Eggs passed into the soil will mature and become infectious in 3 weeks. *T. trichiura* eggs are distinctive with dark bile staining, barrel shape, and the presence of polar plugs in the egg shell (Figure 35-6).

Epidemiology

Similar to *Ascaris lumbricoides,* distribution is worldwide and prevalence is directly correlated with poor sanitation and use of human feces as fertilizer. No animal reservoir is recognized.

Clinical Syndromes

Most infections are with small numbers of *Trichuris* and are usually asymptomatic, although secondary bacterial infection may occur because the heads of the worms penetrate deep into the intestinal mucosa.

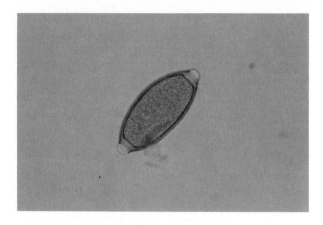

FIGURE 35-6   *Trichuris trichiura* egg. The eggs are barrel shaped, measuring 50 by 24 μm, with a thick wall and two prominent plugs at the ends. Internally, an unsegmented ovum is present.

Infections with many larvae may produce abdominal pain and distention, bloody diarrhea, weakness, and weight loss. Appendicitis may occur as worms fill the lumen, and prolapse of the rectum is seen in children because of the irritation and straining during defecation. Anemia and eosinophilia are also seen in severe infections.

Laboratory Diagnosis

Stool examination reveals the characteristic bile-stained eggs with polar plugs. Light infestations may be difficult to detect because of the paucity of eggs in the stool specimens.

Treatment, Prevention, and Control

The drug of choice is mebendazole. As for *Ascaris lumbricoides,* prevention depends on education, good personal hygiene, adequate sanitation, and avoiding the use of human feces as fertilizer.

HOOKWORMS

Ancylostoma duodenale and *Necator americanus*

Physiology and Structure

The two human hookworms are *Ancylostoma duodenale* (Old World hookworm) and *Necator ameri-*

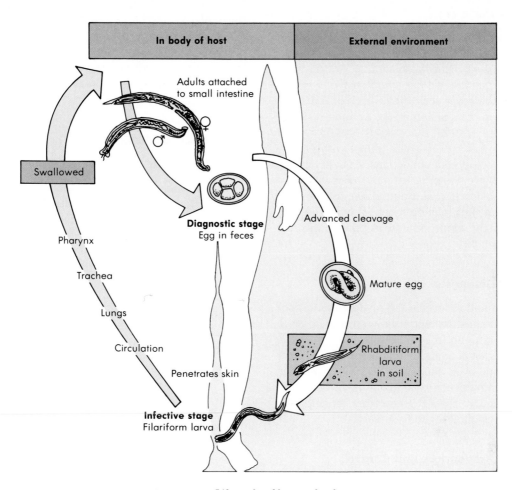

FIGURE 35-7   Life cycle of human hookworms.

canus (New World hookworm). Differing only in geographic distribution, structure of mouthparts, and relative size, these two species will be discussed together as agents of hookworm infection. The human phase of the hookworm life cycle is initiated when a **filariform** (infective-form) larva penetrates intact skin (Figure 35-7). The larva then enters the circulation, is carried to the lungs and, like *Ascaris lumbricoides,* is coughed up, swallowed, and develops to adulthood in the small intestine. The adult *N. americanus* has a hooklike head, which accounts for the name commonly used. Adult worms lay as many as 10,000 to 20,000 eggs per day, which are released into feces. Egg laying is initiated 4 to 8 weeks after the initial exposure and can persist for up to 5 years. On contact with soil, the **rhabditiform** (noninfective)

larvae are released from the eggs and within 2 weeks develop into filariform larvae. The filariform larvae can then penetrate exposed skin (e.g., bare feet) and initiate a new cycle of human infection.

Both species have mouth parts designed for sucking blood from injured intestinal tissue; *A. duodenale* has chitinous teeth and *N. americanus* has shearing chitinous plates.

Epidemiology

Hookworm infections are reported worldwide but are primarily in warm subtropical and tropical regions, and in southern parts of the United States where direct contact with contaminated soil can lead to human disease.

Clinical Syndromes

Skin-penetrating larvae may produce an allergic reaction and rash at sites of entry, and larvae migrating in the lungs can cause pneumonitis. Adult worms produce the gastrointestinal symptoms of nausea, vomiting, and diarrhea and, as blood is lost from feeding worms, a microcytic hypochromic anemia develops. Blood loss with *N. americanus* is 0.03 ml per worm per day; loss 5 to 10 times greater with *A. duodenale*. In severe, chronic infections, emaciation and mental and physical retardation may occur related to anemia from blood loss and nutritional deficiencies. Also, intestinal sites may be secondarily infected by bacteria when the worms migrate along the intestinal mucosa.

Laboratory Diagnosis

Stool examination reveals the characteristic non-bile-stained, segmented eggs shown in Figure 35-8. Larvae are not found in stool specimens unless the specimen was left at ambient temperature for a day or more. The eggs of *A. duodenale* and *N. americanus* cannot be distinguished. The larvae must be examined to identify these hookworms specifically, although this is clinically unnecessary.

Treatment, Prevention, and Control

The drug of choice is mebendazole; pyrantel pamoate is an alternative. In addition to eradication of

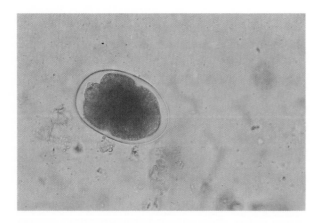

FIGURE 35-8   Human hookworm egg. The eggs are 60 to 75 μm long and 35 to 40 μm wide, thin-shelled, and enclose a developing larva.

the worms to stop blood loss, iron therapy is indicated to raise hemoglobin levels to normal. Education, improved sanitation, and controlled disposal of human feces are critical. The simple practice of wearing shoes in endemic areas helps reduce the prevalence of infection.

Ancylostoma braziliense

Physiology and Structure

This species of hookworm is naturally parasitic in the intestines of dogs and cats and accidentally infects humans. It produces a disease properly called **cutaneous larva migrans** but also called "ground itch" and "creeping eruption." The filariform larvae of this hookworm penetrate intact skin but can develop no further in humans. The larvae remain trapped in the skin of the wrong host for weeks or months, wandering through subcutaneous tissue and creating serpentine tunnels.

Epidemiology

Similar to the situation with ascaris worms, the threat of infection is greatest among children coming into contact with soil or sandboxes contaminated with animal hookworm eggs. Infections are prevalent throughout the year on beaches in subtropical and tropical regions; in summer infection is reported as far north as the Canadian-U.S. border.

Clinical Syndromes

The migrating larvae may provoke a severe erythematous and vesicular reaction. The pruritus and scratching of the irritated skin may lead to secondary bacterial infection.

Laboratory Diagnosis

Occasionally, larvae are recovered in skin biopsy or following freezing of the skin, but most diagnoses are based on the clinical appearance of the tunnels and a history of contact with dog and cat feces.

Treatment, Prevention, and Control

The drug of choice is thiabendazole. This zoonosis, as with animal ascaris, can be reduced by educating pet owners to treat their animals for worm infections and to pick up pet feces from yards, beaches, and sandboxes. In endemic areas shoes or sandals should be worn to prevent infection.

STRONGYLOIDES STERCORALIS

Physiology and Structure

Although the morphology of these worms and epidemiology of their infections is similar to the hookworm, the life cycle of *S. stercoralis* (Figure 35-9) differs in three aspects: (1) eggs hatch into larvae in the intestine and before they are passed in feces; (2) larvae can mature into filariforms and cause autoinfec-

tions; and (3) a free living, nonparasitic cycle can be established outside the human host.

In direct development, like the hookworm, a skin-penetrating larva enters the circulation and follows the pulmonary course, it is coughed up and swallowed, and adults develop in the small intestine. The fertilized female worm then produces eggs that hatch in the intestine into larvae, which are passed into the soil through feces, or that are recognized in the examination of a stool specimen. Egg laying begins 1 month after the initial infection and can persist for many years. The larvae in the soil may reinitiate the direct cycle by changing from a rhabditiform (noninfective) to a filariform (infective) larva to penetrate skin again.

In indirect development the larvae in the soil devel-

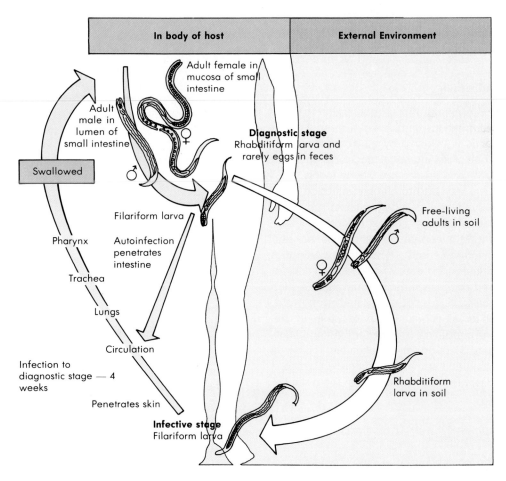

FIGURE 35-9   Life cycle of *Strongyloides stercoralis*.

op into free-living adults that produce eggs and larvae. Several generations of this nonparasitic existence may occur before new larvae again become skin-penetrating parasites.

Finally, in **autoinfection** rhabditiform larvae in the intestine do not pass with feces but become filariform larvae. These penetrate the intestinal or perianal skin and follow the circulation-pulmonary-cough-and-swallow course to become adults, producing more larvae in the intestine. This cycle can persist for years and can lead to hyperinfection and massive, or disseminated—often fatal—infection.

Epidemiology

Similar to hookworms in its requirements for warm temperatures and moisture, *S. stercoralis* demonstrates low prevalence but a somewhat broader geographic distribution, including parts of the northern United States and Canada. Sexual transmission also occurs. Animal reservoirs such as domestic pets are recognized.

Clinical Syndromes

Pneumonitis from migrating larvae, abdominal discomfort, and diarrhea are common in the early stages of infection. This is followed by malabsorption syndrome, ulceration of the intestinal mucosa suggesting ulcerative colitis, and passage of bloody stools.

Autoinfection may lead to chronic infections that last for years even in non-endemic areas. Immunosuppressed patients are at greatest risk of autoinfection. The resulting hyperinfection entails severe pulmonary involvement, with wheezing, coughing, fever, and weakness. Patients who have been severely compromised or those suffering from malnutrition may experience massive or disseminated stronglyoidiasis, in which larval worms migrate to the heart, liver, kidneys, and central nervous system. A complicating factor seen postmortem is secondary bacterial infection of blood and spinal fluid produced because the migrating larvae carry bacteria from the intestine to other sites.

Laboratory Diagnosis

Examination of concentrated stool sediment reveals the larval worms (Figure 35-10) but, in contrast with hookworm infections, eggs are generally not

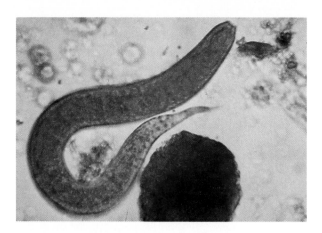

FIGURE 35-10   *Strongyloides stercoralis* larvae. Larvae are 180 to 380 μm long and 14 to 24 μm wide. These are differentiated from hookworm larvae by the length of the buccal cavity and esophagus, as well as the structure of the genital primordium.

seen. Three stools, one per day for 3 days, as for *Giardia lamblia* are recommended because *S. stercoralis* larvae may occur in "showers," with many present one day and few or none the next. Duodenal aspiration has been used to make the diagnosis, serologic procedures are available, and larvae may be found in sputum in cases of massive or disseminated infection.

Treatment, Prevention, and Control

The drug of choice is thiabendazole, with mebendazole as an alternative. Similar to hookworm, *S. stercoralis* control requires education, proper sanitation, and prompt treatment of existing infections. Patients in endemic areas, about to have immunosuppressive therapy, should have at least three stool examinations to rule out *S. stercoralis* infection and avoid the risks of autoinfection.

TRICHINELLA SPIRALIS

Physiology and Structure

Human infection with *T. spiralis* (**trichinosis**) is an accidental occurrence, because this organism is primarily a parasite of carnivorous animals. Figure 35-11 illustrates the simple, direct life cycle, which

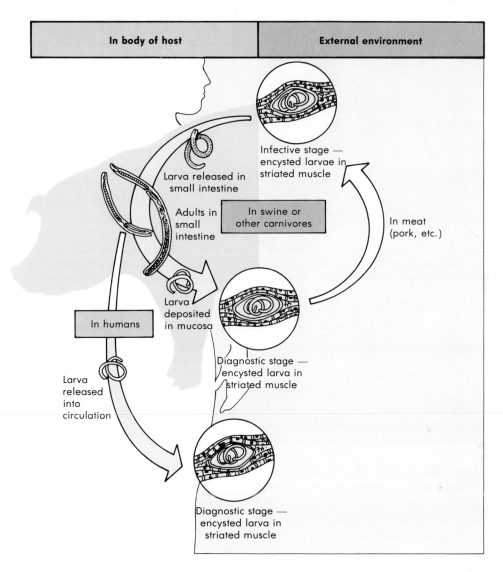

FIGURE 35-11   Life cycle of *Trichinella spiralis*.

terminates in the musculature of humans where the larvae eventually die and calcify. The infection begins when meat containing encysted larvae is digested. The larvae leave the meat in the small intestine and within 2 days develop into adult worms. A single fertilized female worm produces over 1,500 larvae in 1 to 3 months. These larvae move from the intestinal mucosa into the bloodstream and are carried in the circulation to various muscle sites throughout the body, where they coil in striated muscle fibers and become encysted (Figure 35-12).

Epidemiology

Trichinosis occurs worldwide in humans, and its greatest prevalence is associated with consumption of pork products. In addition to its transmission from pigs, many carnivorous and omnivorous animals harbor the organism and are potential sources of human infection. Notably, polar bears and walruses in the Arctic account for outbreaks in human populations—especially with a strain of *T. spiralis* that is more resistant to freezing than the *T. spiralis* strains found in the continental United States and other temperate regions.

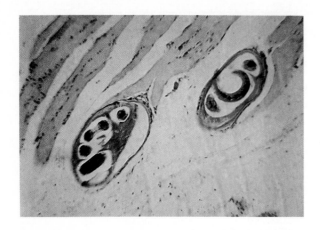

FIGURE 35-12   Encysted larva of *Trichinella spiralis* in biopsied muscle. (From Finegold SM and Baron EJ, editors: Bailey and Scott's Diagnostic Microbiology, ed 7, St Louis, 1986, The CV Mosby Co.)

Clinical Syndromes

In mild infections with few migrating larvae, patients may only experience a flu-like syndrome with slight fever and mild diarrhea. With more extensive larval migration, persistent fever, gastrointestinal distress, marked eosinophilia, muscle pain, and periorbital edema occur. "Splinter" hemorrhages beneath the nails are also a common finding, thought to be caused by vasculitis resulting from toxic secretions of the migrating larvae. In heavy infections severe neurologic symptoms, including psychosis, meningoencephalitis, and cerebrovascular accident, may occur.

Patients who survive the migration, muscle destruction, and encystment of larvae in moderate infections experience a decline in clinical symptoms in 5 or 6 weeks. Lethal trichinosis results when myocarditis, encephalitis, and pneumonitis combine; the patient dies 4 to 6 weeks after infection. Respiratory arrest often follows heavy invasion and muscle destruction in the diaphragm.

Laboratory Diagnosis

Diagnosis is usually established with clinical observations, especially when an outbreak can be traced to consumption of improperly cooked pork or bear meat. The laboratory may confirm this if the encysted larvae are detected in the implicated meat or biopsied muscle from the patient. Marked eosinophilia is characteristically present in patients with trichinosis. Serologic procedures are also available for confirmation.

Treatment, Prevention, and Control

Steroids are recommended for severe symptoms, along with thiabendazole or mebendazole. Education regarding pork and bear meat transmission is essential, especially the recommendation that pork and bear meat be cooked until the interior is gray. Microwave cooking and smoked or dried meat do not kill all larvae.

Laws regulating the feeding of garbage to pigs help control transmission, as may regulations controlling the foraging of bears in garbage pits and public parks. Freezing pork, as conducted in federally inspected meat packing plants, has reduced transmission. Quick freezing of pork at −40° C effectively destroys the organisms, as does low-temperature storage at −15° C for 20 days or more.

WUCHERERIA BANCROFTI AND *BRUGIA MALAYI*

Physiology and Structure

Because of their many similarities, *Wuchereria bancrofti* and *Brugia malayi* will be discussed together. Human infection is initiated by introduction of infective larvae, present in the saliva of a biting mosquito, into a bite wound (Figure 35-13). Various species of *Anopheles, Aedes* and *Culex* mosquitoes are known to be vectors of Bancroft's and Malayan filariasis. The larvae migrate from the bite site to the lymphatics, primarily in the arms, legs, or groin, where growth to adulthood occurs. Nine to 12 months after the initial infection, the adult male worm fertilizes the female, who in turn produces the larval microfilariae that find their way into the circulation. The presence of microfilariae in blood is diagnostic for human disease and is infective for feeding mosquitoes. In the mosquito the larvae move through the stomach and thoracic muscles in developmental stages and finally migrate to the proboscis. There they become an infective third-stage larva and are transmitted in the saliva of the feeding mosquito. The adult form in humans can persist for years.

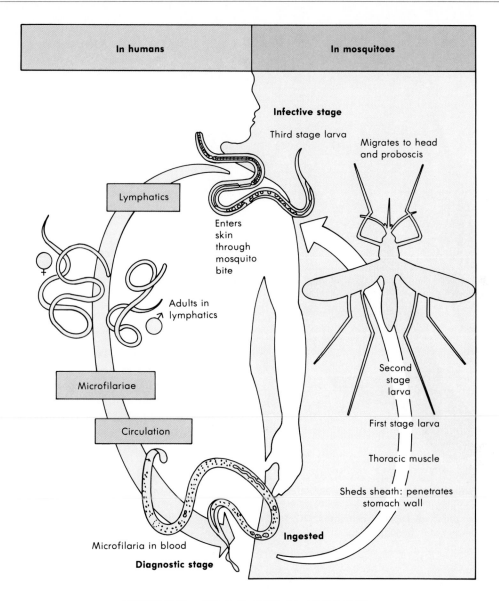

FIGURE 35-13   Life cycle of *Wuchereria bancrofti.*

Epidemiology

W. bancrofti occurs in both tropical and subtropical areas and is endemic in central Africa, along the Mediterranean coast, in many parts of Asia (including China, Korea, and Japan), and in the Philippines. It is also present in Haiti, Trinidad, Surinam, Panama, Costa Rica, and Brazil. No animal reservoir has been identified. *B. malayi* is found primarily in Malaysia, India, Thailand, Vietnam, and parts of China, Korea, Japan, and many Pacific Islands. Animal reservoirs such as cats and monkeys are recognized.

Clinical Syndromes

In some patients there is no sign of disease, even though blood specimens may show many microfilariae present. In other patients early symptoms are fever, lymphangitis and lymphadenitis with chills, and recurrent febrile attacks. As the infection progresses, the lymph nodes enlarge, possibly involving many parts of the body, including the extremities, the scrotum, and the testes with occasional abscess formation. This results from the physical obstruction of lymph in the vessels caused by the presence of adult

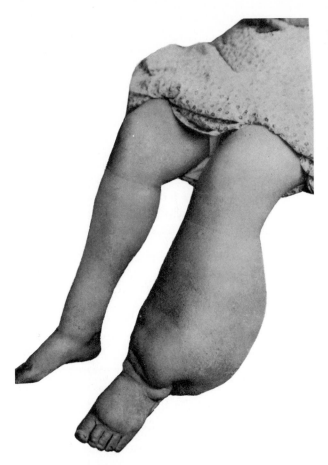

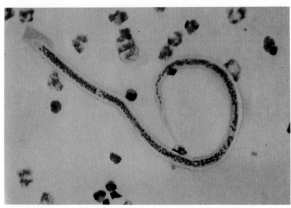

FIGURE 35-15   *Wuchereria bancrofti* microfilaria in blood smear. (From Finegold SM and Baron EJ, editors: Bailey and Scott's Diagnostic Microbiology, ed 7, St Louis, 1986, The CV Mosby Co.)

greater numbers of microfilariae in blood at night. It is recommended that blood specimens be taken between 10 PM and 4 AM to detect infection.

W. bancrofti, as well as *B. malayi* and *Loa loa,* demonstrate a sheath on their microfilariae. This can be the first step in identifying the specific types of filariasis. Further identification is based on study of head and tail structures (Figure 35-16).

FIGURE 35-14   Elephantiasis of left leg caused by *Wuchereria bancrofti.* (From Binford CH and Conner DH: Pathology of tropical and extraordinary diseases, Washington, DC, 1976, Armed Forces Institute of Pathology.)

worms and host reactivity in the lymphatics. The thickening and hypertrophy of tissues infected with the worms may lead to the enlargement of tissues, especially the extremities, progressing to **filarial elephantiasis** (Figure 35-14). Filariasis of this type is thus a chronic, debilitating, and disfiguring disease requiring prompt diagnosis and treatment.

Laboratory Diagnosis

Microfilariae can be demonstrated in blood films as for malaria (Figure 35-15). Concentrations of anticoagulated blood specimens and urine specimens are also valuable procedures. Both *W. bancrofti* and *B. malayi* have a periodicity in production of microfilariae—**"nocturnal periodicity."** This results in

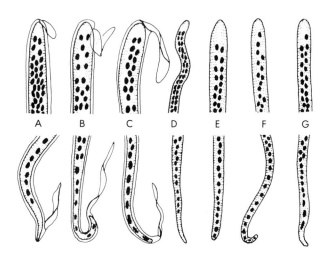

FIGURE 35-16   Differentiation of microfilariae. Identification of microfilariae is based on the presence of a sheath covering the larvae, as well as the distribution of nuclei in the tail region. **A,** *Wuchereria bancrofti;* **B,** *Brugia malayi;* **C,** *Loa loa;* **D,** *Onchocerca volvulus;* **E,** *Mansonella perstans;* **F,** *M. streptocerca;* and **G,** *M. ozzardi.*

Clinically, an exact species identification is not critical, because treatment for all the filarial infections is identical, except for *Onchocerca volvulus.*

Serologic testing is also available through reference laboratories as a diagnostic procedure.

Treatment, Prevention, and Control

The drug of choice for treatment of *W. bancrofti* and *B. malayi* infections is diethylcarbamazine. Education regarding filarial infections, mosquito control, use of protective clothing and insect repellents, and treatment of infections to prevent further transmission are essential. Control of *B. malayi* infections is more difficult because of the presence of disease in animal reservoirs.

LOA LOA

Physiology and Structure

The life cycle of *Loa loa* is similar to that illustrated in Figure 35-13, except the vector is a biting fly call *Chrysops,* the mango fly. Approximately 6 months after infection, production of microfilariae will start and can persist for 17 years or more. Adult worms can migrate through subcutaneous tissues, muscle, and in front of the eyeball.

Epidemiology

Loa loa is confined to the equatorial rain forests of Africa and is endemic in tropical West Africa, the Congo basin, and parts of Nigeria. Monkeys in these areas serve as reservoir hosts in.the life cycle with mango flies as vectors.

Clinical Syndromes

Symptoms usually do not appear until a year or so after the fly bite, because the worms are slow in reaching adulthood. One of the first signs of infection is the so-called fugitive or **Calabar swellings.** These swellings are transient and usually appear on the extremities, produced as the worms migrate through subcutaneous tissues, creating large nodular areas that are painful and pruritic. Because eosinophilia (50%-70%) is observed, Calabar swellings are believed to be the result of allergic reactions to the worms or their metabolic products.

Adult *Loa loa* worms can also migrate under the conjunctiva, producing irritation, painful congestion, edema of the eyelids, and impaired vision. The presence of a worm in the eye can obviously cause anxiety in the patient. The infection may be long lived, and in some cases symptomless.

Laboratory Diagnosis

The clinical observation of Calabar swellings or migration of worms in the eye, combined with eosinophilia, should alert the physician to consider infection with *Loa loa.* The microfilariae, as for *W. bancrofti* and *B. malayi,* can be found in blood. In contrast with the other filariae, *Loa loa* is primarily present during the daytime. Serologic testing can also be useful for confirming the diagnosis.

Treatment, Prevention, and Control

The drug of choice is diethylcarbamazine, as for *W. bancrofti* and *B. malayi.* Surgical removal of worms migrating across the eye or bridge of the nose can be accomplished by immobilizing the worm with instillation of a few drops of 10% cocaine. Education regarding the infection and its vector, especially for persons entering the known endemic areas, is essential. Protection from fly bites by using screening, appropriate clothing, and insect repellents, along with treatment of cases, is also critical in reducing the incidence of infection. However, the presence of disease in animal reservoirs (e.g., monkeys) limits the feasibility of controlling this disease.

MANSONELLA SPECIES

Filarial infections caused by *Mansonella* species are less important than those previously discussed, but physicians should be aware of the names because they may encounter patients with these infections.

All of the *Mansonella* species produce nonsheathed microfilariae in blood and subcutaneous tissues, and all are transmitted by biting midges (Culicoides) or black flies (Simulium). All of these species are treatable with diethylcarbamazine as for previous filarial infections. Species indentification, if desired, can be accomplished with blood smears, noting the structure of the microfilariae. Serologic tests are also available.

Prevention and control require measures involving insect repellants, screening, and other precautions as for all insect-transmitted diseases.

Mansonella perstans

Mansonella perstans occurs primarily in parts of tropical Africa and Central and South America. It may produce allergic skin reactions, edema, and calabar swellings similar to *Loa loa*. Reservoir hosts are chimpanzees and gorillas.

Mansonella ozzardi

Mansonella ozzardi is found primarily in Central and South America and the West Indies. It may produce swelling of the lymph nodes and occasional hydrocele. There are no known reservoir hosts.

Mansonella streptocerca

M. streptocerca occurs primarily in Africa, especially in the Congo basin. It may produce edema in the skin and, rarely, a form of elephantiasis. Monkeys serve as reservoir hosts.

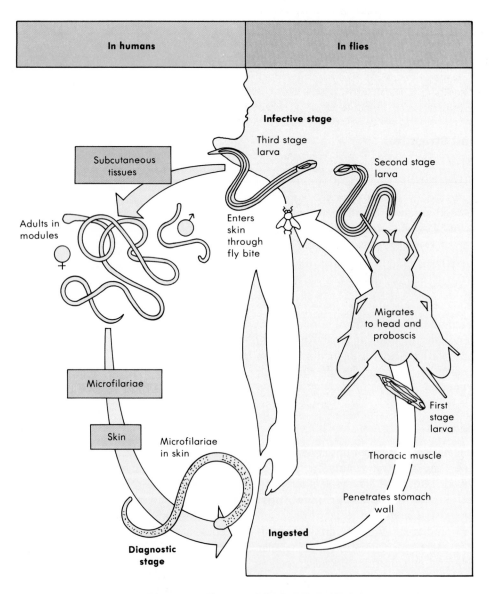

FIGURE 35-17   Life cycle of *Onchocerca volvulus*.

ONCHOCERCA VOLVULUS

Physiology and Structure

Infection occurs following introduction of larvae of *O. volvulus* through the skin during the biting and feeding of the vector, simulian or black fly (Figure 35-17). The larval worms migrate from skin to subcutaneous tissue and develop into adult male and female worms in subcutaneous nodules. The female worm, after fertilization by the male, begins producing non-sheathed microfilariae that migrate to the skin and other body tissues. These nonsheathed microfilariae appear in skin tissue are infective for biting or feeding black flies.

Epidemiology

Onchocera volvulus is endemic in many parts of Africa, especially in the Congo basin and the Volta River basin. In the Americas it occurs in many Central and South American countries. It is primarily a parasite of humans, but some domestic animals (horses and cattle) may serve as reservoir hosts.

Several species of the black fly genus *Simulium* serve as vectors, but none so appropriately named as the principal vector, *Simulium damnosum* ("the damned black fly"). These black flies, or buffalo gnats, breed in fast-flowing streams, which makes control or eradication by insecticides almost impossible because the chemicals are rapidly washed away from the eggs and larval flies.

There is a greater prevalence of infection in men than women in endemic areas because of their work in or near the streams where the black flies are breeding. It has been shown by studies in endemic areas in Africa that 50% of men will be totally blind before they are 50 years of age. This accounts for the common term, **"river blindness,"** applied to the disease onchocerciasis (Figure 35-18). This fear of blindness has created an additional problem in many parts of Africa, because whole villages leave the area near streams and farmland that could produce food. The migrating populations then find themselves in areas where starvation faces them.

Clinical Syndromes

The incubation period from infectious larvae to adult worms is from several months to a year. Disease initially is seen with fever, eosinophilia, and urticaria. As the worms mature, copulate, and begin producing microfilariae, subcutaneous nodules begin to appear

FIGURE 35-18   Onchocerciasis is appropriately called "river blindness." Men are commonly infected with *O. volvulus* while working near rivers where the vector, black flies, breed. Blindness, following chronic infection of ocular tissues, is common. These blind men are guided by young children. (From Binford CH and Conner DH: Pathology of tropical and extraordinary diseases, Washington, DC, 1976, Armed Forces Institute of Pathology.)

and can occur on any part of the body. These nodules are most dangerous when they are present on the head and neck because the microfilariae may migrate to the eyes and cause serious tissue damage, leading to blindness.

A number of skin conditions are related to the presence of this parasite, including pruritus, hyperkeratosis, myxedematous thickening, and a form of elephantiasis called "hanging groin" when the nodules are located near the genitalia.

Laboratory Diagnosis

The method of choice is aseptic removal of a skin snip of the subcutaneous nodule, and microscopic examination of this tissue in a cover-slipped drop of saline to detect the nonsheathed microfilariae. *O. volvulus* microfilariae are not found in blood specimens but may be found in urine specimens because nodules are often located near the genitalia. Serologic tests are also available through reference laboratories.

Treatment, Prevention, and Control

Surgical removal of the encapsulated nodule is often performed to remove the adult worms and stop production of microfilariae. In addition, treatment with diethylcarbmazine followed by suramin is recommended, with an alternative drug mebendazole. The drug ivermectin also holds promise for eradication of migrating microfilariae.

Education regarding the disease and its black fly transmission is essential. Protection from black fly bites by use of protective clothing, screening, and insect repellents, as well as prompt diagnosis and treatment of infections to prevent further transmission, are critical.

The control of black fly breeding is difficult because insecticides wash away in the streams, but some form of biologic control of this vector may reduce fly reproduction and disease transmission.

DIROFILARIA IMMITIS

Several mosquito-transmitted filaria infect dogs, cats, raccoons, and bobcats in nature and occasionally are found in humans. *Dirofilaria immitis,* the dog heartworm, is notorious for forming a lethal worm bolus in the dog's heart. This nematode may also infect humans, producing a subcutaneous nodule or a so-called "coin lesion" in the lung. Only very rarely have these worms been found in human hearts.

The "coin lesion" in the lung presents a problem to both the radiologist and the surgeon because it resembles a malignancy requiring surgical removal. Serologic tests are not available to preclude the surgical intervention.

Transmission of the filarial infections can be controlled by mosquito control and a prophylactic use of the drug ivermectin in dogs.

Dracunculus medinensis

The name *Dracunculus medinensis* means "little dragon of Medina"; this is a very ancient worm infection thought by some scholars to be the "fiery serpent" noted by Moses with the Israelites at the Red Sea.

Physiology and Structure

D. medinensis is not a filarial worm but is a tissue-invading nematode of medical importance in many parts of the world. The worms have a very simple life cycle, depending on fresh water and a microcrustacean (copepod) of the genus *Cyclops* (Figure 35-19). When *Cyclops* harboring larval *D. medinensis* are ingested in drinking water by humans and other mammals, the infection is initiated with liberation of the larvae in the stomach. These larvae penetrate the wall of the digestive tract and migrate to the subcutaneous and connective tissues, usually in the extremities. These larvae are not microfilariae and do not appear in blood or other tissues. When the fertilized female worm becomes gravid, a vesicle is formed in the host tissue, which will ulcerate. When the ulcer is completely formed, the worm protrudes a loop of uterus through the ulcer. On contact with water, the larval worms are released. The larvae are then ingested by the *Cyclops* in fresh water, where they are then infective for humans or animals drinking the water containing the *Cyclops.*

Epidemiology

D. medinensis occurs in many parts of Asia and equatorial Africa. Reservoir hosts include dogs and many fur-bearing animals that come into contact

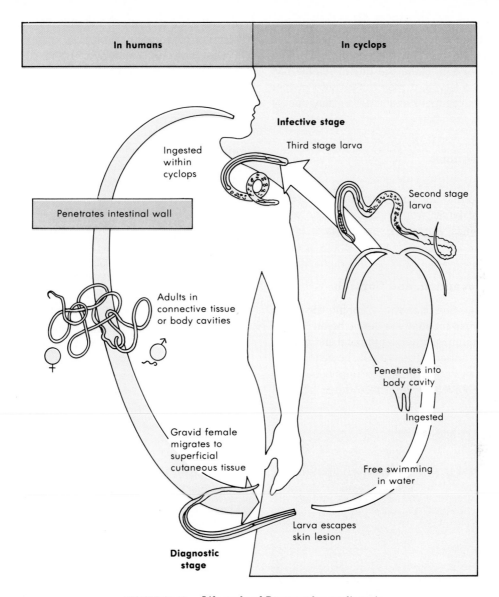

FIGURE 35-19   Life cycle of *Dracunculus medinensis.*

with drinking water containing infective *Cyclops,* including animals found in North America and China.

Human infections are usually the result of ingestion of water from so-called "step wells" where people stand or bathe in the water—at which time the gravid female worm discharges larvae from lesions on the arms, legs, feet, and ankles to infect *Cyclops* in the water. Ponds and standing water are occasionally the source of infection when humans use these for drinking water.

Clinical Syndromes

Symptoms of infection usually do not appear until the gravid female creates the vesicle and the ulcer in the skin for liberation of larval worms. This occurs usually 1 year after the initial exposure. At the site of the ulcer there is erythema and pain, as well as an allergic reaction to the worm. There is also the possibility of abscess formation and secondary bacterial infection, leading to further tissue destruction and inflammatory reaction, with intense pain and sloughing of skin.

If the worm is broken in attempts to remove it, there may be toxic reactions, and if the worm dies and calcifies, there may be nodule formation and some allergic reaction. Once the gravid female has discharged all the larvae, it may retreat into deeper tissue where it is gradually absorbed, or may simply be expelled from the site.

Laboratory Diagnosis

Diagnosis is established by observation of the typical ulcer and flooding the ulcer with water to recover the discharge of larval worms. Occasionally, x-ray examination reveals worms in various parts of the body.

Treatment, Prevention, and Control

The ancient method of slowly wrapping the worm on a twig is still used in many endemic areas (Figure 35-20). Surgical removal is also a practical and reliable procedure for the patient. The drug of choice for treatment is niridazole, with alternative drugs being metronidazole and thiabendazole. These drugs have an antiinflammatory effect and will either eliminate the worm or make surgical removal easier.

Education regarding the life cycle of the worm and avoidance of water contaminated with *Cyclops* is critical. Protection of drinking water by prohibiting bathing and washing of clothing in wells is essential. Persons who live in or travel to endemic areas should boil water before consuming it. Treatment of water with chemicals and the use of fish that consume *Cyclops* as food are also helpful.

Prompt diagnosis and treatment of cases also limit further transmission.

BIBLIOGRAPHY

Beaver PC, Jung RC, and Cupp EW: Clinical parasitology, ed 9, Philadelphia, 1984, Lea & Febiger.

Brown HW and Neva FA: Basic clinical parasitology, ed 5, Norwalk, Connecticut, 1983, Appleton-Century-Crofts.

Chandrasoma PT and Mendis KN: *Enterobius vermicularis* in ectopic sites, Am J Trop Med Hyg 26:644-649, 1977.

Chernin E: The disappearance of bancroftian filariasis from Charleston, South Carolina, Am J Trop Med Hyg 37:111-114, 1987.

Davidson RA et al: Risk factors for strongyloidiasis: a case control study, Arch Intern Med 144:321-324, 1984.

Gilman RH et al: The adverse effects of heavy *Trichuris* infection, Trans R Soc Trop Med Hyg 77:432-438, 1983.

Howard BJ et al: Clinical and pathogenic microbiology, St. Louis, 1987, The CV Mosby Co.

Kumaraswami V et al: Ivermectin for the treatment of *Wuchereria bancrofti* filariasis: efficacy and adverse reactions, JAMA 259:3150-3153, 1988.

Lynch NR et al: Seroprevalence of *Toxocara canis* infection in tropical Venezuela, Trans R Soc Trop Med Hyg 82:275-281, 1988.

Markell EK, Voge M, and John DT: Medical Parasitology, ed 6, Philadelphia, 1986, WB Saunders Co.

Schmidt GD and Roberts LS: Foundations of parasitology, ed 4, St. Louis, 1989, The CV Mosby Co.

Sullivan JJ and Long EG: Synthetic-fibre filters for preventing dracunculiasis: 100 versus 200 micrometers pore size, Trans R Soc Trop Med Hyg 82:465-466, 1988.

Walzer PD et al: Epidemiologic features of *Strongyloides stercoralis* infection in an endemic area of the United States, Am J Trop Med Hyg 31: 313-319, 1982.

Warren KS: Hookworm control, Lancet II (October 15), 897-898, 1988.

Washington JA II, editor: Laboratory procedures in clinical microbiology, ed 2, New York, 1985, Springer-Verlag.

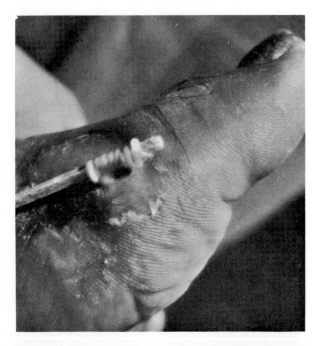

FIGURE 35-20   Removal of a *Dracunculus medinensis* adult from an exposed ulcer by winding the worm slowly around a stick. (From Binford CH and Conner DH: Pathology of tropical and extraordinary diseases, Washington DC, 1976, Armed Forces Institute of Pathology.)

CHAPTER 36

Trematodes

Most of the trematodes **(flukes)** are flat, fleshy, leaf-shaped worms. In general, they are equipped with two muscular suckers: one, an oral type, is the beginning of an incomplete digestive system and the other, the ventral sucker, is simply an organ of attachment. The digestive system consists of lateral tubes that do not join to form an excretory opening. The flukes are hermaphroditic, with both male and female reproductive organs in a single body. Schistosomes are the one exception—they have cylindric bodies (like the nematodes), and separate male and female worms exist.

All flukes require intermediate hosts for completion of their life cycles, and without exception the first intermediate hosts are mollusks (snails and clams). In these hosts an asexual reproductive cycle occurs that is a type of germ cell propagation. Some flukes require various second intermediate hosts before reaching the final host and developing into adult worms. This variation is discussed with the individual species.

Fluke eggs are equipped with a lid at the top of the shell, called an **operculum,** that opens to allow the larval worm to find its appropriate snail host. The schistosomes do not have an operculum, but rather the egg shell splits open to liberate the larva.

The primary medically significant trematodes are summarized in Table 36-1.

Table 36-1 Medically Significant Trematodes

Trematode	Common Name	Intermediate Host	Biologic Vector	Reservoir Host
Fasciolopsis buski	Giant intestinal fluke	Snail	Water plants (e.g., water chestnuts)	Pigs, dogs, rabbits, humans
Fasciola hepatica	Sheep liver fluke	Snail	Water plants (e.g., watercress)	Sheep, cattle, humans
Opisthorchis (Clonorchis) sinensis	Chinese liver fluke	1. Snail 2. Freshwater fish	Uncooked fish	Dogs, cats, humans
Paragonimus westermani	Lung fluke	1. Snail 2. Freshwater crabs or crayfish	Uncooked crabs, crayfish	Pigs, monkeys, humans
Schistosome species	Blood fluke	Snail	None	Primates, rodents, domestic pets, livestock, humans

FASCIOLOPSIS BUSKI

A number of intestinal flukes are recognized, including *Fasciolopsis buski, Heterophyes heterophyes, Metagonimus yokogawai, Echinostoma ilocanum,* and *Gastrodiscoides hominis. F. buski* is the most prevalent and important intestinal fluke. The other flukes are similar to *F. buski* in many respects (epidemiology, clinical syndromes, treatment) and are not discussed further. It is only important that physi-cians recognize the relationship among these different flukes.

Physiology and Structure

This large intestinal fluke demonstrates a typical life cycle (Figure 36-1). Humans ingest the encysted larval stage **(metacercaria)** when they peel the husks from aquatic vegetation (for example, water chestnuts) with the teeth. The metacercariae are scraped from the husk, swallowed, and develop into

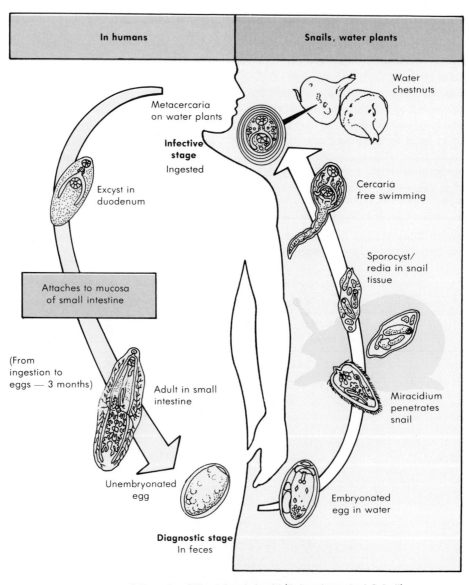

FIGURE 36-1   Life cycle of *Fasciolopsis buski* ("giant intestinal fluke").

immature flukes in the duodenum. The fluke attaches to the mucosa of the small intestine with two muscular suckers, develops into an adult form, and undergoes self-fertilization. Egg production is initiated 3 months after the initial infection with the metacercariae. The operculated eggs pass in feces to water, where the operculum at the top of the egg shell pops open, liberating a ciliated, free-swimming larval stage **(miracidium).** Glands at the pointed anterior end of the miracidium produce lytic substances that allow the penetration of the soft tissues of snails. In the snail tissue the miracidium develops through a series of stages by asexual germ cell propagation. The final stage **(cercaria)** in the snail is a ciliated, free-swimming form that, after release from the snail, encysts on the aquatic vegetation, becoming the metacercariae, or infective stage.

Epidemiology

Because it depends on the distribution of its appropriate snail host, *F. buski* is found only in China, Vietnam, Thailand, parts of Indonesia, Malaysia, and India. Pigs, dogs, and rabbits serve as reservoir hosts in these endemic areas.

Clinical Syndromes

The attachment of the flukes in the small intestine can produce inflammation, ulceration, and hemorrhage. Severe infections produce abdominal discomfort similar to that of duodenal ulcer, as well as diarrhea. Stools may be profuse, a malabsorption syndrome that is similar to giardiasis is common, and intestinal obstruction can occur. Marked eosinophilia is also present. The severity of disease is related to the number of adult flukes in the intestine. Although death can occur, it is rare.

Laboratory Diagnosis

Stool examination reveals the large, golden, bile-stained eggs with an operculum on the top (Figure 36-2). Large (approximately 1.5 cm by 3.0 cm) adult flukes can rarely be found in feces or specimens collected at surgery.

Treatment, Prevention, and Control

The drug of choice is praziquantal, and the alternative is niclosamide. Education regarding safe consumption of infective aquatic vegetation (particularly water chestnuts), proper sanitation, and control of human feces will reduce the incidence of disease. In addition, the snail population may be eliminated with molluscacides. When infection occurs, treatment should be initiated promptly to minimize its spread. Control of the reservoir hosts will also reduce transmission of the worm.

FASCIOLA HEPATICA

A number of liver flukes are recognized, including *Fasciola hepatica, Opisthorchis sinesis, O. felineus,* and *Dicrocoelium dendriticum.* Only *F. hepatica* and *O. sinesis* are discussed in this chapter, although eggs from the other flukes will rarely be detected in the feces from diseased patients.

Physiology and Structure

Commonly called the **sheep liver fluke,** this is a parasite of herbivores (particularly sheep and cattle) and humans. Its life cycle (Figure 36-3) is similar to that of *Fasciolopsis buski,* with human infection resulting from the ingestion of watercress that harbors the encysted metacercariae. The larval flukes then migrate through the duodenal wall, across the peritoneal cavity, and into the bile ducts via the liver

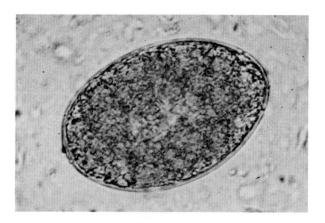

FIGURE 36-2   *Fasciolopsis buski* egg. The eggs are oval, large (75 to 100 nm by 130 to 150 nm), and surrounded by a thin shell. Although an operculum is present, it is rarely seen. (From Koneman EW, Allen SD, Dowell VR, and Sommers HM: Color atlas and textbook of diagnostic microbiology, ed 2, Philadelphia, 1979, JB Lippincott Co.)

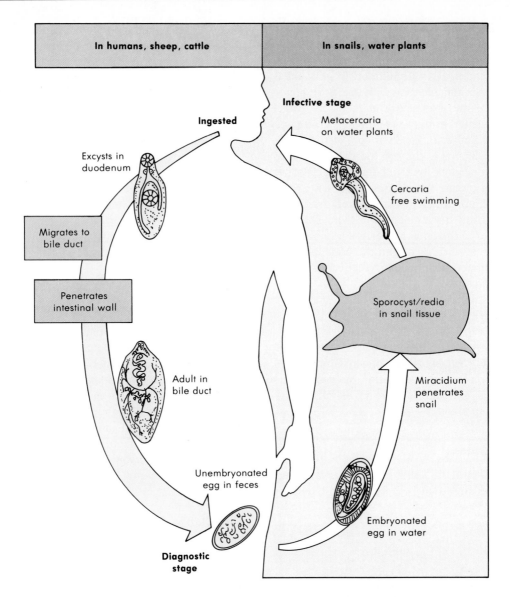

FIGURE 36-3   Life cycle of *Fasciola hepatica* ("sheep liver fluke").

to become adult worms. Approximately 3 to 4 months after the initial infection the adult flukes start producing operculated eggs identical to those of *F. buski* as seen in stool examination.

Epidemiology

Infections have been reported worldwide in sheep-raising areas, with the appropriate snail as an intermediate host, including the Soviet Union, Japan, Egypt, and many Latin American countries. Out-breaks are directly related to human consumption of contaminated watercress in areas where infected herbivores are present. Human infection is rare in the United States, but several well-documented cases have been reported in travelers from endemic areas.

Clinical Syndromes

Migration of the larval worm through the liver produces irritation of this tissue, tenderness, and hepatomegaly. Right upper quadrant pain, chills, and fever with marked eosinophilia are commonly observed. As the worms take up residence in the bile ducts, their mechanical irritation and toxic secretions produce hepatitis, hyperplasia of the epithelium, and biliary obstruction. Some worms penetrate eroded areas in the ducts and invade the liver to produce necrotic foci referred to as "liver rot." In severe infections secondary bacterial infection can occur, and portal cirrhosis is common.

Laboratory Diagnosis

Stool examination reveals operculated eggs indistinguishable from the eggs of *F. buski.* Exact identification is seldom a therapeutic problem because treatment is the same for both. In any instance, where exact identification is desired, a sample of the patient's bile differentiates the species; if the eggs are present in bile, they are *Fasciola hepatica* and not *F. buski,* which is limited to the small intestine. Eggs may appear in stool samples from persons who have eaten infected sheep or cattle liver. The spurious nature of this finding can be confirmed by having the patient refrain from eating liver and then rechecking the stool.

Treatment, Prevention, and Control

The drug of choice is praziquantel, with bithionol as an alternative. Preventive measures are similar to those for *F. buski* control—especially avoiding ingestion of watercress and other uncooked aquatic vegetation in areas frequented by sheep and cattle.

OPISTHORCHIS SINENSIS

Physiology and Structure

This trematode, also referred to as *Clonorchis sinensis* in older literature, is commonly called the **Chinese liver fluke.** Figure 36-4 illustrates this fluke's life cycle, which involves two intermediate hosts. *O. sinensis* differs from other fluke cycles because the eggs are eaten by the snail and then reproduction begins in the soft tissues of the snail. *O. sinensis* also requires a second intermediate host, freshwater fish, where the cercariae encyst and develop into infective metacercariae. When uncooked freshwater fish harboring metacercariae are eaten, flukes develop first in the duodenum and then migrate to the bile ducts where they become adults. The adult fluke undergoes self-fertilization and begins producing eggs. These eggs pass with feces and are once again eaten by snails, reinitiating the cycle.

Epidemiology

O. sinensis is found in China, Japan, Korea, and Vietnam. It is one of the most frequent infections seen among Asian refugees and can be traced to the consumption of raw, pickled, smoked, or dried freshwater fish harboring viable metacercariae. Dogs, cats, and fish-eating mammals also serve as reservoir hosts.

Clinical Syndromes

Mild infections are usually asymptomatic. Severe infections with many flukes in the bile ducts produce fever, diarrhea, epigastric pain, hepatomegaly, anorexia, and occasionally jaundice. Biliary obstruction may occur, and chronic infection can result in adenocarcinoma of the ducts. Invasion of the gallbladder may produce cholecystitis, cholelithiasis, and impaired liver function, as well as liver abscesses.

Laboratory Diagnosis

Examination of stool or duodenal aspirate reveals the small, golden, bile-stained eggs with the distinct operculum (Figure 36-5).

Treatment, Prevention, and Control

The drug of choice is praziquantel. Prevention of infection is accomplished by not eating uncooked fish and implementing proper sanitation policies, including disposal of human, dog, and cat feces in adequately protected sites so that they cannot contaminate water supplies with the intermediate snail and fish hosts.

PARAGONIMUS WESTERMANI

Physiology and Structure

Paragonimus westermani, commonly called the **lung fluke,** is one of several species of *Paragonimus*

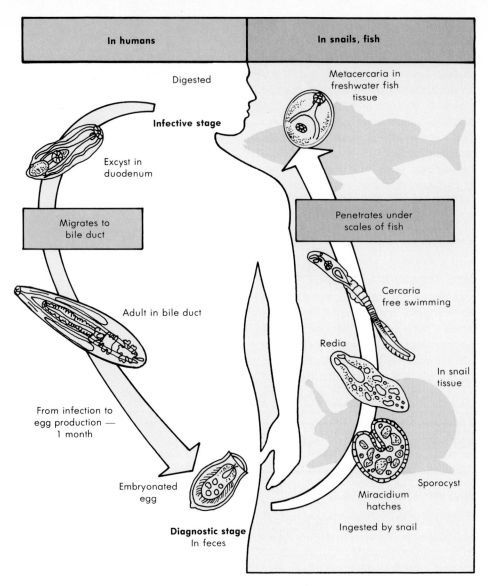

FIGURE 36-4 Life cycle of *Opisthorchis sinesis* ("Chinese liver fluke").

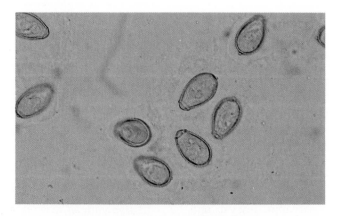

FIGURE 36-5 *Opisthorchis senesis* egg. These ovoid eggs are small (22 to 30 μm long by 12 to 19 μm wide) and have a yellowish-brown thick shell with a prominent operculum at one end and a small knob at the other end.

that infect man and many other animals. Figure 36-6 shows a familiar fluke life cycle from egg to snail to infective metacercaria. The infective stage occurs in a second intermediate host—the muscles and gills of freshwater crabs and crayfish. In humans who ingest infected meat, the larval worm hatches in the stomach and follows an extensive migration through the intestinal wall to the abdominal cavity, then through the diaphragm, and finally to the pleural cavity. Adult worms reside in the lungs and produce eggs that are liberated from ruptured bronchioles and appear in sputum or, when swallowed, in feces.

Epidemiology

Paragonimiasis occurs in many countries in Asia, Africa, India, and Latin America. It can be seen in refugees from Southeast Asia. Its prevalence is directly related to the consumption of uncooked freshwater crabs and crayfish. A wide variety of

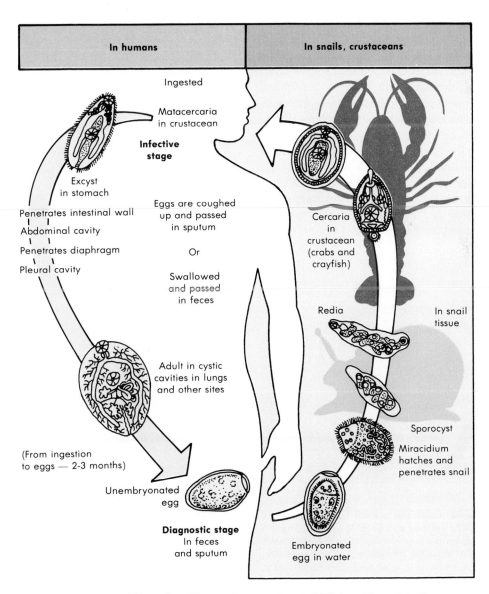

FIGURE 36-6   Life cycle of *Paragonimus westermani* ("Oriental lung fluke").

shore-feeding animals (e.g., wild boars, pigs, and monkeys) serve as reservoir hosts, and some human infections result from ingestion of meat containing migrating larval worms from these reservoir hosts. Human infections endemic to the United States are usually caused by a related species, *Paragonimus kellicotti,* which is found in crabs and crayfish in eastern and midwestern waters.

Clinical Syndromes

The adult flukes in the lungs first produce an inflammatory reaction that results in fever, cough, and increased sputum. As destruction of lung tissue progresses, cavitation occurs around the worms, sputum is blood-tinged and dark with eggs (so-called rusty sputum), and patients experience severe chest pain. Dyspnea increases and chronic bronchitis, bronchiectasis, and pleural effusion may be seen. Chronic infections lead to fibrosis in the lung tissue. The migration of larval worms may cause disease in sites other than the lungs, in particular invasion of the spinal cord and brain, producing severe neurologic disease (visual problems, motor weakness, and convulsive seizures) referred to as **cerebral paragonimiasis.** Migration and infection may also occur in subcutaneous sites, in the abdominal cavity, and in the liver.

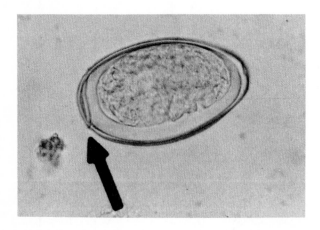

FIGURE 36-7 *Paragonimus westermani* egg. These large ovoid eggs (80 to 120 μm long and 45 to 70 μm wide) have a thick yellowish-brown shell and a distinct operculum (*arrow*). (From Koneman EW, Allen SD, Dowell VR, and Sommers HM: Color atlas and textbook of diagnostic microbiology, ed 2, Philadelphia, 1979, JB Lippincott Co.)

Laboratory Diagnosis

Examination of sputum and feces will reveal the golden-brown, operculated eggs (Figure 36-7). Chest x-ray films often show infiltrate, nodular cysts, and pleural effusion. Marked eosinophilia is common. Serologic procedures are available through reference laboratories and can be helpful.

Treatment, Prevention, and Control

The drug of choice is praziquantel, with bithionol as an alternative. Education regarding the consumption of uncooked freshwater crabs and crayfish, as well as the flesh of animals found in endemic areas, is critical. Pickling and wine-soaking of crabs and crayfish will not kill the infective metacercarial stage. Proper sanitation and control of the disposal of human feces are essential.

SCHISTOSOMES

General Characteristics

The three schistosomes most frequently associated with human disease are *Schistosoma mansoni, S. japonicum,* and *S. haematobium.* They collectively produce the disease called **schistosomiasis,** also known as **bilharziasis** or "**snail fever**." As discussed earlier, the schistosomes differ from other flukes in body structure, in having male and female worms rather than a hemaphroditic existence, and in having eggs that do not have an operculum. They also are obligate intravascular parasites and are not found in cavities, ducts, and other tissues. The infective forms are skin-penetrating cercariae liberated from snails, and these differ from other flukes in that they are not eaten on vegetation, in fish, or in crustaceans.

Figure 36-8 illustrates the life cycle of the different schistosomes. It is initiated by ciliated, free-swimming cercaria in fresh water that penetrate intact skin, enter the circulation, and develop in the intrahepatic portal blood. The female has a long, slender, cylindric body, whereas the shorter male, which appears cylindric, is actually flat. The cylindric appearance derives from folding the sides of the body to produce a groove, the gynecophoral canal, in which the female resides for fertilization. Both sexes

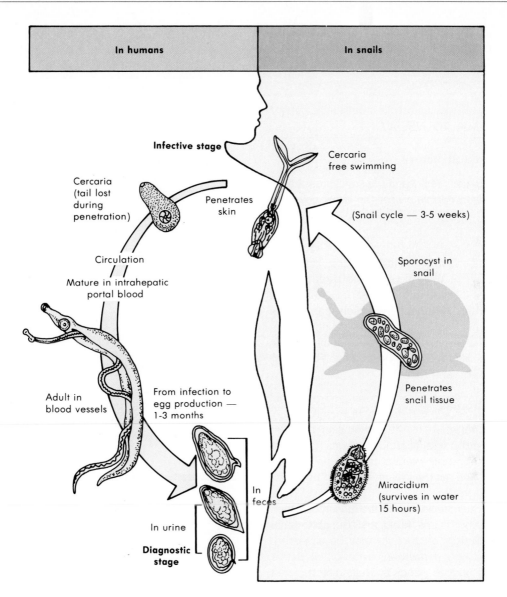

FIGURE 36-8   Life cycle of schistosomes.

have oral and ventral suckers and an incomplete digestive system, typical of a fluke.

As the worms develop in the liver, they elaborate a remarkable defense against host resistance. They coat themselves with substances that the host recognizes as itself; consequently, there is little protective response directed against their presence in blood vessels. This protective mechanism accounts for chronic infections that may last 20 to 30 years or longer.

After 2 to 3 weeks in the liver the maturing worms migrate to their final locations, where fertilization and egg production begin. *S. mansoni* and *S. japonicum* are found in mesenteric veins and produce **intestinal schistosomiasis;** *S. haematobium* occurs in veins around the urinary bladder and causes **vesicular schistosomiasis.** In these locations the worms ingest erythrocytes and the female lies in the male gynecophoral canal, constantly being fertilized and laying eggs. The male drags the female through the vessels, attaching and releasing its ventral sucker.

Because of differences in some aspects of disease and epidemiology, these worms are discussed as separate species.

SCHISTOSOMA MANSONI

Physiology and Structure

S. mansoni usually resides in the small branches of the inferior mesenteric vein near the lower colon.

Epidemiology

S. mansoni is endemic in Africa, Arabia, and Malagasy. Following of the African slave trade, it has became well established in the Western Hemisphere, particularly in Brazil, Surinam, Venezuela, parts of the West Indies, and Puerto Rico. Cases originating in these areas occur in the United States. In all of these areas there are also reservoir hosts, specifically primates, marsupials, and rodents.

Clinical Syndromes

Cercarial penetration of the intact skin may be seen as a dermatitis with allergic reactions, pruritus, and edema. Migrating worms in the lungs may produce cough and, as they reach the liver, hepatitis may appear.

As the flukes take residence in the mesenteric vessels and egg laying begins, fever, malaise, abdominal pain, and tenderness of the liver may be observed. With continued egg deposition, irritation of tissue ensues, and the body responds by walling off the area as if it were a foreign body. The tissue becomes fibrotic, abscesses form, and the patient may experience dysentery with bloody stools.

Chronic infection with *S. mansoni* shows a dramatic hepatosplenomegaly with large accumulations of ascitic fluid in the peritoneal cavity. Grossly, the liver is studded with white granulomas (**pseudotubercles**). Although *S. mansoni* eggs are primarily deposited in the intestine, eggs may appear in the spinal cord, lungs, and other sites. A similar fibrotic process occurs at each site. Severe neurologic problems may follow when eggs are deposited in the spinal cord and brain. In fatal schistosomiasis caused by *S. mansoni,* fibrous tissue surrounds the portal vein in a thick, grossly visible layer (''clay pipestem fibrosis'').

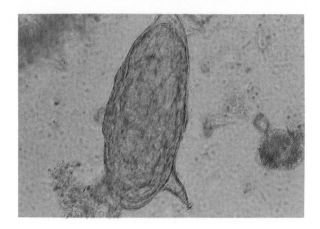

FIGURE 36-9   *Schistosoma mansoni* egg. These eggs are 115 to 175 μm long and 45 to 70 μm wide, contain a miracidium, and are enclosed in a thin shell with a prominent lateral spine.

Laboratory Diagnosis

Stool examination reveals the large, golden eggs with a sharp lateral spine (Figure 36-9). Rectal biopsy is also helpful to see the egg tracks laid by the worms in rectal vessels. Serologic tests are also available.

Treatment, Prevention, and Control

The drug of choice is praziquantel, and the alternative is oxamniquine. Education regarding the life cycles of these worms is essential, as well as control of snails using molluscacides. Improved sanitation and control of human fecal deposits are critical. Mass treatment may one day be practical, and the development of a vaccine may be forthcoming.

SCHISTOSOMA JAPONICUM

Physiology and Structure

S. japonicum resides in branches of the superior mesenteric vein around the small intestine and in the inferior mesenteric vessels. *S. japonicum* eggs (Figure 36-10) are smaller, almost spherical, and possess a tiny spine. These eggs are produced in greater numbers than those of *S. mansoni* and *S. haematobium.* Because of the size, shape, and numbers of these eggs, they are carried to more sites in the body (liver,

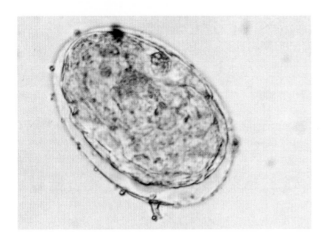

FIGURE 36-10 *Schistosoma japonicum* egg. These eggs are smaller than those of *S. mansoni* (70 to 100 μm long by 55 to 65 μm long) and have a spine that is inconspicuous.

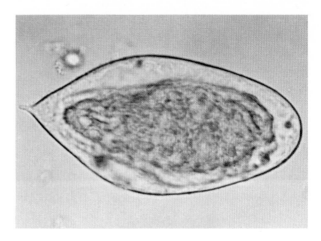

FIGURE 36-11 *Schistosoma haematobium* egg. These eggs are similar in size to those of *S. mansoni* but can be differentiated by the presence of a terminal spine.

lungs, brain), and infection with a few *S. japonicum* adults can be more severe than infections involving similar numbers of *S. mansoni* or *S. haematobium*.

Epidemiology

This **Oriental blood fluke** is found only in China, Japan, the Philippines, and the Celebes. Epidemiologic problems correlate directly with a broad range of reservoir hosts, many of which are domestic (cats, dogs, cattle, horses, and pigs).

Clinical Syndromes

The initial stages of infection with *S. japonicum* are similar to those of *S. mansoni,* with dermatitis, allergic reactions, fever, malaise followed by abdominal discomfort and diarrhea. In chronic *S. japonicum* infection, hepatosplenic disease, portal hypertension, bleeding esophageal varices, and accumulation of ascitic fluid are commonly seen. Granulomas that appear as pseudotubercles in and on the liver are common, along with the ''clay pipestem fibrosis'' as described for *S. mansoni.*

S. japonicum frequently involves cerebral structures when eggs reach the brain and granulomas develop around them. The neurologic manifestations include lethargy, speech impairment, visual defects, and seizures.

Laboratory Diagnosis

Stool examination demonstrates the small golden eggs with tiny spines; occasionally, rectal biopsy is similarly revealing. Serologic tests are available.

Treatment, Prevention, and Control

The drug of choice is praziquantel. Prevention and control may be achieved by measures similar to those for *S. mansoni,* especially education of populations in endemic areas regarding proper water purification, sanitation, and control of human fecal deposits. Control of *S. japonicum* must also involve the broad range of reservoir hosts and consider the fact that people work in rice paddies and on irrigation projects where infected snails are present. Mass treatment may offer help, and a vaccine may be developed someday.

SCHISTOSOMA HAEMATOBIUM

Physiology and Structure

Following development in the liver, these blood flukes migrate to the vesicular, prostatic, and uterine plexuses of the venous circulation, occasionally the portal bloodstream, and only rarely in other venules.

Large eggs with a sharp terminal spine (Fig. 36-11) are deposited in the wall of the bladder, and occasionally in the uterine and prostatic tissues. Those deposited in the bladder wall can break free and are found in urine.

Epidemiology

S. haematobium occurs throughout the Nile Valley and in many other parts of Africa, including islands off the eastern coast. It also appears in Asia Minor, Cypress, southern Portugal, and India. Reservoir hosts include monkeys, baboons, and chimpanzees.

Clinical Syndromes

Early stages of infection with *S. haematobium* are similar to those of infections involving *S. mansoni* and *S. japonicum,* with dermatitis, allergic reactions, fever, and malaise. Unlike the other two schistosomes, *S. haematobium* produces hematuria, dysuria, and urinary frequency as early symptoms. Associated with hematuria, bacteriuria is frequently a chronic condition.

Patients with *S. haemotobium* infections involving many flukes frequently demonstrate a squamous cell carcinoma of the bladder. It is commonly stated that the leading cause of cancer of the bladder in Egypt and other parts of Africa is *S. haematobium.*

The granulomas and pseudotubercles of *S. haematobium* in the bladder may also appear in the lungs. Fibrosis of the pulmonary bed from egg deposition leads to dyspnea, cough, and hemoptysis.

Laboratory Diagnosis

Examination of urine specimens reveals the large, terminally spined eggs, and occasionally bladder biopsy is helpful. *S. haematobium* eggs may appear in stool examination if worms have migrated to mesenteric vessels. Serologic tests are also available.

Treatment, Prevention, and Control

The drug of choice is praziquantel. At present, education, possible mass treatment, and development of a vaccine are the best approaches to control of *S. haematobium* disease. The basic problems of irrigation projects, (e.g., dam building) migratory human populations, and multiple reservoir hosts make prevention and control extremely difficult.

CERCARIAL DERMATITIS ("SWIMMER'S ITCH")

Several nonhuman schistosomes have cercariae that penetrate human skin, producing a severe dermatitis, but these cannot develop into adult worms. The natural hosts of these schistosomes are birds and other shore-feeding animals from freshwater lakes throughout the world, and a few marine beaches. The intense pruritus and urticaria from this skin penetration may lead to secondary bacterial infection from scratching the sites of infection.

Treatment consists of oral trimeprazine and topical applications of palliatives. When indicated, sedatives may be given. Control is difficult because of bird migration and transfer of live snails from lake to lake. Molluscacides such as copper sulfate have produced some reduction in snail populations. Immediate drying of the skin when leaving such waters offers some protection.

BIBLIOGRAPHY

Beaver PC, Jung RC, and Cupp EW: Clinical parasitology, ed 9, Philadelphia, 1984, Lea & Febiger.

Brown HW and Neva FA: Basic clinical parasitology, ed 5, Norwalk, Connecticut, 1983, Appleton-Century-Crofts.

Burton K et al: Pulmonary paragonimiasis in Laotian refugee children, Pediatrics 70:246-248, 1982.

Capron A et al: Role of anaphylactic antibodies in immunity to schistosomes, Am J Trop Med Hyg 29:849-857, 1980.

Healy GR: Trematodes transmitted to man by fish, frogs, and crustacea, J Wildl Dis 6:255-261, 1970.

Howard BJ et al: Clinical and pathogenic microbiology, St. Louis, 1987, The CV Mosby Co.

James SL: *Schistosoma* spp: progress toward a defined vaccine, Exp Parasitol 63:247-252, 1987.

Markell EK, Voge M, and John DT: Medical parasitology, ed 6, Philadelphia, 1986, WB Saunders Co.

Schmidt GD and Roberts LS: Foundations of parasitology, ed 4, St. Louis 1989, The CV Mosby Co.

Washington JA II, editor: Laboratory procedures in clinical microbiology, ed 2, New York, 1985, Springer-Verlag.

Wongratanacheewin S, Bunnag D, Vaeusorn N, and Sirisinba S: Characterization of humoral immune response in the serum and bile of patients with opisthorchiasis and its application in immunodiagnosis, Am J Trop Med Hyg 38:356-362, 1988.

37

Cestodes

The bodies of cestodes, **tapeworms,** are flat and ribbonlike, and the heads are equipped with organs of attachment. The head, or **scolex,** of the worm usually has four muscular, cup-shaped suckers, and a crown of hooklets. An exception is seen with *Diphyllobothrium latum,* the fish tapeworm, whose scolex is equipped with a pair of long, lateral muscular grooves and lacks hooklets.

The individual segments of tapeworms are called **proglottids,** and the chain of proglottids is called a **strobila.**

All tapeworms are hermaphroditic, with male and female reproductive organs present in each mature proglottid. The eggs of tapeworms are nonoperculated with one exception, *Diphyllobothrium latum,* which has an operculated egg similar to fluke eggs.

Tapeworms have no digestive system, and food is absorbed from the host intestine through the soft body wall of the worm. Most tapeworms found in the human intestine have complex life cycles involving intermediate hosts, and in some instances (cysticercosis, echinococcosis, sparganosis) humans serve as a form of intermediate host that harbors larval stages. These extraintestinal larval infections are at times more serious than the presence of adult worms in the intestine.

The most common cestodes of medical importance are listed in Table 37-1.

TAENIA SOLIUM

Physiology and Structure

After a person ingests pork muscle containing a larval worm called a **cysticercus** (''bladder worm''),

Table 37-1   Medically Important Cestodes

Cestode	Common Name	Reservoir for Larvae	Reservoir for Adults
Taenia solium	Pork tapeworm	Hogs	Humans
	Cysticercosis	Humans	—
Taenia saginata	Beef tapeworm	Cattle	Humans
Diphyllobothrium latum	Fish tapeworm	1. Fresh-water crustaceans	Humans, dogs, cats, bears
		2. Fresh-water fish	
Echinococcus granulosus	Unilocular hydatid cyst	Herbivores, humans	Canines
Echinococcus multilocularis	Alveolar hydatid cyst	Herbivores, humans	Foxes, wolves, dogs, cats
Hymenolepis nana	Dwarf tapeworm	Rodents, humans	Rodents, humans
Hymenolepis diminuta	Dwarf tapeworm	Insects	Rodents, humans
Dipylidium caninum	Pumpkin seed tapeworm	Fleas	Dogs, cats, humans

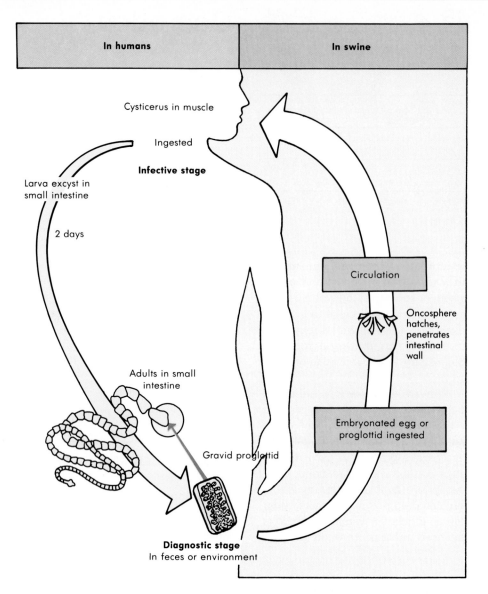

FIGURE 37-1   Life cycle of *Taenia solium* ("pork tapeworm").

attachment of the scolex with its four muscular suckers and crown of hooklets initiates infection in the small intestine (Figure 37-1). The worm then produces proglottids until a strobila of proglottids is developed that may be several meters in length. The sexually mature proglottids contain eggs, and as these proglottids leave the host in feces they can contaminate water and vegetation ingested by swine. The eggs in swine become a six-hooked larval form called an **oncosphere** that penetrates the pig's intestinal wall, migrates in the circulation to the muscles, and becomes a cysticercus to complete the cycle.

Epidemiology

T. solium infection is directly correlated with eating insufficiently cooked pork and is prevalent in Africa, India, Southeast Asia, China, Mexico, Latin American countries, and Slavic countries. It is seen

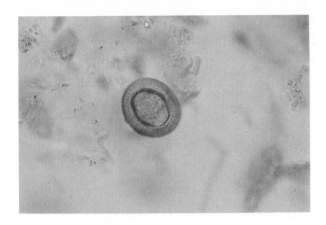

FIGURE 37-2 *Taenia* egg. The eggs are spherical, 40 μm in diameter, and contain 3 pairs of hooklets internally. The eggs of the different *Taenia* species cannot be differentiated.

infrequently in the United States, with most cases acquired in the regions just noted.

Clinical Syndromes

Adult *Taenia solium* in the intestine seldom causes appreciable symptoms. The intestine may be irritated at sites of attachment, and abdominal discomfort, chronic indigestion, and diarrhea may occur. Most patients become aware of the infection only when they see proglottids or a strobila of proglottids in their feces.

Laboratory Diagnosis

Stool examination may reveal proglottids and eggs, and treatment may produce the entire worm for identification. The eggs (Figure 37-2) are identical to those of *Taenia saginata* (beef tapeworm), so eggs alone are not sufficient for species identification. Critical examination of the proglottids reveals their internal structure, which is important for differentiation of *T. solium* and *T. saginata*.

Treatment, Prevention, and Control

The drug of choice is niclosamide; praziquantel, paromomycin, or quinacrine are effective alternatives. Prevention of pork tapeworm infections requires that pork either be cooked until the interior of the meat is gray or frozen at −20°C for at least 12 hours. Sanitation is critical; every effort must be made to keep human feces containing *T. solium* eggs out of water and vegetation ingested by pigs.

CYSTICERCOSIS

Physiology and Structure

Cysticercosis involves infection of individuals with the larval stage of *T. solium,* the cysticerci, which normally infects pigs (Figure 37-3). Human ingestion of water or vegetation contaminated with *T. solium* eggs from human feces initiates the infection. The larval worm penetrates the intestinal wall, enters the circulation, and may be carried to muscle, connective tissue, brain, lungs, and eyes. The immune response to the presence of these larvae in human tissues results in their death and calcification.

Epidemiology

Cysticercosis is found in the areas where *Taenia solium* is prevalent and is directly correlated with human fecal contamination. Autoinfection in humans may occur if eggs from adult worms hatch in the intestine, penetrate the wall, and migrate to other structures.

Clinical Syndromes

A few cysticerci in nonvital areas (subcutaneous tissues, among others) may provoke no symptoms, but serious disease may follow as the cysticerci lodge in vital areas such as the brain or eyes. In the brain they may produce hydrocephalus, meningitis, cranial nerve damage, seizures, hyperactive reflexes, and visual defects. In the eye, loss of visual acuity may occur, and if the larvae lodge along the optic tract, visual field defects result. When the cysticerci eventually die and begin to calcify, these symptoms may exacerbate because of cytotoxic effects.

Laboratory Diagnosis

The presence of cysticerci is usually established by radiographic studies, surgical removal of subcutaneous nodules, and visualization of cysts in the eye.

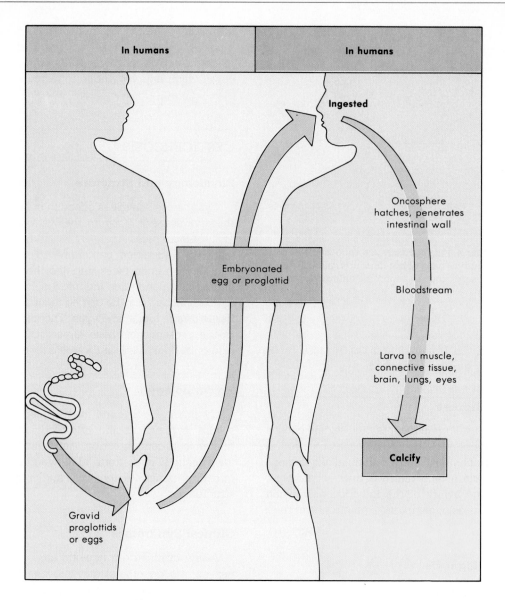

In humans   In humans

Ingested

Oncosphere
hatches, penetrates
intestinal wall

Embryonated
egg or proglottid

Bloodstream

Larva to muscle,
connective tissue,
brain, lungs, eyes

Calcify

Gravid
proglottids
or eggs

FIGURE 37-3   Development of human cysticercosis.

Brain scans are of considerable value in locating the cysts, and serologic tests are available from reference laboratories.

Treatment, Prevention, and Control

The drug of choice is praziquantel; surgery may be necessary. Critical to prevention and control of human infection are the treatment of human cases harboring adult *T. solium* to reduce egg transmission and the controlled disposal of human feces. These measures will also reduce infection in pigs.

TAENIA SAGINATA

Physiology and Structure

The life cycle of *T. saginata,* the **beef tapeworm,** is similar to that of *T. solium* (Figure 37-4), with infection resulting after cysticerci are ingested in insufficiently cooked beef. Following excystment, the larvae develop into adults in the small intestine and initiate egg production in maturing proglottids. A major difference, in contrast with *T. solium* infections, is that

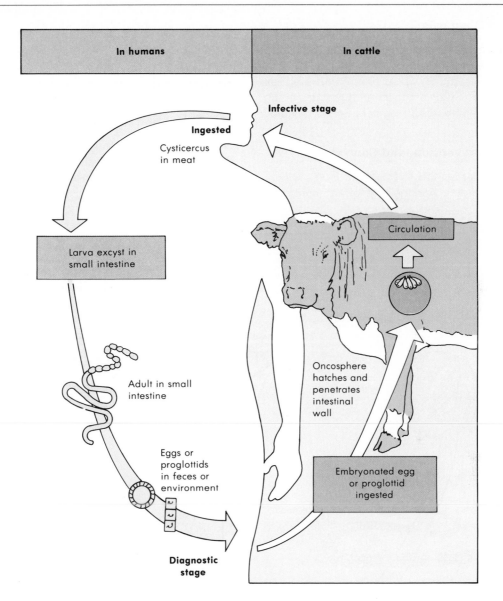

FIGURE 37-4   Life cycle of *Taenia saginata* ("beef tapeworm").

cysticercosis produced by *T. saginata* is extremely rare in humans. The adult *T. saginata* worm also differs from *T. solium* because it lacks a crown of hooklets on the scolex, and has a different proglottid uterine branch structure. These facts are important in differentiating between the two tapeworms but do not affect the selection of therapy.

Epidemiology

T. saginata occurs worldwide and is one of the most frequent causes of cestode infection seen in the United States. Humans and cattle perpetuate the life cycle: human feces contaminate water and vegetation with eggs, which are then ingested by cattle. The cysticerci in cattle produce adult tapeworms in humans when rare or insufficiently cooked beef is eaten.

Clinical Syndromes

The syndrome that results from *T. saginata* infestation is similar to that of *T. solium,* with vague abdominal pains, chronic indigestion, and hunger pains.

Laboratory Diagnosis

The diagnosis of *T. saginata* infection is similar to that of *T. solium,* with recovery of proglottids and eggs, or recovery of the entire worm with a scolex lacking hooklets. Study of the uterine branches in the proglottids differentiates *T. saginata* from *T. solium.*

Treatment, Prevention, and Control

Treatment is identical to treatment for *T. solium.* Education regarding cooking of beef and control of the disposal of human feces are critical measures.

DIPHYLLOBOTHRIUM LATUM

Physiology and Structure

One of the largest tapeworms (20 to 30 feet long), *Diphyllobothrium latum* (the **fish tapeworm**) has a complex life cycle involving two intermediate hosts—fresh-water crustaceans and fresh-water fish (Figure

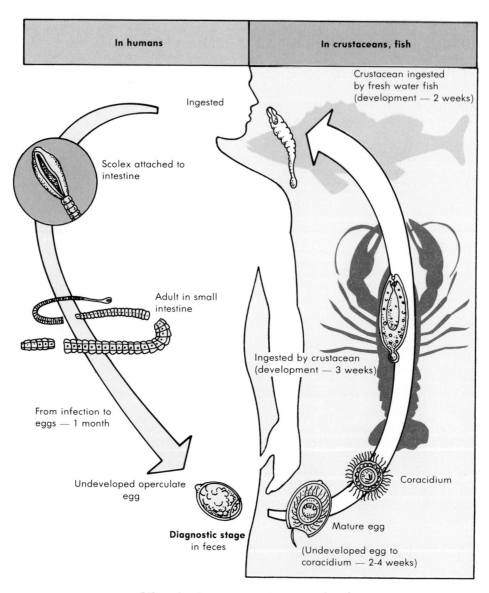

In humans

In crustaceans, fish

Ingested

Crustacean ingested by fresh water fish (development — 2 weeks)

Scolex attached to intestine

Adult in small intestine

Ingested by crustacean (development — 3 weeks)

From infection to eggs — 1 month

Coracidium

Undeveloped operculate egg

Mature egg

Diagnostic stage in feces

(Undeveloped egg to coracidium — 2-4 weeks)

FIGURE 37-5   Life cycle of *Diphyllobothrium latum* ("fish tapeworm").

37-5). The ribbonlike larval worm in the flesh of fresh-water fish is called a **sparganum.** Ingestion of this sparganum in raw or insufficiently cooked fish initiates infection. The scolex of *D. latum* is lance-shaped and has long, lateral grooves, which serve as organs of attachment. The proglottids of *D. latum* are broad, have a central uterine structure resembling a rosette, and produce eggs having an operculum, like fluke eggs, and a knob on the shell at the bottom of the egg. The adult worms may produce eggs for months or years. From 2 to 4 weeks after the eggs in feces reach water, a larval form called a **coracidium** leaves the egg via the open operculum. This free-swimming, ciliated coracidium is then ingested by tiny crustaceans called copepods (e.g., *Cyclops* and *Diaptomus*). The crustacean harboring the larval stage, which develops in about 3 weeks, is then eaten by fish in which the final larval stage (sparganum) develops in fish flesh.

Epidemiology

D. latum occurs worldwide, most prevalently in cool lake regions where raw or pickled fish is popular.

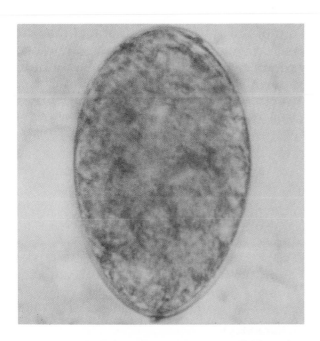

FIGURE 37-6 *Diphyllobothrium latum* egg. Unlike other tapeworm eggs, *D. latum* eggs are operculated. The eggs are 45 μm × 70 μm in size. (From Lennette EH, Balows A, Hausler WJ, and Shadomy HJ: Manual of clinical microbiology, ed 4, Washington, DC, 1985, American Society for Microbiology.)

Insufficient cooking over campfires and tasting and seasoning "gefüllte fish" account for many infections. A reservoir of infected wild animals, such as bears, minks, walruses, and members of the canine and feline families that eat fish, are also sources for human infections. The practice of dumping raw sewage into fresh-water lakes contributes to the propagation of this tapeworm.

Clinical Syndromes

Usually no symptoms are present, and patients first become aware of the infection by finding proglottids in feces. In some individuals who are genetically predisposed to **pernicious anemia,** tapeworm anemia may occur, resulting when the scolex attaches to the jejunum and absorbs significant amounts of vitamin B_{12} from the host.

Laboratory Diagnosis

Stool examination reveals the bile-stained, operculated egg with its knob at the bottom of the shell (Figure 37-6). Typical proglottids with the rosette uterine structure may also be found in stool specimens.

Treatment, Prevention, and Control

The drug of choice is niclosamide; praziquantel and paromomycin are acceptable alternatives. The prevalence of this infection is reduced by avoiding the ingestion of insufficiently cooked fish, controlling the disposal of human feces, especially the proper treatment of sewage prior to disposal in lakes, and promptly treating infections.

SPARGANOSIS

Physiology and Structure

The larval forms of several tapeworms closely related to *D. latum* can produce human disease in subcutaneous sites and in the eye. In these cases humans act as the dead-end host for the larval stage, or **sparganum.** Infections are acquired primarily by drinking pond water and ditch water that contains crustaceans (copepods) that carry a larval tapeworm. This larval form penetrates the intestinal wall and migrates to various sites in the body, where it develops into a sparganum. Infections may also occur if tadpoles, frogs, and snakes are ingested raw, or if the flesh of these animals is applied to wounds as a poultice. The larval worm leaves the cold, dead flesh of the animal and migrates into the warm human flesh.

Epidemiology

Cases have been reported from various parts of the world, including the United States, but the infection is most prevalent in the Orient. Regardless of location, drinking contaminated water and eating raw tadpole, frog, and snake flesh lead to infection.

Clinical Syndromes

In subcutaneous sites sparganosis can produce painful inflammatory tissue reactions and nodules. In the eye the tissue reaction is intensely painful, and periorbital edema is common. Corneal ulcers may develop with ocular involvement. Ocular disease is frequently associated with the use of frog or snake flesh as a poultice over a wound near the eye.

Laboratory Diagnosis

Sections of tissue removed surgically show characteristic tapeworm features: highly convoluted parenchyma and dark staining calcareous corpuscles.

Treatment, Prevention, and Control

Surgical removal is the customary approach; the drug praziquantel has been used when surgery was not feasible. Education regarding possible contamination of drinking water with crustaceans harboring larval worms is essential. This is most likely to occur in pond water and ditch water. Ingestion of raw frog and snake flesh or their use as poultices over wounds also should be avoided.

ECHINOCOCCUS GRANULOSUS

Physiology and Structure

Infection with *Echinococcus granulosus* is another example of accidental human infection, with humans serving as dead-end intermediate hosts in a life cycle naturally occurring in other animals. *E. granulosus* adult tapeworms are found in nature in the intestines of canines (dog, fox, wolf, coyote, jackal, dingo); the cyst stage is present in the viscera of herbivores (sheep, cattle, swine, deer, moose, elk; Figure 37-7). Adult tapeworms in the canine intestine produce infective eggs that pass in feces. When these eggs are ingested by humans, a six-hooked larval stage called an **onchosphere** hatches. The onchosphere penetrates the human intestinal wall and enters the circulation to be carried to various tissue sites, primarily the liver and lungs, but also the central nervous system and bone. This same cycle occurs in the viscera of herbivores. When the herbivore is killed by a canine predator, or viscera is fed to canines, the ingestion of cysts produces adult tapeworms in the canine intestine to complete the cycle and initiate new egg production. Adult tapeworms do not develop in the intestines of either herbivores or humans.

In humans the **unilocular hydatid cyst** is a slow-growing, tumorlike and space-occupying structure enclosed by a laminated germinative membrane. This membrane produces structures on its wall called **brood capsules,** where tapeworm heads **(protoscolices)** develop. **Daughter cysts** may develop in the original mother cyst, and these also produce brood capsules and protoscolices. The cysts and daughter cysts also accumulate fluid as they grow. This fluid is potentially toxic; if spilled into body cavities, anaphylactic shock and death can result. Spillage and the escape of protoscolices can lead to the development of cysts in other sites, because the protoscolices have the germinative potential to form new cysts. Eventually, the brood capsules and daughter cysts disintegrate within the mother cyst, liberating the accumulated protoscolices. These become known as **"hydatid sand."**

This type of echinococcus cyst is called a unilocular cyst to differentiate it from related cysts that grow differently. The unilocular cyst is generally about 5 cm in diameter, but some as large as 20 cm, containing almost 2 liters of cyst fluid, have been reported. Over long periods of time the cyst may die and become calcified.

Epidemiology

Human infection with *E. granulosus* unilocular cyst is directly correlated with raising sheep in many countries in Europe, South America, Africa, Asia, Australia, and New Zealand. It occurs in Canada, Alaska, and in the United States, with cases reported

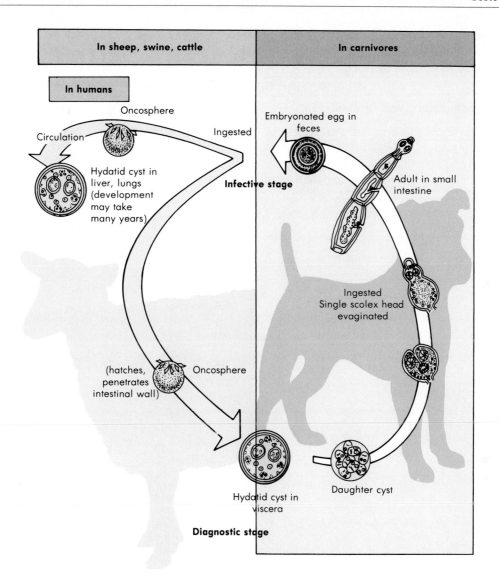

FIGURE 37-7   Life cycle of *Echinococcus granulosus*.

from Utah, New Mexico, Arizona, California, and the lower Mississippi valley. Human infection follows ingestion of contaminated water or vegetation, as well as hand-to-mouth transmission of canine feces carrying the infective eggs.

Clinical Syndromes

Because the unilocular cyst grows slowly, 5 to 20 years may pass before any symptoms appear. In many instances it appears that the cyst is as old as its host. The pressure of the expanding cyst in an organ is usually the first sign of infection. In the liver the cyst may exert pressure on both bile ducts and blood vessels and create pain and biliary rupture. In the lungs cysts may produce cough, dyspnea, and chest pains. In bone the cyst is responsible for erosion of the marrow cavity, as well as the bone itself. In the brain severe damage may occur as a result of the cyst's tumorlike growth into brain tissue.

Laboratory Diagnosis

Radiologic examination, scanning procedures, tomography, and ultrasound techniques are all valuable and may provide the first evidence of the cyst's presence. Aspiration of cyst contents may demonstrate the protoscolices (''hydatid sand'') and is diagnostic. Serologic procedures are also available.

Treatment, Prevention, and Control

Surgical resection of the cyst is the treatment of choice. In some instances the cyst is first aspirated to remove the fluid and hydatid sand, then instilled with formalin to kill and detoxify remaining fluid, and final-ly rolled into a marsupial pouch and sewn shut. If the condition is inoperable because of the cyst's location, the drug mebendazole is used; an alternative is alben-dazole. The most important factor in preventing and controlling echinococcosis is education regarding the transmission of infection and the role of canines in the life cycle. Proper personal hygiene and washing of hands and cooking utensils in environments inhabit-ed by dogs is critical. Dogs should not be allowed in the vicinity of animal slaughter and should never be

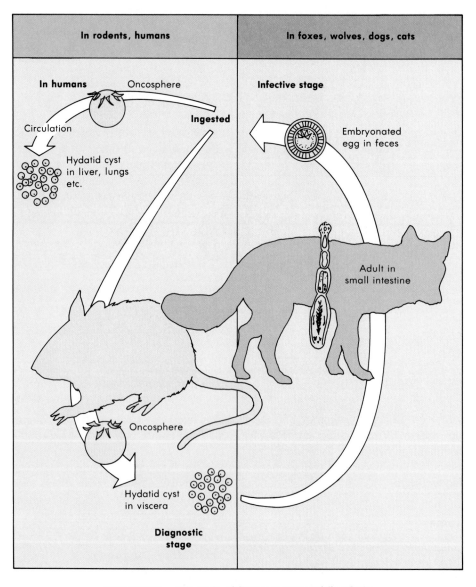

FIGURE 37-8   Life cycle of *Echinococcus multilocularis*.

fed the viscera of slain animals. In some areas the killing of stray dogs has reduced the incidence of infection.

ECHINOCOCCUS MULTILOCULARIS

Physiology and Structure

Like infection with *E. granulosus*, human infection with *E. multilocularis* is accidental (Figure 37-8). Adult *E. multilocularis* tapeworms are primarily found in foxes and wolves, although farm dogs and cats harbor them in some rural environments. The intermediate hosts that harbor the cyst stage are rodents (mice, voles, shrews, and lemmings). Humans become infected with the cyst stage as a result of contact with fox, dog, or cat feces contaminated with eggs. Trappers and workers who handle fur pelts may become infected by inhaling fecal dust carrying eggs.

Infective eggs hatch in and penetrate the intestinal tract to become onchospheres. These forms enter the circulation and take up residence primarily in the liver and lungs, but possibly also in the brain.

The **alveolar hydatid cyst** develops as an alveolar or honeycombed structure that is *not* covered by a unilocular limiting mother-cyst laminated membrane. The cyst grows via exogenous budding, eventually resembling a carcinoma. In humans individual cysts are said to be ''sterile'' and rarely produce protoscolices (hydatid sand).

Epidemiology

E. multilocularis is found primarily in northern areas such as Canada, the U.S.S.R., northern Japan, Central Europe, and Alaska, Montana, North and South Dakota, Minnesota, and Iowa in the United States. There is evidence that the life cycle may be extending to other mid-American states, where foxes and mice transmit the organism to dogs and cats and eventually to humans.

Clinical Syndromes

E. multilocularis, because of its slow growth, may be present in human tissues for many years before any symptoms appear. In the liver, cysts eventually mimic a carcinoma, with liver enlargement and obstruction of biliary and portal pathways. Often the growth will metastasize to the lungs and brain. Malnutrition, ascites, and portal hypertension produced by *E. multilocularis* create the appearance of hepatic cirrhosis.

Among all of the worm infections of humans, *E. multilocularis* is one of the most lethal. If left untreated, mortality is approximately 70% of infected individuals.

Laboratory Diagnosis

Unlike *E. granulosus*, the tissue form *E. multilocularis* presents no protoscolices, and the material so resembles a neoplasm that even pathologists mistake it for carcinoma. Radiologic procedures and scanning techniques are helpful, and serologic methods are available.

Treatment, Prevention, and Control

Surgical removal of the cyst is indicated, especially if an entire hepatic area can be resected. The same surgical approach applies to lesions in the lung, wherein a lobe can be resected. Mebendazole and albendazole, as used for treatment of *E. granulosus*, have shown clinical cures. As with *E. granulosus*, education, proper personal hygiene, and deworming of farm dogs and cats are critical. It is extremely important to treat these animals if they have contact with children.

HYMENOLEPIS NANA

Physiology and Structure

Hymenolepis nana, the **dwarf tapeworm,** is unlike *Taenia*, which measures several meters; this tapeworm is only 2 to 4 cm in length. The life cycle is also simple and involves no intermediate hosts (Figure 37-9).

Infection begins when the embryonated eggs are ingested and develop in the intestinal villi into a larval cysticercoid stage. This cysticercoid larva attaches its four muscular suckers and crown of hooklets to the small intestine, and the adult worm produces a strobila of egg-laden proglottids. Eggs passing in the feces are then immediately and directly infective, initiating another cycle.

H. nana also can cause autoinfection, with a sub-

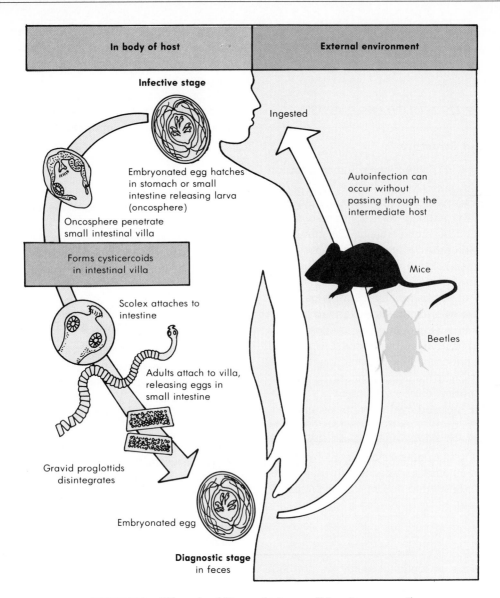

FIGURE 37-9   Life cycle of *Hymenolepis nana* ("dwarf tapeworm").

sequent increased worm burden. Eggs are able to hatch in the intestine, develop into a cysticercoid larva, and then grow into adult worms without leaving the host. This can lead to hyperinfection, with very heavy worm burdens and severe clinical symptoms.

Epidemiology

H. nana occurs worldwide in humans and is also a common parasite of mice. It occasionally develops its cysticercoid stage in beetles, and both humans and mice may ingest these beetles in contaminated grain and flour. Children are especially at risk of infection, and, because of the simple life cycle of the parasite, families with children in day care centers experience problems in controlling the transmission of this organism.

Clinical Syndromes

With few worms in the intestine, there are no symptoms. In heavy infections, especially if autoin-

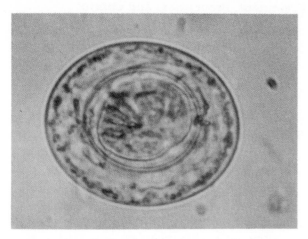

FIGURE 37-10   *Hymenolepis nana* egg. The eggs are 30 to 45 μm in diameter, with a thin shell, containing a six-hooked embryo. (From Lennette EH, Balows A, Hausler WJ, and Shadomy HJ: Manual of clinical microbiology, ed 4, Washington, DC, 1985, American Society for Microbiology.)

fection and hyperinfection occur, patients experience diarrhea, abdominal pain, headache, anorexia, and other vague complaints.

Laboratory Diagnosis

Stool examination reveals the characteristic *H. nana* egg with its six-hooked embryo and polar filaments (Figure 37-10).

Treatment, Prevention, and Control

The drug of choice is praziquantel, with niclosamide an alternative therapy. Treatment of cases, improved sanitation, and proper personal hygiene are essential, especially in the family and institutional environments.

HYMENOLEPIS DIMINUTA

Physiology and Structure

H. diminuta, closely related to *H. nana,* is primarily a tapeworm of rats and mice, but it is also found in humans. It differs from *H. nana* in size, measuring 20 to 60 cm, the scolex lacks hooklets, and the egg is larger, bile stained, and has no polar filaments. The life cycle of *H. diminuta* is more complex and requires

larval insects ("meal worms") to reach the infective cysticercoid stage.

Epidemiology

Infections have been found all over the world, including the United States. Larval beetles and other larval insects become infected when they feed on rat feces carrying *H. diminuta* eggs. Humans are infected by ingesting the larval insects ("meal worms") in contaminated grain products (flour, cereals, etc.).

Clinical Syndromes

Mild infections show no symptoms, but heavier worm burdens produce nausea, abdominal discomfort, anorexia, and diarrhea.

Laboratory Diagnosis

Stool examination demonstrates the characteristic bile-stained egg that lacks polar filaments.

Treatment, Prevention, and Control

The drug of choice is niclosamide, with praziquantel an alternative. Rodent control in areas where grain products are produced or stored is essential. Thorough inspection of uncooked grain products to detect meal worms is also important.

DIPYLIDIUM CANINUM

Physiology and Structure

This small tapeworm, averaging about 15 cm in length, is primarily a parasite of dogs and cats, but it can infect humans—especially children whose mouths are licked by infected pets. The life cycle involves development of larval worms in dog and cat fleas. These fleas, when crushed by the teeth of the infected pet, are carried on the tongue to the child's mouth in kissing or licking. Swallowing the infected flea leads to intestinal infection.

Because of the size and shape of the mature and terminal proglottids, *D. caninum* is often called the **"pumpkin seed tapeworm."** The eggs are distinctive because they occur in packets covered with a

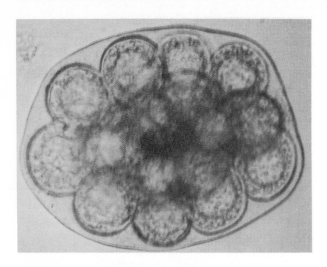

FIGURE 37-11   *Dipylidium caninum* eggs. Free eggs are rarely seen. Instead, egg packets containing 8 to 15 six-hooked onchospheres enclosed in a thin membrane are most commonly found in fecal specmens. (From Lennette EH, Balows A, Hausler WJ, and Shadomy HJ: Manual of clinical microbiology, ed 4, Washington, DC, 1985, American Society for Microbiology.)

tough, clear membrane. There may be as many as 25 eggs in a packet, and a single egg free of the packet is seldom seen.

Epidemiology

D. caninum occurs worldwide, especially in children, and its distribution and transmission are directly correlated with dogs and cats infected with fleas.

Clinical Syndromes

Light infections are asymptomatic; heavier worm burdens produce abdominal discomfort, anal pruritus, and diarrhea. Anal pruritus results from the active migration of the motile proglottid.

Laboratory Diagnosis

Stool examination reveals the colorless egg packets (Figure 37-11), and proglottids may be in feces and brought to physicians by patients.

Treatment, Prevention, and Control

The drug of choice is niclosamide; praziquantel or paramomycin are alternatives. Dogs and cats should be dewormed and not be allowed to lick the mouths of children. Pets should be treated to the eradicate fleas.

BIBLIOGRAPHY

Beaver PC, Jung RC, and Cupp EW: Clinical parasitology, ed 9, Philadelphia, 1984, Lea & Febiger.

Blenkharn JI, Benjamin IS, and Blumgart LH: Bacterial infection of hepatic hydatid cysts with *Haemophilus influenzae,* J Infect 15:169-171, 1987.

Brown HW and Neva FA: Basic clinical parasitology, ed 5, Norwalk, Connecticut, 1983, Appleton-Century-Crofts.

Coltorti E et al: Field evaluation of an enzyme immunoassay for detection of asymptomatic patients in a hydatid control program, Am J Trop Med Hyg 38:603-607, 1988.

Cook GC: Neurocysticercosis: parasitology, clinical presentation, diagnosis, and recent advances in management, Q J Med 68:575-583, 1988.

Desowitz RS: New Guinea tapeworms and Jewish grandmothers, New York, 1983, Avon/Discus.

Groll E: Praziquantel for cestode infections in man, Acta Trop 37:293-296, 1980.

Howard BJ et al: Clinical and pathogenic microbiology, St Louis, 1987, The CV Mosby Co.

Kammerer WS and Schantz PM: Long-term follow-up of human hydatid disease *(Echinococcus granulosus)* treated with a high dose mebendazole regimen, Am J Trop Med Hyg 33:132-137, 1984.

Leiby PD and Kritsky DC: *Echinococcus multilocularis:* a possible domestic life cycle in Central North America and its public health importance. J Parasitol 58:1213-1215, 1972.

Markell EK, Voge M, and John DT: Medical parasitology, ed 6, Philadelphia, 1986, WB Saunders Co.

Saimot AG et al: Albendazole as a potential treatment for human hydatidosis. Lancet 2:652-656, 1983.

Schmidt GD and Roberts LS: Foundations of parasitology, ed 4, St Louis, 1989, The CV Mosby Co.

Washington JA II, editor: Laboratory procedures in clinical microbiology, ed 2, New York, 1985, Springer-Verlag.

VIROLOGY

PART IV

General Medical Virology and Pathogenesis

Viruses were first described as "filterable agents." Their small size allowed them to pass through filters designed to retain bacteria. Viruses range from the structurally simple and small parvoviruses and picornaviruses to the large and complex pox viruses and herpesviruses. Their names may describe certain characteristics, the diseases they are associated with, or even the tissue or geographic locale where they were first identified. Names such as picornavirus (*pico*, meaning small; *rna*, ribonucleic acid [RNA]; *virus*) or togavirus (*toga*, Greek for mantle, referring to a membrane envelope surrounding the virus) describe the structure of the virus, whereas the name papovaviruses describes the members of its family (*papilloma, polyoma,* and *vacuolating viruses*). The name retrovirus (*retro,* reverse) refers to the virus-directed synthesis of deoxyribonucleic acid (DNA) from an RNA template, whereas the poxviruses are named for the symptoms caused by one of its members, smallpox. The adenoviruses (*adeno*ids) and the reoviruses (*r*espiratory, *e*nteric, *o*rphan) are named for the body site from which they were first isolated. *Reovirus* was discovered before it was associated with a specific disease, and thus it was designated an "orphan" virus. Coxsackievirus is named for Coxsackie, N.Y., and many of the toga-, arena-, and bunyaviruses are named after exotic African places where they also were first isolated.

CLASSIFICATION

Until recently no single system for classifying viruses was used. Viruses are now grouped by char-acteristics such as morphology, type of genome, and means of replication.

The units for measurement of viral size are in nanometers (nm). The clinically important viruses range from 20 or 30 nm (picornaviruses) to 300 nm (poxviruses). The latter may just be seen with a light microscope and are approximately one-fourth the size of a staphylococcus.

The electron microscope shows that viruses consist of a protein coat (capsid) and a core of nucleic acid. The larger viruses (e.g., poxvirus, herpesvirus, myxovirus) have an envelope composed of lipid, proteins, and glycoproteins. This lipid envelope can be disrupted by detergents and solvents such as ether, which will then inactivate the virus. Some viruses also contain enzymes required for replication. The viral capsid is composed of many protein subunits called **capsomeres**, which appear as projections.

Table 38-1 DNA Viruses	
Family	**Important Human Viruses**
Parvoviridae	Parvovirus B19
Papovaviridae	Papovavirus (JC, BK, SV40), papillomavirus
Hepadnaviridae	Hepatitis B virus
Adenoviridae	Adenovirus
Herpesviridae	Herpes simplex types 1 and 2, varicella-zoster virus, Epstein-Barr virus, cytomegalovirus, herpes 6 virus
Poxviridae	Smallpox virus, vaccinia virus, molluscum contagiosum virus

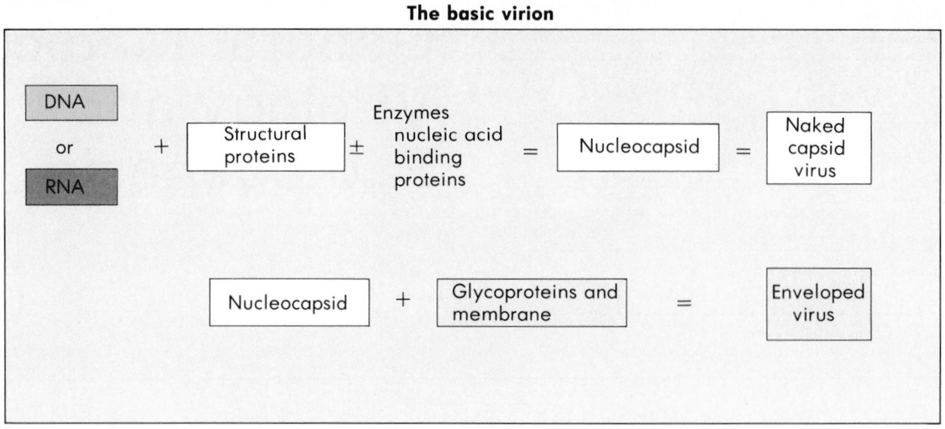

The basic virion

FIGURE 38-1   Schematic structure of the basic virion.

The number of capsomeres differs in each virus family, and they usually appear in one of two geometric arrangements, **cubical (icosahedral)** or **helical**. All true viruses have a nucleic acid that is either **DNA** or **RNA**. The genome may consist of double-stranded or single-stranded DNA or RNA and may be linear, circular, or segmented into pieces, with each piece encoding an individual gene (Figure 38-1).

Nucleic acid composition, as well as capsid arrangement, presence or absence of an envelope, and virus size are the basis for classifying viruses of clinical importance. The DNA viruses associated with human disease comprise six families (Tables 38-1 and 38-2), and the RNA viruses may be divided into 13 such groups (Tables 38-3 and 38-4).

VIRUS COMPONENTS

Viruses are biochemical complexes consisting of an **RNA** or **DNA genome** packaged into a protein **capsid**, which may or may not be surrounded by a membrane **envelope** (Figures 38-1 and 38-2). The protein-covered genome is referred to as the **nucleocapsid**. The proteins on the surface of the capsid and envelope determine the interaction of the virus with the host and elicit the major protective immune response against the virus. Some virion particles also contain enzymes required to facilitate the replication of the virus.

The size and structure of the genome are major

Table 38-2   Properties of Virions of Families of DNA Viruses

Family	Genome* Molecular Weight ($\times 10^6$ daltons)	Nature†	Shape	Virion Size	Polymerase
Parvoviridae	1.5-2.0	ss, linear	Icosahedral	18-26	—
Papovaviridae	3-5	ds, circular	Icosahedral	45-55	—
Hepadnaviridae	1.8	ds, circular‡	Spheric	42	+§
Adenoviridae	20-25	ds, linear	Icosahedral	70-90	—
Herpesviridae	100-150	ds, linear	Icosahedral, enveloped	Capsid, 100-110 envelope, 120-200;	—
Poxviridae	85-140	ds, linear	Brick shaped, sometimes enveloped	$300 \times 240 \times 100$	+

*Genome invariably a single molecule.
†ds, Double stranded; ss, Single stranded.
‡Circular molecule is ds for most of its length but contains an ss region.
§Possibly a reverse transcriptase present.

Table 38-3   RNA Viruses

Family	Important Human Viruses
Paramyxoviridae	Measles, mumps, respiratory syncytial, parainfluenza, Sendai
Orthomyxoviridae	Influenza A, B, and C
Coronaviridae	Coronavirus
Arenaviridae	Lymphocytic choriomeningitis (LCM), Lassa fever virus, Tacaribe virus complex (Junin and Machupo)
Rhabdoviridae	Rabies
Filoviridae	
Bunyaviridae	California encephalitis, sandfly fever, Crimean Congo hemorrhagic fever, etc.
Retroviridae	Human T cell leukemia I and II, human immunodeficiency, etc.
Reoviridae	Rotavirus, reovirus, California tick fever
Picornaviridae	Rhinovirus, poliovirus, ECHO virus, coxsackievirus
Togaviridae	Rubella; western, eastern, and Venezuelan equine encephalitis; Sindbis, Semliki Forest, etc.
Flaviviridae	Yellow fever, dengue, St. Louis encephalitis, etc.
Caliciviridae	Norwalk agent

Table 38-4   Properties of Virions of Families of RNA Viruses

	Genome			Virion			
	Molecular Weight ($\times 10^6$ daltons)	Nature*	Envelope	Shape†	Size (nm)	Polymerase	Symmetry of Nucleocapsid
Paramyxoviridae	5-7	ss, −	+	Spheric	150-300	+	Helical
Orthomyxoviridae	5	ss, −	+	Spheric	80-120	+	Helical
Coronaviridae	6	ss, +	+‡	Spheric	60-220	−	Helical
Arenaviridae	3-5	ss, −	+‡	Spheric	50-300	+	Helical
Rhabdoviridae	4	ss, −	+	Bullet shaped	180 × 75	+	Helical
Filoviridae	4	ss, −	+	Filamentous	800 × 80	+(?)	Helical
Bunyaviridae	4-7	ss, −	+‡	Spheric	90-110	+	Helical
Retroviridae	2 × (2-3)§	ss, +	+	Spheric	80-110	+‖	Helical
Reoviridae	11-15	ds	−	Icosahedral	60-80	+	Icosahedral
Picornaviridae	2.3	ss, +	−	Icosahedral	25-30	−	Icosahedral
Togaviridae	4	ss, +	+	Spheric	60-70	−	Icosahedral
Flaviviridae	4	ss, +	+	Spheric	40-50	−	Icosahedral
Caliciviridae	2.6	ss, +	−	Icosahedral	35-40	−	Icosahedral

*All molecules linear; ss, single stranded; ds, double stranded; + or −, polarity of ss nucleic acid.
†Some enveloped viruses are very pleomorphic (sometimes filamentous).
‡No matrix protein.
§Genome has two identical ss RNA molecules.
‖Reverse transcriptase.

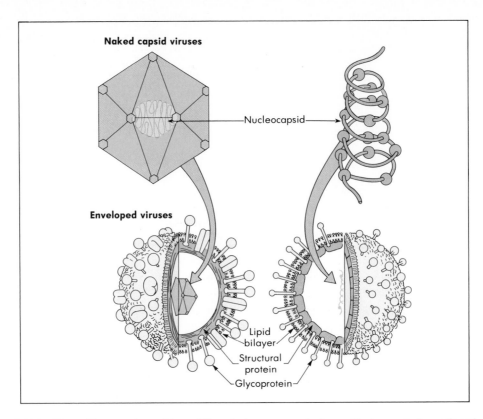

FIGURE 38-2   The structure of two different viruses as naked capsid viruses and enveloped virions, with either icosahedral (left) or helical (right) symmetry.

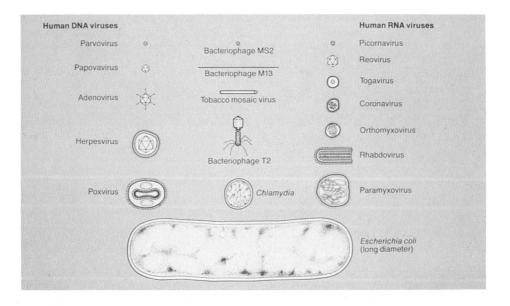

FIGURE 38-3   Sizes of human and other viruses and bacteria. (Courtesy The Upjohn Company; reproduced from Microbiology 1980.)

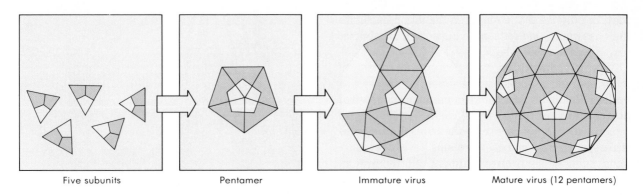

Five subunits   Pentamer   Immature virus   Mature virus (12 pentamers)

FIGURE 38-4   Capsid assembly of picornavirus. Note that this is a simple icosahedral virus with 12 pentamers. (Courtesy Dr. Michael Rossman.)

determinants for distinguishing and classifying the viruses (Figure 38-3). The size correlates with the size of the capsid or envelope. Larger virions can hold a larger genome that can code for more proteins. The RNA of the genome can be either a positive-strand (+) RNA and function as a messenger (mRNA) or a negative-strand (−) RNA and function as a template for the production of mRNA. The capsid is a shell made of building blocks of the capsomeres, which are usually relatively small proteins that self-assemble into the larger capsomere structures and then into the virion capsid (Figure 38-4). Individual capsomeres can often be distinguished in electron micrographs. Each of the components of the capsid must have the chemical features that allow them to fit together and assemble into the larger unit. The simplest structures that can be built in this manner are symmetric and include helical and icosahedral structures. Helical structures form rods, whereas the icosahedron is an approximation of a sphere assembled from symmetric subunits. Nonsymmetric capsids are complex forms and are associated with certain bacterial phages.

The envelope has a membrane structure similar to cellular membranes and consists of lipids, proteins, and glycoproteins (see Figure 38-2). Most of the enveloped viruses are round or pleomorphic. The poxvirus, which has a complex internal and a bricklike external structure, and the rhabdovirus, which is bullet shaped, are two exceptions. As a lipid membrane, the envelope can be disrupted by detergents and solvents (e.g., ether), which will then inactivate the virus. Cellular proteins are rarely found in viral envelopes. The inside surface of some enveloped viruses, especially

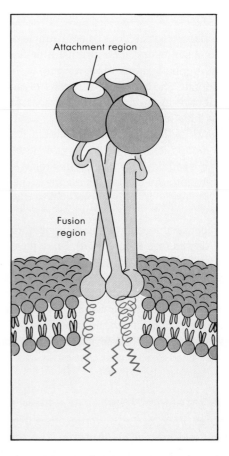

FIGURE 38-5   Diagram of spikes on surface of viral particle. The region for attachment to the cellular receptor is exposed on the spike's surface. (Redrawn from Schlesinger MJ and Schlesinger S: Adv Virus Res 33:1, 1987.)

the negative-strand RNA viruses, is lined with **matrix** proteins that strengthen the virion, facilitate their assembly, and for some viruses (e.g., rhabdovirus) determine their shape.

Viral **glycoproteins** extend from the surface of the virion and for many viruses can be observed as **spikes** (Figure 38-5). Most glycoproteins act as **viral attachment proteins** (VAPs) capable of binding to structures on target cells or erythrocytes (**hemagglutinins,** or HAs). Some glycoproteins have other functions, such as the **neuraminidase** of orthomyxoviruses (e.g., influenza), the Fc receptor and the C3 receptor associated herpes simplex virus type 1 and paramyxoviruses, or the fusion proteins of paramyxoviruses. The glycoproteins are also the major antigens for protective immunity.

STRUCTURE

The tobacco mosaic virus is an example of a virus with helical symmetry. Its capsomeres self-assemble on the RNA genome into rods extending to the length of the genome. The capsomeres cover and protect the RNA. Orthomyxoviruses, paramyxoviruses, and rhabdoviruses also form helical nucleocapsids but with components of the viral RNA-dependent RNA polymerases bound to the RNA (Figure 38-6). These enzymes are required to initiate virus replication and are brought into the cell in this manner. The nucleocapsids of these RNA viruses are enclosed within an envelope.

The picornaviruses have the simplest form of icosahedral capsid (see Figure 38-4). They have a small genome and capsid. Each capsid is composed of 12 identical units that form the 12 adjacent vertices of the structure. Each unit comprises five identical capsomeres composed of four separate proteins. X-ray crystallography has defined the structure of the picornavirus capsid to the 1.5 Å level and shown that each vertex has a keyholelike cleft that binds to a receptor on the surface of the target cell.

Larger icosahedral virions are constructed by inserting structurally distinct capsomeres between the vertices. Each of the 12 vertices has five nearest neighbors and are called **pentons**. The other capsomeres have six nearest neighbors and are referred to as **hexons**. This extended icosahedron is called an **icosadeltahedron**, and its size is determined by the

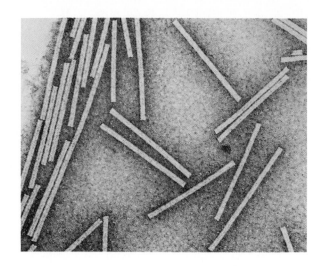

FIGURE 38-6   Tobacco mosaic virus rods, negative staining. (From Fraenkel-Conrat H: The chemistry and biology of viruses, New York, 1969, Academic Press, Inc.)

number of hexons inserted along the edges and planes between the pentons. The adenovirus capsid is composed of 252 capsomeres, with 12 pentons and 240 hexons. Unlike most other icosahedral viruses, however, each penton of adenovirus carries a long fiber that serves as the VAP to target cells and contains the type-specific antigen. The herpesvirus capsid has 162 capsomeres. The reoviruses have an icosahedral capsid that is more complex than those just described. Each virion is composed of a double capsid with glycoproteins at each vertex.

The toga-, myxo-, herpes-, and poxviruses are examples of enveloped viruses, differing in shape, size, and complexity. Togaviruses consist of an icosahedral nucleocapsid containing a positive-strand RNA genome surrounded by an envelope. The envelope contains spikes consisting of two or three glycoprotein subunits anchored to the virion's icosahedral capsid. This causes the envelope to adhere tightly and conform to its icosahedral structure.

Influenza A, an orthomyxovirus, is an example of an enveloped, negative-strand RNA virus. Its envelope has two glycoproteins, the hemagglutinin, which is the viral attachment protein, and a neuraminidase. It is lined by matrix proteins that associate with the ribonucleocapsid segments, promote the assembly of the virus, and confer some structure to the virus.

The herpesvirus envelope is a baglike structure

that encloses the icosadeltahedral nucleocapsid. Depending on the specific herpesvirus, the envelope may contain five or more glycoproteins. The interstitial space between the nucleocapsid and the envelope is called the **tegument** and contains enzymes and proteins that facilitate the virus infection.

The poxviruses are enveloped viruses with a large, complex, bricklike shape. The envelope encloses a dumbbell-shaped, DNA-containing nucleoid structure; lateral bodies; fibrils; and many enzymes required for replication of the virus.

VIRUS REPLICATION

Viruses must be able to recognize and enter appropriate target cells, replicate, and then infect other cells. The cell acts as a factory, providing the substrates, energy, and machinery for replication of the viral genome and synthesis of the viral proteins. The virus conforms to and competes for the same machinery used by the cell to synthesize the mRNA and protein required for its own structure and function. Processes not provided by the cell must be encoded in the genome of the virus. The outcome of the competition with cellular metabolic processes determines the resulting type of infection.

The major steps in virus replication are the same for all viruses and are listed here and shown in Figure 38-7. The manner in which each virus accomplishes these steps and overcomes the cell's biochemical limitations is determined by the structure of the genome and the virion that must be replicated. This is indicated in the examples in Figures 38-8 to 38-10.

A single cycle of virus replication (Box 38-1) proceeds through an **early phase**, with attachment of the virus to the host cell, penetration, and initial uncoating of the viral particle, and a **late phase** from macromolecular synthesis to viral assembly and release. The period before virus assembly is called the **eclipse period**, and the period before virus release is the **latent period** (Figure 38-11). The yield of virus per cell, or **burst size**, can be determined by measuring the amount of infectious virus produced and then dividing by the number of infected cells. The properties of the virus and the cell determine both the length of time a virus requires to traverse these different phases and the burst size.

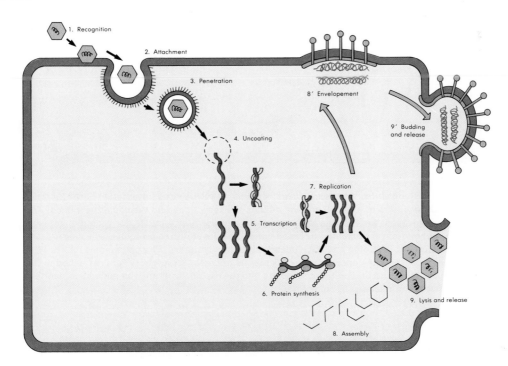

FIGURE 38-7   A general scheme of virus replication.

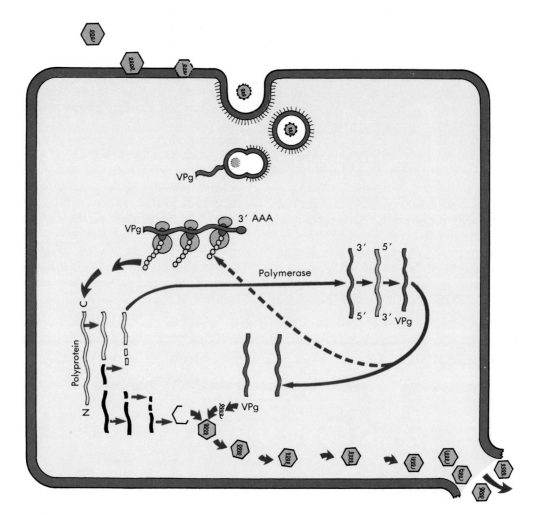

FIGURE 38-8   Replication of picornaviruses: a simple (+) RNA virus. Interaction of the picorna-viruses with receptors on the cell surface defines the target cell and weakens the capsid. The virion is endocytosed, the genome is released and is utilized as a mRNA for protein synthesis. One large polyprotein is translated from the virion genome and then proteolytically cleaved into individual proteins, including an RNA-dependent RNA polymerase. The polymerase makes a (−) strand template from the genome and replicates the genome. A protein *(VPg)* is attached to the viral genome. The structural proteins associate into the capsid structure, the genome is inserted, and the virions are released upon cell lysis.

Box 38-1   Steps in Virus Replication

1. Recognition of the target cell
2. Attachment .
3. Penetration
4. Uncoating
5. Macromolecular synthesis
   a. Early mRNA and protein synthesis: nonstructural genes for enzymes and nucleic acid–binding proteins
   b. Replication of genome
   c. Late mRNA and protein synthesis: structural proteins
6. Posttranslational modification of proteins
7. Assembly of virus
7a. Budding of enveloped viruses
8. Release of virus

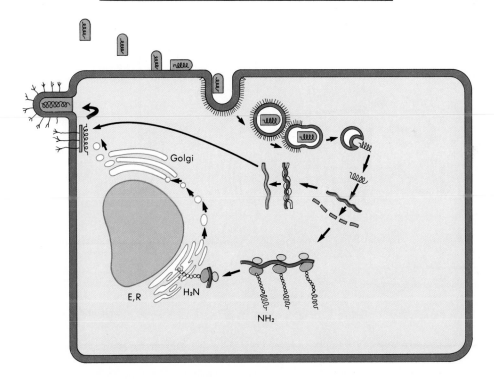

FIGURE 38-9   Replication of rhabdoviruses: a simple enveloped (−) RNA virus. Rhabdoviruses bind to the cell surface, are endocytosed and the envelope fuses with the endosome vesicle membrane to deliver the nucleocapsid to the cytoplasm. The virion must carry a polymerase and produces five individual mRNA and a full length (+) RNA template. Proteins are translated from the mRNAs, including one glycoprotein *(G),* which is co-translationally glycosylated in the endoplasmic reticulum, processed in the golgi apparatus and delivered to the cell membrane. The genome is replicated from the (+) RNA template and the N, L and NS proteins associate with the genome to form the nucleocapsid. The matrix protein associates with the G protein modified membrane, followed by the nucleocapsid and the virion buds from the cell in a bullet-shaped virion.

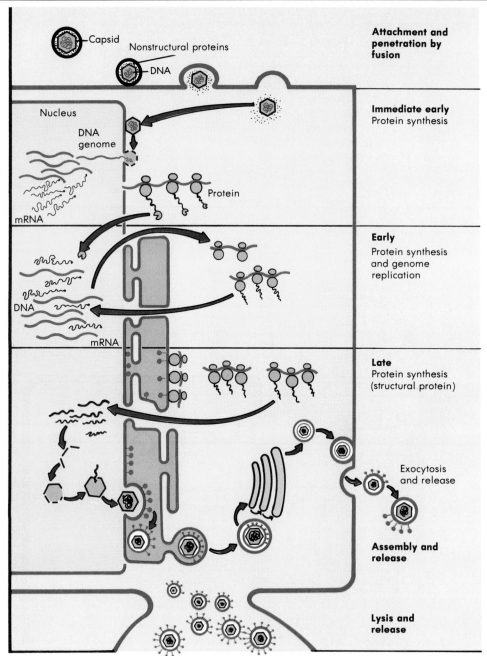

FIGURE 38-10   Replication of herpes simplex virus (HSV), a complex enveloped DNA virus. HSV binds to specific receptors and fuses with the plasma membrane. The nucleocapsid then delivers the DNA genome to the nucleus. Transcription and translation occur in three phases, immediate early, early, and late. Immediate early proteins promote the takeover of the cell; early proteins consist of enzymes, including the DNA dependent DNA polymerase and the late proteins are structural proteins, including the viral capsid and glycoproteins. The genome is replicated before transcription of the late genes. Capsid proteins migrate into the nucleus, assemble into icosadeltahedral capsids and are filled with the DNA genome. The viral glycoproteins are co-translationally glycosylated in the endoplasmic reticulum and diffuse to the contiguous nuclear envelope. The genome containing capsids bud through these modified membranes and are transferred to the golgi apparatus, the glycoproteins are processed, and the virus is released by exocytosis. Alternatively, the virus may be released upon cell lysis.

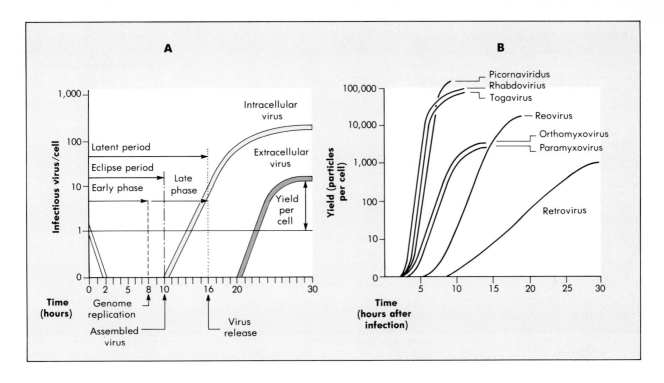

FIGURE 38-11   A, Single cycle growth curves of model virus. **B,** Growth curve and burst size of representative viruses. (**B** redrawn from White DO and Fenner F: Medical virology, ed 3, New York, 1986, Academic Press, Inc.)

Recognition of and Attachment to Target Cell

To infect a cell, the virus must first recognize cells that permit virus replication. Specific capsomeric structures usually found at the vertices of a naked virus or the glycoproteins of enveloped viruses act as viral attachment proteins (Table 38-5) and bind to **receptors** (Table 38-6) on the cell.

The receptors on the cell may be proteins or carbohydrates. Viruses that bind to receptors expressed on a limited number of cells have a restricted **host range**, in which the cell type determines their **tropism**. For example, Epstein-Barr virus binds to the C3d (CR2) receptor, which is expressed on B lymphocytes of humans and New World monkeys. The virus is therefore lymphotropic. Other viruses bind to receptors expressed on many cells within an individual of a species, such as the sialic acid–binding orthomyxoviruses (e.g. influenza). The alpha togaviruses and the flaviviruses are able to bind to receptors and infect cells of many animal species, including arthropods, reptiles, amphibians, birds, and mam-

mals. This allows them to infect and be spread by mosquitos and other insects.

Multiple interactions between the viral attachment proteins and the cellular receptors firmly attach the virus to the cell. These interactions may trigger mechanisms that deliver the viral nucleocapsid into the cell.

Penetration

The mechanism of internalization of a virus depends on its structure. Most nonenveloped viruses enter the cell by **receptor-mediated endocytosis**, or **viropexis**. Endocytosis is a normal process used by the cell for uptake of receptor-bound molecules such as hormones, low density lipoproteins, and transferrin.

Enveloped viruses can be internalized by endocytosis, or their envelope can fuse with the plasma membrane to deliver the nucleocapsid into the cytoplasm. The pH optimum for fusion of an enveloped

Table 38-5   Examples of Viral Attachment Proteins

Virus Family	Virus	Viral Attachment Protein
Picornaviridae	Rhinovirus	VP1-VP2-VP3 complex
Adenoviridae	Adenovirus	Fiber protein
Reoviridae	Reovirus	Sigma 1
	Rotavirus	VP$_7$
Togaviridae	Semliki Forest virus	E1, E2, E3 complex
Rhabdoviridae	Vesicular stomatitis Virus	G
Orthomyxoviridae	Influenza A	Hemagglutinin
Paramyxoviridae	Measles	Hemagglutinin
Herpesviridae	Herpes simplex virus	gD and gB
	Epstein-Barr virus	gp350 and gp220
Retroviridae	Murine leukemia virus	gp70
	Human immunodeficiency virus	gp120

Table 38-6   Virus Receptors of Known Cellular Function

Virus	Target Cell	Function
Epstein-Barr virus (herpes)	B lymphocyte	C3d receptor (CR2)
Reovirus	Neuron	β-Adrenergic receptor
Rabies	Neuron	Acetylcholine receptor (?)
Human immunodeficiency	Helper T lymphocyte, macrophage	CD4 molecule

virus determines whether penetration occurs at the cell surface at neutral pH or in an endosome at acidic pH.

The fusion activity may be provided by the viral attachment protein or another protein. The hemagglutinin of influenza A binds to sialic acid receptors on the target cell and then undergoes a dramatic conformation change under mild acidic conditions to expose hydrophobic portions capable of promoting membrane fusion. Paramyxoviruses have a fusion protein that is active at neutral pH to promote virus-to-cell fusion. This protein can also promote cell-to-cell fusion to form multinucleated giant cells (**polykaryocytes** or **syncytia**). Some herpesviruses and retroviruses can also fuse with cells at neutral pH and induce syncytia.

Uncoating

Once internalized, the nucleocapsid must be delivered to the site of replication within the cell and the capsid or envelope removed. The genome of DNA viruses, except for poxviruses, must be delivered to the nucleus, whereas most RNA viruses remain in the cytoplasm. Poxviruses remain in the cytoplasm and undergo replication in individual subcellular factories.

The uncoating process may be initiated by attachment to the receptor or promoted by the acidic environment or proteases found in an endosome or lysosome. Four major methods of virus uncoating exist:

1. Enveloped viruses that fuse with the cell membrane are uncoated on internalization. Most naked virus capsids disperse in the cytoplasm to release the genome.

2. Picornavirus capsids are weakened when the VP4 protein is released by the insertion of the receptor into the keyholelike attachment site.

3. The herpesvirus nucleocapsid docks with the nuclear membrane and delivers its genome directly to its site of replication.

4. In contrast, the reovirus and poxvirus are only

partially uncoated, leaving an intact core. The poly-merases in the virion are required for viral mRNA synthesis. Poxvirus must initiate synthesis of immediate early proteins, including an uncoated enzyme, to release the DNA-containing core into the cytoplasm.

Macromolecular Synthesis

Transcription and **replication** of the viral genome are probably the most important steps in virus multiplication. The means by which each virus accomplishes these steps depends on the structure of the genome (e.g., single-stranded DNA, double-stranded DNA, positive-strand RNA, negative-strand RNA, or double-stranded RNA) and the site of replication (Figure 38-12).

The machinery for transcription and mRNA processing is found in the nucleus. Most DNA viruses can take advantage of the cell's DNA-dependent RNA polymerase II and enzymes used to make mRNA. Poxviruses, which replicate in the cytoplasm, must provide all these functions. The cell has no means of replicating RNA so that RNA viruses must encode the necessary enzymes for transcription and replication.

DNA Viruses Most DNA viruses require protein synthesis before the genome can be replicated, and their mRNAs will be produced in sequence. mRNA for nonstructural proteins are produced first; after the infection becomes more established, the mRNAs for the structural proteins are made. **Early gene products** (nonstructural proteins) are often DNA-binding proteins and enzymes, such as a DNA-dependent DNA polymerase. These proteins are catalytic, and only a relative few are required. **Late gene products** are structural proteins, and many copies of these proteins are required to package the virus.

Replication of the simple DNA viruses (e.g. parvoviruses, papovaviruses) uses the host DNA-dependent DNA polymerases, whereas the larger, more complex viruses (e.g. adenoviruses, herpesviruses, poxviruses) encode their own polymerases. Viral polymerases are usually faster but less precise than host cell polymerases, causing a higher mutation rate in viruses as well as providing a target for antiviral drugs.

Replication of viral DNA follows the same biochemical rules as for cellular DNA. DNA synthesis is **semiconservative**, and both the viral and the cellular DNA polymerases require a primer to initiate the

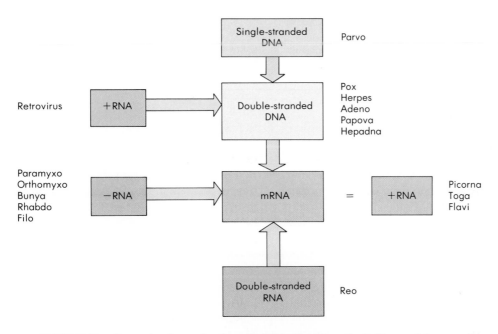

FIGURE 38-12   Strategies for production of mRNA. (Based on the Baltimore Schema; redrawn from Bacteriol Rev 35:235, 1971.)

Different regions of either DNA strand are marked with specific promoters to control transcription of adenoviruses and herpesviruses.

Posttranscriptional processing of the mRNA is used by papovaviruses and adenoviruses to produce several mRNAs from the same larger transcript **(splicing)**. The viral mRNA acquires a 3′ polyadenylated (poly/A) tail and a 5′ methylated cap and then is exported to the cytoplasm.

DNA viruses may enhance the transcription of their mRNA by modifying cellular DNA-binding proteins or by speeding up the growth of the cell. For example, a component of the herpes simplex virion is delivered to the nucleus and binds to a host cell DNA-binding protein to promote the transcription of the immediate early genes of the virus. Transcription will also speed up as the number of viral genomes in the cell increases, providing more templates for mRNA synthesis.

Hepadnavirus replication is unique in that a circular positive-strand RNA intermediate is first synthesized by a DNA-dependent RNA polymerase. An RNA-dependent DNA polymerase (reverse transcriptase) in the core makes a negative-strand-DNA, whereas the RNA is degraded. As the virion is enveloped, a partial positive-strand DNA is synthesized.

A major limitation to the replication of a DNA virus is the availability of deoxyribonucleotide substrates. Most cells in the resting phase of growth are not undergoing DNA synthesis, and deoxythymidine pools are limited. Some DNA viruses (e.g. papovaviruses, adenoviruses) stimulate cell growth, whereas others provide DNA scavenging enzymes to promote substrate availability. It has been demonstrated that the T antigen of SV40, the E7 of papillomavirus, and the E1a protein of adenovirus bind to and alter a growth-controlling DNA-binding protein (the retinoblastoma gene), which may be the mechanism for this process. Herpes simplex virus encodes scavenging enzymes such as a DNAse, ribonucleotide reductase, and thymidine kinase to generate the necessary deoxyribonucleotide substrates for replication of its genome.

RNA Viruses Replication and transcription of RNA viruses are difficult to separate, since the viral genomes are usually either an mRNA (positive-strand RNA) or a template (negative-strand RNA). The RNA viruses must code for **RNA-dependent RNA polymerases (replicases** and **transcriptases)** because the cell has no means of replicating RNA. During rep-

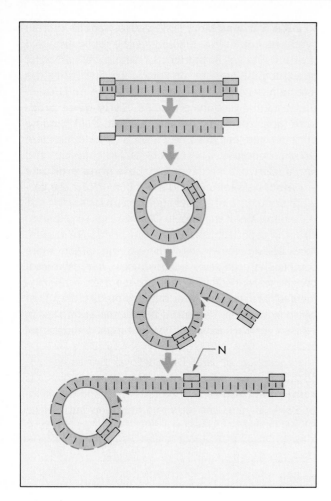

FIGURE 38-13   Model for the replication of herpesvirus double-stranded DNA. *N* is the site for endonuclease cleavage of the genome. Multiple copies of the genome can be replicated and then cleaved for assembly in the virion. (Redrawn from Dulbecco R and Ginsberg HS: Virology, ed 2, Philadelphia, 1986, JB Lippincott Co.)

DNA chain (Figure 38-13). Many DNA viruses have inverted genetic repeats that allow the DNA to fold back and hybridize with itself to provide a primer. The polymerase can then extend the complementary strand and continue DNA synthesis to regenerate the inverted segment and the new genome copy.

Transcription is regulated by specific promoter and enhancer elements similar in structure to those of the host cell. Genes may be transcribed from either DNA strand of the genome and in opposite directions. For example, the early and late genes of the papovaviruses are in opposite, nonoverlapping DNA strands.

lication and transcription a double-stranded RNA replicative intermediate is formed, a structure not normally found in uninfected cells.

Since RNA is degraded relatively quickly, an RNA-dependent RNA polymerase must be provided or synthesized soon after uncoating to generate more viral RNA, or the infection will be aborted. As with the DNA viruses, replication of the genome provides templates for production of more mRNA, which amplifies and accelerates virus replication.

Some RNA viruses (e.g., togaviruses) can control the temporal expression and the extent of production of nonstructural and structural genes. This is accomplished by differences in (1) the efficiency of transcribing the mRNAs, (2) the template used for producing the mRNA, or (3) the efficiency of translating the particular proteins.

The positive-strand RNA virus genomes, (e.g., picornaviruses, flaviviruses, togaviruses) act as mRNA, bind to ribosomes, and direct protein synthesis. The naked positive-strand RNA viral genome is sufficient to initiate infection by itself. After an RNA-dependent RNA polymerase is produced, a negative-strand RNA template is synthesized. The template can then be used to generate more mRNA and replicate the genome. The mRNA is not capped, but the genome does have a short polyA sequence. To promote translation of its mRNA, poliovirus inactivates the 200,000-dalton cap-binding protein of the ribosome with a viral-coded protease.

The genomes of negative-strand RNA viruses are the templates for production of mRNA. The negative-strand RNA genome is not infectious by itself, and a polymerase must be carried into the cell with the genome to make mRNA and initiate viral protein synthesis. The mRNA or another positive-strand RNA species can then act as a template to generate more copies of the genome. Transcription of negative-strand RNA viruses occurs in the cytoplasm, except for influenza. The influenza transcriptase requires a primer to produce mRNA. It uses the 5′ ends of cellular mRNA in the nucleus as primers for its polymerase and in the process steals the 5′ cap from the cellular mRNA. Interestingly, replication of the influenza genome does not require primers and occurs in the cytoplasm.

The reoviruses have a segmented, double-stranded RNA genome and undergo a more complex means of replication and transcription. The reovirus RNA polymerase is part of the inner layer, or core, of the capsid. mRNA units are transcribed from each of the 10 or more segments of the genome while they are still in the core. The mRNA is released into the cytoplasm, where it directs protein synthesis or is sequestered into new cores. The reoviruses provide their own capping enzymes as part of the inner capsid. The mRNA is not polyA. The positive-strand RNA in the cores acts as templates for negative-strand RNA, producing the progeny double-stranded RNA. In this manner the negative-strand RNA portion of the genome is never found free in the cytoplasm.

Although the retroviruses have a postive-strand RNA genome, the virus provides no means for replication of the RNA in the cytoplasm. Instead, the retroviruses deliver two copies of the genome to the nucleus of the cell with an RNA-dependent DNA polymerase (reverse transcriptase) in the virion. A DNA copy of the genome is synthesized and integrated into the host chromatin. The viral genes are then transcribed as cellular genes. Promoters in the viral genes enchance the extent of their transcription.

Viral Protein Synthesis

All viruses use the host cell ribosomes, transfer RNA (tRNA), and mechanisms for posttranslational modification for their protein synthesis. Unlike bacterial mRNAs, the eucaryotic ribosome binds only once to each mRNA, producing only one polypeptide. Each virus deals with this limitation differently, depending on the structure of the genome. For example, the entire genome of (positive-strand) RNA viruses binds to the ribosome and is translated into one giant **polyprotein**. The polyprotein is subsequently cleaved by cellular and viral proteases into functional proteins. DNA viruses, retroviruses and most negative-strand RNA viruses transcribe mRNA for smaller polyproteins or individual proteins. Each segment of the orthomyxovirus and reovirus genome codes for a single protein.

Viruses employ different tactics to promote preferential translation of their viral mRNA. As described, poliovirus alters the ribosome, preventing binding and translation of 5′ capped cellular mRNA. Togaviruses and many other viruses increase the permeability of the cell's membrane, which decreases the ribosomal affinity for most cellular mRNA. Adenovirus infection blocks the egress of cellular mRNA from the nucleus. Herpes simplex virus inhibits cellular macromolecular synthesis and induces degradation of the cell's DNA and mRNA. All these actions also contribute to the cytopathology of the virus infection.

Some viral proteins require **posttranslational modifications** such as phosphorylation, glycosylation, acylation, or sulfation. These processes are performed by the same systems as for cellular proteins. Protein phosphorylation is accomplished by cellular protein kinases and is a means of modulating, activating, or inactivating proteins. Several herpesviruses also encode their own protein kinase, which is carried into the cell in the virion.

Viral glycoproteins are synthesized on membrane-bound ribosomes and have both peptides to allow insertion into the rough endoplasmic reticulum and the necessary amino acid sequences to promote *N*-linked glycosylation. The high mannose form of the glycoproteins is processed through the Golgi apparatus and then expressed on the plasma and other membranes of the cell. O-glycosylation, acylation, and sulfation of the proteins can also occur during progression through the Golgi apparatus.

Assembly

Once all the components of the virus have been synthesized, the virion can assemble. The virion assembly process is analogous to a three-dimensional interlocking puzzle that puts itself together in the box. The virion is built from small, easily manufactured parts that enclose the genome into a functional package. How and where the parts interact are structurally predetermined.

Each part of the virion has recognition structures that allow the virus to form the appropriate protein-protein, protein–nucleic acid, and for enveloped viruses, protein-membrane interactions needed to assemble into the final structure. The assembly process begins when the necessary pieces are synthesized and the concentration of structural proteins in the cell is sufficient to thermodynamically drive the process, much as a crystallization reaction. Assembly may be facilitated by scaffolding proteins or other proteins degraded to provide energy.

The nucleocapsids may be assembled as empty structures, to be filled with the genome, or they may be assembled around the genome. Empty capsids (procapsids) of the picornaviruses and herpesviruses assemble first, and then the genome binds to a capsid protein and is reeled into the shell. Temperature-sensitive mutants of herpes simplex virus have been isolated, which shows that empty capsids form at the nonpermissive temperature and become filled with prescribed lengths of DNA at the permissive temperature. Nucleocapsids of the negative-strand RNA viruses, retroviruses, and togaviruses assemble around the genome and are subsequently enclosed in an envelope.

Assembly of the DNA viruses, other than poxviruses, occurs in the nucleus and requires transport of the virion proteins into the nucleus. These proteins must have the appropriate basic amino acid peptide signal for nuclear transport as part of their structure. RNA virus and poxvirus assembly occurs in the cytoplasm.

Acquisition of an envelope occurs after association of the nucleocapsid with regions of membrane modified with viral glycoproteins. Matrix proteins for negative-strand RNA viruses line and promote the adhesion of nucleocapsids with the glycoprotein-modified membrane. As more interactions occur, the membrane surrounds the nucleocapsid and buds from the membrane in a process analogous to exocytosis.

The site of budding is determined by the type of genome and the protein sequence of the glycoproteins. Most RNA viruses bud from the plasma membrane, but flaviviruses, coronaviruses, and bunyaviruses bud from endoplasmic reticulum and Golgi membranes. Release of the virus from the cell occurs simultaneously with budding from the plasma membrane.

The herpes simplex virus nucleocapsid assembles in the nucleus and buds at the nuclear membrane, taking the high mannose precursor form of the viral glycoproteins. The virion is then transported to the Golgi apparatus in vesicles, the glycoproteins are processed, and the virus is released by exocytosis. The route of envelopment and release of herpes simplex virus is similar to the route of synthesis, processing, and release of secretory glycoproteins. Herpes simplex virus is released after cell lysis.

Reoviruses pose additional problems for assembly because of their segmented genome and double capsid. The inner capsid of segment. Assembly sites appear to be associated with microtubules. The outer capsid assembles on the inner capsid core. For rotaviruses, a reovirus, the virion acquires the glycoprotein, which acts as its viral attachment protein and a transient envelope, by budding into the endoplasmic reticulum. The envelope is then removed before release to produce infectious virus.

For all viruses, errors are made during assembly and empty or defective virions are produced. These are referred to as **defective interfering particles.** Also, the assembly process cannot discriminate between intact or functional and defective genomes. As a result, the particle/infectious unit ratio, also called **particle/plaque-forming unit ratio,** is usually greater than 100 and during rapid viral replication can even be 10^4.

Release

Viruses can be released from cells after lysis of the cell, by exocytosis, or by budding from the plasma membrane. Naked capsid viruses are generally released after lysis of the cell. Release of some enveloped viruses occurs after budding from the plasma membrane. Lysis and plasma membrane budding are efficient means of release. Viruses that bud or acquire their membrane (e.g., poxviruses) in the cytoplasm generally are more cell associated. Viruses that have a neuraminidase (e.g., orthomyxoviruses, certain paramyxoviruses) facilitate their own release by inactivating potential sialic acid receptors on the glycoproteins of the virion and the host cell.

Reinitiation of Replication Cycle

The virus released to the extracellular media is usually responsible for initiating new infections; however, cell-to-cell fusion or **vertical transmission** of the genome to daughter cells can also spread the infection. Some herpesviruses, retroviruses, and paramyxoviruses can induce cell-to-cell fusion to create multinucleated giant cells, (syncytia). The retroviruses can transmit their integrated copy of the genome vertically to daughter cells on cell division.

VIRUS GENETICS

As with other genetic systems, mutations spontaneously occur in viral genomes, creating new virus strains with different properties from the **parental,** or **wild-type, virus.** These variants can be identified by their genetic sequence, antigenic differences (serotype), or differences in functional or structural properties. Mutations in essential genes inactivate the virus, but mutations in other genes can produce antiviral drug resistance or alter the antigenicity of the virus.

The mutation rate is influenced by the integrity of reproduction by the polymerase. The rates of mutation for DNA viruses are usually lower than for RNA viruses. The poor fidelity of the reverse transcriptase of human immunodeficiency virus and other lentiviruses actually works in favor of the virus to induce mutations in the viral antigens, allowing the virus to evade immune detection. Mutagenesis can also be induced chemically or by X-irradiation. In some cases, portions of the genome may be lost or selectively removed to inactivate a particular gene, creating a deletion mutant.

Mutants are identified by the properties induced by the mutation. Mutations in essential genes are termed **lethal mutations.** These mutants are difficult to isolate because the virus cannot replicate. Other mutations may produce **plaque mutants,** which differ from the wild type in the size or appearance of the plaque; **host range mutants,** which differ in the target cell that can be infected; or **attenuated mutants,** which are variants that cause less serious infections in animals or humans. **Conditional mutants,** such as **temperature-sensitive** (ts) or **cold-sensitive mutants,** express a normal phenotype or grow at only certain temperatures. Ts mutants generally grow well or relatively better at 30° to 35° C, whereas elevated temperatures of 38° to 40° C will inactivate the essential gene product.

Genetic interactions can occur between viruses or between the virus and the cell to create new virus strains (Figure 38-14). Intramolecular genetic exchange between viruses or the host and the virus is termed **recombination.** Recombination can occur readily for DNA viruses. For example, co-infection of a cell with the two closely related herpesviruses, herpes simplex virus types 1 and 2, will yield intertypic recombinant strains. These new hybrid strains will have genes from both types 1 and 2. Exchange of genes of a segmented genome, as for influenza or reoviruses, is termed **reassortment.** Hybrid strains expressing randomly chosen functions of each parent are readily produced in this manner.

In some cases a mutant can be rescued by the replication of another mutant or by the wild-type virus. The replication of the other virus provides the missing function required by the mutant **(complementation).** Rescue of a lethal or conditional lethal mutant with a defined genetic sequence, such as a restriction endonuclease DNA fragment, is called **marker rescue.** Marker rescue has been used exten-

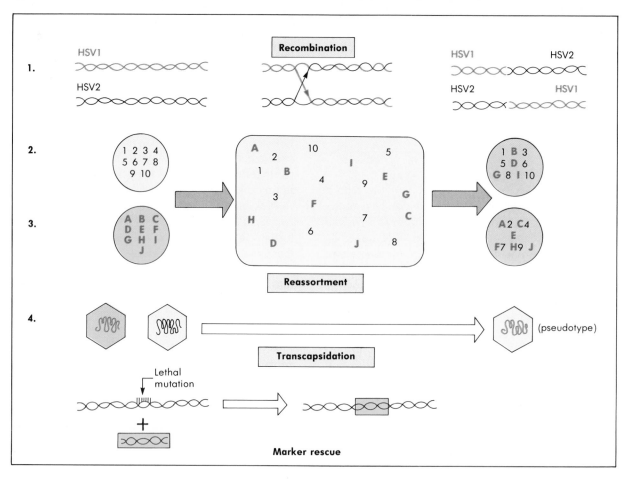

FIGURE 38-14   Genetic exchange between viral particles can give rise to new viral types, as illustrated here.

sively to map the genomes of viruses such as herpes simplex virus. Virus produced from cells infected with different virus strains may be **phenotypically mixed** and have the proteins of one strain but the genome of the other **(transcapsidation). Pseudotypes** are generated when transcapsidation occurs between different types of virus, but this is rare.

Individual virus strains or mutants are selected by their ability to use the host cell machinery and withstand the environment. Cellular properties that can act as selection pressures include the growth rate of the cell and the expression of particular enzymes required by the virus. The body, with its elevated temperature, natural and immune defenses, and tissue structure, selects for those viruses that can evade the defenses and reach the appropriate target cell.

Growth of virus under benign laboratory conditions lacks these selective pressures and allows weaker strains to survive. This process is used to develop attenuated virus strains for use in vaccines.

VIRAL PATHOGENESIS

The outcome of a viral infection is determined by the nature of the virus-host interaction and the host's response to the infection. The immune response limits a virus infection but is also a major factor in pathogenesis. The disease and its symptoms are defined by the target tissue. A particular virus may cause several different diseases or no observable symptoms. For example, herpes simplex virus type 1 can cause gingivostomatitis, pharyngitis, herpes labialis (cold sores), genital herpes, encephalitis, and keratoconjunctivitis. In many individuals, however, this virus remains latent with no observable clinical impact. Also, a particular disease may be caused by several viruses that have a common tissue tropism (e.g., hepatitis, common cold, encephalitis).

Transmission of viruses

Viruses are transmitted by direct contact; respiratory or gastrointestinal secretions; mucous secretions; injection, including insect bites; or tissue transplant, including blood (Table 38-7). Contaminated objects, termed **fomites,** can foster the spread of viruses provided the structure of the virus is sufficiently stable to survive until contact with the host is made. Rhinoviruses and many other viruses can be spread by contact with contaminated objects such as handkerchiefs or toys. During the French and Indian War of 1763, the British used blankets contaminated with smallpox to spread the disease and weaken the Indian enemy. Animals can also act as **vectors** to spread viral disease to other animals and humans. Arthropods, including mosquitos, ticks, and sandflies, can act as vectors for togaviruses, flaviviruses, bunyaviruses, arenaviruses, and reoviruses. These viruses are often referred to as **arboviruses** because they are *a*rthropod *b*orne. Rabies is an example of a virus spread by the bite of an animal.

The mode of transmission available to a particular virus is determined by the tissue producing the virus and its stability under such environmental conditions as drying, pH, temperature, and detergents. Nonenveloped viruses are generally resistant to these environmental stresses, whereas enveloped viruses are labile to them. As a result, enveloped viruses cannot withstand the acidic environment of the stomach and the detergent-like bile of the intestines. Enveloped viruses are generally spread by wet means such as respiratory droplets, blood, injection, organ transplants, mucous, saliva, and semen. Respiratory viruses replicate in the lung and are released in aerosol droplets, whereas gastrointestinal viruses are passed by the fecal-oral route.

Types of Viral Infection

Viral infections can occur differently, depending on the properties of both the virus and the cell (Table 38-8). If the cell is incapable of supporting the replication of a virus, an abortive infection may occur. If the cell supports viral replication, the virus may kill the target cell in a **cytolytic** infection or **persist** in the cell. The persistent infection may be **productive,** or the virus may remain **latent.** Some DNA viruses and retroviruses establish persistent infections, with alterations of the growth characteristics of the cells and **immortalization** or **transformation** of the cell. Retroviruses can continue to produce virus in transformed cells, but production of a DNA virus generally kills the cell.

Viral infection may be **apparent** or **inapparent**

Table 38-7 Modes of Transmission for Representative Viruses

Route of Transmission	Virus
Respiratory	Rhinoviruses, adenoviruses, paramyxoviruses, orthomyxoviruses, coronaviruses
Fecal-oral	Picornaviruses (except rhinoviruses, including hepatitis A virus), adenoviruses, reoviruses
Close contact, injection, and tissue transplants	Herpesviruses, retroviruses, hepadnaviruses
Arthropod or animal bite	Togaviruses, flaviviruses, bunyaviruses, arenaviruses, rhabdoviruses, reoviruses (Colorado tick fever)

Table 38-8 Types of Viral Infections, Cellular Level

Type	Virus Production	Fate of Cell
Abortive	−	No effect
Cytolytic	+	Death
Persistent		
Productive	+	Senescence
Latent	−	No effect
Transforming		
DNA viruses	−	Immortalization
RNA viruses	+	Immortalization

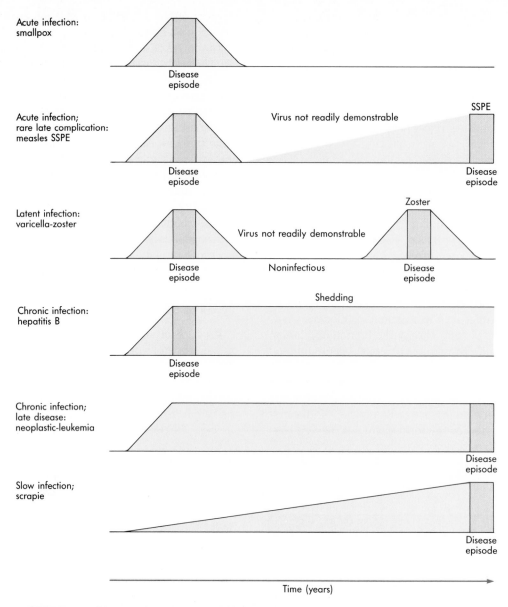

FIGURE 38-15   Diagram depicting acute infection and various types of persistent infection, as illustrated by the diseases in the column at left. *Vertical box* (green) illustrates disease episode. (Redrawn from White DO and Fenner F: Medical virology, ed 3, New York, 1986, Academic Press, Inc.)

and cause **acute** or **chronic disease** (Table 38-9). Inapparent infections occur if no damage occurs to the infected tissue, the tissue is expendable, the damage is rapidly repaired, or the level of damage is below a functional threshold for that particular tissue. For example, most infections by picornaviruses, including poliovirus, are inapparent because the infections are generally restricted to the lining of the gut, which is sloughed off and regenerated. Many infections of the brain are either inapparent or below the threshold of severe loss of function. Encephalitis results when loss of function becomes significant. Inapparent infections are frequently detected by the presence of virus-specific antibody in an individual.

Table 38-9 Types of Viral Infections, Host Level	
Type	**Outcome**
Apparent	Acute
	Chronic or persistent
Inapparent	Subclinical
	Latent

The ability of an individual's immune system to resolve a viral infections usually determines whether acute or chronic disease will result (Figure 38-15).

Stages in Viral Infection

The stages of viral infection and viral disease are shown in Figure 38-16. In the least complicated sequence the virus is acquired from the environment or direct contact, initiates the infection, replicates, initiates an immune response, and then is controlled. The virus gains entry through the mucous membranes of the body, including the eye, nose, mouth, genitalia, gastrointestinal tract, and breaks in the skin. These areas are generally protected by mucus and IgA. Inhalation is probably the most common route of viral infection.

Primary infection usually occurs at the site of entry if susceptible cells are present. The primary infection may result in symptoms but often is only a source of virus for infection of other body sites.

The initial period before detection of symptoms is termed the **prodrome** or **incubation period.** The length of the incubation period is relatively short if infection of the primary site produces the characteristic symptoms of the disease. The incubation periods for many common viral infections are given in Table 38-10.

Following the primary infection, the virus may spread to other sites. The lymphatics, the reticuloen-

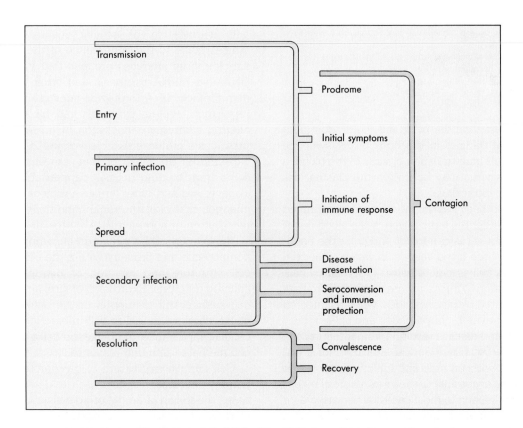

FIGURE 38-16   The stages of viral infection, clinical manifestations, and contagion.

Table 38-10 Incubation Periods of Common Viral Infections

Disease	Incubation Period* (Days)
Influenza	1-2
Common cold	1-3
Bronchiolitis, croup	3-5
Acute respiratory disease (adenoviruses)	5-7
Dengue	5-8
Herpes simplex	5-8
Enteroviruses	6-12
Poliomyelitis	5-20
Measles	9-12
Smallpox	12-14
Chickenpox	13-17
Mumps	16-20
Rubella	17-20
Mononucleosis	30-50
Hepatitis A	15-40
Hepatitis B	50-150
Rabies	30-100
Warts	50-150
AIDS	1-5 years

Modified from White DO and Fenner F: Medical virology, ed 3, New York, 1986, Academic Press, Inc.
*Until first appearance of prodromal symptoms. Diagnostic signs (e.g., rash, paralysis) may not appear until 2 to 4 days later.

dothelial system, and the blood are the predominant means of viral transfer in the body. Most systemic infections result from transport of virus in the blood, a **viremia**. The virus may be free in the plasma, cell associated, or intracellular.

Access to other sites in the body is determined mainly by whether the virus is free in the blood or cell associated. Free virus can infect and cross the endothelial lining of the blood vessels as well as other tissues, whereas cell-associated virus is spread by cell-to-cell interaction. Viruses in macrophages circulate to sites populated by macrophages, whereas infected lymphocytes travel to lymph nodes and spleen.

Symptons and disease resulting from infection of a **secondary** target tissue appear later than for a primary site. As with primary site infection, the target tissue must be accessible to the virus, express receptors for the virus, and in most cases be permissive for replication of the virus. Infection of a specific tissue by a virus is tropism.

Virus replication may be asymptomatic but may provide a source for viremia or for release to the environment via bodily secretions or excretions. The viremia can be maintained by continued production of virus at the primary infection site or at a secondary site, such as the endothelial cell lining of blood vessels, liver, spleen, bone marrow, or even circulating leukocytes. Production of virus by the secretory glands, testes, lung, and gastrointestinal cell lining results in the release of virus.

CYTOPATHOGENESIS

Virus replication initiates changes in cells that can lead to cytolysis or alterations in the appearance, functional properties, or antigenicity of the cell. The effects on the cell may result from virus takeover of macromolecular synthesis, accumulation of viral proteins or particles, or modification of cellular structures, such as incorporation of glycoproteins in cell membranes (Table 38-11).

Virus replication is often incompatible with essential cell functions and viability. Inhibition of synthesis and degradation of cellular macromolecules by viruses leads to the senescence of the cell. Herpes simplex virus produces proteins that inhibit cellular DNA and mRNA synthesis and other proteins that degrade host DNA to provide substrates for genome replication. Protein synthesis can be subverted by altering ribosomal recognition of mRNA. Poliovirus produces a protease that cleaves the 200,000-dalton, cap-binding ribosomal protein, preventing the ribosome from binding normal, capped cellular mRNA. Togaviruses and other viruses promote permeability changes that alter the ionic conditions required for recognition of normal mRNA by the ribosome.

Replication of the virus and accumulation of viral components and progeny within the cell can disrupt cell structure and function or disrupt lysosomes, causing autolysis. Expression of viral antigens on the cell surface and disruption of the cytoskeleton can cause changes in cell-to-cell interactions as well as cellular appearance and make the cell a target for the cell-mediated immune response.

Cell surface expression of glycoproteins of some paramyxoviruses, herpesviruses, and retroviruses initiates the fusion of neighboring cells into multinucleated giant cells called syncytia. Cell-to-cell fusion may occur in the absence of new protein synthesis

Table 38-11 Mechanisms of Viral Cytopathogenesis

Mechanism	Representative Viruses
Inhibition of cellular protein synthesis	Polioviruses, herpes simplex virus, togaviruses, poxviruses
Inhibition and degradation of cellular DNA	Herpes simplex virus
Alteration of cell membrane structure	
Glycoprotein insertion	All enveloped viruses, reoviruses
Syncytia formation	Herpes simplex virus, varicella-zoster, paramyxoviruses, human immunodeficiency virus
Disruption of cytoskeleton	Capsid viruses (accumulation), herpes simplex virus
Permeability changes	Togaviruses, herpesviruses
Inclusion bodies	
Negri bodies (intracytoplasmic)	Rabies
Owl's eye (intranuclear)	Cytomegalovirus
Cowdry's type A (intranuclear)	Herpes simplex virus, subacute sclerosing panencephalitis (measles) virus
Intranuclear basophilic	Adenoviruses
Intracytoplasmic acidophilic	Poxviruses
Perinuclear intracytoplasmic acidophilic	Reoviruses
Toxicity of virion components	Adenovirus fibers

(fusion from without), as for Sendai and other paramyxoviruses, or may require new protein synthesis **(fusion from within)**, as for herpes simplex virus. Syncytia formation allows the virus to spread from cell to cell and escape antibody detection.

Some viral infections cause characteristic changes in the appearance and properties of the target cells. These changes are especially apparent after infection of cells grown in tissue culture. The presence of new, stainable structures within the nucleus or cytoplasm, called **inclusion bodies,** may result from virus-induced changes in the membrane or chromosomal structure or may represent accumulations of viral capsids or the sites of virus replication. The nature and location of these inclusion bodies are characteristic of particular viral infections and are listed in Table 38-11. Vacuolization, rounding of the cells, and other nonspecific histologic changes indicative of sick cells may also result from viral infection.

HOST DEFENSES AGAINST VIRAL INFECTION

Humans are protected from viral infection by natural and immune defenses. On penetration of the natural barriers, the virus faces other defenses that are activated by the infection but are not specific for a particular infecting agent. These include fever, macrophages and other phagocytic cells, interferon, and natural killer cells. Immune defenses are the last to be activated and are specific for the infecting agent. Once the viral infection has penetrated the natural barriers, the immune response is the best and in most cases the only defense. Both the humoral and the cellular immune responses are important for antiviral immunity (Box 38-2).

Box 38-2 Host Defenses Against Viral Infections

NATURAL BARRIERS

Physical defenses: skin, tears, mucus, ciliated epithelium, stomach acid/bile

Fever

Nonspecific response to antigen: interferon, natural killer cells, macrophages

SPECIFIC IMMUNE RESPONSE

Antibody-producing B lymphocytes

T lymphocytes: CD4 helper cells, CD8 killer cells

Natural Defenses

Barriers to Infection Epithelium is the best barrier to infection. Epithelial infections usually occur through abrasions or breaks in the epithelium or mucosa or as a result of a secondary spread of the virus within the body.

The major routes of entry into the body are the eyes, nose, and mouth, urogenital tract, gastrointestinal tract, and breaks in the epithelium. These portholes to infection are protected by tears, mucus, ciliated epithelium, or stomach acid and bile.

Fever Body temperature and fever can limit the replication of or destabilize some viruses. For example, rhinoviruses, do not replicate at 37° C and are therefore restricted to the upper respiratory tract, where the temperature is cooler.

Interferon Interferon comprises a family of proteins made by many mammalian cell types, including lymphocytes, nonlymphoid leukocytes, and fibroblasts, in response to viral infections. Interferon was first described by Isaacs and Lindemann as a factor that "interferes" with virus replication. The interferons have multiple antiviral activities, including prevention of virus replication and activation of cells of the immune response.

One of the best interferon inducers is double-stranded RNA, such as the replicative intermediates of RNA viruses. One double-stranded RNA molecule per cell is sufficient to induce interferon. Several other substances can also induce interferon production, including intracellular microorganisms (e.g., mycobacteria, fungi, protozoa), immune stimulators or mitogens (e.g., endotoxins, phytohemagglutinin), double-stranded polynucleotides (e.g., poly I:C, poly dA:dT), synthetic polyanion polymers (e.g., polysulfates, polyphosphates, pyran), antibiotics (e.g., kanamycin, cyclohexamide), and low molecular weight synthetic compounds (e.g., tilorone, acridine dyes). This wide array of agents suggest that several different interferon-inducing mechanisms may exist. These interferons can be subdivided by several properties, including size, stability, and cell of origin (Table 38-12).

Following induction by a viral infection, interferon is secreted and binds to interferon receptors on other cells to produce an **antiviral state.** Infection of the second cell triggers a cascade of biochemical events to inhibit protein synthesis and initiate mRNA degradation. This essentially puts the cellular factory on strike. The cells do not allow virus multiplication, and the infection is restricted (Figure 38-17). It must be stressed that interferon does not directly block virus replication.

Although originally described as an antiviral agent, interferon has widespread regulatory effects on cell growth, protein synthesis, and the immune

Table 38-12 Physicochemical Properties of Human Interferons

	Interferon		
Property	α	β	γ
Previous designations	Le-IFN	F-IFN	Immune IFN
	Type I	Type I	Type II
Subtypes	>20	Two	Three
Molecular weight (daltons)*			
Major subtypes	16,000-23,000	23,000	20,000-25,000
Cloned†	19,000	19,000	16,000
Glycosylation	No‡	Yes	Yes
pH 2 stability	Stable‡	Stable	Labile
Induction	Viruses	Viruses	Mitogens
Principal source	Epithelium, leukocytes	Fibroblast	Lymphocyte
Introns in gene	No	No	Yes
Homology with Hu-IFN-α	80%-95%	30%-50%	<10%

From White D.O: Antiviral chemotherapy, interferons and vaccines, Basel, 1984, Karger.
*Molecular weight of monomeric form. Interferons often occur as polymers.
†Nonglycosylated form, as produced in bacteria by recombinant DNA technology.
‡Most subtypes, but not all.

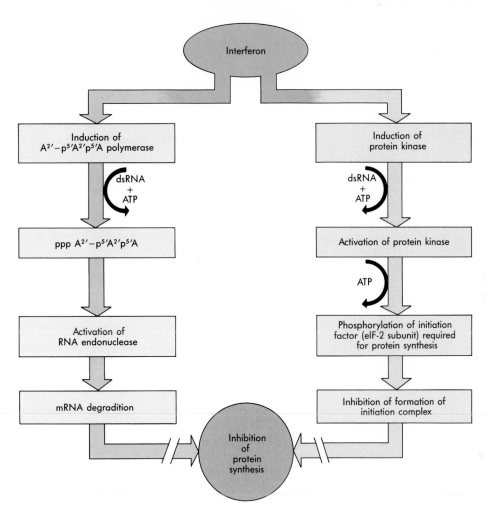

FIGURE 38-17   Two suggested mechanisms of action of interferon on viral protein synthesis. One mechanism involves induction of an enzyme that is activated by double-strand RNA and adenosine triphosphate (ATP). The activated enzyme synthesizes an unusual adenine trinucleotide with a 2′, 5′-phosphodiester linkage. The trinucleotide activates an RNA nuclease. The other mechanism involves induction of a protein kinase that inactivates an initiation factor, eIF-2, by phosphorylating one of the subunits. Both mechanisms inhibit protein synthesis.

response (Table 38-13). All three interferon types block cell proliferation. Interferons have been used in clinical trials for the treatment of cancers. The treatment is tolerated by many patients with advanced cancers but is accompanied by flulike symptoms (chills, fever, fatigue), decreased white blood cell count, and abnormal liver function. The response is different for different cancers, and protocols must be developed empirically for each.

Immune cells (B cells, T cells, natural killer cells, macrophages, basophils, bone marrow stem cells) are sensitive to interferon and also make interferon. Macrophages are one of the best α- interferon producers, whereas T cells are the only cell to produce γ interferon. Interferons can stimulate cell-mediated immunity by enhancing recognition of the target and by activating cellular responses. All interferon types stimulate the expression of human lymphocyte class I antigens (HLAs) on the cell surface, promoting antigen presentation to T cells.

Interferons stimulate pre–natural killer cells to differentiate to natural killer cells and become active.

Table 38-13 Some Biologic Effects of
Interferons

Effect	Comment
Inhibition of multiplication of viruses	
Inhibition of cell division(?)	
Immunomodulation	
All cells	Increased expression of major histocompatibility antigens and Fc receptors
B cell	Antibody production decreased or increased
T cell	Proliferation suppressed; lymphokine release enhanced
Td cell	Delayed hypersensitivity decreased or increased
Tc cell	Cytotoxicity increased
Natural killer cell	Maturation, recycling, and cytoxicity increased
Macrophage	Activated

This provides a rapid natural defense against an infection to limit the virus spread. Interferon also induces major histocompatibility (MHC) class II antigen expression on macrophages. This enhances the cell's ability to present antigen to helper T cells and B cells. The activated macrophages may be a major source of interferon in these cases. Interferon (macrophage activation factor, or MAF) promotes the follow-up procedures after infection. Activation of macrophages would also initiate recruitment, inflammation, and secretion of other biologic response modifiers. The concentration of interferon determines whether an immune activity is enhanced or depressed.

Interferon's broad-spectrum, multifocal activity and potency suggested its potential for treating viral infections soon after its discovery. However, these same attributes have made it difficult to incorporate interferon into clinical use. Effective treatment requires use of the correct interferon subtype, delivered promptly and at the appropriate concentration. Its multiple effects on nonspecific and immune defenses can be detrimental as well as helpful. Incorrect dosage can enhance the side effects and depress immunity. Studies are underway to develop appropri-

ate protocols for treatment using natural and genetically engineered interferons.

Antiviral Immunity

Unlike most bacteria, viruses are intracellular much of the time and hidden from antibody. Protective immune responses control virus infections by limiting virus replication and the spread of the virus. Antibody acts mainly on the virus, whereas cell-mediated immunity is directed at the virus-producing cell (Figure 38-18).

Immune protection to a nonenveloped virus infection depends mainly on antibody, but both the cellular and the humoral responses are important for protection against enveloped viruses. Cell-mediated immunity is especially important for syncytia-forming viruses, which can spread from cell to cell without exposure to antibody, (e.g., measles, herpes simplex virus, varicella-zoster virus).

The immune response can be both protective and destructive. The host's ability to recognize the antigen and mount a protective immune response are major determinants of the infection's outcome. However, hypersensitivity and inflammatory reactions initiated by antiviral immunity can be the major cause of disease-related pathology.

Humoral Immunity The major anti-viral roles of antibody are **neutralization** and **opsonization** of the virus. Antibody can neutralize the virus by binding and blocking the viral attachment proteins and preventing their interaction with target cells or by destabilizing the virus, initiating its degradation. Binding of antibody to the virus also opsonizes the virus, promoting its uptake and clearance by macrophages. Antibody recognition of infected cells can also promote antibody-dependent cellular cytotoxicity.

IgM can neutralize virus in a viremia but is generally too large to enter extravascular tissue spaces (in the absence of inflammation). IgG can diffuse into tissue and through the placenta to the fetus. IgA is an extremely important defense to reinfection at the portals of entry into the body (respiratory, alimentary, and urogenital mucosal surfaces).

Cellular Immunity Cell-mediated immunity usually refers to both nonspecific (macrophage and natural killer cells) and antigen-specific cellular responses. This immunity is generally involved in clearance of virus, lysis of infected cells, or promotion

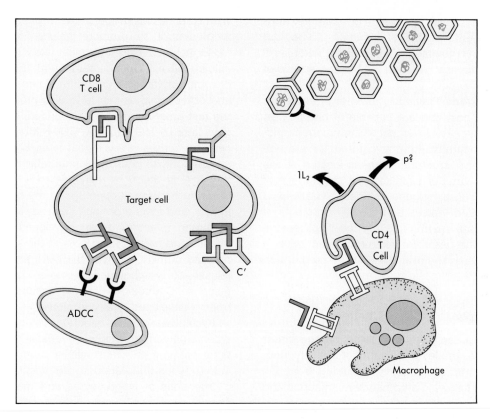

FIGURE 38-18   Antiviral immunity. Refer to the text for a detailed description of immune response to viral infection.

of antibody or inflammatory responses. Macrophages and related cells, including dendritic cells of the skin, Kupffer cells of the liver, and alveolar macrophages, are important for phagocytizing, processing, and presenting antigens to T cells. Natural killer cells (CD16) respond and lyse infected cells. Activated macrophages can distinguish and lyse infected cells.

Specific cellular immunity is conferred by T cells. The different T cells are distinguished by cell surface antigens denoted with CD notations. Helper T cells (CD4) produce lymphokines on activation to stimulate the growth and differentiation of other cells. CD4 cells are also responsible for the T-cell delayed-type hypersensitivity (DTH) response. Unlike other T-cell responses, the DTH response can be initiated by soluble antigen from live or killed virus through macrophage presentation of antigen. Cytotoxic T killer (CD8) cells are activated later in infection and attack cells expressing both viral antigen and the appropriate major histocompatibility complex (MHC) antigens.

The major role of cell-mediated immunity is elimination of abnormal cells. Alterations in cell surface antigens activate the cells. T cells recognize antigens on cell surfaces in the context of MHC antigens. CD4 helper and DTH T cells will respond to antigen presented in context with class II MHC products, whereas CD8 cytotoxic and suppressor T cells respond to antigen presented in context with class I MHC products. The trigger structures for natural killer cells and macrophage are not known.

Cells replicating the virus process and present to T cells viral antigens bound in the hydrophobic cleft of a class I antigens (HLA A and B). This is likely to be a normal degradative pathway of the cell. Phagocytic cells and B cells also degrade antigen and display it on the cell surface bound to class II antigens (HLA DR). The T-cell receptor recognizes the combined unit and, together with the appropriate lymphokine (interleukin 1 for helper cells and interleukin 2 for all T cells), becomes activated. The T-cell receptors are less discriminatory than antibody and may not distinguish between similar epitopes. For example, antibodies distinguish strains of influenza A–bearing variants of the H1 hemagglutinin protein, whereas T cells only distinguish H1- from H2-bearing strains of influenza A.

Viral Antigens Practically all viral proteins are foreign to the host and are capable of eliciting an immune response. However, not all immunogens elicit protective immunity. Protective immunity is developed against the structural antigens located at the surface of the virus or expressed on the surface of infected cells. These antigens include the viral capsid proteins of naked viruses and the glycoproteins of enveloped viruses. An immune response to other viral antigens may be useful for serologic detection of a virus infection and the evaluation of the progression of the disease.

Immune Response to Viral Challenge

The time required to initiate protection, the extent of the response, the level of control of the infection, and potential immunopathology resulting from the infection differ following a primary viral infection and a rechallenge.

Primary Viral Challenge Following replication at a primary site, virus and viral components are phagocytosed by resident macrophages. Alternatively the virus may establish a viremia and be picked up by the reticuloendothelial system in the liver, spleen, or lymph nodes. Swollen lymph nodes can result from activation and growth of lymphocytes in response to the viral challenge. The macrophages degrade the virus, process the viral antigens, and express appropriate fragments on the cell surface in conjunction with HLA class II antigens. Macrophages also release interleukin 1 to promote helper-T-cell activation. CD4+ helper T cells (and most likely DTH cells) specific for these antigens interact with the macrophage; become activated; secrete interleukin 2, γ-interferon, and other lymphokines; and also proliferate. IgM is the first antiviral antibody produced, followed 2 to 3 days later by IgG and IgA. CD8+ killer/suppressor T cells are activated somewhat later. Cytotoxic T cells respond only to viral antigen on infected cells. Secretory IgA is made in response to a viral challenge through the natural openings of the body: the eyes, mouth, and respiratory and gastrointestinal systems.

During infection by an enveloped virus, macrophages and natural killer cells at the site of infection respond quickly and lyse the cells. DTH T cells and release of lymphokines promote an inflammatory response. Cytotoxic T cells are also important in controlling enveloped viral infections.

Resolution of the infection occurs much later, when sufficient antibody is available to neutralize all virus progeny or when cellular immunity has been able to reach and eliminate the infected cells.

Infections of lymphocytes and macrophages are special cases for the immune response. Virus replication may inactivate or promote the elimination of an important effector cell (as for human immunodeficiency virus infection of helper T cells) or initiate an abnormal or excessive response because of viral antigen on lymphocyte cell surfaces (e.g., the overabundance of "atypical lymphocytes" in response to Epstein-Barr virus infection of B cells during mononucleosis). Infection of macrophages can promote the persistence of the virus in the host and their transport through the body (Table 38-14).

Table 38-14 Human Viruses Infecting Cells of the Lymphocyte Macrophage Lineage

Virus	Comment
B lymphocytes	
Epstein-Barr virus	Transformation and polyclonal B-cell activation
T lymphocytes	
Measles	Infects T lymphocytes and proliferates in activated cell
Human T lymphotropic virus 1 and 2	Retroviruses associated with T-cell lymphoma/leukemia
Human immunodeficiency virus 1 and 2	Kills helper cells, induces acquired immunodeficiency syndrome (AIDS)
Macrophages	
Yellow fever, dengue	Viral hemorrhagic fever
Rubella	Facilitates spread of virus
Human immondeficiency virus	Persistence and infection in various tissues, including brain

Secondary Viral Challenge Just as in any war, it is easier to eliminate an enemy if its identity and origin are known and establishment of its foothold can be prevented. Similarly, prior immunity is effective at preventing disease following a rechallenge with a virus. Antibody is the primary defense. Antibody is especially important in preventing viremic spread from the primary site of infection. **Secretory IgA** provides an important defense to reinfection through the natural openings of the body. In addition, **memory B** and **T cells** are present to generate a more rapid, anamnestic response to the virus if the virus escapes or overpowers this initial defense.

Escape From Immune Response

Many mechanisms allow viruses to escape immune clearance. Persistent or lethal infections result when the virus evades immune control. The virus can hide from the immune response in a latent infection and recur when the immune response is weakened, occurs with the herpesviruses. The virus may replicate at a site inaccessible to the immune response, or it may not produce sufficient antigen to elicit a response until it reaches its target tissue, too late for the victim, as occurs with rabies virus. The virus may remain intracellular by passing directly from cell to cell, as with herpes simplex virus and paramyxoviruses, and not be exposed to antibody. During a chronic hepatitis B virus infection, so much viral antigen is produced that it can saturate the individual's antibody. A change in the antigen structure of the virus prevents immune clearance. Human immunodeficiency virus undergoes antigenic changes during the infection of a single individual and also incapacitates the immune system by infecting helper T cells. Some virus infections preferably stimulate T suppressor cells, which may turn down the response too soon (e.g., cytomegalovirus, reovirus 3).

IMMUNOPATHOLOGY

Many disease symptoms associated with viral infection are actually caused by the immune response to the virus. The release of interferon and lymphokines by activated cells initiates effects on multiple body systems to produce the symptoms usually associated with mild viral infections (fever, histamine release, malaise, etc.). Cell-mediated immunity, required for control of many enveloped viruses, also initiates an inflammatory response that is difficult to control and damaging to tissue. For example, the inflammatory and hypersensitivity responses to measles and mumps probably cause more damage to cells than the virus. The presence of large amounts of antigen in the plasma during viremias or chronic infections initiates classic immune complex hypersensitivity reactions. Immune complexes containing virus or viral antigen can activate the complement system, promoting inflammatory responses and tissue destruction. These immune complexes often accumulate in the kidney and cause renal problems.

For dengue virus, prior immunity to a related strain can initiate a more severe disease on challenge with a new strain. Antigen-specific inflammatory and hypersensitivity responses are enhanced and inflict significant damage to endothelial cells (dengue hemorrhagic fever). In addition, the dengue and yellow fever viruses replicate in macrophages, and antibody can facilitate the uptake of virus into a macrophage through Fc receptors.

Children mount effective immune responses to viral infection but generally do not suffer as many of the troublesome sequelae produced by the overzealous adult cell-mediated immune and inflammatory responses. This is illustrated by the more benign nature of enveloped viral infection in children than in adults (e.g., measles, mumps, Epstein-Barr virus, varicella-zoster virus).

Thus viral disease results when the extent of damage produced by virus replication and/or the immune and inflammatory responses becomes so great as to produce symptoms. The nature of the disease is determined by which tissue is affected and the extent of the effect or damage to the tissue. The time course of the disease depends on the initial dose and strain of virus, the capacity of the immune response to clear the infection, and the general health of the individual. The severity of the disease is usually related to the importance of the tissue, how fast the damage is being caused, and how long the damage occurs before resolution.

CLINICAL SYNDROMES

The major sites of viral disease are the respiratory tract; the gastrointestinal tract; the epithelial, mucosal, and endothelial linings; lymphoid tissue; the liver

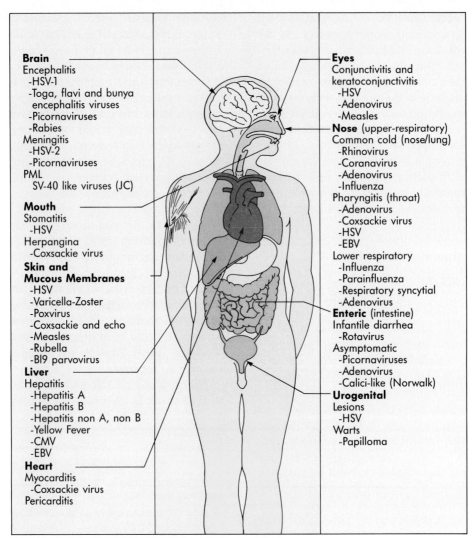

Brain
Encephalitis
 -HSV-1
 -Toga, flavi and bunya
  encephalitis viruses
 -Picornaviruses
 -Rabies
Meningitis
 -HSV-2
 -Picornaviruses
PML
  SV-40 like viruses (JC)

Mouth
Stomatitis
 -HSV
Herpangina
 -Coxsackie virus
Skin and
Mucous Membranes
 -HSV
 -Varicella-Zoster
 -Poxvirus
 -Coxsackie and echo
 -Measles
 -Rubella
 -B19 parvovirus
Liver
Hepatitis
 -Hepatitis A
 -Hepatitis B
 -Hepatitis non A, non B
 -Yellow Fever
 -CMV
 -EBV
Heart
Myocarditis
 -Coxsackie virus
Pericarditis

Eyes
Conjunctivitis and
keratoconjunctivitis
 -HSV
 -Adenovirus
 -Measles
Nose (upper-respiratory)
Common cold (nose/lung)
 -Rhinovirus
 -Coranavirus
 -Adenovirus
 -Influenza
Pharyngitis (throat)
 -Adenovirus
 -Coxsackie virus
 -HSV
 -EBV
Lower respiratory
 -Influenza
 -Parainfluenza
 -Respiratory syncytial
 -Adenovirus
Enteric (intestine)
Infantile diarrhea
 -Rotavirus
Asymptomatic
 -Picornaviruses
 -Adenovirus
 -Calici-like (Norwalk)
Urogenital
Lesions
 -HSV
Warts
 -Papilloma

*Progressive multifocal leukoencephalopathy

FIGURE 38-19   Major target tissues of viral disease.

and other organs; and the central nervous system (Figure 38-19).

Certain specific syndromes may be caused by several different viruses. This is especially true of respiratory diseases, such as bronchiolitis, which may be caused by either respiratory syncytial or parainfluenza virus (see Chapter 48). Similarly, the spectrum of disease potential for any one virus type is quite broad. For example, influenza virus may cause illness ranging from upper respiratory tract infection (URI) to pneumonia.

Many viral infections initiate a primary infection of the respiratory tract that remains localized. Rhinitis, pharyngitis, and pneumonia are the classic diseases associated with respiratory viruses. Other viruses spread from the respiratory tract to other target organs without significantly damaging or altering the function of the lung. The site of infection and therefore of disease is determined by the point of contact with the virus, the tissue expression of receptors, or conditions required for virus replication.

Infections of the gastrointestinal tract can result in

gastroenteritis and diarrhea but frequently produce few or no symptoms. In adults the infected tissue can be sloughed off and replaced very quickly. Reovirus and rotavirus infections of children are often symptomatic because of the tissue damage and consequent loss of fluids and electrolytes. Enteroviruses may cause little or no gastrointestinal symptoms, but clinical disease results when these viruses establish a viremia and spread to other target organs.

Virus-induced skin disease can result from infection through the mucosa or small cuts or abrasions in the skin (herpes simplex virus), as a secondary infection following establishment of a viremia (varicella-zoster virus and smallpox), or through the inflammatory response mounted against the viral antigens (measles rash). The major forms of skin lesions are classified as maculopapular, vesicular nodular, and hemorrhagic rashes. **Macules** are flat, colored spots. **Papules** are slightly raised areas of the skin and may result from immune or inflammatory responses rather than the virus. **Vesicular lesions** are blisters and are likely to contain virus. Papilloma viruses cause **nodular rashes** typically known as warts. Infection of the endothelial cell lining of the vasculature may serve as a source of a viremia but may also compromise the structure of the blood vessel. Viral or immune cytolysis may lead to permeation or rupture of the vessel, producing a hemorrhagic rash and leading to internal bleeding, loss of electrolytes, and shock.

Infections of the eye result from direct contact with virus or viremic spread. Conjunctivitis is a normal feature of many childhood infections and is a characteristic of measles and some adenovirus infections. Keratoconjunctivitis is most often caused by herpes simplex virus, involves the cornea, and can cause severe damage.

Lymphocytes and macrophages are not very permissive for virus replication. Epstein-Barr virus infects and immortalizes B lymphocytes, but the infection is abortive (i.e., viable progeny virus is not produced). The infection does alter the cell surface sufficiently to induce a large, protective, cell-mediated immune response to produce the symptoms of infectious mononucleosis. Cytomegalovirus may induce an abortive infection of T lymphocytes. However, human immunodeficiency viral infection of CD4 helper T cells is initially latent. Activation of the T cell leads to a productive, cytolytic infection. The loss of CD4 cells incapacitates the cell-mediated immune response, leading to immunodeficiency. Macro-

phages are infected by many viruses, including human immunodeficiency virus, which persist in the macrophage or use it to gain access to other tissues, including the brain.

Infection of visceral organs may cause significant disease or result in further spread or secretion of the virus. The disease may cause tissue damage or be limited to discomfort because of tissue swelling. The liver is a prominent target for many viruses, either through viremic infection or the reticuloendothelial system. The liver acts as a source for secondary viremia but can also be damaged by the infection. The classic symptoms of the hepatitis result from infections with hepatitis A, B, and non-A, non-B virus; yellow fever virus; and other viruses. The heart and pancreas are also susceptible to viral infection and damage. Coxsackievirus can cause either myocarditis or pericarditis of the newborn or diabetes mellitus via infection of the beta cells of the pancreas.

Not all infections of the central nervous system cause disease, but neurologic viral diseases are very severe. Herpes simplex virus, varicella-zoster virus, a papovavirus cause asymptomatic, latent infections of the central nervous system but can also cause significant damage. The severity of neurotropic infections results from the importance of the tissue and its inability to repair damage. The actual presentation depends on the tissue infected (neuron, encephalitis; meninges, meningitis) and the extent of viral pathology and immunopathology.

Infection of the secretory glands, accessory sexual organs, mammary glands, and so on results in contagious spread of the virus.

Chronic Infections

Chronic infections occur when the immune system has difficulty resolving the infection. During chronic hepatitis B infection, virus and viral antigen are produced for long periods. Patients with chronic hepatitis B infection have liver damage and problems caused by immune complexes formed between viral antigen and antibody.

Human T-lymphotropic virus 1 (HTLV-1), Estein-Barr virus, and papilloma infections establish chronic infections and have been associated with human cancers. HTLV-1 infection of T lymphocytes and Epstein-Barr infection of B lymphocytes alter the growth pattern and immortalize the cell unless controlled by the immune response. Under unusual situations the immortalized growing cell continues to change and become a cancer. Epstein-Barr virus normally causes infectious mononucleosis but is associated with African Burkitt's lymphoma, and HTLV-1 is associated with human T-cell leukemia. Papilloma viruses induce a simple hyperplasia characterized by a wart. Several strains of papilloma have been associated with human cancers (e.g. type 16, 18, and 31 with cervical carcinoma).

Recurrent Infections

Members of the herpes virus family can cause recurrent infections. Herpes simplex virus, for example, initiates a primary infection of the skin and then spreads to the neuron innervating that dermatome. The virus remains latent in the neuron until an appropriate stress triggers its replication, migration down the neuron, and clinical infection of the dermatome.

Congenital Infections

Development and growth in the fetus are so ordered and rapid that viral infection may damage or prevent the appropriate formation of important tissues, leading to miscarriage or congenital abnormalities. The fetus and newborn receive passive antibody immunity from the mother but still are at risk for severe disease from viruses capable of escaping antibody neutralization.

Rubella and cytomegalovirus are examples of teratogenic viruses that can cause congenital infection and defects. Mothers infected with human immunodeficiency virus pass the virus to their children, but the full consequences for the child are less well known. Herpes simplex virus can be acquired on passage through an infected birth canal. Serious disseminated disease can result in the newborns if they lack maternal immunity and/or cell-mediated immune protection.

Infections of the Immunocompromised Host

Patients deficient in cell-mediated immunity are generally more susceptible to enveloped viral infections, especially the herpesviruses, measles, and even the vaccinia virus used for smallpox vaccinations. Severe T-cell deficiencies also affect the antiviral antibody response. Cell-mediated immunodeficiencies can be congenital or acquired. They may result from genetic defects (e.g., Duncan's disease, DiGeorge syndrome, Wiskott-Aldrich syndrome), infections (e.g., AIDS), or immunosuppressive chemotherapy for a tissue transplant or cancer chemotherapy. Neonates depend on the mother's immunity to protect them from viral infections. The cell-mediated immune system is not mature at birth, and newborns are susceptible to syncytia-forming viruses such as herpes simplex virus or other viruses that spread by cell-to-cell contact, (e.g., cytomegalovirus). Viruses that establish viremia (e.g., rubella) can also infect the fetus unless the mother has developed prior immunity. Neonates receive maternal antibodies through the placenta and then through the mother's milk. This type of passive immunity can remain effective for 6 months to a year after birth.

In immunosuppressed individuals, viruses can cause atypical and more severe presentations. Herpesvirus (herpes simplex virus, cytomegalovirus, varicella-zoster virus) infections, normally benign and localized, can progress locally or disseminate and cause visceral and neurologic infections that may be life threatening. Measles infection can cause a giant-cell (syncytial) pneumonia rather than the characteristic rash in the immunocompromised individual. Hypogammaglobulinemic individuals (antibody deficient) are more susceptible to clinically apparent infection by the live polio vaccine, echovirus, and varicella-zoster virus. These are examples of viruses controlled by secretory IgA and serum IgG.

BIBLIOGRAPHY

Belshe RB, editor: Textbook of human virology, Littleton, Mass., 1984, PSG Publishing Co, Inc.

Fields BN: Virology, New York, 1985, Raven Press.

39

Viral Diagnosis

It is becoming increasingly important for the practitioner to perform viral diagnostic studies because prognostic, epidemiologic, and therapeutic considerations may be greatly influenced by knowledge of the specific virus causing a given illness. Even if no therapy is available, the establishment of a definite diagnosis of viral infection is often beneficial in (1) epidemiologic monitoring, (2) educating physicians and patients, (3) defining the disease process, and (4) evaluating therapeutic implications, both positive and negative. Moreover, identification of a virus as the cause of a patient's illness may be cost-effective, since expensive diagnostic procedures and antibiotic therapy may be avoided or discontinued.

The laboratory can confirm the suspected diagnosis by cytologic examination of clinical specimens; attempting to isolate and grow the virus; detecting the presence of viral particles, antigens, or nucleic acid; or evaluating the patient's immune response to the virus (serology) (Box 39-1).

CYTOLOGY

The most readily available rapid technique for viral diagnosis is cytologic examination of specimens for the presence of characteristic viral inclusions. These intracellular structures may represent aggregates of virus within an infected cell or may be abnormal accumulations of cellular material resulting from the viral-induced metabolic disruption. Papanicolaou (Pap) smears may show these inclusions in single cells or in large syncytia (aggregates of cells containing more than one nucleus), as in a patient with herpes simplex infection of the cervix (Figure 39-1). Cytology is most often used to detect infections with herpes simplex virus or cytomegalovirus, but examination of urine sediment for cytomegalic inclusion-bearing cells is not a sensitive method for diagnosing cytomegalovirus infections. Viral isolation is at least fourfold more sensitive in both infants and adults. Rabies infection may also be detected by finding Negri bodies (rabies virus inclusions) in brain tissue (Figure 39-2).

Box 39-1 Virus Laboratory Procedures
Cytologic examination Virus isolation in culture Detection of viral antigens and genetic material Serology

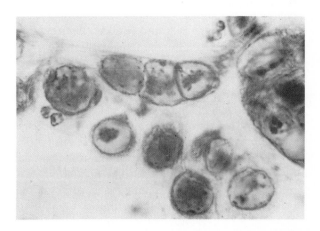

FIGURE 39-1   Papanicolaou smear showing herpes simplex virus.

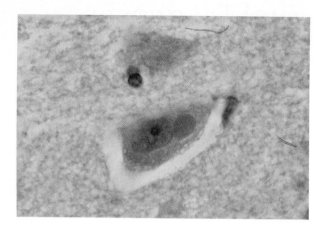

FIGURE 39-2   Negri bodies caused by rabies.

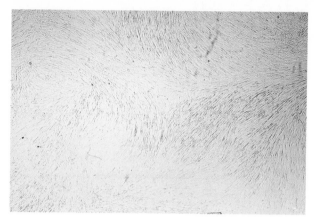

FIGURE 39-3   Tissue culture of human diploid fibroblasts.

VIRAL CULTURE

The "gold standard" for proving the etiology of a viral syndrome is recovery of the agent in tissue culture, embryonated eggs, or experimental animals. Embryonated eggs are still used for the growth of virus for some vaccines but have been replaced by cell cultures for routine virus isolation in clinical laboratories. Likewise, the use of experimental animals rarely occurs in most clinical laboratories. Just as multiple media are used in bacteriology, several different types of tissue culture cells (e.g., monkey kidney, human fetal lung, human amnion, human cancer cells) are inoculated with each viral specimen (Figure 39-3).

Primary monkey kidney cells are excellent for recovery of myxoviruses, paramyxoviruses, and many enteroviruses. These cells also support the growth of some adenoviruses. Human fetal diploid cells are fibroblastic cells that support the growth of a broad spectrum of viruses (e.g., herpes simplex virus, varicella-zoster virus, adenoviruses, picornaviruses) and are the only cells in which cytomegalovirus is recovered. Hep-2 cells are a continuous line of epithelial cells derived from a human cancer. These are particularly excellent for recovering respiratory synctial virus, as well as adenoviruses and herpes simplex virus. Most clinically significant viruses can be recovered in at least one of these cell cultures. Specimens submitted for the isolation of viruses such as human immunodeficiency virus, coxsackie A virus, and togaviruses require respectively co-cultivation with normal human peripheral blood mononuclear cells, inoculation of suckling mice, or the use of cell cultures, which are not generally available. Therefore infections caused by these viruses are most frequently diagnosed serologically or by detection of viral-specific antigens.

Detection of the growth of a virus is by observation of changes in the cell culture monolayer (**cytopathic effect,** or **CPE**) (Figure 39-4). Characteristic CPEs include changes in cell morphology, cell lysis, vacuolation, syncytia formation, and presence of inclusion bodies. **Inclusion bodies** are histologic changes in cells caused by the presence of viral components or changes in cell structures. With experience a technologist can distinguish CPE characteristics of the major virus groups. The observation of which cell culture exhibits CPE and the rapidity of viral growth can be used for the presumptive identification of many clinically important viruses. This approach for identifying viruses is similar to bacterial identification based on growth and morphology of colonies on selective, differential media. Some viruses do not readily cause CPE in cell lines typically used in clinical virology laboratories. However, these can be detected by other techniques, such as (1) interference with the replication of other viruses (e.g., picornaviruses cannot replicate in cells previously infected with rubella virus; this is known as **heterologous interference**) or (2) adsorption of erythrocytes onto cells infected with paramyxoviruses or mumps (Figure 39-5).

Some common viruses are not isolated by standard culture methods, including the human immunodeficiency virus, rubella, measles, rotaviruses, coxsackie

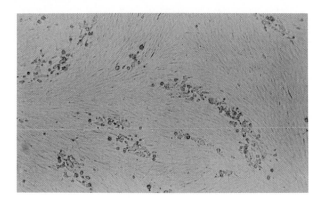

FIGURE 39-4   Changes in the cell culture monolayer, which constitute the cytopathic effect, result from growth of the virus.

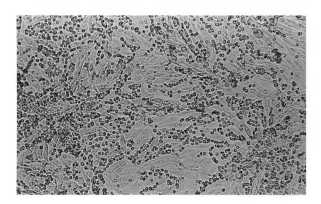

FIGURE 39-5   Erythrocyte adsorption onto infected cells.

A virus, and togaviruses. Specimens from patients suspected to be infected with these agents should be sent to reference laboratories for special culture procedures or, in the case of rotaviruses, for antigen detection or direct electron microscopy.

In contrast with the intrinsic delay of serologic studies, the results of viral culture can be surprisingly rapid. Almost 50% of all viral isolates can be reported within 3 to 4 days of culture, with herpes simplex and influenza A viruses usually detected within 2 to 3 days (Table 39-1).

The selection of the appropriate specimen for viral culture is complicated because several different viruses may cause the same clinical disease (Table 39-2). For example, several types of specimens may be submitted from patients with viral meningitis to enhance the recovery of the possible etiologic agents: cerebrospinal fluid (enteroviruses, mumps virus, herpes simplex virus), throat swabs and washings (enteroviruses), and stool or rectal swabs (enteroviruses). Also, serum should be collected as an acute-phase specimen in case subsequent serologic tests are indicated (e.g., acute and convalescent sera for mumps virus or arbovirus infections). Many considerations, however, allow the physician to select the most appropriate specimens (Table 39-3). For example,

Table 39-1   Frequency of Occurrence and Average Detection Time for Commonly Isolated Viruses

Virus	Isolates		Average Detection Time in Days
	Number	Isolation	
Herpes simplex	416	42	2.7
Influenza A	66	7	3.8
Enteroviruses (echo, coxsackie A and B)	79	8	4.2
Cytomegalovirus	213	21	5.8
Respiratory syncytial virus	35	3	6.1
Varicella-zoster	41	4	6.1
Adenoviruses	80	8	6.4
Parainfluenza 1 and 3	29	3	6.4
Other	41	4	
Total/Overall Average	1,000	100	4.1

From Drew WL and Stevens GR: Lab Med, December 1979, pp 741-746

Table 39-2 Most Common Causative Agents of Viral Respiratory Disease

Disease	Agents
ADULT	
Upper respiratory tract infection (URI)	Rhinoviruses, coronaviruses, adenoviruses, parainfluenza virus
Pneumonia	Influenza virus, adenoviruses
INFANTS AND CHILDREN	
URI	Parainfluenza virus, adenoviruses, influenza virus, rhinoviruses
Pharyngitis	Adenoviruses, parainfluenza virus, Coxsackie virus
Croup	Parainfluenza virus, respiratory syncytial virus
Bronchitis	Parainfluenza virus, respiratory syncytial virus
Bronchiolitis	Respiratory syncytial virus, parainfluenza virus
Pneumonia	Respiratory syncytial virus, parainfluenza virus, adenoviruses

Table 39-3 Specimen Considerations for Viral Agents

Source	Applications and Advantages/Disadvantages*
Throat, nasopharyngeal swab, and aspirate	Nasopharyngeal aspirates are superior to swabs, but swabs are more convenient. Throat swabs are probably adequate for enteroviruses, adenoviruses, and HSV; nasopharyngeal specimens preferred for respiratory syncytial virus and most other respiratory viruses. Nasal specimens are best choice for rhinoviruses.
Rectal swabs and stool specimens	Fecal specimens from patients with viral gastroenteritis show nonculturable agents such as rotaviruses or adenoviruses when submitted for electron microscopy or antigen dectection. In suspected enteroviral disease (aseptic meningitis, myopericarditis, hand-foot-and-mouth disease), positive cultures from rectal cultures support but do not prove enteroviruses are cause of disease. Stool specimens are more productive than rectal swabs. NOTE: Infectivity of viruses surrounded by lipid membrane (HSV, CMV) is destroyed by gastric acidity so they will not be excreted in viable form in feces.
Urine	Used for CMV, mumps, and adenoviruses; in mumps-related central nervous system disease, virus can be isolated from urine even when specimens from other sites are negative.
Dermal lesions	Fluid and cells from vesicles are superior to specimens from ulcers or crusts for both cultures and direct stains.
Cerebrospinal fluid	"Sterile" fluids (pleural, peritoneal, pericardial, joint) can be inoculated directly into tissue culture.
Eye	Secretions are obtained by swab from the palpebral conjunctiva. NOTE: Eye scrapings should be obtained by an ophthalmologist or other trained person.
Blood	Leukocytes are collected by centrifugation to obtain a "buffy coat" layer or on Ficoll Hypaque. Leukocyte culture is especially useful for detecting CMV and HIV in peripheral blood. Anticoagulated blood or a clot can be used for isolation of arboviruses; serum is suitable for enteroviruses.
Tissue	Lung is used for CMV, influenza virus, and adenoviruses. Brain is used for HSV.

*HSV, Herpes simplex virus; CMV, cytomegalovirus; HIV, human immunodeficiency virus.

Box 39-2   Summary of Procedures for Obtaining and Transporting Specimens for Viral Studies

General: Obtain specimens as early in the patient's illness as possible. Inoculate tissue cultures at patient's bedside if possible.

Throat: Use culturette swab (Marion Scientific Corp., Rockford, Ill.) as for bacteriologic culture.

Nasopharynx: Obtain a nasopharyngeal swab or a nasal wash specimen using a bulb syringe and buffered saline.

Stool: Obtain as for bacteriologic culture. If a specimen cannot be passed, a rectal swab of feces may be obtained with a Culturette.

Cerebrospinal fluid: Obtain 1 ml as for bacteriologic culture.

Urine: Obtain as for bacteriologic culture.

Skin or mucosal scrapping: Obtain with Culturette.

Biopsy material: Use sterile technique and submit in a sterile container (e.g., urine culture container).

Blood for culture: Submit at least 3 ml of heparinized blood (green-top Vacutainer tube, Becton-Dickinson Co., Rutherford, N.J.).

Blood for serologic studies: Submit at least 5 ml of clotted whole blood (red-top Vacutainer tube). In certain viral syndromes (e.g., lower respiratory) an acute-phase specimen should be submitted. If a virus is not isolated, a convalescent specimen should be obtained at least 7 days after the acute specimen. Certain viral illnesses (e.g., rubella, rubeola, hepatitis, arbovirus encephalitis) are diagnosed most readily by serologic studies.

Transportation: Transport specimens as rapidly as possible, using a messenger service. Specimens should be stored and transported at refrigerator temperature (4° C). **Do not freeze.** If transportation of swabs will be delayed, place them in a tube of buffered bacteriologic broth medium (e.g., trypticase soy yeast broth) rather than in the Culturette.

---

during the summer, when enteroviral meningitis is prevalent, cerebrospinal fluid, throat, and stool specimens should be submitted. On the other hand, the development of encephalitis in children after being bitten by mosquitos in wooded areas endemic for California encephalitis virus suggests that a serum specimen for antibody testing would be preferred. Central nervous system disease following parotitis would suggest collection of cerebrospinal fluid and urine for the isolation of mumps virus. A focal encephalitis with a temporal lobe localization preceded by headaches and disorientation would suggest herpes simplex virus and the need for biopsy of brain tissue for culture. The specimens that should be collected for other viruses are summarized in Box 39-2.

Timing of Specimen Collection

Proper timing of specimen collection is essential for adequate recovery of viruses. Specimens should be collected early in the acute phase of infection. Studies with respiratory viruses indicated that the mean duration of viral shedding may be only 3 to 7 days. Also, herpes simplex virus and varicella-zoster virus may not be recovered from lesions beyond 5

days after onset. Isolation of an enterovirus from the cerebrospinal fluid may be possible for only 2 to 3 days after onset of the central nervous system manifestations.

Transport to Laboratory

The shorter the interval between collection of a specimen and its delivery to the laboratory, the greater is the potential for isolating an agent. When feasible, all specimens other than blood, feces, and tissue, which need special processing, should be inoculated directly onto cell cultures at the patient's bedside. These should then be transported to the laboratory promptly.

For specimens that cannot be inoculated onto cell cultures immediately, several types of transport media have been used traditionally. It is generally believed that protein (serum, albumin, gelatin) incorporated into a transport medium enhances survival of viruses, although herpes simplex virus and respiratory viruses can survive in some protein-free media, at least for short periods.

Improper storage of specimens before processing can also adversely affect viral recovery. Significant

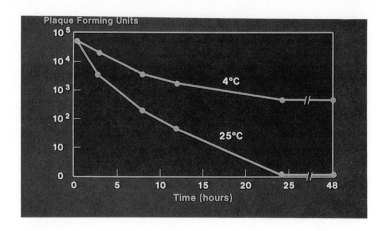

FIGURE 39-6   Stability of herpes simplex type 2 virus in Culturette medium at 4° C and 25° C (room temperature).

losses in infective titer occur with enveloped viruses (e.g., herpes simplex virus, varicella-zoster virus, influenza virus) after specimens have been frozen. This is not observed with nonenveloped viruses (e.g., adenoviruses, enteroviruses). Therefore, when it is impossible to process a specimen immediately, it should be refrigerated but not frozen and packed in shaved ice for delivery to the laboratory if delays in transit are anticipated. Storage of specimens for the recovery of viruses at ambient temperature is far less satisfactory than storage at 4° C (Figure 39-6).

Interpretation of Culture Results

In general the detection of any virus in host tissues, cerebrospinal fluid, blood, or vesicular fluid can be considered highly significant. Recovery of viruses other than cytomegalovirus in urine may be diagnostic of significant infections. For example, both mumps virus and adenovirus type 11 (associated with acute hemorrhagic cystitis) may be recovered in urine. However, the presence of cytomegalovirus in urine is difficult to interpret because this may reflect active, asymptomatic virus replication or indicate a significant infection in the patient. In the newborn, viruria (isolation of virus in urine) in the first 4 weeks of life establishes a diagnosis of congenital cytomegalovirus infection, whereas the onset of viral excretion after 4 weeks of life reflects intrapartum or postpartum infection. Diagnosis of acquired cytomegalovirus infections in older patients usually requires a combi-

nation of findings, including positive cultures, illness compatible with cytomegalovirus disease, reasonable exclusion of other potential etiologic agents, and support by specific serologic or histologic data.

The significance of viruses isolated in upper respiratory tract, vaginal, or fecal specimens varies greatly. At one extreme, isolates such as measles, mumps, influenza, parainfluenza, and respiratory syncytial virus are significant because asymptomatic carriage and prolonged shedding of these viruses are unusual. Conversely, other viruses can be shed without symptoms for periods ranging from several weeks (enteroviruses in feces) to many months or years (herpes simplex virus or cytomegalovirus in the oropharynx and vagina; adenoviruses in the oropharynx and intestinal tract). Varicella-zoster virus, herpes simplex virus, cytomegalovirus, adenoviruses, and Epstein-Barr virus may remain latent for long periods and then become reactivated in response to a variety of stressful stimuli, including other infectious agents. In this setting their detection may not be significant (varicella-zoster virus is an exception) or may merely represent a secondary problem complicating the primary infection (e.g., herpes simplex virus "cold sores" in patients with bacterial sepsis).

Adenovirus isolates typically are found in infants and young children. Based on the epidemiology of adenovirus infections and observed serologic responses, the simultaneous isolation of these viruses from throat and feces is significantly associated with febrile respiratory disease. Isolation of viruses from the throat alone is less frequently associated with disease, and isolates from feces alone have the lowest diagnostic significance.

Enteroviruses are also often found in infants and children, particularly during the late summer and early autumn. A knowledge of the relative frequency of virus shedding among various age-groups in a particular locale is extremely helpful in assessing the significance of results of throat or stool cultures. For example, the peak prevalence of enteroviruses in the stools of toddlers during the late summer may range from 5% in temperate zones to more than 20% in subtropical climates. Even in temperate areas, rates may approach 30% in infants during periods of enterovirus activity. Shedding of enteroviruses in the throat is usually 1 to 2 weeks, whereas fecal shedding may last 4 to 16 weeks. Thus, in a clinically compatible illness, isolation of an enterovirus from the throat supports a stronger temporal relationship to the disease than does an isolate from only the feces.

Herpes simplex virus is unusual in a fecal culture. In such cases it usually represents either severe disseminated infection or local infection of the perianal areas. Detection of herpes simplex virus in the upper respiratory tract may mean nothing other than nonspecific reactivation of virus caused by stress unless typical vesicles or ulcers are also present. Because of the stress-related phenomenon, isolation of HSV in the throat or mucocutaneous lesions of patients with encephalitis cannot be interpreted as causing the central nervous system disease. Currently the definitive way to establish a diagnosis of herpes simplex encephalitis is by direct demonstration of the virus in a brain biopsy. In neonates, however, isolation of the virus from any site should raise the possibility of severe infection.

Isolation of adenoviruses, herpes simplex virus, varicella-zoster virus, and some enteroviruses from the cornea and conjunctiva in patients with inflammatory disease at these sites usually establishes the etiology of the infection.

The laboratory frequently is asked to perform studies to rule out specific viral agents. The ability to exclude the presence of a virus depends greatly on the sensitivity of the detection systems used. The most important prerequisites include:

1. Proper communication with the laboratory before initiating studies
2. Procurement of the appropriate specimens in the early phase of acute illness
3. Careful attention to the procedure of collection, transportation, and processing of specimens
4. Avoidance of fungal or bacterial contamination of processed cultures

When all these criteria are fulfilled, a negative result may be valid unless the virus in question is one that does not grow in the tissue cultures used.

DETECTION OF VIRAL ANTIGENS AND GENETIC MATERIAL

Viral replication produces enzymes, proteins, antigens, and new genomes that can be detected by biochemical, immunologic, and molecular biologic means. Many of the same procedures can be used to detect viral particles directly in clinical samples or viral infections in tissue culture cells.

Electron microscopy can be used to detect and identify some viruses. However, sufficient numbers of virus particles, such as those found in the stools of patients with rotavirus gastroenteritis, must be present to allow detection by electron microscopy. Antibody can facilitate detection (immunoelectron microscopy) by identifying and clumping the viral particles together.

Influenza, parainfluenza, and togaviruses produce a glycoprotein that binds erythrocytes. This property allows detection of the virus produced in cell culture by agglutination of erythrocytes, a process termed **hemagglutination.** The infected cells also adsorb erythrocytes to the surface by **hemadsorption.**

Detection and assay of characteristic enzymes can identify and quantitate specific viruses. For example, reverse transcriptase in cell culture is used as an indicator of infection by retroviruses.

Molecular maps or fingerprints of viruses are used to distinguish different strains or types of virus. Herpes simplex virus types 1 and 2 can be distinguished by protein patterns obtained by electrophoresis, and individual strains of either virus can be distinguished by the restriction endonuclease patterns of their deoxyribonucleic acid (DNA). DNA-DNA hybridization with commercially available probes is becoming a rapid, sensitive means of detecting viruses.

Antibodies can be used as sensitive tools to detect, identify, and quantitate the presence of viral antigen in clinical specimens or cell culture. Monoclonal or polyclonal antibodies prepared in animals or obtained from convalescent patients may be used.

Viral antigens on the cell surface, within the cell, or released from infected cells can be detected by immunofluorescence (IF), enzyme immunoassay (EIA), radioimmunoassay (RIA), and latex agglutination (LA). IF and EIA detect and locate cell-associated antigens, whereas RIA or different variations of enzyme-linked immunosorbent assay (ELISA) are used to detect and quantitate soluble antigens. LA is a rapid, easy assay for soluble antigen. Viruses or viral antigens in a sample cause the clumping of latex particles coated with specific antibody.

Virus-infected tissue or cell cultures can be analyzed by IF or EIA. By attaching a fluorescent signal to an antiviral antibody, the site of antibody binding can be located; this is called **direct immunofluorescence** (DFA). A modification of this technique is detection of unlabeled antiviral antibodies by use of a second antibody with a fluorescent label that will bind to IgM or IgG antibodies; this is called **indirect immunofluorescence** (IFA). EIA uses a second antibody conjugated to an enzyme, such as horseradish peroxidase or alkaline phosphatase, which releases a chromophore to mark the antigen.

Soluble antigen can be quantitated by ELISA, RIA, and LA. The basis for these procedures is the separation and quantitation of antibody-bound and free antigen. Many of the ELISA and RIA techniques use an antibody immobilized to a solid support to capture soluble antigen and a labeled antibody to detect captured antigen.

The detection of viral nucleic acids in clinical specimens can also be accomplished. Cytomegalovirus has been successfully identified in urine by nucleic acid hybridization, but this method is less sensitive than culture. Rotaviruses have also been detected by nucleic acid hybridization of fecal samples.

VIRAL SEROLOGY

The patient's humoral immune response provides a history of infections. Serology can be used to identify the virus and its strain or serotype or to evaluate an infection to determine if it is primary or a reinfection and if acute or chronic. The first antibodies to be produced by the immune system are directed against antigens on the virion or infected cell surfaces. Later in the infection, when cells have been lysed by the infecting virus or the cellular immune response, antibodies are directed against the cytoplasmic viral proteins and enzymes. **Seroconversion** is characterized by a fourfold or greater increase in antibody titer between serum taken during the acute phase of disease and that taken at least 2 to 3 weeks later; it indicates a recent infection. Detection of viral-specific IgM is also usually associated with recent infection.

For many viruses, culture or antigen detection is the best diagnostic test. However, certain viruses (e.g., human immunodeficiency virus, hepatitis A and B, rubella, Epstein-Barr virus, measles, coronaviruses, togaviruses) are difficult to isolate in cell culture, and infections are diagnosed most easily by serologic techniques. When a virologic workup is planned for a patient, it is generally useful to obtain at least 2 to 3 ml of serum during the acute phase of disease and store it at $-20°$ C. This may become valuable, particularly if virus detection subsequently fails or if the significance of an isolate is uncertain. In these instances a convalescent-phase serum specimen may be requested 2 to 3 weeks later, and both the acute and the convalescent sera may then be tested for appropriate viruses. In general, if a virus is isolated, the antibody titers need not be measured to determine whether the patient is infected or colonized by the virus. If a patient has aseptic meningitis and an enterovirus is recovered from the throat, that agent is probably responsible for the illness. Similarly, if influenza virus is recovered from the throat of a patient who has clinical influenza, no serologic confirmation of the etiology is necessary.

Serologic Test Methods

Complement fixation is a standard but technically difficult serologic test. The sample is reacted with antigen or antibody and complement, and the residual complement is assayed by lysis of indicator antibody–coated red blood cells. Antibodies measured by this system generally develop slightly later in the course of an illness than those measured by other techniques. This delayed response is useful for documenting seroconversion when the initial serum specimen is collected late in the clinical course. Members of some virus groups (e.g., adenoviruses, influenza A and B viruses) possess common antigens demonstrable by complement fixation. Unfortunately, members of other large virus groups (e.g., enteroviruses) do not possess group-specific antigen and must be tested individually.

The **neutralization** test is essentially a protection test. When a virus is incubated with homologous type-specific antibody, the virus is rendered incapable of producing infection in an indicator cell culture system. A neutralization antibody response is virus-type specific and develops with the onset of symptoms, with titers rising rapidly and persisting for long periods.

The **hemagglutination inhibition** test can be performed with a variety of viruses that can selectively agglutinate erythrocytes of various animal species (e.g., chicken, guinea pig, human). The hemagglutination capacity of a virus is inhibited by specific immune or convalescent sera. Hemagglutination-inhibiting antibody develops rapidly after the onset of symptoms, plateaus, declines slowly, and may last indefinitely at low levels. This test is useful for both the detection of acute rubella infection and the determination of immunity. In the **passive hemagglutination** procedure, certain viruses can be chemically coupled to the surface of erythrocytes, which then serve as indicators of the presence of homologous antibody in sera.

For the **indirect fluorescent antibody** test, virus-infected cells are placed in prepared wells on microscope slides, then fixed in cold acetone and dried. Patient serum is applied and, following incubation, antihuman globulin conjugated with fluorescein is added to delineate, by fluorescence, the sites of antigen-antibody reaction.

Interpretation of Serologic Results

Virus-specific IgM antibody usually rises during the first 2 to 3 weeks of infection and persists for several weeks to months, eventually being replaced with IgG antibody (Figure 39-7). Thus an elevated titer of specific IgM antibody suggests a recent primary infection, and this may be further supported by demonstrating a fall in IgM antibody in subsequent sera. Detection of specific IgM has been used with success in the diagnosis of infections from Epstein-Barr virus, cytomegalovirus, and rubella and is currently the procedure of choice to establish a recent or acute infection from hepatitis A or B.

Several limitations of interpretation must be remembered. It is now recognized that IgM-specific antibody responses are not always restricted to primary infections. Reactivation or reinfection may result in IgM responses, particularly in herpes virus

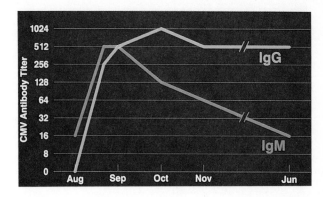

FIGURE 39-7   Antibody response to cytomegalovirus infection.

infections. In addition, patients may continue to produce IgM-specific antibody to rubella or cytomegalovirus for many months after a primary infection. Heterotypic IgM responses may also occur. For example, antibody responses to cytomegalovirus may develop in Epstein-Barr virus infections, and vice versa. Other pitfalls include falsely low or negative IgM titers caused by competition from high-titer IgG antibody for antigen-binding sites and false positive reactions resulting from rheumatoid factor. Both types of errors appear to occur most frequently in solid-phase assays employing immunofluorescence.

A serum specimen can also be used for screening an infant's blood for certain antibodies of the IgG class. Antibodies to *Toxoplasma,* rubella, cytomegalovirus, and herpes simplex virus (known as the **TORCH** screen) may be measured in an effort to determine possible congenital infection with these agents. However, the value of these tests has been misunderstood. They are useful in excluding a possible infection but not in proving an etiology. For example, if rubella antibody is absent, an infant almost certainly does not have congenital rubella infection. To diagnose active rubella infection in such a baby, viral cultures are required. Screening of blood supplies for cytomegalovirus antibody is used to eliminate transmission of antibody-positive blood to seronegative babies and other immunocompromised patients.

The serologic diagnosis of most viral infections is based on demonstration of a seroconversion. However, apparently significant antibody titer rises may result from cross-reactions to related antigens; for example, an antibody rise to parainfluenza virus may

Table 39-4 Serologic panels

Clinical Syndrome	Viruses Under Consideration (Panel)
Myocarditis/pericarditis	Group B coxsackievirus types 1 through 5, influenza A and B, cytomegalovirus
Hepatitis A	Hepatitis A virus
Exanthem	Rubella virus, measles virus, Varicella-Zoster virus
Central nervous system infections	Western and eastern equine encephalitis virus, California encephalitis virus, St. Louis encephalitis virus, lymphocytic choriomeningitis virus, measles virus, Epstein-Barr virus, rabies virus
Heterophile-negative mononucleosis syndromes	Cytomegalovirus, Epstein-Barr virus
Respiratory infections	Influenza A and B; respiratory syncytial virus; Parainfluenza types 1, 2, and 3; adenoviruses

Modified from Chernesky MA, Ray CG, and Smith TF: Laboratory diagnosis of viral infections, Washington, DC, 1982, American Society for Microbiology.

actually result from infection with mumps virus. Furthermore, seroconversions may not be seen with some patient populations (e.g., infants, immunocompromised patients) or when the initial serum is collected late in the course of disease.

Serologic Panels

Selection of several antigens for testing with paired sera in cases where a virus is suspected but not detected can sometimes be made on the basis of clinical syndrome, the known local epidemiology of particular viruses, and the patient's age. This has led to the concept of serologic **batteries** or **panels.** Some examples of possible panels are included in Table 39-4.

Antigens from herpes simplex virus, mumps, western equine encephalitis, eastern equine encephalitis, St. Louis encephalitis, and California encephalitis viruses—and perhaps lymphocytic choriomeningitis virus, Epstein-Barr virus, and human immunodeficiency virus—may be included in a panel of tests for central nervous system diseases. Although herpes simplex antigen is sometimes included in such a panel, a rise in antibody titer is not sufficient to diagnose herpes encephalitis. Many viral central nervous system illness, especially aseptic meningitis, are caused by the enteroviruses; however, the many serotypes and the cumbersome serologic methods necessary for their diagnosis usually make it impractical to include them in a panel. When one or two enteroviruses have been shown to be epidemic in an area in one summer, one can pick up some additional cases by performing neutralization tests on paired sera employing only those specific enteroviruses endemic in the community.

Depending on the patient's age, the viral antigen panel for testing respiratory syndromes might include influenza A and B; respiratory syncytial virus; parainfluenza types 1, 2, and 3; and adenoviruses. For example, respiratory syncytial virus and parainfluenza virus might be routinely tested for infants and young children but not for adults.

To test for viral causes of exanthems, the panel would include measles and rubella. If the disease is vesicular, herpes simplex virus and varicella-zoster virus should be included.

Antigens from group B coxsackievirus types 1 to 5 and perhaps influenza A and B viruses would make up the panel for myocarditis and pericarditis. Although numerous viruses have been implicated in inflammatory diseases of the heart and its covering membranes, the group B coxsackieviruses are believed to account for almost one half of the cases. Unfortunately, much of the clinical illness is expressed at the time when standard methods of virus detection are likely to fail and serologic diagnosis must be attempted.

BIBLIOGRAPHY

Chernesky MA, Ray CG, and Smith TF: Laboratory diagnosis of viral infections, Washington DC, 1982, American Society for Microbiology.

Hsiung GD: Diagnostic virology, ed 3, New Haven, Conn, 1982, Yale University Press.

Lennette DA, Specter S, and Thompson KD, editors: Infections: the role of the clinical laboratory, Baltimore, 1979, University Park Press.

Lennette EH, editor: Laboratory diagnosis of viral infections, New York, 1985, Marcel Dekker.

40

Antiviral Vaccines and Chemotherapy

The control of viral diseases is accomplished by two major approaches: vaccination to prevent disease in an individual and in the community, and treatment of infected patients with specific chemotherapeutic agents. Unfortunately the number of antiviral agents is not as extensive as those agents used to treat bacterial, fungal, or parasitic infections. Thus the most successful approach to control viral diseases has been prevention through vaccination.

TYPES OF IMMUNIZATION

Injection of concentrated antibody for temporary protection of an individual is termed **passive immunization.** Another example is neonates and newborns receiving natural passive immunity when immunoglobulin G (IgG) crosses the placenta or IgA is present in the mother's milk. Stimulation of an immune response by challenge with an immunogen is termed **active immunization.** Natural immunization occurs after each exposure to an infectious agent. Immunization can also occur with exposure to microbes or their antigens in vaccines. Viral vaccines can be subdivided into two general groups, live and inactivated (Figure 40-1.)

Passive Immunization

Passive immunization with immune gamma globulin or specific antibodies is effective for prophylaxis or attenuation of disease. Commercially prepared antiviral antibody for passive immunity is available against hepatitis A and B, varicella zoster (varicella zoster immunoglobin, VZIG), rabies, and measles.

Active Immunization

Immunization with a **live vaccine** is similar to being infected with the infectious agent but under conditions that limit the resultant pathologic condition, thus yielding a subclinical infection. The virus used for the vaccine has **attenuated** (i.e., limited) capacity to cause disease in normal individuals but can be dangerous for immunosuppressed or pregnant individuals. Both humoral and cellular immunity are initiated; the immunity is generally long-lived; and depending on the route of administration, immunity can mimic the normal immune response to the infecting agent.

Attenuated viruses are mutants that cause reduced pathology. They can be obtained by serial passage (7 to 10 passages) of the virus in culture or by genetic engineering (Figure 40-1). These mutants are stable and can grow well under cell culture conditions but poorly in the harsh conditions of the body.

Most of the live vaccines in use are obtained by growth in embryonated eggs or tissue culture cells. Host-range mutants, temperature-sensitive strains (Schwartz measles vaccine), and cold-adapted strains are selected by the cell culture conditions. Examples of attenuated live virus vaccines currently in use are presented in Table 40-1.

The first approaches to development of live vaccines was the use of virulent viruses from other species that share antigenic determinants with the human analog. Jenner developed the first vaccine in this manner. He discovered that people could be immunized against smallpox by infection with cowpox or vaccinia. These animal viruses share antigenic determinants with smallpox but cause benign infections in humans. Similarly the vaccine for the first

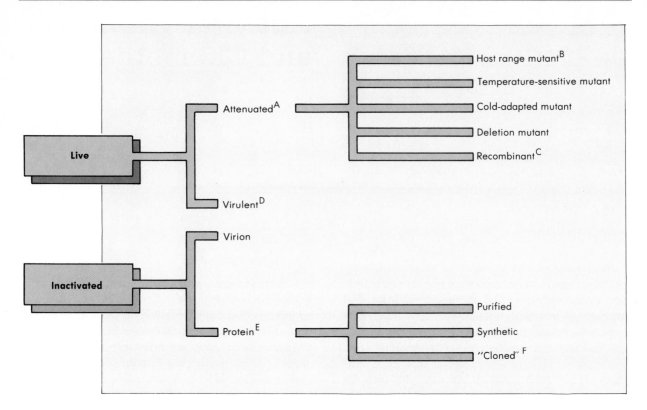

FIGURE 40-1   Types of viral vaccine. **A,** Vaccine that includes genetic reassortants incorporating mutant genes. **B,** Vaccine occurring naturally in a different animal species or selected by passage in cultured cells. **C,** Vaccine incorporating gene from another virus. **D,** Vaccine delivered via an unnatural route or at a safe age (in veterinary medicine). **E,** Vaccine that may be incorporated in a liposome or an adjuvant. **F,** Vaccine developed in prokaryotic or eucaryotic cells through recombinant DNA technology.

Table 40-1   Attenuated, Live Viral Vaccines*

Disease	Vaccine Strain	Cell Substrate†	Attenuation	Route
Yellow fever	17D	Chick embryo	+	Subcutaneous
Poliomyelitis	Sabin 1, 2, 3 (Oral poliovirus vaccine, OPV)	HEF	+	Oral
Measles	Schwarz	CEF	+	Subcutaneous
Rubella	RA27/3 or Cendehill	HEF or RK	+	Subcutaneous
Mumps	Jeryl Lynn	CEF	+	Subcutaneous

Modified from White and Fenner, Medical virology, ed 3, New York, 1986 Academic Press,
*A variety of different viral strains and cell substrates are used in different countries; the selection here is not comprehensive.
†HEF, Diploid strain of human embryonic fibroblasts; CEF, chick embryo fibroblast cultures; RK, rabbit kidney cultures.

tumor virus to be controlled by vaccination, Marek's disease virus of chickens, is the turkey herpesvirus. Use of bovine rotavirus has been suggested as a vaccine to protect infants against human rotavirus.

Sabin developed the first live oral polio vaccine in the 1950s. The vaccine is obtained by multiple passage of the three poliovirus types. At least 57 mutations are accumulated in the polio type 1 vaccine strain. Oral administration of this vaccine elicits both secretory IgA in the gut and IgG in the serum. This provides protection along the normal route of infection by the wild virus.

The vaccinia virus can also be used as a vector to immunize with antigens from other infectious agents. A hybrid virus can be genetically engineered that contains genes from those infectious agents that cannot be attenuated or are inappropriate for live vaccine administration. These antigens are then expressed during replication of the vaccinia virus. This promising approach allows development of a polyvalent vaccine to many viruses in a single, safe, inexpensive, relatively stable vector. A human immunodeficiency virus (HIV) and vaccinia hybrid vaccine is currently in clinical trials.

Inactivated Vaccines

Inactivated vaccines are used for viruses that may be too virulent or when live vaccine use would be too risky, such as viruses with the potential for recurrence or oncogenicity. Inactivated vaccines are composed of chemically killed or heat-killed virus or subunits of the virus. The immunity is usually not lifelong, requires booster shots, and demands a larger dose of vaccine (Table 40-2).

The Salk polio and the influenza vaccines are examples of killed virus vaccines prepared by formaldehyde inactivation of virions. In the past rabies vaccine was prepared by formalin inactivation of infected rabbit neurons and duck embryos. The vaccine is currently prepared by chemical inactivation of virions grown in diploid tissue culture cells.

Subunit vaccines are prepared from viral components obtained from virions, infected cells, sera of chronically infected people (hepatitis B), or growth of cloned viral genes in bacteria or eucaryotic cells. The surface antigen of hepatitis B, the hemagglutinin of influenza, the G antigen of rabies, and the gD antigen of herpes simplex virus (HSV) have all been cloned and grown in bacteria or eucaryotic cells for use as

Table 40-2 Advantages and Disadvantages of Live vs. Dead (Inactivated) Vaccines

Property	Live	Inactivated
Route of administration	Natural* or injection	Injection
Dose of virus; cost	Low	High
Number of doses	Single†	Multiple
Need for adjuvant	No	Yes‡
Duration of immunity	Many years	Generally less§
Antibody response	IgG, IgA∥	IgG
Cell-mediated immune response	Good	Poor
Heat lability in tropics	Yesπ	No
Interference	Occasional; OPV only#	No
Side effects	Occasional mild symptoms**	Occasional sore arm
Reversion to virulence	Rarely; OPV only††	No

From White and Fenner, Medical virology, ed 3, New York, 1986, Academic Press.

*Oral or respiratory, in certain cases.

†Except oral poliovirus vaccine (OPV), to preempt possibility of interference. For some other live vaccines a single booster may be required by law (yellow fever) or desirable (rubella?) after about a decade.

‡However, no satisfactory adjuvants yet licensed for human use.

§However, satisfactory with some inactivated vaccines.

∥IgA if delivered via oral or rspiratory route. OPV can thereby prevent wild poliovirus from multiplying in the gut and thus facilitate near eradication of the virus from the community.

πMgCl$_2$ and other stabilizers, maintenance of refrigeration or cold storage assist preservation.

#Especially in Third World countries.

**Especially rubella and measles.

††10^{-6} vaccinees.

subunit vaccines. The vaccine for hepatitis B was initially prepared from surface antigen obtained from human sera of carriers negative for infectious virus, but this vaccine is now purified from bacteria replicating the gene for the antigen. The antigen is purified, inactivated, and absorbed onto alum for immunization.

Other Approaches to Vaccination

Portions of viral proteins containing specific epitopes and new means of genetically engineering proteins for vaccines are being developed. Anti-idiotypes, which are antibodies that resemble an epitope of a virus, are also being investigated as potential vaccines.

Considerations for Vaccine Use

Vaccines cannot and should not be prepared for all virus diseases. The cost and risk-to-benefit ratio must be considered for each virus. The properties of a good virus candidate for vaccine development are listed in Box 40-1. Smallpox has been eliminated by an effective vaccine program because the virus exists as only one serotype, symptoms are always present, and the vaccine is relatively benign and stable. Measles virus is another good prospect for elimination by vaccine. Rhinovirus is an example of a poor candidate for vaccine development because so many serotypes exist.

Problems with Vaccine Use

The ideal vaccine should elicit dependable, lifelong immunity to infection without serious side effects. Vaccine development has been an empirical science, and unfortunately, limitations and problems occur (Box 40-2). Factors that influence the success of an immunization program include not only the composition of the vaccine, but also the time, site, concentration, and conditions of administration (see Table 40-2). For example, vaccination of infants with the live vaccine for measles may not elicit lifelong protection because of the presence of maternal antibodies and the immature immune response of babies. Inadvertent inactivation of the vaccine by heating or exposure to light, improper dosage, or simultaneous administration of gamma globulin may also compromise the immunization.

ANTIVIRAL AGENTS

In general, the development of antivirals has occurred at a disappointingly slow pace. Interferon was discovered in 1957 and gave promise as a broad-spectrum antiviral, but it has not found its niche as a practical or frequently used antiviral. In part this results from difficulties in producing this biologic product in large quantities. The most effective and available antiviral is **acyclovir,** which selectively inhibits certain herpesviruses with minimal cellular toxicity. This agent is one of the latest in a succession of inhibitors of viral deoxyribonucleic acid (DNA) synthesis. Earlier compounds such as iododeoxyuridine did inhibit herpes viral DNA synthesis but were toxic

---

Box 40-1 Properties of a Good Candidate for Vaccine Development

Virus causes significant illness.
Virus exists as only one serotype.
Antibody will block infection or systemic spread.
Disease, when it occurs, is symptomatic; thus carriers can be detected and prevented from spreading the disease.
Virus does not have oncogenic potential.
Virus used in the vaccine should be heat stable to allow transport to endemic areas.

Box 40-2 Problems with Vaccine Use

Live vaccines can occasionally revert to virulent forms.
Interference by other viruses may prevent infection by the vaccine.
Immunization of an immunocompromised person with a live vaccine can be life threatening.
Side effects to vaccination can occur, such as hypersensitivity or allergic reactions to the viral components, to nonviral material in the vaccine, or to contaminants.
Vaccine development and liability insurance for the manufacturer are very expensive.
Viruses with many serotypes are difficult to control by vaccination.

because of inhibition of host cell DNA synthesis. Acyclovir is only phosphorylated in virus-infected cells, and once phosphorylated, it inhibits viral DNA polymerase. In uninfected cells no phosphorylation occurs, and thus there is no active drug, no inhibition of cellular DNA synthesis, and little or no toxicity.

Other identified antivirals block specific viral replicative steps. For example, **amantadine** blocks the cellular fusion or uncoating of influenza A. Since uncoating is a unique step in viral replication, minimal cellular toxicity from this compound occurs.

Another agent that blocks a specific step in viral replication is **azidothymidine,** which inhibits the reverse transcriptase of retroviruses such as HIV. No reverse transcriptase exists in normal mammalian cells, so this inhibition is selective.

At present the newer antivirals are relatively narrow in their spectrum of activity. For example, acyclovir is only active against herpesviruses, but even within this group, acyclovir is ineffective against cytomegalovirus. Amantadine is only active against influenza A virus and is ineffective against either paramyxoviruses or other myxoviruses.

The development of these newer antivirals gives

Table 40-3   Antiviral Agents

Generic Name	Mechanism of Action	Indications
AVAILABLE ANTIVIRAL AGENTS		
1. Systemic		
Amantadine hydrochloride	Inhibition of influenza A fusion with host cell membrane	Prophylaxis of influenza A; lesser but definite therapeutic effect on influenza A
Acyclovir sodium (ACV)	Inhibition of herpesvirus DNA polymerase	Primary genital HSV; less effective for recurrent genital HSV; for zoster in nonimmunocompromised patients (?)
		Recurrent genital HSV
		Varicella zoster in immunocompromised patients; HSV encephalitis; disseminated HSV infection in neonate or immunocompromised patients
Adenine arabinoside (Ara A)	Inhibition of herpesvirus DNA polymerase	Systemic therapy of neonatal herpes simplex infection
	Incorporation into viral DNA to produce "fraudulent" viral DNA	
Azidothymidine (AZT)	Inhibition of retrovirus reverse transcriptase	AIDS (PCP)
Dihydroxypropoxymethyl guanine (DHPG)	Inhibitor of herpesvirus DNA polymerases	Severe cytomegalovirus infection, especially retinitis
2. Topical		
a. Skin		
Acyclovir		Primary genital herpes simplex infection in patients with normal immunity; limited mucocutaneous herpes simplex infection in immunocompromised patients
b. Eye		
Idoxuridine	Inhibition of herpesvirus DNA polymerase	Topical therapy of conjunctivitis or superficial keratitis caused by herpes simplex virus
Adenine arabinoside	Same as for idoxuridine	Same as for idoxuridine
Trifluridine	Same as for idoxuridine	Same as for idoxuridine
MAJOR EXPERIMENTAL ANTIVIRAL AGENTS		
Ribavirin (Virazole)	Inhibition of RNA virus RNA polymerase	Possibly effective for Lassa fever, severe respiratory syncytial virus pneumonia, influenza A and B
	Depletion of hose cell guanosine	

promise for the future availability of compounds active against entirely different viruses.

The remainder of this chapter reviews the available antiviral agents. A summary of these drugs is presented in Table 40-3.

Acyclovir

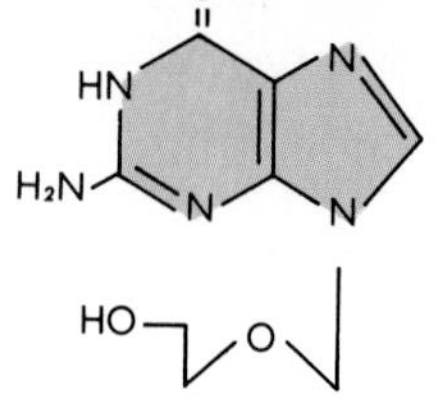

Acyclovir (ACV) differs from the nucleoside guanosine by having an acyclic (hydroxyethoxymethyl) side chain (Figure 40-2). This unique compound must be phosphorylated to be active, and this phosphorylation only occurs in cells infected by certain herpesviruses. This makes the compound essentially nontoxic because it does not adversely affect the metabolism of uninfected host cells. ACV has selective action against herpesviruses because certain members of this virus family induce a thymidine kinase (TK) in cells they infect (Figure 40-3). TK catalyzes the phosphorylation of ACV to a monophosphate. From that

FIGURE 40-2   Structure of acyclovir.

point, host cell enzymes complete the progression to the diphosphate and finally the triphosphate. Activity of ACV against herpesviruses directly correlates with the capacity of the virus to induce a TK. Herpes simplex virus types 1 and 2 (HSV-1 and HSV-2) are the most active TK inducers and are the most readily inhibited by ACV. Cytomegalovirus induces little or no TK and is not inhibited. Varicella zoster and Epstein-Barr viruses are between these two extremes

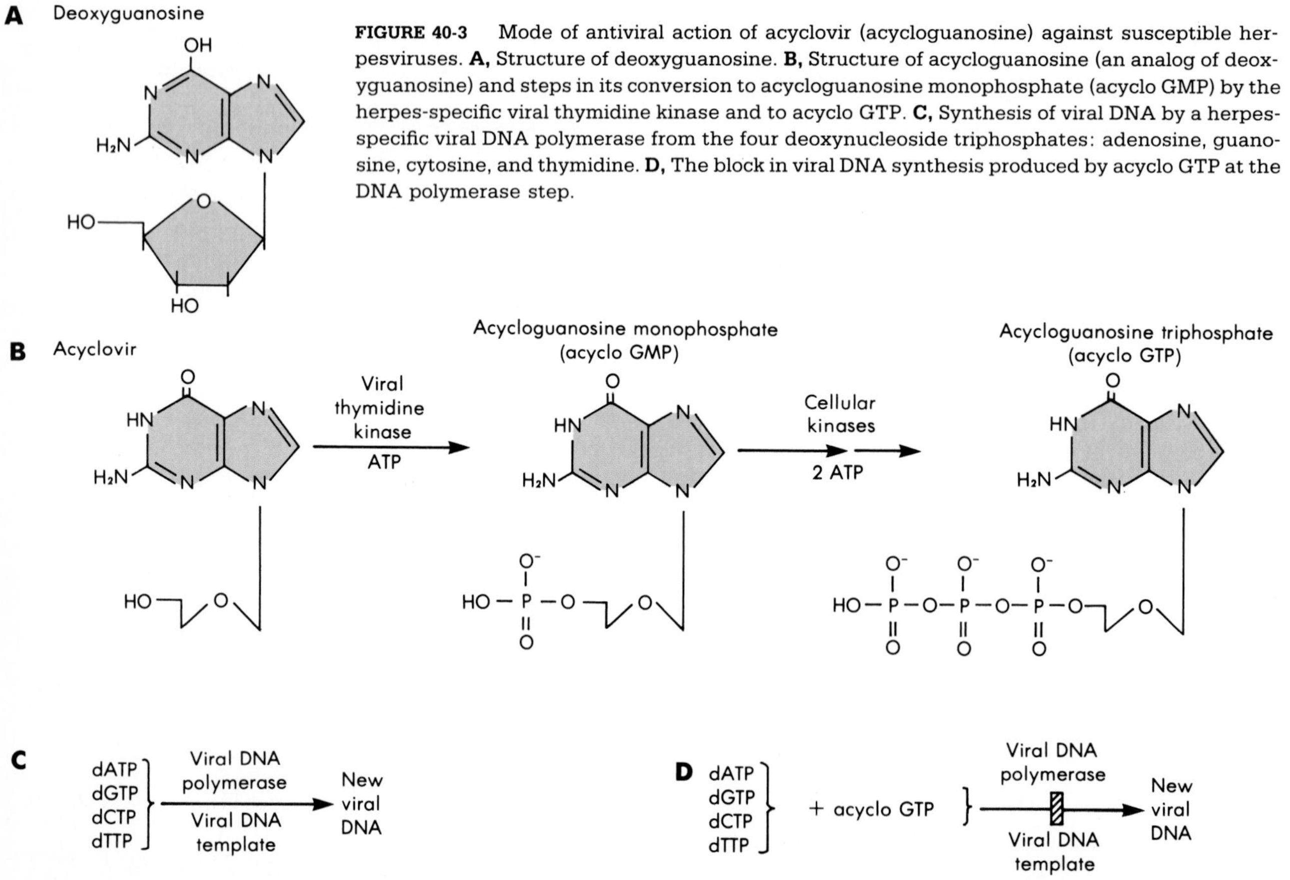

FIGURE 40-3   Mode of antiviral action of acyclovir (acycloguanosine) against susceptible herpesviruses. **A,** Structure of deoxyguanosine. **B,** Structure of acycloguanosine (an analog of deoxyguanosine) and steps in its conversion to acycloguanosine monophosphate (acyclo GMP) by the herpes-specific viral thymidine kinase and to acyclo GTP. **C,** Synthesis of viral DNA by a herpes-specific viral DNA polymerase from the four deoxynucleoside triphosphates: adenosine, guanosine, cytosine, and thymidine. **D,** The block in viral DNA synthesis produced by acyclo GTP at the DNA polymerase step.

Table 40-4 Susceptibility of Herpesviruses to Acyclovir	
Virus	**Mean ID_{50} (μM/ml)**
HSV-1	0.02-0.2
HSV-2	0.2-0.4
Epstein-Barr	0.075-1.5
Varicella zoster	0.5-4.64
Cytomegalovirus	25-90

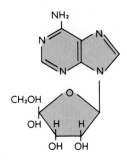

FIGURE 40-4   Structure of adenine arabinoside (Ara A).

in both TK induction and ACV susceptibility (Table 40-4).

ACV triphosphate inhibits viral replication by competing with guanosine triphosphate for the virally induced DNA polymerase. The selectivity and minimal toxicity of ACV is further aided by its 100-fold or greater affinity for viral DNA polymerase than for cellular DNA polymerase. Viral DNA polymerase is able to incorporate ACV triphosphate into the growing viral DNA chain. However, this causes termination of chain growth, since no 3'-hydroxy group on the ACV molecule exists to provide attachment sites for additional nucleotides. Resistant strains of HSV have been recovered from immunocompromised patients, especially patients with AIDS.

Clinical Use

Prophylaxis In patients with frequent severe genital herpes, oral ACV is effective in preventing recurrences. It does not eliminate the virus from the host and therefore must be taken daily to be effective. When the drug is discontinued, the same pattern of recurrences may return, but in many patients the frequency of reactivated infections is reduced compared to patients who did not receive the drug. Because of the uncertain potential for drug toxicity in patients taking ACV chronically, it is recommended that it be discontinued, or at least interrupted, after 6 to 12 months of use. After discontinuation the patient and physician can determine whether the pattern of recurrences has been altered.

Therapy ACV is most effective in the treatment of primary HSV infections or for severe recurrences in immunocompromised patients. It is also useful in herpes encephalitis. Treating recurrent genital herpes in otherwise healthy individuals is only minimally effica-

cious. ACV is also recommended for varicella zoster virus infection in immunocompromised or immune competent persons if the eye is affected. Intravenous treatment is usually necessary, however, since this virus is less sensitive to ACV and higher blood levels are necessary.

Pharmacology and Toxicity

ACV is available in three forms: topical, oral, and parenteral, with parenteral administration required for treatment of severe infections in immunocompromised patients. The drug is remarkably free of bone marrow toxicity, even in patients with hematopoietic disorders. Rapid intravenous infusion may cause impaired renal function, presumably because of crystallization of the drug in the renal tubules. Central nervous system toxicity has been reported in patients treated with high doses for 10 to 14 days or longer.

Adenine Arabinoside

Adenine arabinoside (Ara A) is a purine nucleoside analog identical in structure to adenosine except arabinose is substituted for ribose as the sugar moiety (Figure 40-4). It is phosphorylated by cellular enzymes, especially adenosine kinase, even in uninfected cells, and thus has a greater potential for toxicity than ACV. Both cellular and viral DNA polymerases are inhibited by Ara A triphosphate, but the viral enzyme is 6 to 12 times more sensitive. This is the basis for a somewhat favorable therapeutic/toxic ratio. Ara A is also incorporated into viral DNA, as well as cellular DNA. In view of the latter, the drug may be oncogenic.

Ara A exhibits activity against some herpesviruses

(HSV and varicella zoster) but not Epstein-Barr virus and cytomegalovirus. Resistance can develop by mutation to an altered and less susceptible viral DNA polymerase.

Clinical Use

Until the advent of ACV, Ara A was the principal antiviral used in the treatment of herpesvirus infections. Comparative trials with ACV have shown Ara A to be less efficacious in all HSV and varicella zoster syndromes, with the possible exception of neonatal HSV infection.

Pharmacology and Toxicity

Only intravenous Ara A is available for therapy. The drug is rapidly deactivated by host enzymes to produce an inactive metabolite. Thus the blood levels are usually less than 1 μg/ml, which is substantially less than the concentration needed for in vitro inhibition of herpesviruses. However, intracellular concentrations may be a more important determinant of antiviral activity. The drug is quite insoluble in physiologic solutions, which limits the amount that can be given intravenously. Toxicity consists of nausea and vomiting, central nervous system reactions, and hematologic abnormalities, which are usually reversible.

Amantadine and Rimantadine

Amantadine and rimantadine are amine compounds with selective activity against influenza A virus but no efficacy against any other virus, including influenza B. They are definitely effective in preventing influenza and also may be useful in treatment of this viral infection.

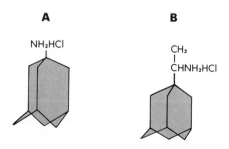

FIGURE 40-5   Structures of amantadine (**A**), rimantadine (**B**).

The structures of amantadine and rimantadine are shown in Figure 40-5. Rimantadine differs from amantadine by the substitution of a methyl group for a hydrogen ion at the alpha position.

These two amines inhibit several early steps in viral replication. It is not clear which of these blocks is responsible for their clinical effectiveness but it is probably at the level of viral fusion with the host cell membrane before uncoating of the virus.

Influenza A strains resistant to these agents rarely have been reported.

Clinical Use

Prophylaxis These acyclics are 50% to 90% effective in preventing influenza A illness when given daily during influenza outbreaks. Although illness is diminished, drug recipients may still develop evidence of infection (i.e., antibody). This is desirable because this antibody may provide some protection against future influenza A exposure.

Therapy In healthy adults or children amantadine shows a slight but statistically significant improvement in symptoms compared to placebos or antipyretics. It is assumed but not proved, that these drugs are efficacious in elderly or other high-risk patients who may have more severe influenza.

Pharmacology and Toxicity

Both drugs are available only as oral preparations, but their pharmacokinetics are quite different. Amantadine is excreted by the kidney without being metabolized. Its half-life is estimated to be 12 to 16 hours. In contrast, rimantadine is metabolized by the liver and then excreted in the kidney; its half-life is estimated at 30 to 36 hours.

The principal toxic effect is upon the central nervous system and consists of nervousness, irritability, and insomnia. These symptoms occur more frequently in patients receiving amantadine than those taking rimantadine; these effects are probably decreased by reduction of amantadine dosage.

Azidothymidine

Azidothymidine (AZT), a nucleoside analog of thymidine, inhibits the reverse transcription of the HIV (Figure 40-6). As with other nucleosides, AZT must be phosphorylated, in this case by host cell enzymes. The basis for the relatively selective therapeutic

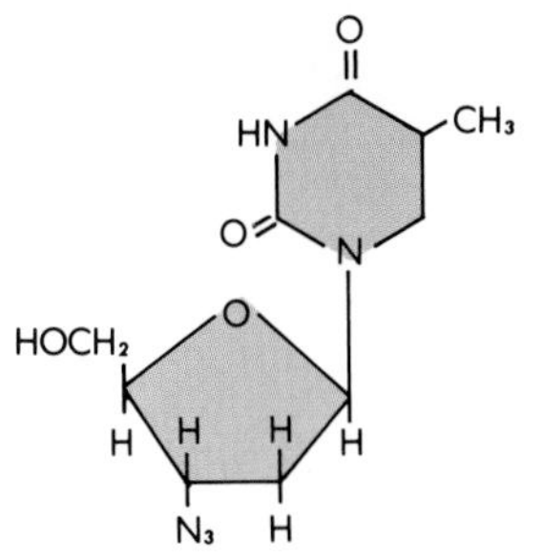

FIGURE 40-6   Structure of azidothymidine.

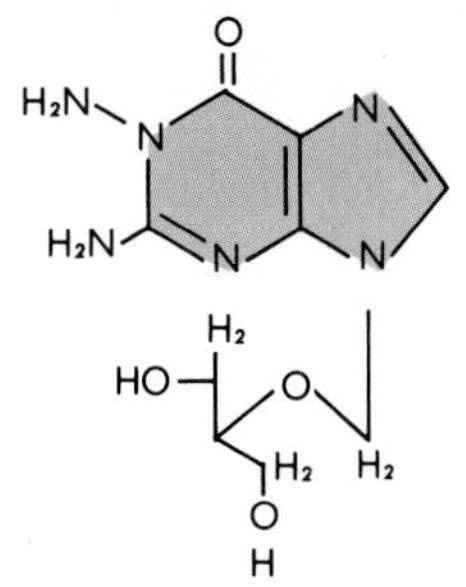

FIGURE 40-7   Structure of dihydroxypropoxymethyl guanine (DHPG).

effect of AZT is that host cell DNA polymerase is more than 100 times less sensitive to AZT triphosphate than the HIV reverse transcriptase. Nonetheless, since host cell enzymes convert it to an active triphosphate, toxicity frequently occurs.

Clinical Use

AZT trials in humans have demonstrated a reduction in life-threatening opportunistic infections and deaths in patients with AIDS receiving AZT compared to those receiving a placebo.

The basis for this efficacy is not entirely clear, since the antiviral effect is incomplete and helper T lymphocytes are only transiently increased. Also unclear is how long the beneficial effects of AZT persist. Nonetheless, AZT represents the first useful therapy for HIV infection.

Pharmacology and Toxicity

AZT is available in oral form only. Toxicity frequently occurs, ranging from nausea to life-threatening bone marrow toxicity. All hematopoietic components may be depressed but are usually reversible with discontinuation of the drug or reduction of the dose.

Dihydroxypropoxymethyl Guanine

Dihydroxypropoxymethyl guanine (DHPG, ganciclovir), differs from ACV by the addition of a single hydroxymethyl group in the acyclic side chain (Figure 40-7). The remarkable result of this single addition to the side chain is that it confers considerable activity against cytomegalovirus (CMV). DHPG is approximately 50 times more active against CMV than ACV. As with ACV, DHPG must be phosphorylated to be

active. Although CMV does not code for a thymidine kinase, DHPG is phosphorylated to a greater degree in CMV-infected cells than in uninfected cells. Presumably viral infection increases endogenous phosphorylating enzymes, but the details of this process are still being investigated. Once phosphorylated, DHPG inhibits all herpesvirus DNA polymerases. The viral DNA polymerases have nearly 30 times greater affinity for the drug than does the cellular DNA polymerase. Strains of CMV resistant to DHPG have been described.

Clinical Use

DHPG is effective in the treatment of CMV retinitis in patients with AIDS. Studies are in progress to determine its efficacy for CMV esophagitis, colitis, and pneumonia.

Pharmacology and Toxicity

DHPG is currently available only for intravenous administration. The limiting factor in DHPG use is the bone marrow toxicity, which may occur in approximately 20% to 40% of recipients. Neutropenia is the most frequent hematologic abnormality, followed by thrombocytopenia. With rare exception, these effects are reversible with discontinuation of drug or reduction of dosage.

Ribavirin

Ribavirin is another analog of the nucleoside guanosine (Figure 40-8). Unlike ACV, which replaces the ribose moiety with an hydroxymethyl acyclic side chain, ribavirin differs from guanosine in that the base ring is incomplete and open. As with other purine nucleoside analogs, ribavirin must be phos-

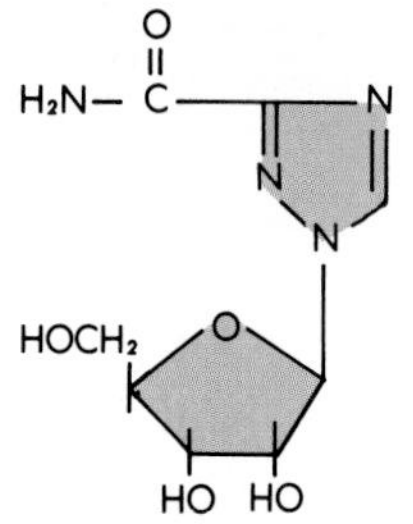

FIGURE 40-8   Structure of ribavirin.

phorylated to mono-, di-, and triphosphate forms. The drug is active against a broad range of viruses in vitro. In vivo ribavirin is somewhat beneficial if given early to persons infected with respiratory syncytial virus (RSV). Some studies have also suggested benefit in measles, influenza A and B, and Lassa fever.

Mechanism of Action

The mechanism of ribavirin's antiviral effect is not as clear as that of ACV. The triphosphate is an inhibitor of RNA polymerase. Another mechanism of action is that ribavirin depletes cellular stores of guanine by inhibiting inosine monophosphate dehydrogenase, an enzyme important in the synthetic pathway of guanosine. Still another mode of action is by decreasing synthesis of the mRNA 5′ cap because of interference with both guanylation and methylation of the nucleic acid base.

These multiple sites of action may explain in part the failure, as yet, to detect ribavirin-resistant mutants of RSV or influenza A.

Clinical Use

Oral and Intravenous Forms   The drug has been reported to benefit patients with Lassa fever, although the studies were not well controlled.

Aerosol Form   In animal studies aerosol administration has been superior to parenteral therapy for respiratory virus infections. In laboratory animals this method enables ribavirin to reach concentrations in respiratory secretions up to tenfold greater than sufficient to inhibit viral replication and substantially higher than those achieved with oral administration. This method has also been advocated for use in RSV infection of infants. In randomized studies of otherwise normal infants with RSV who received aerosol for 20 hours per day, there were statistically signifi-

cant effects on speed of recovery and virus shedding; however, these were detectable only after 4 days of treatment. Aerosol administration has also been used with moderate success to treat influenza in college students, but oral therapy provided only uncertain benefit.

Pharmacology and Toxicity

A reversible anemia has been associated with oral administration of ribavirin. Problems encountered with aerosolized ribavirin include precipitation of the drug in tubing used to administer drug and environmental exposure of health care personnel.

BIBLIOGRAPHY

Bryson YJ et al: Treatment of first episodes of genital herpes simplex virus infection with oral acyclovir: a randomized, double-blind controlled trial in normal subjects, N Engl J Med 308:916-921, 1983.

Burns WH et al: Isolation and characterisation of resistant herpes simplex virus after acyclovir therapy, Lancet 1:421-423, 1982.

Collaborative DHPG Study Group: Treatment of serious CMV infection using DHPG in patients with AIDS and other immunodeficiencies, N Engl J Med 314:801-805, 1986.

Crumpacker CS et al: Resistance to antiviral drugs of herpes simplex virus isolated from a patient treated with acyclovir, N Engl J Med 306:343-346, 1982.

Felsenstein D et al: Treatment of cytomegalivirus retinitis with 9-[2-hydroxy-1-(hydroxymethyl) ethoxymethyl] guanine, Ann Intern Med 103:377-380, 1985.

Fischl MA et al: The efficacy of azidothymidine (AZT) in the treatment of patients with AIDS and AIDS-related complex: a double-blind, placebo-controlled trial, N Eng J Med 317:185-191, 1987.

Reichman RC et al: Topically administered acyclovir in the treatment of recurrent herpes simplex genitalis: a controlled trial, J Infect Dis 147:336-340, 1983.

Straus SE et al: Suppression of frequently recurring genital herpes, N Engl J Med 310:1545-1550, 1984.

Wade JC et al: Intravenous acyclovir to treat mucocutaneous herpes simplex virus infection after marrow transplantation, Ann Intern Med 96:265-269, 1982.

Whitley RJ, et al: Vidarabine therapy of neonatal herpes simplex virus infection, Pediatrics 66:495-501, 1980.

Mills J and Corey L: Antiviral chemotherapy: new directions for clinical application and research, New York, 1986, Elsevier Science Publishing Co.

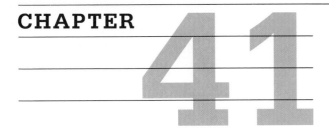

Adenoviruses

Adenoviruses were first isolated in 1953 in human adenoid cell culture. Since then approximately 100 serotypes, at least 42 of which infect humans, have been recognized. All human serotypes are included in a single genus within the family Adenoviridae. Based on homology studies and hemagglutination patterns, the 42 serotypes belong to 1 of six subgroups. Because of their oncogenic potential in animals and their replicative cycle, adenoviruses have been extensively studied by molecular biologists. These studies have elucidated many viral and eucaryotic intracellular processes. Common disorders caused by the adenoviruses include respiratory tract infection, conjunctivitis, hemorrhagic cystitis, and gastroenteritis.

STRUCTURE AND REPLICATION

Adenoviruses are double-stranded DNA viruses with a genome molecular weight of 20 to 25 × 10⁶

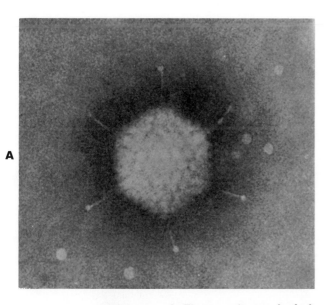

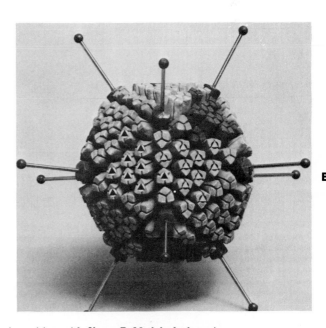

FIGURE 41-1   **A,** Electron micrograph of adenovirus virion with fibers. **B,** Model of adenovirus virion with fibers. (**A** from J Molecular Biol 13:13, 1965, **B** from Ginsberg HS: The adenoviruses, New York, 1984, Plenum Publishers.)

Table 41-1 Major Adenovirus Structural Proteins

Location	Number	Molecular weight	Name	Comment
CAPSID				
	II	120 K	Hexon protein	Contains family antigen and some serotyping antigens
	III	85 K	Penton base protein	Toxic to tissue culture cells
	IV	62 K	Fiber	Responsible for attachment and hemagglutination; contains some serotyping antigens
	VI	24 K	} Hexon-associated proteins	
	VIII	13 K		
	IX	12 K		
	IIIa	66 K	Penton-associated protein	
CORE				
	V	48 K	Core protein 1 linked to viral DNA	
	VII	18 K	Core protein 2	

daltons. The virions are nonenveloped icosahedrons with a diameter of 70 to 90 nm. Projections, or *fibers,* originate from each of the 12 vertices of the protein capsid (Figure 41-1). The capsid is composed of 240 capsomeres, which consist of hexans and pentons. The term **hexons** is used because each is surrounded by 6 identical capsomeric proteins. The 12 pentons, located at each of the vertices, contain a penton base and a fiber. The term **penton** is derived from the five hexons that surround it. The core complex, within the capsid, includes viral DNA and at least two major proteins.

There are at least 11 polypeptides in the adenovirus virion, 9 of which have an identified structural function (Table 41-1). Three structural proteins include the hexon protein, the largest of the capsomeric proteins, which contains a common family antigen and some of the serotyping antigens. The fiber protrudes from the penton base protein and is toxic to cultured cells. The fiber protein is responsible for

Table 41-2 Map of Adenovirus Genome

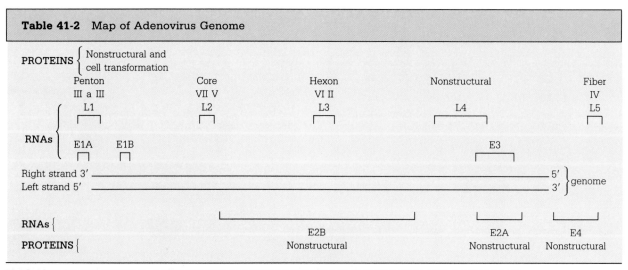

Modified from Jawetz E, Melnick JL, Adelberg EA: *Review of medical microbiology,* ed 17, Norwalk, Conn, 1987, Appleton & Lange.
E, Early MRNA; *L,* late MRNA.

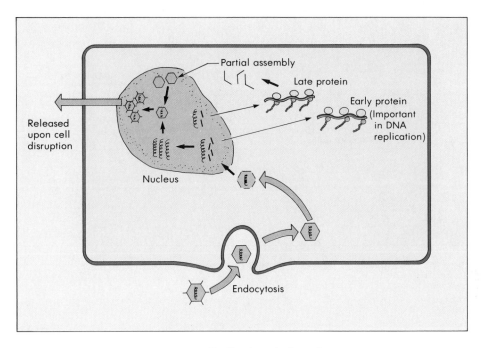

FIGURE 41-2   Replication of adenoviruses.

hemagglutination and also contains some of the serotyping antigens.

A map of the adenovirus genome (Table 41-2) shows the separate locations, on both DNA strands, of early gene transcription. Known structural and nonstructural gene products are shown in Table 41-2, and structural proteins are labeled with roman numerals as in Table 41-1. Early proteins function in cell transformation and DNA replication. Late proteins, synthesized after the onset of viral DNA replication, are primarily structural. Studies done with adenovirus hexon mRNA discovered that eucaryotic mRNAs are usually not colinear with their genes but are spliced products of separate coding regions.

Replication of adenoviruses has been studied extensively in HeLa cell culture. One virus cycle takes approximately 32 to 36 hours and produces 10,000 virions. The replicative cycle is divided into attachment, penetration and uncoating, and early and late events (Figure 41-2). Attachment is probably mediated by the viral fiber protein and a receptor on the host cell membrane. There are approximately 100,000 fiber receptors per cell. The virus enters the cell by endocytosis, is uncoated in the cytoplasm, with subsequent replication in the nucleus. This entire process requires 2 hours. Early transcriptional events,

following shut-down of host cell macromolecular synthesis, lead to gene products involved in cell transformation and viral DNA replication.

Viral DNA replication in the nucleus signals the beginning of the late phase. In addition, late transcripts encode viral structural proteins at a maximum rate at 20 hours after infection. Virion capsomeres are produced in the cytoplasm and then transported to the nucleus for viral assembly. DNA possibly enters the capsid through an opening at one of the vertices. The mature particle is now stable and infectious. Infected host cells release the virus upon degeneration and cell disruption.

PATHOGENESIS

Adenoviruses infect epithelial cells lining respiratory and enteric organs. Following local replication, viremia may occur with spread to visceral organs (Figure 41-3). This dissemination is more likely to occur in immunocompromised patients. Variations in target cell specificity among adenovirus serotypes result from differences in viral fiber proteins. The virus has a propensity to become latent in lymphoid tissue, such as adenoids, tonsils, or Peyer's patches,

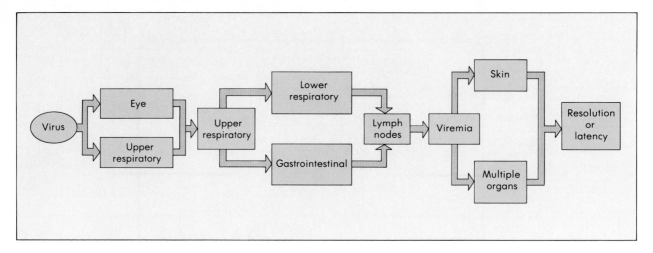

FIGURE 41-3   Mechanisms of adenovirus spread within the body.

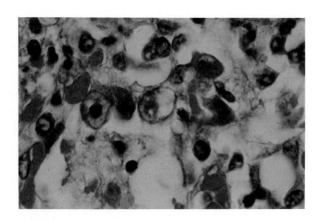

FIGURE 41-4   Histology of adenovirus infection.

and can be reactivated by immunosuppression or infection with other agents. The histologic hallmark of adenovirus infection is a dense central intranuclear inclusion within an infected epithelial cell (Figure 41-4). Superficially, these may resemble inclusions seem with cytomegalovirus, but one distinguishing feature is the absence of cellular enlargement (cytomegaly) in adenovirus infected cells. These inclusions are concentrations of viral DNA and protein. Histologically, adenovirus infections are characterized by mononuclear cell infiltrate and epithelial cell necrosis, as well as the characteristic inclusions. Type-specific neutralizing antibody is associated with protection from reinfection (Box 41-1). The time-course of adenovirus infection is shown in Figure 41-5.

EPIDEMIOLOGY

From 5% to 10% of pediatric respiratory disease can be attributed to adenoviruses. Adenoviruses spread by either respiratory or fecal-oral contact. The virus has also been cultured from semen and therefore may be spread by sexual transmission. Adenoviruses may be shed intermittently and over long periods of time from the pharynx and especially the feces. Most infections are symptomatic, which greatly facilitates their spread in the community. Adenovirus serotypes 4 and 7 seem especially able to spread among military recruits, since antibody prevalence to these serotypes is low in young adults.

Box 41-1   Pathogenic Mechanisms of Adenoviruses

Infects epithelial cells of mucous membranes in respiratory tract, conjunctiva/cornea, and gastrointestinal tract.

Viremia may occur with distribution of virus to oth-er susceptible tissues.

Persists in lymphoid tissue, such as tonsils, adenoids, and Peyer's patches indefinitely and may be reactivated during immunosuppression.

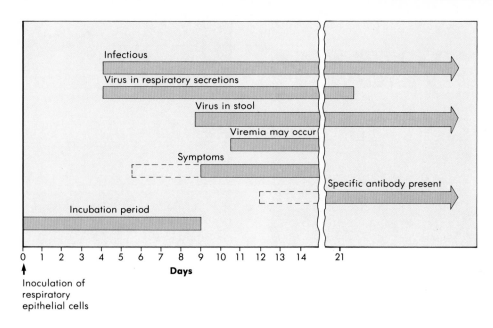

FIGURE 41-5   Time course of adenovirus respiratory tract infection.

CLINICAL SYNDROMES

Adenoviruses cause primary infection in children and, less commonly, adults. Infection by reactivated virus occurs in immunocompromised children and adults. Several district clinical syndromes are associated with adenovirus infection (Table 41-3).

Acute Febrile Pharyngitis and Pharyngoconjunctival Fever

Adenovirus may cause pharyngitis alone or coupled with conjunctivitis (pharyngoconjunctival fever). Pharyngitis alone occurs in young children, particularly those less than 3 years of age, and may mimic streptococcal infection. Pharyngoconjunctival fever occurs more often in outbreaks involving older children.

Table 41-3   Clinical Syndromes Associated With Adenoviruses

Clinical Syndrome	Virus Serotypes	Typical Patient
Acute febrile pharyngitis	1, 2, 3, 4, 5, 7, 7a, 14, 15	Young children
Acute respiratory disease, including fever, cough, pharyngitis, cervical adenitis, and skin rash	2, 3, **4,** 5, **7,** 8, 11, 14, 21	Military recruits
Other respiratory disease	1, 2, 3, 4, 5, 7	Children and adults
Pharyngoconjunctival fever	1, 2, 3, 4, 5, 6, 7, 7a, 8, 14, 37	Older children
Epidemic keratoconjunctivitis	2, 3, 4, 5, 7, 8, 10, 11, 13, 14, 15, 16, 17, 19, 29, 37	Children and adults
Infant diarrhea	3, 7, 11, 15, 17, 32, 33, **40, 41**	Infants
Acute hemorrhagic cystitis	**11, 21,**	Young children
Systemic infection in immunocompromised patients	1, 2, 4, 5, 6, 7, 7a, 11, 31, 32, 34, 35	Children and adults

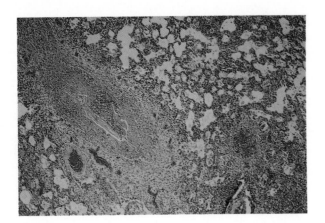

FIGURE 41-6   Hematoxylin and eosin stained sample of adenovirus pneumonia, showing inflammation, necrosis, and exudate.

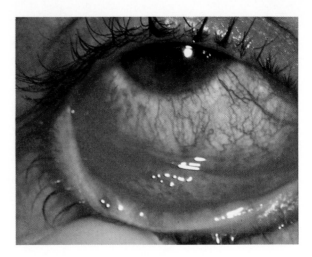

FIGURE 41-7   Conjunctivitis due to adenovirus.

Acute Respiratory Disease

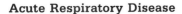

Acute respiratory disease is a syndrome of fever, cough, pharyngitis, and cervical adenitis seen primarily in outbreaks among military recruits because of serotypes 4 and 7.

Other Respiratory Diseases

Adenoviruses are definite but infrequent causes of true viral pneumonia in both children and adults (Figure 41-6). Laryngitis, croup, and bronchiolitis may also occur. Pertussis-like illness with a prolonged clinical course has been associated with adenoviruses.

Conjunctivitis and Epidemic Keratoconjunctivitis

Adenoviruses cause a follicular conjunctivitis in which the mucosa of the palpebral conjunctiva becomes pebbled or nodular, while both conjunctivae (palpebral and bulbar) become inflamed (Fig. 41-7). Conjunctivitis may occur sporadically or in outbreaks that can be traced to a common source, e.g., swimming pools. Corneal involvement may occur with mechanical irritation of the eye and is most striking when is spreads in epidemic form, e.g., "shipyard" conjunctivitis, clinically described as epidemic keratoconjunctivitis.

Infant Diarrhea

Adenovirus serotypes 40 and 41, which are very difficult to isolate in tissue culture, appear to be responsible for episodes of diarrhea in infants. Adenoviruses of several serotypes have been cultivated and have also been recovered from many infants with intussusception. The significance of these agents in this syndrome is not clear.

Acute Hemorrhagic Cystitis

Acute hemorrhagic cystitis with dysuria and hematuria is associated with serotypes 11 and 21. This entity occurs predominantly in young boys.

Systemic Infection in Immunocompromised Patients

Immunocompromised patients are at risk of serious adenovirus infections, although not as often as from infections caused by the herpes viruses. Diseases include pneumonia and hepatitis. Infection appears to be from exogenous or endogenous (reactivation) sources.

LABORATORY DIAGNOSIS

Fluorescent antibody immunoassays have been used with partial success to identify rapidly adenovirus in clinical samples, such as those from the respi-

Table 41-4 Laboratory Diagnosis of Adenovirus Infections

Electron microscopy, immunofluorescence, and enzyme immunoassay	Detection of adenovirus antigens or virions in clinical specimens
HEK cells and HEP-2 cells	Growth and detection of adenoviruses in cell culture
Complement fixation, enzyme immunoassay, hemagglutination inhibition, and neutralization	Detection of seroconversion or fourfold rise in virus-specific antibody titer

ratory tract. Enzyme immunoassay and electron microscopy are used to identify enteric adenovirus serotypes 40 and 41, which do not grow in readily available heteroploid cell cultures but may be responsible for infant diarrhea (Table 41-4). Characteristic intranuclear inclusions can be seen in infected tissue during histologic examination. Inclusions, however, are rare and must be distinguished from those resulting from cytomegalovirus.

Isolation of the virus is best accomplished in cell cultures derived from epithelial cells, for example, primary human embryonic kidney (HEK) cells or continuous (transformed) lines such as HeLa or human epidermal carcinoma (Hep-2) cells. Recovery in cell culture requires an average of 6 days. Isolation of adenovirus in culture has variable significance. If the isolate is from sites not frequently colonized by adenovirus, isolation may be diagnostic of infection (e.g., recovery from conjunctiva, urine, or viscera such as the lung). However, recovery from stool but not the respiratory tract of a patient with pharyngitis provides little diagnostic help. Adenoviruses may be shed in feces for weeks to months after infection. Isolation of adenovirus from the throat of a patient with pharyngitis is usually diagnostic, if laboratory findings eliminate other common etiologies such as *Streptococcus pyogenes*.

Complement fixation, hemagglutination inhibition, enzyme immunoassay, and neutralization techniques have been used to detect specific antibodies following adenovirus infection. A fourfold rise in antibody level or seroconversion between acute and convalescent serum specimens is necessary before the result can be considered diagnostic of active infection. Serologic diagnosis is rarely used except occasionally to confirm the significance of a fecal or upper respiratory isolate.

TREATMENT, PREVENTION, AND CONTROL

There is no known treatment for adenovirus infection. Live, oral enteric-coated vaccines have been used to prevent adenovirus 4 and 7 infections in military recruits, but they are not used in civilian population. Because the virus may be oncogenic, it is unlikely that live virus vaccines will be widely used. However, genetically engineered subunit vaccines could be prepared and used in the future.

BIBLIOGRAPHY

Balows A, Hausler WJ Jr, and Lennette EH, editors: Laboratory diagnosis of infectious diseases: principles and practice, vol II, New York, 1988, Springer-Verlag.

Fields BN, editor: Virology, New York, 1985, Raven Press.

Ginsberg HS: The adenoviruses, New York, 1984, Plenum Publishers.

42

Herpesviruses

Six human herpesviruses exist (Box 42-1). Additionally, herpes B virus of monkeys can also infect humans. The human herpesviruses are ubiquitous, and many individuals become infected at an early age. Unlike most other virus families, the herpesviruses can cause lytic, persistent, latent, and transforming infections. Latent infection, with subsequent recurrent disease, is a trademark of the herpesviruses.

Herpes simplex (HSV) and varicella zoster (VZV) viruses are neurotropic herpesviruses, whereas Epstein-Barr (EBV), cytomegalovirus (CMV), and human herpesvirus 6 (HH6) are lymphotropic herpesviruses. The tropism refers to the cell type in which these viruses establish latent infection (Table 42-1).

STRUCTURE

The herpesviruses are large, enveloped, double-stranded viruses containing deoxyribonucleic acid (DNA). The virion is approximately 150 nm in diameter and has the characteristic morphology shown in Figure 42-1. The DNA core is surrounded by an icosahedral capsid containing 162 capsomeres. This is enclosed by a glycoprotein-containing envelope. The space between the envelope and the capsid is called the tegument and contains viral proteins and enzymes. As enveloped viruses, the herpesviruses are sensitive to acid, solvents, detergents, and drying.

Herpesvirus genomes are linear, double-stranded DNA but differ in size and gene orientation (Figure 42-2). Direct or inverted repeats of DNA bracket regions of the genome, allowing circularization and intragenomic recombination. Inverted repeats in HSV and VZV allow inversion of large portions of DNA and the formation of genomic isomers (Figure 42-3).

The genome of HSV has two sections, the unique long (U_L) and the unique short (U_S), each of which is bracketed by two sets of inverted repeats of DNA. The inverted repeats facilitate the replication of the genome but also allow the U_L and the U_S regions to invert independently of each other to give four different genome configurations or isomers. VZV has only one set of inverted repeats and can form two isomers. EBV and CMV have direct repeats and exist in only one configuration.

REPLICATION

The molecular events of herpesvirus replication are regulated by viral and cellular factors. Viral protein synthesis occurs in three phases: (1) synthesis of early proteins required for subsequent viral protein and

---

Box 42-1   Human Herpesviruses

Human herpesvirus 1	Herpes simplex type 1 (HSV-1)
Human herpesvirus 2	Herpes simples type 2 (HSV-2)
Human herpesvirus 3	Varicella zoster virus (VZV)
Human herpesvirus 4	Epstein-Barr virus (EBV)
Human herpesvirus 5	Cytomegalovirus (CMV)
Human herpesvirus 6	Herpes lymphotropic virus (HH6)

Table 42-1   Properties Distinguishing the Human Herpesviruses

Virus	Primary Target Cell	Means of Spread	Recurrent Disease
NEUROTROPIC HERPESVIRUSES			
HSV-1	Mucoepithelial	Close contact	+
HSV-2	Mucoepithelial	Close contact	+
VZV	Mucoepithelial	Respiratory and close contact	+
LYMPHOTROPIC HERPESVIRUSES			
EBV	B lymphocyte	Close contact	?
CMV	Monocyte, lymphocyte, epithelial cell	Close contact, tissue transplants, blood	+
HH6	B and T lymphocytes	?	?

nucleic acid synthesis, (2) synthesis of specific proteins such as thymidine kinase and DNA polymerase, and (3) synthesis of late structural proteins. The viral and cellular factors determine whether the herpesvirus will establish a lytic, persistent, or latent infection.

HERPES SIMPLEX VIRUS

Herpes simplex virus was the first of the human herpesviruses to be recognized. The name **herpes** is derived from a Greek word meaning to creep. ''Cold sores'' were described in antiquity, and their viral etiology was established in 1919. The term **simplex** may have been derived to distinguish the more benign-appearing lesions caused by this virus from the clinical entity of herpes zoster.

The two types of HSV, 1 and 2, share many common antigens but also have specific glycoproteins that distinguish them. They also differ from each other in many biologic characteristics. The types are at least partially neutralized by antisera produced to the heterologous type.

Structure

The HSV genome is large enough to encode approximately 80 proteins. Among these are DNA-binding proteins that coordinate and promote the transcription and replication of the DNA. The HSV genome also encodes enzymes, including a DNA-dependent DNA polymerase and scavenging enzymes such as deoxyribonuclease, thymidine kinase,

ribonucleotide reductase, as well as other enzymes, among them a protein kinase. Ribonucleotide reductase converts ribonucleotides to deoxyribonucleotides, and thymidine kinase phosphorylates the deoxyribonucleotides to provide substrates for replication of the viral genome. This facilitates the replication of virus in nongrowing cells that lack sufficient substrates for viral DNA synthesis. The substrate specificities of these enzymes and the DNA polymerase differ significantly from their cellular analog and thus represent good targets for the development of antiviral chemotherapy.

HSV virions contain at least five major glycoproteins (gB, gC, gD, gE, and gH) that have *N*-linked and *O*-linked glycans. gB, gD, and gH are important for attachment and entry into the cell. gD is essential for infection and varies little in structure and antigenicity between HSV-1 and HSV-2. Mutants lacking gB can bind to cells but cannot penetrate into the cell to initiate infection. gB is also important in promoting cell-to-cell fusion, a characteristic activity of certain strains of HSV-1. gC is not essential for replication in tissue culture cells but may play a role in the pathogenicity of the virus. gC from HSV-1 (gC1) is a receptor for the C3 component of the complement system, and the gE is, or associates with, an Fc receptor. gC differs extensively between HSV-1 and HSV-2, and antisera to gC is generally type specific.

Replication

HSV can infect most types of human cells and even cells of other species. The virus generally causes lytic infections of fibroblast and epithelial cells and latent infections of neurons (Figure 42-4).

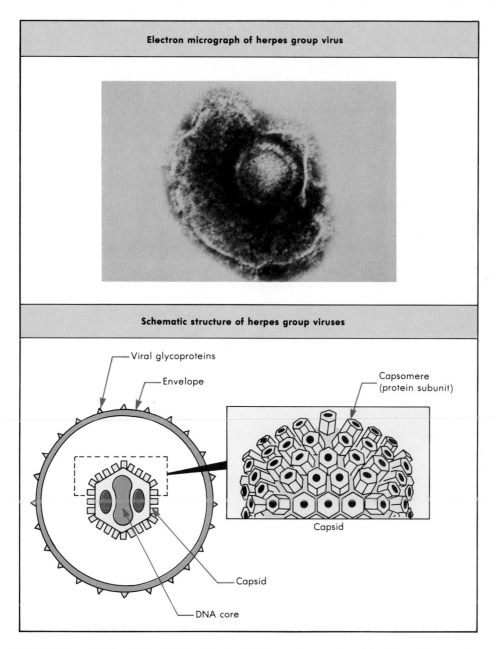

Electron micrograph of herpes group virus

Schematic structure of herpes group viruses

Viral glycoproteins

Envelope

Capsomere
(protein subunit)

Capsid

Capsid

DNA core

FIGURE 42-1   Electron micrograph and general schematic structure of the herpesviruses. The DNA genome of the herpesvirus in the core is surrounded by an icosahedral capsid and a membrane envelope. Glycoproteins are inserted into the envelope.

The broad host range of HSV indicates that the virus receptor must be expressed on many different types of cells. Although receptors for both types of HSV are expressed on similar cells, the two viruses bind to different structures. A candidate for the HSV-1 receptor is heparin sulfate, a proteoglycan found on the outside of many adherent cells.

The major route of HSV penetration is fusion at the cell surface membrane, but virions can also enter cells by endocytosis. On delivery of the nucleocapsid into the cytoplasm, the capsid docks with the nuclear membrane and delivers the genome into the nucleus, where transcription and replication of the genome occur. Virion proteins carried in the tegument are also

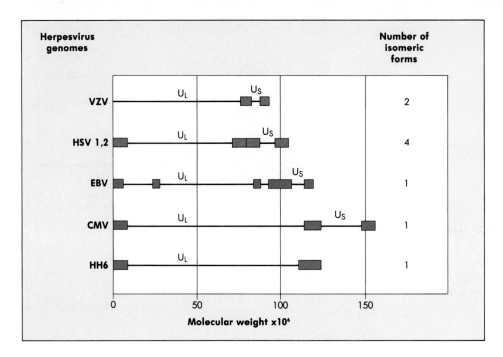

FIGURE 42-2   Herpesvirus genomes. The genomes of the herpesvirus are double stranded DNA. The length and complexity of the genome differ for each of the viruses. Inverted repeats in VZV and HSV allow the genome to recombine with itself to form isomers.

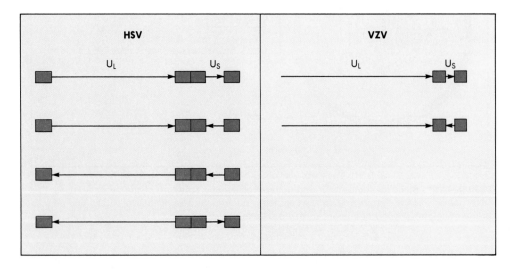

FIGURE 42-3   Isomers of HSV and VZV. Inverted repeats in the HSV and VZV genome promote circularization, inversion, and intragenomic recombination.

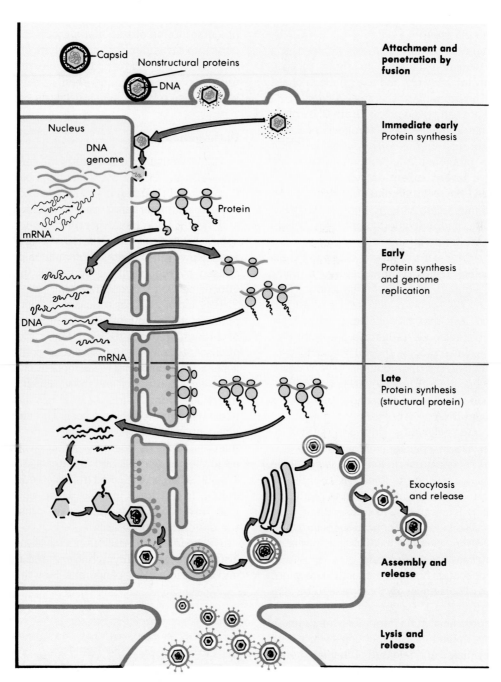

FIGURE 42-4   Replication of herpes simplex virus.

delivered into the cell when fusion of the virus envelope with the plasma membrane occurs. These proteins include a viral encoded protein kinase and a transcriptional regulatory protein that promote the initiation of the infection.

Transcription and protein synthesis proceed in a coordinated, regulated manner in three phases: immediate early (α), early (β), and late (γ). Each phase is required for progression to the next.

The immediate early gene products are DNA-binding proteins that stimulate DNA synthesis and promote the transcription of the early viral genes. Their transcription is promoted by a protein carried in the virion. During a **latent infection,** virus replication does not proceed beyond the immediate early phase. The role of the immediate early proteins in establishing, maintaining, and activating latent infections in neurons is still being determined.

The early proteins consist mostly of enzymes. As catalytic proteins, relatively few copies of these enzymes are required to promote replication. The DNA-dependent DNA polymerase and scavenging enzymes such as deoxyribonuclease, thymidine kinase, and ribonucleotide reductase are early proteins. These enzymes provide the substrates for replication, even in cells not undergoing DNA synthesis. Early proteins also inhibit the production and initiate the degradation of cellular messenger ribonucleic acid (mRNA) and DNA.

Production of the polymerase during the early transcription phase allows replication of the genome. Circular, end-to-end concatameric forms of the genome are made initially. Later in the infection, the DNA is replicated by a rolling-circle mechanism to produce a linear string of genomes in a fashion resembling a roll of toilet paper. Cleavage of the concatamers into individual genomes may occur immediately before or as the DNA is sucked into a procapsid.

Following replication of the genome, the late genes are transcribed. The earliest glycoproteins to be synthesized are gB and gD. Production of gB and gD, which are required for synctia formation, can promote the intracellular spread of the infection even before the virus is assembled and released. Other late genes code for structural proteins, such as the capsid proteins gC and gE, but also include enzymes and other proteins to be packaged in the virion. Many copies of these proteins are required. The capsid proteins are transported to the nucleus, where they are assembled into empty procapsids. and filled with DNA. These

DNA-containing capsids associate and bud from viral glycoprotein-modified portions of the nuclear membrane. The nuclear membrane is contiguous with the endoplasmic reticulum and contains high mannose precursor forms of the viral glycoproteins. The virus buds into the endoplasmic reticulum, is transferred in a vesicle to the Golgi apparatus, where the glycoproteins are processed, and then is exocytosed. The virus is also released when cell lysis occurs.

Pathogenesis

HSV-1 is usually associated with infections above the waist and HSV-2 with infections below the waist (Figure 42-5). Differences in receptor specificity, growth characteristics, and antigenicity of HSV-1 and HSV-2 have been identified, and these explain some distinctions between HSV-1 and HSV-2 disease. However, the site of infection is predominantly caused by the means of spread of the virus. Both viruses cause recurrent infection.

HSV usually causes a localized infection. The virus enters the body by infection of mucosal membranes or through breaks in the skin. Replication of the virus at that site may be inapparent or produce vesicular lesions. The virus undergoes replication in the cells at the base of the lesion, and the vesicular fluid contains infectious virions (Figure 42-6). The lesion generally heals without producing a scar. The virus spreads to adjacent cells and to the innervating neuron. After infection of the neuron, the nucleocapsid is transported to the cell nucleus and initiates a latent infection. The virus can be activated from the neuron by various stimuli (e.g., stress, trauma, fever, or sunlight) and then travels back down the nerve, causing lesions at the dermatome (Box 42-2).

HSV can cause lytic infections of most cells, persistent infections of lymphocytes and macrophages, and latent infections of neurons. HSV replication gen-

Box 42-2	Stresses That Can Trigger Herpes Recurrences
Emotional stress	Upset stomach
Trauma	Fever
Cold	Menstrual cycle
Sunlight	Immune suppression

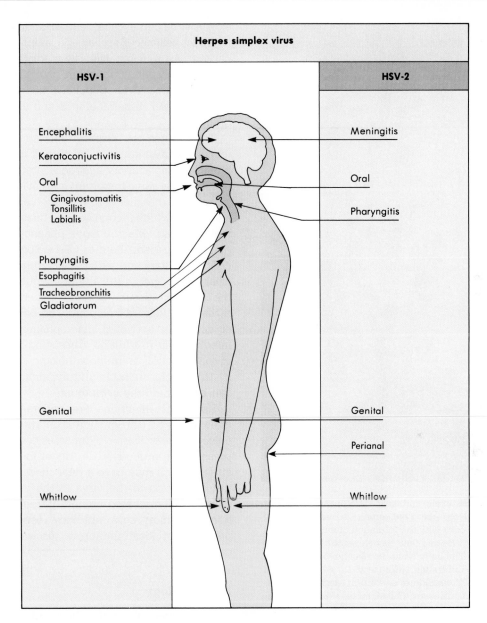

FIGURE 42-5   Disease syndromes of herpes simplex viruses. HSV-1 and HSV-2 can infect the same tissues and cause similar diseases but have a predilection for the sites and diseases indicated.

erally leads to cytolysis as a result of virus-induced inhibition of cellular macromolecular synthesis, degradation of host cell DNA, membrane permeation, cytoskeletal disruption, and senescence of the cell. Changes in nuclear structure and margination of the chromatin occur, along with the production of Cowdry's type A acidophilic intranuclear inclusion bod-

ies. Many strains of HSV also initiate syncytia formation. In tissue culture HSV rapidly infects and kills cells. The cells become rounder and fall off the monolayer, producing plaques.

Latent infection occurs in neurons and results in no detectable damage. These cells may only contain one copy of the viral genome, and only some of the

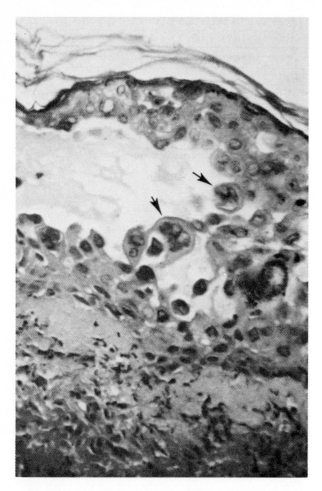

FIGURE 42-6   Hematoxylin and eosin stain of early vesicle of herpes simplex. Note the "ballooning degeneration" of infected epithelial cells with the formation of multinucleated giant cells *(arrows)*. Edema fluid has elevated the overlying stratum corneum. The underlying dermis shows edema and mononuclear cell infiltration. (Courtesy Dr PJ Barr; from Braude AI, Davis CE, and Fierer J, editors: Medical microbiology and infectious diseases, Philadelphia, 1986, WB Saunders Co.)

immediate early proteins are synthesized in these cells. Activation of the genome results in production of a limited number of virions but does not kill the cell.

Immunology   Control and resolution of the HSV infection requires both humoral and cellular immunity. Antibody directed against the glycoproteins of the virus neutralize extracellular virus, limiting its spread. Antibody to HSV-1 also protects against future challenge by other strains of HSV-1 and to some extent against infections by HSV-2, and vice versa. Howev-

er, the virus can escape neutralization and clearance by direct cell-to-cell spread and latent infection of the neuron. As a result the cell-mediated response is essential for controlling and resolving HSV infections. In the absence of functional cell-mediated immunity, HSV infection disseminates to the vital organs and brain.

The course of disease and the immunologic responses to an initial, vs. recurrent, infection are different. During primary infection, interferon and natural killer cells serve important roles in limiting the progression of the infection. Interferon activates natural killer cells, which recognize HSV-infected targets and lyse the cells. Macrophages phagocytize viral antigens and present them to CD4 helper T cells and B cells to activate antigen-specific immunity. Mononuclear cells travel to the site of infection and infiltrate the infected tissue. Delayed hypersensitivity and cytotoxic T killer cell responses assist the natural killer cells and activated macrophages in killing the infected cells. In addition to assisting in the resolution of the infection, immunopathologic changes caused by the cellular immune and inflammatory responses can exacerbate the symptoms.

Recurrent infections are generally less severe, more localized, and of shorter duration than primary episodes. Reactivation of the virus occurs despite the presence of antibody. The trigger for reactivation is stress, which may have a dual effect: (1) to promote the replication of the virus in the nerve and (2) to depress cell-mediated immunity transiently. Activation of memory cells, antibody responses, and the presence of local immunity resolve the infection quickly.

Epidemiology

The epidemiology of HSV is best understood by examining rates of seropositivity to the two types of virus. Seroprevalence studies are much better markers of past HSV infection than is a clinical history, since most infections with either type 1 or type 2 virus are asymptomatic.

HSV-1   In underdeveloped parts of the world, the prevalence of antibody to HSV-1 is greater than 90% by 2 years of age, which reflects crowded living conditions. In a country such as the United States, type 1 antibody increases slowly so that by entry in college, approximately 35% of students are positive. Acquisition continues, and 4 years later 45% of the same stu-

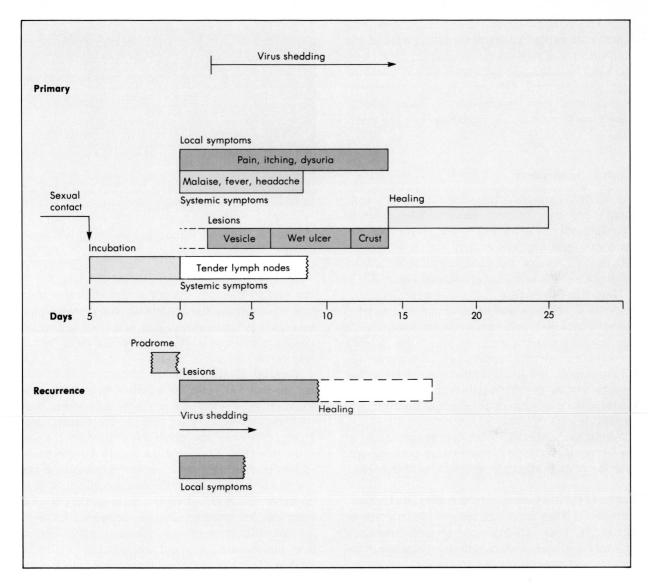

FIGURE 42-7   Clinical course of genital herpes infection. The time course and symptoms of primary and recurrent genital infection with HSV-2 are compared. (Data from Corey L et al: Ann Intern Med 98:958-973, 1983.)

dents exhibit antibody. The virus is present in oropharyngeal secretions, spreads by direct person-to-person contact, and presumably requires salivary exchange or at least salivary contact with a susceptible body surface. The most common spread is mouth to mouth; however, mouth-to-skin contact, resulting in herpetic whitlow or type 1 genital herpes infection, for example, does occur.

HSV-2 Type 2 HSV antibody is much less prevalent than type 1 antibody. In the United States

approximately 1% of freshman college students are positive, but by their senior year 7% may be infected. This rate of increase may continue, depending on the sexual activity of the study group. HSV-2 is transmitted by sexual intercourse or at least by the transfer of genital secretions from one individual to another. Depending on sexual practices, HSV-2 may infect the genitalia, anorectal tissues, or oropharynx. The virus may then be associated with symptomatic or asymptomatic primary infection or recurrences. Asymptomatic recurrences are especially likely in females, who may shed the virus in cervical secretions without any clinical illness. Excretion of HSV-2 in the cervix may occur during pregnancy and lead to vertical transmission during vaginal delivery. Vertical transmission may also result from maternal viremia during primary HSV-2 infection or by an ascending, in utero infection.

Clinical Syndromes

Both HSV-1 and HSV-2 are common human pathogens. A primary infection in either children or adults is usually mild or unnoticeable but may be severe. Recurrent episodes may or may not be associated with lesions, but in either case they provide a source for infection of susceptible individuals (Figure 42-7).

Herpetic Stomatitis Primary herpetic gingivostomatitis in toddlers and children almost always is caused by HSV-1, whereas young adults may be infected with either type 1 or type 2. HSV-2 may result from oral-genital contact. Lesions begin as vesicles but rapidly become ulcerated, whitish areas that may be widely distributed throughout the mouth, involving the palate, pharynx, gingivae, buccal mucosa, and tongue.

Adults may experience recurrent mucocutaneous herpes simplex infection (cold sores) even though they never had clinically apparent primary infection.

Severe HSV stomatitis infection may also occur in patients who are immunosuppressed by disease or therapy. In these patients recurrent mucocutaneous HSV infection may resemble primary gingivostomatitis.

Herpetic Keratitis This disorder may be either superficial or deep. In deep keratitis less lacrimation and irritation are present, but the patient has a greater disturbance in visual acuity. Herpetic keratitis is almost always unilateral and may result in permanent corneal damage and visual impairment.

Herpetic Whitlow This is an infection of the finger. Herpetic whitlow usually occurs in nurses or physicians who attend patients with HSV infections, in thumb-sucking children (Figure 42-8), and in individuals who have genital HSV infection.

Eczema Herpeticum Children with active eczema or a history of this disorder may develop eczema herpeticum. The disease usually involves known eczematous areas, with the sudden onset of numer-

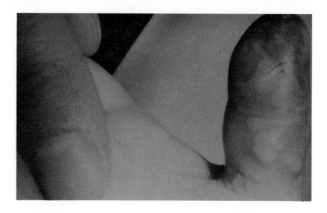

FIGURE 42-8   Herpetic whitlow in 2-year-old child's thumb.

ous vesicular lesions. Bacterial superinfection often occurs, obscuring the underlying viral disease. Viral vesicles may emerge over days, and visceral dissemination to the adrenal glands, liver, and other organs may occur.

Genital Herpes The manifestations of genital herpes partly depend on whether an individual is experiencing a first attack or a recurrent episode. Furthermore, preexisting antibody to the heterologous type of HSV may ameliorate the symptoms in a primary infection. The incubation period for primary HSV infection in an individual with no previous exposure to HSV-1 or HSV-2 is approximately 2 to 20 days (average of 6 days). However, most primary genital infections are asymptomatic. No difference exists in the manifestations of true primary genital herpes infection caused by type 1 (responsible for 10% of genital infections) vs. type 2 virus if the patient does not have preexisting antibody. On the other hand, the course is considerably milder if an individual is experiencing an initial HSV-2 infection in the presence of heterologous HSV-1 antibody.

In primary infection lesions begin as small, multiple papules or vesicles that coalesce into larger pustular or ulcerative lesions after 5 days. The lesions then begin to develop a dry crust 10 to 15 days after their first appearance. The lesions and clinical illness in a patient with primary HSV infection usually resolve within 2 to 4 weeks, averaging 16 days in men and 20 days in women.

Lesions vary in number, but they are often bilateral and usually painful. In males lesions typically are found on the glans or shaft of the penis (Figure 42-9)

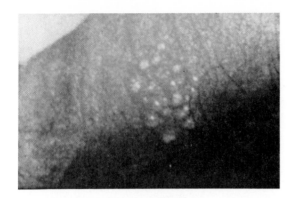

FIGURE 42-9   Primary HSV infection of penis.

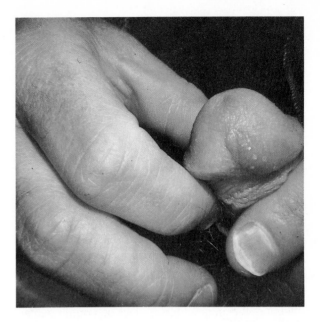

FIGURE 42-10   Recurrent HSV infection of penis.

and occasionally in the urethra. In women itching and mucoid vaginal discharge frequently occur. Lesions may be seen on the vulva, vagina, cervix, perineal area, or inner thighs. In both sexes a primary infection may be accompanied by fever, malaise, myalgia, and inguinal adenitis.

Autonomic dysfunction may develop in patients with primary genital HSV infection associated with sacral anesthesia, impotence, urinary retention, and constipation. These sensory and tone disturbances may be present for 4 to 8 weeks.

Episodes of recurrent genital HSV infection are shorter and less severe than initial episodes. In approximately half the patients recurrences may be preceded by a characteristic prodrome of burning or tingling in the area in which the lesions will erupt. Lesions last an average of 7 to 8 days. They begin as several small vesicles that are unilateral and often coalesce to form one or two ulcers that later crust over. Pain is mild and lasts only 4 days in men and 6 days in women. The vesicles are somewhat less severe than those of a primary infection (Figure 42-10).

Episodes of recurrence may be as frequent as every 2 to 3 weeks or may occur just two or three times per year. Approximately 50% of patients infected with HSV-2 never have recurrence and may never be aware of the primary infection. Unfortunately, any infected patient, including the patient with no history of lesions, may shed virus asymptomatically. These individuals may be important vectors for the spread of this virus. Recurrences are presumed to result from reactivation of latent virus in a sacral ganglion.

On reactivation the virus migrates along the nerve and usually reappears in the same skin or mucosal site with each recurrence. Interestingly, almost all genital recurrences are caused by HSV-2.

Infection in the Newborn HSV infection in the newborn (Figure 42-11) is a devastating disease and most often is caused by HSV-2. It may be acquired in utero but more frequently is contracted during passage through the genital canal when the mother is shedding herpesvirus at the time of delivery.

The life-threatening aspects of this HSV infection result from disseminated involvement of viscera, particularly liver and lung, as well as the central nervous system, which often results in herpes encephalitis.

Proctitis HSV is the leading cause of nongonococcal proctitis in male homosexuals. It is associated with fever, inguinal adenopathy, anorectal pain, tenesmus, constipation, rectal discharge, and blood in the feces. The pain is more severe than in proctitis caused by other agents involved in the ''gay bowel syndrome.''

Perianal or anal lesions, vesicular or pustular rectal lesions, and diffuse ulceration of the distal rectum seen on sigmoidoscopy are characteristic, but lesions are not seen above the lower rectum.

Meningitis Viral or aseptic meningitis caused by HSV is most often a complication of genital HSV-2

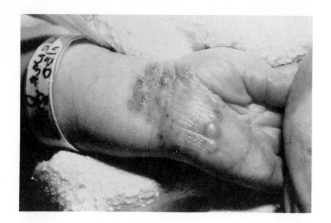

FIGURE 42-11   Vesicular HSV lesion of 10-day-old infant infected in utero.

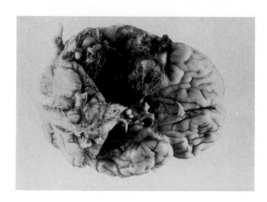

FIGURE 42-12   Right temporal lobe necrosis caused by HSV-1 encephalitis.

infection. It occurs within 10 days of a primary infection that may have been asymptomatic. Patients develop extreme nuchal rigidity as well as other signs of meningitis, such as headache, photophobia, and nausea. Symptoms resolve promptly.

Herpes Encephalitis Aseptic meningitis caused by HSV-2 must not be confused with HSV encephalitis. The latter is an acute, febrile disease that occurs at all ages, is frequently associated with seizures, and results from HSV-1 (Figure 42-12). However, the neonate with herpes encephalitis is more likely to be infected with HSV-2.

Herpes encephalitis is the most common fatal, sporadic encephalitis and is lethal in 50% of patients. It may be part of a primary herpetic infection or may arise in patients with a history of recurrent mucocutaneous herpes simplex infection (cold sores). The virus is thought to migrate via the olfactory nerve to the temporal lobe of the brain, its site of predilection. During viral replication, brain cells are destroyed, and a progressively enlarging necrotic mass may develop. This may give rise to erythrocytes in the cerebrospinal fluid, seizures, focal neurologic abnormalities, and the other expected abnormalities of viral encephalitis.

Diagnosis (Table 42-2)

Cytology/Histology Cytologic examination is a relatively insensitive technique, and intranuclear inclusions are often not detected in tissues that are positive for HSV virus on culture. Furthermore, cyto-

logic changes in HSV infection cannot be distinguished from VZV infection. The **Tzanck test** has been used to examine cells scraped from the base of herpeslike lesions. The cells are smeared onto a slide, fixed, and stained with Wright or Giesma preparations. Giant cells (syncytia), "ballooning" cytoplasm, and Cowdry's type A intranuclear inclusions may be found in cells infected with either HSV or VZV.

Fluorescent Antibody Techniques A rapid 1½ to 2 hours), definitive determination of HSV can be made by demonstrating virus antigen in cells with immunofluorescence or immunoperoxidase techniques. Although more rapid than culture tests, cytologic and antigen detection techniques are, respectively, only 50% and 70% to 80% as sensitive as viral isolation.

Culture Recovery of HSV in culture is related to the stage of the lesion; the highest rates of recovery are from, in order, vesicular lesions, pustular lesions, ulcers, and crusted lesions. Specimens should be collected by aspiration of the lesions or with a cotton swab and then inoculated directly into cell cultures. Delays in inoculation of the cell cultures will adversely affect recovery of HSV. Following incubation of the cell cultures, HSV produces a cytopathic effect (CPE) in heteroploid cells such as HeLa and HEP-2, human embryonic fibroblasts, or rabbit kidney cells. With specimens containing high concentrations of HSV, the CPE may be seen in tissue cultures within 24 hours, but more time is required with lower virus inocula. In general, the mean time for detection of CPE is 1 to 3 days. HSV CPE begins with cytoplasmic

Table 42-2   Laboratory Diagnosis of HSV Infections

Approach	Test/Comment
Direct microscopic examination of cells from base of lesion	Tzanck test
Detection of HSV by cell culture	Most cell cultures support multiplication of HSV
Assay of vesicular fluid and cells for HSV antigen	Enzyme imunoassay; immunofluorescent stain
Detection of antibody to HSV	Immune status testing for patients undergoing immunosuppression HSV-specific IgM tests for the diagnosis of primary infections Detection of primary infection by seroconversion of HSV-specific IgG antibody

granulation, after which the cells become enlarged and appear ballooned. The enlarged cells become rounded and take on a refractile appearance. Less often the virus induces fusion of neighboring cells and the nuclei of the cells aggregate, giving rise to multinucleated giant cells. Although not absolute, the CPE induced by HSV-2 tends to be focal, whereas HSV-1 produces CPE scattered throughout the monolayer. However, in critical situations (e.g., brain or other visceral isolates and isolates from newborns) the isolate should be confirmed as HSV immunologically and the type definitively identified.

Isolates of HSV-1 may be distinguished from isolates of HSV-2 by biochemical, biologic, or immunologic methods. The restriction endonuclease cleavage patterns of the DNA of HSV-1 and HSV-2 are unique and allow unequivocal typing of isolates. In addition, minor variations in patterns between isolates of either type of virus exist and may be useful in epidemiologic studies.

Recovery of HSV in culture indicates active infection, but a positive viral culture from the throat must be interpreted with some caution because HSV may reactivate during any febrile episode. Therefore the pertinence of virus isolation with respect to disease rests primarily on the clinical circumstances associated with the particular patient.

Serologic Procedures Seroconversion in paired sera may provide evidence of a primary HSV infection. However, a significant rise in antibody titers does not usually accompany recurrent disease.

Accordingly, serologic tests should not be used in an attempt to diagnose recurrent HSV infection.

Treatment

Many therapies have been suggested for HSV infections. In virtually all cases, subsequent double-blind, randomized trails revealed the treatments to be ineffective. Acyclovir (ACV) is the exception and is a most effective anti-HSV drug (Figure 42-13).

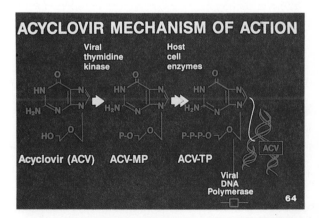

FIGURE 42-13 Acyclovir's mechanism of action. Acyclovir inhibits viral DNA polymerase, but not until it is phosphorylated by viral thymidine kinase to acyclovir monophosphate (ACV-MP). Host cell enzymes further the phosphorylation to acyclovir triphosphate (ACV-TP), and the compound then binds viral DNA polymerase and may also be incorporated into viral DNA, terminating further synthesis.

Both oral and topical ACV are effective in treating first episodes of genital herpes but are only marginally useful in recurrent HSV infection. Parenteral ACV is indicated for severe HSV syndromes such as neonatal disease, encephalitis, and extensive disease in the immunosuppressed patient. HSV resistance to ASV has now been noted. Resistant viruses are defective in thymidine kinase, thereby reducing conversion of the drug to its active form. Fortunately, resistant strains appear to be less virulent.

Adenine arabinoside (Ara A) is also approved for treatment of HSV infection. Ara A reduces mortality in patients with HSV encephalitis from 70% to 40% but cures only 20% of patients and must be given before coma develops.

Iododeoxyuridine is used for treatment of herpetic keratitis. It must be instilled topically every 1 or 2 hours or used as a gel. The drug appears to shorten the course of herpetic keratitis but does not prevent the characteristic recurrences.

Prevention and Control

HSV-1 is transmitted most often from an active mucocutaneous lesion, so avoidance of direct contact with these will reduce infection. Unfortunately, this virus may also be present in the oropharyngeal secretions of an asymptomatic individual, and thus the virus can be transmitted unknowingly.

Physicians, nurses, dentists, and dental hygienists must be especially careful when handling potentially infected tissue or fluids. Gloves can prevent acquisition of infections of the fingers (herpetic whitlow). Individuals with recurrent herpetic whitlow disease are very contagious and can spread infection to patients.

Patients who have a history of genital HSV infection must be instructed to refrain from sexual intercourse when prodromal symptoms or lesions occur. They can resume sexual intercourse only after lesions are completely reepithelialized because virus can be isolated from lesions even when crusted. Although asymptomatic excretors may transmit virus, it is suspected that most new cases of genital HSV infection result from sexual contact with individuals who have active lesions. Use of a condom prevents transmission of the virus in laboratory studies, but these evaluations do not duplicate natural situations. Therefore, although condoms are useful and undoubtedly better than nothing, they may not be fully protective.

If a pregnant woman has active genital HSV at term and the membranes are not ruptured, cesarean section provides a means of preventing contact of the infant with the virus-infected lesions. Unfortunately, as mentioned previously, the virus may be asymptomatically present in genital secretions, and a vaginally delivered infant may be exposed to virus during the delivery.

Vaccines have been proposed for genital HSV for many years. Clearly, vaccination of individuals with recurrent disease would not eliminate the virus but could ameliorate recurrences. Vaccination of an individual before the first contact with HSV-2 might prevent acquisition of "wild" virus and the establishment of latency. Killed vaccines have been available for years but have not been evaluated rigorously. These and nonviable subunit vaccines may not induce the durable immunity necessary to prevent primary HSV-2 infection.

VARICELLA ZOSTER VIRUS

VZV causes chickenpox (varicella) and with recurrence causes herpes zoster, or shingles. VZV shares many characteristics with HSV, including (1) the ability to establish latent infection of the ganglia and recurrent disease at the innervated dermatome, (2) the importance of cell-mediated immunity in controlling and preventing serious disease, (3) the characteristic blisterlike lesions of the disease, and (4) the production of a thymidine kinase. Unlike HSV, the predominant means of spreading VZV is by the respiratory route. After local replication in the respiratory tract, viremia occurs, leading to skin lesions over the entire body. The genome of VZV is smaller than that of HSV, but each gene in VZV is represented by a counterpart in HSV. VZV lacks the potential to transform cells.

Viral Structure

VZV is a member of the herpesvirus family, 150 to 200 nm in diameter, and has the typical morphology of these agents. Although morphology of VZV is similar to that of HSV, the G/C (guanine/cytosine) ratios differ considerably (VZV, 46%; HSV, 67% to 69%). Restriction enzyme profiles of the DNA from large numbers of VZV isolates from varying geographic locations have indicated only minor differences in patterns. Importantly, these studies have concluded on a molecular basis that no detectable differences exist between viruses recovered from patients with varicella or herpes zoster.

Pathogenesis

Primary varicella infection begins in the mucosa of the respiratory tract and then progresses via the bloodstream and lymphatics to cells of the reticuloendothelial system (Figures 42-14 and 42-15). Viral infection proceeds with involvement of a dermal vesiculopustular rash that develops in successive crops (Box 42-3).

As with other herpesviruses, the virus becomes latent following primary infection. The site of latency appears to be dorsal root or cranial nerve ganglia. Reactivation occurs in aged persons or in patients with impaired cellular immunity, suggesting that latency is maintained by intact T lymphocyte–mediated responses. Humoral factors (e.g., antibody) also possibly contribute to maintenance of latency. On reactivation the virus is thought to migrate along neural pathways to the skin, where a vesicular rash

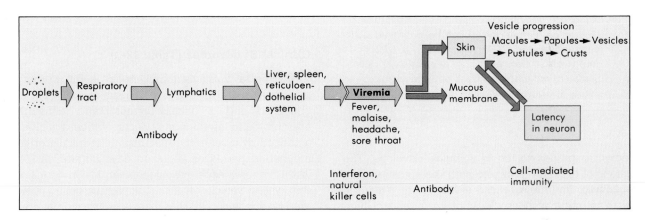

FIGURE 42-14   Mechanism of spread of VZV within the body. VZV initially infects the respiratory tract and is spread by the reticuloendothelial system and by viremia to other parts of the body. The spread can be blocked by the immune response, as indicated.

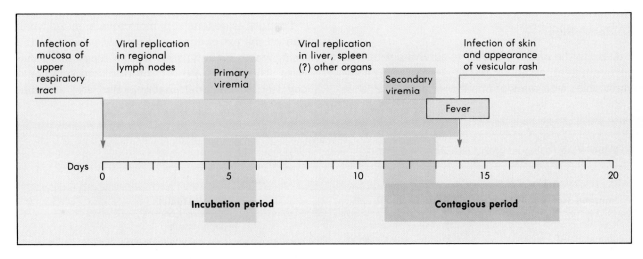

FIGURE 42-15   Time course of varicella (chickenpox). The course of disease in young children, as presented in this figure, is generally shorter and less severe than in adults.

Box 42-3   Pathogenics Mechanisms of VZV

1. Initial replication in respiratory tract occurs.
2. Broad tissue tropism includes most epithelial and fibroblast cells.
3. VZV can form syncytia and spread directly from cell to cell.
4. Latent infection of neurons occurs, usually dorsal root and cranial nerve ganglia.
5. Systemic spread of virus through viremia to the skin causes lesions in successive crops.
6. VZV can escape antibody clearance, and cell-mediated immune response is essential to control infection. Desseminated life-threatening disease can occur in immunocompromised individuals. Recurrence (herpes zoster) can be prompted by immunosuppression.
7. Herpes zoster may result from depression of cell-mediated immunity, although other mechanisms of viral activation may come into play.

---

among susceptible household contacts. The disease is spread principally by droplet transmission through the respiratory tract, although skin vesicles also contain virus particles. Primary VZV infection occurs throughout childhood, and more than 90% of adults in developed countries such as the United States have VZV antibody. Herpes zoster results from reactivation of a patient's latent endogenous virus; therefore outbreaks of zoster do not occur. Approximately 10% to 20% of the population infected with VZV develops zoster; the incidence parallels advancing age and presumably results from declining cellular immunity in older populations. Zoster lesions do contain viable virus and therefore may be a source of varicella infection in a nonimmune individual.

Clinical Syndromes (Table 42-3)

Varicella Varicella (chickenpox) is the primary infection caused by VZV; it is usually a disease of childhood and is usually symptomatic, although asymptomatic infection may occur. Varicella is characterized by fever and a maculopapular rash after an incubation period of about 14 days (Figure 42-16). Within hours, each maculopapular lesion forms a thin-walled vesicle on a maculopapular base ("dew drop on a rose petal") that measures approximately 2 to 4 mm in diameter and is the hallmark of varicella. Within 12 hours the vesicle becomes pustular and begins to crust, and then scabbed lesions appear. Successive crops of lesions appear for 3 to 5 days, and at any given time all stages of skin lesions can be observed.

The rash is generalized, more severe on the trunk than on the extremities, and notably present on the scalp, which distinguishes it from many other diseases. The lesions are very pruritic, and scratching may lead to bacterial superinfection and scarring.

known as herpes zoster, or shingles, develops. The lesions of the rash terminate with the appearance of leukocytes and the synthesis of interferon. When the latter immune responses do not occur and cellular immune responses are impaired, a reactivation with resulting dissemination of the virus may occur via the bloodstream. In such patients the localized dermatomal cutaneous herpes zoster infection is followed by a generalized rash resembling varicella.

Epidemiology

Despite the inability of VZV to survive deleterious environmental conditions, the virus is extremely communicable, with rates of infection greater than 90%

Table 42-3   Clinical Manifestations of VZV

Immune Status of Patient	Infection	
	Primary	Reactivated
Normal	Varicella	Herpes zoster
Compromised	Progressive varicella with complications, such as hemorrhagic rash caused by thrombocytopenia	Disseminated herpes zoster with cutaneous and/or visceral spread of virus

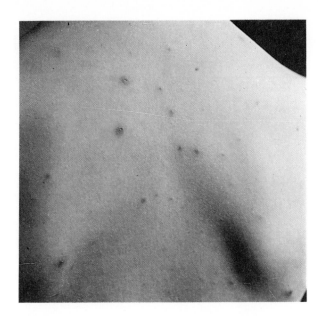

FIGURE 42-16   Characteristic skin rash of varicella.

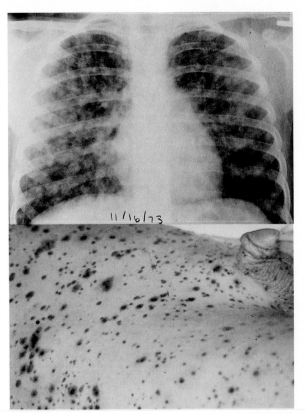

FIGURE 42-17   A, Chest x-ray film showing extensive bilateral nodular infiltrates. **B,** This child had Wiskott-Aldrich syndrome, a form of hereditary immunodeficiency affecting both cellular and humoral immunity and causing thrombocytopenia. Varicella contracted from a playmate proved fatal. In addition to the hemorrhagic skin lesions seen here, the virus affected the brain, lungs, and liver. The child died of a combination of varicella encephalitis and pneumonia.

Mucous membrane lesions in the mouth, conjunctivae, and vagina typically occur. Thrombocytopenia may also develop occasionally; in this case the rash may be hemorrhagic in nature.

Primary infection of adults is usually more severe than for children. Interstitial pneumonia may occur in 20% to 30% of adult patients and may be fatal. Extremely severe, disseminated infection occurs in immunocompromised individuals (Figure 42-17).

Herpes Zoster Herpes zoster occurs sporadically among those who already have experienced varicella infection. Zoster is usually unilateral and occurs in one or more adjacent dermatomes (Figure 42-18). The appearance of chickenpox-like lesions is usually preceded by severe pain in the area innervated by the nerve. The rash, which is usually limited to a dermatome, appears as small, closely spaced maculopapular lesions on an erythematous base, in contrast to the more diffuse pattern of vesicles characteristic of varicella. The lesions of zoster vesiculate rapidly and often coalesce. Following herpes zoster infection, patients may experience a chronic pain syndrome referred to as **postherpetic neuralgia.** This can persist for months to years. It occurs in up to 30% of patients who develop zoster after age 65 years. The neuralgia results from chronic nerve irritation by VZV.

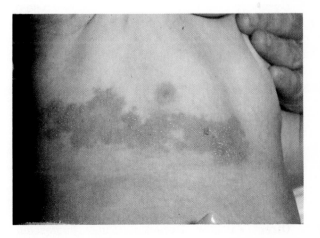

FIGURE 42-18   Herpes zoster in a thoracic dermatome.

Table 42-4 Laboratory Tests for Diagnosis of VZV Infections

Method	Sensitivity	Specificity	Comments
Cytology	80%	40%	Does not distinguish lesions caused by VZV from those resulting from HSV
Culture	50%	100%	Very sensitive for varicella but not for herpes zoster, particularly lesions lasting 5 or more days
Immunofluorescence	95%	95%	Procedure of choice for herpes zoster
Serology	90%	90%	May take several weeks for antibody to develop; antibody rise (i.e., reactivation) possible in the absence of clinical herpes zoster

VZV in Immunocompromised Patients VZV generally follows a benign course in immunologically normal individuals but may be complicated by bacterial superinfection of affected dermal areas, encephalitis, and pneumonia. In contrast, patients with immunodeficiencies, either congenital or caused by drugs or disease, risk serious, progressive disease with VZV. Defects in cell-mediated immunity especially place the patient at risk for dissemination to the lungs, the brain, and the liver, which may be fatal. This may occur with primary exposure to varicella or as disseminated herpes zoster.

Laboratory Diagnosis (Table 42-4)

Cytology VZV infection induces Cowdry's type A intranuclear inclusions in infected cells, which are identical to those caused by HSV. These cells may be seen in skin lesions, respiratory specimens, or organ biopsies. Syncytia may also be seen in Tzanck smears of scrapings of a vesicle's base.

Culture Dermal swabs of fresh vesicles are the most productive source for recovering the virus in cell cultures. VZV has rarely been recovered from specimens of the respiratory tract. Cultures for VZV are performed in the same manner as for HSV but are either negative when lesions have been present longer than 5 days or are slow to turn positive (average of 5 to 6 days for isolation in tissue culture).

Direct Antigen Detection Direct fluorescent antibody (FA) examination of skin lesions is a more sensitive diagnostic technique for VZV infection than culture because the virus is labile during transportation to the laboratory and replicates poorly in most cells in vitro. Once skin lesions are crusted (5 or more days after onset), cultures are usually negative, but antigen can still be detected by FA or other immunologic tests.

Serology Serologic tests to detect antibodies to VZV are used in two ways: screening for immunity to VZV and documenting active VZV infection. Sensitive tests, such as immunoflourescence or enzyme-linked immunosorbent assay (ELISA), must be used to detect low levels of antibody in immune patients. Active VZV infection is documented by seroconversion during primary VZV infection. Virtually any antibody assay can be used for this purpose. Apparent seroconversion or significant antibody rise can also be detected in individuals experiencing herpes zoster or reactivation infection, although it may require several weeks for an antibody rise to occur. The presence of IgM antibody provides an indication of primary VZV infection; however, IgM has also been demonstrated in more than 50% of patients with clinical manifestations of herpes zoster.

Treatment, Prevention, and Control

No treatment is indicated for varicella in children. ACV is being investigated as a possible means for preventing postherpetic neuralgia in immunocompetent adult patients with herpes zoster, but results so

Table 42-5 Treatment and Prevention of VZV

Group	Drug
Antiviral drugs	Acyclovir (ACV)
Passive antibody	Varicella zoster immuno-globulin (VZIG or ZIG)
	Zoster immune plasma (ZIP)
Live vaccine	Attenuated Oka strain VZV*

*In clinical trials.

Box 42-4 Pathogenic Mechanisms of EBV

1. Virus establishes productive infection in epithelial cells of nasopharynx.
2. Virus is mitogenic for B cells, which induces IgM production.
3. Infected B cells activate T cells.
4. Latent infection of B cells occurs with potential for recurrence.
5. Virus immortalizes and induces proliferation of B cells in the absence of T-cell suppression.

far are not encouraging. Varicella in adults may be treated with ACV (Table 42-5). Primary or reactivation infection in the immunocompromised patient should be treated with intravenous ACV.

Since VZV spreads by the respiratory route, its transmission may be reduced by isolation of infected individuals and avoidance of contact with susceptible persons.

VZV disease may be prevented or ameliorated in patients susceptible to severe disease by administration of varicella zoster immunoglobulin (VZIG), which is prepared by pooling plasma from seropositive individuals. VZIG is ineffective in the therapy of patients with either varicella or herpes zoster.

A live attenuated vaccine for VZV has been developed and induces protective or ameliorating antibody. The vaccine is effective as a prophylactic treatment even after exposure to VZV and promotes protection in immunodeficient children. However, questions regarding latency and the duration of immunity have delayed its approval.

EPSTEIN-BARR VIRUS

EBV was first discovered by electron microscopic observation of the characteristic herpes virion morphology in biopsies of an African Burkitt's lymphoma. Discovery of its association with infectious mononucleosis was accidental when serum collected from a laboratory technician during her convalescence from infectious mononucleosis demonstrated antibody to African Burkitt's lymphoma biopsies. This finding was later confirmed in a large serologic study performed on college students.

EBV has been established as the major cause of infectious mononucleosis worldwide. Burkitt's lymphoma in equatorial Africa and New Guinea and nasopharyngeal carcinoma in southeast China also appear to be caused by this virus. EBV has been associated with several other clinical entities, including B-cell lymphomas in patients with acquired or congenital immunodeficiencies. Considerable controversy surrounds the role of this virus as a cause of chronic disease, especially a chronic fatigue syndrome.

Viral Structure

The morphologic structure of EBV cannot be distinguished from other members of the group.

Pathogenesis (Box 42-4)

EBV has a very limited host range and tissue tropism. The virus infects only B lymphocytes of humans and New World monkeys in vitro. Epithelial cells of the oropharynx and nasopharynx may also be infected during the course of the disease. The receptor for EBV has been identified as the receptor for the C3d component of the complement system. This receptor is mainly present on B cells but is also found on epithelial cells of the oropharynx and nasopharynx. The specific attachment protein on the virus is a surface glycoprotein.

The virus establishes a productive infection in the epithelial cells (Figure 42-19), leading to the production of EBV proteins, including the early antigens (EA), viral capsid antigen (VCA), and membrane antigen (MA) glycoproteins (Table 42-6). The EA exists as the EA-R form, which is restricted to the cytoplasm

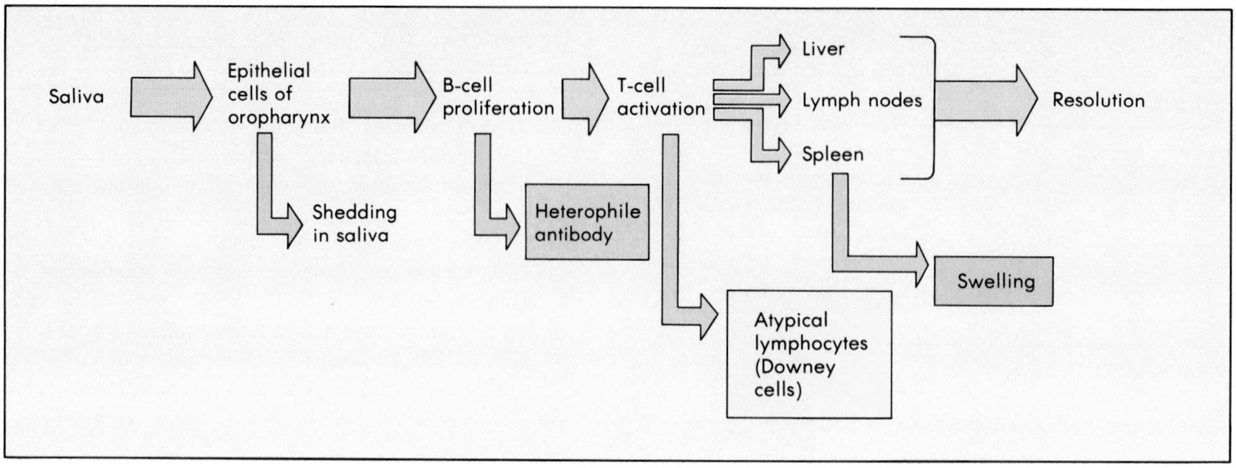

FIGURE 42-19 Pathogenesis of EBV. EBV is acquired by close contact between persons through saliva and infects the B lymphocytes. The resolution of the EBV infection and many of the symptoms of infectious mononucleosis result from activation of T lymphocytes in response to the infection.

Table 42-6 Cellular Antigens Associated With EBV-Infected Cells

Antigen				
Name	**Abbreviation**	**Cellular Location**	**Biologic Association**	**Clinical Association**
EBV nuclear antigen	EBNA	Nuclear antigen	Nonstructural antigen; first antigen to appear; seen in all infected and transformed cells; binds to cell DNA	Anti-EBNA develops late in infection
Early antigen	EA-R	Restricted to cytoplasm	EA-R appears before EA-D; first sign that infected cell has entered lytic cycle	Anti–EA-R seen in Burkitt's lymphoma; anti–EA-D seen in infectious mononucleosis
	EA-D	Diffuse in cytoplasm and nucleus		
Viral capsid antigen	VCA	Cytoplasmic antigen	Late antigen; found in virus producer cells	Anti–VCA IgM transient; anti–VCA IgG persistent
Lymphocyte-determined membrane antigen	LYMDA		Not found on Burkitt's lymphoma cells; found on cells infected in vitro and transformed; found on nonproducer cells	

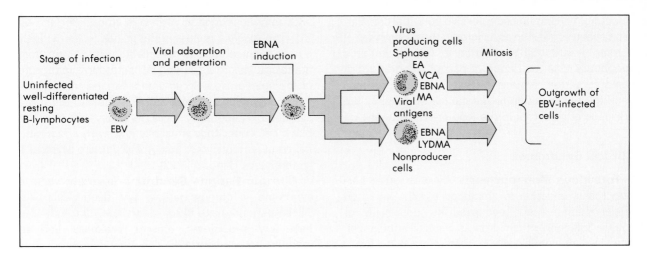

FIGURE 42-20   Progression of EBV Infection. EBV infection may result in lytic or immortalizing infection, which can be distinguished by the production of virus and the expression of different viral proteins and antigens. T lymphocytes are activated and limit the outgrowth of the EBV-infected cells.

and is the first indication of a lytic infection, and the EA-D, which is diffuse in the cytoplasm and nucleus. A lymphocyte-defined membrane antigen (LYDMA), which generates a cell-mediated but not humoral immune response, is also expressed.

Infection of B cells in cell culture generally causes no visible change in the histology of the cell. EBV is a B-cell mitogen that will immortalize the cells and convert them into lymphoblasts that express only EBV nuclear antigen (EBNA) and lymphocyte-defined membrane antigen. B cells isolated from a patient with infectious mononucleosis are also immortalized and grow into cell lines in the absence of T-cell suppression. The presence of virus can be detected by immunologic methods or by hybridization with nucleic acid probes.

Suppressor (CD8) T cells are activated and proliferate in response to the EBV-infected B cells. The large number of atypical lymphocytes (also called **Downey cells**) noted in the blood of persons with infectious mononucleosis are CDB lymphocytes. They increase in the peripheral blood during the second week of infection, accounting for 10% to 80% of the total white cells. The large T-cell response causes the swollen lymph glands, spleen, and liver that occur later in the disease.

T cells limit the proliferation of EBV-infected B cells in tissue culture and are essential in controlling the disease (Figure 42-20). Individuals with impaired cell-mediated immunity may have difficulty resolving an EBV infection, and chronic infections may result. An inability to resolve the EBV infection is indicated serologically by the lack of antibody production to the EBNA. T cell–deficient individuals suffer severe immunoproliferative disorders when infected with EBV.

Epidemiology

The virus infects and is shed by epithelial cells in the oropharynx into the blood and saliva. EBV infection is transmitted in virus-containing saliva. More than 90% of EBV-infected individuals intermittently shed the virus for life, even when totally asymptomatic. Children can acquire the virus at an early age by sharing contaminated glasses and generally undergo subclinical disease. Saliva sharing between adolescents and young adults often occurs by kissing, thus the nickname the ''kissing disease.'' Disease in these individuals may be unnoticed or present in varying degrees of severity as infectious mononucleosis. Even in a developed country such as the United States, approximately 70% of the population has been infected by age 30 years.

The geographic distribution of EBV-associated neoplasms suggests an association with potential cofactors. The immunosuppressive potential of malaria has been suggested as a factor in the progression of chronic or latent EBV infection to African Burkitt's lymphoma. The restriction of nasopharyngeal carcinoma to certain regions of China has suggested a genetic predisposition to the cancer or possibly the presence of cofactors in the food or environment.

Clinical Syndromes

Infectious Mononucleosis As with other herpesviruses, infection in childhood by EBV is much milder than infection occurring in adolescents or adults. Infection of children is usually subclinical. Infectious mononucleosis is characterized by fever, malaise, pharyngitis, lymphadenopathy, and often hepatosplenomegaly. The major complaint of persons with infectious mononucleosis is fatigue (Figure 42-21). The disease is rarely fatal in normal individuals but can cause serious complications resulting from neurologic disorders, laryngeal obstruction, or rupture of the spleen. Neurologic complications can include meningoencephalitis and probably the Guillain-Barré syndrome. Laboratory evaluation reveals a white blood cell count that is usually low, with a predominance of lymphocytes, many of which are atypical T lymphocytes (Figure 42-22).

Chronic Fatigue Syndrome In recent years a syndrome of chronic fatigue has been described. These patients have chronic tiredness and may also have low-grade fever, muscle weakness, and an inability to concentrate. The only laboratory test abnormality is an occasional elevation of the EBV antibody titers. Most patients do not exhibit a classic

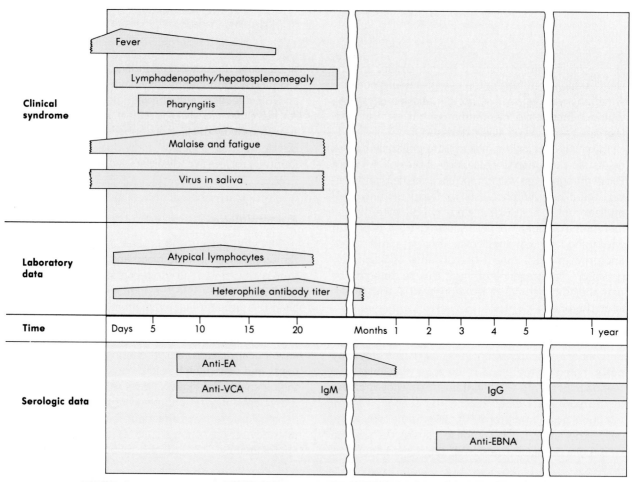

FIGURE 42-21   Clinical course and laboratory findings of infectious mononucleosis. EBV infection may be asymptomatic or produce the symptoms of mononucleosis.

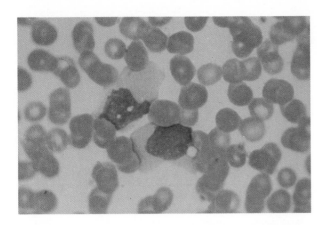

FIGURE 42-22   Atypical T-lymphocyte characteristic of infectious mononucleosis. The cells have a more basophilic and vacuolated cytoplasm than normal lymphocytes, and the nucleus may be oval, kidney shaped, or lobulated. The cell margin may seem to be indented by neighboring red blood cells.

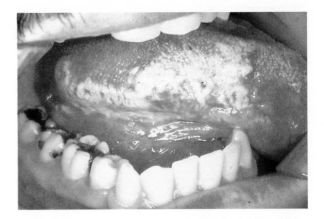

FIGURE 42-23   Hairy leukoplakia caused by EBV.

pattern of EBV infection or recurrence, and serious doubt surrounds the role of EBV in this syndrome.

EBV-Induced Lymphoproliferative Disease in Immunocompromised Individuals Increasing evidence suggests that some immunocompromised patients, especially those with congenital deficiency of T-lymphocyte function, may experience malignant forms of EBV infection leading to invasive lymphomas and other forms of invasive tumors. It also appears that patients with human immunodeficiency virus (HIV) infection may experience EBV-induced lesions of the mouth referred to as hairy leukoplakia (Figure 42-23).

Association With Human Neoplasms

Burkitt's Lymphoma Because EBV transforms or immortalizes lymphocytes in culture, it has been a major suspect in the search for viruses that cause human cancers. Burkitt's lymphoma, a malignant tumor of the jaw and face, contains EBV DNA sequences and EB virions, as seen by electron microscopy. In view of these findings, a cofactor role for EBV in the pathogenesis of Burkitt's lymphoma is suspected. This same tumor occurs rarely outside Africa and in these instances is usually not associated with EBV. Children in the affected regions of Africa are infected with EBV early in life and have high exposure to malaria. EBV may immortalize cells,

which continue to grow in the individual because of malarial immunosuppression.

Nasopharyngeal Carcinoma EBV is also associated with another tumor, nasopharngeal carcinoma. This tumor is endemic to the Orient, occurs in adults, and also contains EBV DNA within tumor cells. Unlike Burkitt's lymphoma, where the tumor cells are derived from lymphocytes, the nasopharyngeal carcinoma tumor cells are of epithelial origin.

Laboratory Diagnosis

Laboratory analysis of an EBV-induced infectious mononucleosis is usually documented by the demonstration of atypical lymphocytes, heterophile antibody, and positive serologic findings (Box 42-45 and Table 42-7). Cell culture is not useful because virus isolation requires the availability of fresh, isolated human B cells or fetal lymphocytes obtained from cord blood and immunofluorescent techniques to identify the EBV virus.

Atypical lymphocytes are probably the earliest detectable indication of an EBV infection, although they are not specific for EBV. These cells are present

---

Box 42-5   Laboratory Diagnosis of EBV
Atypical lymphocytes Heterophile antibody Serology

Table 42-7 Serologic Profile for EBV Infections

Patient's Clinical Status	Heterophile Antibodies	EBV-Specific Antibodies				Comment
		VCA-IgM	VCA-IgG	EA	EBNA	
Susceptible	−	−	−	−	−	
Acute primary	+	+	+	±	−	
Chronic primary	−	−	+	+	−	
Past infection	−	−	+	−	+	
Reactivation infection	−	−	+	+	+	EA restricted or diffuse
Burkitt's lymphoma	−	−	+	+	+	EA restricted only
Nasopharyngeal carcinoma	−	−	+	+	+	EA diffuse only

Modified from Balows A, Hausler WJ Jr, and Lennette EH, editors: Laboratory diagnosis of infectious diseases: principles and practices, New York, 1988, Springer-Verlag New York, Inc.

with the onset of symptoms and disappear with resolution of the disease.

Heterophile antibody is produced by the nonspecific mitogen-like activation of B cells by EBV. Heterophile antibody is an IgM that recognizes the Paul-Bunnell antigen on sheep and bovine erythrocytes but not guinea pig kidney cells. The titer of a patient's antibody is determined as the last dilution that will agglutinate sheep erythrocytes. The monospot test is a more rapid adaptation of the test just described and is more widely used (Figure 42-24). Heterophile antibodies can usually be demonstrated by the end of the first week of illness but may occasionally be delayed until the third or fourth week (Figure 42-25). The test should therefore be repeated weekly in patients suspected to have infectious mononucleosis but who have a negative heterophile antibody test. These antibodies are present for approximately 3 weeks but may last from 1 week to many months. Approximately 5% to 15% of EBV-induced cases of infectious mononucleosis in adults and a much greater proportion in young children and infants fail to induce detectable levels of heterophile antibodies.

When EBV infections cannot be diagnosed by the usual clinical criteria and heterophile antibody tests, specific antibody tests can be performed (Table 42-7 and Figure 42-25). These include (1) IgG antibody to VCA, which appears early in infection and usually persists for life; (2) antibody to EA, which appears in most patients and persists only during the active phase of infection (weeks to months); and (3) antibody to EBNA, which appears 2 to 4 weeks after onset and usually persists for life. In the early phase of

acute infection, IgM-specific VCA titers are usually elevated; however, reactivation of infection may also provoke a similar response. Presence of *both* VCA and EBNA antibody in an acute- or convalescent-phase serum suggests a past infection.

Recent EBV infection is suggested by any of the following: (1) IgM antibody to VCA, (2) presence of VCA antibody and absence of EBNA, (3) rising titer of EBNA, or (4) presence of elevated VCA and EA antibodies.

Heterophile-negative "mononucleosis" is likely to be caused by CMV if the patient is 25 years of age or older. Patients with CMV mononucleosis will have a high percentage of atypical lymphocytes in peripheral

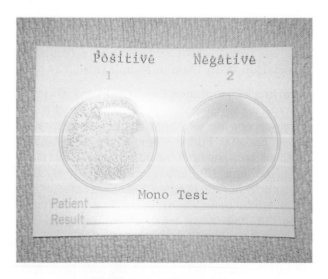

FIGURE 42-24   Monospot test.

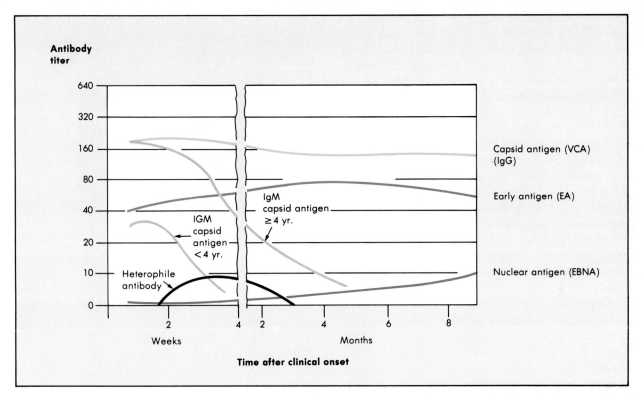

FIGURE 42-25   Serology of infectious mononucleosis. Antibodies to virion antigens and the early antigen are detected during the onset of symptoms, whereas antibodies to EBNA are present on resolution of the infection.

blood, but the heterophile and other monotests are negative. These patients are much less likely to have severe pharyngitis and disseminated lymphadenitis.

Culture   EBV culture is not routinely performed because the virus does not grow in the usual cells used in diagnostic virology laboratories.

Treatment, Prevention, and Control

No effective treatment is available for EBV disease. Acyclovir decreases viral exertion but makes little impact on clinical symptoms. The ubiquitous nature of EBV makes control of infection difficult. Quarantine of infected individuals is also difficult because the disease is often subclinical. Infection elicits lifelong immunity. The best means of preventing infectious mononucleosis is exposure to the virus early in life because the disease is more benign in children. An anti-EBV vaccine might be developed to prevent African Burkitt's lymphoma, but the risk-to-benefit ratio

for prevention of infectious mononucleosis does not warrant its application for general use.

CYTOMEGALOVIRUS

CMV is a common human pathogen, infecting 0.5% to 1.0% of all newborns and approximately 50% of the adult population in developed countries. It becomes particularly prominent as a pathogen in immunocompromised patients.

Viral Structure

CMV is morphologically similar to other members of the herpesvirus group.

Pathogenesis

In general the pathogenesis of CMV is similar to that of other herpesviruses. CMV shares the capacity for (1) cell-to-cell spread in the presence of circulating antibody, (2) establishment of a latent state in the host, (3) reactivation under conditions of immunosuppression, and (4) induction of transient immunosuppression in the recipient.

Cell-to-Cell Spread in Presence of Antibody CMV is highly cell associated and spreads via coalescing cells. This close cell association protects the virus from antibody-mediated inactivation.

Latency Following primary CMV infection, the virus becomes latent. The exact site(s) of latency and the mechanisms of persistence are not completely understood, but leukocytes, especially mononuclear, are suspected to contain latent virus and account for transmission of the virus via blood and leukocyte transfusions. Also, organs such as the kidneys and heart harbor the virus, but the exact cell is not known.

Reactivation Latent CMV infection appears to be reactivated by immunosuppression (e.g., corticosteroids, HIV infection) and possibly by allogeneic stimulation (i.e., the host response to transfused or transplanted cells).

Induction of Immunosuppression Primary CMV infection, even when asymptomatic, induces an extreme reversal of the ratio of helper to suppressor T-lymphocyte subsets. This is primarily the result of an increase of the suppressor cell population, but a reduction in the number of helper T lymphocytes also occurs. Over months these ratios return to or near preinfection levels. Lymphocyte function (e.g., proliferative responses to CMV antigens and mitogens) is also diminished during the acute infection but returns to normal during convalescence.

On entry into the body, CMV infects lymphocytes and leukocytes and is spread throughout the body in these cells (Figure 42-26). Cells may be infected with

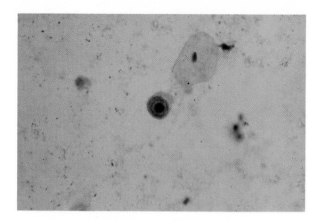

FIGURE 42-26   CMV-infected cell with basophilic nuclear inclusion.

free virus or through cell-to-cell spread. In most cases the virus replicates without causing symptoms and establishes a latent infection in the individual. The virus can be activated, replicate, and be shed for long periods without causing symptoms (Table 42-8).

Epidemiology

CMV has been isolated from urine, blood, throat washings, saliva, tears, milk, semen, stool, amniotic fluid, vaginal and cervical secretions, and tissues taken for transplantation. The major means of CMV transmission are by the congenital, oral, and sexual routes and by blood transfusion or tissue transplantation.

Congenital Infection Almost 1% of all newborns in the United States are infected with CMV at birth (Figure 42-27). Congenital CMV infection is best documented by isolating the virus from the infant's urine during the first week of life.

Table 42-8   Asymptomatic Shedding of CMV

Source	Neonates	Children	Adults
Urine	0.5% to 2.5%	10% to 29%	0% to 2%
Oral secretions	0.5% to 2.5%	10% to 29%	0% to 2%
Cervical secretions			10% to 28%†
Semen			5% to 10%
Breast milk			13% to 27%

†Potential for CMV secretion increases in the third trimester of pregnancy.
*S, Shedding occurs during symptomatic stage.

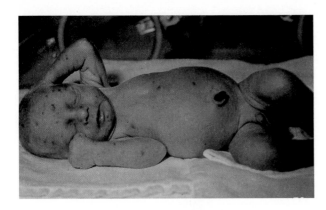

FIGURE 42-27   Congenital CMV infection.

Fetuses are infected by virus ascending from the cervix or by virus in the mother's blood. The cervix is probably the route in immune mothers and the bloodstream the route in mothers with primary infection. Primary infection in the mother causes more severe disease in the infant, including deafness and mental retardation.

Perinatal Infection In the United States up to 20% of pregnant women at term harbor CMV in their cervix. Essentially all these women experience reactivation rather than primary infection. Approximately half the neonates born through an infected cervix acquire CMV infection and become excretors of the virus at 3 to 4 weeks of age. Neonates may also acquire CMV from maternal milk or colostrum. In healthy full-term infants perinatal infection causes no clinically evident disease.

Another means of CMV acquisition by neonates is through blood transfusions. If seronegative babies are exposed to blood from seropositive donors, 13.5% acquire CMV infection in the immediate postnatal period. If premature infants acquire CMV from transfused blood, significant clinical infection may occur, with pneumonia and hepatitis the major manifestations.

Infection in Adults In low socioeconomic populations and in underdeveloped countries, postneonatal CMV infection typically occurs, apparently as a result of crowded living conditions. If such conditions are not present, only 10% to 15% of adolescents are infected. During young adulthood, however, the rate of seropositivity rapidly increases so that by age 35 years approximately 50% show serologic evidence of past infection (Figure 42-28). Evidence suggests that many of these infections are sexually transmitted or at least require very close personal contact. The evidence for sexual transmission of CMV infection follows:

1. The virus does not spread readily among adults by ordinary personal contact, even during prolonged exposure to people who are excreting the virus.
2. CMV has been isolated from the cervix of 13% to 23% of women at venereal disease clinics.
3. CMV has also been isolated from semen and is present there in the highest titer of any body secretion.
4. Antibody prevalence is high among homosexual men (e.g., 93% of homosexuals attending a venereal disease clinic had serologic evidence of past infection).

Posttransfusion Infection Posttransfusion mononucleosis (PTM) infection was described frequently in the past as a complication of surgery.

Infection after Transplantation CMV may also be transmitted by organ transplantation (e.g., kidneys or bone marrow). On the other hand, transplant recipients who are seropositive may reactivate CMV during periods of intense immunosuppression, but these episodes are less apt to be associated with severe clinical disease.

Infection in Immunocompromised Host If a patient becomes immunocompromised by either a disease or its therapy, latent CMV may reactivate and cause clinical illness. Primary infection with CMV may also occur in these patients, and this form is more likely to cause symptomatic illness.

Clinical Syndromes (Table 42-9)

Congenital and Neonatal Infection Approximately 10% of newborns infected with CMV show clinical evidence of disease, such as microcephaly, intracerebral calcification, hepatosplenomegaly, and rash. From 6,000 to 7,000 newborns each year may have unilateral or bilateral hearing loss and mental retardation resulting from congenital CMV infection. Mothers of almost all infants with these stigmata have primary infection during pregnancy.

Full-term and otherwise healthy infants who acquire CMV at birth usually show no ill effects. Infants requiring hospitalization for premature birth or other conditions may develop a recognizable complex of clinical findings. The most frequent sign is hepatosplenomegaly, then sepsis, respiratory deterio-

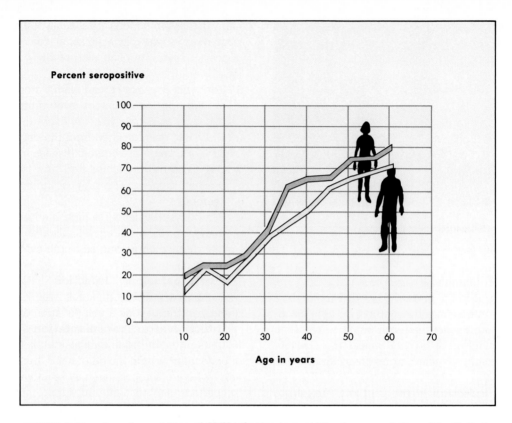

FIGURE 42-28   Age of acquisition of CMV infection in a middle-class population of the United States. Serologic indication of CMV infection was determined by evaluation of the complement-fixing anti-CMV antibody obtained from 4,824 individuals.

Table 42-9 CMV Syndromes

Tissue	Children/Adults	Immunosuppressed Patients
Eyes		Chorioretinitis
Lungs	—	Pneumonia
Gastrointestinal tract	—	Esophagitis, colitis
Nervous system	Polyneuritis, myelitis	Meningitis/encephalitis, myelitis
Lymphoid system	Mononucleosis syndrome, postperfusion syndrome	Leukopenia, lymphocytosis
Major organs	Carditis,* hepatitis*	Hepatitis
Predominant nature of disease	Inapparent infection	Disseminated disease, severe

*Complications of mononucleosis or postperfusion syndrome.

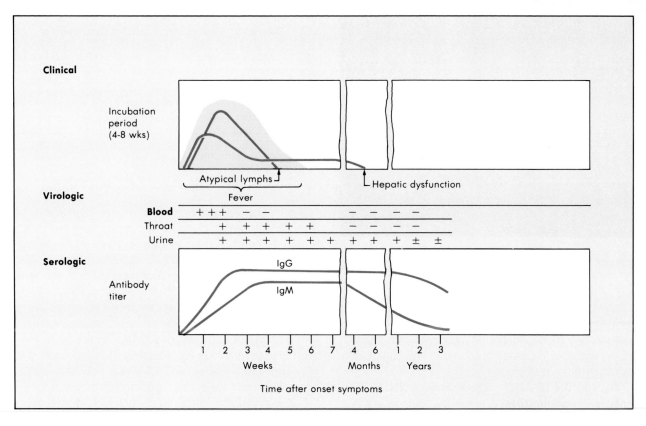

Clinical

Incubation
period
(4-8 wks)

Atypical lymphs

Fever

Hepatic dysfunction

Virologic

	Blood	+++	−	−						−	−	−	−
	Throat		+	+	+	+	+		−	−	−	−
	Urine		+	+	+	+	+	+	+	+	+	±	±

Serologic

Antibody
titer

IgG

IgM

1 2 3 4 5 6 7   4 6   1 2 3
Weeks   Months   Years

Time after onset symptoms

FIGURE 42-29   Clinical course of CMV infection. CMV infection is usually asymptomatic but may produce a mononucleosis syndrome, which can be distinguished from EBV infection.

ration, and fever. A peculiar gray pallor may also occur.

Infection in Adults Although most CMV infections acquired in young adulthood are asymptomatic, patients may develop clinical illness resembling infectious mononucleosis caused by EBV, but with less severe pharyngitis and lymphadenopathy (Figure 42-29). When this syndrome is encountered in patients over age 25 years, it is more likely caused by CMV than by EBV. Although CMV-infected patients develop atypical lymphocytosis similar to those with EBV infection, they have a negative heterophile antibody test.

Posttransfusion Infection Transmission of CMV by blood most often results in an asymptomatic infection; if symptoms are present, they typically resemble infectious mononucleosis. Fever, splenomegaly, and atypical lymphocytosis usually begin 3 to

5 weeks after transfusion. Pneumonia and mild hepatitis may also occur.

Infection in Immunocompromised Patients Retinitis caused by CMV occurs only in patients with severe immunodeficiency (e.g., in up to 10% to 15% of patients with AIDS). Blurring of vision or visual loss are common complaints. Examination of the fundus reveals hemorrhage and exudates. (Figure 42-30).

Pulmonary disease in immunocompromised patients appears as an interstitial pneumonia. Some patients with CMV infection and pulmonary disease have no other pathogens present on diagnostic bronchoscopy, and CMV presumably causes their pneumonia. Others may harbor organisms such as *Pneumocystis carinii,* and the presence of CMV may be only incidental.

CMV is not considered a neurotropic virus, but the

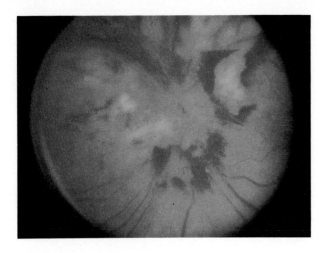

FIGURE 42-30   CMV retinitis.

virus occasionally may produce **central nervous system lesions** in immunocompromised patients. Diagnosis can be confirmed only by brain biopsy, with histologic evidence of the virus and isolation in cell culture.

Up to 10% of AIDS patients may develop CMV **colitis** or **esophagitis.** A lesser percentage of other immunocompromised patients may experience CMV infection of the gastrointestinal tract. Patients with CMV colitis usually have diarrhea, weight loss, anorexia, and fever. CMV esophagitis may mimic candidal esophagitis.

Laboratory Diagnosis (Box 42-6)

Cytology Cytologic examination of urine has been used to diagnose congenital CMV infection but is negative in up to a third of symptomatic congenitally infected infants with culture-positive urine.

Histology The histologic hallmark of CMV infection in vivo is the cytomegalic cell, which is enlarged (25 to 35 μm) and contains a dense, central "owl's eye" basophilic intranuclear inclusion (see Figure 42-26). These cells may be found in any tissue of the body and are thought to be epithelial in origin. These inclusions are seen well with Papanicolaou or hematoxylin and eosin (H and E) stains. However, not all CMV-infected cells are cytomegalic, and cytomegalic cells are rarely numerous in any one organ. It thus appears that typical inclusion-bearing cytomegalic cells are a definitive but insensitive measure of CMV disease.

Histologic examination of a small piece of tissue obtained (e.g., by transbronchial lung biopsy) may

Box 42-6   Laboratory Tests for Diagnosis of CMV Infection*

Cytology	Antigen detection
Histology	Molecular probes
Cell culture	Serology

Samples taken for analysis include urine, saliva, blood, bronchoalveolar lavage, and tissue biopsies.

also be less sensitive than culture, partly because of sampling error.

In an effort to provide a rapid, more sensitive histologic diagnosis, antibodies, especially monoclonal, have been used for direct detection of CMV antigens in tissues by immunofluorescence. From 80% to 100% of the culture-positive biopsies are also positive using this technique.

Both culture and antigen detection are more sensitive than histology for detecting CMV infection, but it is unclear whether patients lacking histologic confirmation truly have CMV pneumonia.

Recent efforts have been directed at detecting CMV by in situ hybridization with molecular probes. However, this procedure is currently less sensitive than culture.

Culture Culture has been generally regarded as the definitive method for detecting CMV infection. It is especially reliable in immunocompromised patients because they often have high titers of virus in their secretions. For example, AIDS patients may have titers with greater than 10^6 viable virus in their semen.

CMV grows only in diploid fibroplast cell cultures and must be maintained for at least 4 to 6 weeks because the characteristic CPE develops very slowly in specimens with very low titers of the virus. More rapid culture results may be achieved by **culture amplification** of a specimen. In this procedure specimens are inoculated by centrifuging them in a shell vial seeded with diploid fibroblast cells. Specimens are examined after 1 to 2 days of incubation by indirect immunofluorescence for either immediate early (IE) antigen or a combination of IE and early antigen (EA).

Serology Seroconversion is usually an excellent marker for primary CMV infection but almost never occurs in AIDS patients, since the overwhelming majority are seropositive before HIV infection. IgG titers to CMV antigen may be very high in AIDS patients, but a high titer level is not diagnostically useful.

Theoretically, CMV-specific IgM antibody develops only during primary infection, but it may reappear during reactivation of CMV. In homosexual men IgM antibody is so prevalent that it is not useful as a positive diagnostic test. The high prevalence of IgM antibody in the sera of homosexual men is presumably a result of reactivation of CMV, although repetitive exposure to differing strains of the virus may account for its presence in some individuals.

Treatment, Prevention, and Control

In the past, idoxuridine, cytosine arabinoside, adenine arabinoside, interferon, and acyclovir (ACV) have all been tried as therapeutic drugs for CMV infection. Little if any benefit was observed, however, and in some instances toxicity was severe.

Some success has been achieved with dihydroxypropoxymethyl guanine (DHPG, ganciclovir, Cytovene). DHPG is an acyclic nucleoside analog of thymidine and is structurally similar to ACV. In vitro, DHPG shows activity toward all human herpesviruses. DHPG itself is inactive and must be converted to the triphosphate to inhibit viral DNA synthesis. The drug is phosphorylated to the monophosphate and converted to the triphosphate (the active form) by viral or cellular enzymes. Although DHPG functions similar to ACV as a selective inhibitor of viral DNA polymerase, unlike acyclovir it is phosphorylated by cellular enzymes and can be incorporated into DNA. Thus DHPG is more toxic than ACV. DHPG can be used to treat life-threatening or sight-threatening CMV infections in immunocompromised patients.

The major preventable routes of CMV spread are the sexual route and the tissue transplant and transfusion routes. Semen is a major vector for sexual spread of CMV for both heterosexual and homosexual contacts. The use of condoms or abstinence from anal receptive intercourse would limit spread. Screening potential blood and organ donors for CMV seronegativity can also reduce the transmission of virus.

Although congenital and perinatal transmission of CMV cannot effectively by prevented, a seropositive mother is least likely to produce a baby with symptomatic CMV disease. Infection before pregnancy and the presence of antibody seem to prevent serious complications of congenital infection. This is what occurs in underdeveloped countries, where almost 100% of women are seropositive at an early age. These women may give birth to infants with high rates of congenital infection because of reactivation of virus, but since they do not experience primary infections during pregnancy, severe congenital infection is largely avoided.

The prime targets for vaccination against CMV would be similar to those for rubella vaccine: children and seronegative women of childbearing age. Live attentuated CMV vaccines have been developed that induce antibody formation as well as cell-mediated immunity. However, the length of immunity, protection from future infection, and possible oncogenesis of the vaccine are not known. Killed, subunit, and vaccinia hybrid vaccines are being considered as alternatives.

HUMAN HERPESVIRUS 6

HH6 is the most recently identified human herpesvirus. As with EBV and CMV, it is lymphotropic and is ubiquitous. At least 45% of children are seropositive for HH6 by age 2 years.

HH6 was first isolated from the blood of AIDS patients and grown in T-lymphocyte cultures. HH6 was identified as a herpesvirus by observation of the characteristic morphology in electron micrographs of infected cells. It is not serologically related and does not hybridize with the other human herpesviruses.

In 1988 HH6 was serologically associated with a common disease of children, exanthem subitum, commonly known as roseola. No other disease association has yet been made for this virus.

Pathogenesis

The finding that HH6 infection occurs very early in life suggests that it must be shed and spread readily. It is most likely spread by close contact or respiratory means.

The only target cell that has been identified is the T lymphocyte. HH6 establishes a latent infection in T cells but may be activated and replicate on mitogen stimulation of the cells. Cells replicating the virus appear large and refractile with occasional intranuclear and intracytoplasmic inclusion bodies. T cell leukemia cell lines also support the replication of the virus. Resting lymphocytes and lymphocytes of normal immune individuals are resistant to infection. A nonlymphocyte cell permissive for HH6 may be identified in the future, as was the case for EBV.

As with EBV and CMV, HH6 replication is controlled by cell-mediated immunity. The virus is likely to become activated in AIDS patients and others with lymphoproliferative and immunosuppressive disorders. Activation of HH6 initiates a lytic infection of T cells, which may contribute to the depression of the number of T cells observed in patients with AIDS. However, no known consequence of HH6 activation has yet been found.

Clinical Syndromes

Exanthem Subitum Exanthem subitum, or roseola, is a common, benign exanthematous disease of children. It is characterized by rapid onset of high fever for a few days, followed by a generalized rash that lasts only 24 to 48 hours. Replication of virus in T cells and a viremia may stimulate interferon production, causing the high fever. The presence of infected T cells or activation of delayed type of hypersensitive T cells in the skin may account for the production of the rash. The disease is effectively controlled and resolved by cell-mediated immunity, but the virus establishes a lifelong latent infection of T cells.

Other Associated Diseases HH6 may also cause a mononucleosis syndrome and lymphadenopathy and may be a cofactor in the pathogenesis of AIDS.

HERPESVIRUS SIMIAE—B VIRUS

B virus *(Herpesvirus simiae)*, which is indigenous to Asian monkeys, can cause a highly lethal central nervous system infection in humans. The virus is transmitted by monkey bites or saliva or even by tissues and cells widely used in virology laboratories. B virus is the simian counterpart of HSV and causes subclinical infections as well as dermal, oral, or eye lesions in monkeys. Once infected, a human may have pain, localized redness, and vesicles at the site of the virus' entrance. Vesicles on the mucous mem-

branes, pneumonia, diarrhea, abdominal pain, and pharyngitis have been reported. However, patients develop an encephalopathy that is frequently fatal; most who survive have serious brain damage.

The diagnosis of B virus infections can be established by virus isolation or by serologic tests. Specimens of vesicular fluids, or more often, biopsy specimens from diseased tissues, are inoculated into cultures of Vero cells or primary rhesus monkey kidney cells. HSV-like CPE will become evident within 7 to 10 days in cultures inoculated with positive specimens. The virus isolate can be identified by neutralization tests using antisera to B virus and HSV. The serologic diagnosis of B virus infection requires demonstration of a fourfold or greater rise in titers to the virus in paired serum samples. Patients with preexisting antibodies to HSV may have sufficient titers to cross-react with B virus. In infected patients a rise in titers to HSV is observed, along with an increase in titers to B virus. Increases in antibodies are observed about 2 weeks after the onset of illness.

BIBLIOGRAPHY

Corey L et al: Genital herpes simplex infections: clinical manifestations, course, and complications, Ann Intern Med 98:958-972, 1983.

Nahmias AJ and Rothman B: Infection with herpes simplex virus 1 and 2, N Engl J Med 289:667-674, 719-725, 781-789, 1973.

Spring SB, Schluederberg A, Allen WP, and Gruber J: Pathogenic diversity of Epstein-Barr virus, J Natl Cancer Inst 81:13-20, 1989.

Thorley-Lawson DA: Basic virologic aspects of Epstein-Barr virus infection, Semin Hematol 25:247-269, 1988.

Weller TH: Varicella zoster: changing concepts of the natural history, control, and importance of a not-so-benign virus, Pt 2, N Engl J Med 309(23):1434-1440, 1983.

Whitley RJ et al: The natural history of herpes simplex virus infection of mother and newborn, Pediatrics 66:489-494, 1980.

Yamanishi K et al: Identification of human herpesvirus-6 as a causal agent for exanthem subitum, Lancet 1:1065-1067, 1988.

CHAPTER 43

Poxvirus

Poxviruses are a large, complex group of viruses that cause disease in humans and other animals. Recent attempts to classify all members of the poxvirus group have resulted in the family Poxviridae. Of the many genera in this family, only species of *Orthopoxvirus* and *Molluscipoxvirus* are associated specifically with humans. The former contains variola virus **(smallpox)** and the later **molluscum contagiosum** virus. Poxviruses in other genera cause incidental infection of humans and naturally infect animals other than man (zoonosis).

Although smallpox has been declared eradicated from the world by the World Health Organization, there is still a need to learn about the poxvirus family. Reasons for understanding this unusual group of viruses include the following: (1) the mechanisms of spread of variola virus within the body represent a model for other virus infections, (2) poxviruses other than variola virus cause human disease, and (3) vaccinia virus (a laboratory altered poxvirus) is under extensive study as a vector for introducing immunizing genes into humans.

STRUCTURE AND REPLICATION

Vaccinia virus morphology differs very little from variola virus and is representative of other poxviruses. Vaccinia virus structure and replication will be used as a prototype poxvirus. Poxviruses are the largest viruses, measuring 230 × 300 nm in size, and are ovoid to brick-shaped (Figure 43-1). They have a capsid that is referred to as complex because it has neither helical nor icosahedral symmetry. An outer membrane and envelope enclose the core and core membrane, which are flanked by two lateral bodies of unknown function. The viral genome is a single strand of large, linear, double-stranded DNA (molecular weight approximately $100\text{-}200 \times 10^6$).

Replication of poxviruses is unique among DNA-containing viruses because the entire multiplication cycle takes place within the host cell cytoplasm (Figure 43-2). Viral penetration occurs within phagocytic vacuoles. Uncoating of the outer membrane occurs in the vacuole. Early gene transcription occurs within the viral core. Among the early proteins produced is an uncoating protein that removes the core membrane, liberating viral DNA into the cell cytoplasm. Replication of viral DNA follows in electron-dense cytoplasmic inclusions, referred to as "factories." Late viral mRNA is translated into structural proteins, which are glycosylated, phosphorylated, and cleaved before assembly. Unlike other viruses, poxvirus membranes form de novo in the cytoplasm, rather than as part of a host cell membrane that is picked up during a "budding" process. About 10,000 viral particles are produced per infected cell.

Live recombinant vaccines can be produced by using vaccinia virus as an expression vector for isolated (foreign) genes (Figure 43-3). The foreign gene, which encodes the immunizing molecule, and specific vaccinia gene sequences are added to a plasmid. This recombinant plasmid is inserted into a host cell simultaneously infected with vaccinia virus. The foreign gene is directed into the "rescuing" vaccinia virus genome because of the homologous vaccinia sequences included on the plasmid. Replication of the recombinant vaccinia genome, followed by maturation and release, results in live vaccinia virus containing a foreign gene which will be expressed when

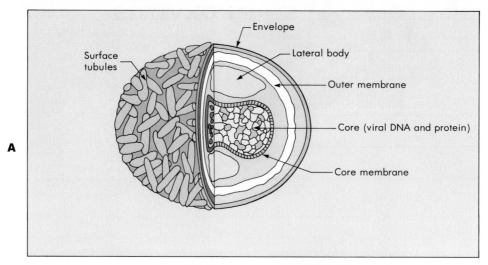

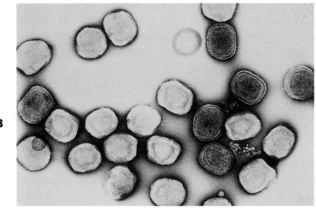

FIGURE 43-1   A, Structure of the vaccinia virus. The viral DNA and several proteins within the core are organized as a "nucleosome." Within the virion, the core assumes the shape of a dumbbell because of the large lateral bodies. Virions released through the cytoplasmic membrane are enclosed within an envelope that contains host cell lipids and several virus-specific polypeptides, including the hemagglutinin; they are infectious. Most virions remain cell-associated and are released by cellular disruption. These particles lack an envelope, so the outer membrane constitutes their surface; like the enveloped particles, they are also infectious. **B,** Electron micrograph of poxvirus. (Courtesy Centers for Disease Control.)

**FIGURE 43-2   Replication of vaccinia virus. (Modified from Fields BN et al, editors: Virology, New York, 1985, Raven Press.)

**FIGURE 43-3   Vaccinia virus as an expression vector for the production of live recombinant vaccines. (Modified from Piccini A and Paoletti E: Vaccinia: virus, vector, vaccine. In Maramorosch K, Murphy FA, and Shatkin AJ, editors: Advances in virus research, vol 34, New York, 1988, Academic Press, Inc)

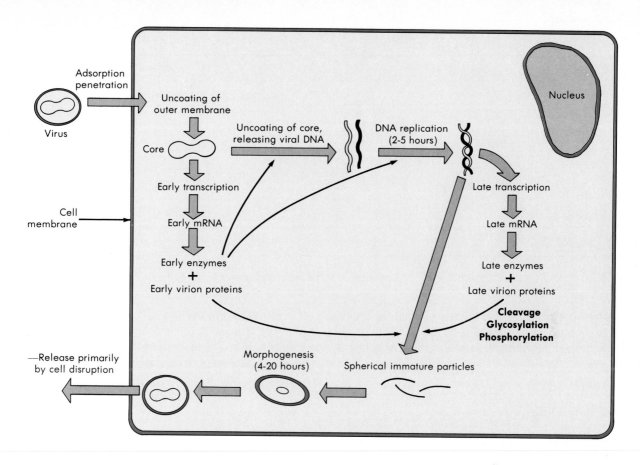

FIGURE 43-2   For legend see opposite page.

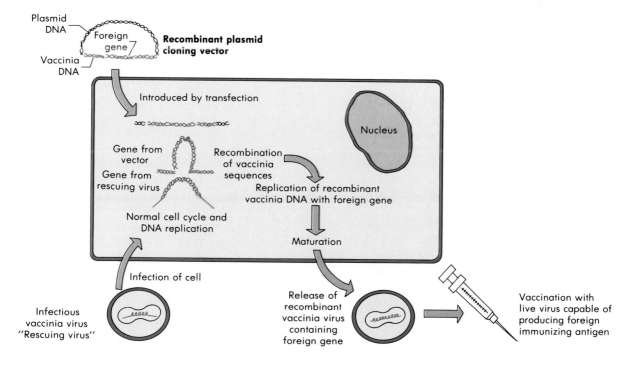

FIGURE 43-3   For legend see opposite page.

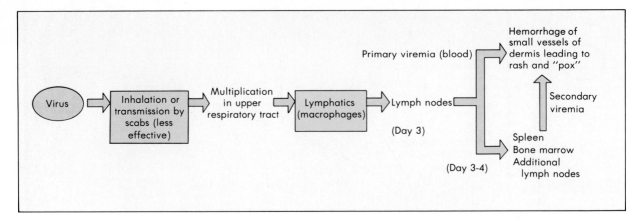

FIGURE 43-4   Mechanisms of spread of poxvirus within the body.

injected as a vaccine. Vaccines for hepatitis B and influenza have been prepared by these techniques. The potential for producing vaccines in the future is unlimited.

PATHOGENESIS

The pathologic hallmark of poxviruses is cell proliferation, manifested by the skin lesions that give this family of viruses their name. Most of the viruses of current human importance are primary pathogens in vertebrates other than humans (cow, sheep, goats) and infect humans only through ''accidental'' occupational exposure (zoonosis). As stated earlier, the exceptions to these are smallpox and molluscum contagiosum.

Smallpox virus is inhaled and replicates in the upper respiratory tract (Figure 43-4). Dissemination occurs via lymphatic and viremic spread. Internal and dermal tissues are inoculated following a second, more intense, viremia. Recovery, when it occurs, presumably follows the appearance of virus-specific immunity (Box 43-1).

EPIDEMIOLOGY

Smallpox (variola) is contagious and spread primarily by respiratory transmission or, less efficiently, by close contact with dried virus on clothes or other materials. Smallpox is now of largely historical interest, because it was declared eradicated from humans

in 1980. The eradication of smallpox is one of the greatest triumphs of medical epidemiology and resulted from a massive World Health Organization (WHO) campaign to vaccinate all susceptible individuals, especially those exposed to any one with the

---

Box 43-1   Pathogenic Mechanisms of Smallpox (Variola) Virus

1. Virus enters through respiratory tract
   Multiplication in respiratory tract produces no symptoms.
   Virus-infected macrophages enter lymphatics and are carried to regional lymph nodes
   Multiplication of virus in lymph node
2. Viremia
   Early viral multiplication occurs in spleen, bone marrow, and lymph nodes
   Virus infection of skin (rash) and visceral organs
   Poor humoral response leads to prolonged viremia and poor clinical outcome
3. Secondary viremia
   Additional lesions throughout host
   Death or recovery with or without sequelae
   Recovery associated with prolonged immunity and lifelong protection

Table 43-1 Diseases Associated With Poxviruses

Virus	Disease	Source	Location
Variola	Smallpox (now extinct)	Humans	Extinct
Vaccinia	Used for smallpox vaccination	Laboratory product	—
Orf	Localized lesion	Zoonosis—sheep, goats	Worldwide
Cowpox	Localized lesion	Zoonosis—rodents, cats, cows	Europe
Pseudocowpox	Milker's nodule	Zoonosis—dairy cows	Worldwide
Monkeypox	Generalized disease	Zoonosis—monkeys, squirrels	Africa
Bovine papular stomatitis virus	Localized lesion	Zoonosis—calves, beef cattle	Worldwide
Tanapox	Localized lesion	Rare zoonosis—monkeys	Africa
Yabapox	Localized lesion	Rare zoonosis—monkeys, baboons	Africa
Molluscum contagiosum	Many skin lesions	Humans	Worldwide

Modified from Balows A, Hausler WJ, Jr., and Lennette EH, editors: Laboratory diagnosis of infectious diseases: principles and practice, vol 2, New York, 1988, Springer-Verlag.

disease, and thereby interrupt the chain of human-to-human transmission. Since there are no animal reservoirs for variola and no chronic carriers (i.e., the virus is not latent), and vaccinated individuals are readily identified by a scar. This campaign, begun in 1967, succeeded with the last case of naturally acquired infection reported in 1977.

At least one fatal laboratory-associated infection with smallpox virus has occurred since the global eradication of the disease. Accordingly, only two World Health Laboratory collaborating centers, where adequate containment facilities are available, now have reference stocks of smallpox virus.

CLINICAL SYNDROMES

See Table 43-1 for a list of diseases associated with poxviruses.

Smallpox

There are two variants of smallpox: variola major, with a mortality of 15% to 40%, and variola minor, with a mortality of 1%. Smallpox is initiated by infection of the respiratory tract with subsequent involvement of local lymph glands, leading to viremia. Viremia is associated with fever, headache, backache, and later seeding of the skin with development of the characteristic vesiculopustular rash (Figure 43-5). The rash (pox) has two characteristics that distinguish it from other exanthems: (1) lesions are virtually all at the same stage of development as they progress from macules to vesicles to pustules to crusting and healing and (2) the rash is centrifugal (i.e., begins centrally on the face, shoulders, chest, and later involves more distal sites.) In addition to skin involvement, visceral organs, especially the spleen, liver, and lungs, are also involved. The incubation period ranges from 5 to 17 days, with an average of 12 days. The time course of smallpox infection is shown in Figure 43-6.

Vaccinia

When smallpox was eradicated, the necessity for vaccination against this disease disappeared. The virus used for vaccination was the vaccinia virus,

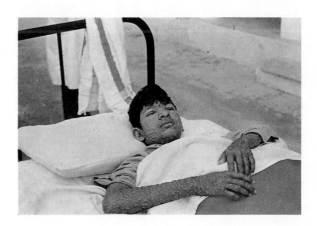

FIGURE 43-5 Child with smallpox. Note characteristic rash.

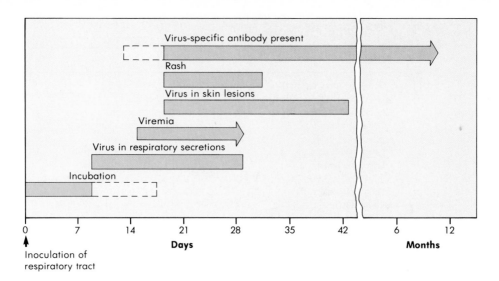

FIGURE 43-6   Time course of smallpox infection.

which was probably derived from an animal poxvirus such as horsepox. The procedure consisted of injecting live virus intracutaneously and observing for the development of vesicles and pustule(s) to confirm a "take." Revaccination at periodic intervals was necessary to maintain immunity. As smallpox waned, it became apparent that there were more complications due to vaccination than there were cases of smallpox. Several of these complications were severe and even fatal, including encephalitis and progressive infection **(vaccinia necrosum);** the latter occurred occasionally when immunocompromised patients were inadvertently vaccinated. The relative ease of person-to-person spread of vaccinia virus infections among unvaccinated individuals posed a threat to immunocompromised contacts of vaccinees.

Orf and Cowpox

The poxvirus of animals such as sheep or goats (orf) and cows (cowpox) can infect humans, usually as a result of accidental direct contact. Nodular lesions produced are usually on the fingers or face and are hemorrhagic (cowpox) or granulomatous (orf or pseudocowpox; Figure 43-7). Vesicular lesions frequently develop and then regress in 25 to 35 days, generally without scar formation. The lesions may be mistaken for anthrax. The etiologic viruses can be grown in culture or seen directly with electron microscopy.

Monkeypox

Over 100 cases of an illnesses resembling smallpox have been attributed to the monkeypox virus. All have occurred in western and central Africa, especially Zaire. There has been concern that this agent might "replace" the smallpox virus and become epidemic in humans, but this has not materialized probably because animal poxviruses seem highly adapted to their particular host and require very close, usually direct, contact for transmission.

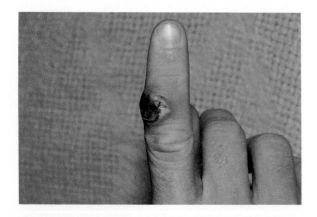

FIGURE 43-7   Orf lesion on finger of taxidermist. (Courtesy Joe Meyers, M.D.)

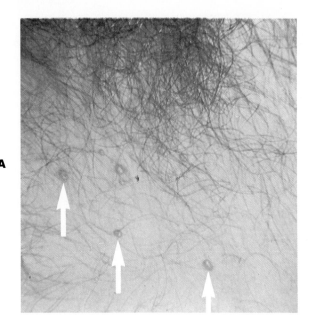

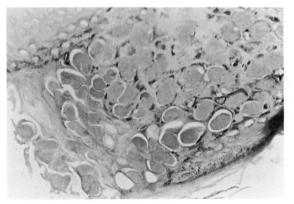

FIGURE 43-8   A, Skin lesion of molluscum contagiosum. **B,** Microscopic view of molluscum contagiosum; epidermis filled with molluscum (M) bodies (×100).

Molluscum contagiosum

The lesions of molluscum contagiosum are caused by a poxvirus that is unclassified because it does not grow in cell cultures. The lesions differ significantly from pox lesions in that they are nodular to wartlike (Figure 43-8). They begin as papules and progress to pearly, umbilicated nodules, 2 to 10 nm in diameter, with a central caseous plug that can be readily expressed. They are most common on the trunk, genitalia, and proximal extremities and usually occur in a cluster of 5 to 20 nodules.

LABORATORY DIAGNOSIS

The diagnosis of smallpox is usually made clinically but is confirmed by growth of the virus in embryonated eggs or cell cultures. Characteristic lesions (pocks) appear on the chorioallantoic membrane of embryonated eggs, and they permit presumptive identification of smallpox virus. Cytologic examination, for cytoplasmic but not nuclear inclusions, is also used, as are electron microscopy, immunofluorescence, and immunoprecipitation.

The diagnosis of molluscum contagiosum is confirmed histologically by the presence of very characteristic large eosinophilic cytoplasmic inclusions **(molluscum bodies)** in epithelial cells. These can be seen in biopsy specimens or in the expressed caseous core. The incubation period for molluscum contagiosum is 2 to 8 weeks, and the disease spreads by direct contact (e.g., sexual contact or wrestling) or fomites (e.g., towels).

TREATMENT, PREVENTION, AND CONTROL

The history of attempts to prevent smallpox is the history of vaccination in general; smallpox was the first disease to be controlled by immunization. The first effort was variolation (i.e., the use of virulent smallpox pus to inoculate susceptible individuals). This practice began in the Far East, was later used in England and then Cotton Mather introduced it to the United States. This practice was associated with fatality in approximately 1%, a better risk than for smallpox itself. Jenner is credited with using a less virulent variant to introduce the practice of vaccination in 1798. This variant virus is designated as vaccinia virus and may have been derived from cowpox or horsepox. Protection conferred by vaccination was considered to last for 3 years and possibly up to 10 years, so reinoculation was required. Controlled studies were never performed to identify the rate of protection against infection, but uncontrolled observations suggested substantial protection. As eradication of smallpox approached, it became apparent that in the developed world the rate of serious reactions to vaccination (see discussion of vaccinia) exceeded the risk of infection. Therefore discontinuation of smallpox vaccination began in the 1970s, and its use was eliminated after 1980.

The lesions of molluscum contagiosum disappear in 2 to 12 months, presumably as a result of acquired immune responses. Treatment, if required, consists of curettage, liquid nitrogen, or iodine solutions.

It should be noted that the antiviral, methisazone, is of more than just historical interest because it was the first antiviral agent that inhibited viral assembly to be used in clinical practice. Since this step in viral replication is unique, it provides an opportunity for selective inhibition of the virus but not normal host cell metabolism. Methisazone was used with moderate success to prevent smallpox in contacts. Its antiviral effect is restricted to the poxviruses, but it gives hope that similar agents can be developed for other viruses.

BIBLIOGRAPHY

Balows A, Hausler WJ Jr, and Lennette EH, editors: Laboratory diagnosis of infectious diseases: principles and practice, vol 2, New York, 1988, Springer-Verlag.

Fenner F: A successful eradication campaign: global eradication of smallpox, Rev. Infect Dis 4(5):916-930, September-October, 1982.

Fields BN, et al, editor: Virology, New York, 1985, Raven Press.

Piccini A, and Paoletti E: Vaccinia: virus, vector, vaccine. In Maramorosch K, Murphy FA, and Shatkin AJ, editors: Advances in virus research, vol 34, New York, 1988, Academic Press, Inc.

44

Papovaviruses

The papovavirus group (Papovaviridae) includes two genera, *Papillomavirus* and *Polyomavirus* (Table 44-1). Human papillomaviruses cause warts. BK and JC viruses, members of the *Polyomavirus* genus, are associated with progressive multifocal leukoencephalopathy and renal disease, respectively. Many viruses in both genera infect animals other than humans and papovaviruses are known to cause chronic latent infections in their natural hosts. Papillomaviruses can induce benign tumors in humans and may cause malignant conversion of some lesions.

Although these two genera are in the same family, they differ in size, antigenic determinants, and biological properties.

HUMAN PAPILLOMAVIRUSES

Structure and Replication

Human papillomaviruses (HPV) are non-enveloped, icosahedral virions measuring 50 nm in diameter. Their DNA is double-stranded and circular. The capsid consists of two structural proteins forming 72 capsomeres (Figure 44-1).

Regions of HPV DNA that encode viral proteins are all located on one strand (the plus strand) and are subdivided into an early region and a late region (Fig-

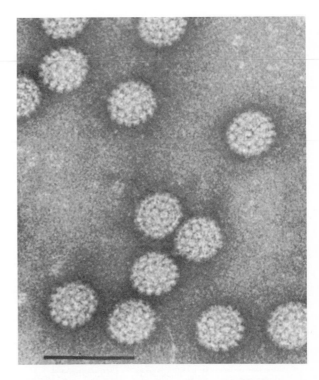

FIGURE 44-1 Human papillomavirus ×240,000 (bar = 0.1 μm). Capsomeres are visible on all particles, but the pattern is obscured by superposition of the images of the front and back surfaces. (From Dalton AJ and Haquenan F: Ultrastructure of animal viruses and bacteriophages: an atlas, New York, 1973, Academic Press.)

Table 44-1 Human Papoviridae and Their Diseases	
Virus	**Disease**
Papillomavirus	Warts
Polyomavirus	
BK virus	Progressive multifocal leukoencephalopathy
JC virus	Renal disease

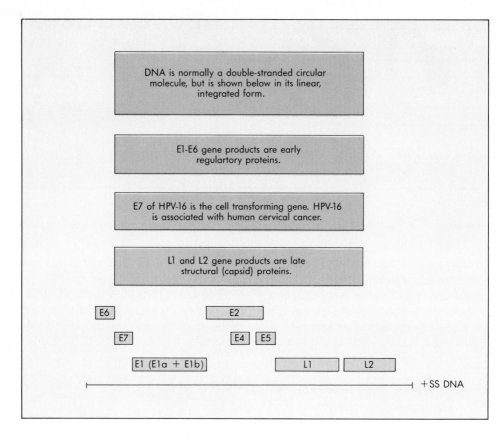

FIGURE 44-2   Genome of human papillomavirus (HPV-16). (Modified from Broker, TR and Botchan M: Cancer cells 4:17, 1986.)

ure 44-2). Early proteins, E1 to E6 and E8, are important in replication, transcription, and oncogenic transformation. Late proteins, L1 and L2, are structural proteins.

Classification of HPV is based on DNA relatedness because these viruses cannot be grown in tissue culture to provide high concentrations of antigens for differentiation. Based on DNA sequence homology, 46 HPV types have been identified and classified in 1 of 16 (A through P) groups. Viruses in similar groups frequently cause similar types of warts.

Pathogenesis

Papillomaviruses infect and replicate in cutaneous and mucosal epithelium, inducing epithelial proliferations. Viral infection remains local and generally regresses spontaneously (Figure 44-3). However, the

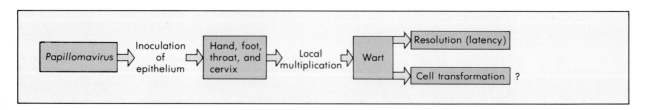

FIGURE 44-3   Mechanisms of spread of papillomavirus within the body.

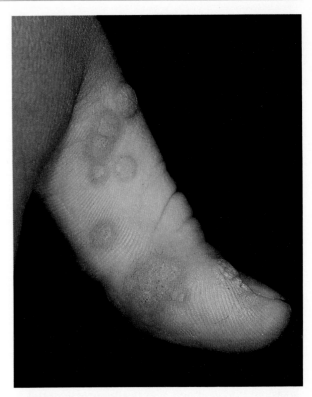

Infect cutaneous and mucosal squamous epithelial
  cells
Induce benign localized tumors
Most regress spontaneously
Some lesions develop into carcinomas
DNA of specific HPV types present in tumors

papillomavirus genome can persist in transformed
cells and may be involved in maintaining tumor
growth by increasing cell proliferation or prolonging
the life of epithelial cells.

Viral DNA has been found in both benign and
malignant tumors. A summary of HPV pathogenic
mechanisms is given in Box 44-1.

Epidemiology

Plantar, common, and flat warts are most common
in children and young adults. The lower incidence of
these warts in adults may result from acquired immunity. Genital warts are most common among sexually
active patients, underscoring their status as a sexually transmitted disease. Recent studies suggest that
genital HPV infections may occur in 20% of females.
Laryngeal papillomas are found most commonly in
young children and middle-aged adults. The annual
incidence is approximately 1 per 1 million population.

HPV infections are transmitted by direct contact.
Virus can be found on bathroom floors and towels.
Inoculation during sexual intercourse, while passing
through an infected birth canal, or by the childhood
habit of chewing warts is known to occur. Transmission is felt to be relatively inefficient.

Clinical Syndromes

Skin Warts Most persons are infected with the
common HPV types (1 through 4), which infect keratinized surfaces usually on the hands and feet (Fig.
44-4) and occur frequently in childhood or early adolescence (Table 44-2).

Benign Head and Neck Tumors Single oral papillomas are the most benign epithelial tumors of the

FIGURE 44-4 Common warts with thrombosed vessels
(black dots) on the surface. (From Habif TP: Clinical dermatology: a color guide to diagnosis and therapy, St Louis, 1985,
The CV Mosby Co.)

oral cavity. They are pedunculated with a fibrovascu-
lar stalk and usually have a rough papillary appearance to their surface. They can occur in any age-
group, are usually solitary, and rarely recur after surgical excision. **Laryngeal papillomas** are commonly associated with HPV-11 and are the most common
benign epithelial tumors of the larynx. Laryngeal pap-
illomatosis is usually considered a life-threatening
condition in children because of the danger of airway
obstruction. Occasionally, papillomas may extend
down the trachea and into the bronchi.

Anogenital Warts Genital warts **(condyloma
acuminata)** occur almost exclusively on the squamous epithelium of the external genitalia and perianal
areas, and about 90% are caused by HPV types 6 and
11. Anogenital lesions infected with these HPV types
rarely progress from benign to malignant conditions
in otherwise healthy individuals.

Cervical Dysplasia, Neoplasia HPV infection of
the genital tract is now recognized as a common sexually transmitted disease; cytologic changes charac-

Table 44-2   Clinical Syndromes Associated With Papillomaviruses

Syndrome	HPV Types	
	Common	**Uncommon**
SKIN WARTS		
Plantar wart	1	2, 4
Common wart	2, 4	1, 7, 26, 29
Flat wart	3, 10	27, 28, 41
Epidermodysplasia verruciformis	5, 8, 17, 20, 36	9, 12, 14, 15, 19, 21-25, 38, 46
BENIGN HEAD AND NECK TUMORS		
Laryngeal papilloma	6, 11	
Oral papilloma	6, 11	2, 16
Conjunctival papilloma	11	
ANOGENITAL WART		
Condyloma acuminatum	6, 11	1, 2, 10, 16, 30, 44, 45
Cervical intraepithelial neoplasia, cancer	16, 18	11, 18, 31, 33, 35, 42, 43, 44

Modified from Balows A et al, editors: Laboratory diagnosis of infectious diseases: principles and practice, vol 2, New York, 1988, Springer-Verlag.

teristic of this viral infection **(keilocytotic cells)** in cervical smears that are Papanicolaou stained are detected in about 5% of all specimens from women (Figure 44-5). Infection of the female genital tract by HPV types 16 and 18 and, rarely, by other types is associated with intraepithelial cervical neoplasia and cancer.The first neoplastic changes noted by light microscopy are termed "dysplasia." Approximately 40% to 70% of the mild dysplasias undergo spontaneous regression.

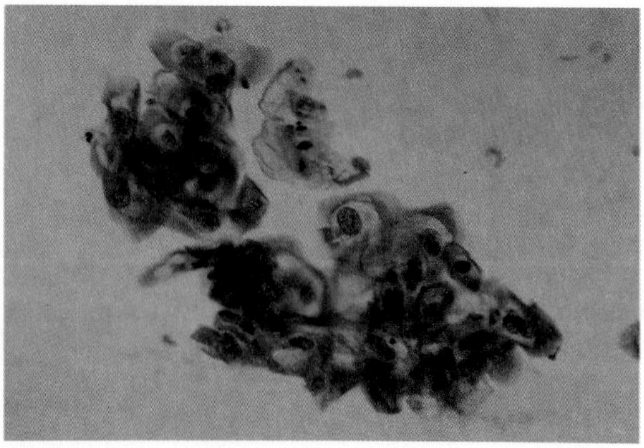

The development of cervical cancer is thought to proceed through a continuum of progressive cellular changes from mild (cervical intraepithelial neoplasia, CIN I) to moderate (CIN II), to severe dysplasia and/or carcinoma *in situ*. This sequence of events has been documented by a number of studies to progress over a period of 1 to 4 years.

FIGURE 44-5   Papanicolaou stain of the exfoliated cervical-vaginal squamous epithelial cells showing perinuclear cytoplasmic vacuolization and nuclear enlargement ("koilocytosis") characteristic of HPV infection (x400).

Laboratory Diagnosis

A wart can be confirmed microscopically by its characteristic histologic appearance, consisting of hyperplasia of **prickle cells** and the production of excess keratin (hyperkeratosis). Papillomavirus infection can be detected by the presence of keilocytotic (vacuolated) squamous epithelial cells which are rounded and occur in clumps (see Figure 44-5). *Papillomavirus* virions can be seen by electron microscopy in lesions as can HPV antigens by immunofluorescent and immunoperoxidase techniques (Table 44-3). Molecular probes are the method of choice for establishing the presence of HPV genomes in cervical swabs and in tissue. Human papillomaviruses do not grow in cell cultures, and tests for HPV antibodies are rarely used except in research surveys.

Table 44-3 Laboratory Diagnosis of Papillomavirus Infections

Test	Detects
Cytology	Keilocytotic cells
Electron microscopy	Virus
Immunofluorescent and immunoperoxidase staining	Viral structural antigens
Southern blot hybridization	Viral nucleic acid
Culture	Not useful

Box 44-2 Pathogenic Mechanisms of Polyomaviruses (JCV and BKV)

Asymptomatic, primary infections in childhood
   Limited replication at primary site of entry, the respiratory tract
   Viremia transports virus to kidney
Virus multiplication in the epithelial cells of the urinary tract
   Transient viremia possible
   Viruses remain latent indefinitely in immunocompetent host
   JCV and BKV genomes found in normal kidneys at autopsy
Reactivation in immunodeficient host
   Impaired T-cell function allows reactivation
   Renewed virus multiplication in urinary tract; viruria occurs
   JC virus in neurotropic hosts results in infection of brain, leading to destruction of oligodendrocytes; causes progressive multifocal leukoencephalopathy
   BK virus infection can lead to ureteral stenosis or hemorrhagic cystitis

Treatment, Prevention, and Control

Spontaneous disappearance of warts is the rule, but this may take many months to years so intervention is sometimes warranted, especially for painful or bulky lesions. Removal by surgical cryotherapy, electrocautery, or chemical means can be effective, although recurrences are common. Application of podophyllin or salicylic acid with formalin or glutaraldehyde is effective. Injection of interferon is also beneficial. Surgery may be necessary for laryngeal papillomas.

Vaccination with formalin-inactivated HPV or autogenous preparations from the patient's own lesions has been at least partly effective, especially for anogenital warts. Neither of these approaches is widely used, but vaccines may assume greater importance if an etiologic relationship between HPV and carcinoma is established. At present the best means of prevention is avoidance of direct contact with infected tissue.

POLYOMAVIRUSES

Although relatively little is known of the human polyomaviruses (BK and JC viruses), other animal polyomaviruses are well-characterized tumor-causing viruses. For example, a simian virus (SV40) causes tumors when injected into hamsters.

Structure and Replication

In comparison with the papillomaviruses, the polyomaviruses are slightly smaller (44 nm in diameter) and contain less nucleic acid (3×10^6 versus 5×10^6 MW). The virion is icosahedral and lacks an envelope. The double-stranded DNA is a circular, supercoiled molecule. The genomes of BKV, JCV, and SV40 are closely related and divided into early, late, and noncoding regions. The early region codes for nonstructural T (transformation) proteins, and the late region codes for three viral capsid proteins (VP1, VP2, VP3). The noncoding region is the origin of DNA replication and the site of the transcription control sequence.

The replicative cycle includes penetration by pinocytosis, followed by uncoating and viral multiplication in the nucleus. Considerable detail is known about the course of cellular infection but will not be discussed completely here because of the relative unimportance of human polyomaviruses.

Pathogenesis

Polyomaviruses have a high specificity for certain hosts and particular cells within that host. In humans both JC and BK viruses probably enter the respiratory tract, spread by viremia, and infect the kidney. In immunocompromised patients, reactivation of virus

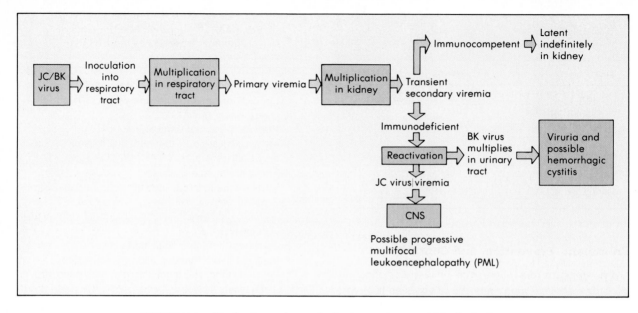

FIGURE 44-6   Mechanisms of spread of polyomaviruses within the body.

in the kidney leads to potentially severe urinary tract infection (BKV) or viremia and central nervous system infection (JCV) (Figure 44-6). Suppressed T-cell function appears most responsible for reactivation. Histopathology results from the cytolytic effects of the virus rather than local immunologic factors (Box 44-2). BKV and JCV cause tumors when injected into hamsters; however, they are not consistently associated with any human tumors.

Table 44-4   Clinical Syndromes Associated With Infections Caused by Polyomaviruses (JKV and BKV)

Agent	Host Status	Syndrome
JCV and BKV	Immunocompetent	Latent asymptomatic infection
JCV	Immunodeficient	Progressive multifocal leukoencephalopathy
BKV	Immunodeficient	Ureteral stenosis and hemorrhagic cystitis

Modified from Balows A et al, editors: Laboratory diagnosis of infectious diseases: principles and practice, vol 2, New York, 1988, Springer-Verlag.

Epidemiology

Polyomavirus infections are ubiquitous, and most humans are infected with both JCV and BKV by the age of 15 years. Respiratory transmission is the probable mode of spread. Immune suppression after organ transplantation or during pregnancy is capable of reactivating latent infections.

Early lots of poliomyelitis vaccine were contaminated with SV40, a simian polyomavirus, that was undetected in the cell cultures of monkey origin used to prepare the vaccine. Although many people were vaccinated with the contaminated lots, no SV40-related tumors have been reported during 25 years of follow-up.

Clinical Syndromes

Primary infection is virtually always asymptomatic, although mild respiratory symptoms might occur and cystitis has been reported (Table 44-4). In up to 40% of immunocompromised patients, urinary excretion of BKV and JCV is commonly observed. Reactivation also occurs in pregnancy, but no effects on the fetus have been established.

Ureteral stenosis in renal transplant patients appears to be associated with BK virus, as does hem-

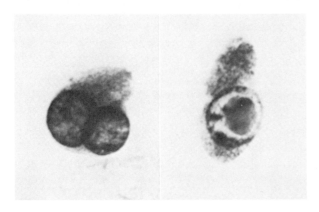

FIGURE 44-7 Urinary tract epithelial cells with dense basophilic intranuclear inclusions due to polyomavirus infection. (From Balows A et al, editors: Laboratory diagnosis of infectious diseases: principles and practice, vol 2, New York, 1988, Springer-Verlag.)

Table 44-5 Laboratory Diagnosis of Polyomavirus (JCV and BKV) Infections

Test	Detects
Papanicolaou smear of urinary epithelial cells	Viral inclusions
Electron microscopy	Virions
Immunofluorescence and immunoperoxidase staining	Viral antigens
Nucleic acid hybridization in clinical specimens	BKV and JCV nucleic acids
Cell culture	
Human diploid lung fibroblasts	Virus isolation—BKV
Primary human fetal glial cells	Virus isolation—JCV

orrhagic cystitis in bone marrow transplant recipients.

Progressive multifocal leukoencephalopathy (PML) is a rare syndrome that occurs in immunocompromised patients, including those with AIDS, and is due to JCV. The virus was first recovered by co-culture of brain tissue from a patient with progressive multifocal leukoencephalopathy and Hodgkin's disease. As the name implies, patients may have multiple neurologic symptoms unattributable to a single anatomic lesion. Impairment of speech, vision, coordination, and/or mentation occurs followed by paralysis of the arms and legs, and finally death. CSF is normal and does not contain antibody to JC virus.

Laboratory Diagnosis

Urine cytologic tests can reveal the presence of JC or BK viruses by showing enlarged cells with dense basophilic intranuclear inclusions resembling those induced by cytomegalovirus (Figure 44-7 and Table 44-5). CMV inclusions, however, are smaller and have a larger halo effect (i.e., a clear zone around the inclusion but within the nuclear membrane). Histologic examination of brain tissue from cases of progressive multifocal leukoencephalopathy reveals similar cytologic changes within the oligodendrocytes. These cells are adjacent to areas of demyelination. There is little if any inflammatory cell response. Electron

microscopy can be used to visualize viral particles in brain tissue.

Culture is the definitive method for documenting active polyomavirus infection, although viral antigen may be demonstrated in tissue or cells by immunofluorescent, immunoperoxidase, and DNA hybridization techniques. Culture of BK virus can be performed in diploid fibroblast cells or Vero monkey kidney cells, both of which are available for use in clinical laboratories. JC virus grows best in primary human fetal glial cells, which are not readily available. Culture of JC virus is therefore performed only in a few research laboratories.

Treatment, Prevention, and Control

No specific treatment is available, but some stabilization or improvement may occur if the immune suppression can be reduced. The ubiquitous nature of polyomaviruses and the lack of understanding of their modes of transmission make preventing primary infection unlikely.

BIBLIOGRAPHY

Arthur RR, Keerti VS, Baust SJ, Santos GW, and Saral R: Association of BK viruria with hemorrhagic cystitis in recipients of bone marrow transplants, N Engl J Med 315:230-234, 1986.

Balows A, Hausler WJ Jr., Lennette EH, editors: Laboratory diagnosis of infectious diseases: principles and practice, New York, 1988, Springer-Verlag.

Fields BN, editor: Virology, 1985, Raven Press.

Houff SA et al: Involvement of JC virus-infected mononuclear cells from the bone marrow and spleen in the pathogenesis of progressive multifocal leukoencephalopathy, N
Engl J Med 318:301, 1988.

Howley PM: On human papillomaviruses, N Engl J Med
315:1089, 1986.

45

Hepatitis Viruses

At least four viruses cause hepatitis (Table 45-1). Although the target organ for each of these viruses is the liver, they differ greatly in their structure, mode of replication, course of disease, and mode of transmission. The hepatitis A and B agents are the best known, but in recent years a non-A, non-B (or C) agent has been described, as has the virus of hepatitis D, the delta agent. Additional non-A, non-B agents may also exist. Although a superficial similarity exists between the symptoms of hepatitis, they can be distinguished clinically to a considerable degree and definitively separated by specific laboratory tests.

Hepatitis A (1) is caused by a picornavirus containing ribonucleic acid (RNA) and is sometimes known as "infectious hepatitis"; (2) is spread by the fecal-oral route; (3) has an incubation period of approximately 1 month; (4) may produce a fulminant and even fatal disease; but (5) is not followed by chronic liver disease.

Hepatitis B, previously known as "serum hepatitis" (1) is produced by a deoxyribonucleic acid (DNA) virus; (2) is spread parenterally by blood or needles, by sexual contact, and perinatally; (3) has a median incubation period of approximately 3 months; and (4) is followed by chronic hepatitis in 5% to 10% of patients. An estimated 300,000 cases of hepatitis B occur in the United States each year, with 4,000 fatalities caused by this agent. Almost the same number of persons probably develop hepatitis A each year, but fatalities are rare.

Hepatitis non-A, non-B resembles hepatitis B in that it is spread by the same routes but differs in that it may give rise to chronic disease in 10% to 50% of patients. Non-A, non-B also differs in having a shorter incubation period. The evidence for more than one non-A, non-B virus derives partly from differences in the incubation period of non-A, non-B hepatitis.

Delta hepatitis, or hepatitis D virus (HDV), is unique in that it occurs only in patients who have active hepatitis B virus infection. HDV replicates only in the presence of actively replicating hepatitis B virus. The B virus acts as a "helper" agent for hepatitis D by providing an envelope for delta virus RNA and its antigen(s). Acute and chronic hepatitis may be caused by the delta agent.

HEPATITIS A VIRUS

Hepatitis A virus (HAV) causes infectious hepatitis and is spread by the fecal-oral route. A common epidemiologic setting for HAV infections is the consumption of contaminated shellfish or other food and water. HAV has the properties of a picornavirus and has been renamed *Enterovirus 72.*

Structure

HAV has the structural characteristics of a picornavirus (Figure 45-1). It has a 27 nm naked icosahedral capsid surrounding a positive-strand RNA genome of 2.5×10^6 daltons. As with the picornaviruses, the HAV genome has a VPg protein attached to the 5' end and polyadenosine attached to the 3' end. The capsid is resistant to acid, detergents, and bile acids and is stable even to the challenges of the gastrointestinal tract, long periods in sewage sludge, and inefficient sewage treatment (Table 45-2).

Table 45-1   Comparative Features: Hepatitis A; B; Non-A, Non-B; Delta

Feature	Hepatitis A	Hepatitis B	Non-A, Non-B	Delta
Incubation period	2 to 6 weeks (average, 4 weeks)	4 to 26 weeks (average, 13 weeks)	2 to 20 weeks	4 to 8 weeks
Virus	27 nm RNA virus	42 nm DNA virus	Hepatitis C (RNA) and possibly other viruses	Incomplete RNA virus
Onset	Abrupt (variable)	Insidious (variable)	Insidious (variable)	Abrupt (variable)
Transmission	Fecal-oral	Parenteral, sexual	Parenteral, sexual	Parenteral, sexual
Severity	Mild	Occasionally severe (up to 25% icteric)	Usually subclinical	Coinfection occasionally severe. Superinfection often severe
Fulminant hepatitis	Rare	Very rare (1% of icteric patients)	Extremely rare	Coinfection occasional
Symptoms	Fever, malaise, headache, anorexia, vomiting, dark urine, jaundice (often asymptomatic)	As with A, but 10%-20% with serum sickness–like reaction	As with A	As with A
Carrier state	None	Yes	Yes	Yes
Chronicity	0%	5%-10%	10%-50%	Coinfection: 2%-5%. Superinfection: >90%
Percentage associated with blood transfusion	Very rare	5%-10%	50%	Occurs, frequency unknown
Serology	Anti-HAV IgM fraction IgG fraction	HBsAg HBeAg Anti-HBs Anti-HBc IgM fraction IgG fraction Anti-HBe	In development	Anti-delta IgM fraction IgG fraction
Postexposure prophylaxis	Immune globulin	HBIG/hepatitis B vaccine	Immune globulin	Prevent B
Mortality rate	<0.5%	1%-2%	0.5%–1%	High
Mechanism of hepatic injury	Cytotoxic antibody response	T lymphocyte–mediated cytotoxicity	Probably direct cytopathic effect of virus	Probably direct viral hepatotoxicity
Association with cirrhosis	No	Yes	Yes	Yes
Association with primary hepatocellular carcinoma	No	Yes	No	No

Replication

Replication of HAV has not been studied as extensively as that of other picornaviruses (see Chapter 50). Laboratory isolates of HAV have been adapted to growth in continuous monkey kidney cell lines, but clinical isolates are very difficult to grow in cell culture. Unlike poliovirus, HAV replication is slow and establishes a steady-state infection with little or no cytopathology in the cells.

HAV interacts specifically with a receptor expressed on liver cells and delivers its genome into the cytoplasm to act as a messenger RNA (mRNA) for the production of a polyprotein. The polyprotein is cleaved to produce a RNA polymerase for transcription and replication and the structural proteins, VP1, 2, 3, and 4.

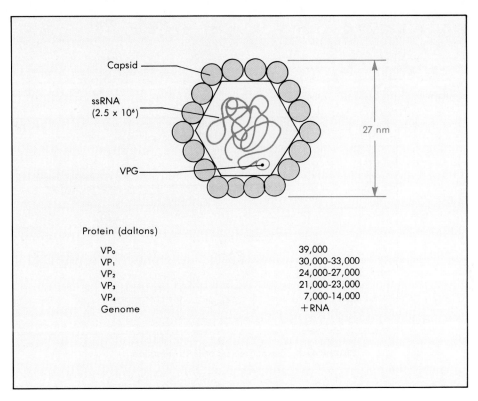

Protein (daltons)

VP_0	39,000
VP_1	30,000-33,000
VP_2	24,000-27,000
VP_3	21,000-23,000
VP_4	7,000-14,000
Genome	+RNA

FIGURE 45-1   Diagram of the proposed structure of the hepatitis A virus (HAV). The protein capsid is made up of four viral polyeptides (VP1 to VP4). Inside the capsid is a single-strand molecule of RNA (molecular weight, 2.5×10^6 daltons), which has a genomic viral protein (VPG) on the 5′ end.

Pathogenesis

The details of the pathogenesis of HAV can only be determined by analogy with other enteroviruses and limited numbers of animal studies. HAV is acquired by ingestion, probably replicates in the oropharynx and epithelial lining of the intestines, initiates a transient viremia, and is targeted to the liver (Figure 45-2). The imperviousness of the poliovirus-like capsid allows the virion to survive passage through the stomach and intestines. The virus crosses this barrier and binds to receptors on parenchymal cells of the liver and replicates. The virus can be localized by immunofluorescence in hepatocytes and Kupffer cells. Virus is produced in these cells and is released into the bile and from there into the stool. Virus is shed into the stool approximately 10 days before symptoms or antibody can be detected.

HAV replicates slowly in the liver without apparent cytopathic effect (CPE). Replication of the virus elicits

Table 45-2 Physical and Biochemical Characteristics of Hepatitis A Virus (HAV)

Characteristic	HAV
Size	25 to 29 mm
Capsid symmetry	Cubic
Virion	Unenveloped
Assembly	Cytoplasm
Nucleic acid type	Single-strand positive RNA
Polypeptides	VP1, VP2, VP3, VP4
Stability at:	
−20° to −70° C	Years
4° C	Weeks to months
50° C (1 hour)	Stable
Boil (5 minutes)	Inactivated
Stability in:	
Formalin (3 days)	Inactivated
Chlorine (30 minutes)	Inactivated

Data from Zuckerman AJ. In Balows A, Hausler W, and Lennette E: Laboratory diagnosis of infectious diseases, vol 3, New York, 1988, Springer-Verlag.

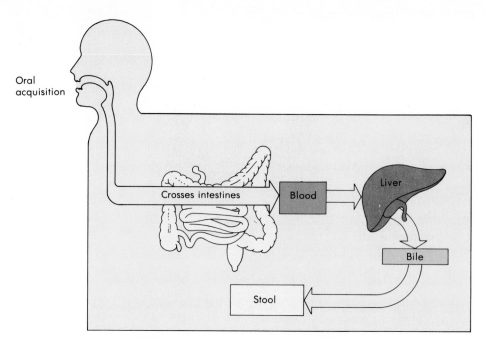

FIGURE 45-2   Pathogenesis of HAV infection.

interferon production, which works to limit virus replication to some extent but also activates natural killer cells. Antibody and complement and antibody-dependent cellular cytotoxicity also facilitate the clearance of the virus and the induction of immuno-pathology. Icterus resulting from damage to the liver occurs simultaneously with detection of antibody to the virus. Mononuclear cell infiltrates are also observed in areas affected by HAV at the same time as the release of liver enzymes.

The liver pathology caused by HAV infection is indistinguishable histologically from that caused by HBV. Liver damage is most likely caused by immuno-pathology instead of virus-induced cytopathology. However, unlike HBV, HAV cannot initiate a chronic infection and is not associated with hepatic cancer.

Epidemiology

Approximately 40% of acute hepatitis cases result from HAV. Spread is by person-to-person contact, usually via a fecal-oral route, or by exposure to contaminated food or water. Close personal contact (e.g., household or sexual) facilitates spread. The virus is present in blood only transiently, but in feces high concentrations of virus are present for several weeks, especially in the 2 weeks before the onset of jaundice. Virus is not present in urine or other body fluids, and a chronic carrier state does not occur.

HAV is distinguished by its ability to survive many months in fresh and salt water. HAV is resistant to detergents, acid, and temperatures elevated as high as 60°C, which promotes its spread into food, even after handwashing.

Raw or improperly treated sewage can taint the water supply and contaminate shellfish. Shellfish are very efficient filter feeders and can concentrate the viral particles, even from dilute solutions.

Day care settings are a major source of spread of the virus. HAV can spread among classmates and to their parents. Since the children and personnel in day care centers are somewhat transient, the number of contacts at risk for HAV infection from a single day care focus can be great.

The incidence of HAV infection is directly related to poor hygienic conditions and overcrowding. Most individuals infected with HAV in developing countries are children who have mild illness and then life-long immunity to reinfection. In more highly developed countries, infection occurs later in life. The sero-

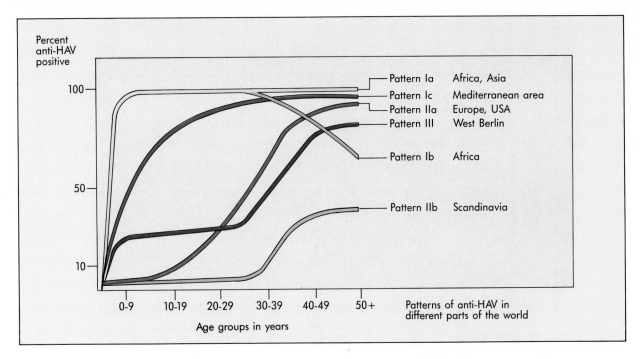

FIGURE 45-3   Different patterns of age-specific prevalence of anti-HAV in different parts of the world. Note that HAV infection has become an adult disease in Western Europe and North America.

positivity rate of adults ranges from a low 13% of the adult population in Sweden to 88% in Taiwan and 97% in Yugoslavia, with a 41% to 44% rate in the United States (Figure 45-3).

Clinical Syndromes

The symptoms caused by hepatitis A are very similar to those for hepatitis B, which are a function of immune-mediated damage to the liver. Infection of children is generally milder than for adults and may be asymptomatic. HAV symptoms occur abruptly 15 to 50 days after exposure and increase for 4 to 6 days before the icteric (jaundice) phase. Jaundice is observed in 2 of 3 adults but only 1 or 2 of 10 children. Symptoms generally wane during the jaundice period. Virus shedding in the stool precedes the onset of symptoms by approximately 14 days but ceases before the cessation of symptoms.

Fulminant hepatitis in HAV infection occurs in 1 to 3 persons per 1,000, with an 80% mortality. Immune complex–related symptoms (e.g., arthritis, rash) occur infrequently in HAV disease.

Laboratory Diagnosis

The diagnosis of hepatitis A is generally made from the time course of the clinical symptoms, identification of a known infected source, and most reliably by specific serologic tests (Figure 45-4). The best means for demonstrating acute HAV infection is the presence of anti-HAV IgM, as measured using enzyme-linked immunosorbent assay (ELISA) or radioimmunoassay (RIA) procedures. Detection of HAV antigen or infectious virus in the stool is possible but difficult, because efficient tissue culture systems for growing the virus are not available, and virus may be shed before symptoms are observed.

Treatment, Prevention, and Control

Spread of hepatitis A is reduced by interrupting the fecal-oral spread of the virus. Potentially contam-

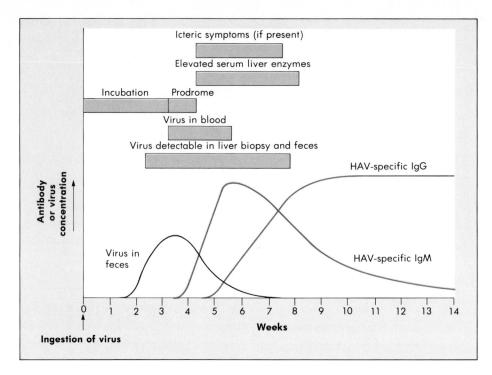

FIGURE 45-4   Time course of HAV infection.

inated water or food, especially uncooked shellfish, must be avoided. Proper handwashing, especially in day care centers, mental hospitals, and other care facilities, is very important. Chlorine treatment of drinking water is generally sufficient to kill the virus. Prophylaxis with immune serum globulin (ISG) given before or early in the incubation period (i.e., less than 2 weeks after exposure) is 80% to 90% effective in preventing clinical illness.

Hepatitis A would seem to be a candidate for vaccine development because only one serotype has been identified. The prospect of a vaccine for HAV has become more of a reality since the development of tissue culture systems for cultivation of the virus and molecular biologic means of subunit vaccine development.

HEPATITIS B VIRUS

HBV is the major member of the hepadnaviruses, a small family of enveloped DNA viruses. Other members of this family were recently discovered and include woodchuck, ground squirrel, and duck hepa-

titis viruses. These viruses have very limited tissue tropisms and host ranges. HBV infects the liver and possibly the pancreas of only humans and chimpanzees. The limited host range and the inability to grow HBV in cell culture has restricted study of the molecular biology of this virus until recently.

Structure and Replication

Hepatitis B is an enveloped DNA virus with several unusual properties (Figure 45-5). The genome is a circular, partly double-strand DNA, encodes a reverse transcriptase, and replicates through an RNA intermediate. The nucleocapsid of HBV also includes a DNA polymerase and a protein kinase surrounded by the core antigen **(HBcAg)** and the hepatitis **HBe** antigen. The HBe antigen is found on the same polypeptide as HBc antigen and is a major constituent of the capsid.

The virion, also called the **Dane particle,** is 42 nm in diameter. Incomplete viral particles in the serum of infected individuals, outnumber the actual virions. These particles can be spheric (22 nm in diameter) or filamentous (22 nm wide and 100 to 300 nm in length)

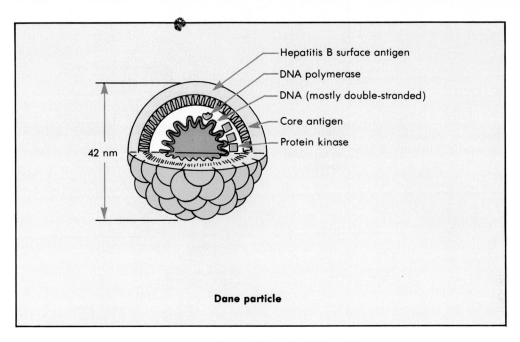

FIGURE 45-5   Hepatitis B virus (Dane Particle) and its components.

(Figure 45-6). The virions are unusually stable despite having an envelope. They resist treatment with ether, low pH, freezing, and moderate heating. This assists their transmission from one individual to another.

The surface antigen **(HBsAg)** originally termed the **Australia antigen,** is found in the envelope and on the surface of the spheric and filamentous particles. The HBsAg elicits protective immunity and is composed of lipid and seven or more polypeptides. The major components are a structurally related polypeptide, p25, and glycopeptide, p29. These polypeptides contain the group (a) and type-specific determinants of HBV (d or y and w or r). Combinations of these antigens (e.g., ady, adw) result in eight subtypes of HBV are useful epidemiologic markers.

HBV interacts with unique receptors found in the

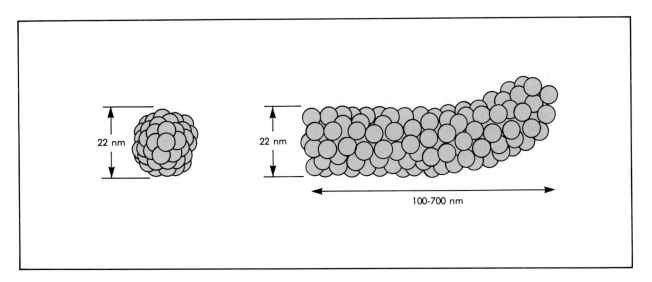

FIGURE 45-6   Structure of incomplete hepatitis B virus particles.

FIGURE 45-7 Proposed pathway for replication of the genome of hepatitis B–like viruses.

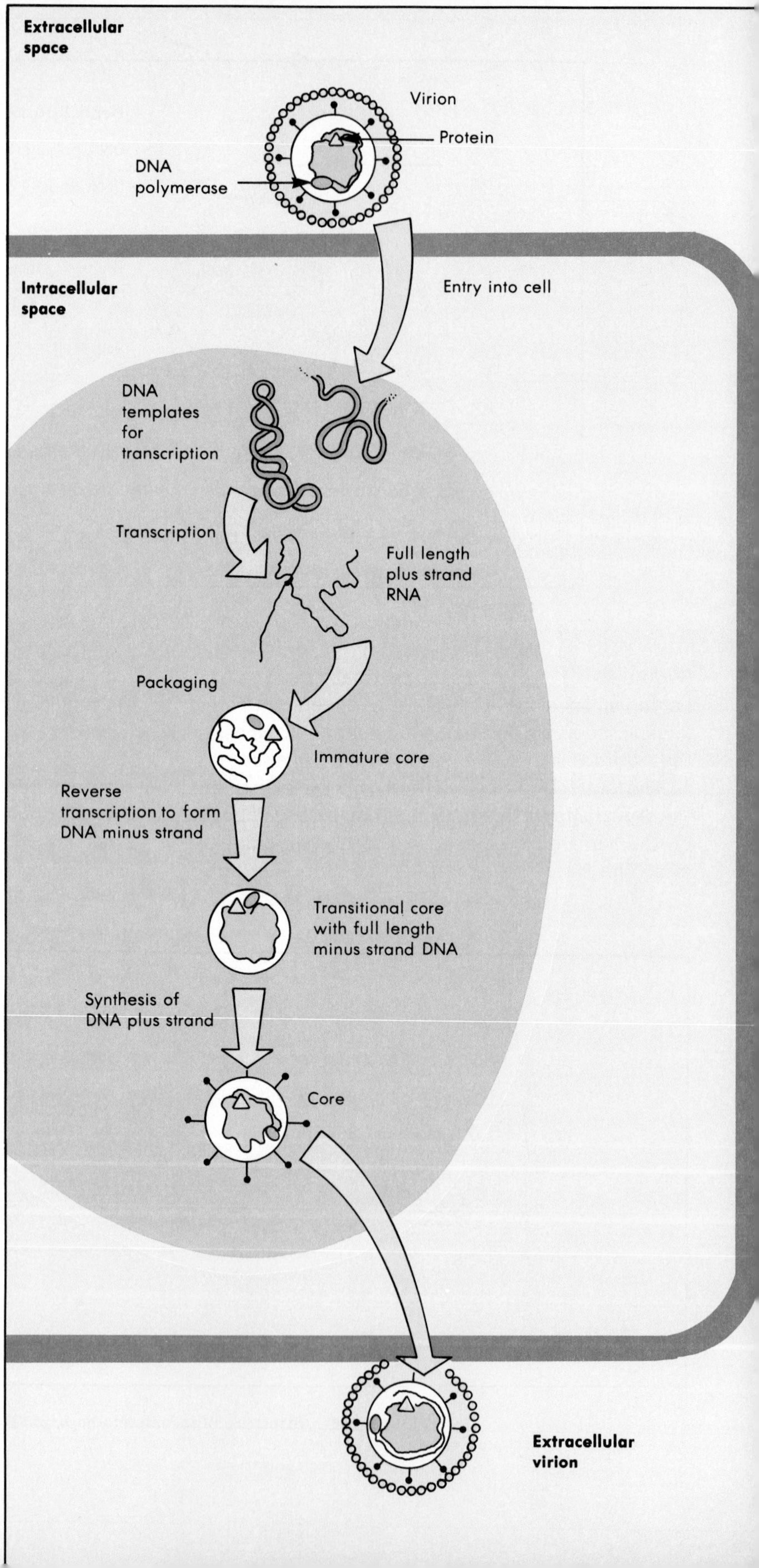

liver but on few if any other tissues. The viral attachment protein is part of the HBsAg complex, as suggested by the neutralizing capacity of anti-HBsAg. HBsAg binds to human and chimpanzee serum albumin, and this interaction may target the virus to the liver.

On penetration into the cell, the nucleocapsid and genome are delivered to the nucleus. The partial strand of the genome is completed to form a complete double-strand DNA circle and then transcribed into individual mRNAs by the host DNA-dependent RNA polymerases. A full-sized (3,200 base pairs) mRNA transcript is also synthesized as a template for replication of the genome (Figure 45-7).

As with the retroviruses, HBV uses a reverse transcriptase to synthesize a complementary DNA (cDNA) from the full-sized positive-strand RNA. Unlike the retroviruses, the cDNA is the negative strand of the virion rather than an intermediate form of the genome to be integrated. The RNA template and polymerase are encapsulated within the nucleocapsid, and negative-strand DNA synthesis occurs on a protein primer, also included in the core. The RNA is degraded as the DNA is synthesized. The positive-strand DNA is then synthesized from the negative DNA template as long as substrates are available. The capsid is enveloped within HBsAg-containing membranes as soon as it is formed, capturing genomes containing RNA-DNA circles with different lengths of RNA. Linear forms of the genome have also been detected. Continued degradation of the remainder of the RNA in the virion yields a partly double-strand DNA genome. The virion is probably released through an exocytic pathway of the hepatocyte and not by cell lysis.

In addition to replication, the entire genome can also be integrated into the host genome. HBsAg, but not other proteins, can often be detected in the cytoplasm of cells containing integrated HBV DNA. The significance of the integrated DNA to replication of the virus is not known, but integrated DNA has been found in hepatocellular carcinomas in humans, woodchucks, and ducks following infection by the corresponding hepatitis virus.

Pathogenesis

HBV can cause acute or chronic hepatitis. Which of these two occurs seems to be determined by the individual's immune response to the infection. Production of virus and a high level of HBsAg continues, and the particles are found in the blood until the infection is resolved. Detection of both the HBsAg and the HBeAg components of the virion in the blood indicates ongoing, active infection (Figure 45-8).

Acquisition of HBV generally requires injection into the bloodstream. The virus must be delivered to the liver to establish infection. The virus replicates, and large amounts of HBsAg are released into the blood in addition to virions. Initiation of virus replication may be as short as 3 days from acquisition, but symptoms may not be observed for 45 days or much longer, depending on the infecting dose of virus, the route of infection, and the individual.

Replication of the virus is not cytopathic and proceeds for relatively long periods without causing liver damage (i.e., elevation of liver enzymes) or symptoms. During this time, copies of the HBV genome integrate into the hepatocyte chromatin and remain latent.

Cell-mediated immunity and inflammation are responsible for both the resolution of the HBV infection and the symptomatology. The actual target for cell-mediated immune protection is not known, but a liver protein (liver-specific protein, or LSP) elicits an autoimmune reaction. Acute cases, with jaundice and other significant symptoms, are generally of short duration, indicating efficient resolution. Patients with chronic hepatitis generally have mild initial symptoms (Figure 45-9). HBsAg production persists in individuals with T-cell deficiencies, but these individuals generally have milder symptoms.

Immune complexes formed by HBsAg and its antibody, anti-HBs, cause damage through hypersensitivity reactions, leading to such problems as vasculitis, arthralgia, rash, or kidney problems.

During the acute phase of infection, the liver parenchyma shows degenerative changes consisting of cellular swelling and necrosis, especially in hepatocytes surrounding the central vein of a hepatic lobule. The inflammatory cell infiltrate is mainly composed of lymphocytes. Resolution of the infection allows the parenchyma to regenerate. Fulminant infections, activation of chronic infections, or coinfection with the delta agent can lead to permanent liver damage and cirrhosis.

Epidemiology

In the United States more than 300,000 persons are infected by HBV each year. This number is impressive, but in underdeveloped nations up to 15% of the

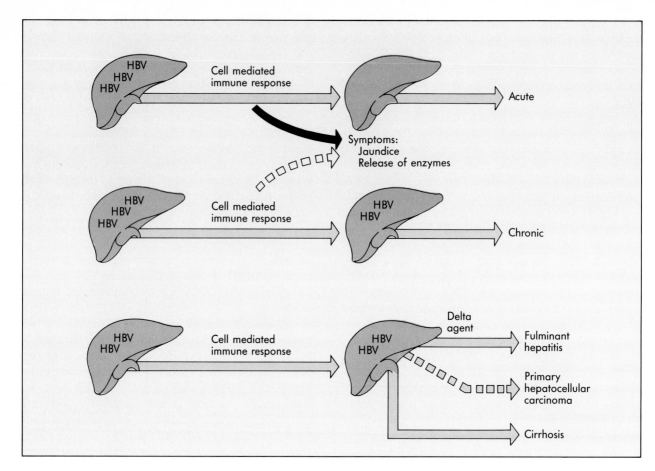

FIGURE 45-8   Pathogenesis of HBV infection.

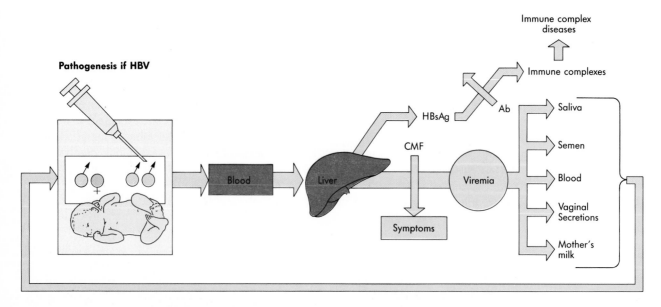

FIGURE 45-9   Major determinants in acute and chronic HBV infection.

Table 45-3 Expected Hepatitis B Virus Prevalence in Various Population Groups

Group	Prevalence of Serologic Markers of HBV Infection	
	HBsAg (%)	All Markers (%)
HIGH RISK		
Immigrants and refugees from areas of high HBV endemicity	13	80-85
Clients in institutions for mentally retarded persons	10-20	35-80
Users of illicit parenteral drugs	7	60-80
Homosexually active males	6	35-80
Household contacts of HBV carriers	3-6	30-60
Patients on hemodialysis units	3-10	20-80
INTERMEDIATE RISK		
Prisoners (male)	1-8	10-80
Staff of institutions for the mentally retarded	1	10-25
Health care workers with frequent blood contact	1-2	15-30
LOW RISK		
Health care workers with no (or infrequent) blood contact	0-3	3-10
Healthy adults (first-time volunteer blood donors)	0-3	3-5

From Hoofnagle JH: Lab Med 14:705, 1983.

population are infected during birth or childhood.

No lower animal reservoir exists for hepatitis B or the other hepatitis viruses; humans are the reservoir and vector. The virus is spread by direct person-to-person contact, especially by chronic carriers of the virus. A chronic carrier is defined as an individual who is HBsAg positive on two occasions at least 6 months apart. Chronic carrier status results partly from impaired immune function. For example, up to 90% of infants infected perinatally are chronic carriers. Overall, in the United States 0.1% to 0.5% of the general population are chronic carriers of HBsAg, and a lesser number are carriers of HBeAg, which is a better marker of infectivity. Approximately 6% of homosexual men and a higher percentage of intravenous drug abusers are chronic carriers. Carrier status may be lifelong. The routes of spread of hepatitis B are percutaneous (e.g., needle sharing, acupuncture, ear piercing, tatooing) or by very close personal contact with exchange of secretions (e.g., sexual, childbirth). The virus can be transmitted by contaminated blood or blood products, but serologic screening of donor units in blood banks has greatly reduced this risk. Given the modes of transmission, groups at high risk for infection have been identified (Tables 45-3 and 45-4).

Table 45-4 Risk of Hepatitis B in Health Care Personnel Ranked by Standardized Prevalences of Serologic Positivity*

Standardized Prevalence	Occupational Subcategory
27	Medical technologists
26	Blood bank and hemodialysis technicians
19	EKG and pulmonary technicians
18	Attending physicians
17	Dentists, nurses, anesthetists, surgical technicians
15	Aides and orderlies; administrative assistants; those employed in laundry, housekeeping, supply departments
13	Laboratory assistants, inhalation therapists, food service workers
12	Chemists, microbiologists, radiation and nuclear medicine technologists
11	Nurses, cytology technologists, those in building maintenance
10	Hygienists, social workers, dieticians, clerical workers
9	Research technicians, dental hygienists, administrators, pharmacists, and their assistants

From Gerety RJ: Hepatitis B, New York, 1985, Academic Press, Inc.
*Compared to control population in whom prevalence was from 4% to 6%. Information provided by the Centers for Disease Control, Atlanta.

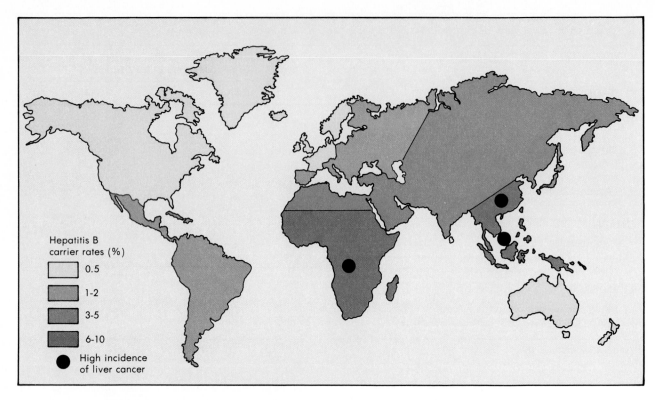

FIGURE 45-10   Worldwide prevalence of hepatitis B carriers and primary hepatocellular carcinoma. (Redrawn from Szmuness W et al: J Med Virol 8:123, 1981.)

High rates of seropositivity are observed in Italy, Greece, Africa, and Southeast Asia (Figure 45-10). In some areas of the world (southern Africa and southeastern Asia) the seroconversion rate of the population is as high as 50%. Primary hepatocellular carcinoma is also endemic in these regions.

One of the major concerns about HBV is its association with **primary hepatocellular carcinoma** (PHC). This type of carcinoma probably accounts for 250,000 to 1 million deaths per year worldwide; in the United States approximately 5,000 deaths per year are attributed to PHC.

Clinical Syndromes

Acute Infection The clinical presentation of HBV in children is less severe than in adults and may be asymptomatic. Clinically apparent illness occurs in up to 25% of those infected with HBV (Figures 45-11 and 45-12).

Hepatitis B infection is characterized by a long incubation period and an insidious onset. Symptoms during the prodromal period may include fever, malaise, and anorexia. This is followed by nausea, vomiting, abdominal discomfort, fever, and chills. The classic icteric symptoms of liver damage—jaundice, dark urine, and pale stools—follow soon thereafter. Recovery is indicated by a decline of fever and renewed appetite.

Fulminant hepatitis occurs in approximately 1% of icteric patients and may be fatal, with much more severe symptoms and indications of severe liver damage, such as ascites and bleeding (see Figure 45-12). Instead of the enlarged liver usually seen in patients with acute hepatitis, the liver shrinks.

Chronic Infection Chronic hepatitis occurs in 5% to 10% of HBV infections, usually after mild or inapparent initial disease. Chronic hepatitis may only be detected by elevated liver enzyme levels on a routine blood chemistry profile, but in up to 10% of patients may cause cirrhosis and liver failure. Chronically infected individuals are the major source for

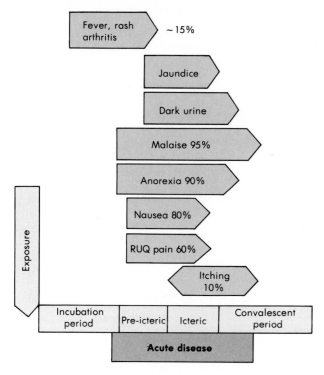

FIGURE 45-11   Symptoms of typical acute viral hepatitis B infection. The timing and incidence of the major symptoms during the four clinical periods of this disease are shown. (Redrawn from Hoofnagle JH: Lab Med 14:705, 1983.)

spread of the virus and are susceptible to fulminant disease because of coinfection with delta virus.

Primary Hepatocellular Carcinoma The World Health Organization estimates that 80% of all cases of PHC can be attributed to chronic hepatitis B infections. PHC is usually fatal and is one of the three most common causes of cancer mortality in the world. In Taiwan at least 15% of the population are HBV carriers, and nearly half of the individuals die of either PHC or cirrhosis.

The mechanisms by which HBV initiates PHC are not known. The HBV genome is integrated into PHC cells, and the cells express HBV antigens. The inability of the immune system to resolve an HBV infection during chronic hepatitis may allow cells with integrated HBV genomes to remain in the liver until a carcinogenic event occurs much later in life. The latency period between HBV infection and PHC may be as short as 9 years or as long as 35 years.

Laboratory Diagnosis

The initial diagnosis of hepatitis can be made from the clinical symptoms and the presence of liver enzymes in the blood. However, identification of the specific viral agent requires quantitation of the HBs

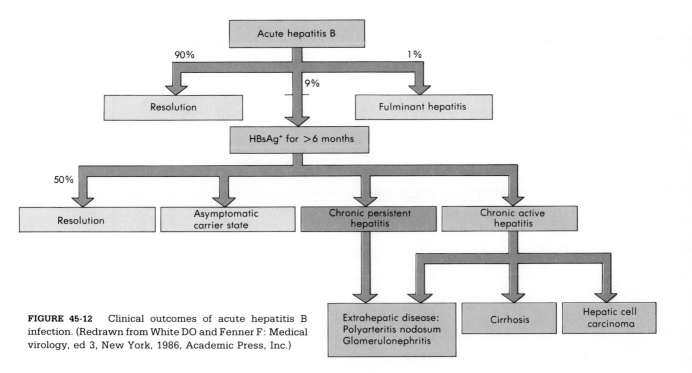

FIGURE 45-12   Clinical outcomes of acute hepatitis B infection. (Redrawn from White DO and Fenner F: Medical virology, ed 3, New York, 1986, Academic Press, Inc.)

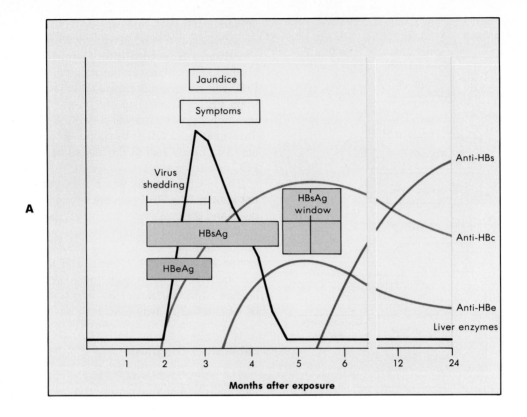

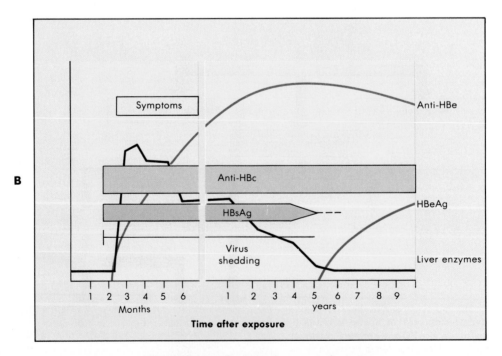

FIGURE 45-13   For legend see opposite page.

Table 45-5   Interpretation of Serologic Markers of HBV Infection

Serologic Reactivity

Pattern	HBsAg	Anti-HBs	Anti-HBc	Interpretations
1	+	−	−	Early (presymptomatic) acute type B hepatitis
2	+	−	+	Acute* or chronic type B hepatitis
3	+	+	+	Acute* or chronic type B hepatitis
4	−	+	+	Recovery from type B hepatitis
5	−	−	+	Long after HBV infection, or immediate recovery phase from type B hepatitis*, or low-level carrier state
6	−	+	−	Long after HBV infection or immunization with HBsAG

From Hoofnagle JH: Lab Med 14:705, 1983.
*Anti-HBc IgM should be present.

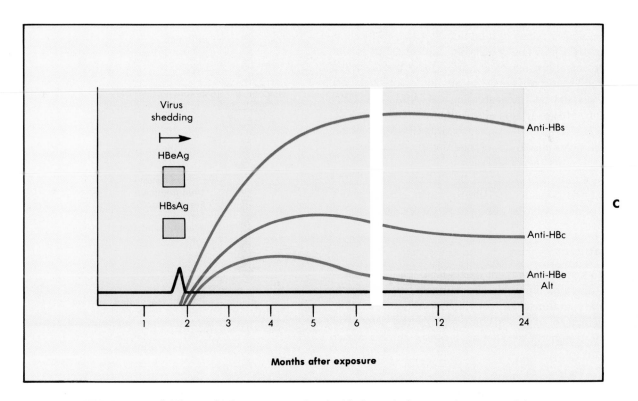

C

FIGURE 45-13   A, The serologic events associated with the typical course of acute type B hepatitis. **B,** Development of the chronic hepatitis B virus carrier state. **C,** Clinical, serologic, and biochemical course of a subclinical, asymptomatic hepatitis B virus infection. *HBsAg,* Hepatitis B surface antigen; *HBeAg,* hepatitis B e antigen; *Alt,* alanine aminotransferase; *anti-HBs,* antibody to HBsAg; *anti-HBc,* antibody to hepatitis B core antigen; *anti-HBe,* antibody to HBeAg. (**A** and **B** redrawn from the Annual Review of Medicine, Vol 32, © 1981, by Annual Reviews, Inc; **C** redrawn from Hoofnagle JH: Lab Med 14:705, 1983.)

antigen and evaluation of the antibody response to HBs and HBc.

Understanding the serology of HBV infection is essential to following the course of the disease (Table 45-5). Acute and chronic HBV infections can be distinguished by the prevalence of HBsAg in the serum and the pattern of antibodies to HBs and Hbc antigens. The serologic pattern of HBV infection (Figure 45-13) can be understood by remembering that B lymphocytes cannot make antibody until they detect antigen and that antibody bound into an antibody-antigen complex cannot be detected.

HBsAg, HBcAg, and HBeAg are secreted into the blood during virus replication. The HBsAg and HBcAg are particulate, whereas the HBeAg is hidden, within the virion. No test is readily available for HBcAg, but HBsAg and HBeAg can be detected in clinical laboratory assays. Detection of HBeAg is the best correlate to the presence of infectious virus.

Antibodies are produced against HBcAg and HBsAg, but so much HBsAg is produced during the symptomatic phase of infection that the antibody is complexed and not detectable. Indication of recent acute infection, especially during the period when neither HBsAg nor anti-HBs can be detected, is best established by measurement of IgM anti-HBc. Release of HBcAg and HBeAg from infected cells on cytolysis allow development of anti-HBeAg to occur and indicate resolution of infection by the cell-mediated immune response.

A chronic infection can be distinguished by the continued presence of HBeAg and/or HBsAg and a lack of detectable antibody to these antigens.

Treatment

No specific treatment exists for hepatitis B infection, although limited studies suggest that interferon, adenine arabinoside, corticosteriods, and azothioprine alone or in certain combinations may have a role in the treatment of different types of chronic active hepatitis.

Prevention and Control

Transmission of HBV by blood or blood products has been greatly reduced by screening of blood donors' serum for the presence of HBsAg and anti-HBc. Additional efforts to prevent hepatitis B consist of avoiding intimate personal contact with a carrier of HBV and avoiding the life-styles that facilitate spread of the virus. Household contacts and sexual partners of HBV carriers are at increased risk, as are hemodialysis patients, recipients of pooled plasma products, health care workers exposed to blood, and babies born of HBV-carrier mothers.

Prophylaxis

HBV Vaccine Two HBV vaccines exist, one derived from human plasma and one genetically engineered. The former is a suspension of alum-adsorbed, 22 nm hepatitis B surface antigen particles obtained

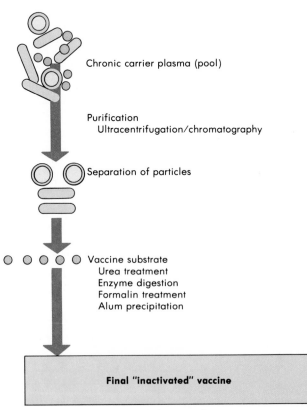

Chronic carrier plasma (pool)

Purification
Ultracentrifugation/chromatography

Separation of particles

Vaccine substrate
Urea treatment
Enzyme digestion
Formalin treatment
Alum precipitation

Final "inactivated" vaccine

Safety tested
Purity tested
Potency tested: In vitro/in vivo

FIGURE 45-14 The manufacture of plasma-derived HBsAg particle vaccine (MSD). The "spheres" are separated from "tubules" and HBV by ultracentrifugation and subjected to procedures to accomplish purification and virus inactivation. The spheres are then adsorbed to alum to yield the final alum-adsorbed, inactivated, plasma-derived HBsAg particle vaccine. (Redrawn from Gerety RJ: Hepatitis B, New York, 1985, Academic Press, Inc.)

from human plasma by ultracentrifugation (Figure 45-14). To inactivate any viable virus in plasma (including retroviruses) the separated antigen suspension is heated and chemically inactivated. The recombinant vaccine is produced by the insertion of a plasmid containing the gene for HBsAg into a yeast, *Saccharomyces cerevisiae*. HBsAg is then harvested by lysing the yeast cells and separating the antigen by chromatography. The purified HBsAg is then filter-sterilized, formalinized, adsorbed to alum, and preserved with thiomersol.

Both vaccines must be given in a series of three injections, with the second and third given 1 and 6 months, respectively, after the first. More than 95% of individuals receiving the full three-dose course will develop antibody.

Preexposure vaccination primarily is recommended for individuals in high-risk groups who are known to be HBsAb negative (see Table 45-3). Screening for existing HBsAb in high-risk groups is probably not cost-effective but is useful in groups with lower risk (e.g., hospital workers). Vaccination is useful even after exposure for two specific groups:

1. Newborns of HBsAg-positive mothers
2. Persons with accidental percutaneous or permucosal exposure to blood or secretions from an HBsAg-positive individual

Newborns should receive both HBV vaccine as well as hepatitis B immune globulin (HBIG; see following section) within 12 hours of birth and subsequent doses of HBV vaccine only at ages 1 month and 6 months.

Hepatitis B Immune Globulin (HBIG) This product is prepared exactly as is ordinary immune globulin (IG), except that the plasma used is selected for a high titer of anti-HBs. HBIG is indicated in postexposure prophylaxis for newborn infants of HBsAg-positive mothers and for persons with accidental percutaneous or permucosal exposure to blood or secretions from individuals who are HBsAg positive.

HEPATITIS C AND NON-A, NON-B

Viruses that cannot be classified as hepatitis A or B or associated with other known viral diseases are termed non-A, non-B hepatitis (NANBH) viruses. Recently a viral agent has been recovered and characterized that may account for 90% of the cases of NANBH infection. It is referred to as hepatitis C, but

Table 45-6	Known Properties of Hepatitis C Virus
Characteristic	**Hepatitis C (NANBH)**
Structure	Enveloped
Genome	Positive-strand RNA, 10,000 nucleotides
Family	Flaviviridae (?)

complete details of the structure, biochemistry, physiology, and pathogenesis of this agent are not yet available (Table 45-6).

Epidemiology

The agent(s) can be spread parenterally or sexually, each resulting in a different time course and severity of disease (Table 45-7). Hepatitis following parenteral distribution of the agents is often seen in transfusion recipients, intravenous drug users, and hemophiliac persons receiving factor VIII or IX. The course of disease classified as NANBH may differ, suggesting that more than one agent may be responsible for NANBH.

Clinical Syndromes

NANBH may have a short (4 to 30 days) or long (3 to 4 months) incubation period. The evidence for more than one NANBH virus derives partly from the differing incubation periods. In its acute form, NANBH is similar to acute hepatitis A and B, but the inflammatory response is less and the symptoms are usually mild. The severity of disease varies depending on the agent causing the disease and host factors, but chronic infection occurs in 10% to 50% of patients.

Laboratory Diagnosis

Until now a diagnosis of NANBH has been made by exclusion of the other hepatitis agents, as determined serologically. Given the isolation of a hepatitis C virus, a specific antibody assay for infection with this virus is expected very soon.

Treatment

Immune globulin has not been shown to be effective definitely in the prophylaxis of NANBH, even

Table 45-7 Epidemiologic and Clinical Features of Hepatitis A and Non-A, Non-B Hepatitis

	Hepatitis A	Non-A, Non-B Hepatitis
Incubation period	2-6 weeks	2-20 weeks
Principal age	Children	?
Seasonal incidence	Throughout year with autumn peak	Throughout year
Route of infection	Fecal-oral	Parenteral, sexual
Virus is in:		
Blood	2 weeks before to 1 week after jaundice	Weeks to years
Stool	2 weeks before to 2 weeks after jaundice	Absent (?)
Urine	Rare	Absent (?)
Saliva	Rare	?
Clinical features		
Onset	Abrupt	Insidious
Fever	Common	Less common
Complications	Uncommon, no chronicity	Chronicity in 10%-50%
Mortality (icteric patients)	<0.5%	0.5%-1%
Laboratory features		
Transaminase elevation	1-3 weeks	1-6 or more months

Modified from Jawetz E, Melnick JL, Adelberg EA, et al: Medical microbiology, ed 18, Los Altos, Calif, 1988, Appleton & Lange.

though it is typically given to healthcare workers for percutaneous exposures to blood from patients.

Prevention and Control

The identification of hepatitis C and a test for its detection may help in screening the blood supply for at least one type of NANBH. Currently the presence of elevated liver enzymes, such as alanine transferase (ALT) and leucine amino transferase (LAT), in donor serum provides nonspecific screening procedures for NANBH among potential blood donors. Also, blood banks are testing for anti-HBc as a "surrogate" test for NANBH on the assumption that individuals who have had hepatitis B in the past are more likely to have had hepatitis NANBH as well.

DELTA AGENT

Structure and Replication

The delta agent, also called hepatitis D virus, is a defective satellite virus that can only replicate in HBV-infected cells. It is a viral parasite, proving that "even fleas have fleas." The delta agent resembles plant virus satellite agents in its size and genome structure (Figure 45-15). These agents also require a helper virus for replication.

The RNA genome of this virus is so small that it cannot encode all the functions necessary for its replication and structure. The delta agent, approximately 35 to 37 nm in diameter, contains a 1.75 kilobase RNA genome that encodes the 68,000-dalton delta antigen but little else.

The actual details of delta agent replication are not known but may be surmised. The HBsAg package around the delta agent should allow it to bind and be internalized into cells in the same manner as HBV. Once inside the cells, the delta antigen protein is translated and the genome replicated. The delta antigen is found in the nucleus of infected cells. The HBV reverse transcriptase or a polymerase activity of the delta antigen must participate in replication of the genome. The genome, delta antigen, and HBsAg associate together and are then released from the cell.

Pathogenesis

The delta agent is spread through means similar to those of HBV, in blood and potentially in semen and vaginal secretions. It can only replicate and cause disease in individuals with active HBV infections. Since

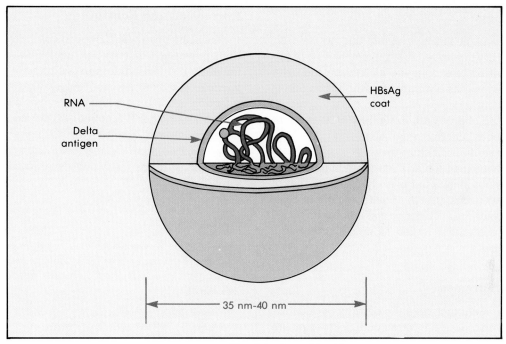

FIGURE 45-15   The delta hepatitis virion.

the two agents are transmitted by the same routes, an individual can be **coinfected** with HBV and delta agent at the same time, or a person with chronic HBV can be **superinfected** with delta agent to cause disease.

On entry into the bloodstream, the HBsAg surrounding the delta agent targets the agent to the liver. Superinfection of an individual allows the agent to replicate in many hepatocytes already infected with HBV. Delta antigen is efficiently produced in these cells and found in the nucleus and cytoplasm. Replication of the delta agent results in cytotoxicity by unknown means. A more rapid, severe progression occurs after superinfection of an HBV carrier than

during an initial challenge with both viruses during coinfection (Figure 45-16). Liver damage occurs in both cases. A persistent delta agent infection is often established in HBV carriers on resolution of the disease.

Antibodies are elicited against the delta agent. However, protection probably relates to the immune response to HBsAg, since it is the external antigen and viral attachment protein. Unlike hepatitis B, hepatocyte necrosis is observed in the absence of mononuclear infiltrates. Histologic examination indicates that damage to the liver occurs by the direct cytopathic effect of the delta agent combined with the underlining immunopathology of the HBV disease.

Consequences of delta virus infection

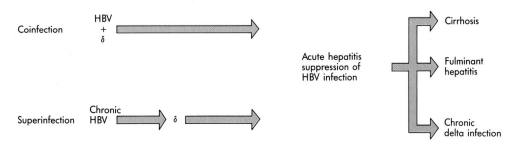

FIGURE 45-16   Consequences of delta virus infection.

Epidemiology

The delta agent infects children and adults with underlying HBV infection. Approximately 40% of fulminant hepatitis infections are caused by the delta agent. The agent has a worldwide distribution and is endemic in southern Italy, the Amazon basin, parts of Africa, and the Middle East. Epidemics of delta agent infection occur in North America and Western Europe, usually in illicit drug users.

The mechanism of spread of delta agent in endemic regions is not known but is likely to mimic that of HBV. Perinatal transmission of delta agent has been observed in one case. Epidemic spread of delta agent is mainly by parenteral routes, as with shared hypodermic needles.

Clinical Syndromes

The delta agent increases the severity of hepatitis B infections. Individuals infected with delta agent are much more likely to develop fulminant hepatitis than with the other hepatitides. Coinfection with HBV intensifies the symptoms of acute hepatitis. Chronic infection with delta agent can occur in those individuals with chronic HBV.

Laboratory Diagnosis

Detection of the delta antigen and antibodies is the only way to determine the presence of the agent. ELISA and RIA procedures are available. The presence of the delta antigen in the blood during the acute phase of disease can be detected on detergent treatment of the serum. HBsAg and HBeAg levels may also drop during delta agent replication.

Treatment

No known specific treatment exists for delta hepatitis.

Prevention and Control

Since the delta agent depends on HBV for replication, prevention of infection with HBV will prevent hepatitis D infection. Immunization with hepatitis B vaccine protects against subsequent delta virus infection, since there would be no viable HBV to perform the "helper" function. If an individual has already acquired hepatitis B, delta agent infection may be prevented by reducing exposure to illicit intravenous drugs or homosexual carriers of this virus.

BIBLIOGRAPHY

Choo QL, Kuo G, Weiner AJ, Overby LR, Bradley DW, and Houghton M: Isolation of a cDNA clone derived from a blood borne non-A, non-B viral hepatitis genome, Science 244:359-362, 1989.

Feinstone SM: non A, non B hepatitis. In Belshe RB, editor: Textbook of human virology, Littleton, Mass, 1984, PSG Publishing Co.

Fields BN, editor: Virology, New York, 1985, Raven Press.

Frosner G: Hepatitis A virus. In Belshe RB, editor: Textbook of human virology, Littleton, Mass, 1984, PSG Publishing Co.

Gerety RJ: Hepatitis B, New York, 1985, Academic Press, Inc.

Hoofnagle JH: Type A and type B hepatitis, Lab Med 14:705-716, 1983.

Lennette EH, Halonen P, Murphy FA, editors: Laboratory diagnosis of infectious diseases: principles and practice, New York, 1988, Springer-Verlag.

London WT, and Buetow K: Hepatitis B virus and primary hepatocellular carcinoma, Cancer Invest 6:317-326, 1988.

Lutwick LI: Hepatitis B virus. In Belshe RB, editor: Textbook of human virology, Littleton, Mass, 1984, PSG Publishing Co.

Robinson W, Koike K, and Will H, editors: Hepadna virus, New York, 1987, Alan R Liss, Inc.

Verme G, Bonino F, and Rizzetto M, editors: Viral hepatitis and delta infection, New York, 1983, Alan R. Liss, Inc.

CHAPTER 46

Parvoviruses

There are three genera within the family Parvoviridae, but only the *Parvovirus* genus contains a human pathogen. The most important pathogenic species in this genus is called *Parvovirus* B19. This virus is also known as the human parvovirus-like agent (PVLA) or the serum parvo-like virus (SPLV). It is thought to be the cause of erythema infectiosum (Fifth disease) and is responsible for episodes of aplastic crises in patients with chronic hemolytic anemias. B19 is also associated with arthritis and intrauterine infection. A second genus, the **dependoviruses** (formerly called **adeno associated viruses**), commonly infects humans but only in association with a second helper virus, usually an adenovirus. Dependoviruses do not appear to cause illness nor do they modify infection by their "helper" viruses. Members of the third genus in the family infect only insects.

STRUCTURE AND REPLICATION

B19 virus is an extremely small (18-26 nm in diameter), non-enveloped, icosahedral virion (Figure 46-1). Growth in an in vitro cell culture system has been unsuccessful, therefore limiting study of physical and chemical properties of the virus. B19 does not appear to agglutinate red blood cells, and only one serotype is known to exist.

B19 virus genome contains one linear single-stranded DNA molecule with a molecular weight of $1.5\text{-}1.8 \times 10^6$ (5.5 kb in length). Plus and minus DNA strands are packaged into separate virions. The viral genome codes for three structural and one nonstructural protein (Box 46-1).

The replicative cycle includes infection of mitoti-cally active cells such as erythroid precursor cells. Transcription, replication, and assembly occur in the host cell nucleus, with release of virus associated with nuclear and cytoplasmic membrane degeneration (Figure 46-2). Steps in DNA replication most likely require host cell functions present only in the late "S" phase, necessitating a mitotically active host cell.

PATHOGENESIS

Intranasal inoculation of volunteers with B19 virus suggests that virus first replicates in the upper respiratory tract, followed by viremia and replication in erythroid precursor cells in the bone marrow (Figure 46-3). Infection of a normal host may result in mild,

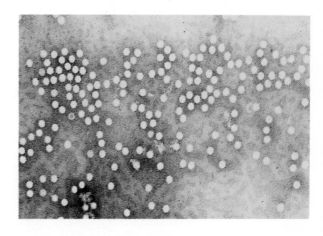

FIGURE 46-1   Electron micrograph of parvovirus. (Courtesy Center for Disease Control, Atlanta.)

Box 46-1   Map of Parvovirus Genome

1. Single-stranded linear DNA genome
   Approximately 5.5 kilobases (kb) in length
   Plus and minus strands packaged into separate B19 virions with approximately equal frequency
2. Coding region for B19 on plus strand
3. Separate coding regions for nonstructural (NS) and structural (VP) proteins

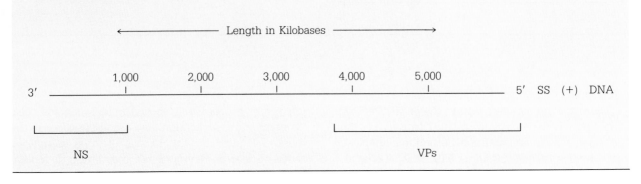

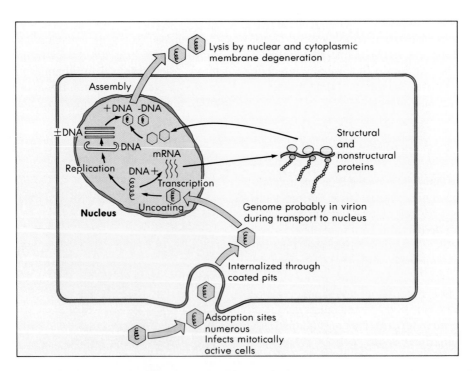

FIGURE 46-2   Postulated replication of parvovirus (B19) based on information from related viruses (minute virus of mice).

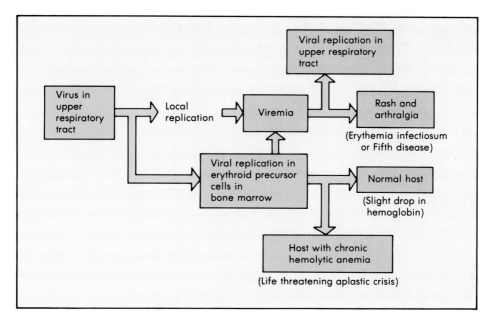

FIGURE 46-3   Mechanism of spread of parvovirus within the body.

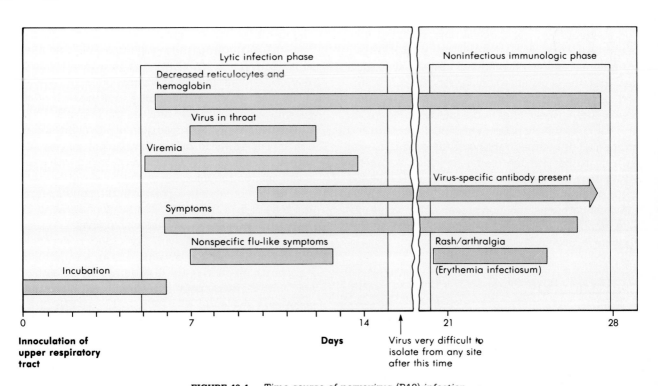

FIGURE 46-4   Time course of parvovirus (B19) infection.

> **Box 46-2** Pathogenic Mechanisms of Parvoviruses (B19)
>
> 1. Infects mitotically active erythroid precursor cells in bone marrow resulting in decreased reticulocyte counts and hemoglobin concentrations
> a. Normal host: Clinically inapparent drop in hemoglobin levels
> b. Host with chronic hemolytic anemia (e.g., sickle cell anemia) Aplastic crisis, which may be life threatening
> 2. Circulating immune complexes, primarily IgM and virion at first, later IgG, that do not fix complement, resulting in erythematous maculopapular rash, arthralgia, and arthritis

> **Box 46-3** Clinical Consequences of Parvovirus (B19) Infection
>
> 1. Mild flu-like illness (fever, headache, chills, myalgia, malaise)
> 2. Erythema infectiosum (Fifth disease)
> 3. Aplastic crisis
> 4. B19 virus crosses placenta, possibly resulting in fetal loss, but does not appear to cause congenital abnormalities

nonspecific symptoms such as sore throat, malaise, and myalgia, and a slight drop in hemoglobin. Infection of hosts with chronic hemolytic anemia (e.g., sickle cell anemia) may result in a life-threatening reticulocytopenia referred to as aplastic crisis. Aplastic crisis is due to B19 infection of red cell precursors and shortened red cell survival time, resulting from the underlying anemia. The time-course of B19 virus infection in healthy children includes the symptoms just described, resulting from lytic virus infection, and a noninfectious, immunologic phase approximately 2 to 3 weeks after initial infection (Figure 46-4). The immunologic phase, which includes rash and arthralgia, is termed erythema infectiosum and appears to be immune mediated, since symptoms accompany the appearance of virus-specific IgM and the disappearance of detectable B19 virus. Antibody produced following infection presumably confers lifelong immunity. A summary of pathgenic mechanisms appears in Box 46-2.

EPIDEMIOLOGY

Up to 65% of the adult population have antibodies to B19. Most infections with this virus occur by 40 years of age. Erythema infectiosum is most common in children from age 4 to 15 years, and it tends to occur in winter and spring. The route of natural trans-

mission is presumably by respiratory droplets. B19 appears to be less contagious than measles or varicella viruses, since erythema infectiosum usually occurs at a later age. Parenteral transmission of B19 by blood-clotting factor concentrate has been described.

CLINICAL SYNDROMES

B19 virus is the cause of **erythema infectiosum** (Fifth disease), which begins with a distinctive rash on the face resembling a cheek that has been slapped. The rash then usually spreads, especially to exposed skin such as the arms and legs (Figure 46-5), and then subsides over a 1 to 2 week period. Relapse of the rash may occur often. In adults, arthritis of hands, wrists, knees, and ankles predominates, and the rash often does not occur or it may precede the arthritis.

The most serious complication of parvovirus infection is the **aplastic crisis** that occurs in patients with chronic hemolytic anemia. In patients with sickle cell anemia the infection is characterized by transient reduction of erythropoiesis in the bone marrow and results in a transient reticulocytopenia (7 to 10 days) and a decrease in hemoglobin levels. Aplasic crisis is accompanied with fever and nonspecific symptoms of malaise, myalgia, chills, and itching. Parvovirus infection in these patients may be associated with a maculopapular rash with arthralgia and some joint swelling.

B19 infection is also associated with fetal infection and stillbirth, although infection of pregnant women may occur without any adverse effect on the fetus. There is no evidence that B19 causes congenital abnormalities (Box 46-3).

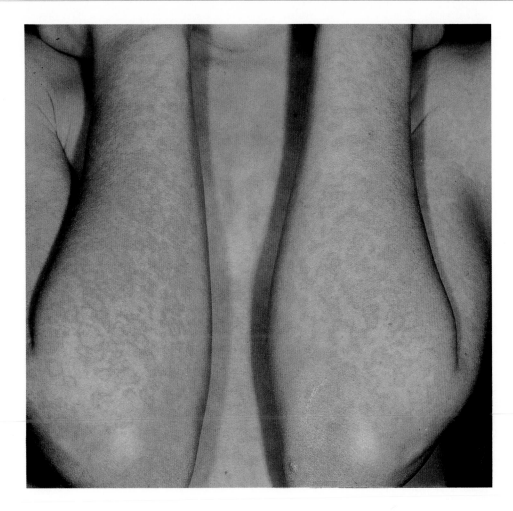

FIGURE 46-5   Rash of erythema infectiosum (Fifth disease). (From Habif TP: Clinical dermatology: a color guide to diagnosis and therapy, St Louis, 1985, The CV Mosby Co.)

LABORATORY DIAGNOSIS

B19 virus can be directly detected in serum or throat washes during the prodromal period or aplastic crisis by immune electron microscopy, enzyme and radioimmunoassay, and nucleic acid hybridization (Table 46-1). However, these assays are available in a limited number of research laboratories only. More readily available, and applicable to patients with erythema infectiosum and aplastic crisis, is the virus-specific IgM test performed on serum. A positive IgM test indicates current or recent infection.

Table 46-1   Laboratory Diagnosis of Parvovirus (B19) Infection

Test	Detects
Immune electron microscopy, enzyme and radioimmunoassay, DNA:DNA hybridization	Detection of virus during prodromal period or aplastic crisis in blood or throat specimens
Enzyme and radioimmunoassay for IgM	Detection of virus-specific IgM antibody during immunologic phase (erythema infectiosum) or aplastic crisis

TREATMENT, PREVENTION, AND CONTROL

No specific antiviral treatment is known. Control of respiratory spread could decrease transmission of B19 virus; however, patients in the clinically apparent immunologic phase (erythema infectiosum) are no longer infectious. Transmission therefore would occur before preventive measures (e.g., isolation) would be enacted.

BIBLIOGRAPHY

Anderson LJ: Role of parvovirus B19 in human disease, Pediatr Infect Dis J 6:711-118, 1987.

Balows A, Hausler WJ Jr, and Lennette EH, editors: Laboratory diagnosis of infectious diseases: principles and practice, vol 2, New York, 1988, Springer-Verlag.

Berns KI: The parvoviruses, New York, 1984, Plenum Press.

Chorba T, et al.: The role of parvovirus B19 in aplastic crisis and erythema infectiosum (fifth disease), J Infect Dis 154:383, 1986.

47

Orthomyxoviruses

The genus *Influenzavirus* is a member of the Orthomyxoviridae family. There are three antigenic types of influenza virus—A, B, and C. Only types A and B have been implicated in causing human disease.

Influenza is a specific clinical syndrome caused by influenza A or B virus. Unfortunately the term "flu" has been used for many other viral infections and even noninfectious diseases (e.g., morning flu in a person who drank excessively the night before). Confusion regarding influenza has also been abetted by terms such as "intestinal flu" to describe gastroenteritis due to one or another of the diarrhea-producing viruses. In reality, gastrointestinal upset is a minor aspect of true influenza.

Epidemics of influenza appear to have been described in ancient times. Probably the most famous influenza pandemic is the one that swept the world in 1918-1919, killing 20 million persons. Pandemics in recent years have occurred in 1947, 1957, 1968, and 1977.

STRUCTURE AND REPLICATION

Influenza virus virions are 80 to 120 nm in diameter and have (1) a segmented RNA genome, (2) helical capsid symmetry, and (3) a lipid envelope. Influenza viruses tend to be very pleomorphic and may appear spherical or tubular (Figures 47-1 and 47-2). Important virally encoded proteins include the hemagglutinin and neuraminidase surface antigens and matrix proteins surrounding the core and forming the inner part of the virus envelope.

The hemagglutinin is a surface polypeptide shaped like a spike and measuring 10 to 40 nm in length. In the laboratory the hemagglutinin can be shown to facilitate the attachment of the virus to a sialic acid–containing glycoprotein receptor on the surface of chicken, guinea pig, and human erythro-

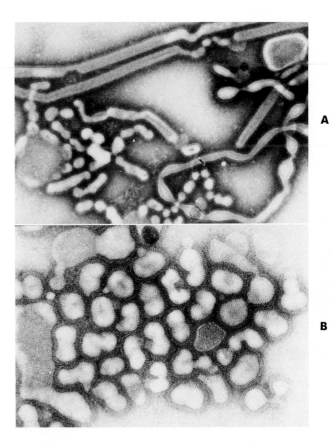

A

B

FIGURE 47-1 **A,** Influenza A, early passage. **B,** Influenza A, late passage. (Courtesy Centers for Disease Control).

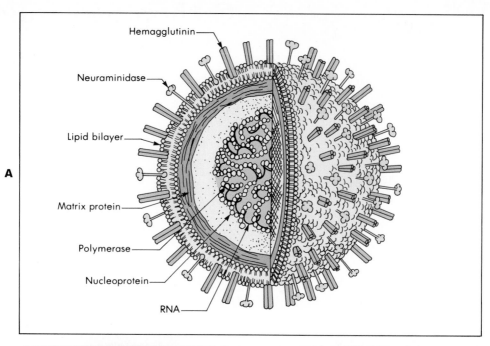

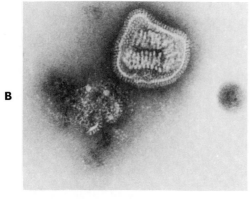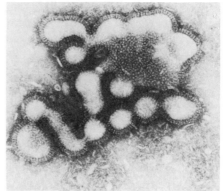

FIGURE 47-2   A, Model of influenza A virus. **B,** Electron micrograph of influenza A virus (left, spherical form; right, filamentous form). (**A** redrawn from page 91, by Bunji Tagawa, from "The epidemiology of influenza," by Martin M. Kaplan and Robert G. Webster, December 1977. Copyright © 1977 by Scientific American, Inc. All rights reserved; Kaplan MM and Webster RG: Sci Am **B** from Balows A, et al, editors: Laboratory diagnosis of infectious diseases: principles and practice, vol 2, Heidelberg, 1988, Springer-Verlag.)

cytes; in the intact animal it probably facilitates attachment of the virus to respiratory epithelial cells. The hemagglutinin undergoes antigenic changes of minor ("drift") or major ("shift") degree. "Shifts" occur only with influenza A, and new hemagglutinins are designated H1, H2, etc.

Neuraminidase is a second glycoprotein spikelike structure on the surface of influenza viruses. In the laboratory this enzyme cleaves the bond between the viral hemagglutinin and the cell receptor. This enzyme appears to play a role in vivo in the release of virus from infected cells. Neuraminidase also undergoes antigenic changes which if major, are designated N1, N2, etc.

The influenza virion contains a segmented single-stranded RNA genome with an aggregate molecular weight of 2 to 4×10^6. Influenza A and B contain eight segments, whereas influenza C, which probably belongs to an as-yet-unnamed second genus, has seven. Each segment has 900 to 2400 bases. Neur-

Table 47-1 Gene Products of Influenza Viruses

Proteins	Function
Polymerase Proteins PA (Polymerase acidic) PB1 (Polymerase basic—1) PB2 (Polymerase basic—2)	Associated with RNA transcriptase activity
Nucleoprotein NP	Internal protein associated with RNA and polymerase proteins
Hemagglutinin HA	Surface glycoprotein responsible for attachment
Neuraminidase NA	Surface glycoprotein with enzyme activity
Matrix proteins M1 M2	Proteins lining inside of virion envelope
Nonstructural proteins N51 N52	Function unknown

Modified from Fields BN, editor: Virology, New York, 1985, Raven Press.

aminidase and hemagglutinin proteins are encoded on separate segments. Other viral proteins are listed in Table 47-1. The matrix and nucleoprotein antigens are type specific and are therefore used to differentiate types A, B, or C. Subtyping is based on hemagglutinin and neuraminidase antigens. Segmentation of the influenza genome facilitates genetic variation through reassortment of genes between two different viruses infecting the same host cell. Reassortment potentially gives rise to antigenically unique strains that can infect a nonimmune population.

Strains of influenza virus are classified by soluble nucleoprotein group antigen (A, B, C), geographic location, date of original isolation, and hemagglutinin and neuraminidase antigens. For example, a current strain of influenza virus might be designated A/Bangkok/1/79 (H_3N_2). This designates an influenza A virus first isolated in Bangkok in January 1979, containing hemagglutinin (H_3) and neuraminidase (N_2) antigens. Strains of influenza B are designated by type, geography, and date of isolation (e.g., B/Singapore/3/64) without specific mention of H or N antigens.

Reassortment of different animal influenza viruses is illustrated in Figure 47-3. Simultaneous infection of a seal with two avian influenza viruses resulted in a highly pathogenic reassortant virus. Although mild infections could be produced in humans and laboratory animals, fatal infections occurred in seals. Such reassortment, resulting in novel strains of new H and N types, is postulated to be the source of pathogenic human strains.

Replication begins with attachment mediated by the hemagglutinin protein spike. Penetration by endocytosis, release of virion core from coated pits following fusion of virion and vesicle membrane, nuclear transcription, and cytoplasmic protein synthesis and replication are highlights of the influenza infectious cycle (Figure 47-4). Release of virions occurs about 8 hours after infection.

PATHOGENESIS

Influenza viruses A and B are uniquely adapted for human infection. Antigenic changes of hemagglutinin and neuraminidase proteins allow inhaled viruses to avoid neutralization by host antibodies. In addition, viral neuraminidase destroys respiratory tract mucous inhibitors of viral attachment. Local upper respiratory tract multiplication and spread to susceptible lower respiratory tract cells cause desquamation of mucus-secreting and ciliated cells. Influenza infection also impairs the T cell–mediated lymphocyte response, chemotaxis of leukocytes, and bactericidal activity of macrophages. As a result, patients in-

Box 47-1 Pathogenic Mechanisms of Influenza A Viruses

1. Antigenic "drift" and "shift" of the hemagglutinin and neuraminidase surface glycoproteins
2. Neuraminidase-mediated destruction of mucous glycoprotein inhibitors of viral attachment
3. Hemagglutinin-mediated fusion of virion with host epithelial cell membrane initiating infectious cycle
4. Mucus-secreting and ciliated epithelial cells of respiratory tract die and desquamate as a result of viral infection
5. Virus-altered "clearing" mechanism in respiratory tract predisposes patient to secondary bacterial infection

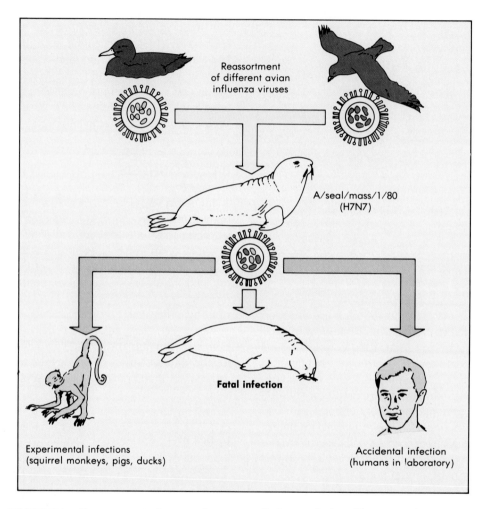

FIGURE 47-3   Reassortment of genome fragments of influenza A virus. Diagrammatic representation of the origin of seal influenza virus A/seal/Mass/1/80 (H7N7). Since all of the eight genes of
seal influenza virus are most closely related to those in different avian viruses, it is proposed that
seals were mixedly infected with influenza viruses by fecal contamination from birds. The resulting reassortant virus had the required gene constellation to replicate in seal tissues and caused
high mortality. This virus has been accidentally transmitted to humans in the laboratory, causing
conjunctivitis. Experimentally, the virus has been shown to replicate preferentially in mammalian species, causing systemic infection in some squirrel monkeys and asymptomatic infection of
pigs. (Modified from Fields BN, editor: Virology, New York, 1985, Raven Press.)

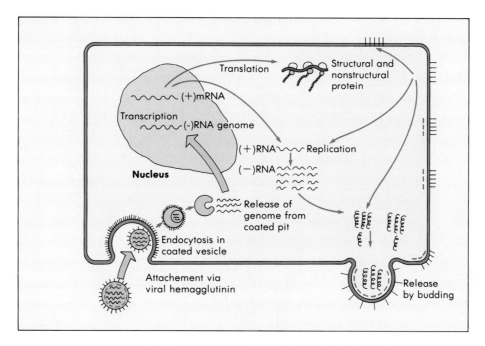

FIGURE 47-4   Replication of influenza A virus. (Modified from Stringfellow DA, et al, editors: Virology: a scope presentation, Kalamazoo, Mich., 1983, The Upjohn Co.)

fected with influenza virus are rendered more susceptible to secondary bacterial infection (Figure 47-5).

Histologically, influenza infection leads to an inflammatory cell response of the mucosal membrane, which consists primarily of monocytes and lymphocytes and few neutrophils. Submucosal edema is present. Lung tissue may reveal hyaline membrane disease, alveolar emphysema, and necrosis of alveolar walls. Viremia may occur but is rarely detected (Box 47-1).

Recovery is associated with cell-mediated immune responses, including interferon production and the generation of cytotoxic T lymphocytes. Recovery is also associated with the development of antibodies to both hemagglutinin and neuraminidase. As long as these antibodies persist at moderate or high levels,

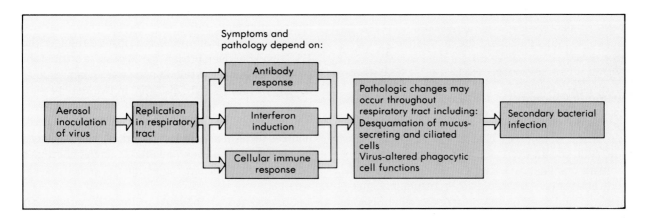

FIGURE 47-5   Mechanisms of spread of influenza A virus within the body.

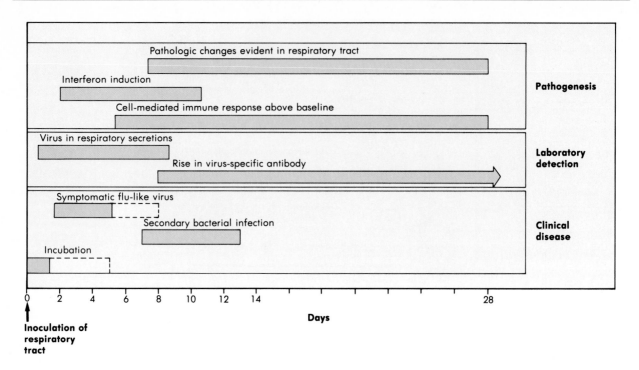

FIGURE 47-6   Time course of influenza A virus infection.

protection against reinfection with the same strain of influenza is conferred. The time course of influenza virus infection is illustrated in Figure 47-6.

EPIDEMIOLOGY

Influenza infection occurs annually, in temperate climates, during the winter months (Figure 47-7). Influenza infection is spread via airborne droplets to susceptible individuals, and the incubation period is 1 to 4 days.

Antigenic Changes

Minor antigenic changes, resulting from mutation of the viral genome, are called antigenic ''drift.'' Antigenic drift occurs every 2 to 3 years, causing local outbreaks of influenza A and B infection (Table 47-2). Major antigenic changes result from genome reassortment (facilitated by the presence of viral RNA in eight different segments), and they occur only with influenza A. These major antigenic changes are

called antigenic ''shift'' and are often associated with pandemics. Antigenic shifts occur infrequently, recently averaging approximately every 10 years (Table 47-3). For example, the prevalent influenza A virus in 1947 was the H1N1 subtype. In 1957 there was a shift in both antigens, resulting in an H2N2 subtype. In 1968 H3N2 appeared, and in 1978 H1N1 reappeared. Presumably, the reappearance of H1N1 reflected the accumulation of a large population of susceptibles, all those less than 30 years of age who lacked protective antibodies. The fact that those over 30 years of age were protected despite the fact that H1N1 had not been present for 20 years is attributable to an anamnestic antibody response. When exposed to any influenza subtype, the individual develops antibodies to that virus, as well as an anamnestic response to the subtype of first exposure. This enables recall of antibody should that original subtype reappear. This has been referred to as the doctrine of ''original antigenic sin,'' although the analogy is a bit weak.

Surveillance of influenza A and B outbreaks is extensive in order to identify the early appearance of new strains that should be incorporated into new vac-

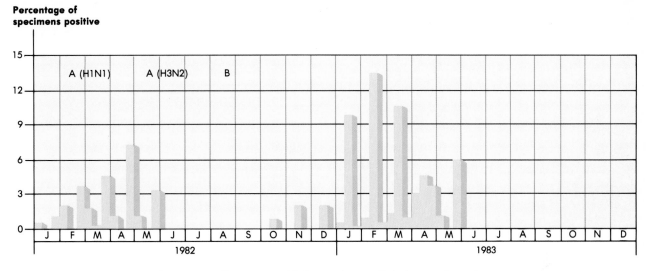

FIGURE 47-7   Occurrence of influenza by month in temperate regions.

cines. Surveillance also extends into animal populations because of the possibility that recombinants responsible for influenza A pandemics arise from animal strains.

CLINICAL SYNDROMES

Clinical illness is caused by influenza A and to a lesser extent by influenza B. The syndrome of influenza is sufficiently typical so that it can often be accurately diagnosed on clinical grounds, especially during a community outbreak. Characteristically, there is a brief prodrome of malaise and headache lasting a few hours. This is followed by the abrupt onset of fever, severe myalgia, and a usually nonproductive cough. The illness persists for approximately 3 days

and then, unless a complication occurs, recovery is complete. Depending on the degree of immunity to the infecting strain of virus, the infection may range from asymptomatic to severe. Patients with underlying cardiorespiratory disease or immune deficiency, even that associated with pregnancy, are more disposed to severe diseases. With influenza B disease respiratory symptoms may be minimal, with gastrointestinal symptoms predominating. In young children influenza resembles other severe respiratory infections, bronchiolitis, croup, otitis media, and, rarely, febrile convulsions (Table 47-4). Complications of influenza include bacterial pneumonia, myositis, central nervous system involvement, and Reye's syndrome.

Primary influenza pneumonia may also occur as a result of progressive influenza virus infection of alve-

Table 47-2 Influenza Epidemics (1968-1977) Resulting From Antigenic Drift

Year of Epidemic	Influenza A Subtype	Strains
1968	H3N2	A/Hong Kong/68*
1972-73	H3N2	A/England/72
1976	H3N2	A/Victoria/75
1977	H3N2	A/Texas/77

*Prototype virus.

Table 47-3 Influenza Pandemics Resulting From Antigenic Shift

Year of Pandemic	Influenza A Subtype
1918	$H_{sw}N1$ Probable swine flu strain
1947	H1N1
1957	H2N2 Asian flu strain
1968	H3N2 Hong Kong flu strain
1978	H1N1

Table 47-4 Diseases Associated With Influenza Virus Infection

Disorder	Symptoms
Acute influenza infection in adults	Rapid onset of fever, malaise, myalgia, sore throat, and nonproductive cough
Acute influenza infection in children	Acute disease similar to adults but having higher fever, gastrointestinal symptoms (abdominal pain, vomiting), otitis media, myositis, and croup more frequently

Complications of influenza virus infection	Primary viral pneumonia
	Secondary bacterial pneumonia
	Myositis and cardiac involvement
	Neurologic syndromes
	  a. Guillain-Barré syndrome
	  b. Encephalopathy
	  c. Encephalitis
	  d. Reye's syndrome

oli. Sputum may be scant, but the process is extensive, leading to hypoxia and bilateral pulmonary densities on x-ray examination. The most common complication, however, is bacterial superinfection of the respiratory tract, leading to bronchitis or pneumonia. The responsible bacteria are usually *Streptococcus pneumoniae, Haemophilus influenzae,* or *Staphylococcus aureus.* In these instances sputum usually becomes productive and purulent.

Although myalgias are the rule in influenza, a true myositis (inflammation of muscle) with release of muscle enzymes is uncommon. It is most apt to occur in children and may be seen with influenza B, as well as influenza A.

Encephalopathy that occurs during the acute influenza illness may be fatal and is associated with cerebral edema, whereas post-influenza encephalitis occurs 2 to 3 weeks following recovery from influenza, is rarely fatal, and is associated with evidence of inflammation.

Reye's syndrome is an acute encephalitis that occurs in children and follows a variety of acute febrile viral infections, including varicella, as well as influenza B and A. Evidence suggests that salicylates increase the likelihood of developing this syndrome. In addition to encephalopathy, there is hepatic dysfunction. The fatality rate may be as high as 40%.

LABORATORY DIAGNOSIS

Immunologic techniques, especially immunofluorescent methods, may be used for the direct detection of viral antigen in respiratory secretions, but they lack sensitivity. Direct detection methods are highly dependent upon the use of antibodies (monoclonal or polyclonal) that are directed against the antigens of the current prevalent influenza strain. This requirement is complicated by antigenic changes in the virus and by the fact that in any given influenza season two or more influenza viruses may be circulating.

Influenza viruses grow rapidly in primary monkey kidney cell cultures, but occasional strains may require chick embryo inoculation for primary isolation. Cytopathic effects (CPE) may be noted in as few as 2 days (average 4 days). Before the development of CPE, the addition of guinea pig erythrocytes may reveal hemadsorption (Figure 47-8) (i.e., the adher-

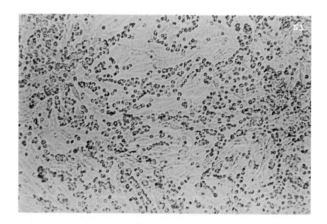

FIGURE 47-8 Hemadsorption of guinea pig erythrocytes.

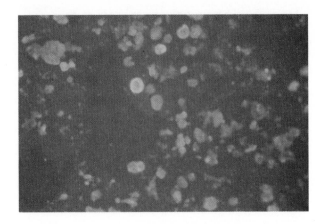

FIGURE 47-9   Immunofluorescent stain of respiratory secretion showing influenza A–infected epithelial cell with fluorescing cytoplasmic inclusions (×200). (Courtesy Richard Thomson, Akron, Ohio.)

ence of these erythrocytes to infected cells). If these erythrocytes are added to the culture as early as 1 or 2 days after inoculation, they may provide evidence of viral presence well before CPE develops. Hemadsorption is not specific for influenza viruses; it may also occur with parainfluenza viruses. Specific identification of the virus causing hemadsorption requires immunologic tests such as immunofluorescence or inhibition of hemadsorption with specific antibody (Figure 47-9).

Serologic diagnosis is a slow method of diagnosing an individual case of influenza, since acute and convalescent sera must be obtained (Table 47-5).

TREATMENT

In general, the treatment of influenza is to relieve symptoms by acetaminophen. In patients with severe progressive disease or with underlying chronic pulmonary disease, therapy with amantadine, or its analog rimantadine, may be useful. These agents block an early step in viral replication and may reduce postinfluenza airway abnormalities. Severity of acute illness can only be lessened if amantadine is started within 24 to 48 hours after onset of illness. Amantadine has the advantage of oral administration in contrast to ribavirin, another drug with marginal effects in treating influenza that must be administered by aerosol inhalation. Bacterial superinfections must be diagnosed and treated appropriately.

PREVENTION AND CONTROL

Influenza virus is present in a community for only a short period (4 to 6 weeks). Its presence is usually evident from the number of sick school children and emergency room visits, as well as from laboratory results and health department bulletins. During periods of influenza activity, it is prudent to reduce interpersonal contact, especially among or with sick individuals. It is also useful to reduce exposure of the elderly, when possible, by omitting elective hospitalization.

Killed (formalin inactivated) influenza vaccine is available each year. Killed vaccines may contain whole viruses, or they may be treated with chemicals to fractionate the virus into subunits. Subunit vaccines are thought to be less immunogenic. Ideally, the vaccine incorporates antigens of the A and B influenza strains that will be prevalent in the community during the upcoming winter. Vaccination is routinely recommended for the elderly and those with chronic pulmonary disease and heart disease.

BIBLIOGRAPHY

Fields BN, editor: Virology, New York, 1985, Raven Press.

Stuart-Harris C: The epidemiology and prevention of influenza, Am Sci 69:166-172, 1981.

Sweet C and Smith H: Microbiol Rev 44:303-330, 1980.

Table 47-5   Laboratory Diagnosis of Influenza Virus Infection

Test	Detects
Immunofluorescence, ELISA	Influenza virus antigens in respiratory epithelial cells
Primary monkey kidney cells (PMK) cell culture	Growth and presence of influenza viruses in cell culture
Complement fixation	Fourfold rise in specific viral antibodies

The family Paramyxoviridae consists of three genera: the *Morbillivirus, Paramyxovirus,* and *Pneumovirus* (Table 48-1). Human pathogens within the morbilliviruses include measles virus, within the paramyxoviruses are parainfluenza and mumps viruses, and within the pneumoviruses is respiratory syncytial virus. The major diseases caused by these agents are well known. Measles causes a potentially serious generalized infection characterized by a maculopapular rash. Parainfluenza viruses cause upper and lower respiratory tract infection, primarily in children, including pharyngitis, croup, bronchitis, and pneumonia. Mumps virus causes a systemic infection, whose most prominent clinical manifestation is parotitis. Respiratory syncytial virus causes local respiratory tract infection in children, ranging from rhinitis and pharyngitis to bronchiolitis and pneumonia. Because of the contagiousness and limited number of serotypes of all these viruses, infections by the paramyxoviruses generally occur in childhood and do not recur, or recur only mildly, in adults.

STRUCTURE AND REPLICATION OF PARAMYXOVIRUSES

The paramyxovirus family consists of negative-sense, single-stranded, RNA viruses with a pleomorphic lipid-containing envelope, a helical nucleocapsid, and 156 to 300 nm virions. (Figure 48-1). They are similar to orthomyxoviruses (influenza A to C) in many respects but are larger and do not have the unique segmented genome of influenza virus.

The important structural components of a typical paramyxovirus are surface glycoproteins forming spike projections called HN and F, a membrane protein called M, a ribonucleoprotein (NP), and a host cell–derived lipid bilayer membrane. The HN glycoprotein has hemagglutinating and neuraminidase activities at different sites on the same molecule, making it responsible for adsorption and lysis of host receptors. The fusion (F) glycoprotein is responsible for virus penetration into the host cell by promoting fusion of viral and host cell membranes. The membrane protein (M) forms the base of the lipid envelope. The NP ribonucleoprotein is the major complement-fixing antigen.

The RNA genome of paramyxoviruses is a single RNA strand approximately 5 to 8×10^6 d. The viral genome must be transcribed into mRNA. Gene products of measles virus are shown in Table 48-2. Measles virus is different from other paramyxoviruses because it does not have neuraminidase activity.

Replication of the paramyxoviruses is represented by the respiratory syncytial virus infectious cycle in Figure 48-2. An important aspect of paramyxovirus infection is viral penetration mediated by the F protein. The membrane-fusing activity is activated by proteolytic cleavage. Cleaved F glycoprotein leads to

Table 48-1 The Paramyxovirus Family	
Genus	**Human Pathogen**
Morbillivirus	Measles virus
Paramyxovirus	Parainfluenza viruses 1 to 4
	Mumps virus
Pneumovirus	Respiratory syncytial virus (RSV)

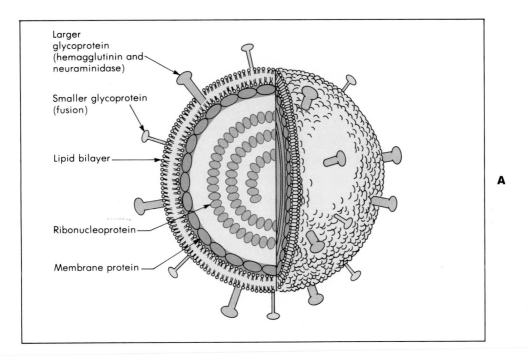

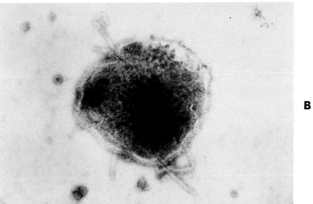

FIGURE 48-1   A, Model of paramyxovirus. The components of paramyxoviruses are as follows: larger virus glycoprotein responsible for hemagglutination; smaller virus glycoprotein, involved in cell fusion by these viruses and probably in the entry of the virus into the cell; lipid bilayer, the lipid is cell-derived; nonglycosylated membrane protein; ribonucleoprotein, the major CF antigen. **B,** Electron micrograph of paramyxovirus. (**A** redrawn; reprinted with permission from Jawetz E, Melnide JL, and Adelberg EA: Review of medical microbiology, ed 17, Appleton & Lange, Norwalk, Connecticut, 1987; **B** Courtesy Centers for Disease Control, Atlanta, Georgia.)

Table 48-2   Virus Encoded Proteins of Measles Virus

Gene Products*	Virion Location	Function
1. Nucleoprotein (NP)	Major internal protein	Protects viral RNA
2. Polymerase phosphoprotein (P)	Associated with nucleoprotein	Possibly part of transcription complex
3. Matrix (M)	Inside virion envelope	Assembly of virions
4. Fusion factor (F)	Transmembraneous envelope glycoprotein	Active in fusion of cells, hemolysis, and virus entry
5. Hemagglutinin (H)	Transmembraneous envelope glycoprotein	Adsorption to nucleated cells and erythrocytes
6. Large proteins (L)	Associated with nucleoprotein	Possibly part of transcription complex

Modified from Fields BN, editor: Virology, New York, 1985, Raven Press.
*In order of transcription.

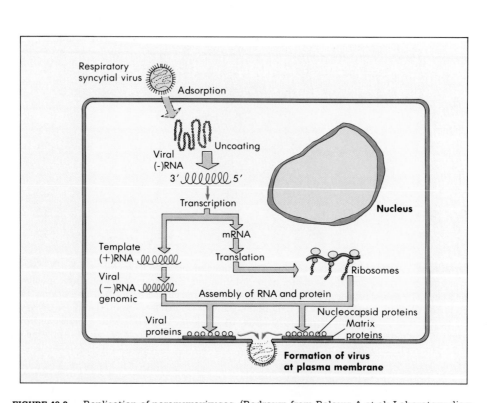

FIGURE 48-2   Replication of paramyxoviruses. (Redrawn from Balows A et al: Laboratory diagnosis of infectious diseases: principles and practice, New York, 1988, Springer-Verlag.)

fusion of viral and host cell membranes. Viral adsorption—mediated by the HN glycoprotein—occurs, but infection does not in the absence of F cleavage. Viral transcription, translation, and replication all occur in the cytoplasm. Mature virions bud from the host cell cytoplasmic membrane, acquiring virally encoded matrix and nucleocapsid proteins, along with a host-derived lipid bilayer envelope.

MEASLES VIRUS

Measles is the sole virus of human importance belonging to the genus *Morbillivirus* within the family of Paramyxoviridae. It has the properties of paramyxoviruses just delineated but is unique for the presence of a hemagglutinin but not a neuraminidase.

Historically, measles has been one of the most common viral infections and one of the most unpleasant. Virtually everyone born before the 1960s can vividly remember the high fever, excruciating headache, and malaise associated with measles infection. Since vaccination was introduced, the yearly incidence of measles has dropped dramatically, from 300 to 1.4 per 100,000 in developed countries. In developing countries, however, it is very common, remains potentially fatal, and in many is the most important cause of death in those 1 to 5 years of age.

Pathogenesis

At the cellular level measles virus is notable for its propensity to cause cell fusion, resulting in giant cells and syncytia. Inclusions occur most commonly in the cytoplasm and are composed of incomplete viral particles. Although infection usually leads to ultimate cell lysis, persistent infection without lysis can occur. This may result from restriction in the synthesis of viral envelope proteins or from the action of antibodies. Persistent infection is more likely to occur in certain cell types (e.g., human brain cells).

Lymphoid hyperplasia is widespread, and characteristic giant cells are seen throughout the reticuloendothelial system. In the intact host, measles is a highly contagious, systemic infection transmitted from person to person by respiratory droplets (Figure 48-3). Local replication of virus in the respiratory tract precedes lymphatic spread and viremia. Wide dissemination is characterized by virus infecting the conjunctiva, respiratory tract, urinary tract, small blood vessels, and the central nervous system. Infection of the endothelial cells lining small blood vessels and the reaction of immune T cells cause the rash that is the hallmark of measles. In most patients recovery follows, with life-long immunity. In some, a postinfectious encephalitis, which is believed to be immune mediated, occurs after the rash. Immunocompromised patients with measles may have continuing

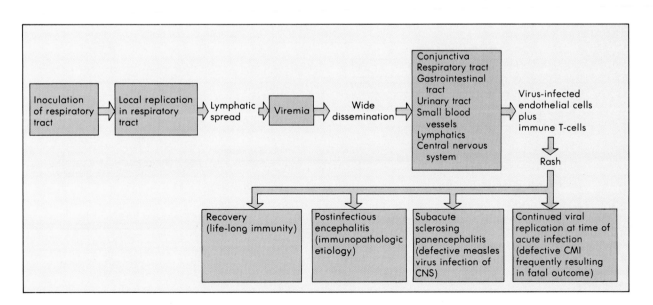

FIGURE 48-3   Mechanisms of spread of measles virus within the body.

Epidemiology

Measles is one of the most contagious infections known. In a household, approximately 85% of exposed susceptible persons will become infected, and 95% of these will develop clinical disease. Epidemics tend to occur in 1- to 3-year cycles when a sufficient number of susceptible individuals has accumulated. Many of these cases occur in preschool-age children who have not been vaccinated and who live in large urban areas. The peak season is in the winter and spring. Control measures among susceptible individuals are difficult because the virus is present in the respiratory tract for up to 4 days before the rash appears.

infection, resulting in death. Years after a measles infection, subacute sclerosing panencephalitis (SSPE) occurs in about 7 in 1,000,000 patients. SSPE results from ongoing replication of defective measles virus in the central nervous system (Box 48-1). Infection spreads directly from cell to cell without mature virus release. The time-course of measles infection is shown in Figure 48-4.

Clinical Syndromes

Measles is a serious febrile illness (Table 48-3). The incubation period is 10 to 13 days. The prodrome starts with high fever and the three "c's"-cough, coryza, and conjunctivitis, in addition to photophobia. At this time the patient is most infectious.

After 2 days of illness, the typical mucous mem-

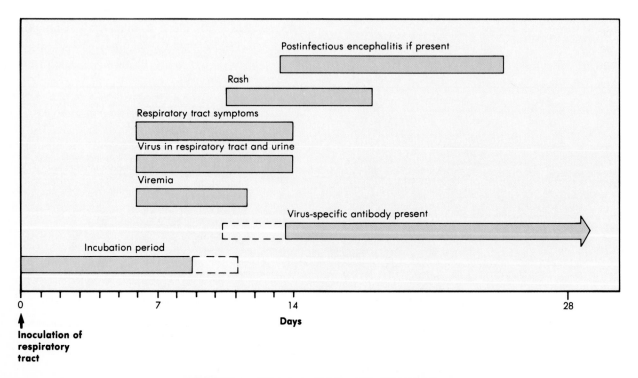

FIGURE 48-4   Time course of measles virus infection.

Table 48-3 Clinical Consequences of Measles Virus Infection

Disorder	Symptoms
Measles	Characteristic maculopapular rash, coryza, cough and conjunctivitis; complications include otitis media, croup, and bronchopneumonia, in addition to encephalitis the most severe complication (principal reason for vaccine)
Atypical measles	Rash is most prominent in distal areas Vesicles, petechiae, purpura, or urticaria may also be present.
Subacute sclerosing panencephalitis	Central nervous system manifestations (e.g., personality, behavior, and memory changes, myoclonic jerks, spasticity, and blindness)

brane lesions, known as **Koplik's spots** (Figure 48-5), appear. As shown, they are seen most commonly on the buccal mucosa across from the molars, but they may appear on any other mucous membrane, including the conjunctiva and the vagina. The lesions, which last 24 to 48 hours, are small (1 to 2 mm) and are best described as grains of salt surrounded by a red halo. Their appearance in the mouth establishes with certainty the diagnosis of measles.

Within 12 to 24 hours after the appearance of Koplik's spots, the exanthem of measles starts below the ears and spreads over the body. The rash is maculopapular, usually very extensive, and often the lesions become confluent. The rash, which takes 1 or 2 days to cover the body, fades in the same order it appeared. The fever is highest and the patient is sickest on the day the rash first appears (Figure 48-6).

One of the most-feared complications of measles is the development of encephalitis, which may be fatal

in 15% of cases. It may occur in up to 0.5% of those infected and usually begins 7 to 10 days after onset of illness.

Pneumonia, which can also be a serious complication, accounts for 60% of reported deaths due to measles virus infection. Mortality with pneumonia, as with other complications associated with measles, increases with malnutrition and is inversely proportional to age. Bacterial superinfection is common in pneumonia caused by measles virus.

Atypical measles occurs in individuals who received the older inactivated measles vaccine and who subsequently were exposed to the wild measles virus. It may also rarely occur after vaccination with attenuated virus vaccine. The illness begins abruptly, with high fever, headache, myalgia, sore throat, coryza, and a nonproductive cough. An erythematous, maculopapular rash begins 3 to 4 days later on the distal extremities and spreads proximally. Vesicles, petechiae, purpura, or urticaria may be seen along with the basic rash. This syndrome probably represents an altered immune response to wild measles virus on the part of a previously sensitized host. The illness is so atypical that it easily can be confused with Rocky Mountain spotted fever, meningococcemia, scarlet fever, and even varicella.

Subacute sclerosing panencephalitis (SSPE) is an extremely serious, very late neurologic sequela of measles. In SSPE the measles virus persists for an unknown reason in the brain and acts as a slow virus. Many months or years after clinical measles, the patient develops changes in personality, behavior, and memory. Myoclonic jerks, blindness, and spasticity follow. Unusually high levels of measles antibodies are found in the blood and spinal fluid. Eosinophilic inclusion bodies (Figure 48-7) are composed of par-

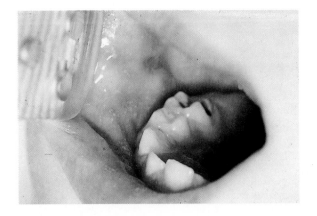

FIGURE 48-5   Koplik's spots in mouth.

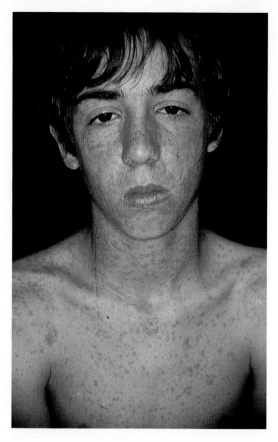

FIGURE 48-6   Measles rash. (From Habif TP: Clinical dermatology: color guide to diagnosis and therapy, St Louis, 1985, The CV Mosby Co.)

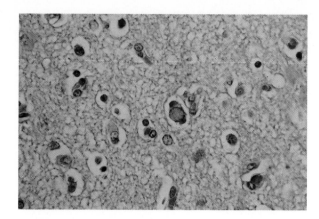

FIGURE 48-7   Brain tissue in SSPE. (hematoxylin and eosin stain ×400)

amyxovirus-like nucleocapsids and are present in the brains of patients with SSPE. The presence of measles virus antigen may be demonstrated by immunofluorescence. The incidence of SSPE has decreased markedly with the success of measles vaccination.

Laboratory Diagnosis

Clinically, measles is usually so characteristic that it is rarely necessary to perform laboratory tests to make a diagnosis (Table 48-4). However, physicians who began training in the United States after 1965 may fail to recognize the infection or may be misled by the appearance of atypical measles.

Culture Measles virus can be grown in primary human or monkey cell cultures, but the virus is not easily recovered and grows slowly. Isolation of mea-

Table 48-4   Laboratory Diagnosis of Measles Virus

Test	Detects
CELL CULTURE	
Isolate from respiratory tract, urine, blood, brain tissue; use primary human or monkey kidney cells	Growth and presence of measles virus; note that isolation is difficult and in postvaccine era is especially unpredictable
DIRECT TEST	
Cytologic testing, electron microscopy, immunofluorescent, or immunoperoxidase stains	Measles virus, cytopathologic changes, or antigens in brain tissue, respiratory secretions or urine
SEROLOGIC TESTS	
Immunofluorescence, enzyme immunoassay, and complement fixation	Fourfold rise in specific virus antibody and virus-specific IgM
Immunofluorescence and enzyme immunoassay	Immune status (presence of antibodies in serum)

sles virus in the postvaccine era has been unpredictable. Recent attempts have been unsuccessful, possibly because of in vivo neutralization of virus by low levels of vaccine-induced antibody. If isolation is attempted, respiratory tract secretions, urine, blood, and brain tissue are recommended specimens. Respiratory and blood specimens are best collected during the prodromal stage up to 1 to 2 days after the appearance of the rash.

Serology Serologic tests demonstrating seroconversion or a fourfold increase in measles-specific antibodies between acute and convalescent sera represent the best method of confirming measles. Single acute serum specimens may be tested for virus-specific IgM antibody. A positive test suggests recent or active infection. Immune status testing to determine past infection or immunization is becoming more common due to vaccine failure and lack of records documenting childhood immunization. Hospitals in areas experiencing endemic measles may wish to vaccinate or check the immune status of their employees to decrease the risk of nosocomial transmission.

Direct Tests Measles antigen can be detected by IFA in pharyngeal cells or urinary sediment, but this test is not generally available. Characteristic cytopathologic results consisting of giant cells occur in cells of the upper respiratory tract and urinary sediment. These are referred to as Warthin-Finkeldey cells and can be identified by the Giemsa method of staining.

Treatment, Prevention, and Control

Measles vaccine has been used since 1963 and has significantly reduced the incidence of measles in the United States. Live, attenuated vaccine is given to all children after 15 months of age, usually along with mumps and rubella vaccine. A childhood immunization initiative in 1977 and a measles elimination program in 1978 contributed to a 1981 record-low incidence of 1.4 cases per 100,000 population. This is a 99.5% reduction from the prevaccination period of 1955 to 1962.

Exposed susceptible individuals who are immunocompromised should be given immune serum globulin to modify their measles infection. This product is most effective if given within 6 days of exposure. Inactivated "killed" vaccines are no longer available, even for the immunocompromised, because of the subsequent effect on naturally acquired measles virus

infection (atypical measles). There is no specific antiviral treatment available for measles.

PARAINFLUENZA VIRUSES

Parainfluenza viruses, discovered in the late 1950s, are members of the Paramyxoviridae family. Four serologic types within the parainfluenza genus are human pathogens, although type 4 causes only mild upper respiratory infection in children and adults. Types 1 to 3 are second only to respiratory syncytial virus as important causes of severe lower respiratory infection in infants and young children. They are especially associated with laryngotracheobronchitis **(croup).**

Parainfluenza viruses have both neuraminidase and hemagglutinin activity, although—unlike influenza viruses—both of these functions are located on the same surface glycoprotein.

Pathogenesis

The precise details of parainfluenza pathogenesis are not understood, in part because few fatal cases occur. The initial site of viral infection is probably the nasopharynx. In many individuals the infection is limited to this site, but in approximately 25% of cases lower respiratory tract involvement occurs and in 2% to 3% this may take the severe form of laryngotracheobronchitis. Viremia is infrequent, so the extension to the lower respiratory tract is probably direct. As with respiratory syncytial virus, the host immune response—in this instance IgE production—may contribute to the pathogenesis. Multiple serotypes and the short duration of immunity following natural infection make reinfection common (Box 48-2), but reinfection is milder, suggesting at least partial immunity.

Box 48-2 Pathogenic Mechanisms of Parainfluenza Viruses

Four serotypes of virus
Infection limited to respiratory tract
Infection induces immunity of short duration

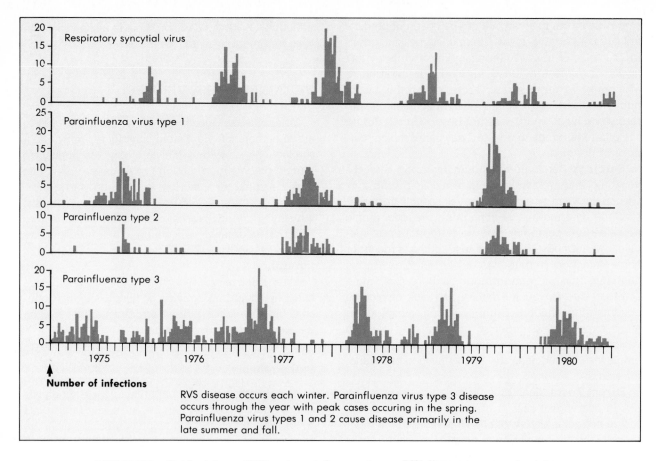

FIGURE 48-8   Epidemiology of RSV and parainfluenza viruses. RSV disease occurs each winter. Parainfluenza virus type 3 disease occurs throughout the year, with peak levels in the spring. Parainfluenza virus types 1 and 2 cause disease primarily in the late summer and fall. (Redrawn from Glezen et al: J Infect Dis 150:851-857, 1984, published by the University of Chicago.)

Table 48-5   Clinical Consequences of Infections by Parainfluenza Viruses

Disorder	Symptoms
Respiratory disorders	Coryza, pharyngitis, and mild bronchitis with fever in young children and adults
	Bronchiolitis and/or pneumonia (similar to RSV disease) in infants
	Croup (laryngotracheobronchitis) in children 6 months to 5 years of age

Table 48-6   Laboratory Diagnosis of Parainfluenza Virus Infection

Test	Detects
Immunofluorescence assay	Parainfluenza virus antigen in respiratory tract secretions
Cell culture	Growth and presence of parainfluenza virus
Use respiratory tract secretions	
Use primary monkey kidney cells	
Serologic examination	IgG antibodies, but may be insensitive or falsely positive because of cross-reactions with other paramyxovirus antigens

Epidemiology

Transmission of parainfluenza is by person-to-person contact and by respiratory droplets. Primary infections usually occur in infants and small children younger than 5 years of age. Reinfections occur throughout life, suggesting that immunity is short-lived. Infections with parainfluenza virus 1 and 2, the major causes of croup, tend to occur in the fall, whereas parainfluenza virus 3 infections occur throughout the year (Figure 48-8). All of these viruses spread readily within hospitals and can cause outbreaks in nurseries and pediatric wards.

Clinical Syndromes

Parainfluenza 1, 2, or 3 may cause respiratory syndromes ranging from upper respiratory infection to pneumonia. Older children and adults generally experience milder infections, although pneumonia may occur in the elderly. The most notable association of parainfluenza viruses with respiratory disease is the association with croup (Table 48-5). Croup results in subglottal swelling, which endangers the airway. Clinically, after a 2- to 6-day incubation period the patient develops hoarseness, a "seal bark" cough, tachypnea, tachycardia, and suprasternal retraction. The principal differential diagnosis is epiglottitis caused by *Haemophilus influenza*.

Laboratory Diagnosis

Cell culture is the method of choice for isolating parainfluenza viruses from respiratory secretions. Primary monkey kidney cells are most sensitive. Early detection of virus is accomplished by hemadsorption using guinea pig erythrocytes. Since parainfluenza viruses have a hemagglutinin similar to influenza, the added erythrocytes attach to the surface of infected cells. Once the presence of a hemadsorbing virus is known, specific immunofluorescent antibody procedures can be used to determine the exact hemadsorbing virus type.

Direct antigen detection has not been very successful and serologic tests may be insensitive or falsely positive due to cross-reactions with other paramyxovirus antigens (Table 48-6).

Treatment, Prevention, and Control

Treatment includes nebulized cold or hot steam and careful monitoring of the upper airway. On rare occasions intubation may become necessary. No specific antiviral agents are available.

Vaccination is ineffective because inactivated vaccines induce antibody production but not protection, possibly due to their failure to induce local secretory antibody. There is no live attenuated vaccine at present.

MUMPS VIRUS

Mumps virus is the cause of acute benign viral parotitis. The disease is recognized because of its obvious clinical manifestation, but it is a systemic infection with the potential for involvement of many organs besides the parotid gland. Mumps virus was isolated in embryonated eggs in 1945 and in cell culture in 1955. Mumps virus shares many properties with parainfluenza viruses, in particular parainfluenza virus 2, but there is no cross-immunity between the two groups. Mumps virus has both a hemagglutinin and a neuraminidase.

Pathogenesis

Mumps virus, of which only one serotype is known, causes a lytic infection of cells (Box 48-3). Initial viral replication occurs in the epithelial cells of the upper respiratory tract. Infection of the parotid gland is probably by direct extension, with initial involvement of ductal epithelium. Dissemination of virus by viremia ensues, with CNS involvement (in up to 50% of those infected) across the choroid plexus and other tissues (Figure 48-9). The time course of human infection is shown in Figure 48-10. Immunity is life-long.

Epidemiology

Mumps is a worldwide infection that—in the absence of vaccination programs—occurs endemically, usually in the winter or spring. In closed populations suffering an outbreak, approximately 90% of susceptible individuals will be infected and 70% of those will experience clinical disease. Infection is acquired in 90% of persons by age 15. The virus is spread by direct person-to-person contact, presumably by respiratory droplets. The virus is present in respiratory secretions for up to 7 days before clinical illness, so control of spread is virtually impossible.

Clinical Syndromes

Usually, clinical illness is manifested as a **parotitis** that is almost always bilateral but may appear first in one parotid gland and then in the other. Onset is sudden. Oral examination reveals redness and swelling of the ostium of Stensen's duct. Submaxillary and submandibular glands may be involved, even in the absence of parotid involvement. Orchitis (usually unilateral), oophoritis, pancreatitis, and meningoencephalitis may occur a few days after the onset of the virus infection but can occur in the absence of parotitis. Mumps virus involves the central nervous system in approximately 50% of patients, and 10% may exhibit clinical evidence of infection. This is usually a

---

Box 48-3   Pathogenic Mechanisms of Mumps Virus
Localized infection of respiratory tract
Capable of viremic spead
Systemic infection, especially of parotid gland, testes, and central nervous system
Inflammation and swelling of parotid gland causes amylase to leak from damaged cells, resulting in significant rise in serum amylase levels

---

syndrome of "aseptic" meningitis, but encephalitis, often with involvement of the cerebellum, also occurs. Because of the latter, children who have mumps encephaltis may be unsteady in standing or walking (Table 48-7).

Laboratory Diagnosis

Virus can be recovered from saliva, urine, pharynx, secretions from Stensen's duct, and cerebrospinal fluid. Virus is present in saliva for approximately 5 days after the onset of symptoms and in urine for up to 2 weeks. Mumps virus grows well in monkey kidney cells and can be recognized by the development of a cytopathogenic effect characterized by syncytia and giant cells. Hemadsorption of guinea pig erythrocytes also occurs on viral-infected cells (Table 48-8).

A clinical diagnosis can be confirmed by serologic testing. As with measles, a fourfold increase in virus-specific antibody level or the detection of mumps-specific IgM antibody indicates active infection. Testing in the past has included measuring antibodies to S (nucleocapsid) and V (viral glycoprotein) antigens. Antibodies to S appear as early as 3 days after the onset of symptoms and disappear in 6 to 8 months. Antibodies to V develop 2 to 4 weeks after symptoms and persist longer than those to S. Relative amounts of S and V antibodies can indicate recent or past infection.

Treatment, Prevention, and Control

Treatment is symptomatic. Antiviral agents are not available.

Given the difficulties of controlling spread, vaccines provide the only effective means for reducing

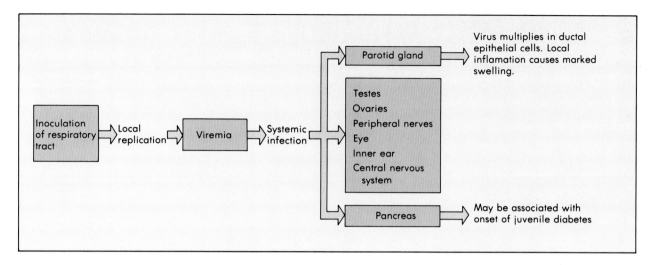

FIGURE 48-9   Mechanism of spread of mumps virus within the body.

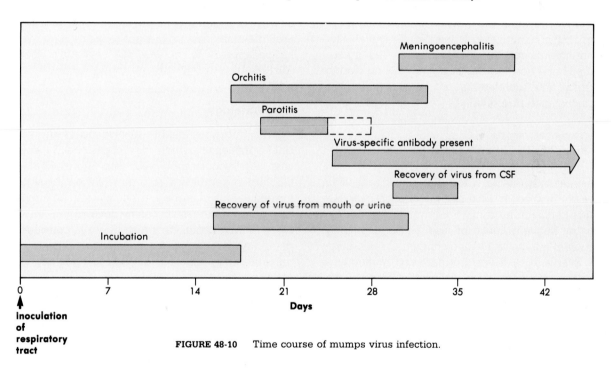

FIGURE 48-10   Time course of mumps virus infection.

Table 48-7   Clinical Consequences of Mumps Virus Infection

Disorder	Symptoms
Subclinical infection	No apparent disease
Mumps	Characteristic parotitis; complications include orchitis, meningitis, and (most seriously) encephalitis

Table 48-8 Laboratory Diagnosis of Mumps Virus Infection	
Test	**Detects**

Cell culture
  Isolate from saliva, urine,
  CSF; use primary or con-
  tinuous monkey kidney
  cells
	Growth and presence of mumps virus

Enzyme immunoassay	Fourfold increase in virus-specific IgG and IgM antibodies

infection. Since the introduction of vaccine in the United States in 1967, the yearly incidence of cases has declined from 76 to 2 per 100,000. Unfortunately, use of this vaccine in the world is very limited. Live attenuated vaccine is the most commonly used preparation, but an inactivated preparation is available for use in immunocompromised patients. The inactivated vaccine does not result in stable, long-lasting immunity and therefore is not recommended for immunocompetent individuals.

RESPIRATORY SYNCYTIAL VIRUS

Respiratory syncytial virus (RSV), first recovered from a chimpanzee in 1956, is a member of the *Pneumovirus* genus in the family Paramyxoviridae. RSV is the most frequent cause of fatal acute respiratory infection in infants and young children. It infects virtually everyone by 4 years of age, and reinfections occur throughout life, even among the elderly.

The major structural differences between RSV and the other paramyxoviruses is the smaller nucleocapsid and the lack of hemagglutinins and neuraminidases for RSV.

Pathogenesis

Respiratory syncytial virus produces an infection localized to the respiratory tract. As the name suggests, RSV induces syncytia, which provides a mechanism for the transmission of the virus to uninfected cells. Most of the pathologic effect of RSV is probably due to direct viral invasion of respiratory epithelium, but an additional component may result from immunologically mediated cell injury. Necrosis of bronchi

and bronchioles leads to "plugs" of mucus, fibrin, and necrotic material within smaller airways. These in turn lead to air trapping and the appearance of hyperinflation that can be seen in chest x-ray films. Narrow airways of young infants are readily obstructed. Natural infection does not prevent reinfection and vaccination appears to increase the severity of disease (Box 48-4).

Epidemiology

RSV infections almost always occur in winter. Unlike influenza, which may occasionally "skip" a year, RSV occurs every year (see Figure 48-8). The incubation period is 4 to 5 days. When the virus is introduced into a nursery, especially into an intensive care nursery, the results can be devastating. Virtually every infant becomes infected, with considerable morbidity; occasional fatalities occur. The virus is transmitted by hand contact and to some degree by

---

Box 48-4 Pathogenic Mechanisms of Respiratory Syncytial Virus

Localized infection of respiratory tract
Pneumonia results from cytopathic effect of virus
Bronchiolitis most likely mediated by host's immune response
Narrow airways of young infants readily obstructed
Maternal antibody does not protect infant from infection
Natural infection does not prevent reinfection
Vaccination increases severity of disease

Table 48-9 Clinical Consequences of Respiratory Syncytial Virus Infection

Disorder	Symptoms
Bronchiolitis and/or pneumonia	Fever, cough, dyspnea, and cyanosis in children less than 1 year old
Febrile rhinitis and pharyngitis	Symptoms occur in children of all ages
Common cold	Upper respiratory infection in older children and adults who are being reinfected against a background of partial immunity

respiratory routes. Virtually all children are infected by RSV by the age of 4, especially in urban centers. Outbreaks may also occur among the elderly, (e.g., in nursing homes). Virus is shed in respiratory secretions for many days, especially in infants.

Clinical Syndromes

Respiratory syncytial virus can cause any respiratory illness, from a common cold to pneumonia. In older children and adults upper respiratory infection with prominent rhinorrhea ("runny nose") is the common manifestation of infection. In infants a more severe lower respiratory illness, bronchiolitis, may occur. Because of inflammation at the level of the bronchiole, there is air trapping and decreased ventilation. Clinically, the patient usually has low-grade fever, tachypnea, tachycardia, and expiratory wheezes over the lungs. Bronchiolitis is usually self-limited, but it can be a frightening disease to observe in an infant. In premature infants, those with underlying lung disease, and in the immunocompromised, it may be be fatal (Table 48-9).

Laboratory diagnosis

Respiratory syncytial virus may be difficult to isolate in cell culture (Table 48-10). It is quite labile during transportation, requires a special cell line (Hep-2) to optimize recovery, and an average of 5 days for detection. For these reasons methods for direct detection of viral antigen in nasal washings have been developed. The best of these methods is detection of viral antigen by immunofluorescence and enzyme immunoassay (Figure 48-11). Reagents for

Table 48-10 Laboratory Diagnosis of Respiratory Syncytial Virus Infection

Test	Detects
Immunofluorescence and enzyme immunoassays; use nasal washings	RSV antigens
Cell culture Use nasal washings or other respiratory tract specimens; use continuous (transformed) cell lines, such as HEP-2 and HeLa cells	Growth and presence of RSV
Serologic tests	Seroconversion; fourfold or greater increase in IgG antibody titer

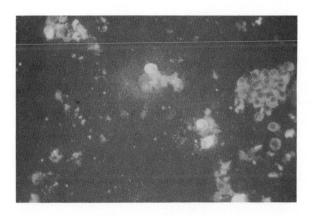

FIGURE 48-11 Immunofluorescent stain of epithelial cells in nasal wash. Note fluorescent RSV cytoplasmic inclusions (×200). (Courtesy Richard Thomson, Akron, Ohio.)

both tests are commercially available. Seroconversion or a fourfold or greater increase in antibody titer can substantiate a diagnosis of RSV infection.

Treatment, Prevention, and Control

In otherwise healthy infants treatment is supportive, with oxygen, intravenous fluids, and nebulized cold steam. In those who are predisposed to a more severe course (e.g., premature or immunocompromised infants), ribavirin administered by inhalation (nebulization) appears to provide benefit. To have a meaningful clinical effect, the drug should be started as early as possible, reinforcing the need for rapid viral diagnosis.

RSV is an important nosocomial pathogen. Infected children must be isolated or cohorted. Since hospital staff caring for infected children are known to transmit virus to other uninfected patients, control measures, including gowns, goggles, masks, and handwashing, are essential.

There is no currently available vaccine for RSV prophylaxis. Indeed, a previous vaccine containing inactivated RSV actually caused recipients to have more severe RSV infection when subsequently exposed to live virus. This is thought to have resulted from a heightened immunologic response at the time of wild virus exposure.

BIBLIOGRAPHY

Balows A, Hausler WJ Jr, and Lennette EH: Laboratory diagnosis of infectious diseases: principles and practice, New York, 1988, Springer-Verlag.

Fields BN, editor: Virology, New York, 1985, Raven Press.

Hinman AR: Potential candidates for eradication, Rev Infect Dis 4:933-939, 1982.

White DO and Fenner F: Medical virology, Orlando, Florida, 1986, Academic Press.

Coronaviruses

The family Coronaviridae is composed of a single genus, *Coronavirus,* which includes strains from humans and other animals. The name "corona" was chosen because the surface projections on the virion have a crownlike appearance when viewed with the electron microscope (Figure 49-1). Coronaviruses are difficult to isolate in cell culture, so infections with this virus rarely diagnosed in clinical practice. Based on serologic studies, coronaviruses cause approximately 10% to 15% of the upper respiratory tract infections and pneumonias in humans. Electron microscopy links coronaviruses to gastroenteritis in children and adults. Only two strains of virus have been isolated; however, other human strains are believed to exist.

STRUCTURE AND REPLICATION

Coronaviruses are enveloped virions, with long helical nucleocapsids that measure 80 to 160 nm. On the surface of the envelope are club-shaped projec-

tions that resemble a solar corona 20 nm in length and 5 to 11 nm wide (Figure 49-2). The genome is composed of plus-stranded, unsegmented RNA with a molecular weight of approximately 7×10^6 d. Coronavirus gene products include four known proteins (Table 49-1). Two peplomeric (spike) proteins, labeled **E2** and **H1,** are responsible for binding to host cells with membrane fusion and hemagglutination, respectively. A core nucleoprotein **(N)** and a transmembrane matrix protein **(E1)** are also encoded.

The replicative cycle is thought to include a unique scheme for producing plus-strand mRNA for translation. Viral attachment, mediated by peplomeric proteins, initiates the cytoplasmic maturation process. A viral-encoded RNA-dependent RNA polymerase generates minus-strand RNA from genomic plus-strand RNA. This RNA, in turn, is the template for synthesis of mRNAs of varying size with a common 3′ terminus, needed for translation and for production of new genomic plus-strand mRNA. The nucleocapsid buds into the lumen of rough endoplasmic reticulum, where envelope and peplomeric components are

Table 49-1 Human Coronavirus Gene Products

Proteins	Molecular Weight (k)	Location	Functions
E2 (peplomeric glycoprotein)	160-200	Envelope spikes (peplomer)	Binding to host cells; fusion activity
H1 (hemagglutinin protein)	60-66	Peplomer	Hemagglutination
N (nucleoprotein)	47-55	Core	Ribonucleoprotein
E1 (matrix protein)	20-40	Envelope	Transmembrane proteins

Modified from Balows A et al, editors: Laboratory diagnosis of infectious diseases: principles and practice, Heidelberg, 1988, Springer-Verlag.

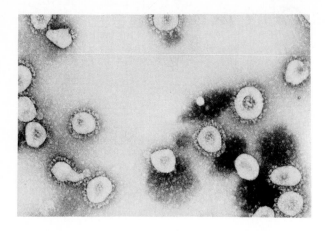

FIGURE 49-1   Electron micrograph of human respiratory coronovirus (×90,000). (Courtesy Center for Disease Control, Atlanta, Georgia).

picked up. Vesicles that contain virus migrate to the cell membrane and are released by exocytosis (Figure 49-3).

PATHOGENESIS

Coronaviruses inoculated into the respiratory tracts of human volunteers infect epithelial cells. Because the optimum temperature for viral growth is 33° to 35° C, infection remains localized to the upper respiratory tract. After infection a rise in coronavirus-specific serum antibody can be detected; however, serum antibody does not prevent reinfection (Box 49-1).

Coronavirus-like particles have been seen by electron microscopy in stools of children with diarrhea. However, the significance of this observation is unknown, and improved cell culture methods and serologic studies are needed to prove the association of coronaviruses with human gastroenteritis.

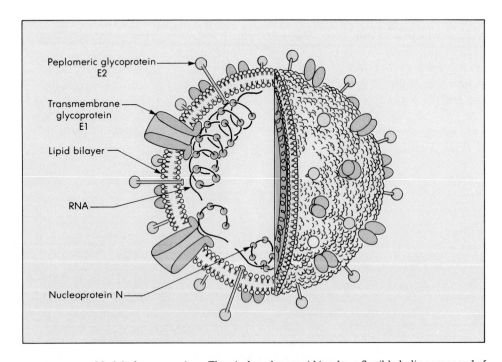

FIGURE 49-2   Model of a coronavirus. The viral nucleocapsid is a long flexible helix composed of the plus-strand genomic RNA and many molecules of the phosphorylated nucleocapsid protein *(N)*. The viral envelope includes a lipid bilayer derived from intracellular membranes of the host cell and two viral glycoproteins *(E1* and E2). (Redrawn from Fields: Virology, New York, 1985, Raven Press, p. 1333)

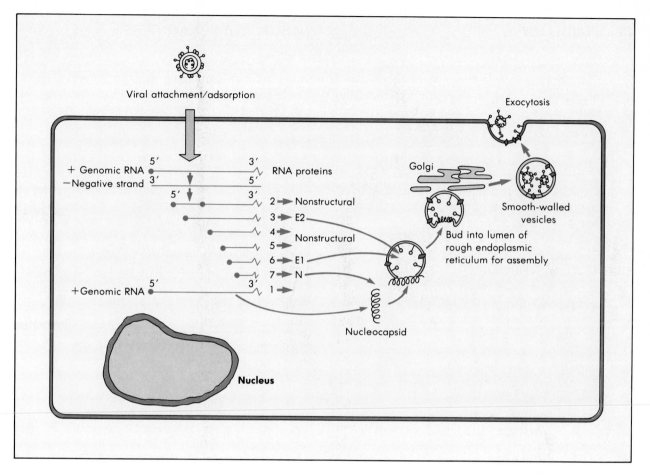

FIGURE 49-3   Replication of human coronaviruses. (Redrawn from Balows A et al, editors: Laboratory diagnosis of infectious diseases: principles and practice, New York, 1988, Springer-Verlag.)

Box 49-1   Pathogenic Mechanisms of Human Coronaviruses

Infects epithelial cells of upper respiratory tract
Grows best at 33° to 35° C
Serum antibodies not protective

Table 49-2   Epidemiology of Human Coronaviruses

Age (Years)	Antibody Prevalence
<1	6%-37%
1-5	54%-80%
5-20	100%
Adult	85%*

Data from Balows A et al, editors: Laboratory diagnosis of infectious diseases: principles and practice, vol 2, New York, 1988, Springer-Verlag.
*Lower antibody prevalence in adults suggests fewer reinfections.

EPIDEMIOLOGY

Coronaviruses may cause 5% to 15% of the "common colds" worldwide. Infections, which occur mainly in infants and children, appear either sporadically or in outbreaks with a winter and a spring pattern. Usually, one strain will predominate in an outbreak. Antibodies to coronaviruses are uniformly present by adulthood (Table 49-2), but reinfections are common despite preexisting serum antibodies. The role of secretory antibodies is unknown. The precise method of spread, whether by aerosol or direct contact, is also unknown (see Box 49-1).

CLINICAL SYNDROMES

Coronaviruses cause upper respiratory infection, similar to "colds" caused by rhinoviruses, but with a longer incubation period (average 3 days). Exacerbations of chronic pulmonary disease such as asthma and bronchitis may accompany the infection. Although the infection is usually mild, pneumonia may occur in children and adults.

Coronaviruses may be associated with gastroenteritis, which occurs year-round. As mentioned previously, confirmation of the etiology of this relationship is needed.

LABORATORY DIAGNOSIS

Reliable isolation of the virus is accomplished using human embryonic trachea organ cultures. These methods are not routinely available, however, and epidemiologic studies have been performed using serologic assays. Many techniques can be used to document increases in virus-specific antibody levels. Serologic tests are not routinely available, but future diagnostic tests using immunofluorescence, enzyme immunoassay, and nucleic acid hybridization to detect virus could make diagnosis routine. Electron microscopy has been used to detect coronavirus-like particles in stool specimens, but this technique is rarely employed. Enzyme immunoassays, similar to those used to detect other viral agents of gastroenteritis (e.g., rotavirus) could be commercially developed when enteric pathogenicity is confirmed.

TREATMENT, PREVENTION, AND CONTROL

Since the method by which coronaviruses spread is unknown, control measures have not been established. Control may not even be necessary because infection is generally mild. However, if coronaviruses are confirmed as agents of infant diarrhea, control would be necessary in hospitals and similar settings. No vaccine or specific antiviral therapy is available.

BIBLIOGRAPHY

Balows A, Hausler WJ Jr, and Lennette EH: Laboratory diagnosis of infectious diseases: principles and practice, New York, 1988, Springer-Verlag.

Fields BN, editor: Virology, New York, 1985, Raven Press.

ter Meulen V, Siddell S, and Wege H, editors: Biochemistry and biology of coronaviruses, New York, 1981, Plenum Press.

Picornaviruses

Picornaviridae is one of the largest families of viruses and includes some of the most important human and animal viruses (Box 50-1). As the name indicates, these viruses are small (pico) ribonucleic acid (RNA) viruses that have a naked capsid structure. The family includes more than 230 members divided into four genera: *Enterovirus, Rhinovirus, Cardiovirus,* and *Apthovirus.* Of these, only *Enterovirus* and *Rhinovirus* cause human disease. The different genera can be distinguished by the stability of the capsid at pH 3, optimum temperature for growth, mode of transmission, and the diseases they cause (Box 50-2).

At least 72 serotypes of human enteroviruses exist, including the polio, coxsackie, and echo viruses. Hepatitis A virus is also classified in this group (enterovirus 72) but is discussed separately in Chapter 45. The capsids of these viruses are very stable in harsh environmental conditions (sewage systems) and in the gastrointestinal tract, which facilitates their transmission by the fecal-oral route. Although

they may initiate their infection in the gastrointestinal tract, the enteroviruses rarely cause enteric disease, and infections are usually asymptomatic. Several different disease syndromes may be caused by a specific enterovirus serotype, and several different serotypes may cause the same disease, depending on the target tissue affected. The most well-known and well-studied picornavirus is poliovirus of which there are three serotypes.

Coxsackieviruses are named after the town of Coxsackie, N.Y., where the first isolation was made. They are divided into two groups, A and B, on the basis of certain biologic and antigenic differences and further subdivided into numeric serotypes by additional antigenic differences.

Box 50-1   Picornaviridae

Enterovirus
  Poliovirus
  Coxsackie A virus
  Coxsackie B virus
  Echovirus (ECHO virus)
  Enterovirus
Rhinovirus
Cardiovirus
Apthovirus

Box 50-2   Properties of Human Picornaviruses

COMMON PROPERTIES

Spheric virion, 25 to 30 nm
Naked icosahedral capsid
Single-strand RNA, linear genome with positive
  polarity
Cytoplasmic replication
Viral RNA translated into polyprotein, which is then
  cleaved

SPECIAL PROPERTIES

pH stability: enteroviruses stable from pH 3 to 9; rhinoviruses labile at acidic pHs.
Optimum growth temperature: enteroviruses, 35° to 37° C; rhinoviruses, 33° C

Modified from Fields BN: Virology, New York, 1988, Raven Press.

The name echovirus is an abbreviation of *enteric* *cytopathic human orphans,* since these agents were not thought to be associated with clinical disease. Now 31 serotypes are recognized. These viruses have a greater tendency than polioviruses to affect the meninges and cause meningitis, but a lesser tendency to infect anterior horn cells.

The human rhinoviruses include at least 89 serotypes and are the major cause of the common cold. The rhinoviruses are adversely affected by acidic pH and replicate optimally at 33° C or colder. This usually limits rhinoviruses to upper respiratory infections.

STRUCTURE

The plus-strand RNA of the picornaviruses is surrounded by an icosahedral capsid approximately 30 nm in diameter. The icosahedral capsid has 12 pentameric vertices, each of which is composed of five protomeric units of proteins. The protomers are made up of four polypepetides, vasopressin $_{1-4}$ (VP_{1-4}). VP_2 and VP_4 are generated by cleavage of a precursor, VP_0. The presence of VP_4 in the virion solidifies the structure. VP_4 is not generated until the genome is incorporated into the capsid and it is released on binding to the cellular receptor. The receptor-binding site is formed by VP_1. The capsids are stable in heat and detergent and, except for the rhinoviruses, are also stable in acid. The capsid structure is so regular that paracrystals of virions often form in infected cells (Figure 50-1).

The genome of the picornaviruses resembles a messenger RNA (mRNA) (Figure 50-2). It is a single

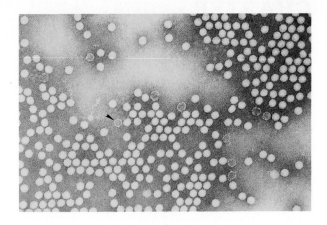

FIGURE 50-1   Electron micrograph of poliovirus. (Courtesy Centers for Disease Control, Atlanta.)

strand of plus-sense RNA of approximately 7,400 bases that has a polyA at the 3′ end and a small protein called VPg (22 to 24 amino acids) attached to the 3′ end. The genome by itself is infectious and can initiate virus replication. The polyA sequence enhances the infectivity of the RNA, and the VPg may be important in packaging the genome into the capsid and initiating viral RNA synthesis.

The genome encodes a polyprotein that is proteolytically cleaved to produce the enzymatic and structural proteins of the virus. In addition to the capsid proteins and VPg, the picornaviruses encode at least one protease and an RNA-dependent RNA polymerase. Poliovirus also produces a protease that degrades the 200,000 d cap-binding protein of eucaryotic ribosomes blocking translation of cellular mRNA.

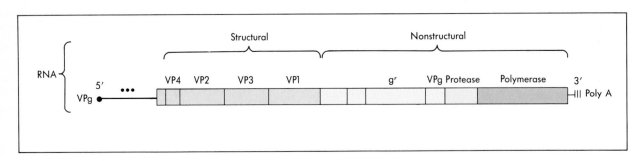

FIGURE 50-2   Structure of the picornaviral genome. Viral protein precursors VP1, VP2, and VP3 are further cleaved into four, three, and four end products, respectively. Most cleavages are carried out by a virus-coded protease. g^r represents the guanidine resistance marker, a genetic locus affecting the action of a drug thought to block initiation of RNA synthesis.

REPLICATION

The specificity of picornavirus interaction with cellular receptors is the major determinant of their target tissue and disease. The VP_1 proteins at the vertices of the virion contain a canyon into which the receptor binds. Arildone and similar compounds bind within this canyon and block the uncoating of the virus.

The picornaviruses have been categorized into several receptor families. The receptors for polioviruses and rhinoviruses have recently been identified as tissue-specific cellular adhesion molecules. These molecules are members of the immunoglobulin superfamily and promote normal and immunologic cell-to-cell interactions. At least 80% of the rhinoviruses and several serotypes of coxsackievirus recognize intercellular adhesion molecule 1 (ICAM-1). ICAM-1 is expressed on epithelial cells, fibroblasts, and endothelial cells. Poliovirus binds to a molecule of similar structure and presumably of similar function. The cells in which the poliovirus receptor is expressed correlate directly with the limited range of poliovirus infection.

On binding to the receptor, the VP_4 is released and the virion weakened. The virus is internalized by receptor-mediated endocytosis, and the virions dissociate in the acidic environment of the endosome, releasing the genome into the cytoplasm.

The genome binds to ribosomes, and a polyprotein is synthesized within 10 to 15 minutes of infection. The polyprotein is initially cleaved by cellular proteases autocatalytically until a viral protease is generated to cleave the rest of the polyprotein. The enzymatic functions of the virus are encoded at the 3′ end and are therefore translated first and most efficiently.

The RNA-dependent RNA polymerase generates a negative-strand RNA template from which the new mRNA/genome and templates can be synthesized. The amount of viral mRNA increases rapidly in the cell, with the number of viral RNA molecules reaching 400,000 per cell. The VPg acts as a primer for the polymerase.

Cellular RNA and protein synthesis are inhibited during infection by several picornaviruses. Cleavage of the 200,000 d cap-binding protein of the ribosome by a poliovirus protease alters the mRNA specificity of the ribosome. Permeability changes induced by picornaviruses reduce the ability of cellular mRNA to bind to the ribosome. Viral mRNA can also outcompete cellular mRNA for the factors required in protein synthesis. These activities contribute to the cytopathic effect of the virus on the target cell.

As the viral genome is being replicated and translated, the structural proteins VP_0, VP_1, and VP_3 are cleaved from the polyproteins and assembled into protomers. The protomers associate into pentamers, and 12 pentamers associate to form the procapsid. After insertion of the genome, VP_0 is cleaved into VP_2 and VP_4 proteins. Assembly of viral RNA into the viral capsid occurs in the cytoplasm, and the virion is released when lysis of the cell occurs.

ENTEROVIRUSES

Pathogenesis

Differences in pathogenesis for the enteroviruses mainly result from differences in tissue tropism and cytolytic capacity of the virus (Figure 50-3). Poliovirus has been studied extensively and is the prototype for the pathogenesis of the enteroviruses. Infections are usually asymptomatic but can range from coldlike symptoms to paralytic disease.

The upper respiratory tract, the oropharynx, and the intestinal tract are the portals of entry for the enteroviruses. The virus initiates replication in the mucosa and lymphoid tissue of the tonsils and pharynx and later infects the gut. The virions are impervious to stomach acid, proteases, and bile and infect lymphoid cells of Peyer's patches and the intestinal mucosa. Primary viremia spreads the virus to receptor-bearing target tissues, where a second phase of viral replication may occur, resulting in symptoms and a secondary viremia. Virus shedding from the oropharynx can be detected for a short time before symptoms begin, whereas virus production and shedding from the intestine may last for 30 days or longer, even in the presence of a humoral immune response.

The nature of the enterovirus disease is determined by the tissue tropism of the virus. Poliovirus has one of the narrowest tissue tropisms, recognizing a receptor expressed on anterior horn cells of the spinal cord, dorsal root ganglia, motor neurons, and few other cells (Figure 50-4). Coxsackieviruses and echoviruses recognize receptors expressed on more cell types and tissues and cause a broader repertoire of

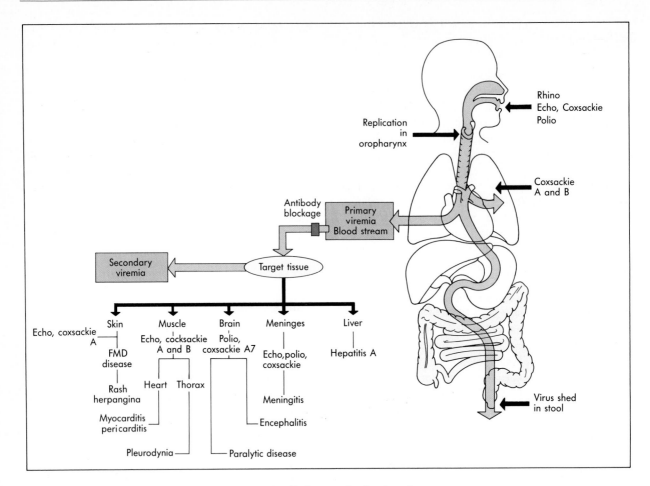

FIGURE 50-3   Pathogenesis of enteroviruses.

Table 50-1   Summary of Clinical Syndromes Associated with Major Enterovirus Groups

Syndrome	Polioviruses	Coxsackie A	Coxsackie B	Echoviruses
Paralytic disease	+	+	+	+
Encephalitis, meningitis	+	+	+	+
Carditis	−	+	+	+
Neonatal disease	−	−	+	+
Pleurodynia	−	−	+	−
Herpangina	−	+	−	−
Rash disease	−	+	+	+
Acute hemorrhagic conjunctivitis	−	+	−	−
Respiratory infections	+	+	+	+
Undifferentiated fever	+	+	+	+
Diarrhea, gastrointestinal disease	−	−	−	+
Diabetes, pancreatitis	−	−	+	−
Orchitis	−	−	+	−
Disease in immunodeficient patients	+	+	−	+
Congenital anomalies	−	+	+	−

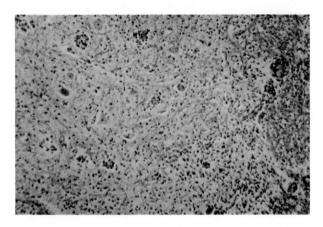

FIGURE 50-4   Poliomyelitis. Anterior horn and adjacent white matter of the spinal cord reveal perivascular mononuclear cell infiltration as well as inflammatory cells in the gray matter around degenerating and dying neurons. Luxol fast blue and hematoxylin and eosin stain were used. (From Lambert HP et al: Infectious diseases illustrated, London, 1982, Gower Medical Publishing Ltd.)

diseases (Table 50-1). In addition to the central nervous system, coxsackievirus and echovirus receptors may be found on heart, lung, pancreatic, and other cells. Differences in the susceptibility to and severity of poliovirus and coxsackievirus infection with age may also be attributed to differences in distribution and amount of receptor expression. Adults are generally more susceptible to serious disease with poliovirus, whereas newborns experience the most serious symptoms from coxsackie B and echovirus infections.

Most enteroviruses are cytolytic, replicating rapidly and causing direct damage to the target cell. Hepatitis A is the exception. It is not very cytolytic, and the kinetics of the immune response to hepatitis A correlate with the appearance of symptoms.

Antibody is the major protective immune response to the enteroviruses. Secretory antibody can prevent the initial establishment of infection in the oropharynx and gut, and serum antibody prevents viremic spread to the target tissue. However, in the individual with poor immune response, antibody production may be too late to block infection of the target tissue. Serum antibody is generally observed 7 to 10 days after infection.

Cell-mediated immunity is not likely to be involved in protection but may play a role in pathogenesis. T cells appear to contribute to coxsackie B virus–induced myocarditis in mice.

Epidemiology

The incubation period for enterovirus disease varies from 1 to 35 days, depending on the virus, the target tissue, and the age of the individual. The shortest incubation periods are for the viruses that affect oral and respiratory sites. Poor sanitation and crowded living conditions foster transmission of enteroviruses (Figure 50-5), and sewage contamination of water supplies can result in poliovirus epidemics. Poliomyelitis occurs throughout the world, and polioviruses are spread most often during the summer and fall. With the success of the polio vaccines, poliovirus infections in the United States now rarely occur. However, in areas where the vaccine is not available or in communities where vaccination is contrary to religious or other teachings, paralytic polio may still occur.

Paralytic polio was once called a middle-class disease. Good hygiene would delay exposure to the virus until late childhood, the adolescent years, or adulthood, when infection would produce the most severe symptoms. Infection during early childhood generally results in asymptomatic or very mild disease.

The coxsackieviruses and echoviruses are primarily spread by enteric routes but may also be spread in droplets to cause respiratory infections. The most frequently isolated enteroviruses are coxsackie A9, A16, B2 to B5, and echo 6, 9, and 11. These viruses are found worldwide. As with poliovirus infection, disease in adults is generally more severe than in children. However, coxsackie B virus and some of the echoviruses can be particularly harmful to infants. Spread of these viruses is enhanced by poor hygiene and sanitation and by crowded living quarters. Summer is the major season for infection with coxsackievirus and echovirus disease.

Clinical Syndromes

The clinical syndromes of the enteroviruses are determined by several factors, including the viral serotype, infecting dose, tissue tropism, portal of entry, age, sex, pregnancy, and state of health.

Poliovirus Infections   ''Wild'' polio infections are becoming rarer because of the success of the polio vaccines (Figure 50-6). However, vaccine-associated

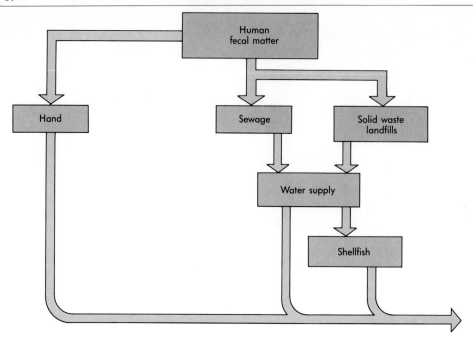

FIGURE 50-5   Transmission of enteroviruses.

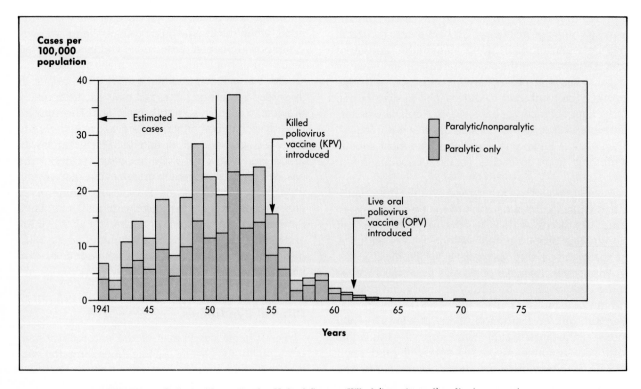

FIGURE 50-6   Polio incidence in the United States. Killed (inactivated) poliovirus vaccine was introduced in 1955 and live (oral) poliovirus vaccine in 1961-1962. Violet, Paralytic polio only; pink, both paralytic and nonparalytic polio. (Reprinted courtesy Centers for Disease Control: Immunization against disease—1972, Washington, DC, 1973, US Government Printing Office.)

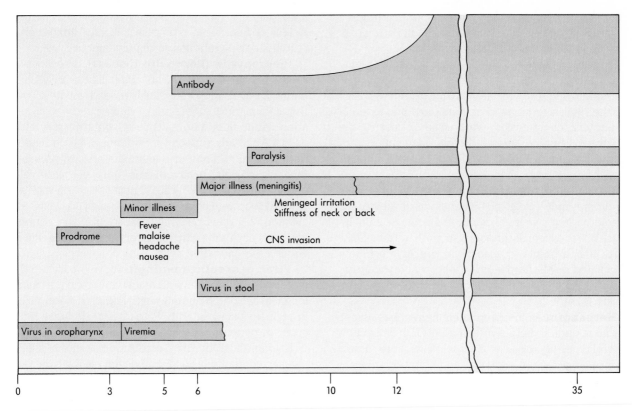

FIGURE 50-7   Progression of poliovirus infection.

cases of polio do occur, and some populations remain unvaccinated.

Poliovirus may cause one of four outcomes, depending on the progression of the infection (Figure 50-7):

1. **Asymptomatic illness** results if the virus is limited to infection of the oropharynx and the gut. At least 90% of poliovirus infections are of the asymptomatic type.
2. **Abortive poliomyelitis,** the **minor illness,** is a nonspecific febrile illness occurring in approximately 5% of infected individuals. Symptoms of fever, headache, malaise, sore throat, and vomiting occur within 3 to 4 days of exposure.
3. **Nonparalytic poliomyelitis** or **aseptic meningitis** occurs in 1% to 2% of patients with poliovirus infections. The virus progresses into the central nervous system and the meninges, causing back pain and muscle spasms in addition to the symptoms of minor illness.
4. **Paralytic polio,** the **major illness,** occurs in 0.1% to 2.0% of persons with poliovirus infections

and is the most severe outcome. Major illness follows 3 to 4 days after minor illness has subsided, thereby producing a biphasic illness. In this disease the virus spreads from the blood to the anterior horn cells of the spinal cord and the motor cortex of the brain. The severity of the paralysis is determined by the extent of the neuronal infection and the neurons affected. Spinal paralysis may involve one or more limbs, whereas bulbar (cranial) paralysis may involve a combination of cranial nerves and even the medullary respiratory center.

Paralytic poliomyelitis is characterized by an asymmetric flaccid paralysis with no sensory loss. Poliovirus type 1 affects 85% of patients with paralytic disease. Types 2 or 3 may cause vaccine-associated disease because of reversion from attenuated virus to virulence.

The degree of paralysis may vary from involving only a few muscle groups (e.g., one leg) to complete flaccid paralysis of all four extremities. The paralysis may progress over the first few days and result in complete recovery, residual paralysis, or death. Most recovery occurs within 6 months, but up to 2 years may be required for complete remission.

Bulbar poliomyelitis can be more severe and may involve the muscles of the pharynx, vocal cords, and respiration and result in death in 75% of patients. During the 1950s iron lungs, chambers providing external respiratory compression were used to assist the breathing of polio patients. Before vaccination programs, iron lungs filled the wards of children's hospitals.

Coxsackievirus and Echovirus Several clinical syndromes may be caused by either coxsackievirus or echovirus (e.g., aseptic meningitis), but certain illnesses are especially associated with coxsackieviruses. For example, coxsackie A viruses are highly associated with herpangina, whereas myocarditis and pleurodynia are most frequently caused by coxsackie B serotypes.

Herpangina is inappropriately named because it has no relation to herpesvirus. Rather, it is caused by several types of coxsackie A virus. Fever, sore throat, pain on swallowing, anorexia, and vomiting characterize herpangina. The classic finding is vesicular, ulcerated lesions around the soft palate and uvula (Figure 50-8). Less typically the lesions may affect the hard palate. The virus can be recovered from the lesions or from feces. The disease is self-limited and requires only symptomatic management.

Pleurodynia (Bornholm disease), also known as the ''devil's grip,'' is an acute illness in which patients have sudden onset of fever and unilateral low thoracic, pleuritic chest pain, which may be excruciating. Abdominal pain and even vomiting may also occur. Although a pleural friction rub may be heard, the physical findings of pneumonia are not present. Muscles on the involved side may be extremely tender. Chest x-ray films are almost always normal, as is the blood leukocyte count. Pleurodynia lasts an average of 4 days but may relapse after the patient has been asymptomatic for several days. Coxsackie B virus is the causative agent.

Viral, or **aseptic, meningitis** is an acute febrile illness accompanied by headache and signs of meningeal irritation, including nuchal rigidity, Kernig's or Brudzinski's sign, or both. Petechiae or skin rash may occur in patients with enteroviral meningitis.

Examination of the cerebrospinal fluid (CSF) reveals a predominantly lymphocytic pleocytosis, but very early in the disease, polymorphonuclear leukocytes may be more numerous. CSF glucose level is usually normal but may be slightly low. CSF protein

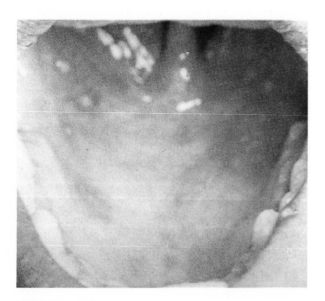

FIGURE 50-8   Herpangina. Characteristic discrete vesicles are seen on the anterior tonsillar pillars. (Courtesy of Dr GDW McKendrick; From Lambert HP et al: Infectious diseases illustrated, London, 1982, Gower Medical Publishing Ltd.)

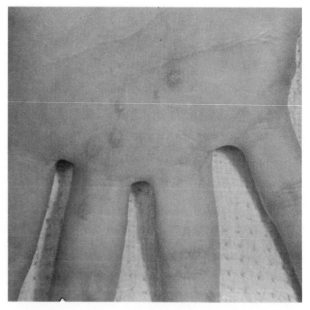

FIGURE 50-9   Hand of patient with enteroviral vesicular eruption caused by coxsackie A virus.

level is normal to slightly elevated. Unless associated encephalitis (meningoencephalitis) exists, recovery is uneventful.

Enteroviral eruptions may occur in patients infected with either echo- or coxsackieviruses. These often cause exanthematous eruptions as well as febrile illness. The eruptions are usually maculopapular but occasionally may appear as petechial or even vesicular eruptions (Figure 50-9). The petechial type of eruption must be differentiated from that of meningococcemia. The child with enteroviral infection is not as ill or as toxic and has a lesser degree of leukocytosis than the child with meningococcemia.

Hand-foot-and-mouth disease is a vesicular exanthem caused by an enterovirus, usually coxsackievirus A16. The colorful name is descriptive, since the main features of this infection are vesicular lesions of the hands, feet, mouth, and tongue (Figure 50-10). The patient is mildly febrile, and the illness subsides in a few days.

Nonpolio enteroviruses, especially coxsackie B viruses, are a cause of **acute benign pericarditis.** This is usually a disease of young adults but may be seen in older individuals, in whom the distinction from myocardial infarction may be difficult. Usually the symptoms are similar, but in pericarditis fever may become more severe than that in the patient with myocardial infarction. Chest pain may be altered by the patient's position and respiration, and a friction rub may be heard.

Myocardial and **pericardial infections** caused by coxsackie B virus occur in older children and adults but are most threatening in newborns. Neonates with these infections have febrile illnesses and sudden, unexplained onset of heart failure. Cyanosis, tachycardia, cardiomegaly, and hepatomegaly occur. Electrocardiographic changes are those found in patients with myocarditis. Mortality is high, and autopsy reveals other involved organ systems, including brain, liver, and pancreas.

Echoviruses may also produce severe disseminated infection in infants.

Enterovirus 70 and a variant of coxsackievirus A24 have recently been associated with an extremely contagious ocular disease, **acute hemorrhagic conjunctivitis.** The infection causes subconjunctival hemorrhages and conjunctivitis. The disease has a 24-hour incubation period and resolves within 1 or 2 weeks.

Respiratory disease, hepatitis, and **diabetes** are some of the syndromes attributed to enteroviruses. Coxsackieviruses A21 and A24 and echoviruses 11 and 20 can cause coldlike symptoms if the upper respiratory tract becomes infected. Enterovirus 72, or hepatitis A virus, causes hepatitis A (see Chapter 45). Coxsackie B infections of the pancreas have been suspected to cause insulin-dependent diabetes because of the destruction of the islets of Langerhans.

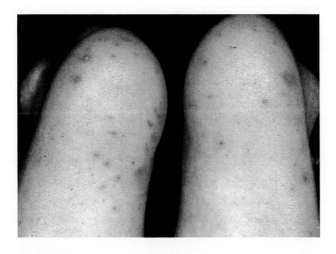

FIGURE 50-10   Hand-foot-and-mouth disease caused by coxsackie A virus. Lesions initially appear in the oral cavity and then develop within 1 day on the palms and, as seen here, soles. (From Habif TP: Clinical dermatology: a color guide to diagnosis and therapy, St Louis, 1985, The CV Mosby Co.)

Laboratory Diagnosis (Table 50-2)

Culture Polioviruses grow well in monkey kidney tissue culture, and the virus may be isolated from the pharynx during the first few days of illness and from the feces for up to 30 days. The CSF is rarely positive for the virus, although a pleocytosis of 25 to 500 leukocytes usually occurs. Neutrophils may predominate early, especially in aseptic meningitis. Protein and glucose levels in CSF are usually normal or only slightly abnormal.

Coxsackieviruses and echoviruses can usually be isolated from the throat and stool during infection and often from CSF in patients with meningitis. Virus is rarely isolated in myocarditis, since the symptoms occur several weeks after the initial infection.

The coxsackie B viruses can be grown on primary monkey or human embryo kidney cells. Many cox-

Table 50-2 Laboratory Diagnosis of
Enterovirus Infections

Test	Detects
Cell culture, e.g. primary monkey kidney cell lines and human diploid fibroblast*	Presence of virus
Serologic tests†	Presence of specific IgM; fourfold rise in serotype specific IgG

*Most coxsackie A viruses cannot be grown in cell culture.
†Not widely available but possibly useful in infections caused by serotypes that cannot be grown in cell culture.

sackie A virus strains do not grow in tissue culture and must still be grown in suckling mice.

Serology Serologic confirmation of enterovirus infection can be made by detection of specific IgM or a four-fold increase in antibody titer between acute illness and convalescence. The many serotypes of echovirus and coxsackievirus makes this approach difficult, but it may be useful in documenting poliovirus infection.

Treatment

No specific antiviral therapy is available for enterovirus infections. Supportive therapy is extremely important for patients with paralytic disease to assist in their potential recovery. Historically the iron lung was used to support patients with bulbar paralysis and impaired breathing.

Prevention and Control

The prevention of paralytic poliomyelitis is one of the triumphs of modern medicine. In the developed world, almost complete control has been achieved by the use of vaccines. Unfortunately, health care delivery systems are not sufficient to provide adequate vaccine administration in underdeveloped countries.

Two types of poliovirus vaccine exist, a formalin-inactivated product known as the inactivated, killed, or Salk vaccine and an attenuated one known as the live, oral, or Sabin vaccine. Both vaccines can induce a protective antibody response (Figure 50-11).

The killed vaccine was proved effective in 1955 but has generally been replaced by the oral preparation because of ease of delivery (Table 50-3). Oral vaccine is attenuated (i.e., rendered less virulent) by passage

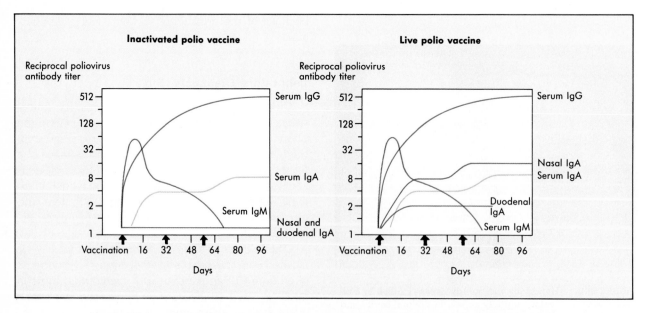

FIGURE 50-11   Serum and secretory antibody response to intramuscular inoculation of inactivated polio vaccine and to orally administered, live attenuated polio vaccine. (Redrawn from Ogra P et al: Rev Infect Dis 2:352-369, 1980.)

Table 50-3 Advantages and Disadvantages of Polio Vaccines

Live	Killed
ADVANTAGES	
Effective	Effective
Lifelong immunity	Can be incorporated into routine immunizations (with DPT)
Induces secretory antibody response similar to that of natural infection	Good stability in transport and storage
Attenuated virus may circulate in community by spread to contacts (indirect immunization)	No risk of poliomyelitis in vaccines or contacts
Easily administered	Safe in immunodeficient patients
Repeated boosters not required	
DISADVANTAGES	
Vaccine-associated poliomyelitis in vaccine recipients or contacts	Does not induce local (gut) immunity
Spread of vaccine to contacts without their consent	Booster vaccine required for lifelong immunity
Unsafe for immunodeficient patients	Injection less acceptable than oral administration
	Must achieve higher community immunization levels than with live vaccine

in cell cultures. Attenuation yields a virus capable of replicating in the oropharynx and intestinal tract and of being shed in feces for weeks. The remote potential for causing paralytic disease is the major drawback of the live vaccine and is estimated to occur in 1 per 4 million doses administered (vs. 1 in 100 of those infected with "wild" poliovirus). The risk of vaccine-associated paralytic poliomyelitis is increased in immunocompromised individuals and is more likely to occur in susceptible adults than susceptible children. Since the live vaccine strain may spread to close, especially household, contacts (a virtue in achieving mass immunization), vaccine-associated poliomyelitis may occur in contacts rather than the actual vaccine recipient.

No vaccines exist for coxsackieviruses or echoviruses. Transmission can presumably be reduced by improvements in hygiene and living conditions.

RHINOVIRUSES

Rhinoviruses are the most important cause of the common cold and upper respiratory infection (URI).

Eighty-nine serotypes have been identified by neutralization with specific antisera, and additional strains have been isolated but not yet typed.

Pathogenesis

In contrast to enteroviruses, rhinoviruses are unable to replicate in the gastrointestinal tract. This phenomenon does not result from degradation of virus by acid gastric contents, but rather probably reflects differences in surface receptors between the two groups of viruses.

At least 80% of the rhinoviruses share a common receptor, which is also used by some of the coxsackieviruses. The receptor has been identified as ICAM-1, as discussed earlier. ICAM-1 is a member of the immunoglobulin supergene family and is expressed on epithelial, fibroblast, and B-lymphoblastoid cells. The function of the molecule is to promote immune interactions, and its expression is stimulated by cytokines released during inflammation.

Secretory and serum antibody are generated to the rhinovirus and can be detected within a week of infection. No antigen is common to all rhinoviruses.

Although secretory IgA is probably more important than serum antibody in preventing and controlling infection, the response dissipates quickly; a better correlate of immunity is serum levels of antibody. Immunity begins to wane approximately 18 months after infection.

Interferon, generated in response to the infection, may both limit the progression of the infection and contribute to the symptoms. As just mentioned, the release of cytokines during inflammation can promote the spread of the virus by enhancing the expression of viral receptors. Cell-mediated immunity is not likely to play an important role in controlling rhinovirus infections.

Rhinoviruses grow best at 33° C, which may partly account for their predilection for the cooler environment of the nasal mucosa. Infection can be initiated by as little as one infectious viral particle. During the peak of illness, titers of 500 to 1,000 infectious virions are reached in nasal secretions. Most viral replication occurs in the nose, and the severity of symptoms correlates with the quantity (titer) of virus in nasal secretions. Biopsies of nasal mucosa taken during a "cold" reveal severe edema of the subepithelial tissue but minimal inflammatory cell response. Infected ciliated epithelial cells may be sloughed from the nasal mucosa.

Epidemiology

Rhinoviruses can be transmitted by two mechanisms: aerosols and direct contacts (e.g., by hands or contaminated inanimate objects). Surprisingly, aerosols are probably not the major route, despite being an apparently efficient mode of transmission. Hands appear to be the most important vector, and direct person-to-person contact is the predominant method of spread. Rhinoviruses can be recovered from the hands of 40% to 90% of persons with colds and from 6% to 15% of inanimate objects around them. The virus can survive on these objects for many hours.

Rhinoviruses produce clinical illness in only half those persons infected. Thus many asymptomatic individuals are capable of spreading the virus even though they have lower viral titers.

Rhinovirus "colds" affect persons most frequently in the early fall and the late spring in temperate climates. These peaks may reflect social patterns (e.g., return to school and day care) rather than any change in the virus itself.

Rates of infection are highest in infants and children. Children under age 2 years are considered the primary vector that introduces colds into a family unit. Secondary infections occur in approximately 50% of family members, especially other children.

Antibody affords incomplete protection, depending partly on the antibody titer. Both nasal secretory (IgA) and serum (IgG) antibody are induced by primary rhinovirus infection. Secretory antibody is probably more important than serum antibody in preventing reinfection. Type-specific immunity begins to wane approximately 18 months after infection.

Although many different rhinovirus serotypes may be found in a given community during many months of sampling, only a few predominate during a specific cold season. The major viruses are usually the newly categorized serotypes, suggesting that a gradual antigenic drift exists, similar to the pattern with influenza viruses.

Clinical Syndromes

URIs caused by rhinoviruses usually begin with sneezing, followed soon by rhinorrhea (Figure 50-12). The rhinorrhea increases and is then accompanied by symptoms of nasal obstruction. Mild sore throat also occurs, along with headache, malaise, and the "chills" (rigors). The illness peaks in 3 to 4 days, but the cough and nasal symptoms may persist for 7 to 10 days or longer. Fever and rigors are not usual features of rhinovirus URIs.

Laboratory Diagnosis

The clinical syndrome of the common cold is usually so characteristic that laboratory diagnosis is unnecessary unless the physician needs to establish which of the many respiratory viruses is causing a specific patient's illness. Although rhinoviruses cause up to one half of URIs, coronaviruses, parainfluenza viruses, and other agents also cause a sizable proportion of colds. Also, at times it may be difficult to distinguish allergic rhinitis from an URI.

The diagnostic methods for rhinoviruses include culture and serology.

Culture Nasal washings are the best clinical specimen for recovering the virus. Rhinoviruses grow in vitro only in cells of primate origin, with human diploid fibroblast cells, (e.g., WI-38) as the optimum system. As already stated, these viruses grow best at 33°

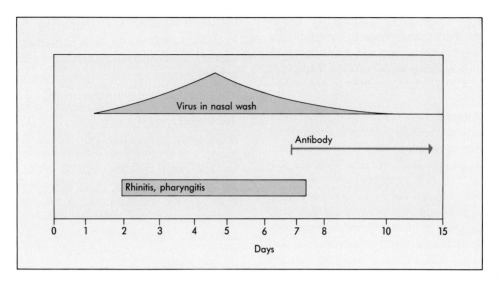

FIGURE 50-12   Time course of rhinovirus infection.

C, which is not the optimum temperature for any other clinically important viruses. Thus their isolation may require a separate incubator. Isolation in tissue culture occurs in 1 to 7 days, with an average of 4 to 5 days. Virus is identified by typical cytopathic effect and the demonstration of acid lability. Serotyping is rarely necessary but is done by using pools of specific neutralizing sera.

Serology Serologic testing to document rhinovirus infection is not practical. No antigen is common to all rhinoviruses; thus it would be necessary to have the patient's viral isolate or a prototype of the specific rhinorvius prevalent in the community in order to perform testing. Multiple serotypes of rhinovirus may circulate simultaneously, increasing the difficulty of serologic testing. If an isolate has been obtained from the patient, it is not necessary to demonstrate seroconversion to prove infection; the isolate itself suffices.

Treatment

Over-the-counter remedies for the common cold abound, but no specific therapy has been found to be effective. Nasal vasoconstrictors may provide relief, but their use may be followed by rebound congestion and worsening symptoms. Antibacterial agents are not beneficial unless a true bacterial infection (e.g., sinusitis) occurs. Even when nasal secretions have

become purulent, a carefully performed Gram's stain will usually show few if any bacteria.

Rigorous studies of vitamin C therapy have not shown efficacy. Experimental antiviral drugs, such as arildone, rhodanine, disoxaril, and their analogs, contain a 3-methylisoxazole group that inserts into the base of the receptor-binding canyon and blocks uncoating of the virus. Enviroxime inhibits the polymerase. A polypeptide receptor analog based on the ICAM-1 protein structure may also have potential as an antiviral drug.

Prevention and Control

Three methods of prevention or control of rhinoviruses are being explored: vaccines, antiviral agents (especially interferon), and interruption of transmission.

Vaccines The multiple serotypes and apparent antigenic drift in rhinoviral antigens pose major problems for the development of vaccines. Formalin-inactivated, parenterally administered vaccines induce antibody in serum but not in nasal secretions. Their use in artificial challenge experiments reduced viral titers and clinical severity of the URI, but neither infection nor illness was prevented. When formalin-inactivated vaccine was given intranasally, infection was significantly reduced and illness during challenge with the homotypic virus was ameliorated.

Multivalent vaccines containing antigens of 10 different rhinoviruses have been tested, but antibody responses occurred to fewer than half the antigens. A new, live attenuated rhinovirus vaccine has been developed but not tested in challenge or natural infection studies.

Antiviral Agents No antiviral drug has been proved therapeutically useful in controlling rhinovirus infections. A series of interesting studies indicate that intranasal interferon is effective for targeted prophylaxis, that is, for short-term use when an index case of URI is detected in a family setting. Long-term use of intranasal interferon (e.g., throughout the "cold season") is associated with unacceptable degrees of local toxicity characterized by nasal congestion and bleeding.

Interruption of Transmission Because hands are the major vector of rhinovirus transmission, handwashing and disinfection of inanimate objects should be beneficial. Impregnation of nasal tissue with antiviral chemicals has been suggested as a means of control, but this is not practical at present.

BIBLIOGRAPHY

Levandowski RA: Rhinoviruses. In Belshe RB, editor: Textbook of human virology, Littleton, Mass, 1984, PSG Publishing Co.

Melnick JL: Live attenuated poliovaccines. In Plotkin SA and Martin EA, editors: Vaccines, Philadelphia, 1988, WB Saunders Co.

Moore M and Morens DM: Enteroviruses including polioviruses. In Belshe RB, editor: Textbook of human virology, Littleton, Mass, 1984, PSG Publishing Co.

Racaniello VR: Poliovirus neurovirulence. In Maramorosch K, Murphy FA, and Shatkin AJ, editors: Advances in virus research, vol 34, New York, 1984, Academic Press.

Robbins FC: Polio-historical. In Plotkin SA and Martin EA, editors: Vaccines, Philadelphia, 1988, WB Saunders Co.

Salk J and Drucker J: Noninfectious poliovirus vaccine. In Plotkin SA and Martin EA, editors: Vaccines, Philadelphia, 1988, WB Saunders Co.

Wilfert CM, Lehrman SN, and Katz SL: Enteroviruses and meningitis, Pediatr Infect Dis 2:333-341, 1983.

CHAPTER 51

Reoviruses

The family Reoviridae contains six genera of viruses, three of which *(Orthoreovirus, Orbivirus,* and *Rotavirus)* infect animals, including humans (Table 51-1). The Reoviridae are nonenveloped viruses with double-layered protein coats and segmented double-stranded ribonucleic acid (RNA) genomes. The name reovirus was proposed in 1959 by Albert Sabin for a group of respiratory and enteric viruses that were not associated with any known disease process, thus the name *r* (respiratory), *e* (enteric), *o* (orphan) -virus. During the 1970s two additional groups of viruses were added. The orbivirus, which were moved from the arbovirus group, and the rotaviruses were both reclassified because of their segmented, double-stranded RNA genomes. The Reoviridae, as with other virus families, have common physiochemical, structural, genomic, and replicative properties (Table 51-2).

ORTHOREOVIRUSES (MAMMALIAN REOVIRUSES)

The orthoreoviruses, also referred to as mammalian reoviruses or simply reoviruses, were first isolated in the 1950s from stools of children. They were initially assumed to be members of the enterovirus group, but were reclassified when their unique double-stranded RNA structure was discovered.

The orthoreoviruses are now known to be ubiquitous, having been isolated in nearly all mammals worldwide. Additionally, viruses have been detected in sewage and in river water. The mammalian reoviruses occur in three serotypes, referred to as reovirus types 1 to 3, based on neutralization and hemagglutination-inhibition tests. All three serotypes share a common complement-fixing antigen. In general, these viruses cause asymptomatic infection.

Table 51-1 Reoviridae Responsible for Human Disease

Virus	Disease
Orthoreovirus*	Mild upper respiratory illness, gastrointestinal illness, biliary atresia
Orbivirus	Febrile illness with headache and myalgia
Rotavirus	Gastrointestinal illness, respiratory illness (?)

*Reovirus is a common name for the family Reoviridae and for the specific genus *Orthoreovirus.*

Table 51-2 Common Properties of the Reoviridae

Parameter	Specific Property
Physiochemistry	Stable at −70°C; little loss of infectivity at room temperature; orthoreoviruses and rotaviruses resistant to lipid solvents and stable over a wide pH range
Structure	Size 60 to 80 nm; icosahedral symmetry; nonenveloped virion
Genome	Double-standed RNA; 10 to 12 segments
Replication	Occurs in cytoplasm

Structure and Physiology

The orthoreovirus is a spheric icosahedron with a double capsid (Figure 51-1). The outer capsid is composed of protein, which surrounds a core composed of a protein shell containing the 10 double-stranded RNA segments with a combined genome molecular weight of 15×10^6 daltons. The virion is known to contain 11 proteins, 9 of which are structural proteins (5 core proteins, 1 core spike protein, and 3 outer capsid proteins). The structure of the core and outer capsid are schematically presented in Figure 51-2. One outer capsid protein is responsible for the hemagglutination property of these viruses. Because the viruses do not have an envelope, they are resistant to lipid solvents. Also, these viruses are stable over wide pH and temperature ranges, characteristics that presumably explain their wide geographic distribution and ability to inhabit mammalian gastrointestinal tracts. Orthoreoviruses survive well in airborne aerosols.

Considerable detail concerning transcription and translation products and specific functions of some of the viral proteins are known. The 10 genome segments, divided into large, medium, and small groups, encode for 11 proteins, also grouped into large, medium, and small. Functions of the outer capsid proteins and core proteins are summarized in Table 51-3.

Viral replication includes rapid adsorption, localization within the cell lysosome, assembly of virions with appropriate genome segments, and release of virus accompanied by host cell destruction. Virus in phagocytic vacuoles fuse with lysosomes, where virus particles remain throughout the replicative cycle. Cytoplasmic inclusions occur as scattered granules that eventually form a collar around the nucleus (Figure 51-3). Mechanisms of assembly and release are not well understood.

Pathogenesis

The orthoreoviruses gain entrance into human hosts through the respiratory and gastrointestinal tracts. Understanding specific pathogenic mechanisms is limited to animal models using gastrointestinal tract inoculation only. Orthoreoviruses are able to reach susceptible intestinal epithelial cells because they can survive the acidic environment of the stomach and proteolytic enzymes and bile within the intestine. The virus then multiplies in the lymphoid tissue of Peyer's patches lining the intestines. The ability to survive in the intestine is determined by the resistance of the virus to proteolytic digestion mediated by the viral outer capsid protein. The virus causes few if any symptoms before entering the circulation and producing infection at a distant site. In the mouse model the outer capsid protein responsible for hemagglutinin activity is also responsible for viral

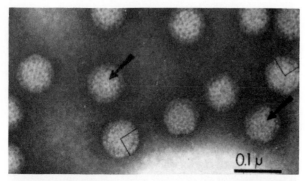

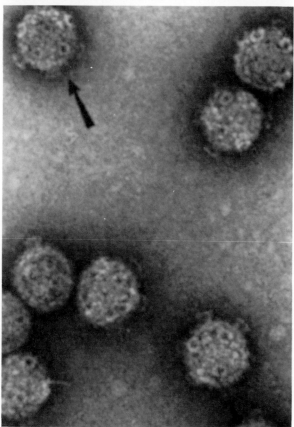

0.1 μ

FIGURE 51-1   Reovirus structure. Note icosahedral capsid and surface projections. (From Luftig RB et al: Virology 48:170, 1972.)

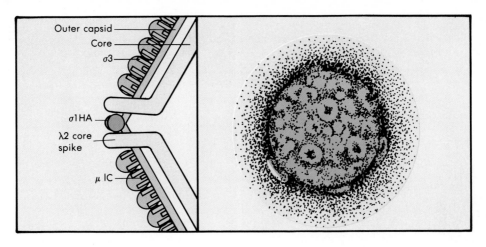

FIGURE 51-2   Structure of orthoreovirus core and outer proteins. (Redrawn from Sharpe AH and Fields BN: N Engl J Med 312:486-497, 1985.)

Table 51-3   Orthoreovirus Genome Segments, Proteins, and Protein Functions

Genome Segments (Molecular Weight)	Proteins	Functions (If Known)
Large segments (2.8×10^6)		
1	λ1 (core)	Not known
2	λ2 (core spike)	Regulate transcription
3	λ3 (core)	Regulate transcription by determining pH optimum of transcriptase
Medium segments (1.4×10^6)		
1	μ1 (core)	Regulates transcription by determining sensitivity of reovirions to proteolytic digestion
2	μ1C (outer capsid)	Resistance to proteolytic digestion and virulence
3	μ2 (core)	Not known
4	μNS	Nonstructural
Small segments (0.7×10^6)		
1	σ1 (outer capsid)	Hemaglutinin: mediates adsorption to host cell; tropism and specific tissue injury; inhibition of host DNA replication
2	σ2 (core)	Not known
3	σ3 (outer capsid)	Inhibition of cellular RNA and protein synthesis; initiation of persistent viral infection in cell cultures
4	σNS	Nonstructural

Modified from Fields BN, editor: Virology, New York, 1985, Raven Press.

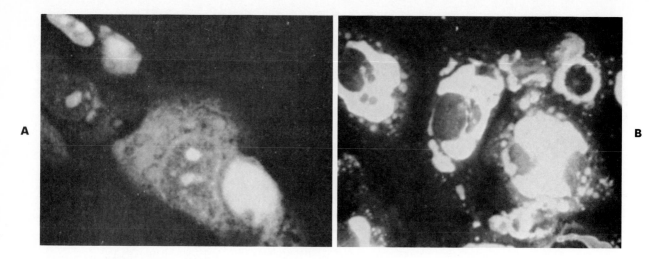

FIGURE 51-3   Orthoreovirus inclusions. **A,** Monkey kidney cells 48 hours after infection with reovirus type 1. **B,** Cells 72 hours after infection. (From Virology 17:342, 1962.)

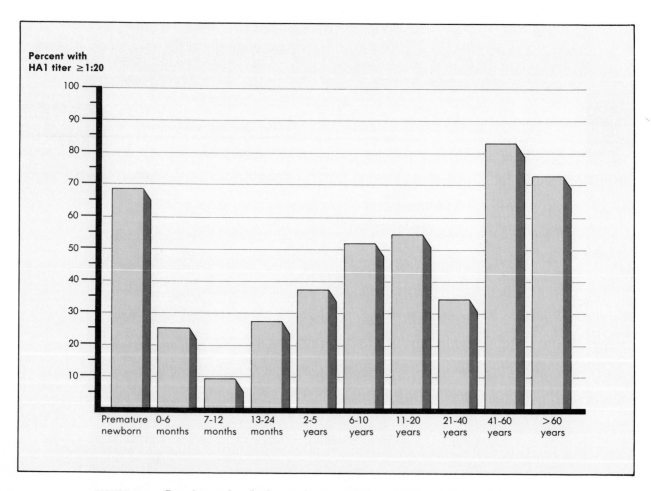

FIGURE 51-4   Prevalence of antibody to orthoreoviruses in children and adults. (Data from Lerner AM et al: N Engl J Med 267:947, 1962.)

spread to the mesenteric lymph nodes and neurotropism. In mice, orthoreoviruses elicit humoral and cellular immune responses to outer capsid proteins, but the role immunity to these viral antigens plays in human infection is not known. Orthoreoviruses, although normally lytic viruses, are also capable of establishing persistent infection in cell culture.

Epidemiology

As mentioned, the orthoreoviruses have been detected worldwide. Seroprevalence studies suggest that most people are infected during childhood, and by adulthood approximately 75% have antibody (Figure 51-4). Immunizing infections are asymptomatic. Most animals, including chimpanzees and monkeys, also have detectable antibody. Whether animals are a reservoir for human infections is unknown.

Clinical Syndromes

Orthoreoviruses infect people of all ages, but linking specific diseases to these agents has been difficult. Most infections are thought to be asymptomatic or so mild that they go undetected. Thus far these viruses have been linked to mild upper respiratory illness (low-grade fever, rhinorrhea, pharyngitis), gastrointestinal disease, and biliary atresia.

Laboratory Diagnosis

Human orthoreovirus infection can be detected using direct detection of viral antigen in clinical material, virus isolation, or serologic assays for virus-specific antibody (Table 51-4). Throat, nasopharyngeal, and stool specimens from patients with suspected upper respiratory or diarrheal disease are used. Detection of virus from tissue, as required for suspected hepatobiliary disease, may also be attempted.

Direct detection of viral antigen has been performed using immunofluorescence and peroxidase immunochemical methods. Availability of these assays, however, is limited to specialized laboratories that have antiserum to the orthoreoviruses. Early in the infectious cycle inclusions occur as small dots in the periphery of the host cell cytoplasm. Later, inclusions are larger and located adjacent to the nucleus. In situ RNA hybridization for the detection of orthoreovirus messenger RNA or double-stranded genomic

Table 51-4   Laboratory Diagnosis of Orthoreovirus Infection

Test	Detects
Immunofluorescence, immunochemical test (peroxidase), RNA hybridization	Viral antigens
Cell culture: mouse L fibroblast cells, primary monkey kidney cells, HeLa cells	Virus
Serologic tests: hemagglutination inhibition, neutralization, complement fixation	Antibody to virus

RNA is another potentially useful technique for the identification of the virus.

Virus isolation can be performed using cell culture. Although many cell lines have been successfully used, mouse L cell fibroblasts, primary monkey kidney cells, and HeLa cells are recommended. Cell cultures are incubated and observed for cytopathic effect (CPE) during a 3-week period. Orthoreovirus CPE includes pronounced cytoplasmic granularity. Hemagglutination and neutralization are used to confirm the presence of virus.

Serologic diagnosis of infection requires the documentation of a fourfold or greater rise in virus-specific antibody between an acute and convalescent specimen, since the presence of antibodies to orthoreoviruses typically occurs in healthy children and adults (Figure 51-4). Techniques used to detect orthoreovirus antibody include hemagglutination inhibition, neutralization, complement fixation, and indirect immunofluorescence.

Treatment, Prevention, and Control

Orthoreovirus disease, as thus far recognized, is mild and self-limited. For this reason treatment has not been necessary, and prevention and control measures have not been investigated.

ORBIVIRUSES

The orbiviruses are a large group of viruses that infect both vertebrates and invertebrates. Disease

Table 51-5 Some Orbivirus Serogroups Known to Cause Human Diseases

Serogroups	Geographic Location
Colorado tick fever	Western United States and Canada
Changuinola	Central and South America
Corriparta	Possibly worldwide
Kemerovo	Europe; possibly worldwide

caused by these viruses include Colorado tick fever of humans, bluetongue disease of sheep, African horse sickness, and epizootic hemorrhagic disease of deer. **Colorado tick fever,** an acute disease characterized by fever, headache, and severe myalgia, was originally described in the nineteenth century and is now believed to be one of the most common tickborne viral diseases in the United States. Although hundreds of infections occur annually, the exact number is not known because it is not a reportable disease. Other human pathogenic orbiviruses have been detected elsewhere in the world (Table 51-5), but little is known about these viruses.

Structure and Physiology

The structure and physiology of the orbiviruses is similar to other Reoviridae, with three major exceptions: 1. The orbivirus life cycle includes both vertebrates and invertebrates (insects). 2. Orbiviruses infect red blood cells, causing viremia. 3. The outer capsid of the orbiviruses has no discernible capsomeric structure, even though the inner capsid is icosahedral (Figure 51-5).

Orbiviruses contain double-stranded RNA with either 10 or 12 segments. Colorado tick fever viruses have 12 segments, and other orbiviruses have 10. The genome molecular weight is approximately 18×10^6 daltons. The virion is nonenveloped, partially resistant to lipid solvents and labile at pH 3.

Replication occurs in the cytoplasm of various cells of insect and mammalian origin. Specifics describing attachment, penetration, assembly, and release during cell lysis are not known.

Pathogenesis

Little is known of the pathogenesis of Colorado tick fever virus or other similar tickborne viruses in

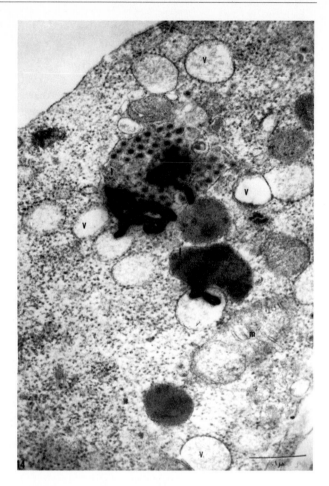

FIGURE 51-5 Electron micrographs of orbiviruses. *I*, Viral inclusion; *V*, vacuole; *m*, mitochondria. (From Fields BN, editor: Virology, New York, 1985, Raven Press.)

humans. Colorado tick fever virus infects hematopoietic cells without severely damaging them. Viremia therefore can persist for weeks or months, even after symptomatic recovery.

Epidemiology

Colorado tick fever has occurred in western and northwestern areas of the United States and western Canada where the wood tick *Dermacentor andersoni* is distributed (Figure 51-6). Ticks acquire the virus by feeding on viremic hosts and subsequently transmit the virus in saliva when they feed on a new host. Many ticks have been shown to be infected; however, *D. andersoni* is the predominant vector and the only proven source of human disease. Natural hosts can be one of many mammals, including squirrels, chip-

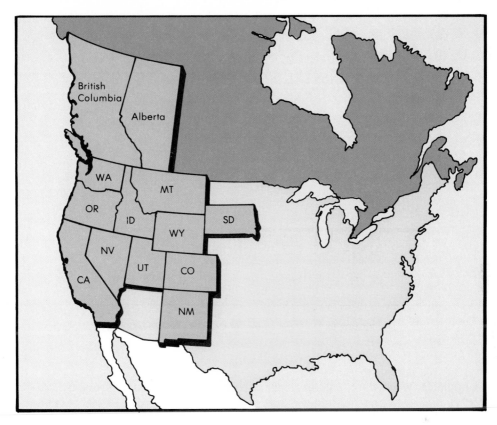

FIGURE 51-6   Geographic distribution of Colorado tick fever.

munks, rabbits, and deer. Exposure to ticks is the major risk factor. Human disease is reported during the spring, summer, and fall months. Colorado tick fever is not contagious. Although the disease is geographically limited, travel makes cases in nonendemic areas possible.

Clinical Syndromes

Acute disease occurs after an incubation period of 3 to 6 days. Although mild or subclinical infections can occur, most infections are symptomatic with fever, chills, headache, photophobia, myalgia, arthralgia, and lethargy (Figure 51-7). Neither respiratory nor gastrointestinal symptoms are prominent features. Hemorrhagic disease, confusion, and meningeal signs are unusual and, when they do occur, are more likely in children. Few physical signs are present on examination but may include fever, conjunctivitis, lymphadenopathy, hepatosplenomegaly, and maculopapular or petechial rash. A leukopenia involving both neutrophils and lymphocytes is an important hallmark of the disease. Leukocyte counts are generally less than 4,500/mm^3, with a relative lymphocytosis. Despite these findings, disease in children and adults is relatively mild, and uncomplicated recovery can be expected.

Colorado tick fever must be differentiated from Rocky Mountain spotted fever, a tickborne bacterial infection characterized by rash.

Laboratory Diagnosis

Specific diagnosis can be made by direct detection of viral antigens, viral isolation, or serologic tests. Because the disease is mild and geographically limited, laboratory tests are not broadly available. Detection of viral antigen in erythrocytes by immunofluorescence staining has been used as a rapid method of diagnosis. Viral isolation can be performed with serum or plasma during the first few days of disease, before the appearance of neutralizing antibody, and later with the blood clot or washed erythrocytes. Viremia is long-lasting. Isolation is best accomplished by

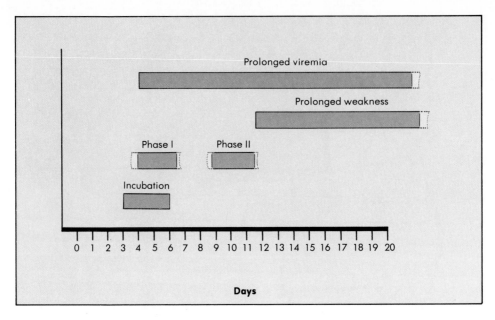

FIGURE 51-7   Time course of Colorado tick fever.

inoculating suckling mice. Infected mice appear ill in 4 to 5 days, and viral antigen can be detected by immunofluorescence. Colorado tick fever virus can also be adapted to cell culture, but primary isolation in cell culture is not sensitive.

Serologic diagnosis of Colorado tick fever must include testing of acute and convalescent specimens because subclinical infections can occur and antibody may persist for a lifetime. Techniques used to measure Colorado tick fever virus antibody include complement fixation, neutralization, indirect immunofluorescence, and enzyme immunoassy. A fourfold rise in antibody between the acute and convalescent specimen or the presence of Colorado tick fever virus–specific IgM in either specimen is presumptive evidence of acute or very recent infection. A sharp decline in IgM antibody occurs about 45 days after onset of illness.

Treatment, Prevention, and Control

No specific treatment is available. The disease is generally self-limited, suggesting that supportive care is sufficient. As mentioned, viremia is long-lasting, implying that infected patients should not donate blood soon after recovery. Prevention includes avoiding tick-infested areas, using protective clothing and tick repellents, and removing ticks before they bite. In contrast with tickborne rickettsial disease where prolonged feeding is required for transmission, the virus from the tick's saliva can enter the bloodstream rapidly. A formalinized Colorado tick fever vaccine has been developed and evaluated but is not practical for use by the general public.

ROTAVIRUSES

The rotaviruses are a large group of gastroenteritis-causing viruses found in many different mammals and birds. As with other viruses in the family Reoviridae, they have a double-layered capsid, a segmented double-stranded RNA genome, and no outer envelope. Although viruses have long been considered as a common cause of childhood diarrhea, it was not until 1973 that a rotavirus was first detected in humans. Since then rotaviruses have been implicated worldwide as common agents of infantile diarrhea.

Structure and Physiology

Rotavirus are icosahedral and 65 to 75 nm in diameter. The name is derived from the Latin *rota,* meaning wheel. Rotavirus particles observed in negative-

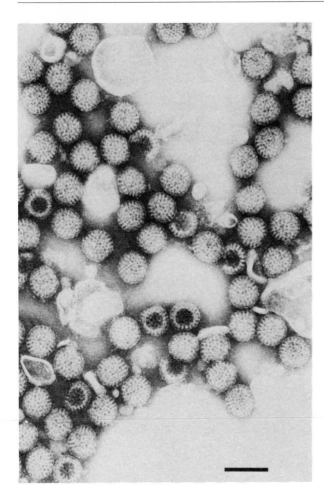

FIGURE 51-8   Electron micrograph of rotavirus. (From Fields BN, editor: Virology, New York, 1985, Raven Press.)

Table 51-6   Functions of Rotavirus Gene Products

Protein	Function
VP1	Inner capsid minor structural protein
VP2	Inner capsid major structural protein
VP3	Inner capsid protein
VP4	Outer capsid protein; enhances infectivity; responsible for hemagglutination
NS	Not known
VP6	Inner capsid major structural protein; contains subgroup antigens
NS	Not known
NS	Not known
VP7	Outer capsid major protein; induces neutralizing antibody
NS	Not known
VP9	Location uncertain; may be outer capsid protein

stain electron micrographs have a sharply defined outer capsid shaped like a wheel, with spokes radiating from the hub (Figure 51-8). Incomplete rotavirus particles are also seen in electron micrographs. These viruses lack the outer layer of the double-layered capsid and are often found in preparations from diarrheal stool and cell culture.

Rotavirus virions are relatively stable at ambient temperature for 24 hours and following treatment with ether and other lipid solvents or with repeated freeze-thawing. Infectivity is maintained at pH 3.5 to 10 and, importantly, is enhanced by proteolytic enzymes such as trypsin.

The double-stranded RNA genome consists of 11 segments, each representing separate genes. Molec-

ular weight of the entire genome is approximately 12×10^6 daltons. Known functions of all gene products are listed in Table 51-6.

Human and animal rotaviruses are divided into serotypes, subgroups, and groups. Serotypes are determined by neutralization reaction and depend primarily on the VP7 outer capsid protein and, to a lesser extent, on the VP4 minor outer capsid protein. Groups are based primarily on the antigenicity of VP6 and on electrophoretic mobility of the genome segments. Subgroups are determined by complement fixation and depend on the VP6 inner capsid protein. Based on these reactions, seven serotypes (1 to 7), two subgroups (I and II), and at least five groups (A to E) have been described for human and animal rotaviruses. The human rotaviruses are summarized in Table 51-7.

Table 51-7   Human Rotavirus Classification

Group	Subgroup	Serotype	Strains
A	I	2	Human
	II	1	Human
		3	Human
		4	Human, pig
B			Human, animal
C			Human, animal

Viral replication follows adsorption to columnar epithelial cells covering the villi of the small intestine. Virus enters the cell by endocytosis and is contained within lysosomes, where uncoating occurs. Approximately 8 hours after infection, cytoplasmic inclusions are seen that contain newly synthesized proteins and RNA. Viral RNA is packaged into core particles, and capsid proteins assemble around the core and bud through the endoplasmic reticulum, acquiring an envelope in the process. This envelope is lost during further maturation but presumably leaves remnants that become the outer capsid.

Pathogenesis

Rotaviruses are able to survive the acidic environment in a buffered stomach or in a stomach after a meal. Studies of biopsies of the small intestine from infants and of experimentally infected animals show shortening and blunting of the microvilli and mononuclear cell infiltration into the lamina propria (Figure 51-9). In addition, disaccharidase activity is decreased. Glucose-coupled sodium transport, however, remains intact. The exact mechanisms that contribute to a profuse watery diarrhea are thought to be lactose intolerance, which leads to a buildup of intes-

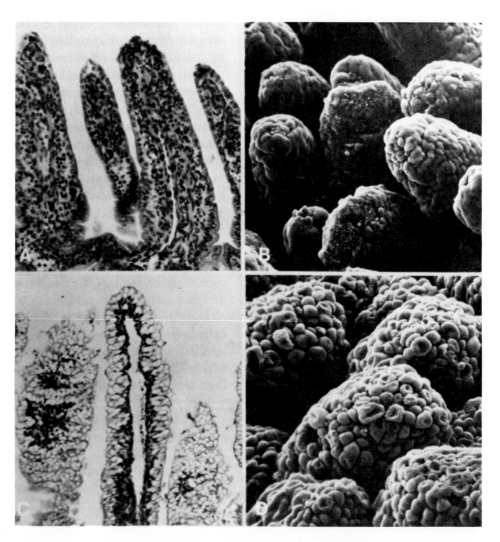

FIGURE 51-9   Microscopic and scanning electron microscopic view of intestinal tissues infected with rotavirus. (From Fields BN, editor: Virology, New York, 1985, Raven Press)

tinal lactose and a resulting osmotic influx of fluid. Loss of fluids and electrolytes can lead to severe dehydration and even death if therapy does not include electrolyte replacement.

Immunity to infection requires the presence of antibody, primarily IgA, in the lumen of the gut. Actively or passively acquired antibody apparently lessens the severity of disease but does not consistently prevent reinfection. In the absence of antibody, inoculation of even small amounts of virus causes infection and diarrhea. Moderate levels of antibody does not prevent infection but prevents diarrhea. High levels of antibody may block both infection, and disease or, conversely, large doses of virus may over-come immune protection. Infection in infants and small children is generally symptomatic, whereas infection in adults is usually asymptomatic, presumably a result of preexisting immunity. A complete understanding of immune protection, including protective effects of breast milk and cross-protection following infection with different serotypes, remains to be elucidated.

Epidemiology

Rotaviruses are found worldwide. Children are infected at an early age, as determined by the early acquisition of antibody (Figure 51-10). Rotaviruses

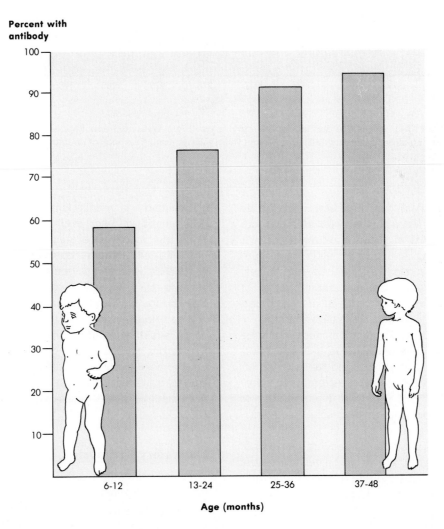

Percent with antibody

Age (months)

FIGURE 51-10   Prevalence of antibody to rotaviruses in children. (From Wyatt RG and Kapikian AZ: Viral gastrointestinal infections. In Feigin RD and Cherry JD, editors: Textbook of pediatric infectious diseases, ed 2, Philadelphia, 1987, WB Saunders Co.)

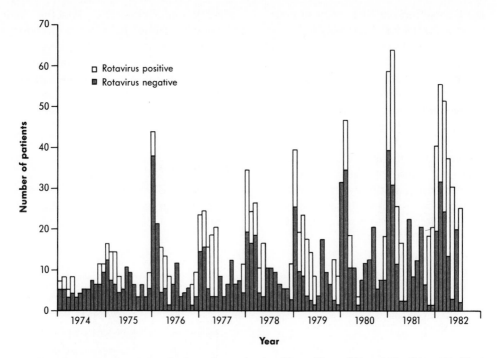

FIGURE 51-11   Seasonal distribution of rotaviruses. (From Wyatt RG and Kapikian AZ: Viral gastrointestinal infections. In Feigin RD and Cherry JD, editors: Textbook of pediatric infectious diseases, ed 2, Philadelphia, 1987, WB Saunders Co.)

are assumed to be passed from person to person by the fecal-oral route. Virus survives well on inanimate objects, such as hands, furniture, and toys, because it can withstand drying. Although domestic animals are known to harbor serologically related rotaviruses, they are not believed to be a common source of human infection. Outbreaks occur in preschool and day care centers and among hospitalized infants.

Rotaviruses are the most common cause of serious diarrhea in young children worldwide. In North America outbreaks occur annually during the fall, winter, and spring months (Figure 51-11). Attempts to correlate these outbreaks with variations in temperature or relative humidity have been unsuccessful. In developing countries rotavirus diarrhea is a severe, life-threatening disease in infants that occurs year-round. Differences in severity appear to be related to the severely malnourished condition of susceptible children.

Clinical Syndromes

The incubation period for rotavirus diarrheal illness is estimated to be approximately 48 hours. The major clinical findings in hospitalized patients include diarrhea, vomiting, fever, and dehydration. Fecal leukocytes and blood in stool are not associated with rotavirus diarrhea. Rotavirus gastroenteritis is a self-limited disease. Recovery is generally complete and without sequelae. In rare instances severe, untreated dehydration can lead to death.

The role of rotaviruses in respiratory infection is controversial. Human studies have detected virus in respiratory secretions of children with respiratory symptoms, and animal studies have documented that aerosol transmission can occur. Currently, however, the exact role these viruses play in respiratory disease is not known.

Laboratory Diagnosis

Clinical findings associated with rotavirus infection are not specific, making laboratory diagnosis essential. Available techniques include direct antigen detection, isolation in cell culture, and serodiagnosis.

Most infections are characterized by large quantities of virus in stool, making the direct detection of viral antigen the method of choice for diagnosis. Originally, the preferred method of diagnosis was the detection of viral particles in specimens by electron microscopy, but the availability of specific antibody has led to the use of enzyme immunoassay and latex agglutination. These methods are preferred because they are quick, easy, relatively inexpensive, and generally comparable to electron microscopy.

Cell culture of rotavirus is possible but is not reliable for diagnostic purposes. Many methods have been used to detect IgG, IgM, and IgA antibodies to rotavirus present in serum, stool, milk, and colostrum. Detection of serum antibody can only be used for diagnosis if acute and convalescent samples are tested. Because so many people have rotavirus-specific antibody, a fourfold rise in antibody titer is necessary for the diagnosis of recent infection or active disease. Serologic tests are primarily used only for research.

Treatment, Prevention, and Control

No specific antiviral therapy is available. Morbidity and mortality in patients with rotavirus diarrhea result from dehydration and electrolyte imbalance. The purpose of supportive therapy is to replace fluids so that blood volume and electrolyte and acid-base imbalances are corrected. Oral solutions used for fluid replacement should initially be low in lactose content because of the lactose intolerance that may accompany rotaviral diarrhea.

Rotavirus vaccines are being developed to protect children, especially those in underdeveloped countries, from potentially fatal disease. Live attenuated vaccines have been prepared using animal rotaviruses because they are better propagated in cell culture. Limited clinical evaluations suggest that oral vaccines are safe and effective. Further evaluation is ongoing.

Acquisition of rotaviruses occurs very early in life. The ubiquitous nature of these viruses and our lack of complete understanding of how they are disseminated make prevention of spread and infection unlikely. Once hospitalized, however, diseased patients must be identified and isolated to limit spread of infection to other susceptible patients.

BIBLIOGRAPHY

Balows A, Hausler WJ, and Lennette EH, editors: Laboratory diagnosis of infectious diseases: principles and practice, New York, 1988, Springer-Verlag New York, Inc.

Christensen ML: Human viral gastroenteritis, Clin Microbiol Rev 2:51-89, 1989.

Feigin RD and Cherry JD, editors: Textbook of pediatric infectious diseases, ed 2, Philadelphia, 1987, WB Saunders Co.

Fields BN, editor: Virology, New York, 1985, Raven Press.

Joklik WK, editor: The Reoviridae, New York, 1983, Plenum Publishing Corp.

Sharpe AH and Fields BN: Pathogenesis of viral infections: basic concepts derived from the Reovirus model, N Engl J Med 312:486-497, 1985.

CHAPTER 52

Togaviruses and Flaviviruses

The togaviruses and flaviviruses are enveloped, positive, single-stranded ribonucleic acid (RNA) viruses. Until recently the flaviviruses were included in the Togaviridae family, but differences in size, morphology, gene sequence, and replication strategy have made it necessary to classify them as an independent virus family. The togaviruses can be classified into four major groups (Table 52-1): *Alphavirus*, *Rubivirus*, *Pestivirus*, and *Arterivirus*. *Alphavirus* and *Flavivirus* are discussed together because of their many similarities. Rubella virus is the only member of the *Rubivirus* group; thus it is discussed separately because its disease presentation (German measles) and its means of spread differ from those of the alphaviruses. No known arteriviruses or pestiviruses cause disease in humans, so these are not discussed further.

ALPHAVIRUSES AND FLAVIVIRUSES

The alphaviruses and flaviviruses are historically classified as arboviruses because they are usually

Table 52-1 Togaviruses and Flaviviruses

Virus Group	Human Pathogens
Togaviruses	
Alphavirus	Arboviruses
Rubivirus	Rubella virus
Pestivirus	None
Arterivirus	None
Flaviviruses	Arboviruses

spread by arthropod vectors. Viruses spread by animals are called zoonoses. These viruses have a very broad host range, including vertebrates (e.g., mammals, birds, amphibians, reptiles) and invertebrates (e.g., mosquitos, ticks). Examples of pathogenic alphaviruses and flaviviruses are given in Table 52-2.

Structure

The alphaviruses are similar to the picornaviruses in that they have an icosahedral capsid and a plus-sense, single-strand RNA genome that resembles messenger RNA (mRNA). They differ by being slightly larger (45 to 75 nm in diameter) than picornaviruses and are surrounded by an envelope (Latin *toga*, meaning cloak). The envelope consists of lipids obtained from the host cell membranes and glycoprotein spikes that protrude from the surface of the virion. Most alphaviruses have two glycoproteins that associate to form a single spike. The COOH-terminus of the glycoproteins is anchored in the capsid, forcing the envelope to wrap tightly and take on the shape of the capsid (Figure 52-1).

The members of the alphaviruses share serologically definable, type-specific antigens. The capsid proteins of the alphaviruses are similar in structure and are the type-common antigens. The envelope glycoproteins express unique antigenic determinants that distinguish the different viruses and also antigenic determinants that are shared by a group, or "complex," of viruses.

The flaviviruses also have plus-strand RNA and an envelope. However, the virions are slightly smaller than those of the alphaviruses (37 to 50 nm in diameter), the RNA does not have a poly A sequence, and

628

Table 52-2 Alphaviruses and Flaviviruses

	Vector	Host	Distribution	Disease
ALPHAVIRUSES				
Sindbis*	*Aedes* and other mosquitos	Birds	Africa, Australia, India	Subclinical
Semliki Forest*	*Aedes* and other mosquitos	Birds	East and West Africa	Subclinical
Venezuelan equine encephalitis (VEE)	*Aedes, Culex*	Rodents, horses	North, South, and Central America	Mild systemic or severe encephalitis
Eastern equine encephalitis (EEE)	*Aedes, Culiseta*	Birds	North and South America, Caribbean	Mild systemic; encephalitis
Western equine encephalitis (WEE)	*Culex, Culiseta*	Birds	North and South America	Mild systemic; encephalitis
FLAVIVIRUSES				
Dengue*	*Aedes*	Humans, monkeys	Worldwide, especially tropics	Mild systemic; dengue hemorrhagic fever/shock syndrome (DHF/DSS)
Yellow fever*	*Aedes*	Humans, monkeys	Africa, South America	Hepatitis, hemorrhagic fever
Japanese encephalitis	*Culex*	Pigs, birds	Asia	Encephalitis
West Nile	*Culex*	Birds	Africa, Europe, Central Asia	Fever, encephalitis hepatitis
St. Louis encephalitis	*Culex*	Birds	North America	Encephalitis
Russian spring-summer encephalitis	*Ixodes* and *Dermacentor* ticks	Birds	Russia	Encephalitis
Powassan	*Ixodes* ticks	Small mammals	North America	Encephalitis

*Prototypical viruses.

a capsid structure is not visible within the virion. All the flaviviruses are serologically related, and antibodies to one virus may cross-neutralize another virus.

Replication

As enveloped, plus-strand RNA viruses, the general scheme for replication of the alphaviruses and flaviviruses is similar and can be discussed together (Figure 52-2). The alphaviruses and flaviviruses attach to specific receptors expressed on many different cell types from many different species. The host range for these viruses include vertebrates, such as humans, monkeys, horses, birds, reptiles, and amphibians, and invertebrates such as mosquitos and ticks. However, the individual viruses have different tissue tropisms, which can account somewhat for their different disease presentations.

The virus enters the cell by receptor-mediated endocytosis (Figure 52-2, *A*). The envelope fuses with the membrane of the endosome on acidification of the vesicle to deliver the capsid and genome into the cytoplasm.

Once released into the cytoplasm, the alphavirus and flavivirus genomes bind to the ribosome as mRNA. Differences in the alphavirus and flavivirus genome structure dictate different translation programs.

The alphavirus genome is translated in early and late phases. The initial two thirds of the alphavirus RNA is translated into a polyprotein, which is subsequently cleaved by proteases into four nonstructural (ns) proteins: ns60, ns89, ns76, and ns72 (cleaved from ns89). These proteins include an RNA-dependent RNA polymerase to transcribe the genome into a full-length 42S minus-sense RNA template and then

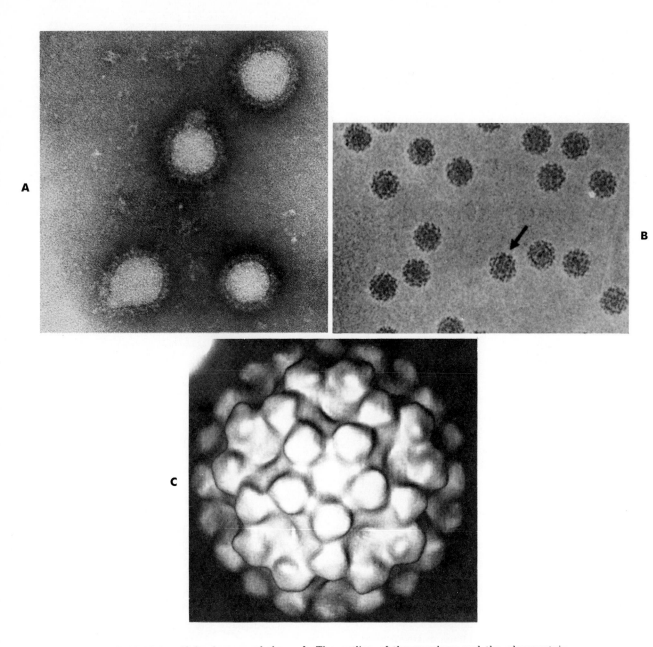

FIGURE 52-1   Alphavirus morphology. **A,** The outline of the envelope and the glycoprotein spikes of the Sindbis virus are visualized by negative staining. **B,** More detail on the morphology of the virion is obtained from cryoelectron microscopy. **C,** Surface representation of the Sindbis virus obtained by image processing of the cryoelectron micrographs indicates that the envelope is held tightly to and conforms to the shape and symmetry of the capsid. (From Fuller SD: Cell 48:923, 1987.)

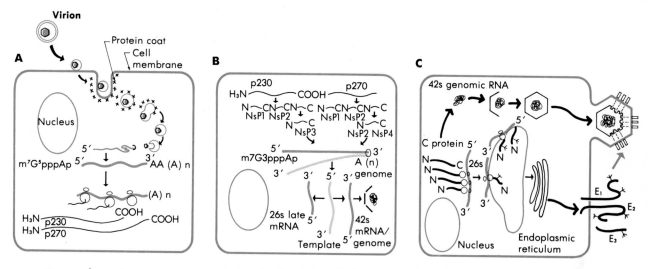

FIGURE 52-2 Replication of Semliki Forest Virus. **A,** Semliki Forest virus binds to cell receptors, is shuttled into a coated pit, and is internalized in a coated vesicle. The virus is transferred to an endosome, and the viral envelope fuses with the endosomal membrane on acidification to release the nucleocapsid into the cytoplasm. Ribosomes bind to the plus-sense RNA genome, and the p230 or p270 (read-through) early polyproteins are made. **B,** The polyproteins are cleaved to produce nonstructural proteins 1 to 4 (NsP_{1-4}), which include a polymerase to transcribe the genome into a minus-sense RNA template. The template is used to produce a full-length 42S plus-sense mRNA-genome and a late 26S mRNA. **C,** The *C* (capsid) protein is translated first, exposing a protease cleavage site and then a signal peptide for association with the endoplasmic reticulum. The E glycoproteins are then synthesized, glycosylated, processed in the Golgi apparatus, and transferred to the plasma membrane. The capsid proteins assemble on the 42S genomic RNA and then associate with regions of cytplasmic and plasma membranes containing the E_{1-3} spike proteins. Budding from the plasma membrane releases the virus.

replicate the genome to produce more 42S plus-sense mRNA. This results in a double-stranded RNA replicative intermediate.

A 26S mRNA corresponding to the other one third of the genome is produced later to code for the structural proteins. The polyprotein translated from the 26S mRNA contains the capsid (C) and envelope (E_{1-3}) proteins. In contrast to the early proteins, many copies of the structural proteins are required for packaging the virus. Late in the replication cycle, viral mRNA can account for as much as 90% of the mRNA in the infected cell. This provides the amplification required to produce the large numbers of structural proteins required to build the nucleocapsid and envelope of the virus.

The structural proteins are produced by protease cleavage of the late polyprotein produced from the 26S mRNA. The C protein is translated first and cleaved from the polyprotein (Figure 52-2, *B*). A signal sequence is then made that associates the nascent

polypeptide with the endoplasmic reticulum. The envelope glycoproteins are then synthesized, co-translationally glycosylated, and then cleaved from the remaining portion of the polyprotein to produce the E_1, E_2, and E_3 glycoprotein spikes. The E_3 remains associated with the Semliki Forest virus spikes but is released from other alphavirus glycoprotein spikes. The glycoproteins are processed by the normal cellular machinery in the endoplasmic reticulum and Golgi apparatus and are also acetylated and acylated with long-chain fatty acids (Figure 52-2, *C*). Alphavirus glycoproteins are transferred efficiently to the plasma membrane.

The C proteins associate with the genomic RNA soon after their synthesis and, for alphaviruses, form an icosahedral capsid. Capsid structures for the flaviviruses may exist but have not been seen. Once completed, the capsid associates with portions of the membrane expressing the viral glycoproteins. The alphavirus capsid has binding sites for the C-termi-

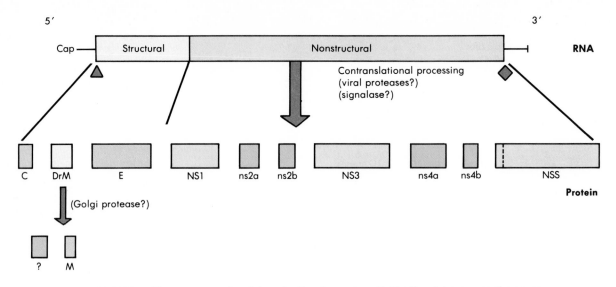

FIGURE 52-3   The genome and proteins of yellow fever virus. Unlike the alphaviruses, the genes for the structural proteins of the flaviviruses are at the 5′ end of the genome/mRNA, and only one species of polyprotein is made, which represents the entire genome. (Redrawn from Rice et al, 1985. Copyright 1985 by the American Association for the Advancement of Science.)

nus of the E glycoprotein spike, which pulls the envelope tightly around itself similar to a form-fitting wrapper (Figures 52-1 and 52-2, *C*). Alphaviruses are released on budding from the plasma membrane.

Attachment and penetration of the flaviviruses occurs as described for the alphaviruses, but the flaviviruses can also attach to the Fc receptors on macrophages, monocytes, and other cells when they are coated with antibody. The antibody actually enhances the infectivity of these viruses by providing new receptors for the virus and by promoting its uptake into these target cells.

The major differences between alphaviruses and flaviviruses are in the organization of their genomes and their mechanisms of protein synthesis. In contrast to the alphaviruses, the entire flavivirus genome is translated into a single polyprotein (Figure 52-3). As a result, no temporal distinction exists in the translation of the different viral proteins. The polyprotein produced from the yellow fever genome contains 4 to 8 nonstructural proteins plus the capsid and envelope structural proteins. Unlike the alphaviruses, the structural genes are at the 5′ end of the genome. As a result, the portions of the polyprotein containing the structural, not the catalytic, proteins are synthesized first and at the highest efficiency. This strategy may allow production of more structural proteins, but it decreases the efficiency of nonstructural protein syn-

thesis and the initiation of virus replication. This may contribute to the long latent period that precedes detection of flavivirus replication.

Another distinction is that budding of flaviviruses occurs predominantly from intracellular membranes into the cytoplasm or into vesicles rather than at the cell surface, as for alphaviruses. The virus can be released by exocytosis, but the most efficient means of virus release requires lysis of the cell.

Pathogenesis

The alphaviruses and flaviviruses can cause lytic or persistent infections of both vertebrate and invertebrate hosts. Infections of invertebrates are usually persistent, with continued virus production. Properties of both the virus and the cell determine whether infection will kill the cell.

The death of the cell results from a combination of virus-induced insults. The large amount of viral RNA produced on replication and transcription of the genome successfully competes with the cellular mRNA for ribosomes. In addition, an increase in permeability of the target cell membrane resulting from the virus infection produces alterations in sodium and potassium ion concentrations, which can alter enzyme activities and favors the translation of viral mRNA over cellular mRNA. The displacement of cellular mRNA from the protein synthesis machinery prevents rebuilding and maintenance of the cell and is a major cause for the death of the virus-infected cell. Some alphaviruses, such as western equine encephalitis (WEE) virus, make a nucleotide triphosphatase that degrades deoxyribonucleotides, depleting even the substrate pool for deoxyribonucleic acid (DNA) production.

The pathogenic characteristics of alphaviruses and flaviviruses are also determined by their route of entry into the host, the concentration of virus within the individual, and specific tissue tropisms of the individual virus type. Since these viruses are acquired from the bite of an arthropod such as a mosquito, the course of infection in both the vertebrate host and the invertebrate vector are important to the understanding of the disease.

Female mosquitos acquire the alphaviruses and flaviviruses on taking a blood meal from a viremic vertebrate host. The virus infects the epithelial cells of the midgut of the mosquito, spreads through the bas-

al lamina of the midgut to the circulation, and then infects the salivary glands. The virus sets up a persistent infection and replicates to high titers in these cells. The salivary glands can then release virus into the saliva. Not all arthropod species support this type of infection. For example, the normal vector for WEE virus is *Culex tarsalis,* but certain strains of virus are limited to the midgut and cannot infect the salivary glands.

After biting a host, the female mosquito regurgitates virus-containing saliva into the victim's bloodstream. The virus circulates free in the plasma and comes into contact with susceptible target cells such as the endothelial cells of the capillaries, macrophages, monocytes, and the reticuloendothelial system (Figure 52-4).

The initial viremia, following replication of the virus in these tissues, produces systemic symptoms such as fever, chills, headaches, backaches, and flu-like symptoms within 3 to 7 days of infection. Some of these symptoms can be attributed to the interferon

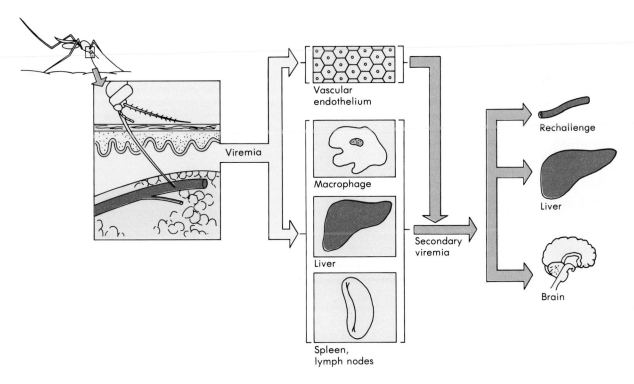

FIGURE 52-4   Spread of alphavirus and flavivirus infection within the host. The female mosquito regurgitates virus into a capillary after taking a blood meal. The virus infects the endothelial cells lining the vasculature and is taken up and replicates in cells of the reticuloendothelial system *(RES).* A secondary viremia can result to allow more extensive infection of the vasculature, the liver, other tissues, and the brain (encephalitis viruses).

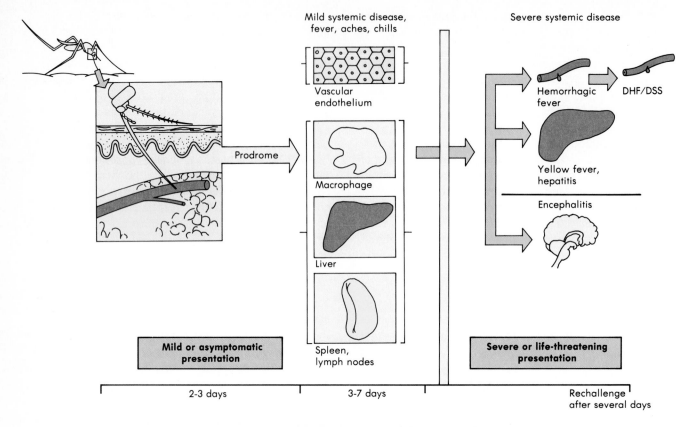

FIGURE 52-5 Disease syndromes of the alphaviruses and flaviviruses. Following a short pro-drome period, a mild systemic disease can result. Most infections are limited to this extent. If sufficient virus is produced during the secondary viremia to escape immune protection and reach critical target tissue, severe systemic disease or encephalitis may result. For dengue virus, rechallenge with another strain can result in severe hemorrhagic disease (dengue hemorrhagic fever, DHF), which can cause shock (dengue shock syndrome, DSS) because of the loss of fluids from the vasculature.

produced following infection of these target cells. This is considered a mild systemic disease, and most virus infections do not progress beyond this point.

Following replication in cells of the reticuloendothelial system, a secondary viremia may result. This can produce sufficient virus to infect target organs such as the brain, liver, skin, and vasculature, depending on the tissue tropism of the virus (Figure 52-5). Access to the brain is provided by infection of the endothelial cells lining the small vessels of the brain or the choroid plexus.

Immune Response Both humoral and cellular immunity are elicited and are important to the control of primary infection and prevention from future infections with the alphaviruses and flaviviruses. Unlike viruses that initially replicate in the lung or periphery,

the primary infection by these viruses is a viremia. This presents the virus immediately to macrophages, the reticuloendothelial system, and the immune response. The immune response to a mild systemic infection with the 17D yellow fever virus vaccine strain is presented in Figure 52-6.

Replication of the alphaviruses and flaviviruses in the macrophage and endothelial cells produces a double-stranded RNA replicative intermediate that is a good inducer of interferon. The interferon produced soon after infection is released into the bloodstream to limit further replication of virus and stimulate the immune response. The interferon also causes the rapid onset of flulike symptoms characteristic of mild systemic disease.

Circulating IgM is produced within 6 days of infec-

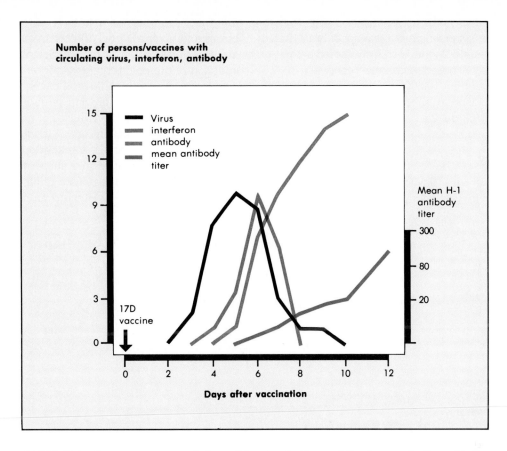

FIGURE 52-6   Immune response during mild systemic disease following vaccination with the 17D yellow fever vaccine. The number of individuals with circulating virus and interferon and their antibody titers (as measured by hemagglutination inhibition) following infection with the vaccine was determined for 15 volunteers. (Redrawn from Wheelock EF and Sibley WA: N Engl J Med 273:194, 1965.)

tion, followed by IgG. The antibody blocks the viremic spread of the virus and subsequent infection of other tissues. Immunity to one flavivirus can protect against other flaviviruses by recognition of the type-common antigens expressed on all members of the viral family. One example of this may be occurring in the Far East, where Japanese encephalitis but not yellow fever virus is endemic despite the presence of the *Aedes aegypti* mosquito vector for yellow fever.

Cell-mediated immunity is also important in control of the primary infection. Natural killer cells, T cells, and macrophages are activated by interferon and can respond to the cell surface antigens displayed on the infected cells.

Immunity to these viruses is a double-edged sword. Inflammation resulting from the cell-mediated immune response can destroy tissue and significantly

contribute to the pathogenesis of encephalitis. Prior immunity can promote hypersensitivity reactions such as delayed-type hypersensitivity, formation of immune complexes with virions and viral antigens, and activation of complement. This can weaken and cause the rupture of the vasculature, leading to hemorrhagic symptoms. A nonneutralizing antibody can also enhance uptake of the flaviviruses into macrophages and other cells that express Fc receptors. Such an antibody can be generated to a related strain of virus in which the neutralizing epitope is not expressed or different. The consequences of such a partial immunity can be devastating.

For example, prior infection by one strain of dengue virus will predispose an individual to dengue hemorrhagic fever when infected by another strain of dengue. The weakening and rupture of the vasculature is believed to result from activation of complement and other hypersensitivity reactions. In 1981 an epidemic of dengue-2 virus in Cuba infected a population previously exposed to dengue-1 (between 1977 and 1980). More than 100,000 cases of dengue hemorrhagic fever/dengue shock syndrome (DHF/DSS) resulted, with 168 deaths.

Epidemiology

Alphaviruses and flaviviruses are prototypical arboviruses (*ar*thropod-*bo*rn viruses). To be an arbovi-rus, the virus must be able to (1) infect both vertebrates and invertebrates; (2) initiate a sufficient viremia in a vertebrate host for a sufficient time to allow infection of the invertebrate vector, and (3) initiate a persistent, productive infection of the salivary gland of the invertebrate to provide virus for infection of other host animals. If the virus is not in the blood, the mosquito cannot pick it up. A cycle of infection occurs in which the virus is transmitted by the arthropod vector and amplified in a susceptible, immunologically deficient host to allow reinfection of other arthropods (Figure 52-7). Humans are usually "dead-

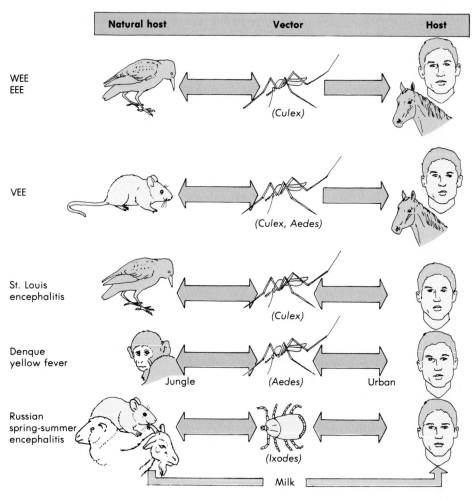

FIGURE 52-7   Patterns of alphavirus and flavivirus transmission. The cycle of arbovirus transmission maintains and amplifies the virus in the environment. Host-vector relationships that can provide this cycle are indicated by the double arrow. "Dead-end" infections with no transmission of the virus back of the vector are indicated by the single-headed arrow. For St. Louis encephalitis, yellow fever, and dengue virus, humans can also act as host, depending on the species of mosquito and the population density of mosquitos and humans. The Russian spring-summer encephalitis virus can be transmitted to humans by a tick bite and in milk from infected goats.

end'' hosts that cannot spread the virus back to the vector because a persistent viremia is not maintained. Table 52-2 lists vectors, natural hosts, and geographic distribution for representative alphaviruses and flaviviruses.

These viruses are usually restricted to the ecologic niche of a specific arthropod vector and its vertebrate host. The most common vector is the mosquito, but some arboviruses are also spread by ticks and sandflies. Even in a tropical region overrun with mosquitos, the spread of these viruses is still restricted to a specific genus of mosquitos. Not all arthropods can act as good vectors for each virus. For example, *Culex quinquefasciatus* is resistant to infection by WEE virus (alphavirus) but is an excellent vector for St. Louis encephalitis virus (flavivirus). As with mosquito-borne viruses, the life cycle of the tick dictates the pattern of spread of Russian spring-summer encephalitis virus and other tickborne viruses (Figure 52-7).

Birds and small mammals are the usual hosts for the alphaviruses and flaviviruses, but reptiles and amphibians can also act as hosts. Development of a large population of viremic animals can occur in these species to continue the infection cycle of the virus.

During the summer months the arboviruses are cycled between a host (bird) and an arthropod (e.g., mosquito). This maintains and increases the amount of virus in the environment. In the winter neither the normal host nor the vector remain to maintain the virus. The virus may persist in arthropod larvae or eggs or in reptiles or amphibians that remain in the locale, or it may be blown by the winds or migrate with the birds and then return during the summer.

Most of the vectors relevant to the alphaviruses and flaviviruses are mosquitos, which are usually found in tropical forests or swamps. When humans travel into these environments, they risk being bitten by virus-bearing mosquitos. For example, up to 50% of the Indians living in the Everglades have antibodies to Venezuelan equine encephalitis (VEE) virus. Pools of standing water, drainage ditches, and sumps in cities can also provide breeding grounds for mosquitos, such as *Aedes aegypti,* the vector for yellow fever. In endemic regions the risk for arbovirus infection increases during the rainy season.

Humans can be hosts for yellow fever, dengue, and Chikungunya viruses (Figure 52-7). These viruses are maintained by *Aedes* mosquitos in a sylvatic or jungle cycle, in which monkeys are the natural host, and also in an urban cycle, in which humans are the host. *A. aegypti* is a vector for each of these viruses and is a household mosquito. It breeds in pools of water, open sewers, and other accumulations of water in cities. St. Louis encephalitis virus can also establish an urban cycle of spread using its vector, *Culex* mosquitos, which prefer stagnant water such as sewage. Large numbers of inapparent infections in high-density populations provide sufficient viremic human hosts for continued spread of these viruses.

Clinical Syndromes

Infection by the alphaviruses usually causes a low-grade disease characterized by flulike symptoms (chills, fever, rash, aches) that correlate with systemic infection during the initial viremia. Eastern equine encephalitis (EEE) virus, WEE virus, and VEE virus infection can progress to encephalitis (as the name implies) in humans and horses, with EEE causing the most severe disease. These viruses are usually more of a problem to livestock than to humans. Other alphaviruses, such as Sindbis and Chikungunya, generally cause only systemic disease.

Most infections with flaviviruses are relatively benign, although encephalitic or hemorrhagic disease can occur. The encephalitis viruses include St. Louis, Japanese, Murray Valley, and Russian spring-summer. West Nile, dengue, and other viruses usually are limited to a mild systemic disease, possibly with a hemorrhagic rash. On rechallenge with a related strain, dengue can also cause severe hemorrhagic disease and shock (DHF/DSS). The hemorrhagic/shock symptoms are attributed to rupture of the vasculature, internal bleeding, and loss of plasma.

Yellow fever infections are characterized by severe systemic disease, with degeneration of the liver, kidney, and heart and hemorrhage of blood vessels. Liver involvement leads to the jaundice, from which the disease obtains its name, but massive gastrointestinal hemorrhages (''black vomit'') may also occur. The mortality following yellow fever was as high as 50% during epidemics.

Laboratory Diagnosis

The alphaviruses and flaviviruses can be grown in both vertebrate and mosquito cell lines but are difficult and dangerous to isolate. When isolation is necessary, the best hosts are suckling mice and mosquito cell lines. In addition to cytopathology, the viruses grown in culture can be detected by immunofluorescence or by hemadsorption of avian erythrocytes.

After isolation the viruses can be distinguished by RNA "fingerprints" of the genomic RNA.

A variety of serologic methods can be used to diagnose infections. A fourfold increase in titer between acute and convalescent sera is used to indicate a recent infection. The serologic cross-reactivity between viruses in a group or complex limits identification of the actual viral species in many cases. Monoclonal antibodies to the individual viruses have become useful in distinguishing the individual species and strains of viruses.

Treatment

No treatments exist for arbovirus diseases other than supportive care.

Prevention and Control

The easiest means to prevent the spread of any arbovirus is elimination of its vector and breeding grounds. Following the discovery by Walter Reed and colleagues that yellow fever was spread by *A. aegypti,* the number of cases was reduced from 1,400 to 0 within 2 years by controlling the mosquito population. Avoidance of the breeding grounds of a mosquito vector is also good prevention.

Vaccination is one of the best means of protection. A live vaccine is available against yellow fever virus and killed vaccines against Japanese encephalitis, EEE, WEE, and Russian spring-summer encephalitis viruses. These vaccines are for individuals working with the virus or at risk for contact. A live vaccine against VEE virus is available but only for use in domestic animals. A vaccine against dengue virus has not been developed because of the potential risk of immune enhancement of disease with subsequent challenge.

The yellow fever vaccine is prepared from the 17D strain isolated from a patient in 1927 and passaged extensively in monkeys, mosquitoes, embryonic tissue culture, and embryonated eggs. The vaccine is administered intradermally and elicits lifelong immunity to yellow fever and possibly other cross-reacting flaviviruses (see Figure 52-6).

RUBELLA VIRUS

Rubella virus shares the structural properties and mode of replication with the other togaviruses dis-

cussed earlier, but it differs in its mode of spread and acquisition. Unlike the other togaviruses, rubella is a respiratory virus.

Rubella infection usually causes a mild exanthematous childhood disease but has serious consequences for the neonate. Rubella, meaning "little red" in Latin, was first distinguished from measles and other exanthems by German physicians, providing the common name for the disease, German measles. An astute Australian ophthalmologist, Norman McAlister Gregg, recognized in 1941 that maternal rubella infection was the cause of congenital cataracts. Maternal rubella infection has since been correlated with several other severe congenital defects. This prompted the development of a unique program to vaccinate children to prevent infection of pregnant women and neonates.

Pathogenesis

Rubella virus has more subtle effects on the cell than the alphaviruses. Lytic infections are only observed in certain cell lines, such as Vero or RK13, but even in these cells the cytopathic effect is very limited. Changes in the cell do occur, and replication of rubella will interfere with the replication of superinfecting picornaviruses. Heterologous interference with picornavirus replication allowed the first isolations of rubella virus in 1962.

As a respiratory virus, rubella differs from other togaviruses in the means of spread within the body, the prodrome period, the time course and symptoms of the disease, and the protective immune response (Figure 52-8). The virus first infects the upper respiratory tract and then spreads to local lymph nodes, which coincides with a period of lymphadenopathy. This is followed by establishment of viremia, which spreads the virus throughout the body. Infection of other tissues and the characteristic mild rash result. The prodrome period is approximately 2 weeks (Figure 52-9). The individual can shed virus in respiratory droplets during the prodrome period and up to 2 weeks after the onset of the rash.

Immune Response The immune response generated against rubella infection is different from the response to alphaviruses and flaviviruses. As a respiratory virus, rubella penetrates the natural defenses of the nasopharynx and lung. Antibody generated soon after initiation of the viremia blocks viremic spread of the virus. Virus replication in the tissues can continue, however, until it can be cleared by cell-mediated

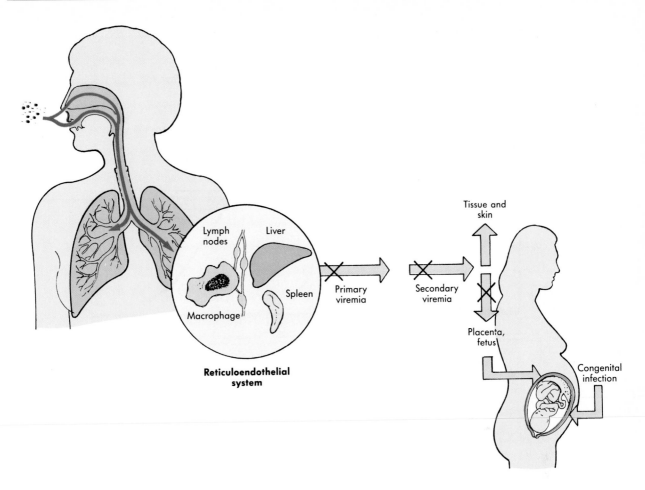

FIGURE 52-8 Spread of rubella virus within the host. Rubella enters and infects the nasopharynx and lung, then spreads to the lymph nodes and reticuloendothelial system. The resulting viremia spreads the virus to other tissues and the skin. Circulating antibody can block transfer of virus at the indicated points. In an immunologically deficient pregnant woman, the virus can infect the placenta and spread to the fetus.

immunity or limited by interferon. Only one serotype of rubella exists, and natural infection produces life-long protective immunity.

Unlike the alphaviruses and flaviviruses, secretory IgA in the respiratory tract plays an extremely important role in protection against a second challenge with rubella. Serum antibody limits the viremic spread and promotes elimination of the virus on reinfection. This prevents infection of other tissues and the onset of disease. Most importantly, serum antibody in a pregnant woman prevents the spread of the virus to the fetus.

Congenital Infection Rubella infection in a pregnant woman can result in serious congenital abnormalities. A primary immune response would generate antibody after the viremic spread of the virus to the placenta. Replication of the virus in the placenta spreads the virus to the fetal blood supply and throughout the fetus. Rubella can replicate in most tissues of the fetus. The virus may not be cytolytic, but the normal growth, mitosis, and chromosomal structure can be altered by the infection. This can lead to improper development, small size of the infected baby, and the teratogenic effects associated with congenital rubella infection. The nature of the disorder is determined by the affected tissue and the stage of development that was disrupted. The virus may persist in tissues, such as the lens of the eye, for up to 3 to 4 years and may be shed for up to a year after birth. The presence of the virus during the devel-

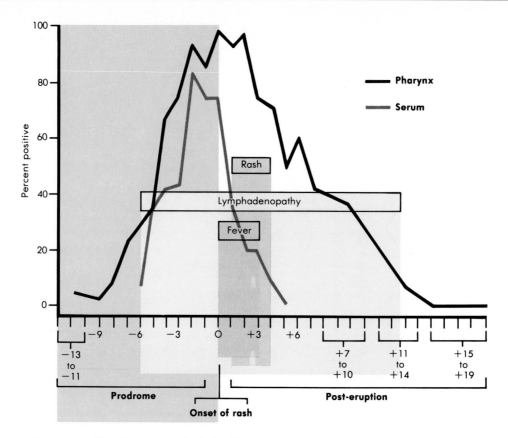

FIGURE 52-9 The time course of rubella disease. Rubella production in the pharynx precedes the appearance of symptoms and continues throughout the course of the disease. Onset of lymphadenopathy coincides with the viremia. Fever and rash are later symptoms. The individual is infectious as long as the virus is produced in the pharynx. (Redrawn from Plotkin SA: Rubella vaccine. In Plotkin SA and Mortimer EA, editors: Vaccines, Philadelphia, 1988, WB Saunders Co.)

opment of the baby's immune response may even have a tolerative effect on the system, preventing effective clearance following birth. Immune complexes may also form in the neonate or the infant to produce further clinical abnormalities.

EPIDEMIOLOGY

Humans are the only host for rubella. The virus is spread by respiratory secretions and is generally acquired in childhood. The virus is less communicable than measles or varicella, but contagion is promoted in crowded conditions such as day care centers.

Approximately 20% of women of childbearing age escape infection during childhood and are susceptible to infection unless previously vaccinated. Programs to ensure that expectant mothers have antibodies to rubella are in effect in many U.S. states.

Before the development and use of the rubella vaccine, cases of rubella in school children would be reported every spring, and major epidemics of rubella occurred at regular 6- to 9-year intervals. The severity of the 1964 to 1965 epidemic in the United States is indicated in Table 52-3. The rate of congenital rubella in such cities as Philadelphia was as high as 1% of all pregnancies during this epidemic. Since the development of the vaccine, the incidence of rubella and congenital rubella is now less than 1 and 0.1 per 100,000 pregnancies respectively.

Clinical Syndromes

Rubella infection is usually not cytolytic, and the disease is normally benign in children. Infection of adults can be more severe, leading to outcomes such as arthralgia, arthritis, and rarely thrombocytopenia

Table 52-3 Estimated Morbidity Associated with the 1964-1965 U.S. Rubella Epidemic

Clinical Events	Number Affected
Rubella cases	12,500,000
Arthritis-arthralgia	159,375
Encephalitis	2,084
Deaths	
Excess neonatal deaths	2,100
Other deaths	60
Total deaths	2,160
Excess fetal wastage	6,250
Congenital rubella syndrome	
Deaf children	8,055
Deaf-blind children	3,580
Mentally retarded children	1,790
Other congenital rubella syndrome symptoms	6,575
Total congenital rubella syndrome	20,000
Therapeutic abortions	5,000

From National Communicable Disease Center: Rubella surveillance, US Department of Health, Education and Welfare, no. 1, June 1969.

Box 52-1 Prominent Clinical Findings in Congenital Rubella Syndrome

Retinitis
Cataracts
Microphthalmia
Glaucoma

Intrauterine growth retardation

Microcephaly
Mental retardation
Autism

Patent ductus arteriosus
Peripheral pulmonic artery stenosis

Hepatosplenomegaly

Cochlear deafness
Central auditory imperception

Thrombocytopenic purpura

Interstitial pneumonitis

Metaphyseal rarefactions

Modified from Alford CA and Griffiths PD: Rubella. In Remington JS and Klein JO, editors: Infectious diseases of the fetus and newborn infant, Philadelphia, 1983, WB Saunders Co. From Plotkin SA: Rubella vaccine. In Plotkin SA and Mortimer EA, editors: Vaccines, Philadelphia, 1988, WB Saunders Co.

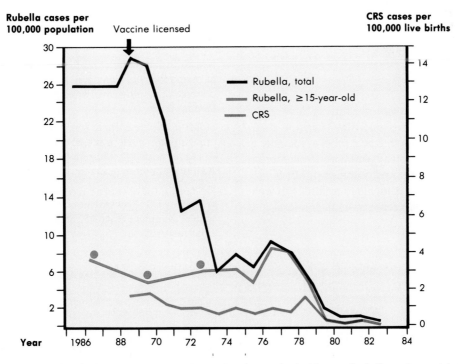

FIGURE 52-10   The effect of rubella virus vaccination on the incidence of rubella and congenital rubella syndrome. (Redrawn from Williams MN and Preblud SR: MMWR 33:1SS, 1984.)

and a postinfectious encephalitis similar to postinfectious measles encephalitis. Immunopathology, resulting from cell-mediated immunity and hypersensitivity reactions, is a major cause of the more severe adult forms of rubella.

Congenital disease is the most serious outcome of rubella infection. The fetus is at major risk until the twentieth week of pregnancy. Maternal immunity to the virus resulting from prior exposure or vaccination prevents spread of the virus to the fetus. The most common manifestations of congenital rubella infection include cataracts, mental retardation, and deafness (Tables 52-3 and Box 52-1). Mortality in utero and within the first year of birth is high for these babies.

Laboratory Diagnosis

Isolation of rubella virus is difficult and rarely done. The diagnosis is usually confirmed by the presence of antirubella-specific IgM. A fourfold increase in specific IgG antibody titer between acute and convalescent sera is also used to indicate a recent infection. Antibodies to rubella are assayed early in pregnancy to determine the immune status of the woman. This test is required in many states.

When isolation is necessary, the virus is usually obtained from urine and is detected by interference with ECHO 11 virus replication in primary African green monkey kidney cell cultures.

Prevention

The best means of preventing rubella is vaccination with the live, cold-adapted RA27/3 vaccine strain of virus (Figure 52-10). The live rubella vaccine is administered subcutaneously alone or in conjunction with the measles and mumps vaccines (MMR-II vaccine). The triple vaccine is included routinely in well-baby care. Vaccination promotes both humoral and cellular immunity.

A major objective of the vaccination program is development of "herd immunity" in the population. Since only one serotype for rubella exists and humans are the only reservoir, vaccination of a large proportion of the population can significantly reduce the chance of exposure to the virus. Elimination of the virus from the community by vaccination is not likely because the virus can escape detection in persons with asymptomatic infections.

BIBLIOGRAPHY

Freestone DS: Yellow fever vaccine. In Plotkin SA and Mortimer EA, editors: Vaccines, Philadelphia, 1988 WB Saunders Co.

Fuller SD: The T-4 envelope of Sindbis virus is organized by interactions with a complementary T-3 capsid, Cell 48:923-934, 1987.

Johnson RT: Viral infections of the nervous system, New York, 1982, Raven Press.

Nathenson N and Miller A: Arbovirus encephalitis. In Nahmias AJ and O'Reilly RJ, editors: Comprehensive immunology, vol 9: Immunology of human infection, part II, New York, 1982, Plenum Medical Book Co.

Plotkin SA: Rubella vaccine. In Plotkin SA and Mortimer EA, editors: Vaccines, Philadelphia, 188, WB Saunders Co.

Russell PK: Immunopathology of dengue hemorrhagic fever: new perspectives. In Ennis FA, editor: Human immunity to viruses, New York, 1983, Academic Press, Inc.

Simizu B: Inhibition of host cell macromolecular synthesis following togavirus infection. In Fraenkel H and Wagner RR, editors: Comprehensive virology, vol 19, New York, 1984, Plenum Press.

Stollar V: Approaches to the study of vector specificity for arboviruses—model systems using cultured mosquito cells. In Maramorosch K, Murphy FA, and Shatkin AJ, editors: Advances in virus research, vol 33, New York, 1987, Academic Press, Inc.

Westaway EG: Flavivirus replication strategy. In Maramorosch K, Murphy FA, and Shatkin AJ, editors: Advances in virus research, vol 33, New York, 1987, Academic Press, Inc.

Bunyaviridae

Bunyaviridae are RNA viruses that were formerly called arboviruses because the usual vectors of infection are arthropods. In most cases the arthropods are mosquitos, but ticks and flies are also implicated.

The Bunyaviridae comprise a ''supergroup'' of at least 200 enveloped, segmented, negative-strand RNA viruses. The supergroup is further broken into four genera based on structural and biochemical features: *Bunyavirus, Phlebovirus, Uukuvirus,* and *Nairovirus* (Table 53-1). *Hantaanvirus* and several other viruses are included with the Bunyaviridae but have not been classified.

STRUCTURE

These viruses are roughly spherical particles 90 to 120 nm in diameter. The envelope of the virus contains two glycoproteins and encloses three unique nucleocapsids (Table 53-2). The nucleocapsids consist of three separate strands of negative-strand RNA of differing length (L, M, S), the RNA-dependent RNA polymerase (L protein), and two nonstructural proteins (NS_s, NS_m) (Figure 53-1). Unlike other negative-strand viruses, the Bunyaviridae do not have a matrix

Table 53-1 Notable Bunyaviridae Genera

Genus	Members	Vector	Pathologic Conditions	Vertebrate Hosts
Bunyavirus	Bunyamwera virus, California encephalitis virus, LaCrosse virus Oropouche virus; 150 members	Mosquito	Febrile illness, encephalitis, febrile rash	Rodents, small mammals, primates, marsupials, birds
Phlebovirus	Rift Valley fever virus, Sandfly fever virus; 36 members	Fly	Sandfly fever Hemorrhagic fever Encephalitis Conjunctivitis, myositis	Sheep, cattle, domestic animals
Nairovirus	Crimean-Congo hemorrhagic fever virus; 6 members	Tick	Hemorrhagic fever	Hares, cattle, goats, seabirds
Uukuvirus	Uukuniemi virus; 7 members	Tick	—	Birds
Hantaanvirus	Hantaan virus; 1 member	None	Hemorrhagic fever with renal syndrome (e.g., Korean hemorrhagic fever)	Rodents

*An additional 35 viruses possess several common properties with Bunyaviridae but are as yet unclassified.

Table 53-2　Genome and Proteins of California Encephalitis Virus

Genome (Negative-Strand RNA)	Proteins	
L　(2.9×10^6)	L	RNA polymerase, 170 kd
M　(1.6×10^6)	G_1	Spike glycoprotein, 75 kd
	G_2	Spike glycoprotein, 65 kd
	NS_m	Nonstructural protein, 15 to 17 kd
S　(0.4×10^6)	N	Nucleocapsid protein, 25 kd
	NS_s	Nonstructural protein, 10 kd

protein. Differences in the sizes of the virion proteins, the lengths of the L, M, and S strands of the genome, and their transcription distinguish the four genera of Bunyaviridae.

REPLICATION

Bunyaviridae follow the rules for replication of enveloped, negative-strand viruses. The virus interacts with cell surface receptors, undergoes endocytosis, and fuses with endosomal membranes upon acidification of the vesicle. A glycoprotein (G1) appears to be the viral attachment protein. Release of the nucleocapsid into the cytoplasm allows mRNA and protein synthesis to begin.

The M strand encodes the NS_m protein and the G1 and G2 glycoproteins, and the L strand encodes the L protein (see Table 53-2). The S strand of RNA encodes two nonstructural proteins, N and NS_s. The S strand of the phleboviruses is "ambisense," meaning that the N protein mRNA is transcribed directly from the genome and the NS_s protein mRNA is transcribed from the replicative intermediate.

Replication of the genome by the L protein also provides new templates for transcription, amplifying the rate of mRNA synthesis. The glycoproteins are synthesized and glycosylated in the endoplasmic reticulum. They are transferred to the Golgi apparatus where budding of the virus occurs. The G1 and G2 are not translocated to the plasma membrane. Virions are released by cell lysis or exocytosis.

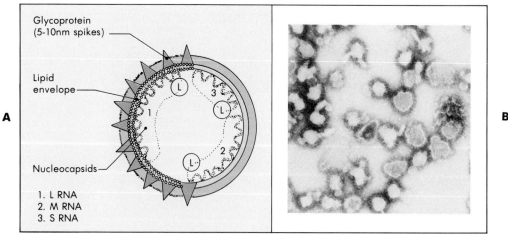

FIGURE 53-1　**A,** Model of *Bunyavirus* particle. **B,** Electron micrograph of La Crosse variant of *Bunyavirus*. (**A,** Redrawn from Fraenkel-Conrat H and Wagner RR, editors: Comprehensive virology, New York, 1979, Plenum Press; **B,** Courtesy Centers for Disease Control, Atlanta, Georgia.)

PATHOGENESIS

As arboviruses, Bunyaviridae share many of the pathogenic mechanisms of the togaviruses and flaviviruses. The virus is spread by the arthropod vector and is injected into the blood to initiate a viremia. Progression past this stage to secondary viremia and further dissemination of the virus can deliver the virus to target organs such as the CNS, major organs, and the vascular endothelium.

Bunyaviridae cause encephalitis by neuronal and glial damage, as well as cerebral edema. Lesions are concentrated in cortical gray matter of the frontal, temporal, and parietal lobes, basal nuclei, midbrain, and pons. In certain of the viremic infections (e.g., Rift Valley fever), hepatic necrosis may occur. In others (e.g., Crimean hemorrhagic fever and Hantaan virus infection), the primary lesion is leakage of plasma and erythrocytes through vascular endothelium. In the latter infection these changes are most prominent in the kidney and are accompanied by hemorrhagic necrosis of the kidney.

EPIDEMIOLOGY

As already stated, these viruses are transmitted from infected mosquitos, ticks, or phlebotomus flies to rodents, birds, and larger animals. Humans are infected only incidentally (Figure 53-2). The viremic vertebrates are a source of virus for arthropods feeding upon them. Unlike the rodent togaviruses and flaviviruses, many members of this family are also transmitted transovarially within the arthropod. The *Hantaanvirus* group does not have an arthropod vector, but spreads from mammal to mammal, usually rodents, and can spread directly to humans.

The viruses in this family are found in the natural environments of their vectors, in South America, Southeast Europe, Southeast Asia and Africa. California encephalitis virus is found in the forests of North America and cause human encephalitis in the United States (Figure 53-3).

CLINICAL SYNDROMES

Bunyaviridae (mosquito borne) cause either a nonspecific febrile illness or encephalitis (Table 53-1). The febrile illnesses caused by Bunyaviridae are nonspecific and indistinguishable from those caused by other viruses. The incubation period for these illnesses is approximately 48 hours, and the fevers last approximately 3 days. Most patients with infections have mild illness, even those infected by agents known to cause severe illness (e.g., Rift Valley fever virus or the LaCrosse virus). Encephalitis illnesses have an incubation period of approximately 1 week. Seizures occur in 50% of patients with encephalitis, usually early in the illness. Signs of meningeal irritation are present in only one third of cases. On an average, illness lasts 7 days. Fatalities occur in less than 1%, but seizure disorders may occur as sequelae in up to 20% of patients. Hemorrhagic fevers such as Rift Valley fever are characterized by petechial hemor-

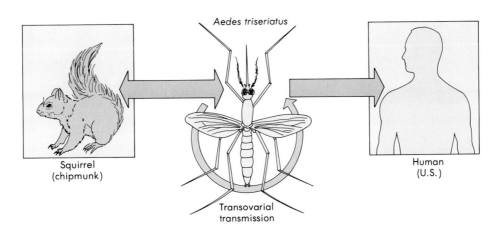

Aedes triseriatus

Squirrel
(chipmunk)

Transovarial
transmission

Human
(U.S.)

FIGURE 53-2   Transmission of La Crosse (California) encephalitis virus.

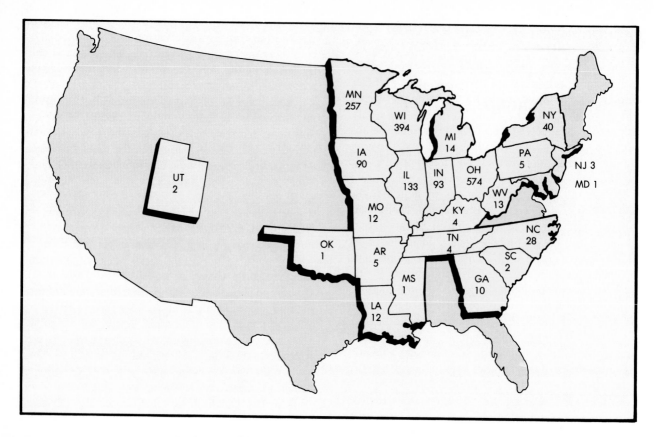

FIGURE 53-3   Distribution of California encephalitis, 1966 to 1981.

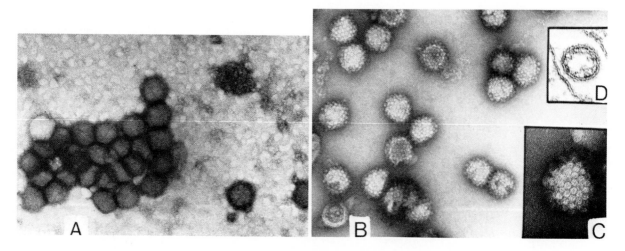

FIGURE 53-4   Electron micrographs of bunyaviruslike viruses. **A,** Karimabad virus negatively stained with phosphotungstate (×75,000). **B,** *Uukuvirus* with 0.7% glutaraldehyde before purification, negatively stained with uranyl acetate. The hexagonal array of subunits and occasional penton groups are evident (×100,000). **C,** *Uukuvirus* virion (×180,000). **D,** *Uukuvirus* virus–infected chick cells showing internal material in juxtaposition to the viral envelope (×60,000). *Uukuvirus* photographs provided by C.-H. Bonsdorff, University of Helsinki. From von Bonsdoroff and Pattersson (1975). Reduced 11% for reproduction. (From Fraenkel-Conrat H and Wagner RR: Comprehensive virology, vol 14, New York, 1979, Plenum Press.)

rhages, ecchymosis, epistaxis, hematemesis, melena, and bleeding of gums. Death occurs in up to 50% of cases with hemorrhagic phenomena.

LABORATORY DIAGNOSIS

See Figure 53-4 for an illustration of bunyavirus-like viruses.

Culture Virus may be recovered by inoculation of suckling mice or cell culture. Blood is usually positive for only 1 to 3 days after the onset of illness.

Direct antigen detection ELISA techniques may detect antigen in clinical specimens from patients with high levels of viremia (e.g., Rift Valley fever, hemorrhagic fever with renal syndrome, and Crimean hemorrhagic fever).

Serology IgM specific assays are useful to document acute infection rapidly. Seroconversion or a fourfold increase in IgG antibody are useful to document recent infection, but cross-reactions within viral genera are common.

TREATMENT, PREVENTION, AND CONTROL

No specific therapy is available for Bunyaviridae infections. Human disease is prevented by interrupting the contact between humans and the vector, whether arthropod or mammal. Arthropod vectors are controlled by insecticide spraying and by netting or window and door screening, protective clothing, and control of tick infestation of animals. Rodent control minimizes transmission of many viruses, especially Hantaan viruses. Rift Valley fever vaccines for use in humans and animals (sheep and cattle) have been developed.

BIBLIOGRAPHY

Bishop DHL and Shope RE: Bunyaviridae. In Fraenkel-Conrat H and Wagner RR, editors: Comprehensive virology, vol 14, New York, 1979, Plenum Press.

Peters CJ and LeDuc JW: Bunyaviruses, phleboviruses and related viruses. In Belshe RB, editor: Textbook of human virology, Littleton, Mass, 1984, PSG Publishing Co.

The members of the family Rhabdoviridae (Greek *rhabdos,* rod) include pathogens for a variety of mammals, fish, birds, and plants. The rhabdoviruses include the genera (1) *Vesiculovirus* (vesicular stomatitis viruses, or VSV), (2) *Lyssavirus* (rabieslike viruses), (3) an unnamed genus comprising the plant rhabdovirus group, and (4) other ungrouped rhabdoviruses of mammals, birds, fish, and arthropods (Box 54-1).

PHYSIOLOGY, STRUCTURE, AND REPLICATION

Rhabdoviruses are bullet-shaped, enveloped virions, 50 to 95 nm in diameter and 130 to 380 nm in length (Figure 54-1). Large projections composed of a glycoprotein (G) cover the surface of the virus (Figure 54-2). This glycosylated protein acts as a hemagglutinin of goose red blood cells and is responsible for the virus's ability to attach to host cells. Within the envelope, the helical nucleocapsid is coiled symmetrically into a cylindric structure, giving the appearance of striations (Figure 54-1). The nucleocapsid is composed of one molecule of single-stranded ribonucleic acid (RNA) and the nucleoprotein (N), large (L), and nonstructural (NS) proteins. The matrix (M) protein lies between the envelope and the nucleocapsid. The N protein is the major structural protein of the virus. It protects the RNA from ribonuclease (RNase) digestion and maintains the RNA in a configuration for transcription. The N protein of rabies virus is phosphorylated, whereas the N protein of VSV is not. The L and NS proteins, also known as M1 protein of rabies, comprise the RNA-dependent RNA polymerase.

The ribonucleoprotein (RNP) is the antigen detected by complement fixation, enzyme-linked immunosorbent assay, and immunofluorescence, whereas the surface glycoprotein G is responsible for eliciting neutralizing antibodies.

The replicative cycle of VSV has been studied in detail and is considered the prototype for the rhabdovirus group (Figure 54-2). Viral attachment to the host

Box 54-1	Rhabdoviridae

Vesiculovirus
Lyssavirus
Plant rhabdovirus group
Ungrouped rhabdoviruses

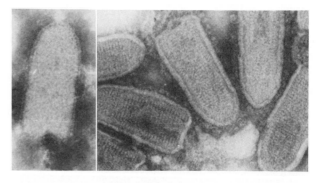

FIGURE 54-1 Rhabdoviridae seen by electron microscopy: rabies virus *(left)* and vesicular stomatitis virus *(right)*. (From Fields BN: Virology, New York, 1985, Raven Press.)

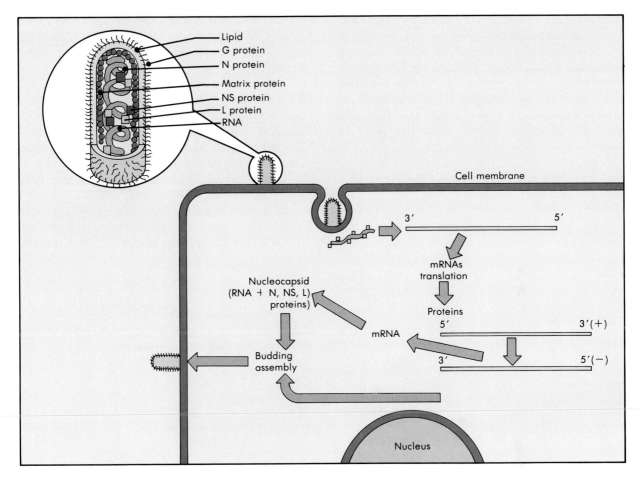

FIGURE 54-2   Model and replication cycle of vesicular stomatitis virus. The nucleocapsid is responsible for the bullet-shaped appearance of the mature virion. It is composed of single-stranded RNA (molecular weight of 3.5 to 4.6 $\times 10^6$ daltons) and three proteins: a nucleoprotein *(N)*, nonstructural protein *(NS)*, and large protein *(L)*. Surrounding the nucleocapsid is a matrix protein and a lipid envelope covered with surface projections (glycoprotein G). (Redrawn from Dulbecco R and Ginsberg HS: Virology, ed 2, Philadelphia, 1988, JB Lippincott Co.)

cell receptor is mediated by glycoprotein G. Entry into the cells appears to occur by endocytosis rather than by fusion with the cell membrane. The acidic pH of the endocytotic vesicle induces fusion of the viral envelope with the lysosomal membrane, which results in the uncoating of the virus and release of the nucleocapsid into the cytoplasm. Viral replication continues in the cytoplasm, where nucleocapsids form cytoplasmic inclusions (**Negri bodies**).

Once the envelope is removed, the RNA-dependent RNA polymerase initiates transcription of the negative strand, viral genomic RNA to produce mRNAs. These mRNAs are then translated into the five viral proteins. The viral genomic RNA is also tran-

scribed into complementary plus strands, then into complementary negative strands. The negative strand RNA is incorporated into new viral particles.

Assembly of the virion occurs in two phases: assembly of the RNP in the cytoplasm and assembly of the envelope in the cell plasma membrane. Assembly of the RNP is initiated by incorporation of the N protein, followed by binding of the polymerase proteins L and NS. The envelope assembly begins with the insertion of the G glycoprotein into the plasma membrane. The final step in virion maturation is the binding of the M protein to the assembled virus in the cell membrane. It appears that M protein functions to bridge the envelope and nucleocapsid and to induce coiling of the RNP into its condensed form. Budding of the mature particle through the plasma membrane or endoplasmic reticular membrane occurs through a poorly understood process. Cell death and lysis follow infection by most rhabdoviruses, with the important exception of rabies virus, which produces little discernible cell damage. In cell culture the replicative cycle of rhabdoviruses is completed in 19 to 24 hours.

PATHOGENESIS

Rabies virus is the rhabdovirus that causes the most significant infection in humans and therefore is discussed here in detail. Viral replication of rabies virus is restricted almost exclusively to neuronal tissue. No significant viremic stage has been documented.

Virus infects the host most frequently after the bite of a rabid animal that carries the virus in the saliva. Inoculation of the infected saliva into subcutaneous tissue or muscle initiates the infection. Other routes of infection, such as inhalation of aerosolized virus, transplantation of infected tissue (e.g., cornea), or inoculation through intact mucous membranes, have been documented.

Virus multiplies at the site of inoculation in striated muscle, remaining localized for days to months (Figure 54-3) before traveling to the peripheral nerves.

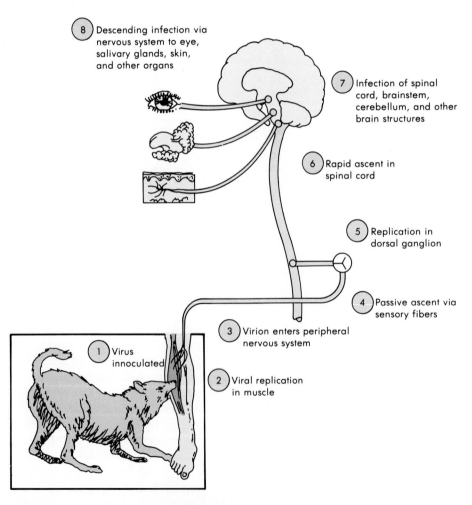

8 Descending infection via nervous system to eye, salivary glands, skin, and other organs

7 Infection of spinal cord, brainstem, cerebellum, and other brain structures

6 Rapid ascent in spinal cord

5 Replication in dorsal ganglion

4 Passive ascent via sensory fibers

3 Virion enters peripheral nervous system

2 Viral replication in muscle

1 Virus innoculated

FIGURE 54-3 Schematic representation of pathogenesis of rabies virus infection. Numbered steps describe sequence of events. (Redrawn from Belshe RB, editor: Textbook of human virology, Littleton, Mass, 1984, PSG Publishing Co, Inc.)

Several factors determine the highly variable incubation period, which ranges from 6 days to 1 year (most cases occur between 30 and 78 days). The progression of the virus to the central nervous system (CNS) is modulated by the virus concentration in the inoculum, the proximity of the wound to the brain, the severity of the wound, the host's age, and the host's immune status. The neuromuscular spindles within the muscle serve as a source of entry of the virus to the peripheral sensory nerves. Subsequently, rabies virus travels by retrograde axoplasmic transport to the dorsal root ganglia before infiltrating the spinal cord. Once the virus gains access to the spinal cord, rapid infection of the brain follows. The affected areas include the hippocampus, brainstem, ganglionic cells of pontine nuclei, and Purkinje's cells of the cerebellum. Virus then disseminates centrifugally from the CNS via afferent neurons to highly innervated sites, such as the skin of the head and neck, salivary glands, retina, cornea, nasal mucosa, adrenal medulla, renal parenchyma, and pancreatic acinar cells. With rare exception, rabies is fatal once clinical disease is apparent.

The mechanisms by which rabies virus evades host defenses are under intense investigation. In naturally acquired infection, neutralizing antibodies are not apparent until after clinical disease is well established. Furthermore, cell-mediated immunity appears to play little or no role in protection from rabies virus infection.

EPIDEMIOLOGY

Rabies is endemic worldwide in a variety of animals. Two forms of rabies occur in animals: urban rabies, with the dog serving as the primary transmitter, and sylvatic rabies, which occurs in many species of wildlife. The principal reservoir for rabies in most of the world is the dog. In Latin America large numbers of stray, unvaccinated dogs and the absence of rabies control programs are responsible for thousands of rabies cases in dogs. Sylvatic rabies in Cuba, Granada, Puerto Rico, and South Africa occurs predominantly in mongooses. In South America vampire bats transmit rabies to cattle, resulting in millions of dollars in losses each year. The two major foci of fox rabies are Western Europe and North America. In the United States in 1987, 4,729 cases of animal rabies were documented, with skunks, raccoons, and bats accounting for 84% (Figure 54-4).

The distribution of human rabies generally follows

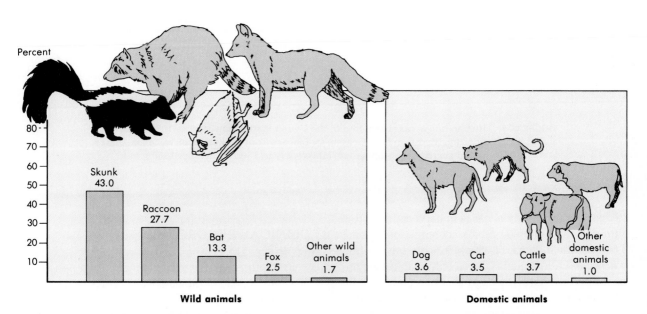

FIGURE 54-4   Distribution of animal rabies in the United States, 1987. (Redrawn from Centers for Disease Control: Rabies surveillance, United States, 1987. In CDC surveillance summaries, MMWR 37, no SS-4, September 1988.)

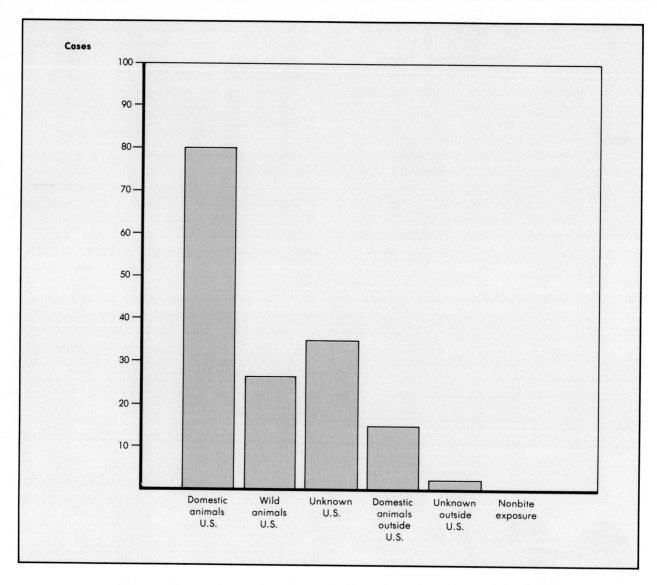

FIGURE 54-5   Human rabies cases reported in the United States from 1950 to 1987 by location and source of exposure. (Redrawn from Centers for Disease Control: Rabies surveillance, United States, 1987. In CDC surveillance summaries, MMWR 37, no SS-4, September 1988.)

the distribution of animal cases in each country. It is estimated that annually rabies accounts for at least 25,000 deaths in India, of which 96% are transmitted by dogs. In Latin America cases of human rabies are transmitted primarily by dogs in urban areas. In the United States, because of effective dog vaccination programs and lack of close contact with skunks, raccoons, and bats, the incidence of human rabies is 0 to 1 case per year (Figure 54-5).

CLINICAL SYNDROMES

Following a highly variable incubation period, the prodrome phase of rabies ensues (Table 54-1). Symptoms such as fever, malaise, headache, pain or paresthesia (itching) at the site of the bite, gastrointestinal symptoms, fatigue, and anorexia typically occur. After a prodrome of 2 to 10 days, neurologic symp-

Table 54-1 Clinical Progression of Rabies

Phase	Duration	Symptoms
Incubation	6 days to 1 year (most 30-70 days)	Asymptomatic
Prodrome	2 to 10 days	Fever, malaise, headache, pain at bite site, gastrointestinal disturbances, fatigue, anorexia
Neurologic	2 to 10 days	Hydrophobia, seizures, disorientation, hallucinations, paralysis
Comatose	1 to 14 days	Coma
Death		

toms specific for rabies appear. Hydrophobia, the most characteristic symptom of rabies, occurs in 20% to 50% of patients. It is marked by violent, jerky contractions of the diaphragm triggered by the patient's attempts to swallow water. Focal and generalized seizures, disorientation, and hallucinations also frequently occur during the neurologic phase. From 15% to 60% of patients exhibit paralysis as the only manifestation of rabies, which may lead to respiratory insufficiency. Following the neurologic phase, which lasts from 2 to 10 days, the comatose phase develops. This phase almost universally leads to death resulting from neurologic and pulmonary complications.

Pathologically, invasion of the virus into the brain and spinal cord causes an encephalitis and neuronal degeneration. Despite extensive CNS involvement, few gross or histopathologic lesions are seen in patients with rabies. Inflammatory lesions normally occur infrequently with wild or ''street'' virus infections.

LABORATORY DIAGNOSIS

Diagnosis of rabies is made by histopathology, detection of viral antigen in the CNS or skin, virus isolation, and serology (Table 54-2). Classically the hallmark of rabies diagnosis has been the detection of intracytoplasmic inclusions (Negri bodies) in affected neurons (Figure 54-6). Negri bodies are found within infected neurons and are matrices of viral nucleocapsids. Generally 2 to 10 μm in diameter, round or oval, and encompassed by a halo, the inclusions are eosinophilic because of the viral nucleocapsid and contain a central mass of basophilic granules (budding virions). Although diagnostic of rabies, Negri bodies are seen in only 70% to 90% of infected human brain tissue.

Antigen detection by immunofluorescence has become the most widely used method for rabies diagnosis. Brain biopsy or autopsy material, impression

Table 54-2 Diagnosis of Rabies Virus Infection

Diagnostic Test	Comments
Histopathology	Detection of Negri bodies in tissue is specific but insensitive (70-90%)
Direct immunofluorescence	High sensitivity and specificity
Culture	High sensitivity and specificity, not readily available; biologically hazardous
Serology	High sensitivity and specificity; antibody may not form until late in the course of disease

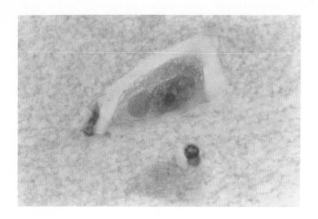

FIGURE 54-6   Negri bodies caused by rabies virus.

smears of corneal epithelial cells, and skin biopsy material from the nape of the neck are the most frequently used specimens to detect viral antigen by direct immunofluorescence. The test has a high degree of sensitivity and specificity in experienced hands.

Rabies in cell culture or by intracerebral inoculation of infant mice. Inoculated cell cultures or brain tissues are subsequently examined by direct immunofluorescence.

Rabies antibody titers in serum and cerebrospinal fluid (CSF) can be measured by a rapid fluorescent focus inhibition test, by mouse infection neutralization, or by plaque reduction. Antibody may not be detectable when symptoms initially develop; titers should be followed for 2 to 3 weeks after onset of symptoms. The presence of high serum antibody titers and any titer of antibody in the CSF suggests infection rather than immunization.

TREATMENT

With the exception of three documented human survivors of rabies, clinical rabies is invariably a fatal infection. Nevertheless, intensive care to support respiratory and circulatory functions, efforts to maintain electrolyte balance, and administration of anticonvulsive and neuroleptic drugs should be applied. Human hyperimmune antirabies globulin and vaccine have been unsuccessful in altering the course of rabies once symptoms have become apparent. The effectiveness of interferon therapy for rabies is currently under investigation.

PREVENTION AND CONTROL

Postexposure prophylaxis currently is the only hope for preventing overt clinical illness in the affected individual. Although human cases of rabies are rare, each year approximately 25,000 individuals receive rabies prophylaxis in the United States alone. Prophylaxis should be initiated for individuals exposed by bite or by contamination of an open wound or mucous membrane to saliva or brain tissue of a suspected animal, unless the animal is tested and shown not to be rabid.

The first protective measure of major importance is local treatment of the wound. Washing the wound with soap and water, detergent, or another substance that inactivates the virus should be done immediately. The World Health Organization Expert Committee on Rabies also recommends the instillation of antirabies serum around the wound.

Subsequently, immunization with vaccine in combination with administration of one dose of equine antirabies serum (ARS) or human rabies immune globulin (HRIG) is recommended. Passive immunization with HRIG provides antibody until the patient produces antibody in response to the vaccine. HRIG is preferred to ARS because ARS has a much higher risk of adverse reactions.

Various vaccines and treatment regimens are used worldwide. Vaccines (**Semple** or **Fermi**) prepared in the brains of adult or suckling animals are used in Latin American and other developing countries. A vaccine prepared with virus grown in Vero cells is available in France. In the United States the human diploid cell vaccine (HDCV) has replaced the duck embryo vaccine. The HDCV is chemically inactivated and administered intramuscularly on days 0, 3, 7, 14, and 28 following exposure. Recently a new cell culture–derived adsorbed rabies vaccine (RVA) has been produced and licensed in Michigan. RVA is prepared in fetal rhesus lung and inactivated with β-propiolactone.

Preexposure vaccination of animal workers, laboratory workers handling potentially infected tissue, and individuals traveling to areas where rabies is endemic should be performed. HDCV administered intramuscularly or intradermally in three doses is recommended.

Ultimately, prevention of human rabies hinges on effective control of rabies in domestic and wild animals. Control of domestic animal rabies depends on removal of strays and unwanted animals and vaccination of all dogs and cats. A variety of attentuated oral vaccines have been used successfully to immunize foxes. A live recombinant vaccinia virus vaccine expressing the rabies virus glycoprotein has been developed and is being tested.

BIBLIOGRAPHY

Anderson LJ, Nicholson KG, Tauxe RV, and Winkler WG: Human rabies in the United States, 1960-1979: epidemiology, diagnosis, and prevention, Ann Intern Med 100:728-735, 1984.

Baer GM, editor: The natural history of rabies, vols I and II, New York, 1975, Academic Press, Inc.

Baer GM: Rabies virus. In Fields BN, editor: Virology New York, 1985, Raven Press.

Blancou J, Kieny MP, Lathe R, Lecocq JP, Pastoret PP, Soulebot JP, and Desmettre P: Oral vaccination of the fox against rabies using a live recombinant vaccinia virus, Nature 322:373-375, 1986.

Centers for Disease Control: Rabies surveillance, United States, 1987. In CDC surveillance summaries, MMWR 37, no. SS-4, September 1988.

Centers for Disease Control: Rabies vaccine, adsorbed: a new rabies vaccine for use in humans, MMWR 37:217-218, 223, 1988.

Emerson SU: Rhabdoviruses. In Fields BN, editor: Virology, New York, 1985, Raven Press.

Ginsberg HS: Rhabdoviruses. In Dulbecco R and Ginsberg HS, editors: Virology, ed 2, New York, 1988, JB Lippincott Co.

Immunization Practices Advisory Committee: Rabies prevention—United States, 1984, MMWR 33:393-402, 407-408, 1984.

Immunization Practices Advisory Committee: Rabies prevention: supplementary statement on the preexposure use of human diploid cell rabies vaccine by the intradermal route, MMWR 35:767-768, 1986.

Robinson PA: Rabies virus. In Belshe RB, editor: Textbook of human virology, Littleton, Mass, 1984, PSG Publishing Co, Inc.

Roumiantzeff M, Ajjan JN, and Vincent-Falquet JC: Experience with preexposure rabies vaccination, Rev. Infect Dis 10(suppl 4):S751-S757, 1988.

Rupprecht CE, Glickman LT, Spencer PA, and Wiktor TJ: Epidemiology of rabies virus variants: differentiation using monoclonal antibodies and discriminant analysis, Am J Epidemiol 126:298-309, 1987.

Steele JH: Rabies in the Americas and remarks on the global aspects, Rev Infect Dis 10, (suppl 4):S585-S597, 1988.

Warrell DA and Warrell MJ: Human rabies and its prevention: an overview, Rev Infect Dis 10 (suppl 4):S726-S731, 1988.

Wiktor TJ, MacFarlan RI, Reagan KJ, Dietzschold B, Curtis PJ, Wunner WH, Kieny MP, Lathe R, Lecocq JP, Mackett M, Moss B, and Koprowski H: Protection from rabies by a vaccinia virus recombinant containing the rabies virus glycoprotein gene, Proc Natl Acad Sci USA 81:7194-7198, 1984.

World Health Organization: WHO Expert Committee on Rabies: seventh report, WHO Tech Rep Ser 709:1-104, 1984.

Wunner WH, Larson JK, Dietzschold B, and Smith CL: The molecular biology of rabies viruses, Rev Infect Dis 10(suppl 4):S771-S784, 1988.

55

Miscellaneous Viruses

CALICIVIRUS

Members of the *Calicivirus* genus are animal pathogens. The Norwalk agent and related viruses are classified as ''calici-like'' because they resemble the group morphologically, but their nucleic acid content has not been defined. The Norwalk agent and antigenically related viruses typically cause outbreaks of gastroenteritis such as occur on cruise ships or in communities as a result of common source contamination (e.g., water). In contrast to infants, who are more often infected by rotaviruses, caliciviruses cause disease primarily in older children and adults. None of these viruses can be isolated in tissue culture, but calici-like agents such as the Norwalk agent can be passed to human volunteers. Their viral nature was established by electron microscopic (EM) examination of feces using antibodies to enhance detection (immune electron microscopy, or IEM).

Physiology and Structure

Caliciviruses are among the smaller viruses, only 27 nm in diameter. They are spheric with 32 cup-shaped surface depressions. No envelope exists, and they exhibit icosahedral symmetry. The nucleoprotein of calicivirus is single-stranded ribonucleic acid (RNA) with a molecular weight of approximately 2.6 $\times 10^6$ daltons. The details of attachment and penetration are uncertain. Replication occurs in the cytoplasm, with release of viral particles accomplished by cell destruction.

Pathogenesis

Jejunal biopsy in human volunteers infected with caliciviruses reveals blunting of villi, cytoplasmic vacuolation, and infiltration with mononuclear cells, but virus particles are not detected by EM of epithelial cells. The virus appears to cause a decrease in brush border enzymes and in turn malabsorption. Although no histologic changes occur in gastric mucosa, gastric emptying may be delayed.

Epidemiology

In developed countries antibodies to Norwalk and related agents are slowly acquired during life so that by age 40 years approximately 50% of adults have been infected (Figure 55-1). In underdeveloped countries infection by these viruses occurs in nearly 100% of children by age 4 years, presumably as a result of poor sanitation. In developed countries outbreaks that occur year-round have been described in schools, resorts, hospitals, nursing homes, restaurants, and cruise ships. A measure of the importance of these agents is that almost 10% of all gastroenteritis outbreaks and almost 60% of those that are nonbacterial are attributed to them. Immunity is generally short-lived at best, and several exposures may be required before it is developed.

Clinical Syndromes

Although caliciviruses are typically associated with diarrheal illness, nausea and vomiting may

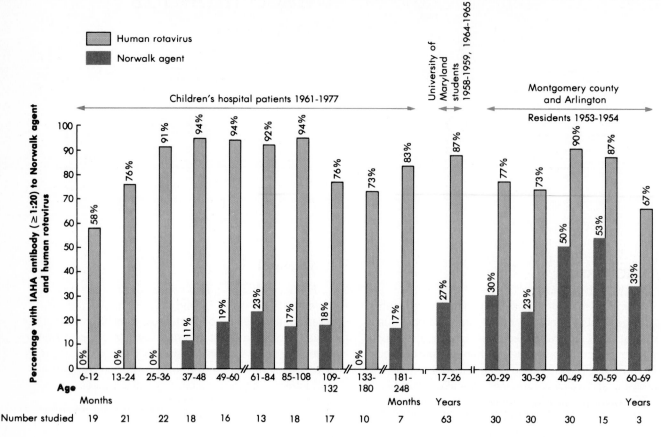

FIGURE 55-1 Prevalence of antibody to Norwalk agent and rotavirus by IAHA in three groups. (Redrawn from Gajdusek DC and Zigas V: Kuru: clinical, pathological and epidemiological study of an acute progressive degenerative disease of the central nervous system among natives of the Eastern Highlands of New Guinea, Am J Med 26:442-469, 1959.)

occur more frequently, especially in children (Figure 55-2). Bloody stoools do not occur. Fever may be present in up to one third of patients. The incubation period is 24 to 48 hours, and the illness lasts 12 to 60 hours.

Laboratory Diagnosis

Culture None of the caliciviruses can be grown in tissue culture.

Direct Detection Standard EM of feces is usually negative for calici-like viruses such as the Norwalk agent because they are present in low titer. Other calici-like viruses may be present in sufficient concentration to permit detection. IEM is the main method of detection of Norwalk type viruses but is rarely employed except in research settings. IEM consists of examining the sample by EM after the addition of an antibody directed against the suspected agent. The antibody causes the virus to aggregate, facilitating recognition. Radioimmunoassay (RIA) detection of Norwalk antigen is an alternative method that may be more sensitive than IEM.

Serology This is the method used to identify most infections. Antibody to the Norwalk agent may be detected by radioimmunoassay (RIA) or immune adherence hemoagglutination assay (IAHA). Antibodies to the other calici-like agents are more difficult to detect.

Treatment, Prevention, and Control

No specific treatment is available. Bismuth subsalicylate may reduce gastrointestinal symptoms. Out-

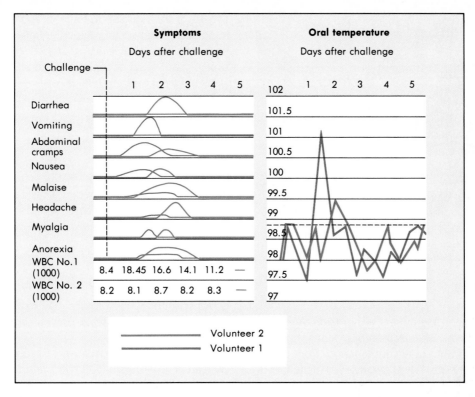

FIGURE 55-2   Response of two volunteers to oral administration of stool filtrate derived from volunteer who received original Norwalk rectal swab specimen. The height of the curve is directly proportional to the severity of the sign or symptom. Volunteer 1 had severe vomiting without diarrhea, whereas volunteer 2 had diarrhea without vomiting, although both received the same inoculum.

breaks may be minimized by handling food carefully and maintaining the purity of the water supply.

ASTROVIRUSES

Astroviruses are small (280 nm) viruses that exhibit five or six points on their surface. They have been seen in fecal specimens by EM, especially in infants and children with mild gastroenteritis. Astroviruses can infect tissue culture, as demonstrated by immunofluorescent assay (IFA) or EM. The nucleic acid is probably RNA, but this has not been proved for human astroviruses. The route of spread is probably fecal-oral; more than 70% of children acquire antibody by age 5 years.

FILOVIRUSES

Two unique RNA viruses, the Marburg and Ebola viruses (Figure 55-3) are members of a new family known as filoviruses. These agents can cause severe or fatal hemorrhagic fevers and are endemic in Africa. Laboratory workers have been exposed to the Marburg agent while working with tissue cultures from African green monkeys. Travelers in or residents of central Africa (e.g., Zaire or the Sudan) may be infected by the Ebola virus.

Physiology and Structure

Filoviruses are single-stranded RNA viruses whose nucleic acid has a molecular weight of approximately

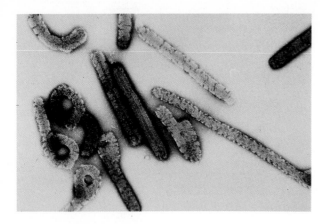

FIGURE 55-3   Electron micrograph of Ebola virus. (Courtesy Centers for Disease Control, Atlanta.)

4.5×10^6 daltons. They appear either filamentous or somewhat circular. The filamentous forms have a diameter of 80 nm but may vary in length from 1,000 to as long as 14,000 nm. Their symmetry is helical and they are enveloped. Virus is replicated in the cytoplasm and released by budding from the cell membrane.

Pathology and Pathogenesis

Cytoplasmic inclusions are seen in cells infected with the virus. In fatal human cases of filovirus infection, the principal histologic finding is necrosis of hepatocytes as well as splenic and lymph node follicular cells. Widespread hemorrhage occurs, presumably because of disseminated intravascular coagulation.

Epidemiology

Marburg virus infection was first detected among laboratory workers in Marburg, Germany, who had been exposed to tissues from apparently healthy African green monkeys. However, it is not clear that these monkeys were or are the reservoir for this virus because inoculation of Marburg virus into these monkeys produces death rather than a carrier state. Rare cases of Marburg virus infection have been reported in Zimbabwe and Kenya.

Ebola virus has only caused disease in Africans in Zaire and the Sudan. In rural areas of central Africa up to 18% of the population have antibody to this virus, suggesting that subclinical infections do occur. The source of the virus and means of transmission are unknown. Rarely, secondary cases of filovirus infections have occurred in health care workers, usually as a result of accidental needle-stick exposure.

Clinical Syndromes

Illness usually begins with flu-like symptoms such as headache and myalgias. Within a few days nausea, vomiting, and diarrhea occur; a rash also may develop. Subsequently, hemorrhage from multiple sites and death occur in as many as 90% of patients with clinically evident disease.

Laboratory Diagnosis

All specimens for filovirus diagnosis must be handled with extreme care to prevent accidental infection.

Immunofluorescence and Electron Microscopy   Direct IFA can detect viral antigens in tissues; virions can be seen by EM in serum or liver tissue.

Culture   Isolation of the virus is the procedure of choice to diagnose filovirus infections. Marburg virus may grow rapidly in tissue culture (Vero cells), although Ebola virus recovery may require animal (e.g., guinea pig) inoculation.

Serology   IgG and IgM antibody to filovirus antigens can be detected by IFA, enzyme-linked immunosorbent assay (ELISA), or RIA. Seroconversion or a fourfold rise of IgG antibody levels may be diagnostic of active infection, as may the detection of specific IgM antibody.

Treatment, Prevention, and Control

No treatment is known. Since the source of filoviruses is unknown, no measures are available for preventing primary infection. Secondary cases in health care workers can be prevented by avoiding exposure to contaminated needles, blood, and so forth.

ARENAVIRUS

The best-known arenaviruses that infect humans are the agents of lymphocytic choriomeningitis (LCM) and Lassa fever. These agents may seem vastly different, but they have a common reservoir in animals, especially small rodents. The infected animals may be asymptomatic or minimally diseased but excreting the virus in their secretions.

Physiology and Structure

Arenaviruses are pleomorphic and enveloped with lipid; the virion has a mean diameter of 120 nm. They contain two-stranded RNA in a linear or circular configuration. The total molecular weight of this RNA is 3.2 to 4.8 × 10^6 daltons. Replication is in the cytoplasm, with budding from the host cell cytoplasm.

Epidemiology

Each arenavirus infects specific rodents; for example, LCM infects hamsters and mice, and the Lassa fever virus infects African rodents. Chronic infection is common in these animals and leads to chronic viremia and virus shedding in saliva, urine, and feces. Infection of humans may occur by way of aerosols, contamination of food, or fomites. Bites are not a usual mechanism of spread. Persistently infected rodents do not usually exhibit illness. Human-to-human infection occurs with Lassa fever virus through contact with infected secretions or body fluids, but this mode of spread rarely if ever occurs with LCM or other hemmorrhagic fevers. The incubation period for arenavirus infections averages 10 to 14 days.

Pathogenesis and Pathology

Arenaviruses are able to infect macrophages and possibly cause the release of mediators of cell and vascular damage. In certain laboratory animals the clinical severity of arenavirus disease appears to be directly related to the host's immunologic response. The greater the immune (especially T lymphocyte) response, the worse is the disease. Whether these mechanisms are operative in human infection is not clear. LCM virus may actually produce an encephalitis as well as meningitis. Perivascular mononuclear infiltrates may be seen in neurons of all sections of brain as well as in the meninges. In patients with the hemorrhagic fevers, petechiae and occasional visceral hemorrhage occur, as do liver and spleen necrosis, but not vasculitis. In contrast to LCM, no lesions are present in the central nervous system (CNS) of patients with hemorrhagic fevers.

Clinical Syndromes

Lymphocytic Choriomeningitis Illness occurs in most infected individuals but is usually nonspecific or influenza-like. Current estimates show that approximately 25% of infected persons exhibit clinical evidence of CNS infection. The name of the virus suggests that meningitis is a typical clinical event, but actually a febrile illness with myalgia occurs more often. The meningeal illness, if it occurs, may be subacute and persist for several months.

Lassa and Other Hemorrhagic Fevers Lassa fever, with its focus of endemicity in West Africa, is the best known of the arenavirus hemorrhagic fevers. Other agents, however, such as Junin and Machupo, cause similar syndromes in different geographic areas (Argentina and Bolivia, respectively). Clinical illness is characterized by fever and coagulopathy. Hemorrhage and shock occur and occasionally cardiac and liver damage. Pharyngitis, diarrhea, and vomiting may be very prevalent, especially in patients with Lassa fever. Death occurs in up to 50% of those with Lassa fever and in a smaller percentage among those infected with the other arenaviruses that cause hemorrhagic fevers. The diagnosis is suggested by recent travel to endemic areas.

Laboratory Diagnosis

The diagnosis of arenavirus infection is usually made through serologic tests, although the virus can be recovered by inoculation of blood or cerebrospinal fluid into suckling mice or Vero monkey cells. Throat specimens can yield arenaviruses; urine is often positive for the Lassa fever virus but negative for the LCM virus. Substantial risk is present for laboratory workers handling body fluids. Therefore, if the diagnosis is suspected, laboratory personnel should be warned and specimens processed only in facilities specialized for the isolation of contagious pathogens.

Treatment, Prevention, and Control

Only supportive therapy for patients with arenavirus infection is currently available. Uncontrolled studies suggest that ribavirin may be useful for those with Lassa fever. Prevention of these rodent-borne infections rests on control of the vector's contact with humans. Most human cases of LCM in the United States have resulted from contact with pet hamsters or in rodent-breeding facilities. If hamsters must be kept as pets, scrupulous handwashing is recommended after contact. In the geographic areas where hemmorhagic fever occurs, trapping of rodents and careful storage of food may decrease contact. Laboratory-acquired cases can be reduced by processing samples for arenavirus isolation in at least P_3 biosafety facilities and *not* in the usual clinical virology laboratory.

BIBLIOGRAPHY

Bishop RF: Other small virus-like particles in humans. In Tyrrell DAJ and Kapikian AZ, editors: Virus infections of the gastrointestinal tract, New York, 1982, Marcel Dekker, Inc.

Blacklow NR and Cukor G: Norwalk virus: a major cause of epidemic gastroenteritis, Am J Public Health 72:1321-1323, 1982.

Caul EO, Ashley C, and Pether JVS: "Norwalk"-like particles in epidemic gastroenteritis in the U.K., Lancet 2:1292, 1979.

CHAPTER 56

Retroviruses

The retroviruses are enveloped, positive-strand RNA viruses with a unique morphology and means of replication. In 1970 Baltimore and Temin showed that the retroviruses encode an RNA-dependent DNA polymerase (reverse transcriptase) and replicate through a DNA intermediate. The DNA copy of the viral genome is integrated into the host chromosome and transcribed as a cellular gene. This discovery, which earned the Nobel Prize, contradicted the central dogma of molecular biology that stated genetic information passed from DNA to RNA and then to protein.

Three subfamilies of retrovirus exist. These subfamilies and their human members are the oncornavirus, or oncovirus (human T-lymphotropic virus 1, 2, and 5; HTLV-1, HTLV-2, HTLV-5), lentivirus (human immunodeficiency virus 1 and 2; HIV-1, HIV-2) and spumavirus (Table 56-1). Spumavirus was the first human retrovirus to be isolated, but no spumavirus has been associated with human disease.

The first retrovirus to be isolated was the Rous sarcoma virus, shown by Peyton Rous to produce solid tumors (sarcomas) in chickens. As with most retroviruses, the Rous sarcoma virus proved to have a very limited host and species range. Cancer-causing retroviruses have been isolated from other animal species and are classified as **RNA tumor viruses** or **oncornaviruses**. Many of these viruses alter cellular growth by expressing analogs of cellular growth–controlling genes **(oncogenes).**

The retroviruses are probably the most studied group of viruses in molecular biology, but not until 1981 was a human retrovirus isolated and associated with human disease. HTLV-1 was isolated from an

Table 56-1 Classification of Retroviruses

Subfamily	Characteristics	Examples
Oncovirus	Associated with cancer and neurologic disorders	
A	Intracytoplasmic precursor of B-type, virus; double-membrane particles	
B	Eccentric nucleocapsid core in mature virion	Mouse mammary tumor virus
C	Centrally located nucleocapsid core in mature virion	Human T-lymphotropic virus (HTLV-1, HTLV-2, HTLV-5), Rous sarcoma virus (chickens)
D	Nucleocapsid core with cylindric form	Mason-Pfizer monkey virus
Lentivirus	Slow disease onset; causes neurologic disorders and immunosuppression; virus with D-type, cylindric, nucleocapsid core	Human immunodeficiency virus (HIV-1, HIV-2), Caprine arthritis/encephalitis virus (goats)
Spumavirus	Causes no clinical disease but characteristic vacuolated "foamy" cytopathology	Human foamy virus

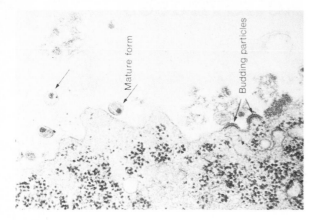

FIGURE 56-1   Electron micrograph of retrovirus. (Courtesy Centers for Disease Control, Atlanta.)

Box 56-1   Characteristics of Retroviruses

Enveloped, spheric virion; 80 to 120 nm
Diploid positive-strand RNA genome, 3.5 to 9 kilobases
Positive-strand RNA, RNA-dependent DNA polymerase carried in virion
Replication through DNA intermediate
Integrates randomly in host chromosome

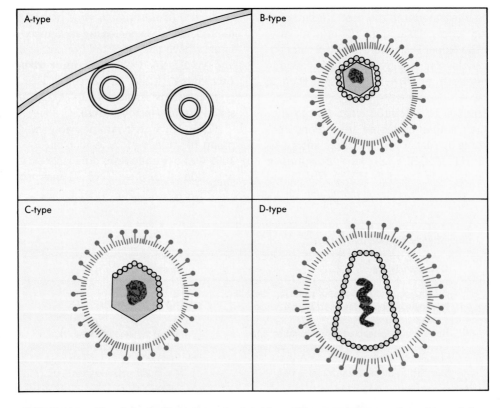

FIGURE 56-2   Morphologic distinction of retrovirions. The morphology and position of the nucleocapsid core are the means of classification. A-type particles are immature, intracytoplasmic forms that bud through the plasma membrane into mature B-type particles. C-type and D-type particles are mature virion forms that are completely assembled at the plasma membrane.

individual with adult human T-cell leukemia by Robert Gallo and associates.

In the late 1970s and early 1980s, the reports that young homosexual men, Haitians, heroin addicts, and hemophiliacs (the initial "4H club" of risk groups) were dying of normally benign opportunistic infections defined a new disease, acquired immunodeficiency syndrome (AIDS). Montagnier and associates in Paris and Gallo and co-workers in the United States reported the isolation of a retrovirus, HIV-1, from patients with lymphadenopathy and AIDS. A variant of HIV-1, designated HIV-2, was isolated later and is prevalent in West Africa.

Our understanding of the retroviruses has paralleled progress in molecular biology. Newer technologies were and are still required to advance our knowledge of the retroviruses and their association with human disease. Their limited host range and tissue tropism has made it difficult to establish appropriate cell systems for their growth. Their ability to integrate into host chromosomes and remain latent has made them difficult to detect, study, and treat. However, the retroviruses have provided a major tool for molecular biology, the reverse transcriptase, and have advanced our understanding of cell growth, differentiation, and oncogenesis through the study of viral oncogenes.

The pathogenic human retroviruses include HTLV-1 and HTLV-2, which are oncoviruses, and HIV-1 and HIV-2, which are lentiviruses. HTLV-1 and HIV-1 both infect CD4 T lymphocytes, are spread by similar means, and have long latent periods, but they cause very different diseases because of the nature of their interaction with the infected host cell.

PHYSIOLOGY AND STRUCTURE

The retroviruses are roughly spheric, enveloped RNA viruses with a diameter of 80 to 120 nm (Figure 56-1 and Box 56-1). The envelope contains viral glycoproteins and is acquired by budding from the plas-

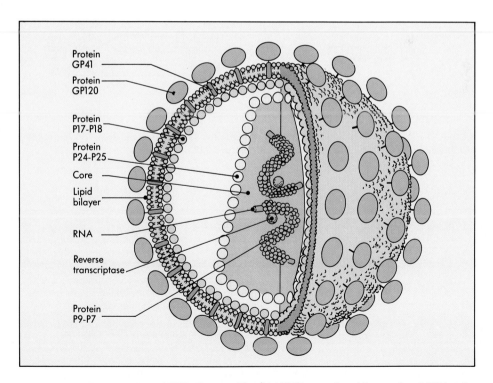

Protein GP41
Protein GP120
Protein P17-P18
Protein P24-P25
Core
Lipid bilayer
RNA
Reverse transcriptase
Protein P9-P7

FIGURE 56-3   Cross section of HIV. The two identical RNA strands with associated RNA polymerase are packaged in a protein core. This is surrounded by proteins and a lipid bilayer. Antibodies to the surface and core proteins are important markers for infection with this virus. (Redrawn from Gallo RC and Montagnier L: Sci Am 259:41-51, 1988.)

ma membrane. The envelope surrounds a capsid that contains two identical copies of the positive-strand RNA genome inside an electron-dense core with 10 to 50 copies of the reverse transcriptase. The morphology of the core differs for different viruses and is used as a means of classifying the retroviruses. The HIV virion core resembles a truncated cone (Figures 56-2 and 56-3).

The retrovirus genome has a 5′ cap and is polyadenylated at the 3′ end (Figure 56-4 and Table 56-2). A cellular transfer RNA (tRNA) is base-paired to each copy of the genome in the virion and acts as a primer for the reverse transcriptase. Although the genome resembles a messenger RNA (mRNA), it is not infectious because it does not encode a polymerase that can directly generate more mRNA. The genome consists of at least three major genes that encode polyproteins for the enzymatic and structural proteins of the virus: gag (group specific antigen), pol (polymerase), and env (envelope). HTLV and the lentiviruses, including HIV, also encode several regulatory proteins.

The viral glycoproteins are produced by proteolytic cleavage of the polyprotein encoded by the env gene. Their size differs for each group of viruses. For HTLV-1, the gp88 is cleaved into a gp46 and p20E, and for

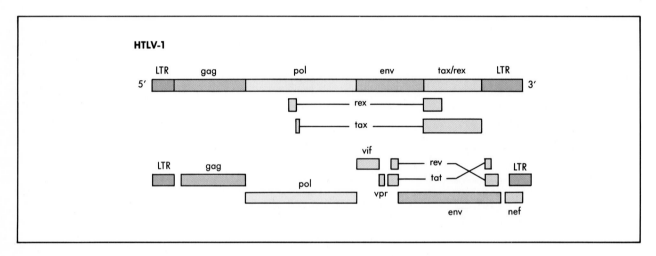

FIGURE 56-4   Genomic structure of human retroviruses. (Redrawn from Gallo R, Wang-Stall F, Montagnier L, Haseltine WA, and Yoshida M: Nature 333:504, 1988.)

Table 56-2   Genes and Their Function in Retroviruses

Gene	Function
gag	Group specific antigen: core and capsid proteins
pol	Polymerase: reverse transcriptase, protease, endonuclease
env	Envelope: spike proteins and glycoprotein
tax/rex	Transactivation
tat	Transactivation, up-regulates HIV replication
rev (art, trs)	Regulation of expression, up-regulates HIV replication, depresses nef
nef (F, 3′ orf, B)	Negative early factor, down-regulates virus expression
vif (Q, sor, A)	Virus infectivity, helps to initiate replication
vpr (R)	Unknown
vpx (X)	Unknown
LTR	Long terminal repeats: promoter, enhancer elements

HIV, the gp160 is cleaved into the gp41 and gp120. These glycoproteins are linked through disulfide bonds to form a lollipop-like spike visible on the surface of the virion. The larger of the glycoproteins is responsible for the tissue tropism of the virus and is recognized by neutralizing antibody. The smaller subunit (gp41 in HIV) forms the lollipop stick and promotes cell-to-cell fusion. The gp120 of HIV is extensively glycosylated, and its antigenicity can drift during the course of a chronic HIV infection, both of which impede immune clearance of the virus. Detection of these glycoproteins is a useful marker of infection.

The capsid is composed of several peptides cleaved from a polypeptide encoded by the gag gene. The reverse transcriptase, an endonuclease, and the protease of HIV are encoded by the pol gene.

CLASSIFICATION

The retroviruses are classified by disease and morphology (Table 56-1). The oncoviruses include the only retroviruses that can immortalize or transform target cells. The lentiviruses are slow viruses associated with neurologic and immunosuppressive disease. The spumaviruses, represented by a foamy virus, cause a distinct cytopathologic effect but do not seem to cause clinical disease.

The retroviruses are further classified by their host range and tissue tropism. Differential expression of viral receptors on cells of different species and tissues, along with their recognition by the retroviral glycoprotein attachment proteins, is one of the major determinants. The oncoviruses are further categorized by the morphology of their core and capsid in electron micrographs as type A, B, C, or D (see Table 56-1).

REPLICATION

Replication of the retroviruses is initiated by the binding of the viral glycoprotein spikes to specific cell surface receptor proteins (Figure 56-5). The gp120 of HIV interacts with a specific epitope of the CD4 surface molecule expressed on T helper lymphocytes, macrophages, and neurons. The virus envelope then fuses with the cellular plasma membrane, delivering the nucleocapsid into the cytoplasm. Some retroviruses may also penetrate by receptor-mediated endocytosis.

Once released into the cytoplasm, the reverse transcriptase uses the virion tRNA as a primer and synthesizes a complementary negative-strand DNA (Figure 56-6). The reverse transcriptase acts as an RNAase, degrades the RNA genome, and then synthesizes the positive strand of DNA. During the synthesis of the DNA, sequences from each end of the genome are duplicated and juxtaposed to create long terminal repeats (LTR) at both ends. This creates sequences that are required for integration and also enhancer and promoter sequences to regulate transcription. The DNA copy of the genome is circular and larger than the original RNA.

The double-strand DNA is then delivered to the nucleus and spliced into the host chromosome at various positions with the aid of a viral-encoded integrase. HIV and other lentiviruses produce a large amount of nonintegrated circular DNA, which is not transcribed efficiently but may contribute to the cytopathogenesis of the virus. Once integrated, the viral DNA is transcribed as a cellular gene by the host RNA polymerase II. Viral transcription is facilitated by host cell–like promoter and enhancer sequences provided by the LTR and, for HTLV and HIV, regulated further by other viral genes. The entire length of the genome is transcribed and processed to produce gag, gag-pol, and env mRNA. The full-length transcripts of the genome can also be assembled into new virions.

The proteins translated from the gag, gag-pol, and env mRNAs are synthesized as polyproteins and are subsequently cleaved to functional proteins. The viral glycoproteins are synthesized, glycosylated, and processed by the endoplasmic reticulum and Golgi apparatus. The glycoprotein is cleaved, the subunits are linked through sulfhydryl bonds, and then they form dimers or trimers that migrate to the plasma membrane.

The gag and the gag-pol polyproteins are acylated and then bind to the plasma membrane. The association of two copies of the genome and cellular tRNA molecules with this aggregate triggers the release of the viral protease and cleavage of the gag polyproteins. This releases the reverse transcriptase and forms the virion core, which remains associated with the virion glycoprotein-modified plasma membrane. The virus buds from the plasma membrane and simultaneously acquires its envelope and is released from the cell. Cell-to-cell spread of HIV is further enhanced

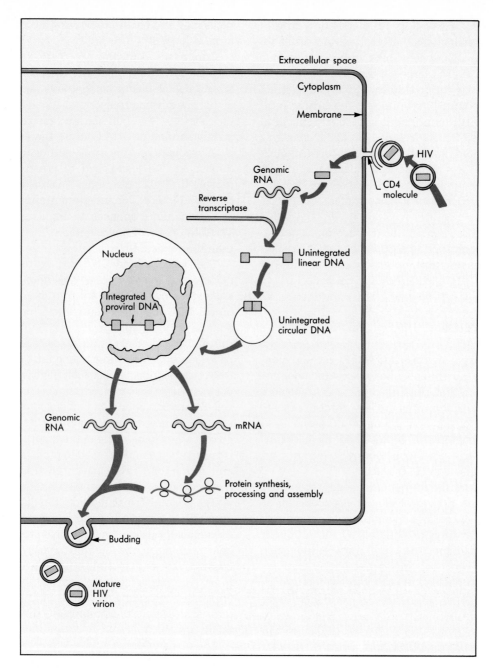

FIGURE 56-5   The life cycle of HIV. (Redrawn from Fauci AS: Science 239:617-622, 1988.)

by its ability to form multinucleated giant cells, or syncytia. This also enhances the cytolytic activity of the virus.

As with other cellular genes, the integrated retroviruses require activation before they are transcribed. The rate of viral genome transcription and whether the virus remains latent depend on the ability of the cell to use the enhancers and promoters encoded in the LTR region, the extent of methylation of the DNA region, and the cell's growth rate. Stimulation of the cell by mitogens or other means, including infection with exogenous viruses (e.g., herpesviruses), may also activate transcription of the virus.

The expression of HIV proteins is regulated by up

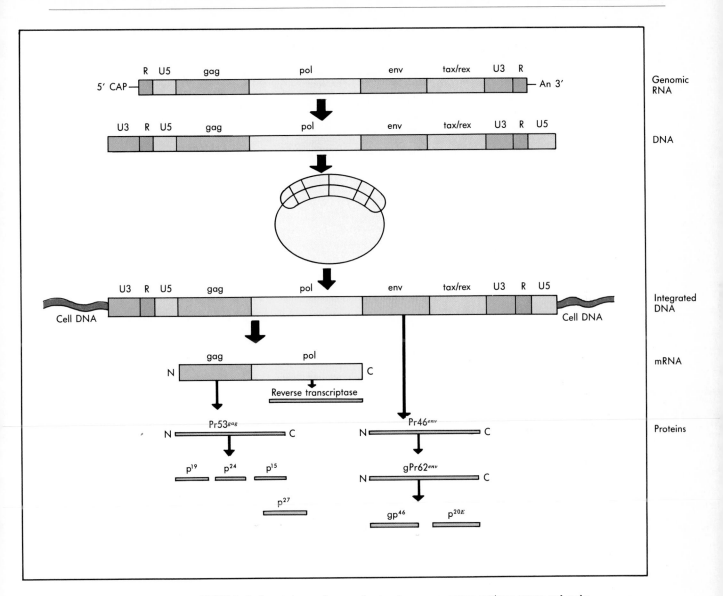

FIGURE 56-6   HTLV-1 viral proteins and genomic structure. *gag*, group antigen gene; *pol*, polymerase; *env*, envelope glycoprotein gene; *tax*, transactivator; *rex*, regulator of expression of virion proteins; *p*, protein; *Pr*, precursor polyprotein; *gp*, glycoprotein; *gPr*, glycosylated precursor polyprotein; *N*, N terminus; *C*, C terminus. (Redrawn from Ng VL and McGrath MS: Cancer Bull 40:276-280, 1988.)

to six genes (see Figure 56-3). The nef, tat, and rev proteins produced by these genes create a network of regulatory factors that control their own synthesis and the synthesis of the virion's proteins. The tat activates all the genes of the virus, the rev activates the structural genes but represses the other regulatory genes, and the nef represses all the viral genes. Nef may play a role in inducing latency (see Table 56-2). HIV is also under cellular control, and activation of the T cell by a mitogen or antigen also activates the virus.

HTLV also encodes transactivator proteins that bind to enhancer sequences in the LTR to stimulate transcription.

HUMAN IMMUNODEFICIENCY VIRUS

Pathogenesis

The major determinant in the pathogenesis and disease caused by HIV is the virus tropism for CD4

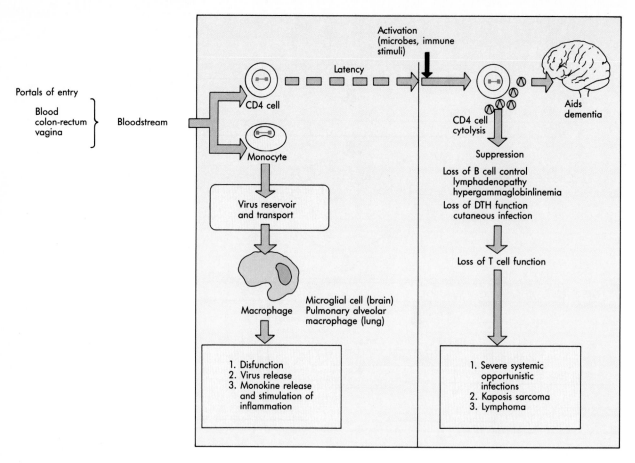

FIGURE 56-7   Pathogenesis of HIV. (Redrawn from Fauci AS: Science 239:617-622, 1988.)

Table 56-3   Immune Function Abnormalities Resulting from HIV Infection

Cells	Abnormalities
T helper (CD4) lymphocytes	Decreased proliferative responses
	Decreased lymphokine production
	Decreased cytotoxic T-cell activity against virus-infected cells
	Decreased DTH response
Monocytes	Decreased chemotaxis
	Decreased IL-1 production
	Decreased microbiocidal activity
Natural killer cells	Decreased cytotoxic activity
B lymphocytes	Decreased antigen-specific humoral responses (antibody production)
	Uncontrolled production of antibody (hypergammaglobulinemia)

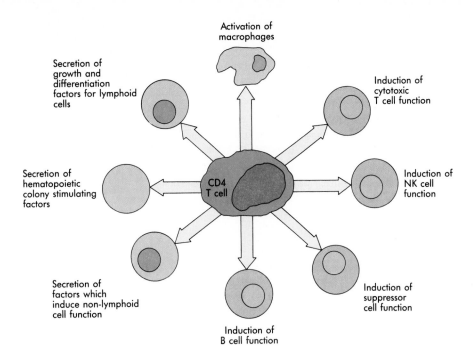

FIGURE 56-8   The CD4 T cell plays a critical role in the regulation of the human immune response by the release of soluble factors. Thus destruction of the cells by HIV has a dramatic impact. (Redrawn from Fauci AS: Science 239:617-622, 1988.)

expressing T cells, macrophages, and brain cells (Figures 56-7 and 56-8; Tables 56-3; Box 56-2). The CD4 cells are the helper and delayed-type hypersensitivity (DTH) T cells. The virus can also infect neurons.

After infection of these cells, the virus replicates but rapidly establishes latency. The virus may remain latent for long periods, but when activated in CD4 T cells, the virus kills the cell. This activation may occur after stimulation of the cell by an antigen or mitogen.

HIV induces several cytopathic effects that may kill the cell. These include accumulation of nonintegrated circular DNA copies of the genome, increased permeability of the plasma membrane, and syncytia formation. The CD4 protein and the envelope glycoproteins play important roles in the cytopathic effects.

Changes in cell function also occur in CD4 T cells latently infected with HIV. HIV infection can reduce the expression of CD4 antigens and production of interleukin 2 (IL-2), reducing the ability of the helper cell to respond to antigen and activate other T cells.

Macrophages express lesser amounts of CD4 than T cells but can still be infected with HIV. The virus does not kill the cells but may alter their function. Monocytes and macrophages are probably the major reservoirs and means of distribution for HIV. Circulating macrophages, microglial cells of the brain, pul-

---

Box 56-2   Mechanisms of HIV-Induced Immune Suppression

DIRECT

HIV cytocidal effect on CD4 lymphocytes
Functional defects in infected CD4 cells
Impaired antigen presentation or monokine production by macrophage

INDIRECT

Generation of suppressor T cells or factors
Induction of autoimmune phenomena
Cytotoxic cell activity against viral or self-proteins
Decreased humoral immune responses

monary alveolar macrophages, and other cells of the monocyte-macrophage lineage can spread the virus and potentially contribute to HIV disease.

The central role of the CD4 helper T cell in the initiation of an immune response and its DTH response is indicated by the extent of immunosuppression induced by HIV infection (Figure 56-8). Activation of the CD4 T cells is one of the first steps in the initiation of an immune response. Helper T cells release lymphokines and γ-interferon required for activation of macrophages, other T cells, B cells, and natural killer cells. When CD4 T cells are unavailable or not functional, antigen-specific immune responses, especially cellular immune responses are incapacitated and humoral responses are uncontrolled. HIV infection also eliminates the CD4 cells responsible for DTH.

In addition to immunosuppressive disorders, HIV can also cause neurologic abnormalities. The macrophage is the predominant cell type of the brain infected with HIV, but neurons and glial cells may also be infected. Infected monocytes and microglial cells may release neurotoxic substances or chemotactic factors to promote inflammatory responses in the brain. Direct cytopathic effects of the virus on neurons are also possible.

The ability of HIV to incapacitate the immune system, remain latent in lymphocytes, and alter their antigenicity allows the virus to escape immune clearance and prevents resolution of the disease (Box 56-3). The virus appears to establish a latent or low-level chronic infection in every infected individual. A slow, progressive decrease in the levels of CD4 cells may precipitate immunodeficiency after long periods.

Epidemiology

AIDS was first described in male homosexuals in the United States and initially was thought to be restricted to this population. We now know that this is not true and that HIV infection is spreading in epidemic proportions throughout the population. As of August 1989, greater than 100,000 people had developed AIDS and greater than 59,000 had died of the complications of the disease. Many more individuals are seropositive for HIV but have not yet exhibited symptoms.

The presence of HIV in blood and semen of infected individuals promotes the spread of the disease to adults by sexual contact and exposure to

---

Box 56-3   Means of Immune Escape by HIV

Infection of lymphocytes and macrophages
Inactivation of CD4 helper cells, incapacitating the system
Antigenic drift of the gp120
Heavy glycosylation of gp120
Latent infection

---

blood and blood products. The virus can also be found in saliva but is not transmitted effectively by this means. The major U.S. adult populations at risk to HIV infections are homosexuals and intravenous drug users. Children acquire the virus perinatally from infected mothers (Table 56-4 and Figure 56-9).

HIV-1 and HIV-2 are much more prevalent in Africa than in the United States. Heterosexual transmission is the major means of spread in Africa, and both men and women are equally affected by the virus.

AIDS was initially observed in young, promiscuous male homosexuals and is still prevalent in the gay community. Anal intercourse may promote efficient transmission of the virus to the bloodstream, but heterosexual transmission by vaginal intercourse can also occur. The major spread of HIV in the heterosexual population is by intravenous drug abusers. Sharing of contaminated syringe needles is common practice in "shooting galleries." In New York City more than 80% of intravenous drug abusers are HIV antibody positive, and these individuals are the major source of heterosexual and congenital transmission of the virus.

Screening of the blood supply has reduced the incidence of transfusion-related transmission of HIV and infection of hemophiliac persons receiving clotting factors. However, as many as 60% of hemophiliac individuals may be HIV antibody positive, reflecting prior exposure to HIV-contaminated factor VIII pooled from thousands of donors.

Interestingly, HIV-seropositive hemophiliac persons are less prone to developing "full-blown AIDS" than are homosexual males and intravenous drug abusers. This has suggested that a cofactor may be required to activate the virus and cause AIDS. Coinfection or reactivation of latent viruses such as the herpesviruses (herpes simplex, Epstein-Barr, cyto-

Table 56-4 Transmission of HIV Infection

Routes	Specific Transmission
Known routes of transmission	
Inoculation of blood	Transfusion of blood and blood products
	Needle sharing among intravenous drug users
	Needle stick, open wound, and mucous membrane exposure in health care workers
Sexual	Homosexuals, male
	Heterosexuals
Perinatal	Intrauterine
	Peripartum
	Breast milk
Routes investigated and shown *not* to be involved in transmission	
Close personal contact	Household members
	Health care workers not exposed to blood

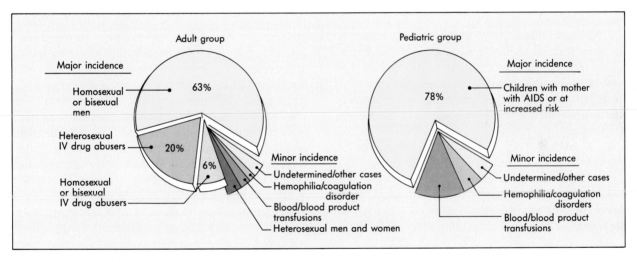

FIGURE 56-9   Incidence of AIDS in adult and pediatric patients. Data collected in 1988. With increased vigilance in certain populations the incidence of new infection with HIV is decreasing in homosexuals and persons receiving blood and blood products. (Redrawn from Redfield RR and Buske DS: Sci Am 259:90-98, 1988.)

megalovirus) or hepatitis B may induce HIV replication.

A small number of health care workers have contracted HIV infection from accidental needle sticks, cuts, or exposure of broken skin to contaminated blood. Prospective studies of needle-stick recipients indicate that fewer than 1% of those exposed to HIV-positive blood experience seroconversion.

HIV-2 is known to infect patients in West Africa and to produce a disease similar to but less severe than AIDS.

Clinical Syndromes—Acquired Immunodeficiency Syndrome

Initial HIV infection is probably asymptomatic in most persons. Acute infection is infrequently recognized but resembles infectious mononucleosis and may be accompanied by "aseptic" meningitis or a rash. This illness subsides spontaneously and is followed by a latent period that may last many years. AIDS may manifest in one of several different ways:

1. **Lymphadenopathy and fever**. This process develops insidiously and may be accompanied by weight loss and malaise. These findings may persist indefinitely or may progress.

2. **Opportunistic infections**. The most common opportunistic infection is pneumonia caused by *Pneumocystis carinii*. Cerebral toxoplasmosis, cryptococcal meningitis, and oral and cutaneous candidiasis often occur, as well as prolonged and severe herpesvirus infections (e.g., oral, anal, and genital herpesvirus; varicella-zoster; and a variety of cytomegalovirus illnesses, especially retinitis and pneumonia). Diarrhea caused by common pathogens (*Salmonella, Shigella, Campylobacter*) and uncommon agents (cryptosporidia, mycobacteria, *Ameoba*) occurs frequently.

3. **Malignancies**. The most notable malignancy is the development of Kaposi's sarcoma, a previously benign skin cancer that disseminates to involve visceral organs in these immunodeficient patients (Figure 56-10). Non-Hodgkins lymphoma and premalignant lesions such as hairy leukoplakia of the mouth also occur.

4. **Wasting**. The progressive development of "slim" disease is especially common in Africa.

5. **AIDS-related dementia**. This may result from HIV infection of the microglial cells and neurons of the brain. Patients may undergo a slow deterioration of intellectual abilities and other signs of neurologic disorder, similar to those of the early stages of Alzheimer's disease.

HIV disease may be viewed as a continuum from asymptomatic infection to profound immunodepression. Some patients may progress through all the syndromes just mentioned, whereas others may have Kaposi's sarcoma or *Pneumocystis carinii* pneumonia without ever having earlier symptoms or signs.

Current understanding is that at least 60% of HIV-infected individuals will become symptomatic, and the overwhelming majority of these will ultimately succumb to AIDS. AIDS is therefore one of the most devastating epidemics ever recorded.

Diagnosis (Table 56-5)

Serology HIV infection is most frequently documented by detection of antibody. Enzyme-linked immunosorbent assay (ELISA) or agglutination procedures are used for routine screening, and more specific procedures, such as Western blot and immunofluorescent assay (IFA), are subsequently used to confirm seropositive results. The ELISA test measures antibody to one or more envelope proteins (e.g., gp120) and can generate false positive data. The Western blot assay determines the presence of antibody to each of the viral antigens, including the core protein (p24). Immunofluorescence detects antibody directed at cell surface antigens. The presence of serum antibody does not label a patient as having AIDS; the latter is a clinical diagnosis that depends on symptomatology, signs, and other laboratory tests. The antibody test is used to confirm the diagnosis of AIDS and to identify carriers who may transmit infection to others, specifically blood or organ donors, pregnant women, and sex partners.

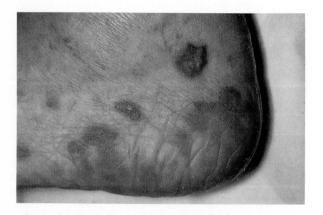

FIGURE 56-10   Kaposi's sarcoma of heel.

Table 56-5 Laboratory Analysis for HIV	
Test	**Detects**
Serology	Serum levels of antibody
ELISA	
Latex agglutination	
Western blot analysis	
IFA	
p24 antigen	Serum levels of p24 antigen
Culture	Isolation of virus
CD4/CD8 ratio	Lymphocyte subset analysis

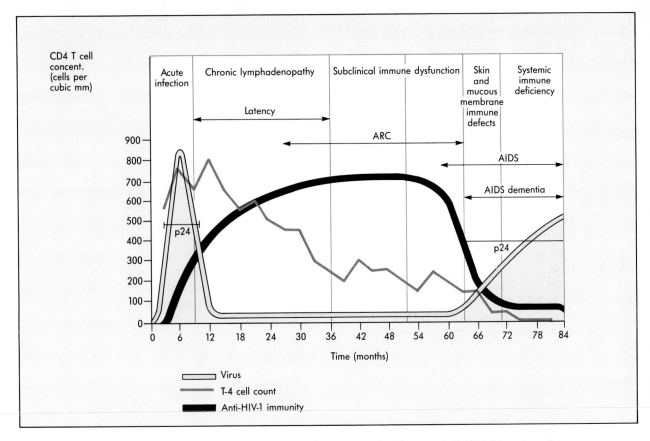

FIGURE 56-11   Time course of HIV disease. Note progressive decrease in T4 (CD4) lymphocytes even during the latency period. (Redrawn from Redfield RR and Buske DS: Sci Am 259:90-98, 1988.)

HIV antibody may develop slowly, requiring 4 to 8 weeks in most patients but 6 months or more in up to 5% of those infected (Figure 56-11). Newer serologic tests using HIV antigens derived by genetic engineering techniques may provide more sensitive measures that are consistently positive shortly after infection. Measurement of specific antibodies to other HIV antigens may be prognostically useful and may assist in staging a patient. For example, early in infection, antibody to p24 appears as HIV enters into latency. Later this antibody diminishes as the p24 antigen reappears, and the process of viral replication resumes.

Antigen Assays The p24 antigen can be detected in the lymphocytes of up to 60% of patients with HIV infection and indicates that active viral replication is probably occurring and is a poor prognostic sign. This antigen is detectable during acute HIV infection and then disappears as the virus enters the latent state (Figure 56-11), only to reappear years after infection.

Culture Culture of HIV is more difficult than for many other viruses of human importance. The virus is cell associated and requires co-culture of patient peripheral blood mononuclear cells (PBMCs) with PBMCs from non-HIV-infected donors in the presence of IL-2 to promote the growth of T cells in culture. Further, these donor cells should be transformed by exposure to mitogens. Although cytopathic effect may occur, detection of the virus in tissue culture is accomplished by measuring the appearance of reverse transcriptase in the medium or detection of viral antigen (p24) in the medium. Culture appears to detect even latent virus, since the procedure will activate provirus from the most antibody-positive individuals.

Immunologic Studies Indication of an HIV infection can be obtained from analysis of T-lymphocyte subsets. The absolute number of CD4 lymphocytes and the ratio of helper to inducer lymphocytes (CD4/CD8 ratio) are abnormally low in HIV-infected individuals.

Treatment

An extensive program to develop antiviral drugs and vaccines against HIV has been initiated worldwide (Table 56-6). Currently, azidothymidine (AZT) is the only anti-HIV drug available approved by the Food and Drug Administration (FDA), but several others are currently in clinical trials.

AZT inhibits the reverse transcriptase and, after incorporation into DNA, causes chain termination. Although it clearly benefits patients at least for many months, its use is limited by bone marrow toxicity. Other nucleotide analogs, including dideoxyinosine and dideoxycytidine, are also being tested against HIV infection. Additional approaches to HIV chemotherapy include CD4 receptor analogs, interferon, and interferon inducers.

Prevention and Control

Education The principal method for control of HIV infection is to alert the population about the methods of transmission and measures that may curtail virus spread. Risk behavior includes promiscuous unprotected sex, intravenous drug abuse, and careless handling of contaminated needles, syringes, and so on. Condoms are useful but imperfect adjuncts if exchange of bodily fluids occurs with a partner who is HIV positive or whose HIV infection status is unknown. Contaminated needles must not be shared or must be disinfected between uses. In some locations efforts are being launched to provide sterile equipment to intravenous drug abusers.

Blood and Blood Product Screening This should be preceded by self-screening to allow HIV-positive potential donors to refrain from donation. Methods to encourage this must be improved. Individuals who anticipate the need for blood, such as those awaiting elective surgery, should consider donating blood beforehand. To contain the worldwide epidemic, blood screening must be initiated in underdeveloped nations.

Vaccine

Several different strategies are being pursued to develop an HIV vaccine. Most approaches have keyed on gp120 and its precursor, gp160. The genes for this protein have been cloned and expressed in different cell systems and used as a subunit vaccine. The gp160 gene has also been incorporated into vac-

---

Table 56-6   Potential Antiviral Therapies for HIV Infection

Drug	Target	Status
ANTIVIRAL DRUGS		
Azidothymidine (AZT), zidovudine	Reverse transcriptase	FDA approved
Dideoxycytidine (ddC)	Reverse transcriptase	Trials
Dideoxyinosine (ddI)	Reverse transcriptase	Trials
Phosphonoformate (foscarnet)	Reverse transcriptase	Trials
Ribavirin	Reverse transcriptase and other steps	Trials
Soluble CD4 peptide analogs	Inhibits attachment of HIV to CD4 cells	Trials
α-Interferon	Many activities	Trials
VACCINES		
gp160/120 subunit vaccines	Genetically engineered, elicits antibody	Trials
gp120/vaccinia hybrid	Genetically engineered, elicits antibody and cell-mediated immunity	Trials
Killed vaccine	Nucleic acid removed	Trials
Anti-idiotype vaccine	Injection of antibody to CD4 to elicit antibody to viral attachment epitope	Experimental

cinia virus to create a hybrid vaccine. Inoculation with this virus mimics vaccination with a live virus and elicits both humoral and cellular responses to the viral antigen. Other approaches include immunization with P17, a protein from the inside of the HIV envelope, a killed HIV vaccine, and anti-idiotypic antibodies to block HIV binding.

The development of a vaccine against AIDS is fraught with several problems unique to the virus. Antibody alone may be insufficient to protect against HIV infection; a cellular immune response may also be necessary. The antigenicity of the virus changes through mutation. The virus can be spread through syncytia and also remains latent in an individual, hiding from antibody. HIV also infects and inactivates those cells required to initiate an immune response.

In addition to the problems of the virus and the vaccine development, the efficacy of the vaccines must be tested in human trials and a proper regimen of vaccination developed to elicit protective immunity. It is not clear which risk group should be vaccinated in any proposed study. Finally, it will be difficult to evaluate the success of the vaccine in limiting the spread and morbidity of HIV infection.

HUMAN T-LYMPHOTROPIC VIRUS AND OTHER ONCOGENIC RETROVIRUSES

The oncoviruses were originally called the RNA tumor viruses and have been associated with leukemias, sarcomas, and lymphomas in many different animals. The members of this family are distinguished by their mechanism of cell transformation and thus the length of the latency period between infection and development of disease (Table 56-7).

A large group of oncoviruses, the sarcoma/acute leukemia viruses, have incorporated cellular genes encoding growth-controlling factors into their genome, such as growth hormones, growth hormone receptors, protein kinases, guanosine triphosphate–binding proteins, and nuclear DNA–binding proteins. These viruses can cause transformation and are highly oncogenic. At least 35 different viral oncogenes have been identified (Table 56-8). Mutations in the viral oncogene from the original cellular proto-oncogene, or the unregulated synthesis of the oncogene, promote the transformation of an infected cell.

Incorporation of the oncogene into many of these viruses replaces coding sequences for the gag, pol, or env genes such that these viruses are defective and require helper viruses for replication. Many of these viruses remain endogenous and are transmitted vertically through the germ line.

The leukemia viruses, including HTLV-1, are replication competent but cannot transform cells in vitro. They cause cancer after a long latent period, at least 30 years. Integration near a cellular growth-controlling gene, or production of a DNA-binding protein (transactivator) capable of activating growth-controlling genes, may promote the outgrowth of that cell. This may be sufficient to transform the cell neoplastically, or the growth may promote other genetic aberrations over a long period. These viruses are also associated with nonneoplastic neurologic disorders and other diseases. HTLV-1 causes adult acute T-cell lymphocytic leukemia (ATLL) and also tropical spastic paraparesis, a nononcogenic neurologic disease.

The human oncoviruses include HTLV-1, HTLV-2, and HTLV-5, but only HTLV-1 has been definitively associated with disease (ATLL). HTLV-2 was isolated from atypical forms of hairy cell leukemia, and HTLV-5 was isolated from a malignant cutaneous lymphoma. HTLV-1 and HTLV-2 share up to 50% homology.

Table 56-7 Mechanisms of Retrovirus Oncogenesis

Disease	Speed	Effect
Acute leukemia/sarcoma	Fast: oncogene	Direct effect Provision of growth-enhancing proteins
Leukemia	Slow: transactivation	Indirect effect Insertion mutagenesis Proximity of viral promoters and enhancers to growth genes

Table 56-8   Representative Examples of Oncogenes

Function	Oncogene	Virus
Tyrosine kinase		
	src	Rous sarcoma virus
	abl	Abelson murine leukemia virus
	fes	ST feline sarcoma virus
Growth factor receptors		
	erb-B (EGF receptor)	Avian erythroblastosis virus
	erb-A (thyroid hormone receptor)	Avian erythroblastosis virus
GTP binding proteins		
	Ha-ras	Harvey murine sarcoma virus
	Ki-ras	Kirsten murine sarcoma virus
Nuclear proteins		
	myc	Avian myelocytomatosis virus MC29
	myb	Avian myeloblastosis virus
	fos	Murine osteosarcoma virus FBJ
	jun	Avian sarcoma virus 17

Modified from Jawetz E, Melnick JL, Adelberg EA, Brooks GF, Butel JS, and Ornston LN: Medical microbiology, ed 18, Los Altos, Calif, 1989, Appleton and Lange.

Pathogenesis

HTLV-1 is cell associated and is spread in cells following blood transfusion, sexual intercourse, or breastfeeding. The virus enters the bloodstream and also infects the CD4 helper and DTH cells. These cells have a tendency to reside in the skin, which contributes to the symptoms of ATLL. Neurons also express a receptor for HTLV-1.

HTLV is replication competent, and the gag, pol, and env genes are transcribed, translated, and processed, as described earlier. HTLV also has a transcriptional regulatory gene, tax/rex, which produces a transactivating protein capable of activating the LTR region in the integrated virus genome and also of activating cellular genes. The virus may remain latent or replicate slowly for many years but may also induce the clonal outgrowth of particular T-cell clones.

Antibodies are elicited to the gp46 and other proteins of HTLV-1. The presence of these antibodies in infected individuals may down-regulate the expression of viral antigens. This would prevent cell-mediated immune clearance of virally infected cells.

Epidemiology

HTLV-1 is endemic in southern Japan and is also found in the Caribbean and among Blacks in the southeastern United States. In the endemic regions of Japan, children obtain HTLV-1 from their mothers in breast milk, and adults are infected sexually. The seroconversion rate on some southern Japanese islands may be as high as 13%, with a mortality resulting from leukemia double that of other regions. Transmission of HTLV-1 by illicit intravenous drug use and blood transfusion is becoming more prominent in the United States.

Clinical Syndromes

HTLV infection is usually asymptomatic but can progress to ATLL in approximately 1 in 20 persons over a 50-year period. ATLL caused by HTLV-1 is a neoplasia of CD4 helper T cells that can be acute or chronic. The malignant cells have been termed "flower cells," are pleomorphic, and contain lobulated nuclei. In addition to an elevated white blood cell count, this form of ATLL is characterized by skin lesions similar to another leukemia, Sézary syndrome. Acute ATLL is usually fatal within 1 year of diagnosis regardless of treatment.

Diagnosis

HTLV-1 infection is detected immunologically by the presence of viral-specific antigens in blood.

Treatment

No treatment has been proved effective against HTLV-1 infection. Presumably AZT and other inhibitors of reverse transcriptase would be effective against HTLV-1, but controlled studies are need to prove the benefit of this approach.

Prevention Control

The means for limiting the spread of HTLV-1 are the same as for HIV. Sexual precautions, screening the blood supply, and increased awareness of the potential risks and diseases will help block transmission of the virus. Routine screening procedures for HTLV-1 in the blood supply are likely to be instituted in the near future.

BIBLIOGRAPHY

AIDS articles, Science 239(4840), Feb 5, 1988.

Allain J-P, Gallo RC, and Montagnier L, editors: Human retroviruses and disease they cause, Symposium Highlights, Chicago, 1988, Excerpta Medica.

Fauci AS: The human immunodeficiency virus: infectivity and mechanisms of pathogenesis, Science 239:617-622, 1988.

Levy JA: The multifaceted retrovirus, Cancer Res 46:5457-5468, 1986.

Ng VL and McGrath MS: Human T-cell leukemia virus involvement in adult T-cell leukemia, Cancer Bull 40:276-280, 1988.

What science knows about AIDS, Sci Am October 1988.

57

Slow Viruses

Slow virus agents are filterable and can transmit disease but otherwise do not conform to the standard definition of a virus. Unlike conventional viruses, no virion structure or genome has been detected; no immune response is elicited; and the agents are extremely resistant to inactivation by heat, disinfectants, and radiation. The slow viruses include the human diseases of kuru, Creutzfeldt-Jacob disease (CJD), and Gerstmann-Straussler syndrome (GSS; a variant of CJD) and the animal diseases of scrapie and transmissible mink encephalopathy (Box 57-1). These agents have long incubation periods and cause damage to the central nervous system (CNS), which leads to subacute spongiform encephalopathies. Most of the information on these agents comes from studies of the scrapie agent, first isolated from sheep but capable of causing disease in hamsters and monkeys.

The long incubation period, which can extend to 30 years, has made study of these agents difficult.

Gajdusek won the Nobel Prize for showing an infectious etiology for kuru and the means for initial analysis of the agent. The major breakthroughs came when the kuru agent was transmitted to monkeys and when it was recognized that the scrapie agent induced a characteristic cytopathology in hamsters considerably before the evidence of disease was evident. These studies have provided a useful means for assessing this agent and its molecular biologic properties.

PHYSIOLOGY AND STRUCTURE

Very little is known about the molecular properties of the agents responsible for these diseases. They are suspected to be viruses because they pass through filters that block the passage of particles greater than 100 nm, but they clearly transmit disease.

These agents can accumulate in high concentrations in organs such as the brain. Isolates diluted as much as 10^{11} can still induce disease in susceptible individuals. The agents are resistant to a wide range of chemical agents and physical treatments, such as formaldehyde, ultraviolet radiation, or heat to 80° C.

The prototype of these agents is scrapie, and the ability to infect and study this agent in hamsters has identified and allowed purification of scrapie-associated fibrils (SAF). The SAF are infectious. The only protein that can be identified with these fibrils is a unique glycoprotein with a molecular weight of 27,000 to 32,000 daltons. Since these structures confer infectivity, the infectious agent of scrapie has been dubbed a **prion,** a small, proteinaceous infectious particle.

Box 57-1   Slow Virus Diseases

HUMAN

Kuru
Creutzfeldt-Jacob disease (CJD)
Gerstmann-Straussler syndrome (GSS)

ANIMAL

Scrapie
Transmissible mink encephalopathy

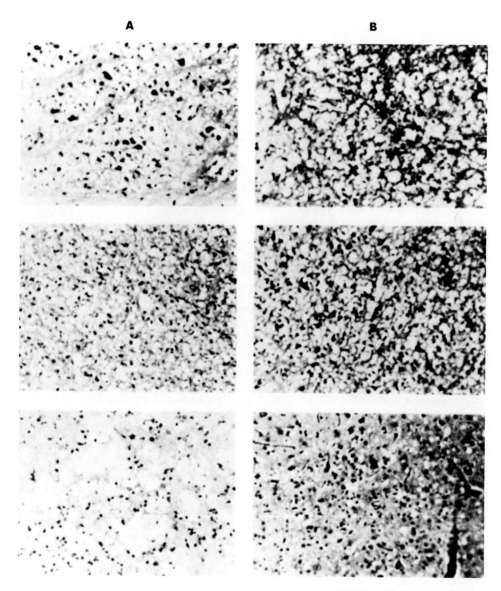

FIGURE 57-1   A, Comparison of spongiform change of scrapie in sheep *(top)*, kuru in humans *(middle)*, and CJD in humans *(lower)*, and **B,** the respective experimentally transmitted diseases in mice for scrapie *(top)* and in chimpanzees for kuru *(middle)* and CJD *(bottom)*; stained with hematoxylin and eosin. (From Fields BN et al, editors: Virology, New York, 1985, Raven Press.)

Although the evidence is strong to indicate that the prion is infectious and can cause scrapie and that similar proteins are responsible for CJD and kuru, other possibilities remain. The protein may be a component of an as yet undiscovered virus or may be induced by a virus infection.

PATHOGENESIS

The term **spongiform encephalopathy** is derived from the characteristic degeneration of neurons and axons of gray matter. Vacuolation of the neurons, amyloid-containing plaques, fibrils, a proliferation and hypertrophy of astrocytes, and fusion between neurons and adjacent glial cells are observed (Fig. 57-1). The agent can be obtained from tissue other than the brain, but only the brain shows any pathology. No inflammation or immune response is generated to the agent, which distinguishes the disease from a classical viral encephalitis.

The incubation period for CJD and kuru may be as long as 30 years, but once the symptoms are evident, the patient dies within a year (Box 57-2).

EPIDEMIOLOGY

The predominant means for transmission of CJD are by injection, transplantation of contaminated tissue (e.g., corneas), or contact with contaminated brain electrodes. However, the sporadic incidence of both CJD and GSS only rarely correlates with acquisition of the agent by these means, and a genetic mechanism for disease has been suggested as more likely for GSS. The incidence of this disease is esti-

> **Box 57-2** Pathogenic Characteristics of Slow Viruses
>
> No cytopathologic effect in vitro
> Doubling time of at least 5.2 days
> Long incubation period
> Strain-dependent pathogenicity
> Vacuolation of neurons (spongiform)
> Amyloid-like plaques, gliosis
> No antigenicity
> No inflammation
> No immune response
> No interferon production

mated to be 1 per 1 million persons, and it occurs worldwide.

In contrast, kuru is limited to a very small area of the New Guinea highlands. Transmission of kuru is related to the cannibalistic ceremonies of the Fore tribe of New Guinea. Kuru means shivering or trembling. Before Gajdusek intervened, the custom of these people was to eat the bodies of their deceased kinsmen. When Gajdusek began his study, he noted that women and children were most susceptible to the disease. The women and children prepared the food and were given the less desirable viscera and brains to eat. Their risk for infection, therefore, was increased by handling the contaminated tissue, with potential acquisition through the conjunctiva or cuts in the skin and also by ingesting the neural tissue, which should contain the highest concentrations of the kuru agent. Cessation of this cannibalistic custom has stopped the spread of kuru.

Table 57-1 Characteristics of Disease Caused by Slow Viruses

Transmission	Disease Syndromes	Incubation Period	Laboratory Diagnosis	Treatment	Control
Transplant of contaminated tissues (e.g., cornea); through cuts in the skin; use of contaminated brain electrodes; ingestion of infected tissue	Loss of muscle control, shivering, tremors, dementia	Up to 30 years or more; fatal within 1 year of symptoms	None	None	5% hypochlorite, 1.0 M NaOH, autoclaving at 15 psi for 1 hour

CLINICAL SYNDROMES

The slow virus agents cause a progressive, degenerative neurologic disease with a long incubation period but with rapid progression to death after onset of symptoms (Figure 57-2). The spongiform encephalopathies are characterized by loss of muscle control leading to shivering, myoclonic jerks, tremors, loss of coordination, and rapidly progressive dementia (Table 57-1).

LABORATORY DIAGNOSIS

There are no methods for direct virile detection in tissue using electron microscopy, antigen detection, or nucleic acid probes. There are also no serologic tests to detect virile antibody. The diagnosis must be made clincially with confirmation by pathologic examination of brain revealing the histologic changes described above in the section on pathogenesis.

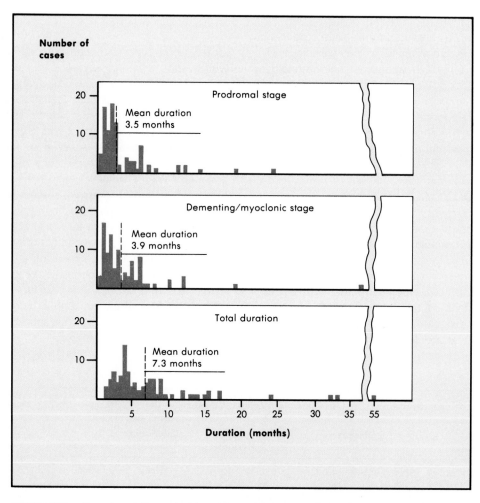

FIGURE 57-2   Duration of illness in 94 patients with transmissible CJD. The mean duration of the prodromal stage was 3.5 months (3.9 standard deviation, SD; range, 0.5 to 24.0), the mean duration of the dementing/myoclonic stage was 3.9 months (4.6 SD; range, 0.5 to 36.0), and the mean total duration was 7.3 months (7.5 SD; range, 1.5 to 55.0) (Redrawn from Fields BN et al, editors: Virology, New York, 1985, Raven Press.)

TREATMENT, PREVENTION, AND CONTROL

No treatment exists for kuru or CJD. The causative agents are impervious to most disinfection procedures. Autoclaving for 1 hour at 15 psi, instead of 20 minutes, or the use of 5% hypochlorite solution or 1.0 M sodium hydroxide, can be used for decontamination. Instruments and brain electrodes are a means of transmission for these agents and should be carefully disinfected before reuse.

BIBLIOGRAPHY

Gajdusek DC. Unconventional viruses causing subacute spongiform encephalopathies. In Fields BN, editor: Virology, New York, 1985, Raven Press.

Prusiner SB: Prions and neurodegenerative disease, N Engl J Med 317:1571-1581, 1987.

Prusiner SB: Molecular structure, biology and genetics of prions. In Maramorosch K, Murphy FA, and Shatkin AJ, editors: Advances in virus research, vol 35, New York, 1989, Academic Press, Inc.

Index

A

Abdominal infection, 63, 212
Abscess
  brucellosis and, 156
  cerebral, 216
  Dracunculus medinensis and, 408
  Entamoeba histolytica and, 356
  Enterococcus and, 83
  Listeria and, 206
  Nocardia and, 214-215, 216
  Streptococcus pneumoniae and, 70
Absidia, 345-346
Acanthamoeba, 357
Acetoin, 16
Acetyl-CoA, 16
N-Acetylmuramic acid
  procaryotic cell and, 6
  Staphylococcus and, 47
Acid
  N-acetylmuramic, 6, 47
  amino, 23-24
    bacteria and, 26
    tricarboxylic acid cycle and, 21
  aspartic
    conversion of, 27
    tricarboxylic acid cycle and, 21-22
  chorismic, 22
  dental caries and, 286
  diaminopimelic, 6
  dipicolinic, 27-28
  keto, 23
  lipoteichoic, 74
  nalidixic, 44
  nucleic, 40, 44
  oxaloacetic, 27
  teichoic, 6
Acid-fast bacillus, 218

Acidovoran group of Pseudomonadaceae, 120
Acinetobacter, 96
  transformation and, 31
Acquired immune deficiency syndrome
  azidothymidine and, 488-489
  cytomegalovirus and, 528
  dihydroxypropoxymethyl guanine and, 489
  human herpesvirus 6 and, 529
  human immunodeficiency virus and, 672-677
  leprosy and, 224
  Pneumocystis carinii and, 380-381
  toxoplasmosis and, 379
Actinobacillus, 146-147
Actinomadura, 320
Actinomyces, 211-214
  subcutaneous infection and, 320
  treatment of, 320
Active immunization, 481, 483
Acute necrotizing gingivitis, 288
Acyclic agents, 488
Acycloguanosine monophosphate, 486
Acycloguanosine triphosphate, 486
Acyclovir, 484-487
  cytomegalovirus and, 529
  herpes simplex virus and, 511-512
Adenine arabinoside, 485, 487-488
  cytomegalovirus and, 529
  herpes simplex virus and, 512
Adeno associated virus, 567
Adenopathy, axillary, 153
Adenosine triphosphate, 256
Adenovirus, 491-497
  clinical presentation and, 495-496
  epidemiology and, 494
  laboratory diagnosis and, 496-497
  pathogenesis and, 493-494
  replication and, 452
  structure and replication of, 491-493